ENCYCLOPEDIA OF

Agricultural Science

Volume 3   M-R

Editor-in-Chief

Charles J. Arntzen

Institute of Biosciences and Technology
Texas A&M University
Houston, Texas

Associate Editor

Ellen M. Ritter

Department of Agricultural
Communications
Texas A&M University
College Station, Texas

Academic Press

San Diego   New York   Boston   London   Sydney   Tokyo   Toronto

Copyright © 1994 by ACADEMIC PRESS, INC.
All Rights Reserved.
No part of this publication may be reproduced or transmitted in any form or by any
means, electronic or mechanical, including photocopy, recording, or any information
storage and retrieval system, without permission in writing from the publisher.

Academic Press, Inc.
A Division of Harcourt Brace & Company
525 B Street, Suite 1900, San Diego, California 92101-4495

United Kingdom Edition published by
Academic Press Limited
24-28 Oval Road, London NW1 7DX

Library of Congress Cataloging-in-Publication Data

Encyclopedia of agricultural science / edited by Charles J. Arntzen,
    Ellen M. Ritter.
            p.   cm.
      Includes index.
        ISBN 0-12-226670-6 (set) -- ISBN 0-12-226671-4 (v. 1)
        ISBN 0-12-226672-2 (v. 2) -- ISBN 0-12-226673-0 (v. 3)
        ISBN 0-12-226674-9 (v. 4)
        1. Agriculture--Encyclopedias.   I. Arntzen, Charles J.
    II. Ritter, Ellen M.

CONTENTS OF VOLUME 3

CONTENTS OF OTHER VOLUMES

CONTENTS OF VOLUME 4

HOW TO USE THE ENCYCLOPEDIA

The *Encyclopedia of Agricultural Science* is intended for use by both students and research professionals. Articles have been chosen to reflect major disciplines in the study of agricultural science, common topics of research by professionals in this realm, areas of public interest and concern, and areas of economics and policy. Each article thus serves as a comprehensive overview of a given area, providing both breadth of coverage for students and depth of coverage for research professionals. We have designed the *Encyclopedia* with the following features for maximum accessibility for all readers.

Articles in the *Encyclopedia* are arranged alphabetically by subject. A complete table of contents appears in each volume. Here, one will find broad discipline-related titles such as "Agroforestry" and "Plant Pathology," research topics such as "Transgenic Animals" and "Photosynthesis," areas of public interest and concern such as "Plant Biotechnology: Food Safety and Environmental Issues" and "World Hunger and Food Security," and areas of economics and policy such as "Macroeconomics of World Agriculture" and "Consultative Group on International Agricultural Research."

Each article contains an outline, a glossary, cross references, and a bibliography. The outline allows a quick scan of the major areas discussed within each article. The glossary contains terms that may be unfamiliar to the reader, with each term defined in the context of its use in that article. Thus, a term may appear in the glossary for another article defined in a slightly different manner or with a subtle nuance specific to that article. For clarity, we have allowed these differences in definition to remain so that the terms are defined relative to the context of each article.

Each article has been cross referenced to other articles in the *Encyclopedia*. Cross references are found at the end of the paragraph containing the first mention of a subject area covered elsewhere in the *Encyclopedia*. We encourage readers to use the cross references to locate other encyclopedia articles that will provide more detailed information about a subject. These cross references are also identified in the Index of Related Titles, which appears in Volume 4.

The bibliography lists recent secondary sources to aid the reader in locating more detailed or technical information. Review articles and research articles that are considered of primary importance to the understanding of a given subject area are also listed. Bibliographies are not intended to provide a full reference listing of all material covered in the context of a given article, but are provided as guides to further reading.

Two appendices appear in Volume 4. Appendix A lists United States colleges and universities granting degrees in agriculture. Appendix B lists United Nations organizations concerned with agriculture and related issues. Both appendices provide address and telephone information for each institution listed.

The Subject Index is located in Volume 4. Because the reader's topic of interest may be listed under a broader article title, we encourage use of the index for access to a subject area. Entries appear with the source volume number in boldface followed by a colon and the page number in that volume where the information occurs.

Macroeconomics of World Agriculture

G. EDWARD SCHUH, *University of Minnesota*

Glossary

Closed economy Economy that has no trade sector, or at least a modest one

Exchange rates Value of one currency expressed in terms of the value of other currencies

Factor markets Markets for inputs or resources used in agriculture such as land, labor, and modern inputs

General equilibrium effect Second-order effect of a change in an economic variable (e.g., the price effect from the increased output due to the introduction of new production technology

Open economy Economy that is connected to the international economy through international trade and capital (financial) markets

The main topics to be considered in addressing the macroeconomics of world agriculture are (1) the monetary and fiscal policies of national governments, (2) currency exchange rate regimes and policies, (3) the international capital (financial) market, (4) international trade policy, (5) international migratory flows of labor, and (6) technological change in

agriculture. Prior to the decade of the 1970s these issues, with the exception of technological change, were largely neglected by authors writing about world agriculture. During the decades of the 1970s and the 1980s, however, writing on this subject has literally burgeoned.

The earlier neglect of some of these issues was rooted in a number of important factors. First, those writing on agriculture tended to take a sectoral approach and to neglect the factor markets and other linkages between agriculture and the rest of the economy. Second, for most analytical problems involving agriculture in that earlier period, national economies could largely be treated as closed and thus many macroeconomic issues did not arise. Third, from the end of World War II through 1973 the world economy operated with the Bretton Woods fixed exchange rate system, which caused the response to domestic monetary and fiscal policies to be channeled to the general economy in ways that did not have macroeconomic implications (see below). Fourth, it was not until a large, well-integrated international financial market emerged in the 1960s and 1970s and the world shifted to a bloc-floating exchange rate system that monetary and fiscal policies began to have a significant effect on agriculture. These two developments caused agriculture, as a tradeable sector, to share in the adjustment to changes in monetary and fiscal policies. Finally, significant international migratory flows of labor are also of comparatively recent vintage.

Important dimensions of the macroeconomics of world agriculture involve certain well-defined general equilibrium effects. These include the consequences of introducing a significant flow of new production technology into the sector and an analysis of how the benefits of that technology are shared in the economy. It also includes a consideration of food as a wage good, the induced effects of technological change in agriculture on real exchange rates, and the role of technological change as the means of dealing with declines in a nation's external terms of trade. These general equilibrium issues will be addressed in this chapter in conjunction with the issues associated with the more narrow, conventional perspective of macroeconomic policy. [See FINANCE.]

I. Increased Openness of the International Economy and Implications

Two developments in the international economy have caused national economies to become increasingly open over the post-World War II period. The first is the growth in international trade at a rate faster than the growth in global GNP. This has been a persistent tendency, with only 5 years of serious economic recession being the exceptions. Popularly, this growth in international trade relative to global trade is referred to as an increased dependence on trade. Economists, however, refer to it as an increased openness of national economies. [See TARIFFS AND TRADE.]

The second development leading to increased openness was the emergence of a large, well-integrated international capital market in the 1960s and 1970s. This capital/financial market on its present scale and scope began to emerge in the 1960s when commercial banks in Europe began to re-lend the ever-larger amounts of U.S. dollars they had on deposit. In the beginning, this was referred to as a Eurodollar market. However, as the banks turned to re-lending the deposits of other currencies they had on deposit as well, it was referred to as the Eurocurrency market.

The largest stimulus to this financial market came in the 1970s, however, and especially after the quadrupling of petroleum prices by OPEC in 1973. Petroleum was both priced and transacted in U.S. dollars. This large increase in price gave rise to a flood of what were called petrodollars. There was much concern at that time that unless these petrodollars were recycled, the international economy would collapse. The commercial banks were "leaned upon" to recycle these petrodollars. They found willing borrowers in the developing countries, who preferred borrowing to devaluations of their currencies and other policy reforms that were more in order. Thus was born the international debt crisis of the 1980s.

The surge of petrodollars was exacerbated by highly stimulative monetary policies in the developed countries. Policy makers in these countries turned to easy monetary policies as the means to facilitate the adjustment of their economies to the shock of higher petroleum prices. Thus, international monetary reserves grew at a rapid rate during this period, increasing the lending power of commercial banks.

The increased openness of national economies has a number of important implications. First, national economies become increasingly beyond the reach of national economic policies as these economies become increasingly open. This loss of "control" can be seen at both the sectoral and national levels. At the sectoral level, changes in exchange rates induced by international financial flows can offset the effects of national sectoral policies. As an example, the rise in the value of the dollar in the first half of the 1980s created serious income problems for U.S. agriculture, with the government responding with greatly increased subsidies for farmers.

At the national level, with a well-integrated international financial market it is not just domestic monetary and fiscal policies that determine the value of a nation's real exchange rate, but the monetary and fiscal policies of other countries as well. For example, the dollar rose significantly in the first half of the 1980s not only because the United States was pursuing contradictory monetary (restrictive) and fiscal (stimulative) policies, but because Japan and the countries of Western Europe were pursuing monetary and fiscal policies that were just the reverse. The interest rate differentials that emerged led to large financial flows to the United States, which in turn drove up the value of the dollar.

The second implication of an increase in the openness of national economies is that the making and implementation of economic policy shifts in a bifurcated way. Some part of economic policy making and implementation shifts upward to the international level and becomes imbedded in international organizations and institutions such as the General Agreement on Tariffs and Trade (GATT) and in systems of economic integration such as the EC-92, the Canada/United States Free Trade Agreement, and the North American Free Trade Agreement. In a parallel fashion, another part of economic policy making and implementation shifts down to the state and local level. The part of policy making that tends to shift upward to the international level involves international trade. The part that shifts downward is that which focuses on domestic resource and income issues.

These shifts in where economic policy making and implementation take place are occurring extensively and significantly around the world. They have important implications for the design of agricultural policy and of policies designed to promote rural development.

II. General Equilibrium Dimensions of Agricultural Development and Policy

Investments in research designed to produce new production technology for agriculture are now widely accepted as the most efficacious means of promoting agricultural development. The development of agriculture by this means has powerful general equilibrium effects which influence who receives the benefits of those investments and which serve as powerful sources of more general economic development.

Consider how this works. The introduction of new production technology into agriculture provides the means of lowering the cost of production. Early adopters of this technology tend to reap its initial benefits, for they will have lower costs of production while the product price is not yet affected. However, as the adoption of the technology spreads, the increase in supply that results tends to drive the product price down (assuming limited or no protectionism). Most of the benefits of the new technology are thus passed to the consumer, especially if the commodity for which the new technology is produced is one that is domestically consumed. The decline in product price puts pressure on the production sector and some producers will be worse off, with some resources, and especially labor, being forced to seek employment in the nonfarm sector. If the commodity is widely consumed in the domestic economy and not traded, the benefits to the consumer can be large. This is one of the reasons why the estimated rates of return to investments in agricultural research are so high. [*See* Prices; Production Economics.]

These distributional effects, largely general equilibrium in nature, are quite significant. Moreover, they have still another dimension. Since poor people tend to spend a larger share of their budget on food than do middle and upper income people, the poor tend to benefit in a relative sense. Moreover, these increases in personal income from the decline in the price of a major consumable good are a powerful source of additional economic growth. They increase the demand for other consumer goods and services and thus stimulate more general economic growth and development.

If the new production technology happens to be for a tradeable good and the country is not important in international markets, there will not tend to be a decline in the price of the commodity as the new technology is disseminated in the sector. Instead, foreign exchange earnings will tend to increase, either because the country will become more competitive in international markets and thus increase exports, or because imports of the commodity (or a competing commodity) will decline and foreign exchange will be saved in that way. In either case, more foreign exchange becomes available to finance a higher rate of growth in the domestic economy. If the effect on the trade side should be sufficiently large, the nation may find the real value of its currency increasing in foreign exchange markets. If that occurs, the benefits will be even more widespread in the economy and the country will experience an increase in national welfare since it will have to give up less in terms of domestic resources to acquire what it wants to acquire from abroad.

A directly related general equilibrium effect occurs through the fact that food is a wage good. Changes in the real price of food can obviously have a significant effect on the real wage workers receive. As the price of food drops, the real wage may rise even though the nominal wage is still unchanged. To put it somewhat differently, the welfare of workers may rise even though the nominal wage has not risen. This helps firms in the economy as a whole become more profitable since workers benefit from higher real wages even though the firms do not have to pay higher nominal wages. It also helps these same firms become, or remain, more competitive internationally because they do not have to continually be raising nominal wages. The positive employment effects of agricultural development by means of introducing new production technology into the sector can thus be quite great, and far reaching in the economy.

Technical change in agriculture can also create general equilibrium effects in the economy by bringing about changes in the real exchange rate. This is most likely to occur when it occurs for a tradeable commodity. If the commodity is important in the trade balance, on either the import or export side, and the technological breakthrough is significant, the real exchange rate can rise either because of a significant increase in export earnings or because of a significant decline in imports of a previously imported commodity or commodities.

This brings us to consideration of a final dimension of general equilibrium effects from agricultural development. Because technological change has been significant in a number of important traded agricultural commodities, the price of agricultural commodities has been experiencing a long-term decline relative to the prices of other goods and services. This is reflected to many countries as a decline in their external terms of trade—the price of their exports relative to the price of their imports. This decline in the external terms of trade has important balance of trade implications, although not—as is widely believed—necessarily negative national income effects.

The classic policy response to such a decline in the external terms of trade is for the country experiencing such a decline to devalue its currency (lower its value relative to other currencies). This will, indeed, result in a decline in national income. However, an alternative policy response is to accelerate the rate of technical change in the domestic sectors experiencing the decline in real prices. By that means the country will remain competitive in international markets, producers will not suffer income losses since their productivity will be increasing in a fashion parallel to the decline in the price of the commodity, and the country will not suffer the loss of foreign exchange that is the root of the balance of payments problem created by the negative shift in the terms of trade. Of course, from a policy standpoint it would be best to anticipate such a decline in the real price of the commodity by keeping the domestic agricultural research establishment up to date with developments in the international economy.

To conclude, these cases demonstrate that the general equilibrium effects of developing agriculture by means of introducing a flow of new production technology into the sector can be very important. Moreover, they have very important implications for what are classically described as macroeconomic policies—exchange rate policy, trade policy, and ultimately monetary and fiscal policy.

III. Exchange Rate Regimes and Exchange Rate Policies

The exchange rate regime prevailing in the international economy has significant implications for the impact monetary and fiscal policies have on world agriculture. The international community entered the post-World War II period with agreement on a fixed exchange rate system—the so-called Bretton Woods Conventions. Under this agreement, nations agreed to fix the value of their currencies relative to each other, and to change them only under dire circumstances and after consultation with their important trading partners and the International Monetary Fund. Imbalances in external accounts were to be eliminated by changes in domestic macroeconomic policies. A deficit on the trade account was to be eliminated by pursuing restrictive monetary policies and a conservative fiscal policy. A surplus on the trade account was to be eliminated by pursuing stimulative monetary and fiscal policies.

This fixed exchange rate regime was chosen because of a widespread belief at the end of World War II that the depth and global scope of the Great Depression had been caused by the extensive use of competitive devaluations which national policy makers used in efforts to dump their domestic problems abroad. The agreement to adhere to a fixed exchange rate regime meant that national policy makers were to deal with problems in their domestic economy by means of domestic economic policies.

This fixed exchange rate regime prevailed until the latter part of the 1960s, when it began to unravel, ultimately coming to an end in 1973 when the United States devalued the dollar for the second time in an 18-month period and declared unilaterally that henceforth the value of the dollar would be whatever the foreign exchange markets said it was. A number of factors led to this turn of events. The first was the decision of the United States to fight the Vietnam War and simultaneously expand the Great Society programs of the Johnson Administration without raising taxes. This caused the U.S. economy to become increasingly out of adjustment with the international economy, and led to a protracted debate between U.S. policy makers and policy makers from Japan and Germany over whether the United States should devalue (reduce the value of) or Germany and Japan should revalue (increase the value of) their currencies. The United States eventually devalued the dollar in February of 1971, for the first time since the Great Depression.

The proximate cause for the United States having forced the world to a system of flexible exchange rates was the failure to re-establish balance in its balance of payments after the 1971 devaluation. However, behind that set of events was a basic development in the international economy that made the fixed exchange rate regime no longer viable. The international capital market had grown enormously during the 1960s and into the early 1970s, at the same time the international communications system had improved substantially. With financial flows tending to dominate foreign exchange markets, it simply was no longer feasible for nations to fix their exchange rates for any extended period of time.

The present exchange rate regime can be best described as a bloc-floating exchange rate system. The world's major reserve currencies, such as the U.S. dollar, the British pound, the Japanese yen, and the German Deutschmark, fluctuate (float) relatively freely in relation to each other. However, a large (but declining) number of less economically important countries fix the value of their currency relative to one of these major reserve currencies.

This bloc-floating exchange rate regime has more flexibility in it than meets the eye. As the major reserve currencies float relative to each other they carry with them the currencies tied to them. Thus, there is a great deal of *implicit* flexibility in the system. This has important policy implications, as will be noted below.

The choice of international exchange rate regime has very significant implications for the management of monetary and fiscal policy and for the effects of these policies on the agricultural sector. Some of these issues will be discussed in the next section. In this section, it is useful to consider two related sets of issues. First, the classic advantage of a flexible exchange rate system has generally been assumed to be that it gives national policy makers more freedom to manage their domestic monetary and fiscal policies as they see fit. However, that assumption fails to recognize the importance of the international financial market and its effects on monetary policy in particular. In point of fact, most countries are not able to completely neutralize the effects of international financial flows on the domestic money supply, with the result that domestic monetary policies become linked one with the other.

Another issue is the problems created with the present bloc-floating system. These problems can be illustrated with the experience of Brazil during the crises of the 1970s and 1980s. When the price of petroleum increased dramatically in 1973 Brazil needed, on classic policy grounds, to undertake a significant devaluation of its currency. It imported approximately 85% of its petroleum at that time and thus the rise in petroleum prices constituted a significant decline in its external terms of trade. It shunned such a devaluation, however, choosing instead to revitalize its import-substituting industrialization policies and to subsidize some of it exports. During the remainder of the 1970s it did exceedingly well, with one of the highest rates of economic growth of any country in the world. Unbeknown to it, seemingly, it was benefitting from the decline in the value of the U.S. dollar, which gave its currency an implicit devaluation relative to many other currencies.

Brazil was again faced with a similar situation in 1979. Petroleum prices took another dramatic jump and it was faced with a similar dilemma. This time Brazil undertook what it described as a "maxi" devaluation, but the failure to pursue a complementary monetary policy that would have forced a realignment in the domestic terms of trade caused the effect of the devaluation to be dissipated quickly. More importantly, the United States changed its monetary policy dramatically to arrest a steep decline in the value of the dollar. The steep rise in domestic interest rates that followed induced a large inflow of capital from abroad and a sharp rise in the value of the dollar.

Interest rates rose generally in the international economy, and Brazil (together with other debtor countries) was faced with refinancing its debt at much higher rates, while at the same time needing to give up increasingly larger amounts of domestic resources to pay for an increasingly expensive dollar. Brazil continued to peg the value of its currency to the value of the dollar in real terms, which caused it to lose its competitive edge in international markets. The result was a decade of stagnant economic growth.

In conclusion, the type of exchange rate regime the international community and national policy makers agree to has a major impact on how domestic macroeconomic policy has to be managed. The shift to a flexible exchange rate regime, or even a bloc-floating exchange rate system, presents important challenges to policy makers. Moreover, as we will see in the next section, the choice of exchange rate regime determines importantly what sectors in the economy bear the adjustments to changes in monetary and fiscal policies. All of this has important implications for global agriculture.

IV. Monetary and Fiscal Policy in an Open International System

During the 1950s and the 1960s and into the early 1970s most countries adhered to the Bretton Woods fixed exchange rate system. During the 1950s and the first half of the 1960s the international financial market was relatively small, and international trade, although growing, was still far less important in a relative sense than it is today. The result was that for many countries macroeconomic policy could be managed as if the national economy were closed and relatively autonomous. Under these circumstances, changes in monetary policy were reflected largely in changes in interest rates, and the effects of these changes were felt rather widely in the economy. Agriculture, for its part, was rather exempt from these changes in policy. The main source of instability for agriculture was the effect of the weather on the supply of agricultural output.

The international system is significantly different today. Exchange rates are no longer fixed on near the extent they were back in that period, and we now have a large, well-integrated international capital market. With flexible exchange rates and a well-integrated international capital market, changes in monetary and fiscal policy in countries that are relatively unimportant in global capital markets will be reflected in changes in the value of a nation's currency instead of in the interest rate. Rather than for interest rates to rise when policy makers pursue restrictive monetary policies, an inflow of capital occurs and the value of the nation's currency rises. This causes the export sector to lose its competitive edge, and imports to enter the country in competition with import-competing sectors. The policy has the same dampening effect on the economy as under the previous configuration of the international economy, but it is the trade sectors that now bear the burden of adjustment. For countries such as Germany and the United States, which *are* relatively important in global capital markets, the effect will be divided between changes in the interest rate and changes in the exchange rate. In these cases the trade effects will be attenuated.

The reverse occurs when policy makers try to stimulate the economy by means of easier monetary policy, with the same distinction between countries that are important and relatively unimportant in global capital markets. In this case, capital tends to flow out of the economy, the value of the currency tends to fall in foreign exchange markets, the export sector becomes more competitive, and so does the import-competing sector. The economy is thus stimulated.

The significance of this change in the way monetary and fiscal policies affect the economy for global agriculture is that in most countries agriculture is a tradeable sector. Most countries either import or export agricultural commodities; many, if not most, do both. Thus, in the new configuration of the international economy agriculture has become a sector that bears the burden of adjustment to changes in monetary and fiscal policy. Consequently, agriculture now experiences unstable demand conditions rooted in changing monetary and fiscal policies, in addition to the traditional supply instability due to instability in the weather. With monetary policy tending to be unstable in the present rudderless international monetary system, agriculture in national economies also has become more unstable. This in turn creates unstable trade flows.

Another consequence of this changed configuration of the international economy is a stronger linkage between financial markets and commodity markets. Changed conditions in financial markets induce shifts in the value of national currencies, and this in turn is reflected in shifts in the demand for agricultural commodities. Under present conditions the international financial markets are the driving force in foreign exchange markets and ultimately in commodity markets.

Over the years the developed countries have developed a package of commodity programs to cope with the instability in agricultural markets associated with weather-related factors and to provide income support. The above changes in the international economy have made these programs increasingly costly in terms of domestic treasury cost. This poses the question of how policy makers should cope with these problems in the future.

A number of solutions are in order, some at the international level and others at the domestic level. At the international level, an important need is for more monetary stability as the means of reducing or eliminating shocks to international commodity markets. Although widely resisted by the U.S. banking community and U.S. policy makers, there seems little alternative to establishing something like an international central bank to provide for a stable increase in monetary reserves to support a stable growth in international money supplies. This can be done with only modest changes in the rules for the present international monetary institutions and while preserving the use of national currencies.

At the domestic level, policy makers need to pursue more stable monetary and fiscal policies, and a flexible exchange rate policy. Given the international linkages which now connect domestic monetary policies through international financial markets, the goal should be to reach agreement on an international code by means of which national governments would commit to pursue neutral monetary and fiscal policies. A neutral monetary policy is one in which policy makers seek to stabilize the price level. A neutral fiscal policy is one that seeks to balance national budgets over, say, a period of 3 years. With such policies in place, monetary shocks to commodity markets will be reduced to a minimum, especially if most of the countries of the world subscribe to the code.

The final element of a national policy to establish more stable commodity markets should be to pursue a flexible exchange rate system. A flexible exchange rate system has the great advantage of starting the adjustment process almost as soon as an external shock begins to affect the economy. Another advantage is that the adjustment is spread widely in the economy. Thus, the onus for adjustment is less on particular sectors and agriculture, for example, would experience less instability due to external monetary disturbances.

Finally, it is widely believed that a significant problem with flexible exchange rates is that they tend to be unstable, especially in the short term. However, that is not the major or key issue. The challenging problem for policy makers is that in the period since we have shifted to a flexible exchange rate system we have experienced wide swings in the values of the major reserve currencies. For example, the value of the U.S. dollar declined almost steadily from where it was at the time of the first devaluation in 1971 to its low point in 1979. By the time the low point was reached in 1979 U.S. producers thought they could compete with almost anybody in the world. From its low point in that year, however, the dollar took off on an almost unprecedented rise, reaching a peak some 6 years later in May 1985. The rise in the value of the dollar made U.S. producers less and less competitive in international markets, and by the time the peak was reached U.S. producers were almost persuaded they could not compete with anyone. After that large swing upward, the value then started on another long-term slide, interrupted only by a period of tight monetary policy in about 1988.

The problem with these large swings in the real value of national currencies is that they mask underlying comparative advantage, and for relatively long periods of time. They also create large pressures for protectionism, precisely because they do mask underlying comparative advantage. Within the sector they lead to large increases in asset values, and then to collapses in those asset values as the value of the currency moves in the opposite direction. It is difficult to imagine commodity policies or programs that could cope with these large swings except at very large treasury costs. The only solution is to create more stable monetary conditions by pursuing policies and institutional changes along the lines outlined above.

V. International Migratory Flows of Labor

Large international migratory flows of labor also can have important macroeconomic implications for agriculture. The potential for large flows of such labor are great at a number of places in the world. They can have significant effects on international comparative advantage, and in turn on international trade. [See LABOR.]

Perhaps the best way to illustrate the issues in such migratory flows of labor is to consider the U.S. case with migrants from Mexico and other Latin American countries. The United States is a significant producer of labor-intensive fruits and vegetables. At the same time, an alternative low-cost source of supply for these commodities is available in Mexico and other countries from which the migrant labor comes. The key to the United States remaining competitive in such commodities is for the migratory labor to continue to enter the country and have gainful employment in these producing sectors. Alternatively, the United States could lower its trade barriers against the imports of these commodities and let the domestic sectors bear the consequences.

In effect the United States has the alternative of "trading" either through the product markets or through the labor market. Whichever way it is, it has important balance of trade and trade policy implications.

VI. Global Liberalization of Agricultural Trade

The global liberalization of agricultural trade can have important macroeconomic implications, as well as important implications for the efficiency of the global agricultural sector. The starting point for this discussion is that there is an important pattern of national policy vis-a-vis the global agricultural sector. For example, the developed countries of the European Community, Japan, and the United States provide high levels of protection for their agricultural sector, with domestic prices substantially above international market-clearing levels. The low-income developing countries, on the other hand, tend to discriminate against their agriculture by overvaluing their currencies (an implicit export tax and import subsidy), implementing export taxes and *confiscos*, and limiting exports by a variety of partial and complete embargoes. The result is that the domestic prices of agricultural commodities in these countries tend to be substantially below international market-clearing levels. This problem is exacerbated because the developed countries, especially the European Community and the United States, tend to dump abroad the excess supplies that accumulate in government hands from the domestic commodity programs.

The net effect of this pattern of trade and exchange rate policies is to create a gross distortion in the use of the world's agricultural resources. Far too much of the world's agricultural output is produced in the high cost developed countries; far too little is produced in the low cost developing countries. There would be much to be gained in terms of increasing the world's total food supply by reducing these government interventions. There would be similar increases in per capita incomes on a global scale since more efficient use would be made of the world's considerable agricultural resources.

Unfortunately, the Multilateral Trade Negotiations have historically tended to ignore these important issues. Back at the beginning of the GATT, the United States insisted that domestic agricultural policies be treated separately from the multilateral mechanisms. Later, when agricultural issues eventually were placed on the table, the European Economic Community insisted that they be negotiated separately from the other issues. More recently, when the United States insisted that agricultural issues be made the center piece of the Uruguay Round, the negotiations were limited to a discussion of limitations of access of foreign producers to domestic markets. No attention was given to barriers imposed by national governments to limit the access of their domestic producers to foreign markets—an equally pernicious restriction on trade.

Similarly, the large and important barriers to trade created by distortions in foreign exchange markets have also been ignored in the trade negotiations. The developing countries have in particular had large distortions in the value of the currencies, with the tendency being to overvalue their currencies, and by a large amount. An overvalued currency is equivalent to a tax on exports and a subsidy on imports. Fewer countries have tended to undervalue their currencies, with Japan being the main exception in the post-World War II period. However, an undervalued currency is equivalent to an export subsidy and a tariff against imports.

Two things are important about these protectionist measures that have been omitted from the negotiating table. First, the share of global agricultural output affected by them is far larger than that affected by barriers to trade that limit access to domestic markets by foreign producers. So are the numbers of people involved, since these policies characterize mostly the developing countries. Second, there is an important synergism between the two patterns of trade distortions. The import subsidies implicit in the overvalued currencies in the developing countries reduce the treasury costs of the protectionist measures practiced by the developed countries and thus make it easier for those governments to sustain those programs. Governments in the developing countries have been so interested in favoring their urban consumers with cheap food that they have not even bothered to impose countervailing duties against the export subsidies of the developed countries—something they would be legitimately entitled to do under the rules of the GATT.

Independently of the trade negotiations, progress is being made on the side of the discriminatory policies the developing countries practice against their agriculture—a positive benefit of the international debt problems these countries experienced during the 1980s. To have the capacity to service their debt these countries have had to become more export oriented and to reduce their imports. This has required that they implement more realistic exchange rate policies and reduce the barriers to exports. These reforms have been aided and abetted by the international community, including the World Bank, the International Monetary Fund, and bilateral development agencies.

The remaining problem is the protectionist policies of the developed countries. Burgeoning Treasury costs may eventually take their toll and bring about the reform of these policies, together with the relative decline in the political power of the farm blocs in those countries. However, that seems unlikely to occur in the immediate future, because as agriculture becomes a smaller and smaller share of GNP the share that subsidy costs make up of the total national budget declines. Whether the GATT will be able to have any significant influence on these policies remains to be seen.

With the reforms in the developing countries alone there are likely to be major shifts in the patterns of global agricultural production and significant shifts in trade patterns. Increased specialization in production is likely to be one of the important consequences of these reforms, with the emergence of increased trade *among* the developing countries. Globally, trade liberalization for the agricultural sector can be an important source of economic growth in the decade ahead, especially among the developing countries. Agriculture will be an even more powerful source of economic growth should the developing countries begin to invest more in agricultural research as their economies recover from the economic difficulties of the 1980s.

VII. Reforms in the Formerly Centrally Planned Economies

The agriculture of the formerly centrally planned economies has been relatively isolated from international agricultural markets, even though some of these countries have at times imported significant quantities of foodstuffs. The shift of these economies to increased dependence on markets to organize their resources, and their increased integration into the international economy, can be expected to have important macroeconomic implications for global agriculture.

The effects of these reforms are not likely to be unidirectional. A contrast of the former Soviet Union and China, two rather large and important economies, illustrates the range of expectations one might have. The former Soviet Union, for example, is a country with a low population density in the aggregate, and one which has a large, unexploited agricultural potential. At one time it was a significant exporter of grains, but centralized planning has lowered the productivity of this important agricultural sector.

As the economy of the former Soviet Union moves toward reform there are likely to be two significant shocks to international commodity markets. First, as the large subsidies on the side of food consumption, which have led to a heavy dependence on livestock products, are removed, the demand for imports of feed grains and soybeans is likely to decline substantially. It is not likely that the former Soviet Union will be importing 40 million tons of grains into the future. Second, reform of the distribution system, the regional shift of production along lines more consistent with comparative advantage, and the reorganization of agricultural production along more efficient lines, will eventually increase available supplies very significantly. Although it may be a decade before these reforms take place and have their effects, it is

very likely that in the future the former Soviet Union will become a net exporter of grains.

Further reform of the China economy is likely to have different consequences for international commodity markets. China is much closer to realizing the potential of its agricultural resources than is the former Soviet Union. Moreover, the population density of the country is much higher, the population totals more than a billion people, and per capita incomes are very low. Further reform of the China economy is likely to result in significant increases in per capita incomes, which will lead to significant increases in demand for food, with a shift toward livestock and livestock products. As reform proceeds in China that country is likely to become a significant importer of agricultural commodities, and especially of feed grains.

VIII. Concluding Comments

Changes in the configuration of the international economy and recent international political developments have brought the macroeconomics of agriculture to the fore. It is now difficult to understand developments in the agriculture of any significant agricultural country without understanding them in the context of a global agricultural sector. Moreover, understanding agriculture in this larger context also requires that one understand the growing importance of monetary and fiscal policy and exchange rate policy as determinants of international trade flows and of the level of income of farm people. The agenda for understanding agriculture has lengthened and broadened in very significant ways.

Bibliography

Centre for International Economics (1988). "Macroeconomic Consequences of Farm-Support Policies (Overview and Seminar Papers)." Canberra, Australia.

Chambers, R. G. (1988). An overview of exchange rates and macroeconomic effects on agriculture. In "Macroeconomics, Agriculture, and Exchange Rates" (P. L. Paarlberg and Robert G. Chambers, eds.), Westview Press, Boulder.

Hayami, Y., and Ruttan, V. W. (1985). "Agricultural Development: An International Perspective," 2nd ed. Johns Hopkins Press, Baltimore.

Huffman, W. E., and Coltrane, R. (1986). "U.S.–Mexican Trade and Immigration." Final Report of Research Agreement between ISU and USDA. Ames, IA.

Johnson, D. G. (1973). "World Agriculture in Disarray." MacMillan and Trade Policy Research Centre, London.

Paarlberg, P. L., and Chambers, R. G. (1988). "Macroeconomics, Agriculture, and Exchange Rates." Westview Press. Boulder.

Peters, G. H. (1991). Agriculture and the macro-economy. *J. Agricult. Econom.* **42** (3).

Schuh, G. E. (1974). The exchange rate and U. S. agriculture. *Am. J. Agricult. Econom.* **56**(1), 1–13.

Schuh, G. E. (1986). "The United States and the Developing Countries: An Economic Perspective." National Planning Association, Washington, DC.

Schuh, G. E. (1991). Open economies: Implications for global agriculture. *Am. J. Agricult. Econom.* **73**(5), 1322–1329.

Measurement of Micro-oxygen Supply and Demand

W. GENSLER, *Agricultural Electronics Corporation*

Glossary

Adsorption Movement of oxygen to the surface of the metal electrode; this is in contrast to absorption which is a penetration of oxygen into subsurface locations

Electrical capacitance Storage of charged ions or electrons due to the presence of a potential difference between two points; capacitance can also be defined as the change in charge stored per unit change in potential; and the units are coulombs/volt or farads

Fick's Laws Two laws in differential equation form describing the movement of a substance due to an inequality in spatial concentration; a solution to the first law equation gives the flow of the substance under a gradient in concentration, and a solution to the second law equation give the space–time values of substance concentration beginning at time equals zero

Path capacitance Storage of neutral oxygen molecules along a diffusion path; the units of path capacitance are moles/unit volume

Reference electrode Electrode in an electrochemical cell which serves to complete the circuit and at the same time maintains a constant potential across the electrode–electrolyte interface

"S" curve of growth Time pattern of growth of biological tissue characterized by a slow beginning, then an accelerated phase, and then a tapering off to a slow phase as the tissue nears maturity

Volumetric diffusion resistance Proportionality constant relating a difference in concentration between source and sink to the flow resulting from the difference; the concentration is expressed in moles/unit volume, the flow is expressed in moles/unit volume/second, and the units of resistance are seconds.

This article is concerned with a method of measuring micro-oxygen supply and demand of the intact whole plant. A measuring electrode is placed in the tissue and a reference electrode is placed in the root zone. Oxygen in the extracellular fluid surrounding the electrode adsorbs on the surface of the measuring electrode thereby creating an electrical potential. As the concentration of the oxygen in the fluid changes, the amount of oxygen adsorbed changes which in turn causes a change in potential. This change in potential is measured across the electrode for indefinite periods.

I. Basic Potential Measurement

Figure 1 shows in schematic form the measuring electrode in the plant and the reference electrode in the soil. The measuring electrode (palladium) penetrates the tissue approximately 3 mm in a radial direction in soft tissue and in a near tangential direction in woody tissue. Electrode diameter is 150 μm. A collar of wound tissue builds up at the electrode entry point thereby sealing the electrode within the tissue. If the electrode is inserted in tissue at the upper shoulder of the "S" curve of growth, there is no scar tissue formation and the electrode is surrounded by a narrow channel (20 μm wide) filled with extracellular fluid.

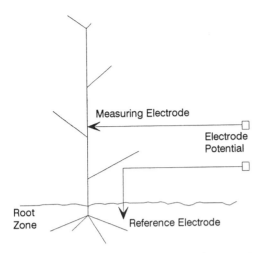

FIGURE 1 External measuring circuit. The measuring electrode is placed in the tissue and a reference electrode is placed in the root zone. The potential at the terminals of these two electrodes is the measured potential. Changes in this potential are attributed to changes at the electrode–extracellular fluid interface.

The other side of the channel is a layer of normally functioning cells.

A reference electrode is placed in the root zone. The location and type of this electrode are not significant. Distances up to 70 m from the plant containing the measuring electrode have yielded normal potential values. For long-term measurements, the major characteristic of this electrode is stability. This is required since changes in the potential measured across the electrode pair are attributed to changes at the measuring electrode–fluid interface assuming all the other potential sources in the path between the reference electrode and the measuring electrode are constant. The validity of this assumption is checked constantly by the use of a second reference electrode in the root zone. The potentials of these two reference electrodes are checked against each other to determine a change in the interfacial potential of either.

In order to compare two noncontiguous sites, an access tube is placed in the root zone as well. The access tube is filled with a salt solution which drains very slowly out of the tube into the soil. A portable electrode is placed in the tube and the potential of the on-site reference electrode is measured against the potential of the electrode in the access tube. By taking the portable reference electrode to a second site and performing the same measurement, the potential of plants at one site can be directly compared to the potentials at the second site. Furthermore, by employing a portable reference electrode with a known relation to the standard hydrogen electrode potential,

an absolute level of potential of the measuring electrode can be achieved valid for any site.

The characteristics of the root zone are not significant. Soils of widely different composition or hydroponic solutions give the same range of potentials.

II. Micro-Anatomical Considerations

The micro-anatomy of the *in situ* electrode in the tissue is shown in schematic form in Fig. 2. Since the cell diameter is smaller than the electrode diameter, there are many cells surrounding the electrode surface. It is the activity of these cells in the vicinity of the electrode surface that determines the oxygen concentration of the extracellular fluid in the channel surrounding the electrode. As these cells increase their oxygen demand, a localized depletion of oxygen occurs. This sets up a gradient of oxygen concentration in the path between the source of oxygen and the cell surfaces. In Fig. 2, the source of oxygen is the aereoplasts. These are gas-filled spaces that maintain a direct gas phase path to the outside. Functionally, they may be considered sources of oxygen whose concentration does not change because of the rapid oxygen transfer through a gas phase.

An increase in oxygen demand by the cell causes a decrease in supply at the cell surface. This sets up a gradient in oxygen concentration in the path between the aereoplast and the cell surface and results in a flow of oxygen along the path. Assuming a constant oxygen concentration in the aereoplast, this flow reaches its maximum value when the oxygen concentration at the surface of the cell goes to zero. An electrode placed in this micro-environment senses these changes in oxygen concentration caused by changes in cellular oxygen demand. In this manner,

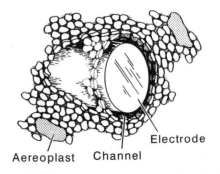

FIGURE 2 Schematic of the electrode in the tissue. The electrode maintains a virtually passive presence in the tissue. As such it can remain in the tissue for indefinite periods. The relative size of the electrode and cells is approximately to scale.

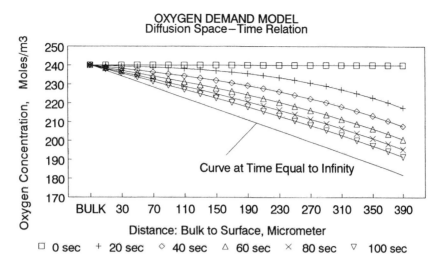

FIGURE 3 Diffusion of oxygen from source to sink. This set of curves shows the oxygen concentration along the path between source and sink following a step function increase in oxygen demand at the sink. At time equal to zero the concentration is uniform in the path. The demand gradually depletes the oxygen along the path until an equilibrium condition is reached whereby the oxygen concentration decreases linearly along the path and the demand is felt completely at the source. Conditions: Source concentration, 240 mmol/m^3; demand, 1.5 mmol/m^3/sec; path length, 400e–6 m; diffusion coefficient: 1e–9 m^2/sec.

the change in oxygen supply can give quantitative insight into the oxygen demand by the cells in the vicinity of the electrode.

Two oxygen concentration calibration points are available at this time: 360 mV corresponds to ambient oxygen concentration and 200 mV corresponds to near-zero concentration. Ambient concentration is the concentration of oxygen in a beaker of water on a table. The near-zero oxygen value can be considered the condition in which there is no dissolved oxygen in the beaker of water.

The amount of oxygen adsorbed on the electrode is in the order of 5e–14 mol. The amount of energy drained from the tissue in the measurement of the potential is in the order of a picojoule. These quantities are insignificant in the energy budgets of the cells involved. The electrode then functions as a "passive" observer of the oxygen activity in the extracellular fluid.

III. Space–Time Analysis of the Path between Oxygen Source and Sink

The method of quantifying the oxygen gradient along the path between the aereoplast and the cell surface is by means of Fick's Laws of Diffusion. Assuming the path length is 300 μm, the diffusion constant is 1e–9 m^2/sec, the initial concentration is 240 mol/m^3, and the oxygen ingestation rate is 2 mol/m^2/sec. Application of Fick's second law yields the space–time values of oxygen along the path shown in Fig. 3.

The curves in Fig. 3 illustrate the basic mechanism of the diffusion process. At time equals zero, the concentration along the path is uniform. When the cells increase their demand, the oxygen is initially drawn from the path volume. As this repository of oxygen becomes depleted, oxygen is increasingly drawn from the aereoplast. At time equals infinity, the path volume has been depleted such that a uniform gradient of oxygen exists between the aereoplast and the cell surface. Oxygen is then drawn entirely from the aereoplast. The linearity of the oxygen gradient along the path would hold regardless of the magnitude of the demand or the path length. Oxygen supply and demand at the cell surface will always move in opposite directions. This opposition is required because the gradient in oxygen concentration is the "cause" of the oxygen movement.

The linearity of the gradient at time equals infinity indicates it is possible to describe the oxygen movement in a form similar to Fick's first law. In this case, the relation between the oxygen concentration gradient and the flow of oxygen into the cell mass is given by

$$J_V = (1/R_{DV}) \times (C_{Source} - C_{Sink}),$$

where J_V is the volumetric flow in mol/m^3/sec, C_{Source} and C_{Sink} are the concentrations of oxygen at the aereoplast and the cell surface, respectively, and R_{DV} is the volumetric diffusion resistance in seconds. In the steady-state, the overall concentration gradient between oxygen source and sink is related to the flow by a constant value. This relation will be used in the model below.

IV. Combined Path–Electrode Interface Model

A quantitative model of oxygen movement along the path and the measurement of oxygen supply by the electrode are shown in Fig. 4. The path between the aereoplast and the extracellular fluid outside the cell can be characterized by a volumetric diffusion resistance and a path capacitance. The oxygen demand on the part of the cells is equivalent to a drain of oxygen from the extracellular fluid. Oxygen flows from the source at the top of the model to the sink at the bottom of the model. The electrode is located at the side, so to speak, as a passive observer, monitoring changes in the oxygen concentration of the fluid. It is functionally equivalent to an electrical capacitor. The potential across this capacitor is a measure of the oxygen concentration in the extracellular fluid.

This model holds for both steady-state and transient conditions. If one is interested in only the steady state wherein there is constant oxygen demand on the part of the cells, the two capacitors will maintain fixed levels.

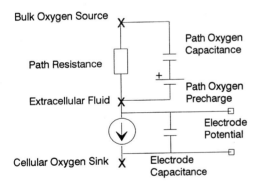

Bulk Oxygen Source

Path Resistance

Extracellular Fluid

Cellular Oxygen Sink

Path Oxygen Capacitance

Path Oxygen Precharge

Electrode Potential

Electrode Capacitance

FIGURE 4 Biophysical and electrochemical model. The physical path between source and sink is represented by a linear resistance to the flow of neutral oxygen molecules across a gradient of concentration. The biophysical capacitance is a storage of neutral oxygen molecules in the volume between source and sink. It is these molecules that are depleted during the transient from one equilibrium demand rate to another. The potential is measured between the extracellular fluid and a base point. The electrical capacitor stores the ionized oxygen adsorbed on the surface of the electrode. The equivalent number of molecules stored in the electrical capacitor is very small compared to the number stored by the biophysical capacitor. The "battery" in series with the biological capacitor is required to precharge the path volume with oxygen.

This model does not take into account oxygen absorbed by cells along the path between the aereoplast and the cells contiguous to the electrode. Oxygen absorption by these cells would yield an increased gradient in the path and can be crudely approximated by a decreased value of the volumetric diffusion resistance. The curves shown in Fig. 3 would have greater slopes if absorption along the path is taken into consideration. However, the basic form of the result would not change.

V. Potential–Oxygen Demand Relation

It is possible to relate oxygen demand to potential changes. The relation between potential change and oxygen supply at a given demand level J_V and constant pH is

(Change in Concentration)/(Change in Potential)
$$= K_1$$

(Change in Concentration)
$$= R_{DV} \times \text{(Change in Demand)}.$$

Therefore,

(Change in Demand)
$$= (K_1/R_{DV} \times \text{(Change in Potential)}.$$

Dissolved oxygen concentration under standard ambient conditions (25°C) is 262 mol/m^3, K_1 then becomes 1 kmol/m^3/V. The value of volumetric diffusion resistance is difficult to determine theoretically or empirically. A approximate method is to approach the value of R_{DV} from the other direction by assuming a demand and introducing empirically observed values of change in potential. If one assumes that the potential change is 0.1 V and the change in demand is 2 mol/m^3/sec, the value of R_{DV} becomes

$$R_{DV} = (1 \text{ kmol/m}^3/\text{V} \times 0.1 \text{ V})/ 2 \text{ mol/m}^3/\text{sec}$$
$$= 50 \text{ seconds}.$$

The relation between potential and oxygen demand is only for changes in these variables. A steady background demand exists over and above these changes. This is not discernible from the measurement of potential.

From the standpoint of a demand assay, the more important relation is between change in demand and change in potential. Using the assumed and empirical values above, this relation is

(Change in Demand)/(Change in Potential)
$$= ((2 \text{ mol/m}^3/\text{sec})/(0.1 \text{ V}) = 20 \text{ mol/m}^3/\text{sec}/\text{V}.$$

These relations are valid only under the assumption that the supply of oxygen is constant.

VI. Typical Patterns

The diurnal pattern of potential change observed in thousands of observations over a wide range of species and conditions is described by a rise in potential in the morning, an afternoon plateau, and then a decrease in potential in the late afternoon and early evening. The night period is characterized by low potentials with the minimum usually in the predawn hours. This is interpreted in terms of a balance between oxygen supply and demand wherein the supply dominates over demand in the morning. The demand dominates over the supply in the later afternoon and evening.

A second very commonly observed pattern is the drop in potential with irrigation or rainfall. This is interpreted as a sudden shift to higher oxygen demand on the part of the plant. It is accompanied by an increase in the stem diameter. The interpretation of this simultaneity is that the plant uses the energy to expand the phloem tissue prior to assimilate translocation. [See PLANT PHYSIOLOGY.]

Bibliography

Armstrong, W. (1979). Aeration in higher plants. In "Advances in Botanical Research" (H. Woolhouse, ed). Vol 7, pp. 226–332. Academic Press, New York.

Bard, A. J., and Faulkner, L. R. (1980). "Electrochemical Methods, Fundamentals and Applications." Wiley, New York.

Bowling, D. J. F. Measurement of the gradient of oxygen partial pressure across the intact root. *Planta* **111,** 323–328.

Crank, J. *et al.* (1981). "Diffusion Processes in Environmental Systems." McMillan, London.

Gensler, W. G., and Diaz-Munoz, F. (1983). Simultaneous stem diameter expansions and apoplastic electropotential variations following irrigation or rainfall in cotton. *Crop Sci.* **23**(5), 920–923.

Gensler, W. G., and Yan, T. L. (1988). Investigation of the causative reactant of the apoplast electropotential of plants. *J. Electrochem. Soc.* **135**(12), 2991–2995.

Hitchman, M. L. (1978). "Measurement of Dissolved Oxygen." Wiley, New York.

Hoare, J. (1968). "The Electrochemistry of Oxygen." Interscience, New York.

Nobel, P. S. (1991). "Physicochemical and Environmental Plant Physiology." Academic Press, San Diego.

times are enormous. The stationary or decline phases are usually brought on by a depletion of nutrients or production of an inhibitory product by the bacteria themselves. In a meat system, the concept of several species of bacteria growing and competing is important.

Several factors affect the growth rate of bacteria on meat. Knowledge of these is essential to controlling growth and, thereby, spoilage of meat. Meat is an excellent source of nutrients so that is rarely a limiting factor. Controlling of temperature is the best method of controlling growth of bacteria. Lowering the temperature slows growth. A good rule to remember is "Life begins at 40;" in other words, if the chiller temperature is above 40°F, growth of bacteria will be rapid. Different bacteria have different optimal temperatures for growth: psychrophiles grow at 20–30°F, mesophiles at 35–40°F, and thermophiles at 55–60°F. Bacteria have differing requirements for oxygen; there are obligate anaerobes, facultative anaerobes, and obligate aerobes. The optimum acidity for growth of most bacteria is near neutrality. The pH of fresh meat is usually in the range of 5.3 to 6.5, so growth can occur easily. Bacteria require moisture for growth, and meat provides a ready source of available water.

All these factors can be manipulated to inhibit growth of given microorganisms. Interrelationship of several factors is an important consideration. The best way to inhibit growth of bacteria on meat is to keep the initial contamination low.

Heat can be used to kill bacteria. Some chemicals, such as chlorine, can be used to kill bacteria. Chemical use is not usually allowed in meat, but is often used to clean and sanitize equipment. Irradiation can be used effectively, but reluctance to use it stems from fear that consumers will not accept food so treated. [See FOOD IRRADIATION.]

Numerous methods can be used to inhibit growth of bacteria. Temperature is a good example and includes heating, chilling, and freezing. Some chemicals normally added to meat during processing are effective. Salt and sugar are good examples, and work by lowering water activity. The nitrite used in curing has a general inhibitory effect and a specific effect against *Clostridium*. Fermentation is inhibitory by lowering pH, and drying is effective by decreasing water activity.

In addition to bacteria, yeasts and molds are also important in meat products. Yeasts are larger than bacteria and reproduce by budding. Molds are filamentous and branched, and produce spores. These psychrophilic obligate aerobes are tolerant to acid and dryness, and can utilize nitrite and nitrate as nitrogen sources. Therefore, they predominate on salted, dried, or fermented meats. Molds can be controlled by heat processing and by vacuum packaging.

Spoilage of meat by bacteria is usually recognized by surface browning, slime, and off-odors or -flavors normally characterized by souring or putrefaction. Yeast growth likewise results in slime, and off-odor or -flavor. Mold growth is characterized by black, white, or green spots, whiskers, and stickiness.

Parasites may be present in muscle. Examples are sarcocystis, tapeworm, and trichinella. Control is through control of infection in the animal and through proper cooking of the product.

Little consideration is normally given to viral contamination of foods. Hepatitis A is one viral disease known to be transmitted in food. Viruses find their way into food via contaminated feces transmitted by water or workers.

The possibility exists for transmission of foodborne pathogens via meat. In recent years, the major interest has centered on *Listeria monocytogenes* and *Escherichia coli* 0157:H7. Both organisms can cause serious illness and are potentially life threatening. The former is usually associated with processed meat and the latter with fresh meat. The best safeguards against a problem developing is prevention of contamination, proper storage and handling, and use of recommended cooking procedures.

The meat processing industry employs quality control programs or HACCP (Hazard Analysis Critical Control Points) procedures to deal with microbiology. Clean-up and sanitation programs are likewise important. A strong trend exists for the industry to employ HACCP as a total program spanning the entire chain from farm production to the consumers table. This is a preventive system rather than an after-the-fact reaction.

IV. Meat Curing and Emulsion Technology

Curing is a procedure that evolved from the process of salting meat to preserve it. By the close of the 19th century, nitrite was known to be the critical ingredient; during the early 20th century, regulations controlling the curing process were established in the United States. Regulations are not discussed here because they vary for different products, are obviously different in different countries, and also change from time to time. Regulations in the United States more or less center on the concept of allowing 0.25 oz sodium nitrite to be added per 100 lb meat.

The ingredients used for curing are salt, sugar, nitrite, ascorbate, seasonings, and possibly other ingredients such as phosphates. Three methods are used to incorporate the curing ingredients into meat. The dry ingredients may be rubbed on the surface of the meat, which is followed by a lengthy time period for penetration and equilibration. This is the so-called dry-curing process and is used now only for specialty items such as country cured hams. The ingredients may be put into solution; then the resulting brine is injected into the piece of meat, or the piece of meat is submerged in the brine so penetration and equilibration can occur. Finally, the curing ingredients can be incorporated into the meat by grinding or chopping. In most cases, the meat is subsequently subjected to a heat processing schedule.

The curing ingredients have functional purposes. Salt is a preservative, affects flavor, and may exert functional properties (especially on the proteins) depending on the processing procedure used. Sugar has a flavor, may influence microbiology, and, in some cases, contributes to browning. Nitrite has a number of effects, the most notable of which is the conversion of the normal red color of meat to the typical pink cured-meat color. The initial result of adding nitrite to meat is an oxidation reaction that produces the brown metmyoglobin color. As time proceeds, this pigment is reduced back to a red color and, with heating, is converted to dinitrosylhemochrome, which is pink. Nitrite also gives a flavor that may be a specific flavor effect or may be indirect through retardation of the off-flavors associated with development of oxidative rancidity. Nitrite-treated meats have a different texture than fresh meats. Finally, nitrite functions in microbiological preservation and is especially important in protecting against outgrowth of spores should they be present. Reducing agents such as ascorbic or erythorbic acid are used to quicken the color development as well as stabilize it. Phosphates may be used to increase water-binding capacity.

Some very popular classes of processed meats, such as wieners, bologna, and loaf products, are known as emulsion products. These are finely chopped, stuffed into casings, smoked, and cooked. The usual procedure to make an emulsion is to begin with the lean meat portion and chop it in the presence of salt. The proteins are thereby extracted and solubilized, and will then act as the emulsifying agents. The remaining ingredients, including fat or fatty meat, are added and chopping is continued. The particle size is further reduced and, as the temperature rises, the fat becomes more plastic. Eventually, the fat is broken down to small globules and the solubilized protein coats the fat particles, forming the so-called emulsion. Heat processing sets or coagulates the protein matrix, trapping the fat particles.

If the emulsion "breaks," the fat is released and leaks out. This failure is most often caused by using meat with insufficient good quality protein to adequately immobilize the fat or by overchopping, which makes the fat particle size too small or raises the temperature too much.

V. Classification of Processed Meats

A categorization and description of the types of processed meat commonly available to the consumer will provide an understanding of the breadth and magnitude of the industry. The several categories of sausage each contain unique types or variations.

Fresh sausage is not cured. It is merely ground meat with seasoning added and may be sold in links or bulk. An example is fresh pork sausage. It is perishable and must be kept refrigerated, and must be cooked thoroughly prior to serving.

Uncooked smoked sausage is invariably cured. It is smoked and must be cooked prior to consumption.

Cooked smoked sausage is cured, smoked, and cooked. It is a very popular item, usually of the emulsion type. Examples are wieners and bologna.

Other cooked sausages are made from fresh meat with or without cure, may be smoked, and are thoroughly cooked and, therefore, ready to eat. Examples are liver sausage and blood sausage.

Dry sausages constitute a large group and, in former times, were often made during the winter months in anticipation of consumption during the summer. Drying is a critical step in the manufacturing procedure and fermentation is very often a part of the process. These sausages have special keeping qualities. Examples are salami and cervelat.

Luncheon meat and jellied products constitute another popular class of processed meats. The former are usually chopped (emulsion type) and possibly contain extenders or other products such as olives. The jellied products contain chunks of meat in a gelatin matrix.

Ham, bacon, and pastrami are examples of a class of product that is made from a muscle or group of muscles. These are invariably cured and heat processed, and often smoked.

A wide variety of canned meat items ranges from the perishable canned ham (which must be kept refrigerated) to a product such as stew, which is heat sterilized and shelf-stable for several years at room temperature.

Ground beef can be considered a value-added processed meat; some companies specialize only in manufacture of hamburger patties.

Finally, a vast and rapidly expanding category is that of precooked meats or prepared meals that contain meat.

VI. Manufacturing Procedures

Although it is impossible to provide detailed procedures or formulations, or to consider all the special or unique products being made, a generalized sausage manufacturing protocol follows that provides an overview of the usual operations.

The first step is a size or particle reduction, which is accomplished by grinding or chopping. Coarse-textured products are ground whereas finer products, especially if an emulsion is to be made, are chopped or subjected to equipment such as an emulsion mill. During or subsequent to particle reduction, the curing ingredients and seasonings are added and distributed. Often an active mixing is employed (sometimes under vacuum), not only to blend all ingredients but also to "work" the meat and thereby permit the protein functionality to result in the desired texture of the final product.

The product is then put into a container, form, or mold that is most often a casing. The casing gives the product a characteristic shape. The product is heat processed to a pasteurizing level. Very often this is accomplished in a smoke house, which permits control of both temperature and humidity. Also, smoke is generated that is a flavor component of many products as well as a contributor to surface color.

A unique type of processed meat is made by incorporating a fermentation step. Appropriate bacteria are encouraged to ferment added sugar with a resulting production of lactic acid. The lactic acid gives a characteristic tangy flavor. Because the pH is lowered, the sausage keeps better; the lactic acid also coagulates the meat proteins, giving a firmer texture. The traditional process was rather lengthy and employed conditions to permit the natural flora of the meat to do the fermentation. Now, it is most usual to add a

specific starter culture, a procedure that is much more rapid and standardized.

Finally, the product is chilled, packaged, and readied for shipment.

VII. Nonmeat Ingredients

Several components other than meat are often used to make processed meats. Care in selecting quality and purity of all ingredients results in a good final product. Obviously, the ingredients are controlled by regulations. Space constraints do not permit discussion of what and how much is permitted in what types of products.

The curing ingredients have already been enumerated and discussed.

Water is an ingredient and is often added in the form of ice. Water helps in distribution of ingredients and the extraction of proteins, and contributes to juiciness in the finished product.

Spices, seasonings, and flavorings are numerous. Flavor enhancers, such as monosodium glutamate, are sometimes used. Antioxidants or mold inhibitors are permitted in some applications. Starter cultures are considered a nonmeat ingredient. Numerous extenders such as those made from milk or various plant sources may be added. Smoke may be applied in the natural vapor form, as described earlier, or as a smoke flavoring.

VIII. Packaging

Packaging protects the product against chemical and physical change and against microbial recontamination, and may be used as a distribution and marketing device. An enormous variety of packaging materials ranges from metal and glass containers, through paper products, to highly complex laminates of plastic films. [See FOOD PACKAGING.]

Quality of processed meats will not be improved by packaging, but only maintained. Optical properties, printability, machinability, and durability are key considerations. The requirements for fresh and cured meat are quite different. For fresh meat, high oxygen permeability and low moisture transmission are ideal to maintain oxymyoglobin and minimize desiccation. Cured meats keep best under vacuum, so the package should be of barrier type to both oxygen and moisture.

Acknowledgments

Muscle Biology Laboratory No. 307.

Bibliography

Anonymous (1983). "Radiation Preservation of Foods." IFT Scientific Status Summary, Chicago.

Buege, D. R., and Cassens, R. G. (1980). "Manufacturing Summer Sausage," Bulletin A3058. University of Wisconsin, Madison.

Cassens, R. G. (1994). "Meat Preservation: Preventing Losses and Assuring Safety." Food and Nutrition Press, Trumbull, Connecticut.

Cliver, D. O. (1988). "Virus Transmission via Foods." IFT Scientific Status Summary, Chicago.

Hotchkiss, J. H., and Cassens, R. G. (1987). "Nitrate, Nitrite, and Nitroso Compounds in Foods." IFT Scientific Status Summary, Chicago.

Kramlich, W. E., Pearson, A. M., and Tauber, F. W. (1973). "Processed Meats." AVI, Westport, Connecticut.

Oblinger, J. L. (1988). "Bacteria Associated with Food-borne Diseases." IFT Scientific Status Summary, Chicago.

Price, J. F., and Schweigert, B. S. (eds.) (1987). "The Science of Meat and Meat Products," 3d Ed. Food and Nutrition Press, Trumbull, Connecticut.

Rust, R., and Olson, D. (1987). "Processing Workshop: 1981–1986. A Collection of Titles from *Meat and Poultry* Magazine." Oman Publishing, Mill Valley, California.

Melons: Biochemical and Physiological Control of Sugar Accumulation

D. MASON PHARR, *North Carolina State University, Raleigh*

NATALIE L. HUBBARD, *E. I. DuPont*

Glossary

Apoplast Space outside the outer limiting membrane of a plant cell; the space is basically constituted by aqueous cell walls

C-3 plant Plants which reduce CO_2 during photosynthesis largely by the reductive pentose phosphate pathway (i.e., the Calvin cycle)

Chloroplast Subcellular organelles which contain the photosynthetic apparatus and chlorophyll

Exocarp Outer tissue or skin of a fruit

Gene promoter Portion of the DNA upstream of a structural gene involved in binding of RNA polymerase to initiate transcription of the gene

Intercalary meristem Region of actively dividing cells between two more or less differentiated tissues

Intermediary cell Specialized cell exhibiting many plasmodesmata and located adjacent to the sieve elements of the minor leaf phloem

Mesocarp Tissue which is the fleshy portion of the muskmelon fruit

Mesophyll Tissue within the leaf which is enriched in chloroplast and which is the major site of photosynthesis in higher plants

Pi translocator Translocator protein located on the inner membrane of the chloroplast envelope and which catalyses the transfer of triose-phosphates and inorganic phosphate between the cytoplasm and the interior of the chloroplast

Phloem A conducting tissue within vascular bundles in which carbohydrates move over long distances in plants

Plasmodesmata Cytoplasmic connections between adjacent plant cells; the plasmodesmata penetrate the walls of adjacent cells and may constitute a major route of metabolite transport between cells

Rubisco Ribulose-1,5-bisphosphate carboxylase/oxygenase, the enzyme of the Calvin cycle which catalyses the addition of CO_2 to ribulose-1,5-bisphosphate; in higher plants the enzyme comprises two different classes of subunits—small and large

Sieve element One of the cells of a longitudinal series of cells comprising sieve tubes within the phloem which are involved in long distance transport

Sink tissue Plant tissue typically exhibiting net import of photosynthates for growth or storage

Source tissue Plant tissue typically exhibiting net export of carbohydrates

Stomata Small pores in the surface of leaves through which gas exchange takes place; stomatal pores are formed by guard cells which regulate the pore aperture

Structural gene Segment of the DNA encoding information for a particular polypeptide which is frequently but not always an enzyme

Symplast Space within a plant cell bound by the outer limiting membrane of the cell

Vacuole Large, membrane-bound, central compartment of plant cells; among other functions, vacuoles may frequently serve as storage sites for soluble sugars

This article focuses attention on the recent advances in understanding carbohydrate metabolism in *Cucumis melo L.* and closely related species. Carbohydrate metabolism is a controlling feature of sugar accumulation (sweetening) in fruits of this species. A discussion is included of the photosynthetic formation of sugars in leaves of these plants, translocation of the sugars, and the biochemical events associated with metabolism and accumulation of soluble sugars in the fleshy fruit mesocarp. Potential strategies for the use of recombinant DNA technology to improve fruit quality are discussed. The effects of selected environmental factors as they influence these specific processes are also addressed.

I. Botany of the Species

It is said by some authorities that *Cucumis melo* may have originated in tropical and subtropical West Africa. The various genotypes of *C. melo* produce a highly varied group of fruits. The variations in fruit size, shape, external skin characteristics, flesh color, and aroma would not immediately suggest that they are all members of the same species. The species is divided into several botanical varieties listed below.

Cucumis melo L.
  var. *cantaloupesis*—cantaloupe (Europe)
  var. *inodorus*—winter or casaba melon
  var. *flexucus*—snake melon, serpent melon
  var. *reticulatus*—netted or nutmeg muskmelon,
    cantaloupe (United States), Persian melon
  var. *conomon*—Oriental pickling melon
  var. *chito*—mango melon, garden melon
  var. *dudaim*—pomegranate melon, Queen Anne's
    pocket melon

Not all botanical varieties produce fruit which have high sugar content. Fruits of var. *reticulatus* accumulate high concentrations of sugar during maturation, whereas fruits of var. *flexucus* do not. Fruit resulting from genetic crosses between botanical varieties have been used to study the fundamental biochemical differences between sugar-accumulating and nonaccumulating melons. Variations among cultivars within botanical varieties have been used for the same purpose.

II. Sugar Concentration, Fruit Quality, and Consumer Acceptance

A. Fruit Sugars and Quality

Sweetness of melons, *C. melo*, is a very important aspect of the quality of these fruits in the United States market. Sweetness in melons is directly determined by sugar concentration. The fruits of citrus, apples, grapes, and other plants contain a high concentration of organic acids such as citric acid or malic acid. These acids strongly diminish the perception of sweetness of fruit sugars. In such fruits, factors controlling sweetness are somewhat complex. For instance, a decline in organic acid concentration without an increase in sugar concentration can increase the perception of sweetness. Other fruits, such as peaches, can contain high concentrations of astringent phenolic compounds which strongly mask the sweet taste of the fruit sugars. Fruits of *C. melo* do not contain a substantial concentration of organic acids or astringent phenols. Thus, sweetness of melon fruits is influenced primarily by the concentration of sugar accumulated in the flesh during maturation. In this sense, the biochemical factors which control sweetness are simpler in fruits of *C. melo* than in fruits of many other species.

Remarkably, a single fruit of *C. melo* may sell for a price of over $60 (U.S.) in Japan. These fruits are given as gifts on occasions of joy or grief in Japanese society in the same manner as flowers are used in western culture. Fruits used for this purpose are produced under very controlled conditions in greenhouses. They must be free of any blemishes and of very high quality. For the mass market in the United States, the fruits are typically produced under field conditions, with about 114,000 acres harvested annually. The current leading centers of production for out-of-state export of the fruit in the United States are California, Texas, and Arizona. They are often sold under the name cantaloupe, muskmelon, casaba melon, or honeydew melon. Sliced green-fleshed and orange-fleshed melons are increasingly evident as an item in the salad-bar trade. Aside from external appearance, sweetness of the fleshy mesocarp is the primary factor that influences consumer acceptability, and thus quality, of the ripe fruit.

United States Standards exist for the marketing of cantaloupes. Grade is based in part upon soluble solids concentration, which can be determined with a hand-held refractometer using juice expressed from the fruit. The refractive index reading is highly correlated with sugar concentration in the fruit flesh. This is not true for juice expressed from fruits of all species, but a high correlation between soluble solids and sugar concentration has been confirmed experimentally many times for cantaloupes. In order to meet the requirements for U.S. Grade Fancy, the fruit must contain at least 11% soluble solids. To meet U.S. Grade No. 1, melons may contain as little as 9% soluble solids.

B. Origin of Variation in Fruit Quality

Unfortunately, quality of fruit in the market is highly variable and therefore often disappointing. This is due most frequently to a lack of adequate sugar concentration which is most often caused by harvest of immature fruit. In certain cultivars, an abscission layer forms where the fruit is attached to the plant, and this layer develops until the fruit slips (detaches) from the plant. At this time the fruit is at optimum maturity for consumption. However, not all of the commercial variants in the species develop abscission layers, so that this criterion is not universally applicable as a maturity guide. Frequently, fruits are intentionally harvested prior to the development of "full slip" and optimum eating quality for the purpose of shipment to distant markets. This is necessary to insure adequate storage life during marketing. As a result, consumers may acquire fruits of inadequate sweetness from the market.

There is a popular misconception among many consumers that fruits of all species, including melons, will sweeten if held at room temperature and allowed to ripen after purchase. Postharvest sweetening does not occur during ripening of fruits of all species. It does occur in fruits such as apples, pears, bananas, and others which contain a significant storage reserve such as starch which is converted to sugar during postharvest ripening. Melons do not contain such a reserve. They will ripen in terms of increased aroma and tissue softening during postharvest holding, but the sugar content of this particular species is set at the time of harvest and can only decrease in the postharvest environment. Thus, from a perspective of marketing high-quality fruit, it is essential to adopt cultivars, production systems, and harvest criteria that ensure adequate sugar content in the fruit at the time of harvest.

C. Dietary Considerations

In melons, rapid stachyose biosynthesis (Fig. 1) and export from leaves to the fruit is needed for fruit sweetening; however, stachyose does not accumulate in the fruit. Sweetness is due to massive accumulation of sucrose, glucose, and fructose in the fruit mesocarp. The fact that stachyose is translocated to the fruit, yet sucrose and hexoses accumulate in the fruit, clearly implies that the predominant translocated sugar, stachyose, is extensively metabolized to sucrose upon or prior to storage in the fruit. From the human perspective it is advantageous that high concentrations of raffinose and stachyose do not accumulate in these fruits. The raffinose sugars are much less sweet than sucrose, and therefore contribute little to the culinary attributes of foods. Also, raffinose sugars are poorly digested due to the absence of an enzyme, α-galactosidase, in the human intestinal mucosa. These large sugars are poorly absorbed through the intestinal wall and subject to gaseous fermentation by intestinal bacteria often causing strong discomfort after their ingestion. At high concentration they can cause an osmotic movement of water into the intestinal lumen and have a cathartic effect causing rapid movement of the digesta through the digestive tract. This rapid passage diminishes the nutritional value of the ingested food. With the exception of beans and peas which are eaten as mature seeds, and a few edible tubers, raffinose sugars do not occur in high concentrations in most foods including melons and other fruits consumed by humans. Both fruits and beans are excellent sources of dietary fiber. Fortunately, such digestive problems are not commonly encountered upon ingestion of the fruit sugars: glucose, fructose, and sucrose.

III. Photosynthetic Formation of Sugars

A. General Photosynthetic Characteristics of Cucumis

Sugar accumulation in melon fruits is dependent upon concurrent leaf photosynthesis and translocation of photoassimilates to the fruit. Sugars produced by leaf photosynthesis must be translocated to the fruit throughout the period of fruit growth as well as during the period of fruit sweetening. Photosynthetic rates within Cucumis are typical of C-3 plants. In high light intensity, net photosynthetic rates may be as high as 30 mg of CO_2 per decimeter of leaf per hour and on cloudy overcast days with low light intensity, around 10 mg of CO_2 per decimeter of leaf per hour. Factors which have a severely adverse impact on leaf photosynthesis and photoassimilate translocation during the final stages of fruit maturation can result in fruit with low sugar concentration. [See PHOTOSYNTHESIS.]

Sucrose is commonly recognized to be a major translocated form of photosynthate in plants. However, in Cucumis and numerous other dicots, stachyose is the major photosynthetic product and the major sugar translocated in the phloem along with smaller quantities of raffinose and sucrose. Stachyose is the tetrasaccharide member of a homologous family of

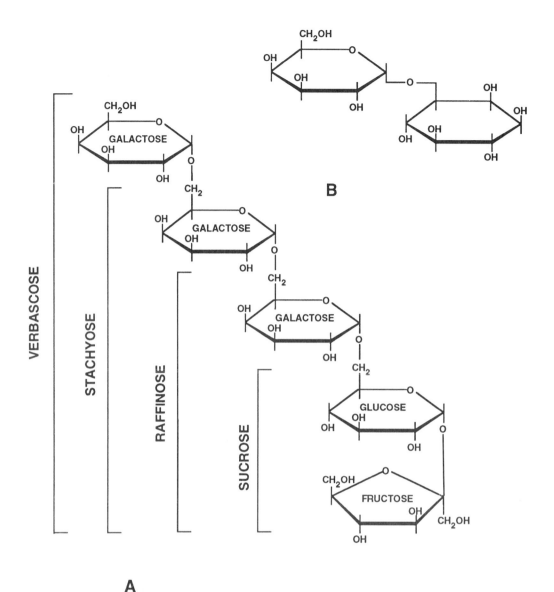

FIGURE 1 Structural representation of sugars of the raffinose family of oligosaccharides. (A) The raffinose sugars. Stachyose and raffinose are the major sugars translocated in the phloem of *Cucumis melo*. (B) Galactinol is an important intermediate in stachyose biosynthesis.

oligosaccharides formed as α-galactosides of sucrose (Fig. 1A). Raffinose, the trisaccharide, and verbascose, the pentasaccharide, are often present in the foliage of stachyose translocating plants including *C. melo*. Galactinol (Fig. 1B) is also present in these leaves as well as in leaves of other stachyose forming plants. This compound serves as the galactosyl donor in the biosynthesis of stachyose.

B. Enzymes of Stachyose Biosynthesis

In foliage in which raffinose saccharides are products of photosynthesis, these sugars label at a rate similar to sucrose when leaves are exposed to $^{14}CO_2$ in the light. In such plants the galactosyl sugars are present as major components in phloem exudate. Further, these sugars have been demonstrated to move along the translocation stream to sink (nonphotosynthetic) tissues where they are metabolized. Stachyose biosynthesis proceeds from sucrose by the following reactions.

1. UDP-galactose + *myo*-inositol ⟶ galactinol + UDP
2. galactinol + sucrose ⟶ raffinose + *myo*-inositol
3. galactinol + raffinose ⟶ stachyose + *myo*-inositol

Recent research clearly indicates the presence of the

enzymes galactinol synthase (1), raffinose synthase (2), and stachyose synthase (3) in cell free extracts from leaves of *C. melo*.

C. Evidence for Spatial Separation of Sucrose and Stachyose Biosynthetic Enzymes

The enzymes of sucrose and stachyose biosynthesis are spatially separated from each other in different cell types within leaves. Studies demonstrating this have also revealed that the fundamental mechanisms by which sugars are loaded into the phloem may be quite different in sucrose and stachyose translocating plants. A simplified scheme showing the probable biosynthesis of sucrose and stachyose in leaves of *Cucumis* from photosynthetic intermediates and the physical path of photoassimilate loading is shown in Fig. 2. Briefly, sucrose is formed from triose-phosphates resulting from CO_2 fixation within the chloroplasts of leaf mesophyll cells. The triose-phosphates are exported into the cytoplasm via the *P*i translocator on the inner membrane of chloroplasts in exchange for *P*i. The triose-phosphates then serve as substrates for the formation of hexose phosphates and UDP-glucose. The latter two compounds are used for the synthesis of sucrose-6-P catalyzed by the enzyme, sucrose phosphate synthase. Sucrose is formed by hydrolysis of the phosphate bond by the enzyme, sucrose-6-P phosphatase. Sucrose then moves through plasmodesmatal pores to the interme-

diary cells where it may be broken down by the enzyme, sucrose synthase, and utilized for stachyose biosynthesis.

This scheme, while tentative, has growing experimental support. It is based upon several different lines of evidence from studies of stachyose translocating plants. In early research on photosynthetic stachyose biosynthesis, it was often assumed that stachyose was made predominantly in the cytoplasm of leaf mesophyll cells. This was known to be true for photosynthetic sucrose formation in other plant species. This early concept about stachyose biosynthesis became less tenable when it was discovered that the incorporation of $^{14}CO_2$ by photosynthetic mesophyll cells or protoplasts, isolated from *Cucumis* leaves, synthesized very little raffinose or stachyose, despite the fact that sucrose became very heavily radiolabeled. This result contrasted very sharply with similar experiments with intact leaves of *Cucumis* in which raffinose and stachyose labeled as major end products of leaf photosynthesis. The dilemma was partially clarified upon the discovery that preparations of minor veins from the leaves of these species, which were contaminated with photosynthetic cells, were able to synthesize raffinose and stachyose very rapidly. This result suggested cooperation between leaf photosynthetic cells and nonphotosynthetic cells of the veins in the overall biosynthesis of stachyose.

Recent work demonstrated that both galactinol synthase and stachyose synthase were located predominantly within intermediary cells of some stachyose translocating plants. Intermediary cells are specialized cells within the phloem tissue of the minor veins of the leaves. The intermediary cells do not contain chloroplasts and are therefore not photosynthetic. Their ability to form raffinose and stachyose is necessarily dependent upon a continuous supply of carbohydrate from the photosynthetic mesophyll cells. All of these observations are accommodated by the model in Fig. 2. Immunocytochemical techniques were used to locate galactinol synthase and stachyose synthase. Specific antibodies were prepared against both purified enzymes. Cytological sections of leaves were challenged with the antibodies followed by protein A–gold staining of the antigen–antibody complex. Specimens were then viewed with the electron microscope to localize the enzyme antibody complexes. From this analysis the major location of both enzymes was found to be in the intermediary cells. These intermediary cells are specialized companion cells, and they are located immediately adjacent to the sieve elements of the phloem. The function of the interme-

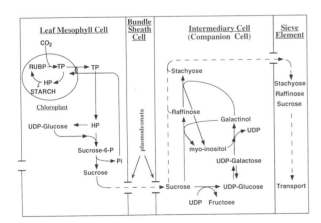

FIGURE 2 Diagram showing the roles of various specialized leaf cells in the overall photosynthetic pathway of raffinose saccharide biosynthesis. Galactinol, raffinose, and stachyose biosynthesis may occur predominantly in intermediary cells, whereas sucrose biosynthesis occurs in the mesophyll cells. Dotted lines indicate movement of metabolites. RUBP, ribulose-1,5-bisphosphate; TP, triose phosphates; HP, hexose phosphates; Pi, inorganic phosphate; UDP, uridine diphosphate.

diary cells seems to include stachyose synthesis and loading into the phloem.

D. Phloem Loading in *Cucumis*

It is the sieve elements of the phloem into which translocated sugars are "loaded" for long-distance transport throughout the plant. The intermediary cells contain extensive plasmodesmatal connections (cytoplasmic connections) to the sieve elements and to the bundle sheath cells which are in turn connected to the mesophyll cells by plasmodesmata (Fig. 2). These connections between cells are thought to function as open pores through which sugars diffuse. Thus, sucrose, formed from photosynthetic intermediates in leaf mesophyll cells, diffuses through plasmodesmata to the intermediary cells where it is rapidly consumed for stachyose synthesis. Bundle sheath cells and intermediary cells have a higher solute concentration than surrounding leaf mesophyll cells, possibly due to the local synthesis of stachyose. This locally high concentration of stachyose may be the driving force for stachyose "loading" into the sieve elements. It has been suggested that this high concentration may be maintained because the bundle sheath and intermediary cell plasmodesmata function as size discriminators which allow the free passage of sucrose but restrict the passage of the larger molecule, stachyose. Accordingly, it is speculated that stachyose formed in the intermediary cells may have only one major route of escape from these cells—through the plasmodesmata connecting the intermediary cells to the sieve elements. These pores are postulated to be nonrestrictive to the passage of stachyose.

It should be noted that some raffinose and stachyose can occur in mesophyll cells of some species, and these sugars can be synthesized to varying extents within mesophyll cells (not depicted in Fig. 2) among different species of plants for temporary storage in those leaf cells. Hence, separate pools of stachyose for storage and for transport can exist in leaves.

This overall mechanism involving movement of photosynthates through the cytoplasm of one cell to the cytoplasm of another via plasmodesmatal pores is known as symplastic movement. This is very different from that of plants such as tomato, tobacco, sugar beet, and others which almost exclusively translocate sucrose. In these plants, sucrose formed photosynthetically in the mesophyll cells leaves the cytoplasm and enters the cell wall environment. This is known as apoplastic movement of sugars. From the apoplast, sucrose is loaded into the sieve elements of the phloem by a specific translocator protein at the companion cell–sieve element interface. A growing body of evidence suggests that these differences in the manner by which sugars enter the sieve elements of the leaf phloem may constitute a fundamental difference between sucrose and stachyose translocating plants.

E. Enzymatic Regulation of Stachyose Biosynthesis

The first enzyme unique to the pathway is galactinol synthase. Galactinol (Fig. 1B) is thought to have no role in carbohydrate metabolism other than as the galactose donor in the raffinose saccharide biosynthetic pathway. Stachyose biosynthesis does not result in a net consumption of *myo*-inositol as this sugar alcohol is regenerated in reactions (2) and (3) above. The enzyme which forms galactinol was first partially purified and kinetically characterized from higher plant leaf tissue in 1982. Subsequently, a survey of plant leaves of 20 species revealed that the activity level of the enzyme, which varied widely among species, controlled the partitioning of leaf carbohydrates between sucrose and leaf raffinose saccharides (raffinose + stachyose). This may be true because carbon biochemically shuttled to the raffinose sugars results in the consumption of sucrose as shown in Fig. 2. Thus, galactinol synthase which provides galactinol for the synthesis of both raffinose and stachyose, controls leaf carbon partitioning between two alternate translocated carbohydrates, sucrose and the raffinose sugars. This phenomenon may be related to differences among species in the ratio of sucrose and raffinose sugars translocated.

Modern developments in recombinant DNA technology and recent advances in understanding carbohydrate metabolism in *Cucumis* and related species may lead to new and exciting research approaches to assure consistent high sugar content in muskmelon fruits. For instance, tomato plants recently have been genetically engineered for overexpression of the enzyme sucrose phosphate synthase in their leaves. This transformation involved the use of the gene promoter of the rubisco small subunit in combination with the sucrose phosphate synthase structural gene from corn. The resulting tomato phenotype has been shown to exhibit higher than normal activity of the enzyme in its leaves and elevated leaf sucrose concentrations. By analogy, it might ultimately be possible to produce, through recombinant DNA technology, a stachyose synthase and/or galactinol synthase overproducer. Such a plant might form leaf stachyose more rapidly and deliver carbohydrates to fruits more rapidly than current cultivars of *C. melo*. Speculatively, such a genotype might be able to produce sweeter fruits or more sweet fruits simultaneously. Experimentally, this may occur in the near future, as both stachyose synthase and galactinol synthase have been purified to homogeneity. A potentially difficult aspect of this strategy is the apparent need for cell-specific overexpression of the enzymes within the leaf intermediary cells in accordance with the scheme shown in Fig. 2.

As discussed below, there are aspects of the enzymology of carbohydrate metabolism involved in the conversion of stachyose to sucrose within the fruit itself which are closely related to the ability of fruits of *C. melo* to accumulate sugars during maturation. Certain of these steps may be subject to favorable modification by genetic engineering. [*See* PLANT BIOCHEMISTRY; PLANT GENETIC ENHANCEMENT; PLANT PHYSIOLOGY.]

IV. Uncertain Role of the Fruit Pedicel

The events associated with the unloading of stachyose from the phloem and the catabolism of this sugar at the fruit are not understood in the same intricate detail as stachyose synthesis and phloem loading in leaves. Of course, such knowledge is actually critical to a more detailed understanding of the biochemical control of sugar accumulation in these fruits. The fact that stachyose is the predominant sugar of transport in the phloem in *Cucumis*, yet is difficult to detect in fruits due to its low concentration, has raised the question as to whether stachyose actually enters the fruit. Some experimental evidence suggested that stachyose may be metabolized rather extensively just prior to entry into the fruit in the fruit pedicel. The fruit pedicel is the small stem which attaches the fruit to the main stem of the plant and has vascular attachments to the main stem and to the fruit. The notion that stachyose is catabolized in this tissue rests in part upon the observation that the fruit pedicel contains an enzymatic pathway capable of degrading stachyose and converting it to sucrose. This pathway does not exist in all tissues of *Cucumis* plants and may be associated primarily with those tissues which actively consume stachyose. These observations led to the hypothesis that stachyose might be converted to sucrose within the pedicel, with subsequent transport of sucrose to the fruit mesocarp tissue.

Certain details which cast doubt on the validity of this idea are presented in Fig. 3 and the following

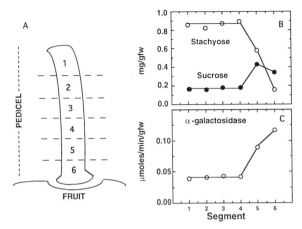

FIGURE 3 Data resulting from the analyses of stachyose and sucrose concentrations and α-galactosidase activity along the length of the pedicel of fruits of *Cucumis sativus*. (A) Schematic illustrating the origin of the fruit pedicel segments which were analyzed. (B) Concentration of stachyose and sucrose in different fruit pedicel segments. (C) α-Galactosidase activity in different fruit pedicel segments. α-Galactosidase was assayed at pH 7.2. gfw, gram fresh weight. [Previously unpublished data of D. M. Pharr, H. N. Sox, and C. E. Anderson, North Carolina State University, Raleigh, NC.]

discussion. When a cucumber (*C. sativus*) fruit pedicel is sectioned into 2-mm cross-sections as depicted in the diagram of Fig. 3A and analyzed, a sharp decline in stachyose and increase in sucrose is observed at the fruit end of the pedicel (Fig. 3B). Alkaline α-galactosidase, an enzyme often associated with stachyose catabolism, is also higher in these pedicel sections than in pedicel sections more distant from the fruit. This fruit end of the pedicel has been examined in thin-section under the light microscope and found to contain an intercalary meristem (a small region of actively dividing and growing cells).

Stachyose consumption within the pedicel is closely related to growth of these meristematic cells at the fruit-end of the pedicel. Stachyose consumption in these cells probably represents catabolism for growth rather than for conversion to sucrose for transport into the fruit as was previously thought. This represents only one of many sites of stachyose utilization for growth throughout the plant. More recently, it has been determined that exudate from the large vascular bundles of the fruit itself does contain stachyose as the major carbohydrate. This indicates that stachyose is transported within the fruit. Furthermore, the fruit mesocarp contains α-galactosidase and other enzymes of stachyose catabolism. Collectively, current evidence favors the idea that stachyose does enter fruits of *Cucumis* in the phloem where it is unloaded and rapidly metabolized to other sugars keeping its concentration low within the fruit mesocarp tissue.

V. Sucrose Synthesis within the Fruit

A. The Enzymes of Sucrose Metabolism

During the final stages of maturation of melon fruits, the stachyose which enters the fruit does not contribute directly to sweetness, but nevertheless is very important, because it is converted by a series of enzymatic reactions to sucrose. The probable scheme for the degradation of raffinose saccharides and subsequent synthesis of sucrose is illustrated in Fig. 4. As is the case for the concentration of any metabolite in cells, sucrose concentration in melon fruit tissue may be regulated in part by the enzymatic capacity of the fruit cells to synthesize and to degrade the molecule. It should be noted here, however, that intracellular compartmentation of enzymes and substrates may functionally and substantially alter the apparent enzymatic capacity for either synthesis or degradation of

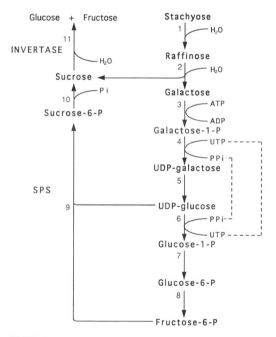

FIGURE 4 Metabolic pathway depicting the conversion of stachyose to sucrose within *Cucumis melo* fruit mesocarp tissue during fruit sweetening. The names of enzymes are : (1) and (2) α-galactosidase, (3) galactokinase, (4) UDP-galactose pyrophosphorylase, (5) UDP-glucose-4-epimerase, (6) UDP-glucose pyrophosphorylase, (7) phosphoglucomutase, (8) phosphoglucoisomerase, (9) sucrose-6-phosphate synthase, (10) sucrose-6-phosphate phosphatase, (11) invertase. [Adapted with permission from Hubbard, N. L., Huber, S. C., and Pharr, D. M. (1989). *Plant Physiol.* **91**, 1527–1534. Waverly, Inc., Baltimore, MD.]

a molecule. As illustrated later, those enzymes immediately associated with sucrose synthesis and sucrose degradation seem to be most important in regulating the concentration of sucrose in melon tissue. These reactions are listed below.

4. UDP-glucose + fructose-6-P → sucrose-6-P + UDP
5. sucrose + UDP → UDP-glucose + fructose
6. sucrose + H_2O → glucose + fructose

The enzymes which catalyze these reactions of sucrose metabolism are (4) sucrose-6-P synthase (SPS), (5) sucrose synthase, and (6) invertase. SPS is the major enzyme in plants which catalyzes the synthesis of sucrose. Reaction (5) is readily reversible but the enzyme is thought to be involved in sucrose degradation in living cells. Reaction (6) is catalyzed by invertase. Two different invertases are known in plant cells which differ in pH optimum. These are known as neutral and acid invertase.

B. Enzymatic Regulation of Mesocarp Sucrose Concentration

All of these enzymes may play an important role during sucrose accumulation in melon fruit. Table I illustrates the pattern of change in soluble sugar concentration in the mesocarp tissue of young, ripening, and ripe fruit of an orange fleshed fruit of *C. melo* (cv Burpee Hybrid). The mesocarp tissue was divided into 1- to 2-cm segments. Segment 1 corresponds to the outermost segment of the mesocarp which is closest to the fruit exocarp. Segment 3 in the case of the unripe and ripening fruits, and segment 4 in the case of the ripe fruit, correspond to the innermost mesocarp tissues (that closest to the seed cavity). Not only is there a distinct increase in sucrose concentration during ripening, but sucrose concentration (and thus sweetness) is highest toward the inner portion of the fruit mesocarp. The outermost portion of the mesocarp, even in fully ripe fruits, does not accumulate sufficient sucrose to result in a sweet taste.

The biochemical mechanisms responsible for the accumulation of sucrose in fruit have been, until recently, poorly understood. This was in large part due to the fact that the enzyme, SPS, which plays a key regulatory role in the biosynthesis of sucrose in plants, was not detected in fruits which accumulate sucrose. This inability to detect SPS in fruits was overcome when more was learned about maintaining activity of this labile enzyme during extraction and activity assays.

SPS activity in *C. melo* fruit was first reported in the late 1980s. As shown in Table I, SPS activity is detectable early in fruit development and increases as

TABLE I

Soluble Sugar Concentration and Sucrose Metabolizing Enzyme Activities in Mesocarp Segments of Young, Ripening, and Ripe Fruit of *Cucumis melo* (cv Burpee Hybrid)

Maturity	Segment	Carbohydrate concentration [mg(g fresh wt)$^{-1}$]		Enzyme activity [μmol hr^{-1} (g fresh wt)$^{-1}$]			
		Sucrose	Hexose	SPS	Acid Inv[a]	Neut Inv[b]	SS[c]
Young, unripe	1 (outer)	0.7	34.5	7.4	40.1	6.3	2.4
	2	1.2	38.0	8.8	27.7	4.2	1.5
	3 (inner)	0.8	41.1	4.4	20.5	4.2	8.1
Ripening	1	2.1	41.2	9.5	20.0	4.1	1.9
	2	13.4	39.5	13.5	6.2	1.9	1.2
	3	22.0	43.5	20.5	6.6	3.2	2.6
Mature, ripened	1	6.6	41.7	24.1	10.8	3.5	2.5
	2	22.1	43.4	23.4	1.0	1.7	3.9
	3	44.3	39.8	22.7	0.1	1.3	0.8
	4	46.7	39.9	28.3	0.0	1.7	5.6

Source: Reproduced with permission from Hubbard, N. L., Huber, S. C., and Pharr, D. M. (1989). *Plant Physiol.* **91**, 1527–1534. Waverly, Inc., Baltimore, MD.

[a] Acid invertase.
[b] Neutral invertase.
[c] Sucrose synthase.

the fruit ripens and accumulates sucrose. There is also a gradient of SPS activity in ripening fruit that correlates with the increase in sucrose from the outer toward the inner portion of the fruit. Early in fruit development, SPS activity is present (indicating the capacity for sucrose biosynthesis) yet there is little or no accumulation of sucrose. This is presumably due to the concomitant biosynthesis and degradation of sucrose. Acid invertase activity is very high early in fruit development and declines with ripening (Table I). Acid invertase is also high in the outer segments of the mesocarp tissue where sucrose concentration is low. Neutral invertase and sucrose synthase activities are relatively low and change little with fruit development or position within the mesocarp.

Studies have also been conducted of sugar accumulation of *C. melo* genotypes which differ markedly in fruit sucrose concentration at maturity. Fruits of these different genotypes exhibit a very similar pattern of decrease in acid invertase with development. However, the extent to which fruit SPS activity increases is much lower in nonsweet fruit than in sweet fruit. This suggests that, whereas changes in activities of both acid invertase and SPS are necessary for sucrose accumulation in the fruit mesocarp, genotypic differences in sucrose concentration may be regulated by genetically determined differences in the activity of fruit SPS.

The capacity for sucrose accumulation in melon fruit tissue can be depicted as the mathematical differ- ence between sucrose biosynthetic capacity (SPS activity) and the sum of the sucrose degrading capacity (acid invertase, neutral invertase, and sucrose synthase activity). The resulting value is negative if the sum of the sucrose degrading enzyme activities exceeds the activity of SPS, and positive if SPS activity exceeds the sum of the sucrose degrading enzyme activities. A negative enzymatic balance would imply that sucrose would not accumulate because of net sucrose degrada- tion, whereas a positive enzymatic balance would im- ply net sucrose synthesis and would lead to sucrose accumulation. In either case sucrose turnover would occur continuously. This relationship between su- crose metabolizing enzymes and sucrose concentra- tion within *C. melo* fruit is illustrated in Fig. 5. These data are from studies of four different genotypes sam- pled at various stages of development and from differ- ent positions within the mesocarp. In any tissue sam- pled, sucrose accumulated only if the capacity for sucrose biosynthesis exceeded the capacity for sucrose degradation. In addition, there was a strong positive linear relationship such that the greater the positive balance for sucrose synthesis, the greater the sucrose concentration in the tissue.

The implication of the studies of sucrose metaboliz- ing enzymes and their relationship to sucrose accumu- lation in melon fruit is that sugar accumulation is strongly influenced by developmentally regulated changes in the activity of sucrose metabolizing en- zymes within the mesocarp tissue of the fruit. Because

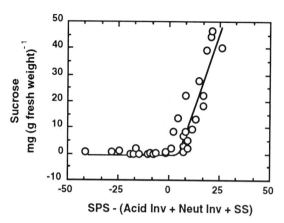

FIGURE 5 Sucrose concentration in muskmelon mesocarp tissue plotted as a function of the enzymatic activity of the tissue for synthesis minus the enzymatic activity for degradation of sucrose. SPS, sucrose phosphate synthase; Acid Inv, acidic invertase; SS, sucrose synthase. [Reproduced with permission from Hubbard, N. L., Huber, S. C., and Pharr, D. M. (1989). *Plant Physiol.* **91**, 1527–1534. Waverly, Inc., Baltimore, MD.]

natural genetic variation exists for these traits in *Cucumis*, traditional plant breeding techniques may be used to enhance fruit sugar concentration. Through the use of recombinant DNA technology, there are many more potential possibilities including the alteration of the developmental timing of the decline in the activity of acid invertase and of the intensity of expression and timing of the increase in SPS activity.

At least two alterations in the developmental patterns of these enzymes might offer an advantage from the standpoint of enhancing fruit sweetness. First, it might be advantageous if the positive balance for sucrose accumulation, which rests primarily with acid invertase and SPS, could be induced to occur at an earlier stage in fruit development, such that fruit would accumulate optimum sucrose prior to the fully ripe stage. Such fruit might be harvested for shipment at a less than fully ripe stage, as is the current commercial practice, but with the assurance of adequate sweetness when full ripeness is achieved in the market channel. Postharvest decline in fruit sugar is not nearly so serious a problem as inadequate sugar resulting from fruit harvest at a premature stage. Second, it would be highly desirable if the fruit mesocarp could be induced to accumulate sucrose uniformly rather than in a developmental gradient from inside to outside as shown by data in Table I.

The genes which encode acid invertase and SPS have been identified and cloned from various plant species. The future success of the genetic engineering strategies suggested above depends upon identifica-

tion and utilization of specific developmentally regulated gene promoters in combination with these structural genes. Continuous and rapid progress in molecular biology of fruits offers hope that such approaches to altering fruit quality by these techniques is probable in the foreseeable future.

VI. Sugar Accumulation in Fruit of Other Species

A. Starch to Sugar Conversion as a Mechanism of Fruit Sweetening

In contrast to fruits of *C. melo*, certain fruits undergo sweetening subsequent to harvest. Bananas are an example of such a fruit for which there is information about the biochemical basis of postharvest sugar accumulation. Commercially, bananas are harvested at an immature stage. At harvest they have a very high starch concentration and an extremely low soluble sugar concentration. After shipment, they are exposed to ethylene gas to induce ripening. Within about 4–6 days after exposure to ethylene, the carbohydrate composition of bananas changes dramatically. Starch concentration declines from approximately 200 mg per gram fresh weight to almost zero. During this time SPS activity increases to extremely high levels and sucrose concentration rises dramatically in the fruit. Once sucrose concentration has peaked and subsequently declined slightly, SPS activity is present, but the enzyme is present in a less active kinetic form corresponding to a diminished rate of sucrose biosynthesis. Concomitant with the high SPS activity and rapid sucrose accumulation, there is a progressive rise in acid invertase activity and an increase in hexose sugar. Sugar accumulation in these fruits can be related in mathematical form to the activities of SPS and acid invertase in a manner analogous to that shown in Fig. 5 for *Cucumis* fruits. This clearly implies the importance of these simple enzymatic reactions of sucrose metabolism in the control of sugar accumulation in banana fruits. [*See* BANANAS.]

High quality of bananas in the market is assured, in part, because their ripening and sugar accumulation can be timed to be optimum for the consumer. This is possible because ripening can be initiated in the postharvest environment with ethylene gas. Thus, sugar accumulation from a stored starch reserve in the fruit, in contrast to concurrent photosynthate as in the melon, may offer a distinct advantage in terms of quality assurance. Genes encoding certain starch biosynthetic enzymes have been cloned, and genetic transformations in higher plants have been achieved. As more detailed knowledge is obtained concerning the enzymatic control of starch accumulation and subsequent conversion to sugar during ripening, it may ultimately become possible to utilize the enzymes of starch biosynthesis and degradation to produce melon fruits which sweeten from a starch reserve when ripened in the postharvest environment, rather than from concurrent photosynthate while attached to the plant.

B. Role of Compartmentation in Fruit Sweetening

By now, investigations over many species of fruits have shown that the relationships between sucrose metabolizing enzyme activities and changes in carbohydrates during fruit ripening are not as straightforward in all species as in the cases of the melon and banana described above. This may not be entirely surprising because the relationships that are described above do not necessitate any potential role of compartmentation within fruit cells of the various enzymes and carbohydrates. For instance, the relationship in Fig. 5 implies equal access to all sugars by each of the sucrose metabolizing enzymes. Soluble acid invertase is believed to be located in the vacuole of cells, while SPS, neutral invertase and sucrose synthase are in the cytoplasm. Thus, if sucrose is synthesized in the cytoplasm, but rapidly compartmentalized into the vacuole, the cytoplasmic sucrose degrading enzymes may play an insignificant role in degrading fruit sucrose. This appears to be true in strawberry fruits, where the activities of neutral invertase and sucrose synthase exceed that of SPS throughout ripening, yet both sucrose and hexoses accumulate. In this way compartmentation may permit sucrose accumulation in fruits without a decline in the activity of all sucrose degrading enzymes. Details of the exact role of cellular compartmentation, as well as that of membrane localized sugar translocators in fruit sugar accumulation, are not as yet fully understood.

Despite the lack of a simple unifying enzymatic scheme to account for accumulation of sugar in fruit across all species, SPS has emerged as a very important enzyme in fruit sweetening. Even in species which accumulate mostly hexose sugars, it has been shown that SPS activity is not only present but increases as sugars accumulate during ripening. Continued research in the regulation of fruit sweetening will, it is hoped, lead to an increased understanding of the exact

mechanisms involved in sugar accumulation in fruit of various species, and the genetic diversity of mechanisms can be exploited.

VII. Interactions between Leaf Canopy and Fruit

A. Influence of Fruit Load on Leaf Canopy Photosynthesis and Carbohydrate Partitioning

Growth and maturation of fruits places a high demand upon the photosynthetic canopy for the manufacture and delivery of carbohydrates. The entire photosynthetic supply from as many as 14 leaves may be continuously and simultaneously required to support the growth of a single fruit of *C. sativus*, and it can be calculated that a similar amount of leaf area is required during the sweetening of fruits of *C. melo*. The growth and sugar accumulation of fruits is very highly competitive with continued vegetative growth. Vegetative growth is much greater in nonfruiting plants, and fruit growth is particularly competitive with root growth.

Plants of *C. sativus*, bearing a single fruit, exhibit twice the photosynthetic rate on a per leaf area basis as plants which are not allowed to bear a fruit. In addition, assimilate export rates from the leaves of fruiting plants are twice as great as that from vegetative plants. This is true for assimilate export both during the daytime as well as at night. Carbohydrate export from the leaves at night is accomplished from the breakdown of a leaf starch reserve which accumulates in mesophyll chloroplasts during the photoperiod of each day. By sunrise each morning the starch reserve within the leaves of fruit bearing plants is almost totally exhausted because of the heavy overnight demands of the fruit, whereas the leaves of vegetative plants still contain a substantial starch reserve. During the daytime, leaves of fruiting plants accumulate starch within their leaves at twice the rate as the leaves of vegetative plants. This is apparently necessary to meet the nightly demands for carbohydrate by the fruit. Associated with the greater photoassimilate export rate from leaves of fruiting plants is a twofold higher activity level of the leaf enzyme, stachyose synthase. This also has been clearly documented in *C. melo* during fruiting.

Thus, an increased rate of leaf stachyose biosynthesis associated with fruiting in these plants may be controlled in part via increased enzymatic capacity for stachyose biosynthesis. From the postulated mechanism of phloem "loading" in Fig. 2, it seems logical to assume that an increased rate of stachyose synthesis per se within the intermediary cells may result in greater assimilate export. Thus, the role of leaf carbohydrate metabolism in supplying assimilates to the growing (sugar accumulating) fruit is not passive. Carbohydrate metabolism within the leaves of these species is very responsive to the demands of the fruit for translocated carbohydrate. Current evidence demonstrates that all steps of leaf photosynthetic carbohydrate metabolism including CO_2 fixation, starch formation within chloroplasts, stachyose synthesis, and assimilate export are up-regulated in response to the strong demands imposed by fruit growth and sugar accumulation. Clearly, the carbohydrate demanding fruit "communicates" with the carbohydrate producing leaf, and brings about a coordinated set of changes in leaf carbohydrate metabolism which are favorable to the development of the fruit. However, the biochemical basis of this "communication" is not understood at this time.

B. Influence of Leaf Canopy Size on Sucrose Metabolism by Fruits

Regulation of carbohydrate metabolism in the leaf canopy (source) and fruit (sink) in *C. melo* is obviously important to sugar accumulation. As implied above, the biochemical activities of sink and source are interdependent. Further evidence that this is true comes from additional studies of genotypes which produce sweet and nonsweet melons. The nonsweet genotype (the same as that characterized for fruit sucrose accumulation and previously described) was found to have approximately half the photosynthetic rate of the sweet genotype on a per plant basis. Partial defoliation of the sweet genotype during fruit sucrose accumulation lowered the canopy photosynthetic activity of this plant on a total canopy basis to that typical of the nonsweet genotype. Even so, the fruits on the partially defoliated sweet genotype accumulated far more sucrose than the nonsweet genotype. Defoliation studies also demonstrated that limiting the photosynthetic activity of the sweet genotype resulted in lower SPS activity of the fruit. Although SPS activity was not as low as that of the nonsweet genotype, the SPS activity in fruit of the sweet genotype was correlated with the extent of defoliation. These observations clearly imply that "communication" between source (leaf canopy) and sink (fruit) tissues does take place and can modify the expression of fruit enzyme

activities which in turn may alter sucrose accumulation within the fruit. The mechanisms of these "communications" are unknown but may be analogous to those alluded to earlier which affect changes in leaf carbohydrate partitioning in response to the presence of a fruit on the plant.

C. Environmental Factors Influencing Leaf and Fruit Metabolism

Perhaps the studies described above demonstrate how some environmental conditions may alter assimilate supply and fruit sweetening. Low night temperatures during fruit growth can increase the time required for melon fruits to mature. An extended maturation period appears to increase soluble solids in melon fruits. Because of the direct dependence of sweetening on translocated photosynthate, the increase in fruit sugar may be a result of the extended time over which the fruit may accumulate photoassimilates.

Other environmental factors can result in fruit which are low in sugar concentration. Extended rainy and cloudy weather prior to harvest causes melons to have less sugar. The low light intensity accompanying these conditions may result in inadequate leaf photosynthesis and photoassimilate delivery to the fruit. In addition, some species, including *C. melo*, often exhibit leaf stomatal closure and markedly reduced photosynthetic rates when they experience excessive soil moisture resulting from heavy rainfall. However, recent experiments have demonstrated that it is possible under certain greenhouse conditions to reduce sugar accumulation in melons in response to root flooding without affecting leaf stomatal resistance or photosynthesis. In these experiments, root flooding during the fruit sucrose accumulation period of *C. melo* resulted in reduced fruit sugar accumulation over a 4-day period but did not result in decreased leaf photosynthesis. Fruits of the root flooded plants exhibited the normal increase in SPS activity and decrease in acid invertase activity, such that failure to accumulate sucrose could not be attributed to altered fruit enzyme expression. Under flooded conditions, root carbohydrate consumption, as measured indirectly by gas exchange, was substantially enhanced as compared to the roots of nonflooded control plants. Adventitious root formation along the base of the stems of the root flooded plants was also observed. Increased metabolism on the part of the flooded root system and newly forming adventitious roots apparently resulted in diversion of photosynthates away from the fruits to the roots. These observations point to the overriding effects at the whole plant level which environmental factors can have on the complex biochemistry and physiology of fruit sugar accumulation.

Much progress has been reported in recent years in understanding the control of the photosynthetic formation and translocation of carbohydrates in *C. melo* and closely related species. Sufficient knowledge of the biochemical use of these translocated carbohydrates within the fruit for sucrose formation exists to postulate certain genetic transformations which might help produce more uniformly sweet fruit for the consumer in future years. During the next decade, consumers may very well see transgenic fruit of higher and more uniform quality than those fruits which have been available in the past. The production of such fruit, using techniques of molecular genetics, clearly represents an exciting challenge for agricultural researchers.

Bibliography

Beebe, D. U., and Turgeon, R. (1992). Localization of galactinol raffinose, and stachyose synthesis in *Cucurbita pepo* leaves. *Planta,* **188,** 354–361.

Burger, Y., and Schaffer, A. A. (1991). Sucrose metabolism in mature fruit and peduncles of *Cucumis melo* and *Cucumis sativus In* "Recent Advances in Phloem Transport and Assimilate Compartmentation" (J. L. Bonnemain, S. Delrot, W. J. Lucas, and J. Dainty, Eds.), pp. 244–247. Quest Editions, Nantes cedex, France.

Gross, K. C., and Pharr, D. M. (1982). A potential pathway for galactose metabolism in *Cucumis sativus* L., a stachyose transporting species. *Plant Physiol.* **69,** 117–121.

Handley, L. W., Pharr, D. M., and McFeeters, R. F. (1983). Relationship between galactinol synthase activity and sugar composition of leaves and seeds of several crop species. *J. Am. Soc. Hortic. Sci.* **108,** 600–605.

Holthaus, U., and Schmitz, K. (1991). Distribution and immunolocalization of stachyose synthase in *Cucumis melo* L. *Planta* **185,** 479–486.

Hubbard, N. L., Pharr, D. M., and Huber, S. C. (1990). Sucrose metabolism in ripening muskmelon fruit as affected by leaf area. *J. Am. Soc. Hortic. Sci.* **115,** 798–802.

Hubbard, N. L., Pharr, D. M., and Huber, S. C. (1990). Role of sucrose phosphate synthase in sucrose biosynthesis in ripening bananas and its relationship to the respiratory climacteric. *Plant Physiol.* **94,** 201–208.

Hubbard, N. L., Huber, S. C., and Pharr, D. M. (1989). Sucrose phosphate synthase and acid invertase as determinants of sucrose concentration in developing (*Cucumis melo* L.) fruits. *Plant Physiol.* **91,** 1527–1534.

Kroen, W. K., Pharr, D. M., and Huber, S. C. (1991). Root flooding of muskmelon (*Cucumis melo* L.) affects fruit sugar concentration but not leaf carbon exchange rate. *Plant Cell Physiol.* **32,** 467–473.

Peirce, L. C. (1987). "Vegetables, Characteristics, Production, and Marketing." Wiley, New York.

Pharr, D. M., Huber, S. C., and Sox, H. N. (1985). Leaf carbohydrate status and enzymes of translocate synthesis in fruiting and vegetative plants of *Cucumis sativus* L. *Plant Physiol.* **77,** 104–108.

Schmitz, K., and Holthaus, U. (1987). Are sucrosyl-oligosaccharides synthesized in mesophyll protoplasts of mature leaves of *Cucumis melo*? *Planta* **169,** 529–535.

Smart, E. L., and Pharr, D. M. (1981). Separation and characteristics of galactose-1-phosphate and glucose-1-phosphate uridyltransferase from fruit peduncles of cucumber. *Planta* **153,** 370–375.

Welles, G. W. H., and Buitrelaar, K. (1988). Factors affecting soluble solids content of muskmelon (*Cucumis melo* L.). *Netherlands J. Agric. Sci.* **36,** 239–246.

Worrell, A. C., Bruneau, J. M., Summerfelt, K., Boersig, M., and Voelker, T. A. (1991). Expression of a maize sucrose phosphate synthase in tomato alters leaf carbohydrate partitioning. *Plant Cell* **3,** 1121–1130.

Yamaguchi, M. (l983). "World Vegetables, Principles, Production and Nutritive Values." Van Nostrand Reinhold, New York.

Meteorology

ROGER H. SHAW, *University of California, Davis*

Glossary

Adiabatic process Process in which no thermal energy is exchanged between the mass of air undergoing the change and its external environment

Coriolis effect Apparent force on a body in motion arising from the rotation of the earth; the effect is that of a force acting to the right of the motion in the northern hemisphere and to the left in the southern hemisphere

Cyclone Large-scale low-pressure system rotating in the same direction as that of the rotation of the earth (anticlockwise in the northern hemisphere); cyclones are classified as either tropical or extra-tropical, the latter normally associated with air masses and fronts

Front Zone of transition between air masses of detectably different temperature and/or moisture content; in the case of a warm front, colder air of poleward origin is being replaced by warm air and in the case of a cold front, colder air behind the front is replacing warmer air

Geostrophic wind Theoretical wind deduced from the balance between the pressure gradient and the Coriolis force and directed parallel to lines of constant pressure such that low pressure (in the northern hemisphere) is to the left

Greenhouse effect Relative warming of the earth as a result of thermal radiation emitted by the surface being largely absorbed by water vapor, carbon dioxide, and other constituents of the atmosphere, and a significant fraction of that energy being radiated back as an additional source of heat for the surface

Lapse rate Rate of decrease of temperature with height; lapse rate can refer to the existing, observable rate of change of temperature with height, or to a hypothetical rate of cooling that would be experienced by a parcel of air undergoing ascent

Potential temperature Temperature that a parcel of air would achieve if brought dry adiabatically to a pressure level of 1000 mb

Vorticity Measure of rotation in a fluid: vorticity is composed of two parts: that due to curvature of streamlines, and that due to shear (rate of change of speed in a direction perpendicular to the streamlines); the vorticity of a fluid is equivalent to twice the local angular velocity of the fluid element in the absence of deformation, and is measured in radians per second

Meteorology is the study of the envelope of air surrounding the planet and of the phenomena associated with the atmosphere. A component of meteorology is the study of weather, which is the condition of the atmosphere mainly with respect to its day-to-day effects on life and human activities. Meteorology also relates to climatology which is concerned with long-term manifestations of the weather represented by a statistical collective of weather conditions over a specified long period of time, usually at least a few decades in duration. Meteorology deals with both the physical and chemical aspects of the atmosphere.

I. Composition and Thermal Structure of the Atmosphere

A. Atmospheric Constituents

The atmosphere extends from the earth's surface to an indefinite height. Density and pressure decrease in an approximately exponential manner and asymptotically approach conditions of interplanetary space. Nevertheless, the atmosphere can generally be considered as a thin shell blanketing the globe with approximately half of its mass to be found in the lowest 5.5 km above the surface and 99% of its mass contained within a layer of thickness equivalent to less than 0.5% of the earth's radius.

The atmosphere is composed of a mixture of gases which, apart from a few variable constituents (mainly water vapor) that make up only a small fraction of its total mass, remains essentially unchanged in fractional content up to a height of about 80 km. To avoid the complication of varying vapor content, it is convenient to consider the composition of "dry" air, in which case the predominant constituents are, in order of decreasing concentration by volume; nitrogen (78.09%), oxygen (20.95%), and argon (0.93%). Carbon dioxide currently comprises only 0.036% by volume [a global mean of 358 parts per million (ppm), in 1993 as based on an extrapolation from observations made in earlier years] but has extremely important properties with regard to the absorption and emission of terrestrial (longwave or far-infrared) radiation. Carbon dioxide concentrations have exhibited a steadily increasing trend (currently 1.8 ppm per year) due primarily to the burning of fossil fuels; roughly 50% of the anthropogenic production of CO_2 remaining in the atmosphere, the rest being absorbed by oceans and incorporated into increased biomass.

There are numerous trace gases in the atmosphere, some of which are meteorologically important, such as ozone (at high altitudes) and methane, and some of which are considered to be air pollutants, such as ozone (at low levels), sulfur dioxide, and various oxides of nitrogen. Ozone is produced by photochemical reaction both near the earth's surface, from the action of sunlight on hydrocarbons emitted both naturally and anthropogenically, and in the upper part of the atmosphere by the photodissociation of molecular oxygen and subsequent recombination. It is responsible for absorbing and largely eliminating biologically harmful uv radiation from the sun.

Water vapor is highly variable in both time and space, occupying from practically zero to as much as 4% of the volume of the atmosphere near the surface in humid tropical regions. The amount of water vapor present in the atmosphere is strongly dependent upon temperature and proximity to sources of evaporation. Hence, water vapor changes rapidly with latitude, season, elevation above the earth's surface, and surface moisture content. There is very little water vapor above altitudes of about 10 km. Water vapor assumes great importance in meteorology for three reasons. First, water vapor condenses to form clouds and rain or other forms of precipitation and, hence, comprises part of the hydrologic cycle; second, water vapor is a strong absorber and emitter of thermal radiation, and is thus a participant in the "greenhouse" effect; and third, the processes of evaporation at the surface and condensation to form clouds consume and release enormous amounts of thermal energy and thus play important roles in the energy balance of the earth/atmosphere system. [See WATER RESOURCES.]

B. Temperature Structure of the Atmosphere

There is a natural division of the atmosphere into four height regions: *troposphere, stratosphere, mesosphere, and thermosphere,* according to the way in which temperature changes with height (Fig. 1). Meteorologically, the most important of these regions is the troposphere, which extends from the surface to a height of about 12 km in mid-latitudes. At low latitudes, because of enhanced vertical convection in the form of thunderstorms or strong cumulus activity, the upper limit of the troposphere achieves altitudes of 16 km or more while, in polar regions, it is limited to depths of the order of only 8 km.

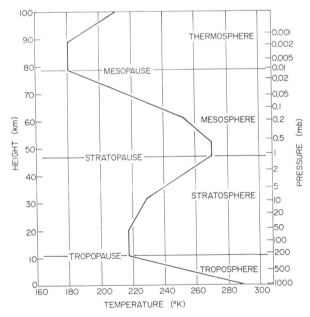

FIGURE 1 Vertical temperature profile for the U.S. Standard Atmosphere. [Reprinted with permission from Wallace, J. M., and Hobbs, P. V. (1977). "Atmospheric Science, An Introductory Survey," Fig. 1.8, p. 23. Copyright © 1977 by Academic Press, New York.]

On the average, temperature decreases with height through the depth of the troposphere. This trend is halted at the *tropopause*. It is then reversed in the stratosphere because of radiative heating as solar ultraviolet radiation is absorbed at high altitudes by nitrogen, oxygen, and, most importantly, ozone to produce a local temperature maximum at an altitude of around 50 km. This temperature maximum terminates the stratosphere and leaves only 0.1% of the mass of the atmosphere to remain at higher levels.

Virtually all weather activity is confined to the troposphere, which contains approximately 80% of the mass of the atmosphere and nearly all of its water vapor. Vertical overturning occurs in the troposphere under appropriate conditions. The stratosphere, by contrast and because warmer air overrides air of lower temperature, is a stable layer and resistive to mixing.

The mesosphere and thermosphere are of no consequence in terms of weather activity but, at altitudes exceeding 60 km, ultraviolet radiation is able to ionize the rarefied atmosphere to form the ionosphere, with free electron density achieving a peak at about 300 km. The ionosphere is commonly divided into D-, E-, and F-regions and has important effects on radiowave propagation.

II. Radiation Processes in the Atmosphere

A. Solar and Terrestrial Radiation

The distinction between spectral compositions of solar and terrestrial radiation is crucial to an understanding of the energy balance of the globe. Effectively, the two radiation streams do not overlap in their spectral qualities, and this is a direct consequence of the difference in temperatures of their radiating surfaces. Planck's law describes the amount of radiation (the monochromatic irradiance $E_\lambda\star$) emitted by a blackbody, a perfect emitter and absorber, at each wavelength λ as a function of the absolute temperature T of the body:

$$E_\lambda\star = \frac{C_1}{\lambda^5[\exp(C_2/\lambda T) - 1]}.$$

Here, $C_1 = 3.74 \ 10^{-16}$ W m^2 and $C_2 = 1.44 \ 10^{-2}$ m °K. Consequences of Planck's law are that (i) the total amount of radiant energy emitted, the blackbody irradiance, is proportional to the fourth power of the absolute temperature (the Stefan-Boltzmann law), and (ii) the wavelength corresponding to the peak in the spectrum is inversely proportional to absolute temperature (the Wien displacement law). [*See* MICROCLIMATE.]

Apart from demonstrating the vast difference between the amount of radiant energy emitted at the surface of the sun (a "color" temperature of roughly 6100°K) and at the surface of the earth (temperatures mostly between, say, 255° and 310°K), Planck's law explains why solar radiation peaks at a wavelength of 0.475 μm, while the peak in the terrestrial radiation spectrum corresponding to these two temperatures appears between 11.4 and 9.3 μm. Solar radiation lies within the visible and near-infrared portions of the electromagnetic spectrum, while terrestrial radiation is in the thermal band and the two are almost devoid of overlap.

The mean flux density of solar radiation at the earth's orbital radius, called the "solar constant," is about 1370 W m^{-2}. This is the energy flux density outside the influence of the atmosphere. Minor changes occur in the "constant" due to changes in the output of the sun (± 0.1%) and, because the earth's orbit is slightly elliptical. The sun and earth are closest in January (the southern hemisphere summer) and farthest apart in July (the northern hemisphere summer), the 1.7% difference in distance yielding a 3.4% difference in irradiance because of the inverse square dependence of irradiance on the distance from the source.

Different parts of the earth are exposed to different amounts of solar radiation depending on the time of day and time of year, which change the zenith angle of the sun (the angular distance from the overhead position). Such changes in exposure account for the daily heating cycle and for the annual change in season. In addition, changes in the path length of the solar beam through the atmosphere directly relate to zenith angle, and influence the depletion by scattering and absorption.

Of great importance is the difference between how the atmosphere transmits, absorbs, and scatters solar and terrestrial radiation streams. A general statement can be made that the atmosphere is relatively transparent to energy from the sun but is relatively opaque to terrestrial radiation. In the absence of dense clouds, a large fraction of the incident solar radiation penetrates the atmosphere and reaches the surface. On the other hand, most of the longwave radiation emitted by the earth's surface is absorbed by the atmosphere and only a small fraction is lost directly to space.

Short ultraviolet wavelengths are almost totally eliminated by O_2 and O_3 in the stratosphere but, otherwise, solar radiation under clear sky conditions is depleted only weakly by water vapor absorption bands in the near-infrared. Within the spectrum of terrestrial radiation, strong absorption by water vapor and carbon dioxide allows little energy to penetrate the atmosphere. However, satellite-borne thermal sensors utilize an "atmospheric window" in the region of 8 to 11 μm to observe cloud patterns both night and day by distinguishing between cloud-top temperature and that of the earth's surface.

B. The Earth's Energy Budget and the "Greenhouse" Effect

Since radiant energy is the only mechanism by which the earth/atmosphere system can exchange energy with outer space, there exists a state of radiation balance in which the amount of solar radiation absorbed equals the amount of terrestrial radiation emitted to space. Satellite observations place the planetary albedo, the reflectivity to solar radiation, at 0.30 and, assuming the earth to radiate as a black body, the radiative temperature of the earth is calculated to be 255°K or -18°C. Obviously, this is a much lower temperature than exists at the earth's surface, where a globally averaged temperature in the region of 288°K or 15°C is observed.

This difference of 33°C is explained by the so-called "greenhouse" effect. Being a good absorber of thermal radiation, the atmosphere is also a good emitter (Kirchoff's law) and thus re-emits thermal energy to the surface to maintain a relatively high surface temperature. Clouds are also very effective absorbers and emitters of thermal radiation, hence the big difference between nocturnal cooling at the ground under clear sky and cloudy conditions, especially with low and relatively warm cloud cover.

The last one or two decades have seen extensive debate over the likely consequences of the well-documented increase in the CO_2 content of the atmosphere, CO_2 being one of the principal greenhouse gases. It is estimated that CO_2 concentrations were of the order of 280 ppm in pre-industrial times (1750–1800) but rose to 316 ppm by 1960 and to 358 ppm by 1993, and it is anticipated that a further doubling could occur over the next 100 years. Other gases such as N_2O, CH_4, and chlorofluorocarbons (CFCs) are also increasing in the atmosphere and some predictions indicate that in combination they will have as large an effect as that of CO_2. Computer simulations of the atmosphere (general circulation models, or GCMs) which attempt to reproduce all pertinent physical processes generally result in predictions of

increases in global temperatures of the order of 2° to 5°C with larger changes occurring at higher latitudes than at lower latitudes. There is much uncertainty, however, because of the numerous feedback effects that are not well understood and which are difficult to predict. For example, GCMs tend to handle cloud formation and precipitation poorly, yet cloud amount and surface moisture both play important roles in the global energy budget.

III. The Behavior of Gases in the Atmosphere

A. The Gas Laws and Adiabatic Processes

The thermodynamic state of the atmosphere or of any of its constituents is specified by three state variables: pressure p, temperature T, and density ρ or its inverse, specific volume α. These are related by the equation of state:

$$p = \rho RT \quad \text{or} \quad p = \frac{RT}{\alpha}.$$

Here, R is the specific gas constant, which for dry air is equal to 287 J/kg/°K.

Pressure is the force per unit area exerted by the molecular motions of the atmosphere and common meteorological units are the kiloPascal (kPa) and the millibar (mb), where 10 mb≡1 kPa. One *standard atmosphere* pressure is equal to 101.325 kPa or 1013.25 mb, and corresponds to a globally averaged mean sea level pressure. Observed sea level pressures deviate considerably from this average value, such that pressures of the order of 105 kPa would correspond to a very strong anticyclone, while a central pressure of 97 kPa might be observed in association with a very strong cyclone.

At any altitude, the air pressure is almost exactly that which is sufficient to support the weight of atmosphere above, a state of *hydrostatic balance*. Thus, pressure decreases with height and at a rate given by the hydrostatic equation

$$\frac{dp}{dz} = -\rho g \quad \text{or} \quad \frac{dp}{dz} = -\frac{pg}{RT},$$

where, in the second form of the equation, density is eliminated by application of the equation of state.

The equation of state alone is not sufficient to define the reaction of the atmosphere to processes which cause a mass of air to change pressure, such as in the rise of warm, buoyant plumes of air over a heated ground surface, or orographic uplift, when air is forced over rising terrain. However, if the process can be assumed to be *adiabatic,* that is, proceeds in such a manner that no thermal energy is exchanged between the mass of air undergoing the change and its external environment, the first law of thermodynamics provides the necessary additional information. In an adiabatic process, the mechanical work of expansion, as a mass of air ascends to a region of lower pressure, comes entirely from the internal energy of the air. A combination of the first law of thermodynamics under this condition and the equation of state yields Poisson's equation which defines how temperature changes for a change in pressure during an adiabatic process:

$$T_1 = T_2 \left(\frac{p_1}{p_2} \right)^{0.286}.$$

where the exponent 0.286 is R/C_p, the ratio of the specific gas constant for dry air to the specific heat at constant pressure. It is convenient to define *potential temperature* θ as the temperature that would be achieved by bringing a parcel of air from any initial altitude dry adiabatically to a reference pressure level of 100 kPa or 1000 mb. Poisson's equation expresses θ in terms of the pressure and temperature of a parcel of air, such that

$$\theta = T \left(\frac{1000}{p} \right)^{0.286},$$

where p is in millibars. Notice that, as a parcel of air rises or sinks in the atmosphere and, if it does so adiabatically, its potential temperature remains fixed. On the other hand, expansion during ascent would cause a reduction in the measured temperature of a parcel, while descent and compression would cause an increase in measured temperature.

This decrease or increase of temperature as a parcel of air ascends or descends adiabatically through the atmosphere defines a reference *lapse rate,* or rate of decrease of temperature with height. Combining the hydrostatic equation and the first law of thermodynamics for such a process, it follows that

$$\frac{dT}{dz} = -\frac{g}{C_p} = -9.8°C\,km^{-1}.$$

This is the *dry adiabatic lapse rate* Γ_d, the rate of decrease of temperature with height in the atmosphere which would match the rate of cooling of a parcel of air lifted dry adiabatically.

B. Moist or Saturated Adiabatic Processes

The occurrence of condensation or evaporation during the formation or dissipation of cloud or fog has a large influence on the reaction of the atmosphere to vertical displacement. Under extreme circumstances, the violence of a severe thunderstorm or of a hurricane can be devastating to human life and property. Both derive most of their energy from the latent heat released during the condensation process.

There are a number of ways that meteorologists express the water vapor content of the atmosphere: (1) *absolute humidity* is the density of water vapor expressed as the mass of water vapor per unit volume of air; (2) *specific humidity* is the ratio of the mass of water vapor to the total mass of the air in any given volume; (3) closely related to this is *mixing ratio* which is the ratio of the mass of water vapor to the mass of the dry components of the atmosphere in any given volume; and (4) *vapor pressure* is the partial pressure exerted by the molecules of vapor in the air.

Saturation is the state in which moist air exists in equilibrium with an associated condensed phase. In equilibrium (at saturation), the rate at which water molecules leave a surface of liquid water or ice matches the rate at which water molecules arrive at that surface. The *saturation vapor pressure* is a unique function of temperature and approximately doubles for each 10°C rise. Below 0°C, the saturation vapor pressure with respect to ice is less than that with respect to liquid water because the crystalline structure of ice more tightly binds water molecules to the surface.

Relative humidity is a useful and common expression for the ''drying power'' of the air because it expresses the proximity to the saturated state. It is defined as the ratio of the existing vapor pressure to the saturation vapor pressure at the existing temperature, and is usually shown as a percentage.

Dew point temperature is the temperature at which saturation would occur if moist air was cooled isobarically (at constant pressure). On further cooling, water vapor normally condenses to form liquid water or solid ice (meteorologists generally consider a direct transition between solid and gas phases, in either direction, to be a process of sublimation). Dew point is useful in identifying *air masses* because its diurnal change is smaller than that of air temperature. Together with air temperature, the dew point indicates the relative closeness to saturation.

In order to evaporate water, large quantities of heat are required but this thermal energy is released again when water vapor condenses. The energy associated with these phase changes, *latent energy,* is of great importance to meteorology. The latent heat of vaporization (or condensation) has a small dependence on temperature but, at 0°C is equal to $2.50 \ 10^6 \ \text{J kg}^{-1}$. Considering the heat capacity of water, one can appreciate the importance of latent heat when it is observed that it takes approximately six times more heat to completely evaporate a given mass of liquid water than it does to raise its temperature from 0° to 100°C, the normal boiling point at sea level.

It is common for adiabatic cooling to be sufficient to bring the air to a state of saturation. If lifting continues, the air will continue to cool but the release of latent heat will offset the temperature decrease to some extent. Thus, during a saturated ascent in which water vapor is condensing in the form of cloud, the rate of decrease of temperature of a parcel of air, the *saturation or moist adiabatic lapse rate* Γ_s, will be less than during a dry adiabatic ascent. The difference in lapse rates between dry and saturated processes will depend on the amount of water vapor available for condensation. Thus, near the earth's surface in warm climates, the saturation adiabatic lapse rate will be considerably smaller than $9.8°\text{C km}^{-1}$ while, high in the atmosphere or wherever the temperature is low and there is little moisture content, the difference will not be great.

C. Atmospheric Stability and Instability

The stability or instability of the atmosphere determines the likelihood of convective activity (development of cumuliform clouds), the likelihood of atmospheric turbulence, and the extent of mixing or the potential for pollution episodes. Stability or instability is a consequence of the actual lapse rate of a layer of atmosphere and of the reaction of a volume or parcel of air from within that layer to a vertical displacement.

A layer of atmosphere is said to be *unstable* if, when vertically displaced upward, a parcel becomes warmer than its new environment, positively buoyant, and would continue its upward drive unassisted. On the other hand, a layer is *stable* if, under the same circumstances, the parcel becomes cooler than its new environment, negatively buoyant, and the displacement is resisted. The stability of a layer of air is classified according to the lapse rate γ that exists in the layer and its comparison with the two reference lapse rates Γ_d and Γ_s. If $\gamma > \Gamma_d$, the layer is said to be *superadiabatic* and is absolutely unstable. Such conditions commonly occur near the earth's surface during periods

of solar heating, and mixing of the lower atmosphere is enhanced. When $\gamma < \Gamma_s$, the layer is absolutely stable and mixing is suppressed. An *inversion* is said to occur when air temperature actually increases with height, as might result from nocturnal cooling near the ground, or when a warm air mass overrides a cooler, denser one. Finally, a layer of atmosphere is said to be in a state of *conditional instability* when $\Gamma_s < \gamma > \Gamma_d$, that is, the existing lapse rate is greater than the saturated adiabatic rate but less than the dry adiabatic rate. The layer is stable if dry, but is unstable if condensation is occurring.

For the purpose of weather forecasting, meteorologists commonly employ stability indices that provide an overall assessment of the state of the troposphere with respect to the likelihood of the development of convective activity. One such index is the *Lifted Index*, which is obtained as the difference between the temperature of the atmosphere at the 500 mb level and that of a parcel of air hypothetically lifted from near the surface dry adiabatically until condensation occurs (the lifting condensation level) and then moist adiabatically to 500 mb. The less positive, or the more negative the index, the more likely is the possibility of thunderstorm or severe weather activity. Stability indices are used in conjunction with other weather information, such as the presence or absence of weather fronts or other triggering mechanisms.

IV. Clouds and Precipitation

A. The Condensation Process

The equilibrium or saturation vapor pressure with respect to a microscopic droplet of water is greater than that over a plane water surface. This is because molecules at the surface of such a droplet are bound less strongly to the droplet than would be the case for a flat water surface. Since molecules can escape more easily, a higher vapor pressure in the surrounding air is needed to maintain equilibrium, and it is necessary to have a certain degree of supersaturation (relative to a flat water surface) to maintain a small droplet in equilibrium. The percentage supersaturation is inversely proportional to the radius of the drop.

This is the *radius effect* and raises the question: how does a droplet form in the first place because, initially, it would be a conglomeration of only a few molecules and would need a very large degree of supersaturation to prevent it from evaporating again? The answer is that *condensation nuclei* (microscopic particles of sea salt, dust, combustion products, and anthropogenic pollution) in the atmosphere provide a surface on which droplets can form. The atmosphere is generally quite heavily laden with condensation nuclei, roughly 10^5 per liter, and condensation occurs readily with only very small degrees of supersaturation.

Condensation is further enhanced by the fact that some aerosols are hygroscopic or wettable and the presence of dissolved solute in a water droplet reduces the saturation vapor pressure below that of pure water. For very small droplets, this *solution effect* is sufficient to allow microscopic droplets to form at relative humidities below 100%, although the creation of a visible cloud requires humidities close to saturation.

B. Cloud Classification

Clouds are generally classified into two main groups: convective clouds and layer clouds. The distinction is based on the presence or absence of instability within the layer of atmosphere in which they form.

Convective or *cumuliform clouds* are the result of local ascent of warm buoyant parcels of air in a conditionally unstable environment. Such clouds exhibit distinct spatial variability, with diameters of individual cells of the order of a few hundred meters to 10 km or more. Updraft velocities within the clouds can be large: tens of cm/sec to tens of m/sec under very strong convection. Convective clouds have lifetimes from minutes to a few hours in extreme cases. Layer or *stratiform clouds* are the result of forced lifting of stable air by large-scale atmospheric processes such as fronts. Their horizontal extent can be very large, of the order of hundreds of kilometers. Updraft velocities are much smaller than in the case of cumuliform clouds and would be measured in centimeters per second. Layer clouds can persist for periods of hours or tens of hours.

Clouds are further categorized in terms of the height of their bases, resulting in 10 main characteristic forms or genera. The 10 cloud genera are: *cirrus, cirrostratus, cirrocumulus, altostratus, altocumulus, stratus, nimbostratus, stratocumulus, cumulus,* and *cumulonimbus.* The first three in this list are high clouds composed mostly of ice crystals. Cirrus has the appearance of wispiness, often referred to as mares' tails. Cirrostratus has a uniform light grey appearance and a characteristic 22° halo around the sun or moon, evidence of the presence of ice crystals. Cirrocumulus shows evidence of convective instability at cloud level, in the form of patchiness. Altostratus and altocumulus

are middle layer clouds and are composed mostly of water droplets. Their stratiform or cumuliform appearance signifies the degree of stability or instability of the layer in which they have formed. The remaining clouds are low clouds or have low bases. Stratus is a uniform grey cloud which produces either no precipitation or only small droplets (drizzle) or ice crystals. Nimbostratus is a dark grey cloud from which precipitation is falling and usually, but not always, reaching the ground. This cloud has large horizontal and vertical extent. Stratocumulus is predominantly stratiform but lumpy in nature with distinct light and dark patches indicating a degree of instability in the layer of air in which it has formed. Cumulus clouds are formed as individual elements in a buoyantly active layer of atmosphere. Thermals which are topped by cumulus normally originate at the ground surface and, over a land surface, commonly exhibit a diurnal cycle, but there are other situations where instability is triggered aloft. The amount of growth of cumulus cells is dependent on the degree of instability in the atmosphere. In an extreme case, cumulus can develop into cumulonimbus with extensive vertical development and electrical discharge.

The only distinction between a cloud and a fog is that a fog has its base at ground level. A variety of processes lead to saturation of the air and to the creation of a fog: cooling of the ground on calm clear nights can result in a *radiation fog;* warm air moved over a cold surface can be brought to its dew point to create an *advection fog;* a *warm frontal fog* is formed when rain falling through colder air beneath supersaturates the air; relatively warm water evaporates into an overriding layer of cold air and the mixture of warm moist air and cold air becomes supersaturated to form *steam fog.*

C. Precipitation Processes and Types

Once initiated on a condensation nucleus, small droplets grow to raindrop size by a combination of mechanisms. In warm, humid environments, the condensation process is sufficient to cause droplets to grow to a size where drops will collide with each other and coalescence creates drops that are large enough that they can fall through the cloud updraft. Droplets smaller than about 20 μm in diameter tend not to collide because they lack sufficient inertia and follow the airstream around other drops. On the other hand, in latitudes beyond the tropics or subtropics, lower temperatures and smaller amounts of water vapor in the air inhibit the growth of droplets by condensation alone. In such "cold" clouds, a third mechanism is critical. This is the *Bergeron–Findeisen three-phase process.*

First proposed in 1933, the theory explains how precipitation can fall from relatively short-lived cumulus clouds. Crucial to its understanding are that, at common cloud temperatures, icing nuclei are needed to initiate droplet freezing or direct vapor to ice particle conversion, and that icing nuclei are far less abundant than condensation nuclei. Thus, clouds composed largely of water droplets exist at temperatures substantially below 0°C. Further, the saturation vapor pressure with respect to ice is less than that with respect to water, allowing relatively rapid growth of a small number of ice particles at the expense of the water droplets. Such ice crystals can quickly gain sufficient size to allow collision and coalescence to precipitable size particles, which may or may not melt, depending on the temperature of the air through which they fall.

Rain is precipitation in the form of drops with a diameter greater than 0.5 mm, while *drizzle* consists of drops smaller than 0.5 mm. Droplet size is dependent on the vertical extent of the cloud producing the precipitation and on the magnitude of the updrafts present in the cloud. A distinction is usually made between continuous rainfall from clouds of large horizontal extent and showers of shorter duration and usually from convective clouds. *Virga* is seen as precipitation falling from the base of a cloud into a sufficiently dry atmosphere that evaporation occurs before the ground is reached.

Snow is composed of white or translucent ice crystals which may agglomerate into flakes. Forms of frozen precipitation created by the freezing of supercooled water droplets include *snow pellets, ice pellets,* and *hail* which falls from cumulonimbus clouds as irregular particles of ice with diameters of 5 mm or more. Deep convection and large updrafts are favorable for the formation of large hailstones. *Freezing rain or drizzle* is precipitation that falls in liquid but supercooled form and freezes upon impact with a cold surface, causing the accumulation of ice on the surface and on exposed objects.

D. Severe Weather Phenomena

If the atmosphere is sufficiently unstable, cumulus clouds can develop into cumulonimbus with lightning discharge, possibly with hail and, in extreme cases, accompanied by a tornado. An *air mass thunderstorm* is generally a summertime event when solar heating of the ground initiates convection into a humid and sufficiently unstable air mass. Other common triggering mechanisms are frontal boundaries, orographic forcing, and atmospheric gravity waves. In the formative, cumulus stage, there are updrafts across all points at the base of the cloud. In the mature stage, the cumulonimbus cloud has appreciable vertical development with an anvil top of cirrus cloud, and has perhaps penetrated into the lower stratosphere. In addition to updrafts within the cloud, strong downdrafts are also formed by the drag of rain and hail falling through the cloud and by evaporative cooling and resulting negative buoyancy as dry air is entrained into the cloud. In the dissipating stage, there are weak downdrafts throughout the cloud. The typical lifetime of an air mass thunderstorm is 1 hr or less.

Severe thunderstorms with large hail and strong wind gusts generally require strong vertical shear in the wind through the depth of the cloud, in addition to the conditions previously specified, and are usually associated with conditions in which there is divergence aloft to enhance the updraft. The vertical shear of the wind causes a tilting of the updraft so that, when precipitation becomes too heavy to be supported by the updraft, it falls into the downdraft region rather than reducing the intensity of the updraft. Violent updrafts keep hailstones suspended in the cloud long enough to grow to considerable size.

Downbursts are intense downdrafts exiting the base of a cumulonimbus cloud, which spread out horizontally on reaching the ground producing a *gust front.* Gust fronts are particularly dangerous to aircraft during take-off and landing because of the possibility of rapid changes in head or tail winds.

Definitive of a thunderstorm is the lightning discharge resulting from electrification within the cloud. Lightning occurs most often, about 80% of the time, within a cloud but also occurs between cloud and ground. The effect of electrification is to create a charge distribution in which positive charges accumulate in the upper part of the cloud and negative charges accumulate in the middle and lower levels of the cloud. This further induces a positive charge on the ground surface which "shadows" the cloud as it moves along.

The electrical discharge, is initiated when the localized electric field exceeds 3 million V/m along a path several tens of meters long. For "cloud-to-ground" lightning, a faint stepped leader approaches the ground and a large luminous *return stroke* surges upward to the cloud along its path. The discharge takes only 100 μsec and our eye cannot discern the direction of the stroke but sees it as an instantaneous flash. The path of the electrical current is only a few centimeters wide but the stroke heats the air to about 30,000°C, five times the surface temperature of the sun. The resulting shock wave from the rapid expansion of the air is what we hear as thunder. While light travels at 10^9 km/hr, sound travels much more slowly, about 340 m/sec (1 km in 3 sec or 1 mile in 5 sec), so the distance between a lightning stroke and an observer can be calculated by timing the difference between the appearance of the flash and the sound of the thunder.

V. Atmospheric Circulations

A. Atmospheric Motions and the Coriolis Effect

On the largest scale, the atmosphere is driven by the imbalance between the absorption of solar radiation and the earth's emission of thermal radiation at different latitudes. In a longitudinally averaged sense and at latitudes below about 35°, the amount of radiation from the sun absorbed by the earth and its atmosphere exceeds the amount of terrestrial radiation lost to space. At higher latitudes, the reverse is true. This differential heating is compensated by atmospheric circulations (and by oceanic currents), which redistribute heat about the globe such that there is a net transfer of thermal energy poleward and a thermal balance is achieved at each latitude.

The atmosphere responds to a great variety of influences over a wide range of scales, from the small turbulent motions responsible for the gustiness of the wind near the ground, to the largest patterns that influence the climate of continents. The larger the scale of motion, the more important is the influence of the rotation of the earth about its axis. Essentially, the atmosphere is in motion in a noninertial (accelerating) frame of reference. Meteorologists compensate for this by introducing an apparent force, the *Coriolis force,* that acts on any object or mass of air in a state of motion. The effect of rotation on motions tangent to the surface of the earth is greatest at the poles and reduces to zero at the equator, such that the Coriolis force F_c is given by the following relationship:

$$F_c = (2\Omega \sin \phi)U,$$

where $\Omega = 7.292 \ 10^{-5}$ radians s^{-1} is the angular velocity of the earth, ϕ is the latitude angle, and U is the wind speed. The quantity $f = 2\Omega \sin \phi$ is commonly referred to as the Coriolis parameter. The Coriolis force acts perpendicularly to the velocity vector: to the right in the northern hemisphere and to the left in the southern hemisphere.

Except at low latitudes where the Coriolis effect has little impact, a first approximation to the balance of forces governing atmospheric flow is the state of *geostrophy* in which forces due to the pressure gradient and Coriolis turning are in balance. This is expressed mathematically in the following form:

$$U_g = -\frac{1}{\rho f}\frac{dp}{dn} \quad \text{or} \quad U_g = -\frac{g}{f}\frac{dz_p}{dn},$$

where U_g is the *geostrophic* wind speed, and n is the horizontal distance measured perpendicularly to the *isobars* (lines of constant pressure) in the direction of decreasing pressure. On weather charts representing conditions aloft, it is normal to show pressure patterns as contours of the topographic surface over which pressure is constant. In this case, the second form of the geostrophic equation is appropriate and z_p is the height above mean sea level of the specified pressure level. On weather maps, common mandatory pressure levels are 850, 700, 500, 300, and 200 mb.

Buys-Ballots law states that the geostrophic wind blows such that, in the northern hemisphere, low pressure or low height is on the left. In the southern hemisphere, it would be on the right. Thus, in the northern hemisphere, the wind flows anticlockwise around a low and clockwise around a high. With lower tropospheric temperatures at higher latitudes, pressure surfaces aloft generally slope downward toward each of the poles. The geostrophic condition then explains the normal westerly winds (winds from west to east) that characterize the middle latitudes. Such winds generally increase as the tropopause is reached and tend to be concentrated over a relatively small latitude band in each hemisphere to form the *jet streams.*

While providing a useful explanation for the primary aspects of mid- and high-latitude wind patterns, geostrophy provides only a crude approximation when air parcels are undergoing acceleration, such as in curved flow, or when friction at the surface decreases the wind speed below its geostrophic value. In this latter circumstance, the wind tends to spiral in toward a low pressure center and out from a high.

B. The General Circulation of the Atmosphere

Within bands extending approximately 20° north and south of the climatic equator, direct, thermally driven

Hadley cells consist of rising air near the equator, poleward motion aloft, sinking air near 20° latitude, and returning equatorward air at low altitudes to complete the circulation. The zone of rising motion is called the *intertropical convergence zone* (ITCZ) and is characterized by extensive convective and thunderstorm activity, and by high rainfall. This zone is frequently seen on satellite images as discontinuous bands of cloud circling the globe in equatorial latitudes, and coincides over land with tropical rainforest ecosystems. The two downwelling portions of the Hadley cells create *subtropical high pressure* regions at the earth's surface; the world's great deserts are found in these latitudes. The return flow near the surface exhibits the persistent *trade winds,* from the northeast in the northern hemisphere and from the southeast in the southern hemisphere.

The increasing strength of the Coriolis force prevents direct thermal circulations from extending further from the equator and, at higher latitudes, heat is directed poleward by wave and circulation patterns that are principally in the horizontal and associated with disturbances in the mid-latitude *westerlies*.

A major influence on the general circulation system is the presence of continents and oceans. Land masses warm up in the summer and cool in the winter much more than large bodies of water which mix and store great amounts of heat. Temperature changes at the ocean surface are much more conservative between seasons than is the case for a land mass. In summer the North Pacific and North Atlantic oceans are relatively cool, decreasing the temperature of the overlying air masses with the result that the cool dense air masses form the Pacific and Atlantic high pressures. Meanwhile, the continents exhibit lower pressure, especially over the extensive Euro-Asian continent. The reverse is true in wintertime. The ocean highs reduce in strength and are pushed southward and the northern oceans assume predominantly low pressure. Higher pressure builds up over the colder land masses at this time.

The predominance of such pressure patterns has a pronounced influence on the climate of certain regions of the globe. For example, the *monsoons* of southern Asia are a major perturbation on the simple view of a global circulation. Low pressure over the continent during the summer months moves the inter-tropical convergence zone onto the land mass and produces persistent heavy rainfalls at that time of year. The dry summers of California are another example, and result from the dominant high pressure over the Pacific.

C. The Westerlies and Extra-tropical Cyclones

To a large extent, the cycle of weather patterns experienced in the middle latitudes is explained by the waves that develop in the westerlies, and the travelling *cyclones* (regions of low pressure at the surface) and *anticyclones* (high pressure) associated with them, as seen on charts constructed from data collected with the global network of upper air and surface synoptic weather stations.

The main band of westerlies in the mid- to upper troposphere is characterized by sinusoidal-like waves of varying amplitude and wavelength, with north to south perturbations of a few degrees to a few tens of degrees latitude. These wavelike perturbations are called *Rossby* waves and propagate toward the east at a rate that is dependent on their wavelength and on the wind speed in mid tropospheric levels; the longer the wavelength of the wave, the slower the rate of travel. The very largest waves may even retrograde, and move toward the west. The overall picture is of long waves progressing slowly eastward, shorter waves rippling through them, and the air itself travelling eastward at even higher speed.

The amount of rotation of the atmosphere as the wind circulates around lows and highs and around the troughs and ridges of an upper air contour map is measured in units of *vorticity*. Rotation in the same sense as that of the earth (anticlockwise in the northern hemisphere, clockwise in the southern hemisphere) is considered to be positive and is called *cyclonic,* while that in the opposite direction is negative and *anticyclonic*. Vorticity can be measured relative to a frame of reference fixed to the earth: relative vorticity; or as the sum of relative vorticity and a component due to the fact that the earth itself is rotating: absolute vorticity.

As the air blows through and exits a mid- and upper tropospheric trough (a region of maximum absolute vorticity), conservation of angular momentum forces the air in the upper troposphere to become horizontally divergent, while in the lower troposphere, the air becomes horizontally convergent. This pattern of divergence and convergence downstream from the trough (to the east) is compensated by ascending air in this region. Ascent causes adiabatic cooling and usually results in condensation in the form of clouds and possible precipitation. On the other hand, upwind (to the west) of a trough, convergence aloft and divergence at lower levels is accompanied by subsidence, warming, drying of the air, and generally cloud-free skies.

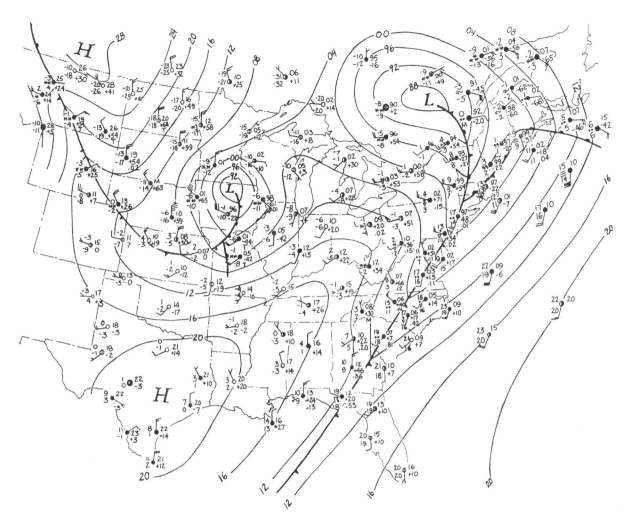

FIGURE 2 Surface synoptic chart at 12 GCT 20 November 1964. Stations report air temperature in Celsius (upper left of station circle); dew point (lower left); sea level pressure in millibars with either 1000 or 900 removed (upper right); pressure change in 10ths of a millibar over the previous 3 hr; direction from which the wind approaches (arrow); wind speed in knots, each long barb indicating a speed of 10 knots and a short barb for 5 knots; weather symbols appear to left of station circle; cloud amount is indicated within the circle. Contour lines are isobars reproduced every 4 mb and labeled with either 1000 or 900 removed. A cold front approaches the Atlantic coast, while a warm front moves northward through Nova Scotia. An occluded front extends from the low center in Quebec to the junction of the cold and warm fronts in Connecticut. [Reprinted with permission from Wallace, J. M., and Hobbs, P. V. (1977). "Atmospheric Science, An Introductory Survey," Fig. 3.8, p. 122. Copyright © 1977 by Academic Press, New York.]

Low pressure at the surface is usually found below the region of upper level divergence ahead of a trough aloft, implying a westward tilt to the system in its developing stage, a feature that disappears as the system matures and eventually dissipates. Convergence at low levels accentuates zones of interchange between air masses, creating or strengthening *fronts*. Fronts are labeled as *cold, warm, stationary,* or *occluded* according to their direction of progress or, in the case of the occluded front to indicate that they have been lifted off the surface by an advancing air mass. Rotation around the low distorts frontal boundaries to create a characteristic wavecrestlike structure; the low pressure is then called a wave- or frontal-cyclone with a warm front advancing ahead of the low as it rotates around the center, and a cold front advancing from behind and bringing cold air from the north (in the northern hemisphere). In the final stages, the cold front has advanced sufficiently that the warm air sector is lifted off the surface and the system is occluded. Figure 2 illustrates a typical frontal-cyclone in the northern hemisphere.

D. Tropical Cyclones

Tropical cyclone is the general name for a cyclone that forms over tropical oceans. Regional names for such cyclones are *hurricanes* for those forming over the Atlantic, Caribbean, and eastern Pacific; *typhoons* for those forming over the western Pacific; and *cyclones* for those forming over the Indian Ocean and those reaching the vicinity of Australia. By international agreement, tropical circulation systems are classified as tropical cyclones if they contain winds of 65 knots (34 m/sec) or greater. Lesser categories are: *tropical storm* (wind speeds between 35 and 64 knots), and *tropical depression* (wind speeds up to 34 knots).

Tropical cyclones are cyclonically rotating systems that originate between approximately 5° and 15° north or south latitude (a Coriolis force is needed to initiate rotation) over oceans with air and water temperatures of at least 27°C. They generally form in association with an easterly wave in the trade winds in an environment of weak vertical wind shear, and in one that sustains deep convection. Because of the absence of contrasting air masses, they are more circular than extra-tropical cyclones, particularly near their centers. Bands of cumulonimbus spiral in toward a central eye which exhibits a weak downdraft and is often cloud free. Total horizontal extent can range in size from diameters of 100 km to more than 1000 km but the highest wind speeds are found near the base of the wall of cumulonimbus surrounding the eye. Tropical cyclones generally travel westward, usually curving poleward as they reach land, and often recurving toward the east if they become caught up in the westerlies.

Tropical cyclones derive their energy from the release of latent heat as humid air from the tropical ocean converges at low level toward the center of the system, and condenses within the convective cloud bands. Catastrophic winds and torrential rainfall can cause extensive property damage and loss of life if a tropical cyclone passes over island communities or reaches shore, but once it moves inland, the system is deprived of its source of energy and gradually dissipates.

Bibliography

Ahrens, C. D. (1991). "Meteorology Today: An Introduction to Weather, Climate and the Environment," 4th ed. West, St. Paul, MN.

Grotjahn, R. D. (1993). "Global Atmospheric Circulations, Observations, and Theories." Oxford Univ. Press, New York.

Holton, J. R. (1992). "An Introduction to Dynamic Meteorology," 3rd ed. Academic Press, San Diego.

Lewis, R. P. W. (1991). "Meteorological Glossary," 6th ed. Her Majesty's Stationery Office, London, Millwood, NY.

Liou, K. N. (1992). "Radiation and Cloud Processes in the Atmosphere." Oxford Univ. Press, New York.

Peixoto, J. P., and Oort, A. H. (1992). "Physics of Climate." American Institute of Physics, New York.

Pielke, R. (1990). "The Hurricane." Routledge, New York.

Microclimate

KENJI KURATA, *University of Tokyo*

Glossary

Albedo Ratio of reflected solar radiation to incident solar radiation; radiation reflectivity of natural surfaces strongly depends on the wavelength of incident radiation; refers to the reflected fraction of incident solar radiation integrated over its whole wavelength range; albedo of the ground surface plays an important role in the energy budget at the ground surface; albedo of most natural surfaces under clear sky conditions is dependent on solar altitude and takes higher values for lower solar altitudes

Bowen ratio Ratio of the sensible heat flux at the underlying surface to the latent heat flux at the same surface and can be determined from measurements of temperature and humidity profiles above the ground; this ratio is often used to estimate the sensible and latent heat fluxes from measurements of the net radiation and the conductive heat flux into the soil

Monin-Obukhov length Parameter of atmospheric stability which has a dimension of length and is a function of fluxes of momentum and heat; many of the turbulence statistics in the surface layer can be expressed as a function of the height divided by the Monin-Obukhov length; length is positive for stable, infinitely large for neutral, and negative for unstable conditions

Penman-Monteith equation Equation formulating actual evapotranspiration from a plant canopy, as a function of net radiation, conductive heat flux into

the soil, temperature and humidity at a height above the canopy, and aerodynamic and canopy resistances; one of the advantages of this equation is that it requires measurements of the air temperature and humidity only at one height, rather than at two or more heights; difficulty in estimating the canopy resistance is one of the limiting factors for wide application

Roughness length Height at which the "concentration" of an entity transferred vertically in a surface layer assumes its surface value, where the zero height is adjusted due to the presence of roughness elements on the surface; in the case of momentum transfer, the roughness length is the height where the horizontal wind velocity assumes a value of zero

Stefan-Boltzmann law Law that states the radiant energy emitted by a black body is proportional to the fourth power of its absolute temperature; this is one of the basic laws which describe the radiation exchanges among materials on the earth's surface and the sky

Turbulent transfer coefficient Coefficient relating turbulent flux of an entity to the gradient of the potential of the entity; turbulent transfer coefficient for momentum is often called eddy viscosity; coefficient for sensible heat, water vapor, or any other scalar is often called eddy diffusivity; in a developed turbulent flow, the turbulent transfer coefficient is orders of magnitude greater than the corresponding molecular viscosity or diffusivity

Microclimate is a research area dealing with the states, fluxes, and transformations of mass, momentum, and energy, taking place in and through the atmosphere near the earth's surface. Interrelations of the atmospheric processes with living organisms' (vegetation, animals, human beings, etc.) activities and the soil surface are particularly important, and hence microclimate can be regarded as an interdisciplinary area of climatology (meteorology), plant and animal physiology, ecology, hydrology, and soil physics.

I. Introduction

Understanding of climatological processes near the ground surface gives basic knowledge in controlling or improving the physical environment of crops, and thus promotes crop growth and higher yield. It also constitutes an important part of the environmental sciences as well. Air pollutant fluxes near the ground surface are governed by microclimatological processes. Understanding the desertification process requires microclimatological knowledge. Formulations of momentum and heat fluxes near the earth's surface give boundary conditions of the global circulation models, which are and have been developed for predicting climate change caused by the increase in greenhouse gas concentrations in the atmosphere. [See DESERTIFICATION OF DRYLANDS; METEOROLOGY.]

Main climatological factors near the ground are irradiance, wind, air temperature, air humidity, and soil surface temperature. To understand the mechanism of microclimate creation, fluxes of relevant entities need to be characterized, and include shortwave and longwave radiation, momentum flux, sensible and latent heat fluxes, and conductive heat flux in the soil. These fluxes (except for momentum flux) are mutually related in the sense that transformations of fluxes (energy forms) take place through interactions with vegetation and soil. For example, incident shortwave radiation is absorbed by a plant canopy and transformed into sensible and latent heat fluxes into the atmosphere. These transformation processes are affected by the surface characteristics (vegetation, topography, soil moisture content, etc.). Momentum flux is not involved in these transformation processes but is closely related to the turbulent transport of sensible and latent heat, carbon dioxide, and other gases.

This article describes microclimate mechanisms focusing on the above-mentioned fluxes and the heat balance at the ground. Measurement techniques or estimation methods of each flux are also addressed. The main concern in this article is the microclimate near and within plant canopies.

II. Fluxes near the Ground

A. Radiation

1. Radiation Laws

Materials emit and absorb electromagnetic radiation. An ideal material which absorbs all incident radiation and emits maximum radiation at all wavelengths is called a black body. The spectral distribution of emitted radiation from a black body is strongly dependent on temperature and is given by Plank's Distribution Law. Examples of spectral distributions of black body radiation are given in Fig. 1, where black body temperatures are assumed to be 6000 K (approximately equal to the sun's surface temperature) and 300 K (approximately equal to the earth's surface temperature). Higher black body temperature gives shorter peak wavelength, λ_{max}, (wavelength at which emitted radiant energy per unit wavelength takes a maximum value) in the spectral distribution. The value of λ_{max} is given by Wien's Displacement Law:

$$\lambda_{max} = 2898/T,$$

where λ_{max} is given in micrometers and T means the absolute temperature (unit: K). Temperatures of 6000 and 300 K result in λ_{max} of 483 nm (1 nm = 10^{-9} m = 10^{-3} μm) and 9.66 μm, respectively.

Figure 1 shows that most of the energy emitted from a black body at a temperature of 6000 K (hence that of solar radiation) is contained within the wavelength range from 0.15 to 3 μm, while most of the radiant energy emitted from bodies at temperatures prevailing on the earth's surface is confined to the wavelengths between 3 and 100 μm. Because the overlap between the spectral distributions is negligible, the regions are distinguished from each other by the terms "shortwave radiation" and "longwave radiation."

Radiant flux emitted from a unit black body surface per unit time (Φ in W m^{-2}) is given by the Stefan-Boltzmann Law:

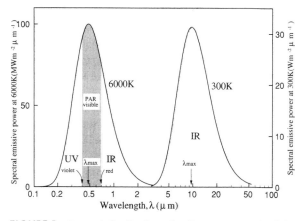

FIGURE 1 Spectral distribution of radiant energy emitted from black bodies at temperatures of 6000 and 300 K.

$$\Phi = \sigma T^4,$$

where σ is the Stefan-Boltzmann constant (5.67×10^{-8} W m^{-2} K^{-4}). Radiant flux emitted from a "grey" body is calculated by multiplying the right-hand side of the above equation by a constant ε ($0 \le \varepsilon \le 1$):

$$\Phi = \varepsilon \sigma T^4,$$

where ε is termed emissivity of the material. Kirchhoff showed from theoretical considerations that the emissivity of a material is equal to the absorptivity of the material, where absorptivity is defined as a fraction of the incident radiation absorbed by the material.

2. Solar Radiation

The solar radiant flux density (solar radiant energy which passes through a unit plane perpendicular to the flux per unit time) at the average distance of the earth from the sun, without atmospheric attenuation, is called the solar constant. It is well known that this "constant" varies by a small amount over time, due to, e.g., solar activity variation. The approximate value of the solar constant is 1.37 kWm^{-2}.

As illustrated in Fig. 1, the solar radiation spectrum can be divided into three portions: ultraviolet (200–400 nm), visible/PAR (400–700 nm), and infrared (>700 nm). Ultraviolet radiation less than 290 nm, which is harmful for living creatures on the earth, is almost completely absorbed by ozone in the stratosphere and does not reach the earth's surface. The visible portion approximately overlaps the region of the spectrum in which photosynthesis is stimulated and therefore is called photosynthetically active radiation (PAR). The PAR fraction of the total solar radiation energy at the earth's surface is approximately 50%. The infrared spectrum in solar radiation is often called near infrared, but the boundary between near infrared and far infrared differs in the literature.

The solar radiant flux is attenuated when passing through the atmosphere due to scattering by air molecules and aerosols and absorption by gases. Both scattering and absorption processes strongly modify the quantity and quality (spectral distribution) of the solar radiation. Scattering by particles much smaller than the radiation wavelength, such as air molecules, is called Rayleigh scattering, in which scattering is identical in forward and backward directions. Aerosols scatter solar radiation in a different mode: forward scattering has more energy than backward. This is called Mie scattering.

The main molecules present in the atmosphere which absorb solar radiation are ozone, water vapor, carbon dioxide, and oxygen. Absorption is a spectrum-selective process and as a result of absorption by these gases, the spectral distribution of solar radiation at the earth's surface has a "zigzag" form. Ozone has a strong absorption band in the ultraviolet portion, while water vapor and carbon dioxide absorb radiation in the infrared wavelengths.

Solar radiation which directly reaches the ground from the direction of the sun is called direct radiation. All other scattered radiation (including radiation reflected by the ground and scattered in the atmosphere) which reaches the ground from the sky or from clouds is called diffuse radiation. Direct radiant flux density (radiant energy passing through a unit plane perpendicular to the radiation per unit time) at the ground, I_g, is given by Bourguer's Law:

$$I_g = I_n \tau^m,$$

where I_n is the extraterrestrial radiant flux arriving outside the atmosphere, τ is the transmissivity of the atmosphere, and m is the air mass through which the radiation traverses. The value of I_n is given as (solar constant/r^2), where r is the distance from the earth to the sun divided by its average. The transmissivity of the atmosphere is highly dependent on the amount of water vapor and aerosol contained in the atmosphere. Average monthly values measured at 14 locations in Japan are given in Fig. 2. Transmissivity takes lower values in summer than in winter due to higher water vapor content in summer. The air mass is defined to be unity when the sun is directly overhead (the solar altitude h is 90°). The value of m is approximately equal to cosec h.

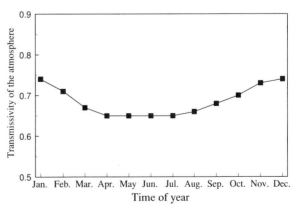

FIGURE 2 Mean monthly transmissivity of the atmosphere (average of 14 locations in Japan).

Incident solar radiation on a surface is partly reflected. The fraction of the total incident radiant energy which is reflected is called albedo. The values of albedo of most natural surfaces under clear sky conditions are dependent on the solar altitude, and hence show diurnal variations as well as seasonal variations. When the solar altitude is below 30°, the albedo increases as the solar altitude decreases. The approximate daily mean albedo value of fresh snow cover is 0.75–0.95, thus reflecting most of the incident solar radiation. The albedo of soil is a function of the moisture content: it takes higher values as the moisture content decreases. Reported values, for example, are as follows: wet sand (0.18–0.24), dry sand (0.37–0.42), wet loam (0.08–0.14), dry loam (0.17–0.30), wet clay (0.08), and dry clay (0.14). The albedo of farm crops ranges between 0.15 and 0.26, while forests or orchards have lower albedo (0.12–0.18) due to stronger "trapping" of the incident solar radiation by taller vegetation.

3. Terrestrial Radiation

Materials on the earth's surface and in the atmosphere have much lower temperatures than the solar surface, and therefore the radiation they emit has much longer wavelengths than solar radiation (longwave radiation, see Fig. 1). Although we are usually not aware of the fact that materials around us emit and receive longwave radiation, longwave radiation exchange plays an important role in microclimate mechanisms. We can distinguish two components of terrestrial radiation: downward radiation from the atmosphere (atmospheric radiation) and upward radiation from the ground.

The downward atmospheric radiation is emitted by some gases (in particular, water vapor and carbon dioxide) contained in the atmosphere. Therefore, the magnitude of the downward radiation is a function of the profiles of the atmosphere's temperature and water vapor content. Because these data are seldom available, and also the downward radiation is not measured at most of the meteorological observatories, empirical formulae for estimating the downward radiant flux density, L_d, have been developed. Under clear skies the value of L_d is approximately given by

$$L_d = \sigma(T_a - 20)^4,$$

where T_a is the air temperature (K) near the ground. This means that under clear skies the effective temperature of the sky, assuming that the sky is a black body, is 20° below T_a. Another formula often used for clear sky conditions is that of Yamamoto-Brunt:

$$L_d = \varepsilon_a \sigma T_a^4,$$
$$\varepsilon_a = 0.51 + 0.066\sqrt{e},$$

where e is the vapor pressure near the ground, given in mb. This formula assumes that the sky is a black body with the same temperature as the air near the ground but the apparent emissivity, ε_a, is a function of the vapor pressure.

The presence of clouds increases downward radiation, because clouds are more perfect emitters than water vapor in the infrared portion. Several formulae have been proposed to estimate L_d under cloudy conditions as a function of the cloud cover fraction of the sky and the cloud type.

The absorbed portion of L_d by the ground surface is $\varepsilon_s L_d$, where ε_s is the emissivity (therefore, by Kirchhoff's principle, absorptivity) of the surface. The rest, $(1 - \varepsilon_s)L_d$, is reflected toward the sky. Therefore, the upward longwave radiation, L_u, consists of two components:

$$L_u = \varepsilon_s \sigma T_s^4 + (1 - \varepsilon_s)L_d,$$

where the first term represents the radiation emitted from the ground, and T_s is the surface temperature (K). The net loss of longwave radiation from the surface, F, is given as

$$F = \varepsilon_s(\sigma T_s^4 - L_d) = L_u - L_d.$$

The number of measurements of ε_s is restricted compared to other factors such as albedo. The data of ε_s for the same type of surface do not always agree among different measurements, presumably because of the different characteristics of the sensors. Most of the measurements of ε_s for natural surfaces vary within 0.85–0.99. In practical applications, it is often assumed that ε_s is unity.

4. Radiation Balance at the Ground Surface

The net gain of radiation by unit ground surface per unit time, R_n, is called net radiation and is given by

$$R_n = (1 - \rho)S_t + L_d - L_u,$$

where ρ means the albedo of the surface, and S_t is the total shortwave radiation received by the horizontal surface. The amount of S_t consists of the direct radiation components, S_b, and the diffuse radiation components, S_d:

$$S_t = S_b + S_d = I_g \sin h + S_d.$$

The above relations are schematically illustrated in Fig. 3, and some measurements on the diurnal varia-

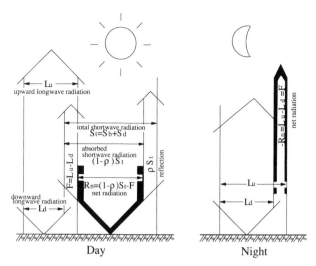

FIGURE 3 Schematic illustration of radiation balance at the ground. [Adapted from Uchijima, Z. (1980). Solar energy absorption and partitioning at cultivated land. *In* "New Energy Application to Controlled Environment Agriculture" (T. Takakura *et al.*, eds.), Fig. 19, p. 59. Copyright © 1980 by Fuji Technosystem, Tokyo.]

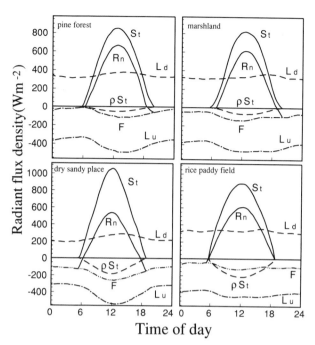

FIGURE 4 Examples of diurnal variations of radiation balance under clear sky conditions. [Adapted from Uchijima, Z. (1980). Solar energy absorption and partitioning at cultivated land. *In* "New Energy Application to Controlled Environment Agriculture" (T. Takakura *et al.*, eds.), Fig. 34, p. 72. Copyright © 1980 by Fuji Technosystem, Tokyo.]

tions of each component under clear sky conditions are given in Fig. 4. The characteristics of the radiation balance at a dry sandy place are much different from those of pine forest, marshland, and paddy field. The sandy area receives less downward atmospheric radiation compared to other places due to less vapor content in the atmosphere, and emits more upward radiation during the daytime due to higher surface temperature. Moreover, the albedo of the sandy surface is larger than that of other surfaces. As a result, the ratio of the net radiation to the total shortwave radiation, R_n/S_t, during the daytime is much smaller at the dry sandy area than at other surfaces.

B. Turbulent Fluxes

1. Turbulent Flux Mechanism

Fluxes of momentum, sensible heat, latent heat (water vapor), carbon dioxide, etc., in the atmosphere near the ground surface have a strong effect on microclimate. These fluxes are governed by turbulent flow. Turbulence is a characteristic of fluid flow that is irregular, random, chaotic, and rotational and has a strong diffusive ability. Figure 5 schematically illustrates temporal variations of vertical velocity w and concentration s of some entity S (amount of the entity S contained in a unit volume air) measured at a fixed point in a turbulent flow. Measured values show strong fluctuations around their mean values.

The entity S is transferred by turbulent air movement. If a parcel of air, whose S concentration is s, moves upward at the vertical velocity of w, the instantaneous flux of S, defined as the amount of S transferred upward through a unit horizontal plane at the measurement height per unit time, is given by the product sw. This product also fluctuates as illustrated in Fig. 5. However, we are not concerned with instantaneous values of flux but with its average value. The mean value of sw during a certain period gives the vertical flux F_s of the entity S: $F_s = \overline{sw}$, where the bar means the temporal average. We decompose w and s into their mean values $\overline{w}$ and $\overline{s}$ and deviations from the mean values w' and s': $w = \overline{w} + w'$, and $s = \overline{s} + s'$. Putting these decompositions into the above equation results in $F_s = \overline{sw} + \overline{s'w'}$, where we utilized the relations $\overline{s'} = 0$ and $\overline{w'} = 0$. If we assume we are measuring above a homogeneous horizontal terrain, the relation $\overline{w} = 0$ holds, and we get

$$F_s = \overline{s'w'}. \tag{1}$$

In the above description, molecular diffusion was neglected, because it is orders of magnitude smaller than that by turbulent air movement.

To parameterize the flux $F_s = \overline{s'w'}$, a simple for-

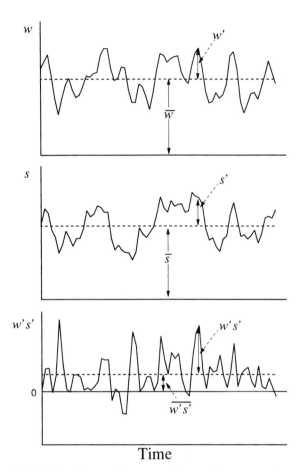

FIGURE 5 Schematic illustration of temporal variations of vertical velocity component w, concentration of an entity s, and product $w's'$ in a turbulent flow. The bar indicates a mean value and the (') indicates deviation from the average.

mulation of flux often used in other areas such as molecular diffusion is also adopted: flux is proportional to the potential gradient. This assumption gives:

$$F_s = -K_s \frac{\partial \bar{s}}{\partial z}, \qquad (2)$$

where z is the height. The minus sign on the left-hand side means that the flux is directed from the higher potential side to the lower potential side. The coefficient K_s corresponds to the diffusivity in the case of molecular diffusion and called turbulent transfer coefficient, which has a unit of $m^2 \ s^{-1}$, irrespective of the entity S.

2. Momentum Flux and Wind Profile

In this and following subsections, vertical fluxes within a surface layer above a homogeneous horizontal terrain are considered. The surface layer is an atmospheric layer immediately above the ground sur-

face, where vertical variations of vertical fluxes can be ignored. The thickness of the surface layer varies within 10–100 m.

Let us first consider the flux of horizontal momentum ρu, where ρ is the density of air and u is the horizontal velocity component in the mean direction of the wind. Substituting ρu for s in Eq. (1) gives the vertical momentum flux, $\rho \overline{u'w'}$. Because wind velocity increases upward, the momentum is transferred downward. The downward momentum flux $\tau = \rho \overline{u'w'}$ has the same dimensions as the shearing stress (momentum/area/time = force/area) and is called Reynolds stress.

From the correlation, $\overline{u'w'}$, a quantity having a dimension of velocity can be defined $u_\star = \sqrt{-\overline{u'w'}}$, where $u_\star$ is called the friction velocity.

Substituting ρu for s in Eq. (2) and equating the upward momentum flux with the negative of the shearing stress gives

$$\tau = \rho K_m \frac{\partial \bar{u}}{\partial z}, \qquad (3)$$

where the coefficient K_m is the turbulent transfer coefficient for momentum and is often referred to as the eddy viscosity. The coefficient K_m increases with the wind velocity, which is proportional to $u_\star$, and also with the height, and therefore the simplest assumption for K_m for neutral conditions (referred to later) is

$$K_m = k \ u_\star \ z,$$

where k is a constant. Putting this equation into Eq. (3) and integrating yields

$$\bar{u} = \frac{u_\star}{k} \ln \left(\frac{z}{z_0} \right), \qquad (4)$$

where z_0 is the height at which $\bar{u}$ is assumed to be zero and is termed roughness length. In the region just above the roughness elements the wind profile deviates from that given by Eq. (4). Therefore, in reality at $z = z_0$, $\bar{u}$ might not be exactly zero. The constant k is called the von Karman constant and is often assigned the value of 0.4.

The more general wind profile above a tall roughness under neutral conditions is given by

$$\bar{u} = \frac{u_\star}{k} \ln \frac{(z - d)}{z_0}, \qquad (5)$$

where d is called the displacement length and represents the adjustment of zero height due to the presence of roughness elements such as plant canopies and houses. As illustrated in Fig. 6, the extrapolated value of $\bar{u}$ is zero at $z = d + z_0$. Typical values of z_0 and d for plant canopies are $z_0 = 0.13h$ and $d = 0.64h$, where h is the canopy height. However, it is well known that these values vary with, e.g., canopy density, canopy structure, flexibility of the plant, and wind velocity. As the canopy density increases, d also increases, while z_0 takes a maximum value at a certain canopy density.

Equation (5) is valid under neutral conditions in which vertical heat flux is negligible. These conditions occur, for example, under very cloudy conditions with strong winds. Under sunny conditions, the ground surface is heated by solar radiation and upward heat flux takes place (unstable condition), while on a clear night with weak wind the ground surface loses energy by longwave radiation and downward heat flux takes place (stable condition). The air temperature decreases with increased height under unstable conditions, and increases with increased height under stable conditions (more exactly, the potential temperature gradient is the criterion for stability, but in this article we do not distinguish temperature from potential temperature, because we consider only a small height difference).

For parameterization of the atmospheric stability, two parameters are often used: the Richardson number and the Monin-Obukhov length, L. Because recently Monin-Obukhov length has been used more commonly, we also adopt it in this article. The Monin-Obukhov length is defined as

$$L = -\frac{\rho c_{\mathrm{p}} \bar{T} u_\star^3}{kgH},$$

where c_{p} is the specific heat of air for constant pressure,

g is the gravitational acceleration, and H is the vertical sensible heat flux (upward positive). The value of L is independent of the height and is negative under unstable conditions, infinitely large under neutral conditions, and positive under stable conditions. We define a nondimensional variable, ζ, by $\zeta = (z - d)/L$. Applying the Monin-Obukhov similarity relations to the wind shear yields

$$\frac{k(z - d)}{u_\star} \frac{\partial \bar{u}}{\partial z} = \phi_{\mathrm{m}}(\zeta), \qquad (6)$$

where ϕ_{m} is a nondimensional function with a smaller or larger value than unity for unstable or stable conditions, respectively. Integration of this equation results in an equation of the form

$$\bar{u} = \frac{u_\star}{k}\left[\ln\frac{z - d}{z_0} - \psi_{\mathrm{m}}(\zeta)\right]. \qquad (7)$$

A number of empirical formulae have been proposed for ϕ_{m}. The most commonly used Businger-Dyer form for unstable conditions is: $\phi_{\mathrm{m}} = (1 - 16\zeta)^{-\frac{1}{4}}$. This yields

$$\psi_m = \ln\left[\left(\frac{1 + x^2}{2}\right)\left(\frac{1 + x}{2}\right)^2\right] - 2\arctan x + \frac{\pi}{2}, \qquad (8)$$

where $x = (1 - 16\zeta)^{\frac{1}{4}}$.

For stable conditions the formula, $\phi_{\mathrm{m}} = 1 + 5\zeta$, is often used, which yields $\psi_{\mathrm{m}} = -5\zeta$. Figure 7 shows nondimensional wind profiles, $ku/u_\star$, for different stability conditions. Some experiments have shown, however, that the Monin-Obukhov similarity theory

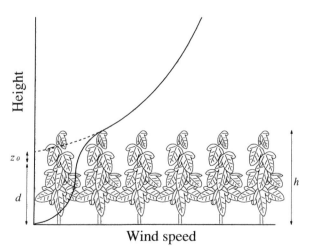

FIGURE 6 Schematic illustration of the wind profile above and within a plant canopy.

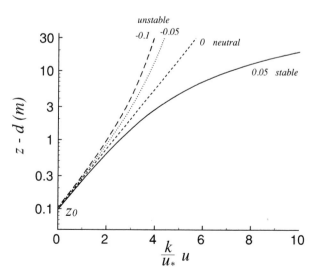

FIGURE 7 Nondimensional wind profile, $ku/u_\star$, for $z_0 = 0.1$ m. Numbers in the figure indicate $1/L$ (m^{-1}), where L is the Monin-Obukhov length.

is not valid near the roughness elements. In particular, over surfaces with large values of z_0, such as forests, this limitation on the validity of the theory restricts its practical applications.

3. Sensible, Latent Heat Fluxes

For vertical fluxes of scalars, such as sensible heat, latent heat (water vapor), carbon dioxide, and other gases over homogeneous terrain, approaches similar to that for estimating the momentum flux can be followed. For sensible heat flux, the concentration s defined earlier can be regarded as sensible heat contained in a unit volume of air, given as $c_p\rho T$. For water vapor flux, s can be regarded as the mass of water vapor per unit volume of moist air (absolute humidity) which is given as ρq, where q is the mass of water vapor per unit mass of moist air (specific humidity). For other fluxes of scalars, the same argument holds as that for the water vapor in this subsection. For example, flux of carbon dioxide can be parameterized in the same way as the water vapor flux parameterization that follows in this subsection, by regarding q as mass of carbon dioxide per unit mass of air.

Substituting $c_p\rho T$ and ρq for s in Eqs. (1) and (2) results in

$$H = c_p\rho \overline{T'w'} = -c_p\rho K_h \frac{\partial \overline{T}}{\partial z}, \qquad (9)$$

for sensible heat flux H, and

$$E = \rho \overline{q'w'} = -\rho K_v \frac{\partial \overline{q}}{\partial z}, \qquad (10)$$

for water vapor flux E, where K_h and K_v are turbulent transfer coefficients for sensible heat and water vapor, respectively.

Defining nondimensional ϕ-functions in a way similar to that in Eq. (6)

$$-\frac{kc_p\rho u_\star(z-d)}{H} \frac{\partial \overline{T}}{\partial z} = \phi_h(\zeta), \qquad (11)$$

$$-\frac{k\rho u_\star(z-d)}{E} \frac{\partial \overline{q}}{\partial z} = \phi_v(\zeta), \qquad (12)$$

and integrating yields formulae similar to those of Eq. (7)

$$\overline{T}_s - \overline{T} = \frac{H}{kc_p\rho u_\star}\left[\ln\left(\frac{z-d}{z_{0h}}\right) - \psi_h(\zeta)\right], \qquad (13)$$

$$\overline{q}_s - \overline{q} = \frac{E}{k\rho u_\star}\left[\ln\left(\frac{z-d)}{z_{0v}}\right) - \psi_v(\zeta)\right], \qquad (14)$$

where T_s and q_s are, respectively, the temperature at $z = d + z_{0h}$ and the specific humidity at $z = d + z_{0v}$, and z_{0h} and z_{0v} are the roughness lengths for sensible heat and water vapor.

The Businger-Dyer form of the above ϕ-functions for unstable conditions is $\phi_h = \phi_v = \phi_m^2 = (1 - 16\zeta)^{-\frac{1}{2}}$, which yields

$$\psi_h(\zeta) = \psi_v(\zeta) = 2\ln\left[\frac{1+x^2}{2}\right],$$

where $x = (1 - 16\zeta)^{\frac{1}{4}}$ as in Eq. (8).

For stable conditions, $\phi_h = \phi_v = \phi_m = 1 + 5\zeta$ is often used and therefore, $\psi_h = \psi_v = \psi_m = -5\zeta$.

C. Conductive Heat Flux at the Ground Surface

A part of the net radiation received at the ground surface during the daytime is transferred into the soil by conduction, while at night, heat thus stored during the daytime is released into the atmosphere from the ground surface. Thus, conductive heat flow also contributes to the energy balance at the ground surface. The vertical conductive downward heat flux, F_g, can be written as

$$F_g = -\lambda\frac{\partial T}{\partial z},$$

where z is the depth, defined as positive downward, and λ is called the thermal conductivity (W m^{-1} K^{-1}). The downward heat flux at the soil surface, $G = (F_g)_{z=0}$ is, therefore, given as

$$G = -\lambda\left(\frac{\partial T}{\partial z}\right)_{z=0}.$$

Assuming λ is independent of the depth, the temperature variation is formulated as

$$\frac{\partial T}{\partial t} = \alpha\frac{\partial^2 T}{\partial z^2},$$

where $\alpha = \lambda/C_s$ is the thermal diffusivity, and C_s is the volumetric heat capacity of the soil (J m^{-3} K^{-1}). Intensive studies have been conducted on two important parameters of the soil, λ and C_s. The volumetric heat capacity is the sum of contributions from the soil components and can be written as follows, using De Vries's data of specific heat and the density of each component:

$$C_s = 10^6 \times (1.94\,x_m + 2.50\,x_o + 4.18\,x_w),$$

where x_m, x_o, and x_w are the volume fraction of mineral, organic matter, and water, respectively.

The thermal conductivity, λ, is also a function of the water content and increases with increasing water content. Measured data of moist soils ranges from 0.5 (peat soil) to 2.5 (sand) W m^{-1} K^{-1}.

III. Heat Balance at the Ground Surface

The net radiation received by materials on the ground surface is transformed into several forms of energy such as sensible heat and latent heat. Different surface types have different characteristics of this transformation. To quantitatively understand the energy transformation, we think of the energy budget for a layer of surface materials. For a layer of plant canopy, the following equation holds:

$$R_n = H + lE + G + Q, \qquad (15)$$

where R_n is net radiative flux density at the upper surface of the canopy, H is sensible heat flux density, lE is latent heat flux density (l: latent heat of vaporization), G is conductive heat flux density into soil, and Q is heat stored in the canopy layer per unit ground area. All fluxes on the right-hand side are positive when fluxes are leaving the layer. The heat storage term, Q, consisting of sensible and latent heat storage in the canopy air and biomass heat storage, is usually negligible compared to R_n. However, some experimental measurements have shown that this storage term also contributes to the energy budget in the case of forest, which has larger air mass and biomass. Energy consumed for photosynthesis can be neglected.

Figure 8 shows a heat budget analysis for 12-m-tall natural forest with prevailing species of red pine in Japan and that of lawn which is adjacent to the forest, on 2 clear days in summer, July 15 and July 30. July 15 was in the middle of Baiu (rainy season) in Japan and therefore the soil underneath the canopies was wet. On July 30, Baiu was over, and during the preceding 14 days no precipitation was recorded. Note that net radiation is smaller above the lawn than above the forest, due to the larger value of albedo of the lawn. On July 15, a larger part of the net radiation was partitioned into latent heat at both canopies. On July 30, the latent heat flux from the lawn into the atmosphere was suppressed and the sensible heat flux was the largest component, while at the forest, the latent heat flux was larger on July 30 than during Baiu. The suppression of evapotranspiration from the lawn on July 30 was due to the water shortage in soil, while promotion of evapotranspiration from the

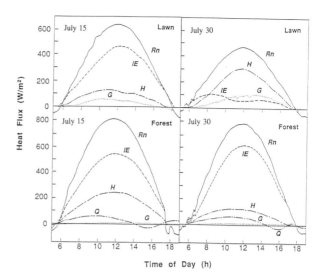

FIGURE 8 Diurnal variations in heat balance components at a natural forest and a lawn on two clear days in summer. [Adapted with permission from Harazono, Y., Kiyota, M., and Yabuki, K. (1992). Characteristics of thermal environment and water budget of a forest and a lawn developed. *J. Agric. Meteorol.* **48** (2), 147–155, Fig. 2, p. 150. Copyright © 1992 by the Society of Agricultural Meteorology of Japan. The Q-component was added to the original figure with permission from the first author.]

forest, which had a larger soil water capacity, was caused by the drier air above the forest (greater driving force of evapotranspiration).

IV. Measurement and Estimation of Sensible Heat and Water Vapor Fluxes

A. Aerodynamics-Based Methods

Various methods have been proposed for measuring sensible or vapor fluxes, based on the statistics of turbulence or wind and relevant profiles. Direct measurement of correlations, $\overline{u'w'}$, $\overline{T'w'}$, or $\overline{q'w'}$ is called the eddy correlation method and has become more widely applied with the progress of instrumentation technology. The sensors must have a short enough response time (minimum of 10 Hz fluctuations must be followed in open field measurements), and very accurate sensor alignment is required (less than 1° error). For measurement of wind fluctuations, w', a sonic anemometer is often used.

From the measurements of mean wind speed, temperature, and humidity at two or more heights, fluxes of the sensible heat and vapor can be calculated based on the flux-profile relationships presented earlier. For neutral conditions, the identity of the turbulent transfer coefficient, K, is often assumed such that $K_m = K_h = K_v$. For nonneutral conditions, integration of Eqs. (7) and (13) from z_1 to z_2 yields

$$\overline{u_2} - \overline{u_1} = \frac{u_\star}{k}\left[\ln\left(\frac{\zeta_2}{\zeta_1}\right) - \psi_m(\zeta_2) + \psi_m(\zeta_1)\right],$$

$$\overline{T_1} - \overline{T_2} = \frac{H}{kc_p\rho u_\star}\left[\ln\left(\frac{\zeta_2}{\zeta_1}\right) - \psi_h(\zeta_2) + \psi_h(\zeta_1)\right],$$

where subscripts 1 and 2 refer to heights z_1 and z_2, respectively. For calculating fluxes, an iterative method is required. First, the Monin-Obukhov length, L, is roughly estimated. This L-value yields ζ_1 and ζ_2. With these ζ-values, $u_\star$ and H are estimated using the above two equations. These H- and $u_\star$-values give a new value for L. This procedure is repeated until estimated values converge. Using thus determined values, the flux of water vapor can be calculated with the integrated form of Eq. (14).

If the surface temperature, $\overline{T_s}$, and the surface humidity, $\overline{q_s}$, can be easily measured or estimated, fluxes can be estimated using bulk transfer coefficients, C_h and C_v, and measurements of wind speed, $\overline{u}$, temperature, T, and specific humidity, $\overline{q}$ at one height. Equations (13) and (14) can be transformed into the forms: $H = c_p\rho\,C_h\,\overline{u}\,(\overline{T_s} - \overline{T})$ and $E = \rho\,C_v\,\overline{u}\,(\overline{q_s} - \overline{q})$. The coefficients, C_h and C_v, are determined either through experiments or theoretically. In the latter case, it is usually assumed that $z_0 = z_{0h} = z_{0v}$ for simplicity, although there is no physical reason for this assumption. This bulk coefficient method is often applied to fluxes above water surfaces.

B. Heat-Balance Based Methods

The ratio of sensible heat flux to latent heat flux, $\beta = H/lE$, is an important microclimatological parameter and called the Bowen ratio. This ratio can be calculated with air temperatures and specific humidities at two heights, using the equation

$$\beta = \frac{c_p(\overline{T_1} - \overline{T_2})}{l(q_1 - q_2)},$$

which can be obtained by integrating Eqs. (9) and (10) with respect to z, and with the assumption that $K_h = K_v$. The fluxes of sensible heat and latent heat can be obtained from measured values of this ratio and $R_n - G$ in Eq. (15):

$$lE = \frac{R_n - G}{\beta + 1}$$

$$H = \beta lE,$$

where the heat storage term Q in Eq. (15) is neglected.

Various formulations of evapotranspiration from the plant canopy have been proposed using resistance analogues. Among them is the well known Penman-Monteith equation:

$$lE = \frac{\Delta(R_n - G) + c_p\rho[e_s(\overline{T}) - \overline{e}]/r_a}{\Delta + \gamma(1 + r_c/r_a)},$$

where $\overline{T}$ and $\overline{e}$ are the measured or estimated air temperature and water vapor pressure at a certain height above the canopy, respectively, $e_s(\overline{T})$ is the saturation vapor pressure at $\overline{T}$, Δ is the change of the saturation vapor pressure e_s with temperature, $\partial e_s/\partial T$, at temperature $\overline{T}$, $\gamma = c_p P/(0.622l)$ is the psychrometer constant (approximately 67 Pa K^{-1}: P is the atmospheric pressure), and r_a and r_c are the aerodynamic resistance and the canopy resistance, respectively. The Penman-Monteith equation is often used because it requires measurement or estimation of $\overline{T}$ and $\overline{e}$ at one height only, in addition to measurement of $R_n - G$. However, difficulties of estimating resistances, r_a and r_c, restricts its application.

V. Microclimate within Plant Canopies

Much effort has been devoted to investigation of solar radiation penetration into plant canopies in relation to canopy photosynthesis. It is well known that solar radiant flux, S, within a plant canopy (either that of the direct radiation, or diffuse radiation, or the sum of both) can be expressed as

$$S = S_o\exp(-k\,F_l),$$

where S_o is the solar radiant flux above the canopy, k is the extinction coefficient, and F_l is the downward cumulative leaf area index. F_l is given as

$$F_l = \int_z^h F_a dz,$$

where F_a is the leaf area density (leaf area per unit volume), and h is the canopy height. The extinction coefficient k is a function of the canopy structure (e.g., leaf inclination) and the solar altitude. For horizontal leaves, k takes a value of unity.

The wind speed decreases rapidly with depth into the canopy due to the momentum absorption by plant elements. The air temperature has a maximum value during the daytime at the layer where most of the solar radiation is absorbed. At night, the minimum air temperature can be found at the layer where the maximum longwave radiation energy loss takes place.

Acknowledgments

The author thanks Masashi Komine and Nancy Okamura for assistance in preparation of this article.

Bibliography

Asrar, G. (guest ed.) (1990). Special issue: Land surface-atmosphere interactions. *Agric. Forest Meteorol.* **45** (1–2).

Aylor, D. E. (guest ed.) (1989). Special issue: Biometeorology. *Agric. Forest Meteorol.* **47** (2–4).

Carlson, T. N. (guest ed.) (1991). Special issue: Modeling stomatal resistance. *Agric. Forest Meteorol.* **54** (2–4).

Monteith, J. L., and Unsworth, M. H. (1990). "Principles of Environmental Physics," 2nd ed. Edward Arnold, London.

Myneni, R. B., Ross, J., and Asrar, G. (1989). A review on the theory of photon transport in leaf canopies. *Agric. Forest Meteorol.* **45**, 1–153.

Russell, G., Marshall, B., and Jarvis, P. G. (ed.) (1989). "Plant Canopies: Their growth, Form and Function," Cambridge Univ. Press, Cambridge.

Schmugge, T. J., and André, J.-C. (ed.) (1991). "Land Surface Evaporation Measurement and Parameterization." Springer-Verlag, New York.

ten Berge, H. F. M. (1990). "Heat and Water Transfer in Bare Topsoil and the Lower Atmosphere," Pudoc, Wageningen.

Minerals, Role in Human Nutrition

YOSHINORI ITOKAWA, *Kyoto University, Japan*

I. Definition and Classification of Minerals
II. Mineral Deficiency Diseases
III. Diseases Caused by the Excessive Intake of Minerals

Glossary

Essential minerals Minerals which must be ingested everyday in order to maintain health and performance
Major minerals Minerals which are present in the body in large amounts and have a daily requirement more than 100 mg
Mineral deficiency Lack of a mineral to the extent that it is severe enough to cause disease
Mineral toxins (or mineral poisoning) Disease caused by excessive amounts of mineral
Trace elements Elements with a daily requirement less than 100 mg which are present in the body of adults at concentrations less than 10 g

Minerals are classified as essential and nonessential on the basis of their physiological requirement for humans. Essential minerals need to be ingested on a daily basis, and can be further divided into major minerals and trace elements. The major minerals include calcium, phosphorus, potassium, sulfur, chlorine, sodium, and magnesium. They are present in the body in large amounts and their daily requirement is also large. In contrast, essential trace elements are present in the body at low levels and their daily requirement is small. The essential trace elements which have been widely recognized include iron, iodine, zinc, copper, manganese, selenium, cobalt, and mollybdenum. Although the other trace elements, including fluoride, silicon, rubidium, lead, cadmium, vanadium, arsenic, nickel, tin, and lithium, have been found to be biologically essential in animal studies, these elements are not recognized as essential for humans since their functions are not yet clearly understood. Specific deficiency diseases are caused by the deficiency of any of these essential minerals; however, excessive intake of minerals can also induce specific diseases. [*See* Food Composition.]

I. Definition and Classification of Minerals

The term "mineral" is not a definite concept but is used arbitrarily to mean many things. In the fields of medicine and nutrition, "minerals" are defined as elements contained in living organisms other than carbon, nitrogen, oxygen, and hydrogen which compose the principal constituents of organic substances in human body such as carbohydrates, fat, and protein. Thus, "mineral" is nearly a synonym for "inorganic substance."

A variety of elements are thus considered to be minerals, and various classifications for them can be devised. First, minerals can be chemically classified as metals and nonmetals. As common features, metals are lustrous, conduct electricity and heat well, and have malleability (the capacity to deform) as well as ductility (capacity to stretch). These properties can be easily understood from the actual observation of gold, silver, iron, and other metals. Minerals which do not have these features are categorized as nonmetals, e.g., phosphorus, selenium, and iodine. The group of minerals that include boron, silicon, germanium, and arsenic has features intermediate between those of metals and nonmetals, and these substances are called metalloids. Metals are also divided into heavy metals and light metals. Those with a specific gravity above 4 are defined as heavy metals (e.g., gold, silver, and lead), while those with a specific gravity below 4 are defined as light metals (e.g., calcium and magnesium).

Minerals may also be classified on a biological basis, and they are divided into essential and nonessential minerals on the basis of their indispensability for humans. A mineral is considered to be essential if certain abnormalities are caused by its deficiency and if these abnormalities resolve following correction of the deficiency. Only a limited number of minerals have actually been confirmed to cause deficiency diseases in humans. However, if the deficiency of a certain mineral causes disease in animals or birds, the mineral is also inferred to be essential for humans even if a specific deficiency disease has not been detected. The 26 minerals listed in Table I have been found to be biologically essential in animal studies. However, the importance of minerals other than those indicated as essential is not proven since their functions are not yet clearly understood. The number of essential minerals may increase in the future, since some of the minerals that are now thought to be nonessential may be shown to be necessary by further studies.

As also shown in Table I, the essential minerals are classified into major and trace minerals. Although the definition of trace minerals is not same as that of trace elements, all the trace elements are actually minerals. The distinction between major and trace minerals is not based on their intrinsic properties but on the amount present in the body or the required daily intake. Trace elements have a required daily intake less than 100 mg. Table II shows the mineral content of an average daily diet, which was calculated at our department. According to the above definition, there are 5 major minerals (sodium, potassium, phosphorus, calcium, and magnesium), and the 13 minerals listed below iron in Table II are categorized as trace minerals.

The minerals contained in the human body are listed in Table I. The five major minerals, as classified on the basis of dietary content, are also considered to be major minerals on the basis of their biological content. The major minerals are made up of nonmet-

TABLE I
Features of Essential Minerals

	Mineral	Essentiality	Chemical feature	Approximate amount in adult human body (mg)
Major minerals	Calcium	★	Light metal	1,160,000
	Phosphorous	★	Nonmetal	670,000
	Potassium	★	Light metal	150,000
	Sulfur	★	Nonmetal	112,000
	Chlor	★	Nonmetal	85,000
	Sodium	★	Light metal	63,000
	Magnesium	★	Light metal	25,000
Trace elements	Iron	★	Heavy metal	4,500
	Fuloride	★★	Nonmetal	2,600
	Silicon		Metalloid	2,300
	Zinc	★	Heavy metal	2,000
	Rubidium		Light metal	360
	Lead		Heavy metal	120
	Copper	★	Heavy metal	80
	Cadmium		Heavy metal	50
	Vanadium	★★	Heavy metal	18
	Arzenic		Metalloid	18
	Manganese	★	Heavy metal	15
	Iodine	★	Nonmetal	15
	Selenium	★	Nonmetal	13
	Nickel		Heavy metal	10
	Molybdenum	★	Heavy metal	9
	Tin	★★	Heavy metal	6
	Chromium	★	Heavy metal	2
	Cobalt	★	Heavy metal	2
	Lithium		Light metal	2

★ established; ★★ possible.

TABLE II
Contents of Minerals in Japanese Food[a]

	Mineral	Urban food	Rural food
Major minerals	Sodium	4541	5584
	Potassium	1865	2348
	Phosphorus	1210	1317
	Calcium	630	505
	Magnesium	206	292
Trace elements	Iron	11.7	10.9
	Zinc	13.8	11.9
	Silicon	21.5	19.9
	Manganese	8.4	3.0
	Copper	2.8	1.4
	Lead	0.12	0.15
	Vanadium	0.14	0.11
	Nickel	0.06	0.11

[a] Three meals and snacks.

als and light metals, while the trace minerals are mainly heavy metals.

II. Mineral Deficiency Diseases

In humans, homeostasis works to prevent the deficiency of major minerals and hence diseases do not usually arise from mineral deficiency unless it persists for a long period. However, once a mineral deficiency disease develops, no effective treatment is available and the consequences are often serious. Some trace minerals such as iron and iodine commonly cause deficiency diseases. In addition, trace element deficiency can be induced by certain types of medical technology. For some trace minerals, it is still unknown what type of disease their deficiency may cause. Thus, many issues remain unsolved concerning mineral deficiency diseases. In this article, the diseases associated with mineral deficiency in humans will be reviewed in detail.

A. Sodium Deficiency

Loss of appetite, nausea, fatigue, myalgia, and hemoconcentration occur in humans fed a sodium-free diet. Such abnormalities resolve when salt and water are administered. If the patient is left untreated, however, coma will develop and death will result. The minimum daily requirement for sodium is 2.5 g on a salt-equivalent basis. Laborers who work in high-temperature environments will become depleted of salt as a result of sweating, so they may require 10–15 g of salt daily. Firemen or miners working under high temperature and humidity may develop heat cramps (myalgia of the four limbs and cramping of the abdominal muscles) due to salt depletion resulting from excessive perspiration. If such symptom appear, saline should be administered.

B. Potassium Deficiency

Hypopotassemia in humans can be caused by a lack of potassium in the diet, urinary potassium loss due to diabetes mellitus, or renal impairment, and potassium loss from the gastrointestinal tract due to diarrhea and vomitting. The symptoms associated with hypopotassemia are weakness, loss of appetite, myasthenia, confusion, hypotension, arrhythmias and tachycardia, ECG abnormalities, and alkalemia. Potassium replacement will correct these symptoms. However, potassium deficiency is influenced by the balance with sodium. Even in a potassium-deficient state, no appreciable symptoms of alkalemia or hypopotassemia will occur if sodium is also deficient. Conversely, if an excessive amount of sodium is retained in the body because of a sodium excretion defect, alkalemia or hypopotassemia will be enhanced.

C. Calcium Deficiency

The adult human body contains about 1 kg of calcium, most of which exists in the hard tissue of the bones and teeth. The bones and teeth provide a reservoir of calcium, and calcium continuously moves between the blood and bones.

When an individual is placed on a calcium-free diet, osseous calcium is consumed and calcium deficiency symptoms do not develop immediately. However, the bones and teeth become weak. A long-term low calcium intake may lead to osteoporosis with aging. In chronic calcium deficiency, the bones become fragile and prone to fracture, although bone pain rarely occurs. Dental disease and osteoporosis are more frequent in women who have given birth, suggesting that they were in a calcium-deficient state during pregnancy and lactation. In children, the teeth are formed during fetal life and infancy so that the children of mothers with a low calcium intake have weak teeth and are prone to dental caries. Such children may also have poor development of the jaw bones as well as a poor set of teeth.

Osteomalacia, a disease with somewhat similar symptoms, is largely due to vitamin D deficiency rather than calcium deficiency.

D. Magnesium Deficiency

Magnesium deficiency may occur in persons who have low magnesium intake and have impaired absorption due to chronic diarrhea and increased excretion from long-term diuretic therapy or excessive alcohol intake.

The symptoms induced by magnesium deficiency are tremor and muscular cramps, as well as mental abnormalities such as depression, anxiety, and confusion. Recent attention has also been paid to the association between chronic magnesium deficiency and heart disease, and it has been speculated that the relationship between magnesium and calcium levels may have an important role in the development of heart disease due to magnesium deficiency. Figure 1 shows the dietary calcium/magnesium (Ca/Mg) ratios and the nation-specific mortality rates for ischemic heart diseases (IHD: myocardial infarction and angina pectoris). In nations showing a high dietary Ca/Mg ratio, such as Finland, the United States, and the Netherlands, the IHD-related mortality is high, while the IHD-related mortality is low in nations with a low dietary Ca/Mg ratio, such as Japan, Yugoslavia, and Greece. Thus, there is a correlation between the dietary Ca/Mg ratio and IHD-related mortality. Figure 2 shows a comparison of Mg and Ca concentrations in the heart between patients with cardiac death and accidental death. The cardiac Mg concentration is lower and the cardiac Ca concentration is higher

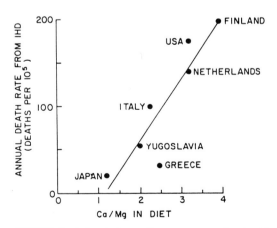

FIGURE 1 Ischemic heart disease rates correlated with dietary calcium/magnesium ratios. [From Karppanen, H., *et al.* (1978). *Adv. Cardiol.* **25,** 9–24.]

in the patients with cardiac death than in those with accidental death.

In a dog model of myocardial infarction, it was found that the infarct size was increased by magnesium deficiency. Thus, more attention may have to be paid to magnesium deficiency in relation to cardiovascular disease.

E. Iodine Deficiency

Goiter is caused by iodine deficiency. This is thought to occur because the two thyroid hormones (triiodothyronine and thyroxine) are not produced effectively in the presence of iodine deficiency, so that the thyroid gland becomes enlarged in an attempt to produce more of the hormones.

Iodine is mostly excreted in the urine and the iodine intake can be estimated by measuring the urinary iodine content. Figure 3 shows the urinary iodine excretion and the incidence of goiter in various nations and localities. The data suggest that iodine deficiency goiter is likely if the daily urinary iodine excretion decreases to below 50 μg.

Although goiter is the most obvious feature of iodine deficiency, the knowledge of iodine deficiency has greatly expanded in recent years and the term iodine deficiency disorders (IDD) is used to describe effects of iodine deficiency which include effects on growth or development as well as goiter.

F. Zinc Deficiency

A genetically transmitted intestinal inability to absorb zinc causes acrodermatitis enteropathica, a congenital disease featuring crust-forming dermatitis in the periorbital and perioral regions as well as on the hands and feet. Hair is lost and the patient eventually becomes bald. If the patient is left untreated, dysgenesis and malnutrition will increase the susceptibility to infectious diseases and may lead to death. Injection of zinc will produce a marked improvement.

Episodes of dermatitis resembling acrodermatitis enteropathica and hair loss have been reported in patients receiving total intravenous alimentation with inadequate zinc. Photographs of dermatitis in such patients are presented in Figs. 4 and 5. For trace elements such as zinc, the daily requirement is very small and deficiency symptoms rarely appear when a normal diet is consumed. However, when patients depend on an artificial intravenous alimentation mixture for their entire nutrition and when no trace elements are added to the solution, disease may eventually be

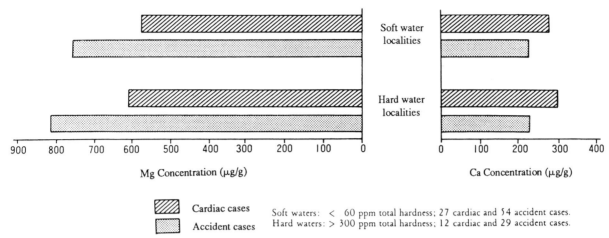

FIGURE 2 Myocardial magnesium and calcium concentrations in autopsy studies. [From Marier, J. R., *et al.* (1979). National Research Council of Canada. 17581, 65–84.]

caused by trace element deficiency. This is one of the issues raised by advanced medical technology that has to be handled with care.

A study of neonates and infants in Japan performed nationwide in major hospitals has shown that children with zinc deficiency have poor weight gain, anemia, and hair loss. Such children were found in large numbers and the majority of them were under 1 year of age. This was considered to be because diseases causing malabsorption such as diarrhea are common in neonates and also because the contents of zinc are low in premature infants.

The symptoms described above are typical manifestations of zinc deficiency, but mild zinc deficiency is

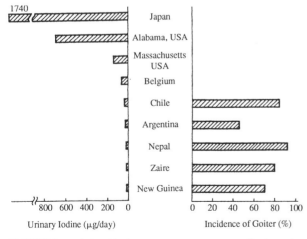

FIGURE 3 Relationship between average amount of urinary iodine excretion and the incidence of goiter in various nations. [From Ermans, A. M. (1978). "The Thyroid," p. 537. Harper & Row, New York; and Katsura, E., *et al.* (1959). *Eiyo to Shokuryo* (in Japanese) **12,** 34–36.]

also associated with the delayed healing of injuries or burns. In addition, low zinc levels in hair or blood have been reported in about half of the individuals with taste blindness, a condition in which food cannot be tasted.

G. Copper Deficiency

Menkee's kinky hair syndrome is a congenital disease related to the intestinal inability to absorb copper, in which characteristic curly hair appears, cramps develop, muscle tone decreases, and mental development is retarded.

Figure 6 shows the fetal copper content in relation to the gestational age. As seen in this figure, the fetal copper content increases rapidly in late gestation, so that premature infants delivered before their copper stores in liver are adequate are likely to develop copper deficiency after birth. When the copper content of mother's milk is low, copper deficiency also develops in young infants. Poor weight gain, leukopenia, anemia, and osteopenia occur in nursing infants with copper deficiency.

In a manner similar to zinc deficiency, copper deficiency also can develop in patients undergoing long-term total intravenous alimentation. Anemia is the most common symptom in such patients. Copper is contained in serum ceruloplasmin, which plays a role in oxidizing bivalent iron to trivalent iron and in the transport of oxidized trivalent iron. Thus, anemia is probably induced by copper deficiency because the function of ceruloplasmin is inhibited and it becomes impossible for iron to be utilized biologically. The induction of bone abnormalities by copper deficiency

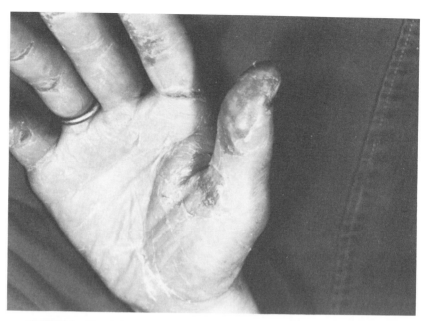

FIGURE 4 Zinc deficient dermatitis in patient receiving total parenteral nutrition.

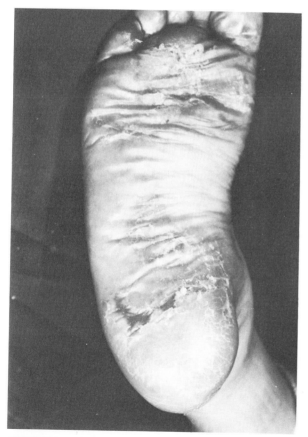

FIGURE 5 Zinc deficient dermatitis in patient receiving total parenteral nutrition.

is thought to be ascribable to the fact that copper is a constituent of lysyloxydase, an enzyme involved in crosslinking of collagen.

H. Manganese Deficiency

When a manganese-free diet was given to young males ages 19 to 22 years, the urinary and fecal manganese levels decreased rapidly while there was no significant change in the serum manganese level. Most of the subjects complained of pruritus in the legs, ankles, and chest, and an erythematous skin rash was observed on the upper abdomen, the inguinal region, and the legs. These symptoms were diagnosed as miliaria crystalline, and the skin rash eventually healed spontaneously.

Although manganese deficiency can be produced experimentally as above, clinical manganese deficiency has not been described except under special circumstances. However, some reports have recently been published on manganese deficiency in patients receiving long-term intravenous alimentation. One of the cases reported was a girl with short intestine syndrome who was receiving intravenous alimentation. Since her gain in height was poor, skeletal examination was performed. This revealed a notched scapula with increased opacity and also showed increased X-ray lucency at the ends of the long bones of the legs. No manganese was detected in the bone biopsy sample and the serum manganese level was low, so

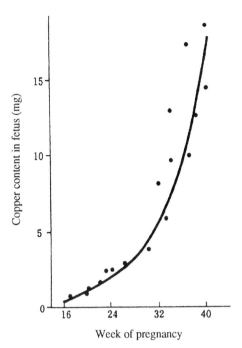

FIGURE 6 Copper content in fetus in relation to the gestational age. [Shaw, J. C. L. (1980). *Am. J. Dis. Child.* **134,** 74.]

manganese deficiency was diagnosed. When manganese was administered, her height increased rapidly and a normal weight was attained after 4 months of manganese treatment along with an increase in the serum manganese level. Manganese activates glycosyltransferase, an enzyme that synthesizes bone matrix mucopolysaccharide. Therefore, decreased activity of this enzyme may have caused the bone abnormalities associated with manganese deficiency in this patient. According to our research, the daily dietary intake of manganese is 3–9 mg in the Japanese population, and manganese deficiency will not occur so long as normal diet is consumed. However, because manganese deficiency is a potential complication in patients who are receiving long-term total intravenous alimentation, manganese deficiency should be monitored. Measurement of manganese level in the lymphocytes can be useful for a diagnosis.

I. Selenium Deficiency

An endemic disease of unknown origin occurred frequently in the Keshan district of Manchuria (Heilungchiang) in northeastern China around 1935. This disease caused many deaths and was known as the "strange disease of Keshan." It was confirmed to be a myocardial disease and was named Keshan disease.

The Health Committee of Manchuria investigated the pathogenesis of this disease and concluded that carbon monoxide poisoning was the cause. However, Chinese researchers recently published a new view on Keshan disease, stating that the disease is caused by selenium deficiency. The soil selenium concentration is low in the area with a high incidence of Keshan disease and therefore the agricultural products of that area contain little selenium and selenium deficiency occurs in the inhabitants. They also reported that blood selenium concentrations were low in the inhabitants of this region. In addition, the incidence of Keshan disease has decreased markedly as a result of the administration of selenium preparations to the people living in the district.

III. Diseases Caused by the Excessive Intake of Minerals

When there is excessive intake of an essential mineral, intestinal absorption is regulated and renal excretion is increased so that toxicity does not usually develop. However, when mineral intake exceeds the capacity of these regulatory systems, mineral hyperalimentation disease may be induced. The excessive intake of a nonessential mineral often causes serious illness and is usually called poisoning. The distinction between essential and nonessential minerals depends on the daily requirement, and some of the minerals which are now considered nonessential may actually have some biologically indispensable activity. In the following section, the typical diseases arising from excessive mineral intake are reviewed.

A. Sodium Hyperalimentation

Sodium is usually available as sodium chloride (salt), in which form it is bound to chloride. Sodium hyperalimentation can therefore be regarded as salt hyperalimentation. There is a close relationship between salt intake and hypertension. Population-based epidemiologic studies have shown that hypertension and stroke are more frequent in areas where the salt intake is higher. The native inhabitants of the Pacific islands as well as the highlanders of New Guinea are known for their low daily intake of salt (1–3 g) and they have a low incidence of hypertension. In contrast, the daily salt intake in the Tohoku district of Japan is very high at 15 g, and a high incidence of stroke has been reported in this district. The correlation between salt intake and hypertension has thus been demonstrated by population-based studies. However, comparison among individuals shows little correlation between salt intake and hypertension, and this may be explained by the large inter-individual variations in salt tolerance. The conclusion of studies in this field is that excessive salt intake is a risk factor for cerebrovascular disease.

Many studies have shown that an increase in blood pressure due to sodium hyperalimentation can be inhibited by the administration of potassium. Excessive potassium administration will increase urinary sodium excretion and lead to sodium loss. Thus, sodium hyperalimentation can be modified by altering the potassium intake.

B. Iodine Hyperalimentation

While iodine deficiency causes goiter, this disease is also induced by excessive iodine intake. The costal area of Hokkaido in Japan is known for a high incidence of goiter. The area is famous for the production of tangle and 16 g of tangle is eaten daily per person. While the average daily intake of iodine in Japan is approximately 1 mg, the intake in this district is 50–80 mg.

C. Arsenic Poisoning

Arsenic is categorized as an essential trace element, but it is also known to be very toxic and has been used for both suicide and murder. In the Okayama Prefecture of Japan, large-scale arsenic poisoning was caused in 1955 by powdered milk which had been contaminated with arsenic. Poisoning developed in about 12,000 infants and 131 of them died. This was reported to be the world's largest food poisoning incident. It turned out that 2–5 mg of arsenic was ingested daily by the infants. Typical symptoms of arsenic poisoning were observed, including fever accompanied by hypokinesis and loss of appetite, sleep disorder, diarrhea, vomitting, coughing, lacrimation, dermatosis, melanoderma, hyperkeratosis, and anemia.

D. Lead Poisoning

Lead ingested in the diet enters the liver, and the majority of it is excreted into the bile while some accumulates in hard tissues such as bone. Lead accumulated in the bone is eluted under various circumstances, and causes toxicity to target tissues such as the bone marrow when the serum lead concentration becomes too high. Hematological disorders such as anemia then develop. Lead poisoning is associated with symptoms such as malaise, headache, fatigue, and anorexia, as well as with more specific symptoms such as constipation, lead colic (periumbilical pain), a lead-pale face (the pallor is due to capillary constriction), and a lead rim (the gingival rim is colored black-purple due to lead sulfide).

E. Mercury Poisoning

Mercury has not yet been established as an essential trace element, but it seems likely to be proven essential in the future. On the other hand, mercury is also very toxic and mercury poisoning has been known since ancient times. Mercury poisoning is manifested primarily as neurological symptoms with tremor, vertigo, irritability, moodiness, depression, stomatitis, and diarrhea. Symptoms of organic mercury poisoning include loss of vision and hearing, and mental deterioration.

The daily intake of mercury per person is 40 μg for organic mercury in Japan, and this level of mercury intake is higher than elsewhere in world. Thus, the Japanese should be concerned about mercury poisoning. Table III shows the mercury content of various foods. As seen in this table, the mercury content is higher in large sea fish such as tuna and shark. The

TABLE III

Total Mercury and Organic Mercury in Various Foods (μg/100 g)

Food	Total mercury	Organic mercury
Large tuna	159.8	139.0
Small tuna	76.0	58.5
Shark	31.9	26.2
Goby	22.0	1.0
Oyster	18.3	0
Mackeral	17.9	0.1
Prawn	12.0	0
Egg	6.0	0
Rice	5.9	0.8
Beef	5.1	0.4
Soybean	4.7	0
Chinese cabbage	3.7	0
Spinach	3.4	0
Sea bream	3.4	2.2
Radish	3.2	0.1
Eel	3.0	1.9
Onion	2.3	0
Sardine	1.6	0
Orange	1.0	0
Milk	0.7	0

mercury content of rice reflects not only the level naturally present in the environment but also residual agricultural chemicals in soil and mercury leached from mines. Thus, the higher mercury intake in Japan is accounted for by the traditional Japanese diet in which fish and rice are staple.

The worst poisoning on record is the Minamata disease in Japan. Two outbreaks of severe methylmercury poisoning occurred in 1956 and 1964, when organic mercury was discharged into an oceanic bay (Minamata) and into the river (Agano river near Niigata). Fish in the bay and river accumulated the organic mercury about 1,000,000 times the water concentration and people who consumed seafood for several years manifested neurological and mental disabilities. Numbers of certified Minamata disease patients by Japanese government reached 1373 in Minamata and 641 in Niigata.

F. Cadmium Poisoning

Like mercury, cadmium is thought to be biologically essential in trace amounts, although it shows strong toxicity as well.

The dietary cadmium intake in Japan is about 50 μg per person daily, and half of the daily intake is ingested as rice. This cadmium intake is higher than that in Europe (10–30 μg). The visceral cadmium concentration is accordingly higher in the Japanese than in Europeans.

About 5% of the cadmium ingested is absorbed from the intestines and 30–50% of this absorbed cadmium accumulates in the liver and kidneys. When renal cadmium accumulation becomes excessive, renal dysfunction is induced. Renal impairment is said to appear when the daily cadmium intake is continuously 250 μg or more. Cadmium first accumulates in the proximal tubular epithelium of the kidneys and

causes degeneration of this epithelium. The lesions then extend to the distal convoluted tubules and the glomeruli. Bone is also affected by cadmium poisoning, and osteomalacia is the typical bone disease.

Sometimes, river water polluted by cadmium from mining is used for irrigation of rice fields in Japan. As a result, the cadmium content of rice in some areas is increased to more than 10 times. Since rice is a staple food for Japanese, the daily cadmium intake is elevated in these area.

Itai-itai disease meaning ouch-ouch disease had prevalence endemically in Toyama prefecture from the early 20th century. The rice fields of this area had been polluted by cadmium in waste from mines. Most of the patients were postmenopausal women and the most characteristic features of the disease are lumbar pains and leg myalgia. Pressure on bones produces further pain. Renal dysfunction is considered by the symptoms of proteinuria, glucosuria, and aminoaciduri. Based on large-scale epidemiological and clinical studies by the Japanese government, the Japanese Ministry of Health and Welfare declared in 1968 that "The Itai-itai disease is caused by chronic cadmium poisoning, on condition of the existence of such inducing factors as pregnancy, lactation, imbalance in internal secretion, aging, deficiency of calcium, etc." The number of Itai-itai disease patients certified by the Japanese government from 1962 to 1975 was 129.

Bibliography

Itokawa, Y., and Durlach, J. (1989). "Magnesium in Health and Disease." John Libbey, London/Paris.

Levander, O. A., and Cheng, L. (1982). "Micronutrient Interactions: Vitamins, Minerals and Hazardous Elements." The New York Academy of Science, New York.

Underwood, E. J. (1977). "Trace Elements in Human and Animal Nutrition." Academic Press, New York.

Natural Rubber Production in Arid and Semiarid Zones

DAVID MILLS, *Ben-Gurion University of the Negev*

Glossary

Apomixis Asexual reproduction by seeds that occurs without fertilization of the ovum cell; facultative apomixis involves some degree of sexual fertilization

Arid and semiarid zones Regions characterized by a hot dry climate; a semiarid zone ranges from 53 to 118 units on Emberger's scale of 8–1000 units for which values below 53 units are considered to represent a humid climate, and those above 118 units an arid climate

Latex Milky colloidal suspension present in about 12,500 species of plant belonging to 20 families; the latex contains a variety of organic compounds, such as carbohydrates (sometimes rubber) and proteins

Natural rubber High-molecular-weight hydrocarbon, *cis*-1 : 4-polyisoprene, which is usually found in latex of some plants and obtained by coagulation of the latex with chemicals, by drying, or by other processes

Resin Mixture of a number of organic compounds such as polyphenols, cinnamyl derivatives, terpenoids, and fatty acids (linoleic, linolenic, palmitic, and stearic acids)

Water use efficiency (WUE) Amount of biomass, fruit, seed, or secondary metabolites produced by a unit of water absorbed by a plant

Rubber production in arid and semiarid zones provides an alternative to the sole commercial source of natural rubber, the tropical rubber tree, *Hevea brasiliensis*. An alternative source of natural rubber is strategically important to the western industrial world, which currently relies on a rubber supply from remote sources that could become unavailable due to political turmoil. Natural rubber is a commodity of critical importance to the defense and economy of many nations. Natural rubber has performance specifications that cannot be met by its synthetic alternatives and is preferred in applications that require good elasticity, flexibility, and resilience, together with low heat build-up. Natural rubber constitutes 30% of the total world rubber production.

This article describes rubber-producing plants native to semiarid zones, with emphasis on guayule, *Parthenium argentatum* Gray, which is the best alternative to the rubber tree in nontropical areas, and which has a long history of cultivation and commercialization.

I. Rubber-Producing Plants of Arid and Semiarid Zones

Rubber is synthesized in over 2000 plant species, representing the families Apocynaceae, Ascepiadaceae, Aseraceae (Compositae), Euphorbiaceae, Loranthaceae, Moraceae, Rubiaceae, Caryophyllaceae, Boraginaceae, Onagraceae, Labiatae, and Sapotaceae. Most species of these families are native to tropical zones, but a number are indigenous to arid, semiarid, and temperate zones. Rubber-producing species of four families native to arid and semiarid zones are listed in Table I.

Most of these species contain a low concentration of rubber, up to 10%, which usually has a low molecular weight. One such rubber-containing species is rabbitbrush (*Chrysothamnus nauseosus*), a perennial plant native to the western United States. The plant is fast

TABLE I

Rubber-Producing Plants Native to Arid and Semiarid Zones

Family	Species	Geographical distribution	Rubber content (%)[a]
Apocynaceae			
	Amsonia palmeri	S.W. USA	2.5
	Apocynum androsaemifolium	USA	0.4–5.1
	Apocynum cannabinum	USA	0.4–5.1
Asclepiadaceae			
	Asclepias albicans	S.W. USA	0.9–5.4
	A. californica	California	4.1 (leaves)
	A. eriocarpa	California	2.4 (leaves)
	A. erosa	S.W. USA	2.5–13
	A. latifolia	C. USA, N. Mex.	2–3.8
	A subulata	S.W. USA	0.5–0.6
Compositae			
	Chrysothamnus nauseosus	W. USA	0.1–6.7
	C. paniculatus	S.W. USA	1.2–3.2
	C. teretifolius	S.W. USA	1.7–4.5
	Ericameria laricifolia	S.W. USA	2–5.2
	E. nanum	W. USA	6–10
	Guardiola platyphylla	S.W. USA	1.5–2.3
	Helianthus occidentalis	TX and AR	1.6 (leaves)
	Parthenium argentatum	Mexico, Texas	Up to 26
	P. tomentosum and *P. schottii*	Mexico	0.1–0.5
	P. fruticosum	Mexico	0.1–0.5
	P. incanum	Mexico, Texas	1–2
	Scorzonera tau-saghyz	Kazakhstan	Up to 30 (roots)
Euphorbiaceae			
	Euphorbia balsamifera	Sahel, Africa	Up to 20 (latex)
	E. resinifera	North Africa	Up to 20 (latex)
	Jatropha cardiophylla	S.W. USA	3 (stems)
	Pedilanthus bracteatus	N. Mexico	28 (latex)
	P. macrocarpus	N. Mexico	6–10

Source: Bowers, J. E. (1990). "Natural Rubber-Producing Plants for the United States." NAL, Beltsville, MD.
[a] Where not mentioned, rubber was analyzed from the whole plant.

growing and salt and cold tolerant, and it can grow in alkaline soils. The rubber content is low, being 1–2% of the plant dry weight (maximum recorded 6%), and the rubber has a low molecular weight of 50,000. *Amsonia palmeri* is a perennial desert herb that contains 2.5% rubber. It is slow growing but can be harvested twice a year. *A. palmeri* also has potential as a feed stock for fermentation and for use as an animal feed. *Apocynum androsaemifolium* and *A. cannabinum* are perennial herbs of the forest, prairie, pasture, and roadside. High-quality rubber of 0.2–5.2% has been reported for the whole plant and 2–3% for latex. Of the two, *A. cannabinum* is a better candidate for rubber exploitation since it produces more biomass. *Asclepias* spp. are native to the deserts of the southwestern United States and northern Mexico. Rubber of 0.4–13%, on a whole plant basis, is found

mainly in the leafless stems in some species of this genus and in the leaves in others. Rubber level is usually higher during the dormant season. An annual rubber yield of 60 kg/ha from *A. subulata* is expected after harvesting a dense cultivation after 3 years, followed by another harvest of the regrowth 3 years later. Of the genus *Parthenium*, *P. schottii*, *P. tomentosum*, and *P. fruticosum*, which are native to the dry zones of northern Mexico, are tall shrubs, 4–10 m high, and contain only traces of rubber of molecular weight 2000–50,000. *P. incanum* (mariola), 1–1.5 m in height, contains a somewhat higher content of rubber, with a higher molecular weight of 150,000.

High rubber producers are *Euphorbia balsamifera*, *E. resinifera*, *Pedilanthus bracteatus*, *Scorzonera tau-saghyz*, and *Parthenium argentatum*. The *Euphorbia* species are latex-bearing, slow-growing succulent trees and shrubs. *E. balsamifera, E. resinifera,* and *P. bracteatus* have a high rubber content in the latex (up to 20–28% rubber of the dry latex). The rubber is, however, of low molecular weight. The feasibility of growing shrubby *Euphorbia* as an agricultural plant has not been determined. *Scorzonera tau-saghyz* is a perennial herb native to the high mountains of Kazakhstan. It tolerates severe winters, dry summers, and low annual precipitation. Rubber of high molecular weight accumulates in the roots (up to 30% of the root), this figure being the highest known in any tissue of rubber-bearing plants. This level is reached, however, after 5 years of growth. This species is very susceptible to fungal and insect infestation, which may limit its commercial exploitation.

Guayule (*P. argentatum* Gray) produces the highest quantity of rubber (per plant or per unit area) of high molecular weight ($1–2 \times 10^6$) among species native to arid and semiarid zones or to nontropical zones. The potential of guayule as a commercial source of natural rubber was realized as long ago as the previous century. The first large-scale commercial exploitation of rubber from guayule was undertaken in 1888 by the New York Belting & Packing Co., with the extraction of rubber in hot water from about 50 tons of guayule shrub. During 1902–1904 the pebble mill extraction method was developed, leading to the establishment of several factories in Mexico, mainly by the Continental–Mexican Rubber Co. Rubber extracted from wild stands of guayule was used mainly in the manufacture of automobile tires, and by 1910 about 50% of U.S. rubber came from guayule. Since natural sources of the shrub rapidly became depleted and factories were forced to close down, cultivation of the species was initiated in Mexico in 1910 and continued in the United States after the 1912 Mexican Revolution. Extensive commercial cultivation took place between 1931 and 1941, mostly in the Salinas Valley of California, where a total of about 1.4×10^6 kg of rubber was processed. In 1942, after the rubber supply from Southeast Asia was cut off, the U.S. government, through the U.S. Department of Agriculture (USDA) and other departments, initiated the Emergency Rubber Project (ERP), a massive project involving more than 1000 scientists and technicians and many more workers. During a brief period of 3.5 years, the ERP produced a total of about 1.4×10^6 kg of rubber from shrubs grown on 13,000 ha in Arizona, Texas, and California. After World War II, due to the development of synthetic rubber, surplus stocks of *Hevea* rubber, and economic and strategic considerations, the project was stopped

at a stage at which only 15% of the planted guayule had been utilized. Plantations were burned or disked, and research gradually slowed down until it came to a stop in 1959. Natural stands of guayule in Mexico were harvested and processed until 1950, about 133×10^6 kg of rubber having been produced from the beginning of the century. Renewed interest in guayule in the United States and other countries commenced with the significant rise in oil prices and the energy crisis in the early 1970s. Since then, research on various aspects of guayule cultivation has been carried out in the southwest United States in universities and by the USDA. Recently, a project has been completed to supply about 50 tons of rubber from 111 ha of plantations in the Gila River Indian Community in Arizona; the rubber was processed in a new prototype pilot plant run by the Firestone Tire and Rubber Co. In addition to the United States and Mexico, where guayule is native, interest in this species has arisen elsewhere in the world: small research programs are being conducted in Australia, South Africa, India, and Israel.

II. Botany of Guayule

Guayule is native to the semiarid plateau, 1200–2100 m, of north-central Mexico and southern Texas. The plant is restricted to outwash fans and slopes of calcareous soils. Temperatures in this high plateau range from $-18°$ to $49°C$, and annual precipitation ranges from 230 to 400 mm, occurring principally from late spring to early autumn. On the scale of the Emberger Aridity Index, guayule's native habitat ranges between 53 and 500 units, corresponding to a semiarid to arid climate. Guayule is a bushy perennial shrub, 60–100 cm high, that can live in nature for up to 30 years. Its narrow 10-cm long leaves are covered with a drought-protecting wax, and dense T-shaped trichomes give the plant its characteristic light-green–gray sheen. Both wax and trichomes slow water loss from the leaves. The plant is thus drought-resistant and can withstand extreme water stress with leaf temperatures approaching the thermal death point of many other species. Guayule is able to recover very rapidly from water stress after it is resupplied with water, the stressed silver-gray folded leaves becoming dark green and turgid in a matter of hours after water is supplied. The root system consists of a long tap root, which can penetrate, depending on the soil type, more than 6 m, and lateral roots that can spread 3 m from the plant. The lateral roots enable the plant to utilize the moisture of short, sporadic rains. Adventitious shoots (retoños) develop on shallow roots exposed by erosion. Adventitious roots that develop at the base of the retoños increase the extent of the root system.

Guayule is a semidormant semideciduous plant. Active vegetative growth takes place during the warm season (when soil moisture is available). In the cold season, which is also the dry season in its habitat, the plant sheds most of its lower leaves, leaving compact terminal clusters of small leaves.

Guayule exhibits a juvenile phase, lasting from seed germination until the plant develops 9–10 nodes, during which flower initiation is not possible. This phase usually lasts for 6 months but can be shortened to 1 month under greenhouse conditions. The plant is a long-day plant, requiring 9.5–11 hr of light for flower initiation, which occurs within 16 days and requires 10 photocycles for full flowering. Flowering in guayule ceases during periods of high moisture stress and cool temperatures and resumes after new vegetative growth begins, when days become warmer and the soil becomes moist. Guayule has a compound inflorescence, the flowers being borne in heads on a common receptacle. The flower head contains five fertile (seed producing) ray-florets, each with two attached sterile disk-florets. Other disk-florets contain fertile stamens which produce pollen, and an abortive pistil. The achene complex, i.e., the seed, when shed, consists of the achene, to the base of which are attached the two sterile disk-florets, a bract, a persistent ligule (corolla), and the two-lobed stigma. The achene contains an embryo invested with two seed coats.

The canopy of guayule may vary considerably in size and shape; some canopies are narrow and erect, some are wide, and some have a few widely spaced branches, while others are compact. Canopy shape is determined by the formation of the inflorescence. The monopodial growth of the seedling, while still herbaceous, is terminated by the first inflorescence. Active growth occurs in a number of the uppermost branches, each ending in an inflorescence, and after several years this branching scheme usually gives rise to a closely branched symmetrical shrub. Since the patterns and intensity of flowering depend upon genetic consistency and environmental factors, such as soil moisture, flowering accounts for some of the morphological variations in canopy shape.

III. Rubber Accumulation

A. Rubber Biosynthesis

The biosynthesis of rubber can be divided into three stages: (1) the production of acetyl-coenzyme A (CoA), (2) the conversion of acetyl-CoA to isopentinyl pyrophosphate (IPP), and (3) the polymerization of IPP into polyisoprene. It is believed that acetyl-CoA is produced from acetate and acetoacetate, the true *in situ* substrates of rubber biosynthesis. The source of acetyl-CoA may, however, be pyruvate produced by a reaction involving the pyruvate dehydrogenase complex, by β-oxidation of fatty acids, or by the metabolism of branched chain amino acids. Via a series of reactions that require ATP and NADPH, including the synthesis of mevalonic acid, acetyl-CoA is converted to IPP, a 5-carbon molecule. IPP is then converted to dimethylallyl pyrophosphate (DMAPP) by the enzyme IPP isomerase. IPP and DMAPP are condensed to nerylpyrophosphate, a 10-carbon molecule, by rubber transferase (polyprenyl transferase), an enzyme that on purification has been estimated to have a molecular weight of 60,000. Other allylic pyrophosphate initiators are the 10-carbon molecules geranyl pyrophosphate, farnesylpyrophosphate, and geranylgeranyl pyrophosphate. Rubber transferase adds IPP repeatedly to the initial 10-carbon chain to produce a long polymer of up to 30,000 units of isoprene.

B. Rubber Localization

In guayule, rubber is synthesized and accumulated within the parenchyma cells, unlike *Hevea*, which produces rubber as a milky latex that accumulates in canals called laticifers. In *Hevea* rubber is collected by cutting into the bark, while the rubber of guayule has to be extracted from the cells subsequent to tissue maceration. In guayule concentric rings of resin canals are located within the cortex of the secondary growth. Initial polymerization of the polyisoprene molecule occurs in the epithelial cells of the canals, and polymerization is then continued in the surrounding cells, while resin is secreted into the canal. The rubber that is synthesized accumulates in the form of a rubber particle, which is surrounded by a dense osmophilic layer, that is probably composed of phospholipids. The rubber particle is initiated as a small irregular particle in the cytoplasm and increases in volume as more rubber is synthesized (Fig. 1). As the plant ma-

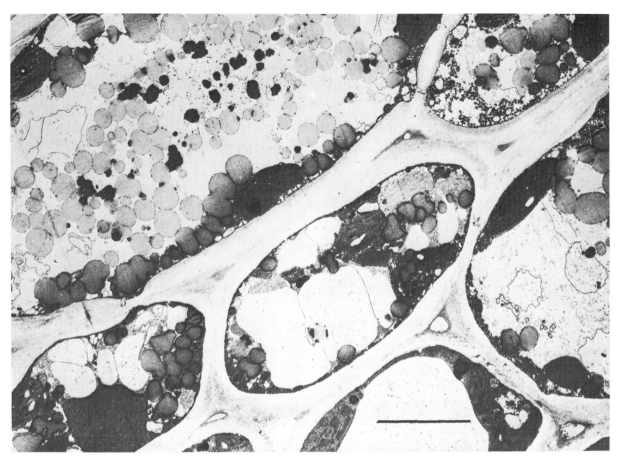

FIGURE 1 Cross-section of cortical parenchyma tissue of guayule stem. Rubber particles are located in the cytoplasm of some cells as well as in the vacuole of others. Bar represents 4.3 μm. [Reproduced with permission from Goss, R. A. *et al.* (1984). *cis*-Polyisoprene synthesis in guayule plants (*Parthenium argentatum* Gray) exposed to low, nonfreezing temperatures. *Plant Physiol.* **74,** 534–537.]

tures, more and more particles appear in the vacuole, where they become spherical in shape. Eventually, the vacuoles become completely filled with rubber particles that may fuse to form a large mass of rubber.

In guayule, rubber is produced and accumulates in most of the parenchyma cells of the stems, branches, and roots of various sizes and developmental stages. Rubber concentration and quantity are lower in the roots than in the branches. Small amounts of rubber, usually of a low-molecular-weight compound, can be found in the leaves, and almost none is found in the peduncles (flower stalks). In young branches most of the rubber is located in the primary cortex, in the pith, and in and around parenchymous resin canals. As the branch matures, rubber also accumulates in the xylem and phloem rays. The bark is the principal rubber-containing tissue, containing up to 80% of the total amount of rubber. The pith cells accumulate

relatively less rubber, and lignified xylem cells and vessels and cambial cells and their immediate derivatives are devoid of rubber.

Collections from Mexico have been found to contain as much as 26% of rubber by weight in old plants ≥10 years old. In cultivated plantations, younger plants 2- to 5-years-old usually contain 5–10% rubber and sometimes even higher amounts, up to 20%, depending on factors such as soil moisture, climate, and age.

C. Biosynthesis Control

Rubber accumulation in guayule is affected by many processes, mainly root and branch biomass production, synthesis and translocation of intermediates for rubber biosynthesis, and induction and activation of enzymes catalyzing reactions in the rubber biosynthesis pathway. These processes are affected and partially regulated by environmental conditions, sometimes in a conflicting manner. Two well-documented environmental factors are low temperatures and water stress. Active growth of guayule occurs mainly in the warm months. In the northern Negev of Israel, for example, an approximately equal increase in biomass of leaves and branches was observed in the summer. In the winter, when the plant is not fully dormant, a considerable number of leaves were shed, but some increase in branch biomass was observed. In winter, the leaves were photosynthetically active (CO_2 fixation), and carbon was translocated along the stems to a larger extent than in summer. CO_2 fixation also occurs in the bark to some extent. Rubber content, as a percentage of dry weight, is often higher during or at the end of winter. This rubber has a higher molecular weight than rubber synthesized in the summer. Young guayule plants exposed to low temperatures (7°C at night) under controlled conditions accumulated twice as much rubber as plants growing at 21–24°C. Incorporation of labeled acetate (substrate in excess) into rubber in detached branch slices—a method that determines the rubber-producing potential—has shown that the rate of rubber biosynthesis may be as much as fivefold higher in winter than in summer, reaching a value of 26% incorporation of labeled acetate into rubber of the total incorporation (4–5% in summer). The incorporation of labeled IPP into rubber in a crude stem homogenate also increased in cold-treated plants, suggesting that low temperatures cause an increase in the activity of rubber transferase.

A number of investigations have shown that guayule plants exposed to water stress accumulate more rubber (Table II). Water stress may develop during the irrigation cycle; its extent and duration depend upon the quantity and frequency of water application. Water stress in guayule has been measured or estimated in terms of the crop water stress index (CWSI), in terms of the relative leaf water content (RLWC), or with a psychrometer. All these methods were able to demonstrate that water stress is correlated with the fraction of available water in the soil. It has been demonstrated that under stress, RLWC falls to values of ≤30%, and that water and osmotic potentials fall to 3.5 MPa, while net photosynthesis is not impaired. These data suggest that a significant level of osmotic adjustment was achieved in the stressed guayule plants. Water stress results in an inhibition of growth and in the production of new stem tissue. While it causes a decrease in carbon translocation into the stem

of guayule, there is a simultaneous increase in the incorporation of translocated carbon into rubber. A twofold increase in labeled acetate incorporation was reported in stressed plants vs irrigated plants. As is the case for cold stress, water stress leads to an improvement in the quality of the rubber, the product having a higher molecular weight.

D. Rubber Quality

The rubber extracted from guayule has a molecular weight of $1–2 \times 10^6$, the molecular weights of the polyisoprene chains being similar to those in *Hevea* rubber. It has been shown spectrometrically that in terms of structure and composition purified guayule rubber is similar to that of *Hevea*. Mechanical properties, such as Mooney viscosity, plasticity retention index, tack, etc., of the two rubbers have been shown to be similar. However, guayule rubber is less viscous and has a shorter stress relaxation time, a result of the linearity of the polymer chain and the low level of crosslinking. In addition, as opposed to *Hevea* rubber, guayule rubber lacks natural antioxidants that preserve the rubber after extraction, preventing oxidization and degradation. The low cure rate and the lack of antioxidants may be easily overcome by applying chemicals during the process of purification. Guayule rubber has been tested in a range of military applications (tank track pad recipes, aircraft tires), and its performance has been found comparable to that of *Hevea* rubber.

IV. General Agromanagement of Guayule

A. Pollination

Guayule is both a wind- and an insect-pollinated plant. Pollinators of the plant are honeybees, ladybugs, lygus bugs, and cucumber beetles among others. Pollination by honeybees increases seed yield significantly. Seeds produced on flowers exposed to bees, as opposed to seeds produced on caged plants (no bees) or bagged plants, exhibit a higher germination rate. The rubber content of plants exposed to bee pollination was found to be significantly higher than that in plants for which pollination by bees was prevented.

B. Seed Yield and Technology

Seed yield varies greatly with plant age and vigor, irrigation, and method of harvesting. Viability and

TABLE II

Effect of Water Status on Biomass and Rubber Production in Clipped 2-Year-Old Guayule Plants Grown under Dry-Land and Irrigated Conditions

Location	Density (plants/m^2)	E.T.[a] (cm)	Dry biomass (ton/ha)	Rubber content (%)	Rubber yield (kg/ha)	Rubber WUE (g/m^3)
Irrigation						
Yuma, Arizona	4.9	162	5.8	8.3	480	30
		285	11.8	7.5	885	31
		409	12.7	7.3	930	23
Hillston, Australia		140	1.7	11.2	200	14
Condobolin, Australia		194	4.7	8.8	410	21
Beer-Sheva, Israel	3.2	105	4.8	10.3	490	47
		159	7.7	9.6	740	47
		205	10.8	9.8	1060	52
Dry land						
Rio Grande, Texas	3.2	124	6	6.0	360	29
Condobolin, Australia	1.9	110	1.7	6.2	103	9
Kingaroy, Australia		152	3.7	9.9	370	24
Hillston, Australia		70	1.3	10.1	155	21

Source: Data taken from Bucks *et al.* (1985). Irrigation water, nitrogen, and bioregulation for guayule production. *Trans. Am. Soc. Agr. Eng.* **28,** 1196–1205; Benzioni *et al.* (1989). Effects of irrigation regimes on the water status, vegetative growth and rubber production of guayule plants. *Exp. Agr.* **25,** 189–197; Whitworth, J. W., and Whitehead, E. E. (eds.) (1991). "Guayule Natural Rubber." Guayule Administration Management Committee and USDA Cooperative State Research Service.

[a] E.T. (evapotranspiration) equals irrigation (when applied) plus rainfall.

yield of seeds decrease with an increase in water stress at flowering. To achieve maximum seed production, plants should be fertilized and irrigated regularly, and exposure to insects should be controlled. Under such conditions a yield of 110 kg/ha was obtained in Salinas, California, in the early 1950s.

Seed harvesting has been attempted by various methods. Revolving-brush-type harvesters were developed during the ERP and found inefficient. In the early 1950s and the 1980s more efficient suction-type harvesters were developed; in these machines a vibrator or a rotating brush that dislodges the seeds is used in combination with a suction blower that collects the seeds from the canopy. A suction device has also been developed to collect scattered seed from the soil surface. An efficient seed cleaning and threshing method was developed by the ERP and comprises the following steps: (a) removal of coarse debris such as leaves by means of a shaker or scalping screen; (b) separation of the components of the remaining coarse plant material into clusters of disk florets, seeds, and fine trash by a vibrating clipper; (c) separation of floral parts from the achene by a thresher, and (d) separation of empty from filled achenes by a gravity separator, with recovery of 95% of the filled achenes. For long-term storage, seeds should be kept in dry conditions. For example, seeds stored immedi-

ately after harvesting in sealed steel drums, at moisture content of 4%, were still fully viable after 16 years of storage. For a short-term storage, e.g., 1 year, seeds may be kept in open containers at 7–8% moisture.

The proportion of filled achenes (each containing an embryo) ranges from 0 to 70% and is most commonly between 20 and 45%. Germination of freshly harvested guayule seed is very low. Research has revealed that there are three types of seed dormancy: (a) embryo dormancy, that persists for about 2 months and can be partially broken by storage at 4°C in a moist atmosphere or completely broken by soaking the seeds in 1 g/liter gibberellic acid for 6 hr; (b) seed coat dormancy, lasting from 6 months to many years, that can be partially broken by scarification or exposure to oxidizing chemicals or completely broken by soaking the seeds in 1 g/liter gibberellic acid for 22 days; and (c) dormancy caused by water-soluble germination inhibitors, released from the floral components that are attached to the achene.

C. Vegetative Propagation

Guayule may be propagated vegetatively both by cuttings and in tissue culture. Eighty percent rooting of cuttings can be obtained after 12 hr of soaking in indolebutyric acid (IBA), followed by 17 days in aerated water at 25°C. Ninety-five percent rooting was obtained in moist, aerated soil after 2 to 3 weeks. Rooted cuttings are easily transplanted in the field. Shoot multiplication *in vitro* may be stimulated with 1 mg/liter benzyladenine, and root initiation with 0.5 mg/liter IBA. Conditions for optimal propagation vary among cultivars. Rooted plantlets should be hardened in the greenhouse before transplanting. Currently, vegetative propagation (by cuttings or tissue culture) cannot compete economically with greenhouse- or nursery-grown seedlings. Nevertheless, despite the low genetic variability in polyploid guayule cultivars (see below), vegetative propagation can facilitate cloning of individual superior plants for experimental purposes, or perhaps for commercial application when means are found to reduce production costs. [*See* PLANT PROPAGATION.]

D. Soil and Field Establishment

Guayule grows best in nonacid, noncompacted, well-drained soils that support better root penetration, higher growth rates, and higher rubber yields. Soils with moderate amounts of clay (10–25%) in the topsoil appear to be most suitable for cultivating guayule. Before planting, the soil should be subsoiled to a depth of ≥ 0.5 m to break up compacted soil. Transplanting and sowing guayule are best when temperatures are warm and the soil is moist, conditions that support vigorous vegetative growth. In irrigated plots, transplanting may be performed during the spring, summer, and autumn months, but should be avoided in the winter when the plant is dormant. Guayule plants should be planted in rows, with the distance between plants within a row being 30–50 cm, and that between the rows being 30–100 cm, resulting in a wide range of densities of 3 to 100 plants/m². Both biomass and rubber yields increase with an increase in density, according to a certain formula, the constants of which depend on factors such as water application.

Two basic methods have been used for establishing field stands: (1) direct seeding, and (2) transplanting seedlings raised in the field or in the greenhouse. Direct seeding is problematic and has not yet been developed successfully. The problems encountered are long seed dormancy, slow growth of the seedlings, and various stresses such as drought, soil salinity, diseases, and weed competition. Stands ranging from 17 to 57% establishment have been obtained, depending on the variety, the time of sowing, and the specific

agricultural practices. Pretreating seeds with polyethylene glycol (25%) and gibberellic acid (100 μM) stimulated germination in the light over a broad range of temperatures (15–33°C) and also stimulated the development of normal seedlings. Fluid drilling, a direct seeding technique that involves sowing pregerminated seeds suspended in a gel carrier, improves direct seeding in guayule. Addition of fungicides to the gel controls damping-off, thus improving this technique even further.

A method for the production of transplants in the field was developed many years ago by the Intercontinental Rubber Co. The method involved the sowing of pregerminated, hypochlorite-treated seeds that were then covered with sand in the spring or early summer. Overhead irrigation was applied several times a day during the summer, and weeding was done by hand. After hardening of the plants during the autumn and winter, machinery was used to prepare bare-rooted, topped seedlings for transplanting. This technology was perfected and simplified in the ERP as follows: seeds were sown dry in sandy loamy soils; furrowing replaced overhead irrigation; and oil emulsion spraying was used for weed control. It was found that seedlings 4–5 mm in diameter at the root crown, with a root length of 18 cm after cutting, were best for transplanting and gave the best survival. Bare-rooted seedlings packed in wax paper could survive for several weeks in cold storage. If kept under low moisture, seedlings could survive for 2 to 3 months after transplanting in a dry field before the soil was wetted. Survival could have been improved to values comparable with transplants with soil if bare-rooted seedlings had received supplemental irrigation. Higher survival rates of transplants with the root system in soil are obtained in rain-fed fields. The methodology of raising seedlings in a nursery is inferior to that in a greenhouse: in the nursery there are problems of weed and climate control, and the methodology is labor and irrigation intensive. Thus, growing transplants in a greenhouse is currently the preferred method of propagation. In the greenhouse, a temperature of 20–30°C enables germination within 2 to 3 days; a light soil mix provides good drainage; and aeration and the use of appropriate containers facilitate easy removal of the seedling with only minor damage to the root system. After about a month, seedlings are transferred gradually to harsher conditions, such as a screen-house, for hardening. Greenhouse-grown seedlings can be transplanted into the field 2 to 3 months after germination. Seeds may be germinated in the winter in a heated greenhouse and transplanted in the spring. Greenhouse-grown seedlings are transplanted with the container soil into a pre-irrigated field. Larger containers that allow the development of a bigger root system give rise to a better plant growth. Thus, as compared to direct seeding, stands obtained from transplanting are much more successful, giving 80 to 100% establishment.

E. Irrigation and Fertilization

The native guayule plant survives and grows under 400 mm and less of precipitation. However, for commercial production, irrigation is needed to increase rubber yield per plant or per unit area. This is particularly true in areas such as the Negev Desert of Israel that lack precipitation in the warm growing season. Biomass production in terms of branch dry matter and rubber and resin production are directly related to water availability (Table II). Biomass production of guayule responds to water application in a linear manner even at high water application levels of 1500–1700 mm/year as long as the soil does not become oversaturated, since guayule is sensitive to water logging. Rubber usually accumulates to somewhat higher levels in plants receiving less irrigation (see earlier). However, since irrigation has much more impact on biomass production, rubber yield, which is the multiplication product of biomass and rubber concentration, is directly correlated with water availability (Table II). For clipped plants, quantities of $2–5 \times 10^3$ and $20–50 \times 10^3$ m^3/ha of water are required to produce a ton of biomass and rubber, respectively (Table II). Water use efficiency (WUE) for rubber usually ranges from 20 to 50 g/m^3/ha for clipped plants and from 40 to 60 g/m^3/ha for whole plants. Higher WUE values recorded in Israel (Table II) are perhaps due to the use of drip irrigation. Data in the literature indicate that $1.2–1.6 \times 10^3$ and $15–25 \times 10^3$ m^3 of water is required to produce a ton of dry shrub and a ton of rubber, respectively, for whole plants (including part of the root system but no leaves). WUE of whole guayule plant is low, ranging from 0.4 to 0.8 tons of biomass/ha. These values are lower than those for wheat or alfalfa grown in semiarid regions but higher than those for cotton. From whole plant harvests of existing cultivars, an annual rubber yield of at least 200 kg/ha may be expected under dry-land cultivation (400–500 mm of rainfall), and a yield of ≥500 kg/ha under irrigation (1000–1300 mm of irrigation + rainfall). The availability of water depends on the soil texture and composition, and as such soil differences are dominant factors in

influencing the yield of rubber. The relationship between rubber yield and irrigation should, therefore, be determined for each location. Based on considerations of the effect of water stress on rubber quality and the cost of water, intermediate quantities of irrigation (i.e., in the range of 400–1000 mm/year) should be applied. The irrigation should be applied in such a way as to allow alternate periods of relatively vigorous vegetative growth with periods of high rubber synthesis. [See Irrigation Engineering: Farm Practices, Methods, and Systems.]

Since saline water is sometimes abundant in arid and semiarid zones, and since the salinization of cultivated areas takes place as a result of poor water management, the effect of salinity on development, growth, and rubber accumulation in guayule has been studied. Emergence of guayule seedlings is very sensitive to salinity: the absence of emergence at 6 dS/m has been attributed to hypocotyl mortality associated with salt accumulation at the soil surface or with direct contact with the saline water. Four-year-old, field-grown guayule plants, established with freshwater, were found to be sensitive to salinity in a study conducted in California. Biomass production and rubber yield decreased at a rate of 11% per dS/m above a threshold of 8 dS/m in the soil (equivalent to 6 dS/m in the irrigation water). Plant mortality, rather than a reduction in growth, appears to limit rubber production under conditions of salinity. After 3 years of salinization nearly all plants irrigated with 9 dS/m had died, as did 40% of those irrigated with 6 dS/m water. Recovery and regrowth of clipped plants were also significantly affected by salinity.

The nutritional requirement of guayule is not considered to be high and is definitely lower than that of other crops. In field-grown plants nutrient deficiency symptoms are difficult to recognize, and the symptoms have not yet been delineated. Nitrogen and calcium applications promote growth of directly seeded guayule. Under irrigation, 20–210 kg/ha nitrogen stimulates biomass and rubber production. Guayule responds better to nitrate nitrogen than to ammonium nitrogen. Other elements, such as potassium, iron, phosphate, and magnesium, usually do not affect rubber yields.

F. Bioregulator Application

A promising approach to improving rubber yields is through bioregulation of rubber synthesis. An effective bioregulator is 2-diethylaminoethyl-3,4-dichlorophenylether (DCPTA), which causes a two-to sixfold increase in rubber levels of young plants. DCPTA also leads to an increase in the incorporation of labeled IPP into polyisoprene in a crude extract. It has been suggested that DCPTA acts by increasing the activity of rubber transferase. Yet, this bioregulator has no effect on rubber accumulation in mature plants.

G. Harvesting

In principle, the guayule shrub can be harvested at any time, as long as the plant is alive (rubber disintegrates in dead plants), which means that it can be considered as a living stock of natural rubber. Both biomass and rubber increase asymptotically with age, with growth parameters depending on environmental and genetic factors, such as cultivar, soil type, and water management. As such, for practical reasons, the optimal age of harvesting should be established for each set of conditions. Harvesting is usually considered after 2 to 4 years of growth, the period of the highest growth rate. The general harvesting procedure is to dig out the plant, removing the crown with 15–20 cm of the root system. An alternative method is harvesting the upper portions of the plant (about two-thirds of the entire plant rubber), namely, clipping or pollarding, leaving the crown to generate new growth, followed after about 2 to 3 years by digging out the entire plant. This latter method is advantageous vs replanting of the seedlings, which is a costly operation. The success of clipping as a harvesting method depends on a high regeneration rate. The latter depends on genotype and the timing of the harvesting. Optimal regeneration occurs when clipping is performed at the end of the dormant season when the rubber level is maximal. Clipping almost to ground level in the summer may be detrimental. Large variability in survival, ranging from 6 to 97%, was observed among different genotypes in two studies. Some studies have shown that plants clipped first and then dug out yielded more rubber than unclipped plants of comparable age. However, other studies have shown that the yield of clipped plants was the same or even lower than that of unclipped plants.

At the final harvest, different strategies are possible. The plants may be forage harvested before lifting the stumps, or the entire plant may be lifted, then foraged or baled. Postharvest delays in processing result in a reduced rubber recovery due to oxidation and deterioration of the rubber. Processing should, therefore, commence as soon as possible after harvest.

V. Improvement of Guayule Germplasm

Existing guayule germplasm occurs in various ploidy levels, including diploid ($2N = 36$), triploid ($2N = 54$), tetraploid ($2N = 72$), and even higher levels. Aneuploid deviations from the primary numbers and the presence of B chromosomes have also been found, resulting in genetic variability within the native populations. The mode of reproduction varies with ploidy level. Naturally occurring diploid plants reproduce sexually, whereas polyploids reproduce by facultative apomixis. Depending upon whether meiosis and/or fertilization takes place in megaspore mother cells of the tetraploid parents, four classes of progeny may be identified: apomictic tetraploids, hexaploids, polyhaploids ($N = 36$), and amphimictic. Pollination is not required for embryo development in polyploids but is needed for normal endosperm development. The classes of progeny of the apomictic plants described above, possessing different ploidy levels and genetic combinations between the parents, demonstrate some potential for genetic variability. However, since the predominant class (70–100%) is the one resulting in offspring identical to the maternal parent, it has been suggested that apomixis does not provide sufficient variation and much more emphasis should be placed on sexual breeding and selection, a system in which the offspring has much more likelihood of being a new genetic recombinant. Nevertheless, apomixis offers the advantage of a high uniformity and relative genetic stability of lines.

A number of approaches have been used to improve guayule. The most widely used is the simple approach of selection among apomictic polyploids. It is based on some plant-to-plant variation in the facultative apomictic populations, on efficient screening, and on a high degree of heritability of selected traits. Genetic variation is found in both native and cultivated apomictic guayule. Native populations of guayule are often isolated in a number of niches distant from each other, a situation that induces genetic variability between populations. Collection and examination of native guayule germplasm was initiated in 1912 by McCallum. The first guayule cultivar, 593, originated from this collection and from the initial cultivation efforts. In 1942, Powers together with McCallum and Olson made 66 collections in Mexico; they were followed by Hammond and Hinton who collected 174 accessions in 1948. These two collections gave rise to the well-known 26 USDA lines or varieties. Fifteen of these lines (11591, 11600, 11604, 11605,

11609, 11619, 11635, 11693, 11701, 12229, N565, N565II, N566, 4265-X, and 4265-XF) are descendants of Powers' collection no. 4265 made from five plants at a single location in the State of Durango, Mexico. Seven other lines (12231, 11633, 11646, N396, N575, N576, and A48118) are from the collection of Hammond and Hinton, made in a relatively small area in the same state. Despite the apparently narrow genetic base of these cultivars, a surprising amount of variability exists among them regarding vigorousness, morphology, and rubber content. The genetic variation is also reflected in the different performances of the cultivars in relation to each other at various locations (Table III), differing in terms of soil type, climatic conditions, and agromanagement techniques. These data emphasize the necessity to select or develop superior germplasm separately for each location. It should be mentioned that selections from open-pollinated populations have recently been successful in California, giving rise to two new high-yielding cultivars, namely Cal 6 (tetraploid) and Cal 7 (triploid). The annual rubber yield of the USDA lines varies from 170 to 740 kg/ha, whereas the new lines yielded 240 to 940 kg/ha annually in three locations in the United States (Table III). [*See* PLANT GENETIC RESOURCES; PLANT GENETIC RESOURCE CONSERVATION AND UTILIZATION.]

Repeated selection within a selected population of diploids is aimed at increasing the frequency of genes affecting rubber content. The procedure consists of screening a large population of young diploids for rubber content and removing low-rubber-yielding plants. The remaining high-rubber-producing plants are allowed or forced to cross-pollinate and to produce seeds for the next cycle. A newly released cultivar, Cal 3, is the result of such procedure. High-rubber-yielding plants may be made polyploid and apomictic either by doubling their chromosome number or by hybridizing them with polyploid plants.

Another approach to breeding consists of interspecific hybridization and backcrosses between guayule and other species of the genus *Parthenium*. These species vary in their growth habits and habitat: some are herbaceous annuals that grow under temperate conditions, and some are tropical trees, while others are desert shrubs. Despite the differences, *Parthenium* species do cross with one another, producing fertile hybrids. *Parthenium* species, with the exception of guayule, contain very small amounts of rubber which is of low molecular weight. Nevertheless, they possess other features that are very attractive for guayule breeding, including high biomass production, leading to high rubber yield (*P. schottii*, *P. tomentosum*, and *P. fruticosum*), cold tolerance (*P. alpinum* and *P. ligu-*

TABLE III

Annual Rubber Yields of 2- to 3-Year-Old Guayule Cultivars in Various Locations in the United States, Israel, and Australia

	Annual rubber yield, kg per hectare						
	Las Cruces, N. Mexico		Rio Grande, Texas		Riverside, California		Beer-Sheva, Israel[c]	Condobolin, Australia
Cultivar	I[a]	II[b]	I	II	I	II		
11604	—	209	—	209	—	738	330	202
11605	266	217	294	217	531	572	276	286
11591	280	—	326	—	464	—	200	177
N576	275	—	275	187	517	395	200	232
11619	305	—	272	—	523	—	250	353
12229	254	—	239	—	427	—	243	340
N565	275	—	226	—	563	—	169	380
Cal 6	—	327	—	324	—	938	—	—
Cal 7	—	239	—	274	—	887	—	—

Source: Data taken from Estilai, A., and Ray, D. T. (1991). Genetics, cytogenetics, and breeding in guayule. *In* "Guayule Natural Rubber" (J. W. Whiteworth and E. E. Whitehead, eds.) Guayule Administration Management Committee and USDA Cooperative State Research Service; Mills, D., Forti, M., and Benzioni, A. (1989). Performance of USDA lines in the northern Negev of Israel. *Econ. Bot.* **43,** 378–385; Milthorpe, P. L. (1984). "Guayule Research and Development in New South Wales." Agricultural Research and Advisory Station, Condobolin. Department of Agriculture New South Wales.
[a] First Uniform Regional Variety Trials (1982–1985).
[b] Second Uniform Regional Variety Trials (1985–1988).
[c] Clipped plants.

latum), and resistance to diseases and pests (*P. tomentosum* and *P. incanum*). For such hybridization, diploids of guayule are crossed with diploids of other *Parthenium* species. Since diploids of *Parthenium* are usually self-incompatible, emasculation is not necessary. F1 hybrids between guayule and *P. schottii*, *P. fruticosum*, and *P. tomentosum* produce annual yields of more than 40,000 kg/ha of biomass, but their rubber content is only 1–3% and the molecular weight of the rubber is about half that of pure guayule. An increase in both rubber level and quality should be the aim of future crosses and backcrosses.

VI. Guayule Processing and By-products

The economic feasibility of producing rubber from guayule depends both on the development of a process for rubber production and on the definition of valuable by-products and efficient processes for their production. To release the rubber it is necessary to break the plant cells by means of pebble mills, hammer mills, single-disk attrition mills or, more efficiently, with roll mills or extruders of the types used for oilseed processing. Rubber and resin may be recovered either by flotation or by sequential or simultaneous extraction. In the flotation process chopped shrub is treated with a dilute caustic solution to create rubber "worms" that are skimmed off the surface and treated with acetone to remove the resin. The rubber is then dissolved in hexane, antioxidants are added, dirt is removed, and finally the hexane is evaporated. This process suffers from two disadvantages: the low efficiency of deresination of the rubber and the large volume of caustic flotation liquid that requires extensive treatment before it can be reused or discarded. In the sequential extraction process, as opposed to the flotation process, the milled shrub is first deresinated by extraction with acetone. Rubber is then removed by extraction with hexane. Both extractions can be carried out by immersion or by gravity or countercurrent percolation. The solvents can then be recovered for reuse. Incomplete desolventization of residual solids results in losses of solvents. In addition, contamination of hexane with acetone complicates the recycling of the hexane. In the more complicated simultaneous extraction process, ground shrub is treated with a solvent that dissolves both the resin and the rubber. A polar solvent is added to the solution or miscella to coagulate the rubber. An example of such a solvent system is toluene as the rubber/resin solvent and methanol as the polar component. Simultaneous extraction is characterized by rapid extraction of resin and rubber, efficient separation of the latter from the bagasse, and selective removal of low-molecular-weight rubber. These features make this process the most suitable technology for commercial guayule processing. Two recent process development programs carried out at Texas A&M University and Bridgestone/Firestone, Inc. based on simultaneous extraction technology yielded rubber of acceptable characteristics.

For each ton of guayule rubber that is extracted, about one ton of resin and about 12 tons of bagasse are produced. About 15% of the total rubber is of low molecular weight. This low-viscosity low-molecular-weight rubber may be used as a plasticizer or as a processing aid in adhesive and molded product manufacture and in other applications. It can also replace depolymerized rubber that is prepared from natural or synthetic rubber by thermolysis.

The resin in guayule is composed of a range of organic components. Since the resin composition is affected by genetic and environmental parameters, it has not yet been fully elucidated. Guayule resin has been found to have promise as a wood protectant against marine and terrestrial wood-destroying organisms, such as fungi and termites, and its potential as such will increase as the use of creosote becomes regulated. The resin can serve as a plasticizer for high polymers and has also been used experimentally as an adhesion modifier in amine-cured epoxy resins. Coatings containing guayule resin have properties for temporary protection of aircraft or land vehicles. The resin can be used directly for combustion or converted to liquid fuel by thermolysis. Since noncondensable gas is produced in the process, under certain conditions the resin may also be converted almost entirely to gaseous fuel. It should be mentioned that guayulin A, a unique component of the resin, has been identified as a potent skin irritant causing dermatitis in animals and apparently affecting workers who processed the resin.

The wax extracted from the leaves is physically similar to carnauba wax, which is used in the manufacture of high-grade polishes, carbon paper, and explosives. The wax may not be economically viable due to its low yields. Guayule bagasse affords the potential for use as biomass fuel. Applications include the production of process steam or use as a cellulosic base for extruded fireplace logs. It can also be used for production of gas that can be converted into liquid hydrocarbon mixtures equivalent to diesel or aviation fuel. Alternatively, the bagasse can be converted into fermentable sugars, but this process requires further development.

VII. Economics and Prognosis for Guayule

The economic feasibility of commercializing guayule has been under consideration since the 1970s in the United States (California, Arizona, and Texas), Israel, and Australia. After the costs of the current technology and prices for natural rubber had been taken into account, it was concluded that a domestic industry for utilizing guayule was not economically viable. The most recent economic feasibility study was conducted in Texas in 1990. This study showed that the on-farm rubber price required to cover costs of guayule production under irrigation or dry land cultivation ranges from $1.3 to $1.7/kg, assuming crop establishment to be by direct seeding and by-products to have no value. If the current price of processed rubber is taken as $1.1/kg and if processing costs are assumed to be $0.5/kg, costs are almost double the market price. A profit is thus envisaged only if the rubber yield is doubled and if the resin can be sold for $0.5/kg. For guayule to offer the farmer an attractive alternative to sorghum or cotton, the processed rubber would have to command prices of 2.8 and $3.8/kg, respectively, prices that are two- to threefold higher than the current market price for rubber.

Despite the history of successful commercial production by the Intercontinental Rubber Co. and the successful production under the ERP during World War II, guayule remains an experimental crop. As such, however, guayule rates very high in its potential for long-term commercial development. It is generally accepted that for guayule to become a commercially feasible crop, a certain combination of improved rubber yields, favorable production costs (such as by implementing direct seeding and improving rubber extraction), exploitation of the by-products, and an increase in the price of rubber are necessary. Currently, no commercial use has been established for any of the above by-products. None has unique chemical or physical properties and thus none represents a high-value use. Recent research has, however, advanced the development of guayule, as can be seen in the production of new high-yielding cultivars. Even

better cultivars, in addition to improved agromanagement and rubber extraction processes, are to be expected in the future. Moreover, other circumstances may facilitate the commercialization of guayule. Prices of synthetic rubber and of natural rubber currently extracted from *Hevea* may rise in the future, and the open-ended supply of petroleum, an essential raw material for the production of synthetic rubber, is not assured. In addition, current social changes in Southeast Asia may result in an increase in labor costs and a shift to more profitable crops such as palm or coconut trees and hence an increase in the prices of rubber from *Hevea*, a labor-intensive crop. Under such circumstances, guayule, as a fully mechanized crop, may be able to compete with *Hevea*. A viable guayule industry in nations capable of growing this species would relieve their dependence on the import of *Hevea* rubber, thus improving their economies and reducing their dependence on the import of a strategically important commodity. Recently, concern is being raised about the allergic reaction elicited by medicinal products such as gloves, enema nozzles, etc., made from the latex of the rubber tree. This allergic reaction is sometimes fatal. It has been suggested that the latex produced from guayule may contain fewer allergens. If proven safer, the use of guayule rubber in products coming into contact with human tissues may promote the commercialization of guayule.

Bibliography

Backhaus, R. A. (1985). Rubber formation in plants—A mini review. *Israel J. Bot.* **34,** 283–293.

Benedict, C. R. (ed.) (1986). "Biochemistry and Regulation of *cis*-Polyisoprene in Plants." NSF Sponsored Workshop, College Station, TX.

Benzioni, A., and Mills, D. (1991). The effect of water status and season on the incorporation of $^{14}CO_2$ and [^{14}C]acetate into resin and rubber fractions in guayule. *Physiol. Plant.* **81,** 45–50.

Bowers, J. E. (1990). "Natural Rubber-Producing Plants for the United States." NAL, Beltsville, MD.

Hammond, B. L., and Polhamus, L. G. (1965). "Research on Guayule (*Parthenium argentatum*): 1942–1959." USDA, Agriculture Research Service, Technical Bulletin No. 1327.

Miyamoto, S., and Bucks, D. A. (1985). Water quantity and quality requirements of guayule: Current assessment. *Agricult. Water Management* **10,** 205–219.

National Academy of Sciences (1977). "Guayule: An Alternative Source of Natural Rubber." National Research Council, Washington DC.

Stewart, G. A., and Malafant, K. W. (1989). "Factors Affecting the Productivity of Cultivated Guayule (*Parthenium argentatum* Gray)." CSIRO. Victoria.

Thompson, A. E., and Ray, D. T. (1989). Breeding guayule plant. *Breeding Rev.* **6,** 93–165.

Wagner, J. P., and Parma, D. G. (1989). Establishing a domestic guayule natural rubber industry. *Polym.-Plast. Technol. Eng.* **28,** 753–777.

Whitworth, J. W., and Whitehead, E. E. (eds.) (1991). "Guayule Natural Rubber." Guayule Administration Management Committee and USDA Cooperative State Research Service.

Nematicides

MICHAEL McKENRY, *University of California, Riverside*

Glossary

Nematode A multicellular, nonsegmented, true roundworm which can be pathogenic to specific agricultural crops
Soil Fumigants Chemicals volatile enough to move through soil pore spaces, dissolve into soil water films, and kill various microflora and fauna of soil
Soil Profile A vertical cross-section of the soil from the surface into the underlying, unweathered material

A nematicide is any agent lethal to nematodes. A nematistat, or nemastat, is any chemical, situation, or phenomenon which holds a nematode population in equilibrium, but this term is frequently used to refer to sublethal dosages of nematicides which disrupt nematode behavior. Since root parasitic nematodes may reside as deep in soil as the deepest root, it is to be expected that any soil-applied nematicidal agent will provide a gradient of lethal to sublethal to nonlethal effects emanating from the point or line of chemical injection. Gradient concentrations of nematicides can also be expected to occur when systemic-type nematicides are applied to plant parts but these gradients can be expected to be affected by plant physiology and "sink" effects rather than distance from the point of application. In crops of lower value one attempts to reduce the nematode population enough to attain a nondamaged crop. For crops of higher value, such as nursery crops or permanent plantings of trees or vine crops, the attempt is to reduce pest populations as low as possible prior to planting. The most effective preplant nematicides are those biocides which also kill old roots and plant pests remaining throughout the soil profile. The most effective postplant nematicides are those which kill nematodes in soil and roots without damaging the existing root system. [*See* PLANT PATHOLOGY.]

I. A Historical Perspective

The decades of the 1940s and 1950s were the years of discovery and development for nematicidal agents. It was the advent of nematicidal agents that permitted field-level confirmation of the usually subtle damage caused by nematodes. By the 1960s the best performing nematicides were widely used especially in crops of higher cash value. From 1975 to 1990 the ability to detect environmental concentrations of these nematicides improved by 10,000-fold. During this period of increased environmental monitoring and increasing concern about the environmental fate of nematicides, total nematicide usage continued to increase but the number of active ingredients decreased from four or five to one or two.

Nematicides have been one of the many inputs which have allowed for "intensive" agriculture. The ability to grow as many as three cash crops in a single year and the ability to sustain permanent tree and vine crops as economic units are testimony to the value of nematicides. For those crops where there is no nematode resistance or where resistance is to one nematode species but not to others, nematicides have been a valuable tool toward providing an inexpensive and diverse food and fiber supply. Nematicides have provided growers with a tool to avoid the expense of crop rotations and scarcity of new appropriate land.

The use of nematicides has also resulted in numerous incidents of off-target contamination. As the level of chemical detection improved, scientific efforts were focused toward improvements in nematicide delivery systems, reductions in treatment rates and improvements in timing of applications. In the 1940s the requirement to obtain a registered nematicide was proof of nematicide performance. In the 1990s there are cancer studies, wild-life studies, groundwater studies, air pollution studies, ozone depletion studies, and at the Federal level no real requirement relative to chemical efficacy. Of course, the cost of these studies is passed on to the grower who constantly tries to lower his costs. For example, in the 1960s treatment rates for 1,3-dichloropropene were as high as 2500 kg/ha as a preplant treatment for grapes where Dagger nematode, *Xiphinema index*, and the grape fan leaf virus were known to occur. By the 1970s treatment rates were down to 350 to 800 kg/ha depending on soil conditions. By 1990 treatments of 1,3-dichloropropene above 120 kg/ha were not permitted in California because of part per trillion air pollution control requirements in areas adjacent to treated fields.

The greatest task in attaining efficient nematode control has been delivery of the nematicidal agent to the site of the pest. Nematicidal agents are commonly lethal at concentrations of 1 to 20 ppm. Formulations have included soil applied gases, liquids, or granules as well as plant systemic and root protecting agents. Gases applied at high rates (e.g., 450 kg/ha methyl bromide) are not easily kept beneath the soil surface. Liquids delivered for nematode control within soil can move deeper at part per billion levels unless there is a mechanism for gradual degradation of the toxicant. Some systemic materials may occur as residues in food if improperly applied. Root protectants, including manures and amendments, also carry salts and nitrates which have the potential to accumulate in groundwater or in the plant.

II. Delivery to the Target Pest

A. Volatile Nematicidal Agents

In a properly prepared soil the fastest movement of a pesticide is by those chemicals which are able to move through air passageways of soil. True soil fumigants are applied either as a gas or a liquid which quickly volatilizes to become a gas. Soil fumigants are usually delivered to soil through a tube just behind a steel shank. The shanks are inserted 15 to 75 cm beneath the soil surface as they are pulled across a field by a tractor. The disrupted soil surface is immediately resettled, smoothed, and packed. To avoid volatization the highly volatile chemicals must either be applied deep, applied through more shanks, or covered with a tarpaulin. In the case of methyl bromide a thin film of polyethylene tarpaulin may be applied to the field surface in order to hold concentrations at the surface for a longer exposure time. One advantage of highly volatile nematicidal agents is that the active ingredient reaches the pest or the woody roots surrounding the pest at high enough concentrations to kill the root and thereby the habitat of the pest. Highly volatile nematicides are useful as preplant treatments because they are also general biocides at their usual delivery rates.

B. Less Volatile Nematicidal Agents

Several chemicals are occasionally referred to as fumigants but have a relatively strong affinity for soil water films and organic matter and therefore may only move distances of centimeters as a gas. An example is the methyl-isothiocyanate liberators (MIT) such as Vapam (Zeneca Co.). If the soil-borne pest is primarily in the surface 30 cm of soil profile such chemicals can be effective when applied to properly prepared soils. Since nematode pests can reside quite deep in soil and can travel up from 1m below, this group of nematicides is usually most effective when applied by mixing into irrigation water and applying enough water to reach sufficient soil depths. The concentrations of nematicide delivered throughout the soil water films are lower than those from highly volatile substances and less apt to kill woody roots or tubers. The fact that water seldom moves uniformly through a soil profile indicates that the chemical must be well mixed into water and the water carrier must be uniformly applied to a properly prepared soil. Although this group of nematicides is termed less volatile it should also be noted that applications through sprinklers which atomize water droplets can result in much chemical loss before the water droplet strikes the soil surface.

C. Systemic Nematicidal Agents

Systemic chemicals may be applied to soil, plant leaves, roots, or the trunk. Waxy surfaces on leaves, the soil or organic mantle surrounding roots, and the bark on the trunk can each reduce successful introductions of the toxicant into the plant and consequently necessitate an increase in treatment rates. Nematodes attacking roots frequently develop specialized feeding cells within the plant and these may not necessarily be a "sink" for the nematicidal agent.

Several carbamate-type nematicides including aldicarb, oxamyl, and carbofuran exhibit acropetal translocation (from roots to leaves or fruits). These carbamates are also associated with increased growth response (IGR) or the "carbamate response" which can occur whether nematodes are present or absent. Growers have liked the carbamates because of the multiple benefits they provide including IGR, insecticidal value translocated to young leaves, and their early season nematicidal value. It should be noted, however, that unless the nematode of concern is a foliar or stem feeder or unless the active ingredient can be translocated into specialized nematode feeding sites within roots, the preponderance of nematicidal effect occurs on soil-dwelling nematode stages. This author did not see the first example of a true systemic nematicide until 1991 when working with fosthiazate, an organophosphate capable of indirectly or directly eliminating well-established nematode populations within 2-year old woody-rooted plants. In certain regions of the world today oxamyl is still recommended as a foliar treatment for soil-dwelling nematodes.

Organophosphate (OP) nematicides do not tend to promote an IGR effect within plants. Phenamiphos- and ethoprop-treated plants grow better after treatment but only if the nematodes are indeed causing growth reductions. The growth benefit occurs as a result of lethal and especially sublethal effects derived from a soil drench treatment and not due to systemic activity within the plant. Small percentages of phenamiphos can be basipetally translocated in the field (from above-ground to below-ground) but effective treatments are largely a result of effects on soil-dwelling stages. For this reason, multiple treatments of OP and carbamate chemicals have tended to be much more effective than single treatments when endoparasitic nematodes are attacking perennial crops. A true basipetally translocated nematicide is not yet commercially available. However, fosthiazate is an interesting candidate.

Since much of the nematicide benefit of OP and carbamate materials occurs outside the root, it follows that other root protectants might also be useful against nematodes.

D. Root Protectants

Root protectants include a mixture of nematode control mechanisms usually associated with increased bi-

ological activity in soil or along the rhizosphere. They are included here because undoubtedly some of the activity is in the form of toxins such as ammonia (NH_3) released by microbes but other methods of microbial competition play a role. These microbes may already be present in soil or they may grow along with the root surface as roots spread through soil. One difficulty with root protectants, in general, is that their performance is so sporadic that most large corporations have steered away from investing in them. Since many root protectants are products of, or the result of, biological control agents there is ever increasing interest and field testing of their worth. There are few success stories for addition of single biocontrol agents per se to soil, but there are nematode suppressive and nematode conducive soils found around the world and these are the focus of contemporary biological control studies.

Root growth stimulators or root protectants include substances such as humic acids, oil crops or oil cakes, naturally occurring plant substances or extracts, fatty acids, manures, amendments, and various organisms or compounds which release nematode antagonistic substances into soil. Many of these substances also promote IGR. It has been stated by numerous promoters of root protectants that root protectants provide increased growth or yield in a nematode infested soil without actually reducing the nematode population. To the contrary, this author would state that with an appropriate number of soil samples, a root protectant, if it is to provide viable nematode relief, must also provide a reduction in nematode population to complement the increase in plant growth if one is to call such a substance nematicidal. As an example, the chaff of sesame, *Sesemum niger*, is federally registered as a nematicidal agent. Even at high treatment rates it seldom exhibits significant nematode reduction but it does exhibit some nematode reduction. These findings indicate that additional studies are needed but sesame chaff is probably more correctly referred to as a root protectant or soil amendment rather than as a nematicidal agent. However, what if sesame chaff only performs well in nematode infested sites? And how about substances like ethroprop which seldom give rise to significant nematode reductions in the field but are referred to as nematicidal?

E. Relative Performance of Various Preplant Soil Fumigants

Visualize a back yard location for summer vegetables in a warm climate. The yard is spotted with trees, their roots criss-crossing beneath the vegetable garden. The root knot nematode has a very large host range and can be expected on the roots of most trees. What treatment does one use to control nematodes in such a garden? In some U.S. states today there is not one effective, legal nematicide treatment for such situations. When orchard and vineyard crops are removed much of the root system remains in the ground and portions of the old roots can be found alive 4 to 10 years later. Live roots result in viable nematode populations capable of attacking the subsequent crop. Planting of resistant crops can help. The best example is from stone fruit and nut crops, *Prunus* spp. Today, 300,000 hectares of *Prunus* spp in the United States are on Nemaguard rootstock with complete resistance to all *Meloidogyne* species. Unfortunately some of that land also has Ring nematode, *Criconemella xenoplax*, or Root lesion nematode, *Pratylenchus vulnus*, to which this rootstock is highly susceptible. It follows that preplant fumigants have been a very important component in the re-establishment of orchard and vineyard land. A useful characteristic of methyl bromide and Telone has been their ability to kill old roots from the surface 1.6 m of soil when properly applied. In order to reduce costs, especially among annual crops the grower does not usually broadcast treat the entire field but only those zones where the plants are seeded.

The data in Table 1 depict the long-term performance of eight different preplant treatments made in the fall following the removal of an 80-year-old vineyard. Strawberries were planted 6 months after treatment and grown for 18 months. Data are shown as percentage nematode control compared to a nontreated check with eight replicates of each treatment.

Numbers of 100% do not indicate that all the nematodes are gone but indicate that the sampling method is not sensitive enough to detect them at that point in time. A "flipped soil" involves a dual application with a 30-cm-deep plowing 10 days after the first treatment.

These data indicate the futility of organophosphate treatments in the presence of heavily infested older root systems. They indicate that MIT in large volumes of water can be effective. They indicate the value of a tarpaulin versus the value of a dual application with MB. We also observe that use of *cis*-1,3-D at treatment rates greater than 150 kg/ha results in a biological vacuum and that when nematodes do reappear this occurs faster at higher treatment rates than at rates below 150 kg/ha.

III. Mechanisms of Nematode Control

A. Direct Lethal Effects

At the laboratory bench every one of the nematicidal or nematistatic agents is lethal to nematodes at some concentration. The LD_{95} (lethal dosage for 95% of test animals) for root knot nematode, *Meloidogyne* spp, is 19 μg/ml of methyl bromide for 24 hr. Dosages of five-fold that value can kill nematodes in 5 hr and MB is frequently delivered to the site of the nematode at dosages 100 times the LD_{95} value, thus killing old roots too. Sublethal effects are an insignificant event with regard to the more volatile nematicidal agents such as methyl bromide. Carbamates and OP compounds, on the other hand, have LD_{95} values in the area of 1 μg/ml for a 72-hr exposure. They

TABLE I

Long-Term Performance of Various Preplant Nematicidal Agents

Treatment	Nematode control expressed as a percentage of nontreated				
	1 mo	7 mo	12 mo	20 mo	24 mo
MB 336 kg/top 30 soil flipped then 174 kg/ha	100	100	100	99.2	95.0
MIT 356 kg in 17 cm water	100	100	100	99.2	89.0
MB 336 kg tarped	100	100	100	99.2	41.0
Telone II 560 kg/top 30 cm soil flipped then 224 kg/ha	100	100	100	99.2	60.0
Telone II 560 kg/ha	100	100	100	70.0	37.0
Cis-1,3-D 140 kg/ha	100	99.4	75.5	68.0	65.0
MB 336 kg, without tarpaulin	99.5	98.2	58.1	76.0	50.0
Ethoprop 22.4 kg/ha in 17 cm water	75.0	0	58.0	0	0

are frequently delivered via water at concentrations of 50 to 100 times that level but are quickly diluted as they move among the soil water films so that direct lethal effects may only occur within centimeters of the application site. Additionally, nematode stages residing deep within roots such as woody perennials or old tubers may be completely unaffected by the nematicide treatment.

B. Sublethal Effects

Carbamates and OP compounds at concentrations and exposure times delivered to the nematode site are frequently too low to achieve the lethal dosage value. With grapes, for example, 1 g of active ingredient is applied per vine throughout the surface 3 feet of soil profile using low volume drippers. This 20 to 50 mg/liter treatment is delivered into soil as a gradient and diluted by subsequent irrigations which occur at 2- to 5-day intervals. This treatment is repeated once or twice on a 28-day cycle and can result in 6 to 8 months of 75 to 90% relief from endoparasitic nematodes. A few weeks after the first treatment one can detect populations two to three times higher than the non-treated population indicating that there is a positive effect on egg hatch. One can calculate the exposure levels and determine that adsorption, biological degradation, and hydrolysis of the active ingredient ensures that lethal dosages are not present within a week of treatment, and yet these treatments are quite effective when timed to the root flush activity of the grapevine, the new roots being the target of the nematode.

C. Comparative Performance of Three Postplant Nematicides

The use of low-volume irrigation results in a concentrating of plant roots and root parasitic nematodes, and allows for the concentration of nematicide theraputants in a field setting with minimal application costs. Plant–nematode systems grown under a dripper system also become a useful tool for characterizing the difference in nematicides and their sites of action. One such model system that this author has used involves the repeated treatment of 2-year-old grapevines grown in individual 60 cm diameter by 1.3-m-deep microplots in the presence of varied root knot nematode species (*Meloidogyne* spp). In one such experiment these nematicide treatments were carried out on four specific *Meloidogyne* populations (see Fig. 1). On grape, *M. hapla* is one of the more nomadic-species of *Meloidogyne* exhibiting small, superficial galls and producing relatively high soil to root ratios

of juveniles which search for new root tips. They tend not to maintain old galls for lengthy periods and do not damage grape unless very high population levels are reached on young root systems. The *M. incognita* is Race 3, the cotton race, having large sized galls with multiple females present and providing high soil population levels. The *M. javanica* population is a pathotype from Thompson Seedless grapes which can maintain large galls but tends to produce low soil population levels relative to root population levels. It is possible that many juveniles of this pathotype seldom migrate through soil but stay within the root, successfully penetrating the giant cells developed by their mother. The fourth population, *M. arenaria*, is a pathotype of Cibola alfalfa near Blythe, California. Little is known about it except that it can provide very large galls on alfalfa and can cause substantial damage to grape.

After one year of nematodes and grapevines together the vines were treated with three sequentially spaced treatments of either Sincocin, a root protectant-type fatty acid, Phenamiphos, an OP with some systemic activity, or Fosthiazate, a basipetally systemic nematicide. Nematodes in roots and soil around the roots were sampled in the fall of the second year. Figure 1 depicts the root population level of juveniles 150 days after the last nematicide treatment. The active ingredient in Sincocin is the fatty acids which appear to promote microbial growth along the rhizoplane which is inhibitory to the nomadic *M. hapla* population but not the other three *Meloidogyne* spp. Phenamiphos exhibits useful nematode reduction in soil but is only moderate in its ability to affect adult females within roots and may be ineffective on juvenile nematode stages unless they migrate through soil. Fosthiazate treatments were highly effective against soil- and root-dwelling populations of *Meloidogyne*. This comparison provides an indication of the specificity of control which may occur depending upon the mode of action of the nematicide and the feeding or root penetrating preference of selected nematodes. Each compound is nematicidal but their performance will be determined by the specific nematode pest present in the field.

IV. Listing of Nematicidal Agents

A. Volatile Nematicidal Agents

Carbon disulfide or carbon bisulfide was first used in the late 1800s to control phylloxera, *Daktylosphaeria vitifoliae*, in French vineyards at application rates of

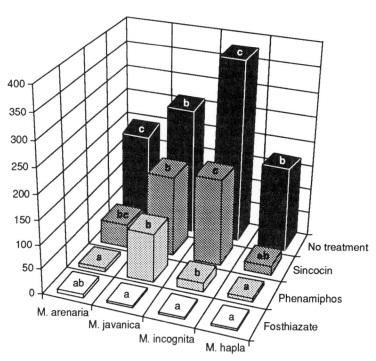

FIGURE 1 Relative performance of three nematicidal agents applied to second-year grapevines infected with each of four root knot nematode species.

2800 to 5000 kg/ha. This gaseous material has also been used to control oak root fungus, *Armillaria mellea*, of citrus. It has nematicidal value but has seldom been used specifically for that purpose. The major problem with its use is its high flash point. The pulling of steel shanks through a rocky soil can result in a spark and explosion. This active ingredient has now been complexed into solutions as Enzone (Gowan) for safer delivery into soil as a liquid.

Methyl-bromide is an odorless gas above 4°C which is delivered to soils with or without the use of a plastic tarpaulin. It is used for control of nematodes, weeds, certain soil fungi, and bacteria. It is an effective nematicide but will not control nematodes if they reside in the surface 15 cm of soil profile unless a tarpaulin is present or there is a dual application. Hydrolysis, methylation, and volatilization are major events in the final fate of methyl-bromide. It has not been a groundwater contaminant. The major contemporary concern about its future use is its volatility and its potential for transport to and degradation of the ozone layer of the stratosphere.

Chloropicrin or teargas is occasionally mixed with methyl bromide to provide control of a broader group of soil organisms. It may also occur at 2% in mixtures with methyl bromide to indicate the presence

of methyl bromide. It is not used as a nematicide per se.

1,3-dichloropropene (1,3-D) consists of a mixture of *cis* and *trans* isomers, each having slightly different nematicidal and dispersion characteristics. Since 1943 there have been various mixtures including D-D (Shell), Telone, and Telone II (Dow). In spring 1990 the use of 1,3-D in California was suspended as a result of its presence at part per trillion levels in the air space within two California communities. This mixture had been the backbone of many nematode control programs for decades. It has been reintroduced as a California nematicide but at treatment rates of 120 kg/ha or less with numerous restraints to avoid atmospheric contamination. Its relatively fast hydrolysis rate precludes it from being a ground water contaminant. However, 1,2-dichloropropane, another component previously in the 1,3-D mixture, was highly persistent in soil and has been detected from groundwater. Dichloropropene was reported in the late 1980s to be a weak carcinogen.

1,2-dibromoethane or ethylene dibromide (EDB) is the least volatile of this group of fumigants but is highly persistent. Its spectrum of control included nematodes, some grasses, and soil insects. Treatment rates were in the order of 20 to 120 kg/ha to soil. In

1984 it was detected in U.S. bakery goods since it was a popular commodity fumigant especially for stored grains. It was also added to leaded gasoline as a lead scrubber. EDB has proven to be a groundwater contaminant in Florida, Washington, California, Hawaii, and elsewhere. EDB is a known mutagen and carcinogen.

B. Less-Volatile Nematicidal Agents

1,2-dibromo-3-chloropropane or *DBCP* was marketed under brand names such as Nemagon(Shell), and Fumazone (Dow), and others. It became popular with growers in the late 1950s and its popularity gradually increased until August 1977 when laborers in its manufacture associated it with a reversible testicular dysfunction. The 1977 suspension by California was about to be lifted when in May 1979 it was found in a series of groundwater samplings from an area which had received heavy use. Over the next decade and a half numerous water wells were found contaminated in various locations worldwide. Some towns in the San Joaquin Valley of California have found all their water wells producing DBCP-contaminated water in excess of 0.2 ppb DBCP (the current action level). No well concentrations in excess of 48 ppb were ever reported from the United States The half-life for DBCP is reported to be 54 to 138 years.

Methyl isothiocyanate liberators (MIT) such as Metam Sodium are sold under a variety of trade names including Vapam (ICI Americas), Soil Prep (Buchmin Labs), Basamid (BASF), and several others. In 1991 this was the only nematicidal agent that could still be purchased by the homeowner in the garden section of the local hardware store. The performance of this product has always been inconsistent, largely due to application methodologies. In 1991 a train car of the active ingredient was accidentally dumped into a northern California river and lake, causing much attention and acute environmental damage. The product is a useful broad spectrum biocide with inconsistent performance attributed largely to problems of improper application and a relatively poor capability to penetrate and kill old woody roots or tubers.

The slow-release liquid formulation of *carbon bisulfide* called Enzone (Gowan) has value because of its short half-life. It is being tested as a postplant treatment but may have greater value as an alternative preplanting treatment.

There have been occasional reports that *ammonia releasing fertilizers* have nematicidal value. Commercial fertilizers such as urea and urea ammonium nitrate are both clearly nematicidal when re-applied at 30- to 45-day intervals via low volume irrigation systems. However, equivalent rates of calcium nitrate can be equally nematicidal when applied in a similar manner. In 1990 it was reported that potassium nitrate and ammonium nitrate had the potential to repel nematodes at relatively low treatment rates. Multiple treatments with these fertilizers can also reduce nematode populations. It may be that repeated interference with the osmoregulatory system of nematodes in soil is a major deterrent to endoparasitic and ectoparasitic nematodes, and the NH_3 release may only be a bonus. Treatments with 145 kg/ha sucrose, which also interferes with osmoregulation, through a low volume dripper in a vineyard setting has also proven nematicidal but reduced grape yield significantly after 2 years of retreatment.

C. Nonvolatile Nematicidal Agents

Aldicarb (Temik) has been a useful nematicide/insecticide for numerous crops. It has appeared in shallow groundwater of Florida and elsewhere and as residue in certain vegetables in fruits causing human sickness. This highly toxic material is currently available for only a few crops in the United States primarily on ornamentals and cotton.

Oxamyl (Vydate) provides an IGR similar to aldicarb and does not provide the breadth of nematode protection that aldicarb does. Multiple treatments increase the spectrum of activity but oxamyl is particularly useful against root lesion nematodes, *Pratylenchus* spp, as well as foliar nematodes.

Carbofuran (Furadan) has been most effective against certain insect pests but at rates two to fourfold that of aldicarb can provide some nematode relief. This product is most noted for its association with bird kills during its application which has severely restricted its use.

Phenamiphos (Nemacur) has proven to be an excellent nematicide for use on perennial crops. It provides relief from endoparasitic nematodes at relatively low treatment rates of 2 kg/ha via low volume irrigation. It is less effective against ectoparasitic nematodes and inconsistent in performance without low volume irrigation. Reports of it being found 10 m deep in soil at ppb levels appear to be erroneous since the parent material simply cannot move that far in soil without degrading.

Ethoprop (Mocap) is an organophosphate highly adsorptive to soil. It has provided inconsistent performance but is useful for specific crops such as potatoes.

D. Root Protectants

Organic matter including manure and composts have weak nematicidal value but the improved water holding capacity they provide to soil tends to reduce nematode damage without reducing nematode population levels. In a similar manner, more frequent irrigation scheduling can halve the damage caused by many nematode species. Manure also contains nitrates, ammonia, and humic acids which can reduce nematode populations to a slight degree.

Fatty acids especially those in the range of 6 to 12 carbons have nematicidal value and are occasionally marketed. One example of such a compound is Sincocin which along with cytokinins and various plant extracted components has been sold as a nematicide outside the United States. These substances may exhibit much of their effect at the root surface and not against those nematodes which live within roots or root galls.

Chitinase-producing microbes are reported to reduce nematode populations and they can be stimulated by high soil supplements of carbon and nitrogen and a source of chitin. Occasionally such products have become commercially available but they have been very expensive. The value of such products is that they show that the addition of a specific nematode controlling organism to soil has a much better chance for success if the proper habitat for the organism is also provided (e.g., food base).

Naturally occurring substances, many from plants, have been tested as nematicides over the last four decades. The author's own work has confirmed the nematicidal properties of some. However, these plant residues or plant extracts can also be allelochemicals resulting in negative plant growth. Under arid conditions this author has reported that a crop of marigolds (*Tagetes* spp.) turned under in November and planted to cherries the following spring resulted in cherry trees of significantly reduced vigor. In wetter climates, marigolds have improved growth of certain nematode-sensitive crops such as potatoes.

Over this last decade the author has been studying a winter legume bred for wide resistance to nematodes. Cahaba White Vetch, developed in Alabama, is a non-host to the nematodes found in arid tree and vine crops—except Ring nematode, *C. xenoplax*—making it a useful winter cover crop. The refuse or the extract of the refuse contains nonfermenting nematicidal properties as well as a low amount of nitrogen when applied via dripper system to a field. As with commercial fertilizers, repeated treatments appear to take their toll on soil-dwelling stages of the nematode. This extract and the extracts of marigold are also strong antioxidants being able to kill a small fish in 5 min at field treatment rates. Sesame chaff is a registered nematicide for the United States. It does appear to have nematicidal properties but performance data are lacking.

V. The Future of Nematicides

Over the last five decades there have been many useful nematicides with specific and general application. Some of them have included highly volatile substances that did not stay in soil, nondegradable substances that transported to groundwater, and highly toxic systemic substances that can translocate to food stuffs. Given the numerous incidents of off target nematicides, the number one priority for nematicides should be a downward translocating nematicidal agent. This agent could be an organism which colonizes along the rhizosphere, or it could be within the roots and effective on endoparasitic and ectoparasitic nematodes. The concept of treating large volumes of soil with large volumes of nematicide has been very useful to growers but the environmental concern of today will not permit continued use unless such chemicals are quick to degrade. The development of low-volume irrigation has provided a ready vehicle for nematicide delivery at relatively low treatment rates. Desperately needed is a procedure or pesticide that can kill the roots of previously planted perennial crops. These old roots provide nutrition for nematodes and other microbes for years after the top portion of the plant has been removed.

The food supply of America is abundant, diverse, and relatively inexpensive. At this point in time the American public is calling for a million-fold safety factor between the cancer-causing level of a pesticide and the action levels which protect our air, water, and food supply. Is this enough? The answer varies but relative to soil applied pesticides we are sure of one thing. Nematicides without a predictable degradation rate have no value in the future.

Bibliography

Brown, R. H., and Kerry, B. R. (1987). "Principles and Practice of Nematode Control in Crops." Academic Press, Orlando, FL.

McKenry, M. V., and Roberts, P. A. (1985). "Phytonematology Study Guide." Publication No. 4045, Univ. of California Press, Oakland, CA.

Sasser, J. N., and Carner, C. C. (1985). "An Advanced Treatise on Meloidogyne," Vol. I. North Carolina State Univ. Graphics, Raleigh, NC.

Stirling, G. R. (1991). "Biological Control of Plant Parasitic Nematodes." CAB International, Tucson, AZ.

Veech, J. A., and Dickson, D. W. (1987). "Vistas on Nematology." Soc. of Nematologists, Hyattsville, MD.

Nitrogen Cycling

DALE W. JOHNSON, *Desert Research and University of Nevada*

Glossary

Denitrification Conversion of nitrate (NO_3^-) and nitrite (NO_2^-) to gaseous dinitrogen oxide (N_2O) and nitrogen gas (N_2); several genera of heterotrophic bacteria (bacteria which obtain energy from organic matter) are capable of denitrification under anaerobic (low oxygen) conditions; nitrite reduction can also occur via chemical reactions (chemodenitrification) under aerobic conditions

Nitrification Two-stage, microbially mediated conversion of ammonium (NH_4^+) to nitrite (NO_2^-) and subsequent conversion of nitrite to nitrate (NO_3^-); organisms most commonly associated with nitrification are chemautrotrophic bacteria (bacteria which derive their energy from chemical reactions), but it may also be accomplished by heterotrophic nitrifiers, which generate NO_2^- and NO_3^- from organic N compounds but obtain energy from organic matter; nitrification is an acid-producing process that can significantly affect soil acidity and nutrient availability

Nitrogen fixation Process by which atmospheric nitrogen gas (N_2) is converted to ammonium (NH_4^+) by bacteria or actinomycetes; nitrogen fixing organisms may be free-living or associated with nodules in plant roots and on a global scale, nitrogen fixation is the most important pathway of N input to natural terrestrial ecosystems

Nitrogen is rather unique among the major plant nutrients in that the primary source is the atmosphere rather than soil minerals. Nitrogen plays a critical role in plant photosynthesis and growth, and is the most commonly limiting nutrient in terrestrial ecosystems. Nitrogen enters terrestrial ecosystems primarily via fixation and atmospheric deposition, whereupon competition among decomposers, nitrifying bacteria, nonbiological processes, and plants ensues within the soil. This competition largely determines whether nitrogen will be retained within the terrestrial ecosystem in either plants or soils, or whether it will be lost via denitrification and leaching. The relative importance of nitrogen cycling processes varies greatly from site to site, depending upon local conditions.

I. Sources of Nitrogen

A. Global Scale

On a global scale, the largest reservoir of nitrogen (N) is in igneous (volcanic origin) rocks within the earth's mantle (Table I). There are also large reserves within both igneous and sedimentary rocks within the earth's crust. However, N concentrations in rocks and minerals are very low and rarely do minerals provide appreciable N to terrestrial biota. The atmosphere, being composed of 79% N , is the ultimate source of most N for terrestrial ecosystems. Nitrogen (N) is unique among nutrients in this respect; minerals provide the primary sources of the other major nutrients (P, S, K, Ca, Mg, Mn).

Compared to the relatively low amounts of N in terrestrial biomass and soils, the atmosphere is a very large reserve from which to draw. However, the predominant form of N in the atmosphere—N_2—is unavailable to plants and most microorganisms because of the strength of the triple bond in the N_2 molecule. Thus, despite the great abundance of N in both the lithosphere and atmosphere, N is the most commonly limiting nutrient in terrestrial ecosystems.

TABLE I

Nitrogen Content of the Atmosphere and Terrestrial Ecosystems

Atmosphere	3.9×10^9
Terrestrial biomass	3.5×10^3
Soils	
Organic matter	1 to 2.2×10^5
Clay fixed NH_4^+	2.0×10^4
Exchangeable[a]	1×10^4
Rocks	
Igneous (in the crust)	1.0×10^9
Igneous (in the mantle)	1.6×10^{11}
Sedimentary (fossil N)	3.5 to 5.5×10^8
Coal	1.0×10^5

Sources: Paul, E. A., and Clark, F. E. (1989). "Soil microbiology and Biochemistry." Academic Press, San Diego; Schlesinger, W. H. (1991). "Biogeochemistry—An Analysis of Global Change." Academic Press, San Diego; Stevenson, F. J. (ed.) (1982). "Nitrogen in Agricultural Soils." Agronomy No. 22, American Society of Agronomy, Madison, WI.
Note. Units are in teragrams, or 10^{12} g.
[a] Assumed to equal 1% of soil organic N.

The numbers in Table I serve to put N cycling processes in perspective on a global scale, but they are necessarily tentative. For instance, estimates of N in the earth's mantle are very uncertain, given the limited access we have for sampling in that geologic strata. The range of values for soil N gives an idea of the magnitude of uncertainty associated with this number, and, although ranges for vegetation N are not available, this pool is quite uncertain as well.

B. Terrestrial Scale

If we define our system boundaries to include the plant–soil system only (Fig. 1), we find that most (normally more than 90%) N resides in the soil, followed by vegetation and/or litter. Table II illustrates the N distribution from representative forest and grassland ecosystems. In these ecosystems, soil solution and exchangeable N—that is, soil ammonium (NH_4^+) and nitrate (NO_3^-) loosely held on the soil colloids by electrostatic forces and immediately available to plants and microbes—are very low relative to soil total N. Thus, even though soil N normally exceeds plant N by many fold, N deficiencies in natural ecosystems are common because only a small fraction (typically 1% or less) of total soil N is available to plants at any given time. Most soil N is tied up in recalcitrant forms of organic matter or is chemically "fixed" within clays. Thus, terrestrial vegetation must rely upon atmospheric as well as soil sources of

N even though total N supplies in most ecosystems are large relative to plant needs.

II. Inputs of Nitrogen to Terrestrial Ecosystems

Nitrogen fixation—the process by which atmospheric N_2 is "fixed" by microorganisms and made available to plants as ammonium (NH_4^+)—is the most important pathway of N input to natural terrestrial ecosystems on a global scale (Table III). Other significant inputs of N to terrestrial ecosystems include atmospheric deposition, N fixation via lightning, and inputs of fertilizer. On a local scale, any of these processes can dominate N inputs; for instance, atmospheric deposition may be the most important N input in polluted sites without N-fixing vegetation, or fertilizer may be the most important N input to agricultural soils. [*See* FERTILIZER MANAGEMENT AND TECHNOLOGY.]

A. Atmospheric Deposition

Atmospheric deposition may be either wet (rain, snow, fog inputs of NH_4^+, NO_3^-, or organic forms) or dry (gaseous inputs of N_2O or NH_3, particulate inputs of all forms, or HNO_3 vapor). Because atmospheric deposition usually impinges first on vegetation, leaf surfaces are often the sites of initial entry into the ecosystem via foliar uptake of N.

Only in recent years has the technology for measurement of dry deposition of N been fully developed and the importance of dry deposition inputs of nitrogen been recognized. Dry deposition is especially important in polluted areas, where it may equal or exceed precipitation (Table IV).

B. Nitrogen Fixation

Nitrogen fixation, or the conversion of atmospheric N_2 into forms usable by plants and microbes, can occur via either abiotic (lightning) or biotic mechanisms. Biotic N fixation is generally of much greater significance. Biotic nitrogen fixation is accomplished by a group of bacteria referred to as diazotrophs which include organotrophic bacteria (bacteria which use organic substrates as a source of energy), phototrophic sulfur bacteria, and cynobacteria (blue-green algae). These organisms may be free-living or associated with nodules in plant roots. Free-living nitrogen fixing organisms obtain energy either from sunlight

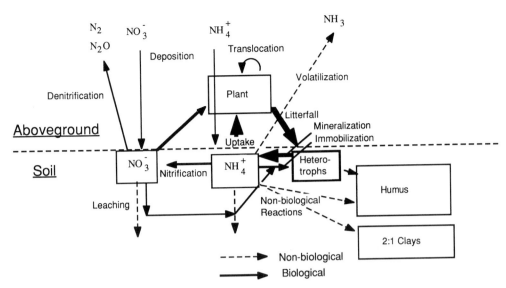

FIGURE 1 Schematic representation of nitrogen cycling in soils. [Revised from Johnson, D. W. (1992). Nitrogen retention in forest soils. *J. Environ. Qual.* **21**, 1–12.]

through photosynthesis (e.g., cyanobacteria) or from organic substrates in soils or leaf surfaces. Nitrogen fixers associated with plant roots reside in nodules where the organisms obtain carbohydrate from plant roots and excrete excess fixed N (N not used by themselves) as NH_4^+. Some of the more common nitrogen fixers include bacteria *Rhizobium* species associated with the roots of legumes, actinomycetes (*Frankia* spp.) associated with the roots of *Alnus* and *Ceanothus* species, and cyanobacteria (e.g., *Nostoc* spp.) associated with lichens and cryptogamic crusts in desert ecosystems. [See SOIL MICROBIOLOGY.]

The overall cost of nitrogen fixation has been estimated to be on the order of 1.1 to 2.4 mol of glucose,

or 2.9 to 6.1 g of carbon per gram of nitrogen fixed. These energy/carbon costs to the host plant must be weighed when considering the benefits of nitrogen fixation vs commercial fertilization.

Both ammonium and nitrate are known to inhibit both nodulation and N-fixation, providing a potential negative feedback mechanism that should come into play to prevent excessive N fixation and nitrate leaching. However, this feedback mechanism does not always operate efficiently. Recent studies have shown

TABLE II

Distribution of Nitrogen in a Representative Forest and Grassland Ecosystem

Component	Deciduous forest	Shortgrass prairie
Vegetation (live and dead)	492 (8%)	169 (6%)
Litter	274 (5%)	60 (2%)
Soil		
Exchangeable	75 (1%)	20 (0.5%)
Total	5080 (95%)	3330 (91.5%)

Sources: Henderson, G. S., Harris, W. F. (1975). An ecosystem approach to characterization of the nitrogen cycle in a deciduous forest watershed. pp179-193 *In* Forest Soils and Land Management, B. Bernier and C. H. Winget (eds). Les Presses de l'Universite Laval, Quebec; and Nitrogen budget of a shortgrass prairie ecosystem. *Oecologia* **34**, 363-376.
Note: Units are in kg ha^{-1}.

TABLE III

Nitrogen Fluxes between the Atmosphere and Terrestrial Ecosystems

Atmospheric deposition	+25
Nitrogen fixation	
Lightning	+<20
Biological	+140 to 250
Fertilizer	+26
Plant uptake	1200 to 1400
Ammonium volatilization	−26 to 53
Denitrification	−130 to 135
Leaching	−93
Runoff, erosion	−25 to 36
Fire	?

Sources: Paul, E. A., and Clark, F. E. (1989). "Soil microbiology and Biochemistry." Academic Press, San Diego; Schlesinger, W. H. (1991). "Biogeochemistry—An Analysis of Global Change." Academic Press, San Diego; Stevenson, F. J. (ed.) (1982). "Nitrogen in Agricultural Soils." Agronomy No. 22. American Society of Agronomy, Madison, WI.
Note. Units are in teragrams, or 10^{12} g.

TABLE IV

Wet and Dry Deposition of Nitrogen (Ammonium + Nitrate) to a Mixed Deciduous and Red Spruce Forest in Western North Carolina

Flux	Deciduous forest[a]	Coniferous forest[b]
Dry deposition		
Vapors	2.2	8.6
Particles	1.9	3.5
Total dry	4.1	12.1
Precipitation	6.9	6.2
Fog/cloud	0	8.7
Total	11.0	27.0

Source: Johnson, D. W., and Lindberg, S. E. (eds). (1991). "Atmospheric Deposition and Forest Nutient Cycling: A Synthesis of the Integrated Forest Study." Ecological Series 91, Springer-Verlag, New York.

Note: Units are in kg ha^{-1} yr^{-1}.

[a] Mixed deciduous at Coweeta Hydrologic Lab, Otto, NC.

[b] Red spruce forest in the Great Smoky Mountains National Park, NC.

that prolonged nitrogen fixation by red alder (*Alnus rubra*) in forests of the Pacific Northwest can lead to high rates of nitrification and nitrate pollution in groundwater. Apparently the symbiotic relationship between red alder and nitrogen-fixing *Frankia* in nodules of its roots does not result in an ideal, efficient exchange of carbohydrate for nitrogen but rather results in considerable "waste" of fixed N over the long term. The extent to which this occurs with other nitrogen-fixing species has not been sufficiently studied.

The "wasted" N fixed by *Frankia* in red alder has significant side effects that bear upon plant succession in that ecosystem type. Nitrogen fixation is often associated with pioneering vegetation, giving them a competitive advantage on very N-poor soils, such as those found on recently deglaciated or eroded sites. It has long been known that a stage of N-fixing vegetation "paves the way" for subsequent vegetation stages by improving the N status of soils. Recent work in red alder (*A. rubra*) has also shown that N-fixation can create soil conditions (increased acidity, lowered phosphorus availability) unfavorable for the continued presence of the N fixing species. Thus, N fixation, at least in the case of red alder, may play an even more significant role in both soil modification and vegetation succession than previously suspected.

III. Plant Uptake, Allocation, and Recycling

A. Uptake

As noted above, nitrogen is the most commonly limiting nutrient in natural ecosystems, and therefore

plants have adapted strategies of efficiently taking up and recycling nitrogen. Nitrate is taken up by nearly all plants, and either NO_3^- or NH_4^+ may be taken up by many plants. Uptake may occur through either roots or foliage. Inside the plant, both forms are converted to amino (-NH_2) groups attached to organic molecules which are then transported throughout the plant. In the case of NO_3^-, conversion to amino groups is accomplished by an enzyme called nitrate reductase and imposes a significantly greater energy cost to the plant than conversion of NH_4^+ to amino groups. It is something of a mystery as to why most plants show a preference for NO_3^- rather than NH_4^+, given the greater energy cost of utilizing NO_3^-. The answer may lie in greater mobility NO_3^- in the soil and greater ease of NO_3^- uptake (because of less competition for uptake sites with ions of similar size and charge). Also, NH_4^+ is toxic to plant roots at lower concentrations than is NO_3^-.

B. Uses of Nitrogen in Plants

Nitrogen is a constituent of proteins, and therefore has a major effect upon plant growth. Nitrogen deficiency, which is often accompanied by a yellowing appearance of foliage, causes an accumulation of carbohydrates such as starch (which cannot be utilized for protein synthesis) and a reduction in growth. One of the most important N-containing enzymes in plants is 1,5-biphosphate carboxylase oxygenase, often referred to as RuBISCO. This enzyme, which occurs in all green plants, plays a critical role in photosynthesis in that it is responsible for the capture of carbon dioxide (CO_2) which has entered the plant chloroplast. [See PHOTOSYNTHESIS.]

C. Nitrogen Allocation in Plants

Plants have several ways to deal with N limitations or deficiencies. Forests growing under N-deficient condition allocate more carbohydrate to roots in order to promote more root growth and exploration of soil reserves than their counterparts growing with adequate N supplies. Internal translocation is also an important mechanism for dealing with N deficiencies. Evergreen coniferous trees often maintain large stores of N in older foliage that can be tapped for translocation to growing tissue during periods of N stress. Deciduous vegetation typically translocates 30–50% of the nitrogen in leaves back into the branches, stem and roots during autumn senescence for use in the following spring growth flush. In this way, plants can efficiently recycle N internally without having to compete with microbes to take up N from decomposing litter.

IV. Biological Transformations in Soils

Once dead plant parts come into contact with the litter layer, or, in the case of roots, directly with the soil, decomposition begins. During the decomposition process, there is first a release of organic C as CO_2, then, at a later stage when decomposer demands for N are satisfied, release of N as NH_4^+. The latter is referred to as N mineralization. After N mineralization, N can undergo a variety of biological and chemical transformations in the soil causing it to either re-enter the terrestrial N cycle or leave the terrestrial ecosystem to enter either the hydrosphere or atmosphere and the global cycle. Detailed descriptions of these transformations are given in the following sections.

A. Heterotrophic Uptake and Mineralization

During the initial stages of decomposition, N is not released from litter but rather is utilized by decomposer microbes (heterotrophs). This is necessary because the carbon to nitrogen ratio (C:N ratio) in microbes is on the order of 5 to 15, whereas that in plant litter is on the order of 100 to 200. Heterotrophs utilize only a small fraction of organic C to build their own bodies, the remainder being used as an energy source and converted into CO_2 in the process. Thus, as litter decomposes, its C:N ratio decreases to a critical threshold before any N is released (as NH_4^+) that can be taken up by plants. This critical C:N ratio is typically on the order of 20-25, somewhat higher than that of decomposer microbes, indicating that microbes are not 100% efficient in N recovery.

A major factor affecting both decomposition and N mineralization from decomposing litter is lignin content. Lignin has a poly-phenolic structure that is very resistant to attack by decomposers and therefore slows the overall decomposition rate. Lignin also chemically combines with NH_4^+ to form very stable humus compounds, and thus retards N release from decomposing litter. Several studies have shown the rates of decomposition and N mineralization rates are more accurately predicted by lignin:N ratios than C:N ratios. More information on chemical reactions of inorganic N with lignin and other soil organic compounds is provided later in the article.

B. Nitrification

Nitrification is the microbially mediated conversion of NH_4^+ to NO_2^- (nitrite) and NO_3^-. The organisms most commonly associated with nitrification are chemautotrophic bacteria, or bacteria which derive their energy from chemical reactions. These bacteria are classified into two categories: those that oxidize NH_4^+ to NO_2^- (such as *Nitrosomonas* species), and those that oxidize NO_2^- to NO_3^- (such as *Nitrobacter* species). The N transformations in the two stages of the nitrification process are shown in Fig. 2 (top). In most cases, the two sets of organisms are found together and NO_2^- seldom accumulates as such in soils. Notice that the first stage is acid (H^+) producing, and that both stages require oxygen. Not shown in Fig. 2 are the transformations associated with heterotrophic nitrification. Heterotrophic nitrifiers generate NO_2^- and NO_3^- from organic N compounds, but obtain energy from organic carbon oxidation rather than from the nitrification process.

Autotrophic nitrification is strongly affected by aeration, moisture, temperature, organic matter, and pH. Oxygen is a requirement for the nitrification reaction (Fig. 2), and thus adequate aeration is necessary. Adequate moisture is also necessary for the reaction, but excessive moisture inhibits nitrification by reducing aeration. Temperature strongly affects nitrification; the optimum is around 30° to 35°C, and nitrification is strongly inhibited at temperatures below 5°C or above 40°C. Organic matter affects nitrification indirectly by causing heterotrophic competition for both O_2 and NH_4^+.

Laboratory studies indicate that optimum pH values for autotrophic nitrification are between 6.6 and 8.0, and that autotrophic nitrification is severely limited at lower pH's. Because nitrification produces acid (Fig. 2, top), it is potentially a self-limiting process. However, several field studies have documented high rates of nitrification at soil pH's of 3.8 to 5.0, clearly showing the danger of overextrapolating laboratory results to field conditions. The reasons for this disparity between laboratory and field results is not known, but may be due to the presence of high pH microsites within acidic soils. It is also possible that nitrification in acidic soils is due to the activities of less pH-sensitive heterotrophic nitrifiers.

Autotrophic nitrifiers are also known to be inhibited by certain organic chemicals, both naturally and synthetically produced. Some studies have shown that nitrification rates decrease during forest succession due to the production of nitrification inhibitors in late successional ecosystems. It is not known how

Nitrification

$$2NH_4^+ + 3O_2 \text{-----------} > 2NO_2^- + 4H^+ + 2H_2O \qquad \text{(\underline{Nitrosomonas})}$$

$$2NO_2^- + O_2 \text{------------} > 2NO_3^- \qquad \text{(\underline{Nitrobacter})}$$

Denitrification

$$4NO_3^- + 2H_2O \text{------------} > 2N_2 + 5O_2 + 4OH^- \qquad \text{(\underline{Pseudomonas})}$$

$$2NO_3^- + H_2O \text{------------} > 2N_2O + 2O_2 + 2OH^- \qquad \text{(\underline{Pseudomonas})}$$

Chemodenitrification

$$RNH_2 + HNO_2 \text{------------} > ROH + H_2O + N_2 \qquad \text{(Van Slyke Reaction)}$$

$$NH_4^+ + HNO_2 \text{------------} > 2H_2O + N_2 \qquad \text{(Reaction with ammonium)}$$

$$3HNO_2 \text{------------} > HNO_3 + H_2O + 2NO \qquad \text{(Spontaneous decomposition of nitrous acid)}$$

FIGURE 2 Nitrogen transformations during nitrification (top), denitrification (middle), and chemodenitrification (bottom).

inhibitors might function under conditions of chronically elevated NH_4^+ inputs.

Nitrification can be accompanied by gaseous N_2O loss (especially following fertilization. Nitrate produced can be taken up by plants and microbes or reduced to gaseous forms (N_2, N_2O) through denitrification and chemodenitrification (described below). Because NO_3^- is poorly adsorbed to most soils, nitrification in excess of plant and microbial NO_3^- uptake in aerobic soils nearly always leads to increased NO_3^- leaching, which may have consequences for both water quality and soil acidification.

C. Denitrification

Denitrification refers to the microbially mediated conversion of NO_3^- and NO_2^- to gaseous N_2O and N_2. Several genera of bacteria are capable of denitrification, including *Pseudomonas*, *Bacillus*, and *Alcaligenes*. Most are heterotrophic (obtain energy from organic matter) and are normally aerobes (grow in the presence of oxygen) without carrying on denitrification. Under anaerobic (low oxygen) conditions, however, denitrifying organisms can grow in the presence of NO_3^- and/or NO_2^-, utilizing these ions as electron acceptors instead of O_2. Figure 2 (middle) shows two reactions that occur during denitrification by *Pseudomonas* bacteria.

Denitrification requires the presence of NO_3^-, limited O_2 availability, and decomposable organic carbon. The requirement for NO_3^- is obvious, since it is a reactant in the denitrification process. Limited O_2 availability is required in order to cause the heterotrophic denitrifying organisms to begin utilizing NO_3^- and NO_2^- as electron acceptors. Limited O_2 is usually associated with high moisture (e.g., waterlogged) conditions. Denitrification occurs most readily at pH's near neutrality (6 to 8), but can occur more slowly at pH's above 4.

Nitrite reduction can also occur via chemical reactions under aerobic conditions. This is sometimes referred to as chemodenitrification. Some of these reactions are depicted in Fig. 2 (bottom). In the Van Slyke reaction, amino groups (RNH_2) are converted to N_2 gas in the presence of nitrous acid (HNO_2). Similar reactions may occur in the presence of NH_4^+. Nitrous acid may also decompose into nitric acid (HNO_3) and nitrous oxide (NO), the latter of which may further react with O_2 and water to form more nitric acid.

D. Competition among Heterotrophs, Nitrifiers, and Plants for Soil Nitrogen

Although NH_4^+ strongly adsorbs to cation exchange sites, large soil NH_4^+ pools seldom occur in undisturbed soils because of competition for NH_4^+ among decomposers (heterotrophs), plants (autotrophs), and nitrifiers (chemautotrophs) for NH_4^+ (Fig. 1). This competition for N among heterotrophs, plants, and nitrifiers for NH_4^+ is thought to play a major role in determining the degree to which N is retained within the ecosystem. Several studies have indicated that heterotrophs are the most effective short-term competitors for N. In that heterotrophic demand for N depends upon the supply of decomposable organic C substrate, high levels of low-N organic matter in litter and soils greatly reduce N availability to plants. Thus, plants can experience severe N deficiency even in the presence of large total soil N pools if soil organic C pools are also very large. High levels of organic C may also facilitate increased activity of denitrifying organisms if anaerobic conditions exist.

Once heterotrophic demand for NH_4^+ is satisfied—for example, when organic C substrates are depleted, depriving heterotrophs of energy sources, or following N fertilization—NH_4^+ supply to both plants and nitrifiers is increased. It has been generally assumed that nitrifiers are the least effective competitors for N, and thus that nitrification proceeds only after both heterotrophic and plant N needs are met. However, this assumption has been largely based upon the lack of NO_3^- in soils; recent studies find significant nitrification and microbial NO_3^- uptake in some grassland soils even when soil NO_3^- pools were very low. There is also evidence that nitrifiers become more effective competitors for NH_4^+ when given a slow, steady supply.

V. Nonbiological Transformations in and Effects on Soils

A. Reactions with Organic Matter

Nonbiological incorporation of N into humus is a much-neglected but potentially important process of soil N cycles (Fig. 1). A substantial proportion of soil organic N is typically associated with humus, and some of the important mechanisms of N incorporation into humus are nonbiological in nature. The inhibitory effect of lignin upon decomposition and N mineralization noted earlier is due in part to the formation of stable nitrogenous compounds from lignin by-products, reducing nitrogen availability to decomposer organisms. Physical condensation reactions of phenols (originating from partially degraded lignin and some fungal pigments) with either amino acids or ammonia result in the formation of "brown, nitrogenous humates." These nonbiological, autocatalytic reactions are important in the production of humus. Nonbiological incorporation of ammonia into humus is enhanced by high pH (because NH_3 is the

reactive form of N), and high NH_3 and/or NH_4^+ concentrations. Substantial amount of N from urea fertilizer can be immobilized by nonbiological reactions with humus because of the high pH and ammonia conditions that urea fertilization creates.

B. Ammonium Fixation in Clays

Unlike other nutrients, N rarely accumulates in readily available inorganic forms (i.e., NH_4^+, NO_3^- or NO_2^-) in soils. Significant amounts can accumulate in unavailable inorganic NH_4^+ "fixed" within certain clay, however. Clay-fixed NH_4^+ can account for anywhere from <1 to over 50% of total soil N, depending upon the clay types and organic N concentration in the soil.[1]

The ammonium ion, like the similar-sized potassium ion (K^+), can become very tightly bound between the layers of 2:1 clays (i.e., clays which contain a layer of aluminum oxide octahedra sandwiched between two layers of silica oxide tetrahedra). These clays normally have a net negative charge due to Al for Si substitution in the Si-layer, giving them the capability of adsorbing cations (positively charged ions). The capacity of soils to retain cations is referred to as cation exchange capacity. Potassium and NH_4^+ are both cations and therefore can occupy cation exchange sites, but both are of such a size that they can also be trapped between clay layers, clamping them together very tightly and "fixing" the trapped NH_4^+ and K^+ between the layers. A clay saturated with interlayer K^+ is referred to as mica. Clay-fixed NH_4^+, unlike exchangeable NH_4^+, is only slowly available to plants and microbes. In some soils, clay-fixed NH_4^+ is nearly totally unavailable to plants and microbes, whereas in other soils, it can be a significant slow-release form of available N.

The presence of K^+ can greatly influence the availability of clay-fixed NH_4^+. Figure 3 shows both K^+ and NH_4^+ fixed between a modified mica where some of the interlayer K^+ has been replaced by NH_4^+ (right-hand side), and Ca^{2+} has replaced interlayer K^+ and NH_4^+ (left-hand side). Here it can be seen that clay-fixed NH_4^+ can be blocked by clay-fixed K^+, making the clay-fixed NH_4^+ completely unavailable for either plant or microbial uptake.

[1] The term, "fixed" which is often used to describe NH_4^+ trapped between layers of certain clays should not be confused with the term "fixed" used to describe the microbial conversion of atmospheric N_2 to NH_4^+ form. For that reason, the term for NH_4^+ between clays will henceforward be referred to as "clay-fixed" NH_4^+.

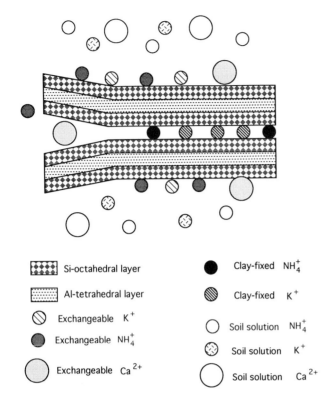

▦ Si-octahedral layer		●	Clay-fixed NH_4^+
▨ Al-tetrahedral layer		◍	Clay-fixed K^+
◎ Exchangeable K^+			
◕ Exchangeable NH_4^+		○	Soil solution NH_4^+
		⊛	Soil solution K^+
◯ Exchangeable Ca^{2+}		○	Soil solution Ca^{2+}

FIGURE 3 Schematic drawing of 2:1 clay with ammonium, potassium, and calcium in solution, on exchange sites, and ammonium and potassium fixed.

VI. Outputs of Nitrogen from Terrestrial Ecosystems

The most important output process for N on a global scale is via denitrification (the microbially mediated conversion of nitrate, NO_3^- to N_2O, and N_2 gases) (Table III). Details of the denitrification process have already been discussed. Most denitrification occurs in wetlands, a consequence of the requirement for anaerobic conditions. Estimates of N leaching—primarily in the mobile NO_3^- form—put this process as nearly as important as denitrification on a global scale, with runoff and erosion being considerably less.

As was the case with inputs, however, the relative importance of N outputs on a local scale may differ considerably from global averages. For instance, leaching is usually of much greater significance than denitrification in well-drained terrestrial ecosystems, whereas the reverse is true of bogs, marshes, and wetlands. Interestingly, both nitrogen fixation and denitrification are important processes in desert soils. Given the requirement for anaerobic conditions, it would not seem likely that denitrification would be important in desert soils. Denitrification is important in desert soils, however, for several reasons. Due to the lack of water for leaching, NO_3^- levels can become quite high in desert soils, providing a substrate for denitrification which occurs in certain microsites in the soil and in low areas during and immediately following precipitation events.

Ammonia (NH_3) volatilization occurs only at relatively high pH (>7) because of the preponderance of the tightly bound ammonium (NH_4^+) form at lower pH. Ammonia volatilization from animal feces and urine can be a very significant N loss pathway in grassland or agricultural ecosystems where grazers are present in relatively large numbers. Ammonia volatilization can also be important in arid soils where pH conditions are appropriately high, but in general, NH_3 volatilization is of less importance than denitrification in these systems.

Global estimates of N release from fire are not available as of this writing, but fire can be a significant N loss pathway from certain ecosystems. Fire converts vegetation, litter, and a portion of soil organic N into N_2O and N_2 gases which escape from the ecosystem. It is noteworthy that N-fixing vegetation is adapted to fire both in terms of ability to thrive on N-poor soils and in terms of life cycle. N-fixing vegetation frequently predominates postfire vegetation for several years until ecosystem N stores are replenished.

Bibliography

Johnson, D. W. (1992). Nitrogen retention in forest soils. *J. Environ. Qual.* **21,** 1–12

Paul, E. A., and Clark, F. E. (1989). "Soil Microbiology and Biochemistry." Academic Press, San Diego.

Schlesinger, W. H. (1991). "Biogeochemistry—An Analysis of Global Change." Academic Press, San Diego.

Stevenson, F. J. (ed.) (1982). "Nitrogen in Agricultural Soils." Agronomy No. 22. American Society of Agronomy, Madison, WI.

Van Miegroet, H., and D. W. Cole. (1984). The impact of nitrification on soil acidification and cation leaching in red alder ecosystem. *J. Environ. Qual.* **13,** 586–590.

Waring, R. H., and Schlesinger, W. H. (1985). "Forest ecosystems—Concepts and Management." Academic Press, Orlando, FL.

Oat

DAVID W. CUDNEY, *University of California, Riverside*

CLYDE L. ELMORE, *University of California, Davis*

Glossary

Caryopsis Kernel or dry, one-seeded, indehiscent fruit common to members of the grass family

Dormancy State of reduced physiological activity which allows seeds to remain viable in the soil for long periods of time

Fatuoid False wild oat that often appears in cultivated oat as a result of hybridization of wild oat (*Avena fatua*) and the cultivated variety

Panicle Inflorescence with a branched main axis, whose branches bear loose flower clusters

Spikelet Unit of inflorescence in grasses which comprises a small group of grass flowers

Sucker mouth Oval-shaped structure at the base of wild oat kernels which is the result of the formation of an abscission zone; the abscission zone causes the wild oat seed to drop from the plant to the soil at maturity

Tiller Grass stem arising from a lateral bud at the base of the plant: tillering is the process of tiller formation

Oat, of the genus *Avena,* is a complex series of diploid, tetraploid, and hexaploid species. The genus contains both useful crop plants and serious weeds. The development of oat as a weed and as a crop parallels that of wheat (*Triticum*) and barley (*Hordium*) species. Weedy members of the genus have worldwide distribution and are found wherever cereal pro-

duction occurs. The domesticated oat ranks sixth in world cereal production and is adapted to cool climates with over 700 mm of rainfall or irrigated conditions. Most oat production is for livestock feed and is harvested for grain, silage, or hay. Oat has also been useful as a cover crop or as companion crop in mixed culture, particularly with legumes. Human consumption of oat grain products have increased in the last few years.

I. History and Origin

The area where oat originated is unknown; however, the consensus is that it may have been in the western Mediterranean region. It is also difficult to demonstrate when oat domestication occurred. Recent findings have shown wild oat to be present with barley in a cave in southern Greece which was dated to 10,500 B.C. Early remains of *Avena sterilis* were found in Syria which was dated to at least 8600 B.C. and in a prepottery Neolithic village in southern Jordan (7000 B.C.), and another site yielded a strong twisted awn that would indicate wild oat in the same period. *Avena sativa* has been mentioned from Switzerland, Germany, and Denmark as early as 2000 to 1000 B.C. There is no mention of oat in the Bible.

Early use of oat was probably as a forage crop with later use for food and as a mixture with barley for the making of beer. There are early descriptions of the use of oat for medicinal purposes, as food for horses, and food for humans in time of need. The use of oat as a porridge was indicated in 415 A.D. and has often been associated with people of the working or poorer classes. Oat for food was planted extensively in northern Europe with long history in Scotland and more recently in England, Spain, Germany, and Ireland. During the famine in Ireland when potatoes were scarce, oat became the food to keep people alive. Food use of oat increased during the 18th and 19th centuries; however, it decreased again in the mid- to late 20th century.

Oat probably was not in the Americas before the Columbus voyages. It was most likely introduced by the English and other Europeans over northeastern America during the 16th and 17th centuries. Oat may also have been introduced from the south via the Spanish expeditions during the 16th century. Oat plantings spread over what is now the United States with greatest concentration in the northern and north-eastern states.

II. Economic Importance

The economics of oat have changed over the years. At one time oat was a mainstay of the diet for many people of European decent. Much of the grain was used for animal food and the forage for animal feed. As time progressed, however, farmers began to grow crops which were more efficient in the production of energy and protein, and oat production decreased. Oat currently ranks sixth in world cereal production behind wheat, maize, rice, barley, and sorghum. As of 1985, the former USSR, followed by the United States, Canada, West Germany, and Poland were the principal producers and utilizers of oat and oat products. In the last 100 years the area that was devoted to oat production has declined by one-half; during the same period, yields have increased about 60%. In the United States it is estimated that 60% of the production is consumed on the farm where it is produced. Although there is little world trade of oat, the United States is an importer.

About 75% of all oat is used for feed for animals. Most of this is fed to milk cows and horses. Oat grain is an excellent conditioning feed for horses and cattle because of the fiber content. Use for human consumption is most often as rolled oat and to a lesser extent as oat bran and oat flour. Rolled oat is prepared in a process where the grain is heated, dried, and hulled. The resulting "groat" is then steamed and rolled to produce the familiar flake. Oat grain does not compete with corn, or oil seed crops for fattening livestock. As the number of working horses declined, there was a concurrent decrease in oat consumption. [*See* FEEDS AND FEEDING.]

Food consumption by humans declined until the early 1980s when research indicated that the water-soluble fiber of oat products was beneficial for health.

Currently the use of oat products for human consumption is increasing.

III. The Oat Genus

Oats are members of tribe, Aveneae, within the grass family (Poaceae). All but one of the common 31 species (*A. macrostachya*) are annuals. Members of the genus have branched, panicle-type, inflorescences. The panicle may be symmetrical (tree-shaped) or one sided. The spikelets are borne terminally on the thickened end of slender, drooping branches which radiate from a central stem or peduncle.

The spikelet (Fig. 1) is subtended by two papery, boat-shaped glumes. Two or more florets are produced per spikelet. Florets are bisexual and are composed of the lemma and palea and the reproductive organs (the ovary and three stamens). The lemma

FIGURE 1 Spikelet of *Avena fatua*. Note two glumes at base of inflorescence and awns which arise midway down the lemmas of the three florets. Awns are bent (geniculate) and twisted below the bend. (Photo courtesy of Jack Kelly Clark, University of California, Davis.

and palea are two bracts that enclose the caryopsis or kernel at maturity. The lemma is the lower of the two bracts and it ranges in color from white, yellow, gray, and red to black at maturity. Lemma color and tip shape are often used in species identification. Tips are usually indented with two points. An awn, or stiff bristlelike appendage, arises in many of the species from the midrib of the lemma at or slightly below the middle of the lemma. The awn is often bent (geniculate) near its middle and coiled below the bend. The kernel or caryopsis is long and narrow, more or less spindle shaped, and is deeply furrowed. Fine hairs cover the caryopsis usually more prominently toward the tip.

In wild types and the progenitors of the cultivated types, the florets break off from the spikelet at maturity and fall to the ground leaving the panicle with the glumes still attached. They do this through the formation of an abscission layer at the base of the first floret. This natural separation of the lower floret or kernel from the spikelet is termed disarticulation. Disarticulation leaves a well-defined oval cavity at the base of the lemma which is called a sucker mouth (Fig. 2). The sucker mouth is usually not found in cultivated oat species because separation of the oat from its axis is by fracture at harvest and results in roughened tissue with no observable cavity.

Leaves of oat are smooth and flat with rounded tips. The leaf is composed of a leaf blade, a ligule (papery extension above the collar), a collar which lacks auricles (clasping appendages), and a sheath that extends down the stem (Fig. 3).

Oat species are often grouped in accordance with their ploidy or chromosome number (diploid, tetraploid, and hexaploid). Of the 31 most common species 16 are diploid (14 chromosomes), 7 are tetraploid (28 chromosomes), and 8 are hexaploid (42 chromosomes).

The most common cultivated oats are hexaploid and include *A. sativa*, the common white or northern

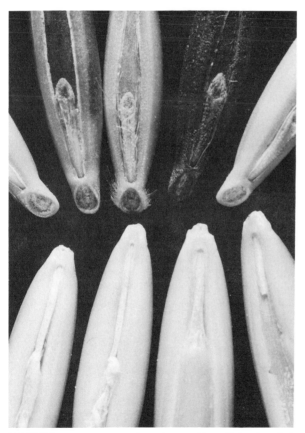

FIGURE 2 Seed of *Avena fatua* above with well-developed abscission zones (sucker mouths), which allow seed to be shed prior to harvest. Seed of cultivated oat, *Avena sativa* var. Ogle, have no developed abscission zone and are broken off at harvest. (Photo courtesy of Jack Kelly Clark, University of California, Davis.

FIGURE 3 Typical leaf of *Avena fatua*. Note papery ligule (collarlike structure) and absence of auricles. (Photo courtesy of Jack Kelly Clark, University of California, Davis.

oat, and *A. byzantina* the cultivated red oat. *Avena byzantina* has sometimes been included in *A. sativa*. *Avena nuda* (also known as naked or hull-less oat) is a diploid oat that is sometimes cultivated. It has papery lemmas which thresh freely from its kernels at harvest. Among the wild species there are 3 that are most often found as weeds in cultivated fields. The common wild oat, *A. fatua*, is hexaploid as is *A. sterilis* (wild red or animated oat). Slender oat, *A. barbata*, is tetraploid.

Self-pollination is the norm in oat but some natural crossing can occur. Crossing varies (from 0 to 10% usually less than 1%) depending on the environment and the species. Fatuoids or false wild oat often appear in cultivated oat varieties. Fatuoids result as a cross between the weedy *A. fatua* and the cultivated variety. They tend to resemble both parents to some extent and often shed their seed prior to harvest like the wild species.

IV. Development and Biology

A. The Seed

When the papery lemma and palea are removed from the oat seed, the kernel or caryopsis is exposed. It consists of an outer coat, a starchy endosperm, and the germ or embryo. The embryo is attached to the endosperm by a shield-shaped structure known as the scutellum (actually the cotyledon). The embryo consists of the plumule (precursors of the leaves and stem tissue of the plant), the scutellum, the hypocotyl (embryonic stem), and the radicle or first root (other root primordia may also be evident).

B. Germination

The minimum temperature for germination of cultivated oats is near 7°C. Wild oat requirements have been shown to be between 10° and 23°C with reduced germination at higher temperatures. Unless the seed is dormant, germination generally occurs when temperature, moisture, and oxygen requirements are met.

C. Dormancy

A dormant seed is one that fails to germinate even though it has imbibed water and temperature and environmental conditions favor germination. Loss of dormancy of mature seed with time is termed "after ripening." In cultivated oat the after ripening period is short, lasting but a few weeks. Wild oat requires a much longer period, often a year or more. Research has shown that dormancy in wild oat is affected by environmental conditions during seed formation including temperature and moisture. The environment under which seed is stored also has an effect on dormancy. Usually seed stored at higher temperatures is lower in dormancy. Light, soil oxygen, and gibberellic acid have also been implicated in the regulation of dormancy. [*See* DORMANCY.]

Imbibition is the process where dry seed absorbs water and the seeds increase in fresh weight. Germination begins after imbibition as nuclear activity, protein synthesis, and metabolic activity increase and growth of the embryo occurs.

D. Growth

The energy for germination and early seedling growth is sustained by the hydrolysis and remobilization of reserve substances such as lipids, proteins, carbohydrates, and mineral nutrients stored in the endosperm of the seed. As the seedling grows it becomes progressively more dependent on the process of photosynthesis taking place in the newly formed leaves for its energy.

During germination the radicle pushes down from the germinating seed elongating from the scutellar node. Two or three additional roots may also emerge from the coleoptilar node. These roots compose the seminal root system. The seminal root system serves the seedling oat plant for several weeks until the main root system of the plant becomes established.

The root system which sustains the oat plant for the major portion of its life is fibrous and develops as advantitious roots from the "crown" of the oat plant at or just below the soil surface. These roots actually arise from shortened stem nodes within the crown region. Tillers or additional stems also arise from this region.

The oat coleoptile is the first leaf to emerge from the germinating embryo. It is a pointed hollow, tubular leaf which pushes up through the soil during emergence. The first true leaf then emerges through an apical pore at the tip of the coleoptile. The oat coleoptile shows both phototrophic (grows toward the light) and geotrophic (grows upward) responses. Oat coleoptiles were the object of much of the early research involving growth responses to plant growth substances (plant hormones).

The oat stem, or culm, is composed of a succession of alternating nodes and hollow internodes. These nodes and internodes originate from tissue in the growing point of the stem tip (apical meristem). The first, or basal, internodes are relatively short. As the stem grows the later internodes become longer, terminating with the inflorescence. Each node contains meristimatic tissue which is capable of producing a new stem (tiller) or roots. Several tillers are generally produced by each plant. The number is dependent on the density of the plant population, availability of moisture and nutrients, and the length of the growing season. The formation of new tillers is stimulated by emergence of the panicle on the first stem, decapitation of the main stem, or lodging (the plant falls over and lays along the ground). The formation of new tillers is thought to be regulated by the balance of plant growth substances (cytokinin and auxin). The meristematic tissue in the upper stem nodes usually remains quiescent unless lodging occurs, whereupon it may become activated.

Oat leaves alternate in two ranks, with one leaf at each node. Each leaf originates successively along the margin of the growing point (apical meristem) of the stem. As the leaf expands it is composed of a blade, a collar, and a sheath (which surrounds the stem and attaches to the stem). As the plant grows, new leaves emerge blade first in a sequential fashion from the leaf sheath of the preceding leaf.

E. Photosynthesis

Photosynthesis takes place in the leaves of the oat plant and to some extent on the green stem tissue. Most photosynthesis takes place in the upper leaves of the plant canopy. The lower leaves become less efficient as they age and eventually senesce. Although wild oat competes in winter cereal plantings for moisture and nutrients, recent research has shown that under optimum growth conditions it competes strongly for light. It does so by extending its flag leaf (the last leaf produced on the stem before the oat panicle) above the crop canopy and intercepting light. Oat, like most cool season grass plants, has a C_3 type of carbon fixation pathway. This makes it well adapted to cool growing conditions, but at a disadvantage during high temperature and light conditions when compared to C_4 species. [See PHOTOSYNTHESIS.]

F. Transpiration

Both photosynthesis and water loss from the plant (transpiration) are regulated by the opening and closing of stomata in the leaf epidermis. Stomata are small openings in the leaf epidermis which are regulated by changes in the turgor of two surrounding guard cells. When stomata are open, they allow the exchange of CO_2 and O_2 which is necessary for photosynthesis and respiration. Water vapor also is exchanged or lost. Thus, transpiration can be regarded as part of the cost the plant pays for doing business. But water movement within the plant necessitated by transpiration, also helps with the transport of mineral nutrients and the maintenance of cell turgor. As soil water becomes less available or when water demand becomes high, the stomata close, reducing water loss. This reduces water loss, but it also slows photosynthesis and if continued for extended periods of time results in reduced plant growth and low yield. Oat, like many grasses, rolls its leaves inward in response to water deficit. This rolling is made possible by the action of bulliform cells in the upper surface of the leaf.

V. Cultural Practices

A. Soil Preparation and Planting

A seedbed of loose soil should be prepared by disking or plowing followed by a shallow cultivation with a harrow or springtooth. Seeding can be done by broadcasting or by drilling with a grain drill. Drilling is generally more satisfactory because the seed can be placed to a uniform depth (25 to 65 mm) and is covered by soil to give a more uniform stand. Drill rows are usually spaced 10 to 20 cm apart. Broadcasting requires a further cultivation to incorporate the seed into the soil. Seeding rate varies depending upon the use of the crop. When oat is produced for grain, planting rates of from 70 to 100 kg per hectare are usually recommended. When oat is produced for forage (grazing or hay), up to twice the planting rate for grain is used. Seeding rate can be adjusted for soil texture, fertility, and moisture conditions, and to affect weed population and disease incidence.

Oat can be planted into an unprepared seedbed left after the previous crop (no-till). This is best done in fields where the soil is left relatively mellow. Yields have been reduced in heavy, poorly drained soils. No-till planting allows earlier planting than conventional methods.

B. Nutritional Requirements

Fertilizer requirements for oat depend on the soil where the crop is to be planted. Many soils may need lime as well as the conventional fertilizers. A soil test will be helpful to determine the need for additions of nutrients. Nitrogen (N) fertility is difficult to determine because there is little knowledge of how much residual N is in the soil available to the plant. Often low rates of nitrogen (no more than 22 kg ha^{-1}) are applied at seeding with the plan of top-dressing more later to achieve a total of from 65 to 100 kg per hectare. Too much nitrogen at planting or as a foliar feeding can injure oat. Too little nitrogen can reduce yield and the competitiveness of the crop with weeds. The amount of phosphorus and potassium can be planned from the soil test. When difficient, at least as much of these fertilizers as the crop yield is expected to remove from the soil needs to be supplied.

C. Water Requirements

Oat grows best in a deep soil with good water holding capacity. Oat should have adequate, but not excessive, soil moisture. Oat requires more water than other cereal grains. It is commonly grown in areas averaging 40 to 110 cm precipitation, with most production in areas with 80 cm or less.

D. Growing Season

Oat is adaptable for use as a winter or spring crop. Winter oat planting dates vary from locale to locale even within a state. Often planting may begin in September in some parts of the United States up until 3 to 4 weeks before an expected frost or into November in the southern United States. Time of planting will depend upon the severity of the winter and whether the crop is intended for grain or forage. Early plantings are needed to allow the crop to establish well before a killing frost. Growers plant early for a good grain yield, and later for a forage yield. Spring oat is generally planted in the early spring after the soil is dry enough to be worked properly. Early plantings generally give the greatest grain yield.

E. Harvesting Techniques

Oat forage is harvested by grazing with livestock or by machine. Oat is harvested at the milk stage of the grain for silage or in the soft dough stage for hay. Oat hay is prized for horse or other livestock feed and has been used historically as bedding. Grain can be swathed at the hard dough stage and when the kernels have lost all green color. The crop is picked up from the windrows with a combine equipped with a pickup attachment and threshed. Harvesting can start sooner with swathing and is feasible in areas that are dry during the harvest season. In recent years, most grain is harvested directly with a field combine. This change is possible because of new varieties that resist lodging and shattering of the grain.

F. Companion Planting and Intercropping

In some regions of the United States oat is planted at reduced seeding rates with legumes to suppress weeds and enhance the total forage yield of the first cuttings. There has also been interest in planting oat in thin alfalfa stands during their last season of production to increase forage production.

Oat is planted as an inter-row crop in newly planted orchards and vineyards in the western United States. It has also been used as a cover crop to reduce erosion and maintain organic matter. As a cover crop, it is planted with a legume such as vetch, winter pea, or annual clovers to be cultivated down as a mulch and nitrogen source in the spring.

G. Oat as a Rotational Crop

In a farm management system, oat is often one of the rotational crops. It can be grown in the cooler months, when other crops are out of season, or as a catch crop for forage or grain. Oat might follow full-season crops such as corn, soybean, potato, cotton, or grain sorghum. In areas where residual chemicals such as dinitroaniline, triazine, or sulfonylurea herbicides have been used in the preceding crop, oat can be injured. There are studies showing increased yields following oat in a rotation.

VI. Pests of Oats

A. Insects

Oat is generally less susceptible to damage from insect pests than barley or wheat, but it can sustain damage at any point in its life cycle. The presence of insects in oats may not result in economic damage to the crop. The population level of an individual pest at and above which economic injury will take place is called the economic injury level. This pest level, if known for a given pest species, should be used as an indicator of need for control measures. There are relatively few key insect pest species that cause yield loss year after year in oat. These include green bug

(*Schizaphis graminum*) and the chinch bug (*Blissus leucopterus*). Most of the insect pests in oats are occasional pests that cause sporadic losses, depending on the year. Examples of occasional pests include armyworm (*Pseudaletia unipuncta*), billbug (*Calendra parvulus*), cutworms, grasshoppers, Morman cricket, several additional aphid species, and cereal leaf beetle (*Oulema melanopus*). Mites, another type of arthropod, can also become a secondary pest if natural predators are lost after the use of certain insecticides.

B. Diseases

There are fungal, bacterial, and viral diseases of oat. The best control measures for diseases, if available, are resistant varieties. The difficulty with many diseases is that new strains of the diseases continue to develop, necessitating the development of still more resistant oat varieties. Additional control measures involve planting disease-free seed or manipulating the environment so that ideal conditions for infection do not occur at a time when the crop is most susceptible. This may require altering planting dates. Crop rotation to reduce disease innoculum can be helpful, if the rotational crops are not alternate hosts and weed hosts are excluded. [*See* PLANT PATHOLOGY.]

Two of the more serious types of fungal diseases include the smuts and the rusts. There are two smut diseases which cause serious damage to oat. They are loose smut (*Usilago avena*) which replaces the oat floret with a loose mass of spores, and covered smut (*Ustilago kolleri*) which replaces the floret with a spore mass covered with a grayish membrane, frequently within the lemma and palea. Both resistant varieties and fungicides have been effective in the control of these diseases. There are also two serious rust diseases of oat. They are stem rust (*Puccinia graminis avenae*) and crown rust (*Puccinia coronata*). Both produce rust-colored pustules and have alternate hosts which participate in the life cycle of the disease. Crown rust is the more serious of the two diseases and both are best controlled by growing resistant varieties. Eradication of the alternate hosts has also been helpful. Other fungal diseases include Septoria leaf blight (*Septoria avena*) which is best controlled by resistant varieties and *Fusarium* diseases which can be reduced by rotation with nonhost crops and can be treated with certain fungicides.

Two of the more serious bacterial diseases include bacterial blight and blade blight. There are two species of bacerial blight and one of blade blight that infect oat, all are species of the genus *Pseudomonas*. The use of uncontaminated seed and avoidance of contaminated debris can help to reduce disease incidence.

There are at least 10 common viruses that can infest oat. The most common of these is barley yellow dwarf virus (BYDV). Like all of the other oat viruses BYVD is vectored (transmitted by a secondary organism), in this case it is transmitted by aphids. Once an aphid feeds on an infected plant it can transmit BYVD to the next plant on which it feeds. Treatment of the aphids with insecticide is usually not effective, because by the time the aphids are noticed transmission has already occurred. Even if one flight of aphids were controlled, new flights would continue to transmit the disease. BYVD is most severe in oat when the plants are infested in their early growth stages. One effective method of reducing the severity of BYVD is to shift planting dates so that the young oat plants develop at a time when aphid activity is low and disease transmission less likely. The best control measure is the use of resistant varieties.

C. Nematodes

Nematodes are small, microscopic, or nearly microscopic roundworms that can parasitize plants and animals. Plant parasitic nematodes can reduce yield severely and can be found in large numbers in the soil after a suitable host crop. Important oat nematode pests include the stem nematode (*Ditylenchus dipsaci*), the cereal root-knot nematode (*Meliodogyne naasi*), and the cereal cyst nematode (*Heterodera avenae*). The most commonly used control measure for nematodes is rotation with a nonhost crops.

D. Weeds

Oat like many cereal crops is fairly competitive. One of the best weed control measures is to establish a dense stand of oat with narrow rows as quickly as possible, before weeds can become established. This goal is often achieved; however, weedy species often do become problems and competition from weedy species can reduce oat yield. Robust, rapidly growing broadleaf weeds such as wild radish (*Raphanus sativus*) readily compete for nutrients, moisture, and light. Control of broadleaf weeds in oat is made more difficult because oat is more sensitive to injury from the common foliar-applied broadleaf herbicides than other cereal crops. There are no herbicides which will selectively control wild oat in cultivated oat. Wild oat must be controlled in rotational crops prior to planting cultivated oat. [*See* WEED SCIENCE.]

Bibliography

Coffman, F. A. (1977). "Oat History, Identification and Classification." USDA Technical Bulletin 1516. U.S. Government Printing Office, Washington, DC.

Hickman, J. C. (ed.) (1993). "The Jepson Manual: Higher Plants of California." University of California Press, Berkeley.

Hoffman, L. A., and Livezey, J. (1987). "The U.S. Oats Industry." USDA-ARS Rep. 573. U.S. Government Printing Office, Washington, DC.

Jones, D. P. (ed.). (1976). "Wild Oats in World Agriculture." Agricultural Research Council, London.

Lawes, D. A., and Thomas, H. (eds.) (1986). "Proceedings of the Second International Oats Conference." Martinus Nijhoff, Dordrecht.

Marshall, H. G., and Sorrells, M. E. (eds.) (1992). "Oat Science and Technology." No. 33. American Society of Agronomy, Inc., Madison, WI.

Martin, J. H., Leonard, W. H., and Stamp, D. L. (1976). "Principles of Field Crop Production," 3rd ed. Macmillan, New York.

Orchard Management: Soil Environment and Resources

DAVID ATKINSON, *The Scottish Agricultural College*

Glossary

Arbuscular mycorrhiza Symbiotic association between a root and a fungus characterized by the development of arbuscules, tree-shaped, double plant–fungal membrane systems within the cortical cells of the root; described as endomycorrhizal because the association between the root and the fungus is "internal"; the fungus within the root is connected to a network of extramatricular hyphae which ramify out into the soil, greatly extending the surface area of the root and potentially promoting the uptake of nutrients; describes most or all fruit species under field conditions

Budget Balance sheet for the nutrient or water use of a production system, which details the needs of the components of the system, i.e., the quantities of nitrogen required by leaves, fruit, wood, and roots during a single production cycle; budgets can also be used to estimate the total needs for nutrients (fertilizers) or water (irrigation)

Cover crop Vegetative cover or permanent sod sown, or allowed to develop by managed natural regeneration, in a fruit plantation specifically to cover the ground surface and provide a suitable surface for

machinery movement or a means of controlling soil condition: the commonest species used are slow-growing grasses, prostrate legumes, or mixtures of these; in some orchard situations, plants may be sown as a "green manure" and incorporated into the soil on an annual basis, usually in the spring

Genetically modified organism Organism where the genotype has been modified by the addition of genetic information from another species by means other than normal crossing (a transgenic organism); genetic changes usually involve the addition of a small number of genes by means of a vector

Irrigation Application of water to supplement natural precipitation and so increase crop growth and production, or to alleviate water stress; most common application is to the soil surface to recharge water stored within the soil profile and as a result maintain transpiration close to the rate dictated by physical environmental conditions

Mist application Variant of irrigation rare in commercial practice which aims to reduce transpiration while maintaining stomatal conductance, thereby minimizing water stress and maintaining growth; leaves are maintained in a "wet" condition and the moisture layer on the leaf surface is evaporated before incoming radiation is available to "power" transpiration

Mulch Natural, e.g., straw, or synthetic, e.g., polythene, cover applied to the soil to minimize the loss of soil moisture, optimize soil temperature, or as an aid to weed control; mulches may be applied over the whole of the soil surface area or more commonly over the area adjacent to the crop plant and may be left in place for much of the cropping cycle or for a brief period

Organic production Production of crops in the absence of synthetic fertilizers or pesticides. Organic systems attempt to harness natural processes and to cycle nutrients within a production unit. Animal production by-products may be used as a source of nutrients. The basis of an organic farming system is the interdependence of individual components and the holistic nature of the enterprise

Resource Material needed for production to occur; usually applied to soil water or to mineral nutrients such as nitrogen or nitrogen-containing materials; an assessment of the amount and availability of soil resources is essential to the estimation of a crop's needs for added resources such as irrigation and fertilizers

Rootstock Lower section of a compound fruit tree onto which a scion variety is budded or grafted; gives rise to the root system and the basal section of stem material and is used to control the amount or form of the growth of the scion variety; may also be used to provide resistance to a soil-borne disease or pest; compound trees are described using the nomenclature scion/rootstock as in Red Delicious/M9

Soil fertility Inherent features of a soil which allow crop growth to proceed at a given rate; influenced by the organic matter and mineralogical contents, which control its potential to release nutrients, its microbial population, including pathogens, symbionts, and microorganisms involved in nutrient transformations, and the texture; "apparent fertility" is also affected by soil structure, the physical arrangement of solid, liquid, and gaseous materials; mineral and organic materials mix to form aggregates which contain air- or water-filled spaces which influence movement, e.g., of water and energy (heat), through the soil and the ability of biological organisms, e.g., roots, to grow

Sustainable System of production in which the use of resources and the ability to produce crops are durable, i.e., where the production of the current crop does not adversely affect the ability to produce future crops; although not synonymous with organic or low input systems or with systems where crop protection involves integrated pest management (IPM), it is often used in this way

Orchard soil management can have two separate purposes. First, soil management should optimize the growth and production of fruiting species by the supply of necessary edaphic resources. Second, it aims to give a surface suitable for the movement of the machinery needed for crop harvesting and crop protection. Soil needs to be maintained as both a road and a growth medium. Soil management aims to provide the best practical solution to these dual aims. The edaphic resources naturally present in the soil,

and needed for growth, are often present in amounts smaller than those needed for growth. Where this is the case, natural resources would commonly be supplemented by the addition of fertilizer (nutrient) or irrigation (water). If additions are not to be in excess of needs, resulting in loss to the environment and an environmental problem, then the rate of addition must be related to crop or soil demand for additional resources. The crop's use of soil resources will be influenced by the activities of other organisms. These may be beneficial, as with arbuscular mycorrhizas, or detrimental, as with pests, weeds, or diseases. The balance between beneficial and harmful organisms and the crop will be affected by the use of crop protection materials (pesticides), the design of the production system, and its management. This balance will change during the life of an orchard. In general, young trees are more susceptible than more established trees to other organisms. The optimum management system for an orchard will thus vary with age, between soil types, and between climatic zones. The appropriate soil management systems for apple production in California and Northern Europe will differ, although the basic principles upon which decision making depends will be common.

All food production currently occurs against an increased awareness of interactions between anthropogenic activities and the environment. The delivery of crop production and protection practices with minimal effects on the natural environment and nontarget organisms are now critical to the design of orchard management systems. In addition, there is a requirement for production to be sustainable. As a consequence, the budgets of orchard systems are being assessed in terms of their energy, carbon, and nitrogen budgets. Soil management is thus primarily about the management of these interactions. This article reviews the basic soil and plant factors influencing the need for nutrients and water and the roles of soil management and the supply of additional resources in optimizing production. [*See* ORCHARD MANAGEMENT SYSTEMS.]

I. The Resource Requirements for Optimum Production

The demand for resources and the ability of the soil to meet these demands will be influenced by a range of factors. Some of the major factors are reviewed here to provide a basis for the discussion of management effects.

A. Orchard Nutrient Budgets

Orchard trees, like other crops, need substantial amounts of nitrogen (N), phosphorus (P), potassium (K), calcium (Ca), magnesium (Mg), and sulfur (S), as well as smaller amounts of trace elements of which those most commonly deficient are iron (Fe), manganese (Mn), and boron (B). N, P, and K are the elements most commonly added as fertilizers. A budget for a mature heavy cropping orchard in a single season is shown in Table I. The annual uptake of Ca and K was similar, greater than that of N and much greater than that of P. Nitrogen was relatively uniformly distributed between the various components. There was little Ca in fruit (2%) but much in leaves (51%). K was principally in and about equally divided between fruit and leaves (87%), while P was concentrated in the fruit (35%). This type of budget is essential to making any predictions of the effects of changes in cultural practice on the demand for particular nutrients. As the major nutrients are unequally divided between vegetative (leaves and wood) and reproductive growth (fruit) an increase in relative crop production will increase the need for P and K relatively more than that for N and Ca. In contrast an increase in vegetative growth will increase relative demand for Ca and N. In most budgets, little or no allowance is made for amounts contained within the root system (or amounts lost by leaching). During a single season new root growth may require up to 21 Kg N ha^{-1}. An annual nutrient budget is, however, an incomplete basis for assessing the ability of soils to supply resources to the trees, because the different components of the tree increase in mass, and so need nutrients, at different times during the season (Fig. 1). Variations in the amounts of shoot, fruit, roots, and leaves, and in the times when they need resources, mean that the soil system must meet both total needs and specific demands at appropriate times during the season. The ability of the soil to supply nutrients and the system's needs for supplementary nutrients will depend upon

1. The characteristics of the root system.
2. The absolute demand for nutrients any given time. This will be affected by the relative balances between fruit, leaves, wood, and roots and between orchard crop and cover crop or weeds.
3. The extent to which nutrients within the tree can be remobilized to supply new growth.
4. The availability of nutrients. The availability of nutrients to the crop root system will depend upon the amount of nutrient element within the soil, its chemical speciation, and the relationship between the amount present and the amount available, which is a major concern for P and for micro-nutrients such as Fe, Zn, Bo, and Mn.

B. Orchard Water Budgets

The needs of the orchard for water depend on transpiration by the crop, the use of water by the cover crop or weeds, and evaporation from the soil. The water budget of an orchard will depend on

1. The age and type of tree. Older trees and trees in intensive plantings use more water than younger or widely spaced trees. For example, in an orchard of trees of Golden Delicious/M9 water use increased from 15 liter m^{-3} for trees with a leaf area index (LAI) of 0.7 m^2 m^{-2} to 55 liter m^{-3} at an LAI of 2.1. The form of the water

TABLE I

Annual Nutrient Uptake by a Mature Heavy-Cropping Red Delicious Orchard (45 Mg ha^{-1} yr^{-1})

Annual uptake into	Nutrient (kg ha^1)				
	N	P	K	Ca	Mg
Fruit	21	6.3	57	4	2
Prunings	12	2.2	4	28	2
Tree framework + roots	18	4.2	14	46	2
Leaves	48	3.3	52	86	18
Dropped fruitlets, etc.	12	1.7	15	4	1
Total	111	17.8	142	168	25

Source: Batjer, L. P., Rogers, B. L., and Thompson, A. H. (1952). *Proc. Am. Soc. Horticult. Sci.* **60,** 1–6.

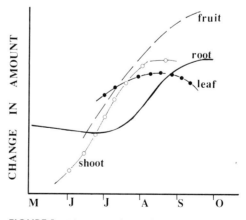

FIGURE 1 The seasonal periodicity of total root length, woody shoot production, leaf growth, and fruit production for a young apple tree. [Redrawn from Atkinson, D. (1985). "Pome Fruit Quality," p.5 University of Bonn.]

use: LAI relationship depends on the configuration of the orchard, i.e., whether the trees are in a square or rectangular arrangement, and whether the soil surface is bare or vegetated.

2. The total amount of water available from the soil profile and the ease of its availability to the crop. This is affected by soil type: more water is available from clay than sandy soils, but water is more easily removed from sandy soils; the depth of the soil profile: in general the amount of available water increases by 12–25 mm for every 15 cm increase in soil depth, and the effective depth of the root system. The recharge of soil moisture content following rainfall also influences the quantity of water available. An example of this is illustrated in Fig. 2. A soil treated with herbicide, following a series of arable cultivations, had a higher water content early in the season than did a soil with a partial grass cover. Penetration of heavy rainfall into the grass-covered soil occurred more rapidly and to a greater extent than with the herbicide-treated soil which subsequently maintained a higher deficit.

The effects of orchard design and management on the water budget of an orchard are complex. To assess the needs of an orchard system, information is needed on

   a. The total water content of the soil profile.

   b. The moisture release characteristics of the soil: the release characteristics of sandy and clay soils differ greatly and herbicide-treated soil usually has a lower percentage moisture available at any water potential than does a cover cropped soil, i.e., at a potential of -40 kPa the water content was 30% for a vegetated soil and 24% for a bare soil.

   c. The contribution of the cover crop or evaporation from the soil surface to the total water budget, i.e., a

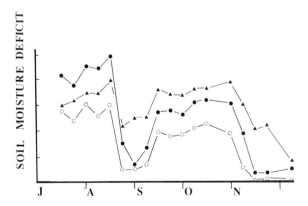

FIGURE 2 The soil moisture deficit (l tree^{-1}) as influenced by soil management. In mid-August a fall of around 40 mm rain reduced the deficit in all treatments. (▲) Herbicide management, bare soil following cultivation; (○) herbicide management, bare soil following grass; (●) a wide herbicide-treated strip in grass. [From Atkinson, D. (1981). *In* "Pests Pathogens and Vegetation" (J. M. Thresh, ed.) p.421. Pitman UK.]

cover crop exposed to radiant energy will transpire and increase the soil moisture deficit. The importance of this may be illustrated for an established orchard of Cox/M26 in the U.K. Here the soil moisture deficit, totaled for the surface 750 mm depth of soil, in the inter-row area in mid-summer was 22 mm for bare soil, but 76 mm under a grassed soil. The rate of soil moisture depletion during the June–July period was 0.05 % v/vd^{-1} under bare soil compared with 0.12 under grass.

   d. The ability of the soil to accept rainfall (Fig. 2).

   e. The size and distribution of the crop root system.

C. Soil Resources

The ability of the soil to meet the needs of the tree for nutrients and water will depend upon its mineral and organic matter content and its physical structure. The extent to which deficiencies in nutrient supply will actually become apparent will depend upon the crop's growth rate and the total amounts needed of the various elements. Although orchard management will aim to influence nutrient release, where the needs of the crop in total or at any point in time are higher than the soil system can supply, the crop will become nutrient deficient unless its demand upon the soil is reduced. This can be achieved by either a reduction in the amount of crop produced or the provision of an additional source of nutrient, i.e., fertilizer. The supply of N from the soil can, however, be substantial. For example, a series of fertilizer experiments in the U.K. have demonstrated the absence of a response to added nitrogen in trees with an annual net removal of nitrogen of 35 kg N over 30 years. [*See* FERTILIZER MANAGEMENT AND TECHNOLOGY; SOIL FERTILITY.]

Similarly the factors detailed previously indicate that orchard management can influence the demand for water. The amount of water needed will, however, be mainly dictated by potential evaporation which is largely a function of wind and radiant energy. The demand for water by the trees, however the orchard is managed, may consequently exceed the water-holding capacity of the soil and in this case, if water stress is to be avoided, then additional water, i.e., irrigation, must be provided. The extent to which supplementary nutrients and water are needed is one of the major themes of this article.

D. Spatial and Temporal Factors Influencing Resource Supply and Demand

Water and nutrients are normally unevenly distributed through the soil volume. This distribution interacts with management. Most nutrients have their highest concentrations near to the soil surface. This is a consequence of the distribution of organic matter, proximity to the site of the application of fertilizer, the higher soil temperature, and the resulting enhanced amount of soil microbial activity. Both nutrient availability and microbial activity will be modified by soil water potential. The concentration of tree and cover crop roots found near the soil surface will result in the water near the soil surface being depleted most rapidly, so that the water potential near the soil surface may be very low for much of the season. As a consequence, deeper, less well-supplied zones can become important for the nutrition of the tree. The extent to which this occurs will be influenced by soil management. In an established orchard in the U.K. on June 18 the soil water potential at 250 mm depth was -1.3 MPa under grass, but only -0.08 MPa under bare soil. In orchards the soil volume around the fruit tree can be divided into two functional areas. The area adjacent to the tree, which is usually maintained with a bare surface, provides a major proportion of the resource needs of the tree. In contrast, the inter-row area functions primarily as an access area and frequently has a vegetative cover. The difference in soil management between these areas introduces further heterogeneity. In addition, the need of the tree for nutrients varies with time during the season (Fig. 1), while tree age also influences the temporal needs of the tree.

The availability of water from the soil is a consequence of past water use and rainfall. However, the link between water use and solar radiation means that soil water potential, in the absence of irrigation, is usually high in early summer and low in autumn. Such factors interact with soil management so that nutrients, e.g., N, which are liberated by the mineralization of organic matter, soil microbial processes sensitive to temperature and water potential, will vary between depths and during the course of a single season. In a mature orchard in the U.K., even in the absence of fertilizer applications the mineral nitrogen content ($NO_3 + NH_4$) under bare soil varied during a single season from a minimum of 20 Kg ha^{-1} in the surface 600 mm to a maximum of 70 kg. [See SOIL-WATER RELATIONSHIPS.]

E. Effects of the Growing System on Resource Demand

The type of orchard system used should in principle not have a large effect on the amounts of nutrients needed when the trees reach maturity or a stable level of production. In trials carried out in New York State, Washington State, East Malling, U.K., and Luddington, U.K., annual values for net nutrient removal (fruit and prunings) of 24, 39, 26, and 33 Kg N ha^{-1} and 43, 71, 41, and 51 Kg K ha^{-1}, respectively, were found. Although these amounts of nutrients represent components which are removed from the orchard, the soil resource must also be able to provide enough nutrient to allow the growth of nonremoved components such as leaves. For nutrients like N, this amount can be several times that removed from the orchard. The mobility of N and the amounts lost annually by leaching make strict N budgeting difficult. While the amounts of nutrient removed from the different orchard types at maturity are relatively similar, the period required to achieve this state will vary. In the more traditional orchard, maturity will normally take 12 years, while plateau production can occur at 5–7 years in intensive systems. Studies of apple trees planted at a range of densities and receiving nitrogen applications within the range 0 to 280 Kg N ha^{-1} have indicated that while trees at the wide spacing increased their leaf N status only in the range to 70 Kg N, trees at the highest density responded to rates up to 210 Kg N. Studies in the U.K. have shown that the development of an optimal leaf canopy (a major sink for nutrients) varies between varieties and systems. Trees of Cox/MM104 achieved an LAI of 1.49 at 9 yr, Cox/M26 an LAI of 1.36 at 7 yr, Golden Delicious/M9a an LAI of 1.5 at 5 yr, and a bed system of Cox/M27 an LAI of 1.4 at 2 yr. As the nutrient content of trees is greatly influenced by the amount present in leaves (Table I), growing systems will affect tree nutrient demand. In addition, there will be parallel affects on water use. In an orchard of Golden Delicious/M9, water use increased from 15 to 42 liter m^{-3} as the LAI increased from 0.7 to 1.5. Intensive orchards will reach peak water and nutrient demand earlier than with more traditional plantings.

II. The Tree Root System

Effects of orchard management are likely to impact initially upon the tree root system. Root system development will be affected by competition from other trees, particularly in high-density plantings, and from cover crops. Most soil management practices are designed to benefit the functioning of the root system. As a result, it is important to understand the growth and dynamics of tree and cover crop root systems.

A. The Development and Dynamics of the Root System

The mature fruit tree root system includes roots which differ in age, diameter, and the extent of the development of woody tissues. The young apple root is white in color and after a period of 1–5 weeks begins to turn brown in color. Browning of the cortical tissue is followed by cortical disintegration which is aided by the soil fauna. Following the loss of the cortex some roots develop secondary tissues, e.g., wood and bark, and become part of the woody root framework. Other roots, usually the higher order laterals disappear. In the early years of a tree's life approximately 25% of roots will survive and 75% decompose. The rate of production of new root is not uniform over the growing season. The periodicity of new growth varies between species in both the form and timing of the cycle, although in most species, root growth begins in the spring, in advance of shoot growth. A bimodal pattern of growth in apple, an early peak in spring prior to shoot growth, and a second more substantial peak in late summer/autumn, is commonly reported. In *Prunus* a single peak of new growth seems most common. In all species, the form and amount of growth are influenced by factors such as pruning and cropping. The vigour of new growth varies with tree age. For example, in the U.K. newly planted apple trees on M9 showed a single peak of root growth which coincided with shoot growth, while older trees showed a pattern similar to that described above. Heavy pruning and cropping both reduce the late season peak of new root growth. This is entirely consistent with the inverse relationship found in perennial species between root and shoot growth, i.e., as shoot growth accelerates root growth slows and vice-versa.

The periodicity of new root production is important because new growth can greatly increase the size of the root system. In addition, as all woody roots begin life as white roots, growth is needed to replace roots lost from the system.

The extent to which roots of different ages are able to function in the absorption of water and mineral nutrients has been debated for many years. It is commonly assumed that uptake occurs exclusively through the younger parts of the root system, often described as "absorbing root." Studies by a number of workers, however, have shown that the older roots of fruit trees are able to absorb both water and some mineral nutrients, e.g., in laboratory trials woody and white roots of cherry absorbed ^{32}P at a similar

rate per unit surface area. A similar situation existed for water. In addition, field studies with apple have shown that commonly in summer, when transpiration is high, the amount of new root present is often small. If uptake occurred only through "absorbing root," rates of absorption would need to be unrealistically high.

The usual pattern of development gives a root system which increases in size with age (Table II). The size of the root system increases rapidly over the initial 3–5 years and then more slowly. The area of soil exploited by the root system of Jonathan/M4 doubled between years 1 and 2, 2 and 3, and 3 and 5 and then remained relatively constant as did fine root length. The number of roots 1–5 mm in diameter was relatively constant after Year 3. The depth of the root system increased rapidly during Year 1 to occupy the surface 1 m of soil and then increased more slowly. Although the total depth of the root system had increased to almost 4 m by Year 6 only 6% was below 1 m depth. The basic tree root system, on sites with no abnormal physical limitations, thus consists of a scaffold of roots running parallel to the soil surface within the top 30 cm of soil. This is supplemented by a series of sinker roots which may proliferate at depth (Fig. 3). The spread of the root system is usually much greater than that of the branch system. The permanent structure of the root system is important as a source of new root growth. An advantage of a woody root system is that a crop of new roots can easily be produced at a range of depths in order to increase the absorptive potential of the system without the need for growth from the trunk, providing that soil conditions remain in the acceptable range.

B. Variation in Root System Morphology between Species and Rootstocks

The size of tree root systems varies between species. At 1 year of age apple trees, Jonathan/M4, occupied 3.2 m^2 compared with 1.1 m^2 in pear and 2.4 m^2 in Sour Cherry. By 5 years of age, the areas filled were 18.2 m^2 in apple, 21.3 m^2 in pear, and 13.4 m^2 in cherry. Comparisons of the overall size of different species are difficult because root system size is affected by a range of cultural and edaphic factors. While apple roots have been found to a depth of 8.6 m a range of 1–2 m is common with around 70% of length in the surface 30 cm of soil. The average density of roots in the soil also varies between species. Typical values for root length density (Lv, mm ml^{-1}) are 0.1–1.1 for apple, 1.2–5.6 for pear, and 2.9–5.6, for peach.

TABLE II

The Development with Time of the Root System of Jonathan/M4 on a Sandy Soil

Tree age (years)	No. roots 1–5 mm diam	No. roots 11–15 mm	Length of roots >1 mm (m)	Area of root system (m²)	Length of vertical roots (m)
1	82	4	38	4.0	3
2	159	5	120	9.5	18
3	354	11	246	14.4	36
4	308	22	358	22.5	69
5	372	14	408	29.1	74
6	444	15	400	28.1	43

Source: Tamari, J. (1986). "Root Location of Fruit Trees and Its Agrotechnical Consequences." *Akademiai Kiada,* Budapest.

The amount of new growth within a season varies between species but seems generally to be higher in *Prunus* than *Malus*.

Rootstock will influence the size and distribution of the root system. Data on root distribution, as affected by rootstock, are often confounded as a consequence of the different rootstocks normally being used for different types of orchard system. For example, dwarfing stocks would be used in the more intensive systems and the more vigorous stocks for extensive plantings. A study where trees of the apple variety Cox, on one of the rootstocks M27, M9, M26, MM106, and MM111, were planted at a common 4 × 4m spacing indicated that root system size was affected by rootstock but that vertical and horizontal distribution and response to grass competition were purely a function of root system size, i.e., the larger trees which were on the more vigorous stocks had more root at a distance from the trunk. The size of the root system did not parallel the size of the shoot system. Trees on M26 were larger than those on M9 but had a smaller root system.

C. Relationship between Root System Form and Functioning

As orchard management will influence the form of the root system, any understanding of the wider effects of orchard design requires an understanding of relationship between root system form and functioning, under field conditions. Soil management may affect

a. The amount of new root growth produced in a season. New growth is often higher under grass than bare soil, an effect enhanced where irrigated. In an orchard of 5-year-old Cox/M26, the estimated root length in July was 66 m m^{-2} under bare soil, 75 m m^{-2} for grass, and 130 m m^{-2} under irrigated grass. The

morphology of the new growth also varied with much growth under grass being short-lived lateral roots. Comparable winter values for the above treatments were 44, 23, and 47 m m^{-2}.

b. The timing of new growth. In the study discussed above, growth occurred earlier with grass management and was sustained longest where the grass was irrigated.

c. The extent of mycorrhizal infection. Soil management treatment can modify the level of infection with arbuscular mycorrhizas (AM). In a trial with trees of Cox/MM106 in the U.K. roots from trees under herbicide management at 0–15 cm depth showed 6% infection at a time when infection under irrigated grass was 26%.

d. Root survival. This seems to be higher under bare soil, probably as a consequence of the production of fewer higher order lateral roots, so trees under bare soil should be able to absorb water/nutrients with fewer resources devoted to the production of new roots.

e. The distribution of new growth within the soil volume. Studies, using injections of ^{32}P at a range of depths, have shown uptake from 30 and 90 cm depth as higher under grass than bare soil management. This is likely to be a consequence of (a) and (c) above.

The combined results of these effects are that the root systems of trees grown in herbicide-treated rows with grassed alleys tend to show a concentration of roots in the herbicided row compared to the situation found under a uniform treatment. This distribution corresponded to the pattern of the uptake of nitrogen. In trees grown in herbicide-treated rows with grassed inter-rows, the uptake of N was predominantly from the herbicide-treated area. In contrast, uptake by herbicide-managed trees was from the whole of the orchard area.

D. Root System Plasticity

It follows from the above that the form and development of the root system will vary with soil manage-

FIGURE 3 The root system of a 26-year-old tree of Fortune/M9 grown with a herbicide-treated square around the trunk and the remainder of the soil under grass. [From Atkinson, D. (1973). *In* "Report of E Malling Research Station for 1972," p.75.]

ment. This can be illustrated by a study of 3-year-old apple trees grown with or without a straw mulch. With the mulch they showed an increased number of roots, 199 compared to 118; an increased length of roots > 1 mm diameter, 182 compared to 90; an increase in the area of soil surface exploited, 25.4 compared to 17.6 m²; and an increase in the percentage of roots in the surface 20 cm, 17% compared to 7%. This root system was associated with an enlarged shoot system and a reduced root:shoot ratio. The inherent flexibility of the tree root system is indicated by the effects of root pruning which results in a substantial and rapid redistribution of resources to the root system at the expense of the shoot system. Root pruning has thus been used as a means of controlling tree vigour.

III. Orchard Management Practices Used to Augment Resources

Information presented above has indicated the factors which influence the crop's need for nutrients and wa-

ter and the mechanisms employed by the plant to maximize its ability to obtain nutrients and water from the soil. This section details the main methods to augment soil resources.

A. The Use of Fertilizers

Fertilizers can increase the total amount of nutrient in the system and the amount available at particular times. Long-term studies of orchards in the U.K. have monitored the effects of the addition of 140 Kg N ha^{-1} yr^{-1} on the level of soil mineral nitrogen (NO_3 and NH_4). In a cultivated soil, the addition was associated with an increase in mineral N of 170, 140, and 180 Kg ha^{-1} in three successive seasons. With comparable trees under grass the same fertilizer additions were associated with increases in N of 40, 20, and 40 kg N ha^{-1}. These differences occurred despite differences in the rate of mineralization under grass, (230–320 Kg ha^{-1} yr^{-1}) and bare soil (160–210 kg ha^{-1}). The precise effect of fertilizer additions on soil nutrient availability and content will thus depend upon the amounts of leaching and fixation (microbe and plant). Effects of fertilizer must be seen in the context of whole orchard nutrient budgets.

B. The Use of Irrigation

Irrigation can maintain the moisture potential of the whole or a part of the orchard soil volume close to field capacity. The main limitations to this are cost, equipment needs, and the availability of the amount of water needed to match potential evaporation. Soil moisture is, therefore, usually allowed to fluctuate between set limits which are defined in terms of either a maximum soil moisture deficit, e.g., 50 mm, or a minimum soil moisture potential at a particular depth, e.g., -70 kPa at 30 cm. Different irrigation methods will result in a different spatial distribution of the water. Using sprinkler or flood irrigation, the soil moisture content will be replenished through the whole volume. Only part of the soil volume is modified with drip irrigation. In either case, the amount of water needed can be estimated in several ways although a water balance approach is among the most common. Here the amount of water added is related to evaporation from a given surface area of water, usually a "Class A pan." This amount of potential evaporation is modified by the use of a crop coefficient, a factor which links potential crop water loss to free water evaporation. In California, this ratio varies from 0.45 in March to 0.75 in August. Values estimated for other sites and species differ from these and may exceed 1. [See IRRIGATION ENGINEERING: FARM PRACTICES, METHODS, AND SYSTEMS.]

The use of irrigation and the particular levels of water application and timing selected will themselves influence root system development and hence influence the ability to use irrigation and increase the risks associated with a failure to irrigate. This is most acute when water is targeted to a limited part of the soil volume, as with trickle irrigation. For example, work in Michigan with peach showed that the application of irrigation for 11 years resulted in a shallower root system. The interaction between irrigation and the plant developed under irrigated conditions is an important consideration when designing an irrigation system.

C. The Need for Weed Control

Unwanted and unmanaged vegetation in the orchard increases water use (Fig. 2), the duration of stress, and the irrigation need, e.g., the soil moisture deficit was 500 liter tree^{-1} under grass management, but only 300 liter tree^{-1} under bare soil. The control of weeds can thus result in the provision of additional water. The amounts of available N, P, and K will also be potentially higher in a bare soil although with a higher risk of leaching, especially for $NO_3 - N$. [See WEED SCIENCE.]

D. The Use of Cover Crops

Cover crops like other plants will use substantial amounts of water but may be needed to stabilize a soil surface against water or wind erosion, to allow machinery access at times of the year when the soil surface would otherwise be unstable, or to facilitate the penetration of rainfall. For example, in an orchard in the U.K. the hydraulic conductivity of the soil surface was 75 mm hr^{-1} for a vegetated surface but only 5 mm hr^{-1} for a bare soil surface (Fig. 2). Additionally, a cover crop can benefit the nutrient budget of an orchard by retaining nitrogen within the soil system. This results in the prevention of leaching and the later release by mineralization. The fixation of nitrogen by legumes within the cover crop and the upward recirculation of leached cations from depth (increasing surface soil pH) are other potential benefits of cover crops. The amounts of nutrients released by these processes are variable but often significant, e.g., under soil covered by a legume, like clover, soil mineral N levels may exceed 49 kg N ha^{-1} compared to 34 kg N under grass.

E. Organic Fruit Production

Nutrient budgets indicate that for fertile soils with significant organic matter contents, the needs of orchard trees for major nutrients are relatively small and thus that some orchards can be maintained without fertilizers for significant periods. Orchard systems thus have the potential to be managed as organic systems; the legume component of the cover crop can be used to augment the nitrogen supply, although the sward will compete for water. The main problem is weed control. The usual means of weed control in arable organic farming systems, i.e., cultivation, presents difficulties in a perennial crop system because the cultivation implements will damage the surface roots of the crop and reduce production. In cultivated orchards, roots are excluded from the surface 15 cm of soil where nutrients are most available. Pest and disease control can sometimes, with difficulty, be managed without pesticides but the nonavailability of herbicides represents a more difficult problem. The use of a mulch, Polythene or natural, in the tree row can aid weed control, but will complicate management and is more difficult to use in the cropping orchard.

F. The Effect of Growth Regulation on the Need for Nutrients and Water

Regulation of tree size is normally achieved through one of the following routes:

1. Mechanical reduction in tree size by pruning.

2. Selection of a rootstock to give a tree of an appropriate size.

3. The use of chemical growth regulators to reduce the amount of vegetative growth.

Commonly all three methods are used together, although concerns about the safety of chemical growth regulation have seen recent decreases in the use of chemical regulation. Table I indicates that vegetative growth is responsible for much of the gross need for nutrients by the tree. Reducing shoot and leaf growth will thus reduce demand. Additionally, reducing leaf area should reduce the need for water unless this water is simply used by the cover crop. However, some growth regulators, e.g., paclobutrazol, seem to reduce water use more than can be accounted for on the basis of a simple reduction in leaf area. Paclobutrazol modifies root growth and distribution, increasing the proportion produced near the soil surface. It may also modify competition between the orchard sward and the crop. An increase in the balance of fruit to vegetative growth may relatively reduce the need for N and Ca, but will correspondingly increase the need for P and K. The extent of effects produced by chemical regulation indicate the scope for genetic modification (breeding or genetic modification) to modify the distribution of growth.

G. Management of Soil Microbial Organisms

As the demand for nutrients by orchard crops is relatively low compared to intensive arable crops, the management of the soil microbial population can have a significant impact on nutrient cycles and so will be important in orchards. Ultimately, the release of nutrients from all organic matter returned to the soil will depend upon microbial activity. In addition, most fruit species are mycorrhizal which will aid their P and micro-nutrient supply. Given that the total needs of fruit trees for nitrogen are relatively low, free-living soil organisms, e.g., blue-green algae, may make a

more significant contribution, $5–10$ kg ha^{-1} yr^{-1}, than is true for more nutrient demanding crops. Pathogenic organisms, fungi, and nematodes will also influence the functioning of the root system. Attempts to modify these processes by inoculation have rarely been successful. Attempts to modify them by changes in soil management have had a significant impact. Some examples of management effects are

a. Bacterial action in soils varies spatially and temporally but "activity," as either O_2 uptake or CO_2 production, is usually higher in soils under grass compared with cultivated soils, usually by a factor of 3–6.

b. The presence of vegetation influences the weight of earthworms present in an orchard. In the spring, in an established orchard in the U.K., earthworm biomass was 76 g m^{-2} under grass and 17 g m^{-2} under bare soil. The addition of organic matter doubled the number of worms present.

c. Microbial activity is usually higher under grass although even under bare soil nitrogen mineralization can be substantial, around 300 kg N ha^{-1} yr^{-1}. [See Soil Microbiology.]

In contrast to the above processes, there have been many attempts to influence mycorrhizal infection by inoculation. Levels of root infection by AM fungi can be low, e.g., 1% under bare soil in mid-summer. Where the availability of nutrients, such as P, is limiting then inoculation with AM has at times benefited growth and survival. Increases in growth of several hundred percent have been reported. Even in the absence of nutrient deficiency AM fungi have improved the survival of *Prunus* rootstocks at the transplanting stage. Fruit plants are infected by a number of different AM species and the magnitude of the growth effect can vary between species, probably as a consequence of variation in the level of infection. Maximum benefit seems to derive from infection with a mixed-species inoculum. In addition, regardless of the soil nutrient status, infection with AM fungi will modify the form of the root system with infection resulting in an increase in the proportion of higher order laterals. Infection of fruit species with AM fungi has been shown to influence both leaf P content and the tree's ability to obtain P from low P soils. For example, citrus plants grown on a sandy soil showed no response to P additions over the range 0–600 mg kg^{-1} soil, approximately equivalent to 0–780 kg P ha^{-1} applied to the surface 100 mm of the profile. In contrast nonmycorrhizal plants responded over the whole of this range. Citrus are among the species which have responded best to inoculation with AM fungi. On most soils infection with AM fungi will increase the percentage P found in leaves. The infection of apple with a mixture of *Glomus fasciculatum*, *G. mosseae*, and *Gigaspora margarita* increased the percentage P in leaves from 0.16 to 0.24%. An application of P at 44 mg liter^{-1}, approximately 44 kg ha^{-1}, gave a leaf P content of 0.23%. Infection with AM fungi may also increase resistance to soil-borne diseases. The mechanism by which this occurs is unclear but both effects may be related to the dynamics of the root system, i.e., a larger number of short-lived roots may reduce the ability of the pathogens to infect while direct effects on the pathogen may also be involved.

A complete review of the effects of soil-borne diseases influencing fruit plants is beyond the scope of this article. The suggested mechanisms for apple specific replant disease, which seems to involve *Pythium sylvaticum* in the U.K., illustrates this type of effect and interactions with management. Replant "disease" is a component of the "short life replant problem." Where apple trees are replanted in an area previously used for apple, then the growth of the replanted tree is poorer than might have been expected on the basis of soil physical and chemical properties. The growth of the root system is particularly poor. This effect seems to be due to *P. sylvaticum*. Given that all woody roots must develop initially from white roots, a reduction in white root production and survival must reduce the potential for the development of the perennial root system and the base for future new root production. This "pruning" of new roots by the pathogen will reduce the tree's flexibility to respond to sudden stresses. The incidence of the pathogen can be reduced by fumigation, using agents like formaldehyde or chloropicrin, but both can be dangerous to health and will adversely affect all soil microbes. In trials in Netherlands the replacement of the soil in the planting hole with compost seems to have been beneficial to the growth of young apples planted into previous orchard sites.

H. The Use of Mulches

A mulch which can be derived from either a living, e.g., straw, or a synthetic source, e.g., polythene, will influence soil moisture and temperature, and the development of weeds. All of these effects will influence the release of nutrients and their availability to the crop and the availability of soil water. The use of mulches has been shown to promote tree growth although they will complicate orchard management

and are more compatible with young than with the mature cropping orchard.

IV. Nutrient Applications

In a perennial crop the relationship between the amount of nutrient in soil and its uptake and movement to particular plant tissues is complex. In addition, nutrient levels in the various plant tissues are often poorly correlated. For example, in a single season in a series of trials in the U.K., established trees of Cox showed correlation between leaf and fruit P which varied between $+0.492$ and -0.815. Despite this there exist standard values for particular fruit species which indicate the need for supplementary nutrient supply.

A. Criteria for Assessing Nutrient Demand

The need for additional nutrients may be assessed by soil analysis, plant tissue analysis, or reference to a nutrient budget. In practice, all are used, although tissue analysis is probably the most common. Soil analysis would be the method of choice for soil pH and also provides useful information on P, K, and Mg. Here emphasis is placed on tissue analysis. The value of tissue, usually leaf, analysis depends upon a number of assumptions, i.e.,

1. That the leaf is the principal site of plant metabolism.
2. That changes in nutrient supply to the plant are reflected in leaf composition (or other tissues analyzed).
3. That these changes are most pronounced at certain developmental stages.
4. That the concentration of nutrients in the leaf (or other tissue), at a specific growth stage, is related to crop performance.

Analytical values are better related to nutrient status than to potential crop yield which is affected by many other factors. Interpretation of the results of leaf analysis is usually based on critical levels (Table III). Val-

TABLE III

Critical Nutrient Levels in Apple (% DW)

Element	Deficient	Insufficient	Optimum	High
N	1.7	1.8–2.0	2.1–2.2	2.3
K	0.8	0.9–1.5	1.5–2.0	2.1
Mg	0.15	0.25	0.35	0.4
P	0.11	0.15	0.22	

Source: Faust, M. (1989). "Physiology of Temperate Zone Fruit Trees." Wiley, New York.

ues below these levels are normally taken to indicate that the soil is unable to supply a particular nutrient in the correct amount or at the correct time. In addition, in fruits, like apple, which are usually stored prior to sale, the concentration of nutrients such as Ca and P in the tissue of the fruit will influence storage potential. Fruit analysis is used for two entirely separate purposes, i.e.,

1. As an indication of the need for adjusted orchard nutrition.
2. As a guide to the length of time that fruit can be stored and the storage regime needed to maximize storage life.

For mature trees the orchard fertilizer regime will be as heavily influenced by probable effects on fruit tissue concentrations and their significance for fruit storage as by concentrations which may influence growth. Leaf nutrient levels, although a valuable guide to growth limiting deficiencies, are a poor guide to fruit concentrations. Applications of N and K, even if indicated by leaf analysis, would usually be restricted because of adverse consequences for fruit storage. In young, precropping trees, these restrictions do not apply and the tree's needs can be more easily related to leaf analysis (Table III). In the young tree, however, the remobilization of nutrients from woody tissues can contribute up to 60% of current need for N, a similar proportion of P, and up to 30% of K. The concentration of nutrients in leaves varies substantially between years and so leaf analytical data for a single year are an imperfect guide to fertilizer needs. Monitoring of concentrations over several years is more helpful.

B. Types of Fertilizer

Three types of nutrient additives are employed in orchards:

1. Chemical fertilizers.
2. Organic manures.
3. Sprays of dilute nutrient solutions.

Most orchards receive nutrients in the form of chemical fertilizers. Organic manures are used where the crop is being produced in an organic system or where organic manures are locally available, e.g., sewage sludge may be used in orchards close to towns. Sprays can be used to enhance the supply of nutrients at particular times or where it is desired to increase the concentration of a specific element in a specific tissue, e.g., Ca in fruit to improve storage. The major

nutrients added as chemical fertilizers are N, P, and K. N is usually added as an NO_3-based material because NH_4-based materials may adversely influence soil acidity. The distribution of tree roots in an orchard grown with herbicide-treated rows and grassed interrows, the commonest system of orchard soil management, suggests that there will be considerable benefit from restricting the application of fertilizer to the herbicided area. Where this is followed, materials resulting in soil acidification, e.g., $(NH_4)_2 SO_4$, must be avoided.

C. Nutrient Status for Optimal Production Storage and Disease Control

The influence of nutrient status can be divided into three types of effect, i.e.,

1. Effects upon vegetative processes. Some aspects of these processes have been discussed previously. Within limits, vegetative growth increases in proportion to N supply so that a large supply of N is necessary for the young tree, but perhaps less needed for the mature tree.

2. Effects on fruiting. N application can increase fruit bud production, probably as a consequence of effects on the size of spur leaves. The growth of spur leaves is, however, more often influenced by N availability in the previous season than by that available in the spring preceding cropping. An oversupply of N will delay bloom although in N-deficient trees flower fertility may be reduced.

3. Effects on fruit characteristics. An excessive supply of N can raise the concentration of N in the fruit and as a result increase susceptibility to storage diseases and thus the need for fungicides Control of N levels in the tree is thus important to the development of an integrated pest management (IPM) system. High N fruit is also more likely to suffer from a range of storage disorders, e.g., internal breakdown. The optimal concentrations of N, P, K, Ca, and Mg needed for the growth and storage of fruit can be defined. For the variety Cox, these are leaf 2.6% N max, 0.24% P min, 1.6% K max, 0.25% Mg min, fruit 50–70 mg N 100 g FW^{-1}, 11 mg P min, 130–160 mg K, 4.5 mg Ca, 5.0 mg Mg. Critical levels vary between different fruit varieties. In the variety Cox, the optimum concentrations are N 50–70, P 11, K 130–170, Mg 5, Ca 4.5 mg 100 g FW^{-1}. The level of Ca is critical for long-term storage while the ratios of Ca:K and Ca:P are also important. The Ca:K ratio shown above would be expected to, on 2 occasions in 10, result in a bitter pit level >10%. To guarantee the total absence of bitter pit at the K levels shown above, the Ca concentration would need to be around 7.8 which is very rare.

The least precise links in the fertilizer/soil nutrient/ nutrient concentration chain are those

a. between fertilizer addition and the resultant concentration of nutrient in soil, and
b. between the concentration in the soil and the concentration in the crop tissue.

Links between soil and fruit concentrations are less good than those between soil and either leaf or wood. This imprecision is due to a number of climate and soil factors. Most of these cannot be predicted in advance, i.e., the amount of rainfall and therefore the amount of leaching. Some of these relationships can be stabilized by soil management.

V. Management of the Orchard Floor

Available tools for the management of the orchard floor fall into three groups:

a. cultivation
b. vegetative covers and mulches
c. chemical weed control agents.

All three groups of methods modify the physical, chemical, and biological environment of the soil and thus change the conditions under which the crop root system must develop, the soil's ability to provide nutrients and water, and the tree's capacity to grow and produce quality fruit.

A. Mechanical Cultivation

In arable crops mechanical cultivation is used to create a seed bed for establishment and to control weeds. Cultivation usually results in a weakening of soil aggregation and thus may render the soil more prone to wind erosion and rainfall-initiated structural degradation. The bulk density of cultivated soil is variable being low immediately after cultivation but similar to that of an herbicide-treated surface after a period of time whose duration usually depends upon rainfall. Those soil properties which are related to bulk density, i.e., total porosity, aggregation, and pore size distribution, usually show similar variation. Cultivation destroys the perennial roots present within the cultivation depth. It normally results in a poor traffic bearing surface and conditions where sprayers cannot be used without causing structural damage. Even with several cultivations weed control is often less good than that achieved with herbicide-based management. The duration of good control tends to be shorter,

so that cultivation is less common than either cover cropping or herbicide management. It does, however, give some of the advantages of herbicide management, but without chemical residues. It is ineffective in controlling rhizomatous grass weeds. In addition, mechanical tillage can be more costly and requires a higher energy input (machine power) than that needed for chemical or biological control.

B. Cover Cropping

In the 1950s, many growers abandoned cultivation for "sod" or cover crop culture. This conserved nutrients within the soil profile and provided a working surface which allowed access by machinery at most times of the year and thus allowed sprays to be applied at the correct time. Following an establishment period, a cover crop provides a relatively stable environment and soil physical conditions which are better than those found under cultivation or herbicide management. The soil bulk density at the surface is normally lower than with herbicide management, often by as much as 0.2 g ml^{-1}. This lower soil bulk density is associated with an increase in total porosity, a higher proportion of the pore space present as small pores ($<30 \mu$M), and enhanced aggregate stability. A vegetative cover, however, normally results in a reduced soil moisture potential for much of the year and in extreme conditions can result in much of the potential soil moisture being depleted before the tree leafs out. If irrigation is unavailable the consequences of this can be severe. Data on the relative water content of grassed and bare soil surfaces are shown in Fig. 2. Under grass management soil organic matter levels are usually greater than in herbicide-treated soil. The presence of the cover crop means that most nutrient levels, but especially available P and NO$_3$, are low and the sowing of a new cover is often associated with nitrogen deficiency in the tree, unless additional N is applied. In the presence of grass, the acidification of the soil surface, and the associated high levels of available Mn, which occur with herbicide management, are usually absent. Cover crop competition can be modified in a number of ways:

1. The species selected will influence the amount of water used. Comparison of the percentage of available soil moisture remaining at 200 mm depth at a site in Michigan showed the lowest amount, 12%, with Ladino clover, and the highest amount, 21%, with Quackgrass.
2. Water use will depend upon cover management. Mowing the grass reduced the early summer water use of a range of grass species by 6–15%. With leguminous

species mowing had little or no effect. The differences in the water content under long and mown grass were principally below 500 mm depth.

3. Orchard covers can be managed using growth regulators. Suitable combinations of paclobutrazol and mefluidide applied at the correct time have been shown to reduce growth, resulting in fewer cuts being needed during the year. Compared to uncut grass, chemically controlled grass appears to use less water although it will still develop a substantial deficit in most seasons.

C. Herbicide-Based Management

Following the development of chemical herbicides, full effects of weed competition on the growth of fruit crops were able to be measured. In perennial crops, weed competition has been shown to cause reductions in growth as high as 15–96%, and reductions in crop yield of 17–66%. Vegetation in the orchard can modify the number of fruits produced, mean fruit size, fruit color, skin finish, mineral composition, and internal quality. In an orchard of Cox/MM106 planted in 1973, the cumulative crop for the period 1976–1981 was 70 kg tree^{-1} under grass with a narrow herbicide strip, and 125 kg for trees grown with herbicide-treated soil. In addition, the grade out of the fruit, i.e., the percentage distribution of fruits, with different given mean diameters, was changed. For example, in one year with the grass treatment 80% of fruits had diameters <65 mm. The comparable figure for the bare treatment was 60%. As large fruits have a higher value this size effect enhances the positive effects on yield. [See HERBICIDES AND HERBICIDE RESISTANCE.]

Effects of soil management on fruit quality are variable. Herbicide managed fruit is usually less well colored, with a better skin finish but with a higher potential for low temperature breakdown as a result of a reduced P content. Effects of soil on bitter pit vary between years. From the perspective of production, herbicide-based management has many advantages over grass management, but many soil parameters are influenced detrimentally. The significance of these adverse changes will vary with the length of time the orchard is under herbicide management, and the soil type. Soil deterioration with herbicide management is also influenced by whether it follows periods under arable cultivation or under grass (Fig. 2). An herbicide-managed surface derived from grass retained the characteristics of a grassed area for a considerable time. The relative advantages of herbicide-based management for tree growth and grass-based management for soil condition have resulted in the compromise solution of a wide herbicide-treated row with a grassed inter-row becoming the fruit industry standard. In many countries concerns over the safety of chemical herbicides and the herbicide residues of in drinking water have resulted in the withdrawal of clearance for some soil-acting herbicides. These materials remain critical for intensive, early cropping systems. In countries with strict environmental legislation, weed-free conditions, where needed, are now being maintained with shallow cultivations, although studies to assess the use of mulches are currently ongoing. Environmental legislation is affecting orchard design. Multirow bed systems, which depend upon chemical control, will disappear to be replaced by single-row systems designed specifically to facilitate control using shallow cultivations or mulches.

VI. Irrigation Needs

The basic water balance of an orchard can be described as

Water + use by tree crop	Water + use by cover crop	Evapo- + ration from soil surface	Leach- = ing	Stored + water in soil profile	Rainfall.

If the amounts of stored water in the soil profile, usually 100–130 mm in 1 m soil depth, and the rainfall, normally 200–300 mm during the growing season in Europe, are inadequate for the needs of the crop and cover crop, then the crop will suffer from water stress which will only be alleviated by the provision of added water, i.e., irrigation. Even in the U.K., which has a relatively wet climate and low potential evaporation, the need for additional water is common, and in much of the United States and in southern Europe irrigation is needed in most if not all years.

A. The Need for and Effects of Water

Comparison in an orchard of Cox/MM106 with trees grown in a narrow herbicide strip either with or without irrigation gave a cumulative 5-year fruit yield of 70 kg tree^{-1} without irrigation and 130 kg with irrigation. In high-density systems where the rapid development of optimum cropping is vital, irrigation has been shown to increase initial crop by as much as 24%, mainly as a consequence of increased fruit numbers. For the maximum production of quality fruit, the maintenance of a soil water potential of > -50 kPa appears to be necessary. Even in northern Europe this cannot usually be achieved just by soil management and even trees under herbicide-based management have been shown to respond to irrigation. Even when the soil water potential is close to field capacity, the trees usually experience water stress for part of the day. At mid-day water loss by transpiration is faster than the rate of water movement through the plant and so water stress results from this imbalance between transpiration and water uptake and transport. This type of stress can only be ameliorated by a reduction in the rate of water loss. With leaf wetting ("mist irrigation") some of the latent heat arriving at the leaf surface is used to evaporate the water deposited on the leaf surface by the "mist" system. While this water is being evaporated transpiration is reduced, resulting in leaves with a water potential of -0.8 MPa, soon after water deposition, compared to control leaves with a water potential of -2.0 MPa. A system of this type has given a 20–30% increase in crop production, the lower figure for an irrigated orchard. The practical use of such a system remains hypothetical because of the complications which exist in relation to disease. The need for irrigation and the amounts of water which need to be added will depend upon both the design of the orchard and its soil management, some of which will reduce the need for irrigation.

B. Methods of Irrigation

With the exception of the "mist" system, all water application systems aim to replenish water stored in the soil profile. This is commonly achieved in one of three ways:

a. The flooding of the orchard surface from pipes at the edge of the orchard.
b. A system of sprinklers either permanently installed over the trees or moved down the tree rows.
c. A permanent system of trickle emitters or minisprinklers installed either at the soil surface or within the canopy.

The management of permanent systems is simpler than that of systems which need to be moved. Flood systems will tend to render the orchard inaccessible to machinery for a period following flooding. In contrast, trickle and permanent sprinkler systems provide the flexibility to deliver smaller amounts of water on a daily basis, and both can be linked to sensors which initiate small or large volume events when a given set of conditions have occurred. Irrigation is, however, more commonly based on measurements from soil moisture sensors, or on the basis of measured evaporation, e.g., from a Class A pan, modified by crop factors, or on the basis of a water budget such as

$$\begin{array}{ccc} \text{Estimated water} & - \text{ Rainfall} & = \text{ Irrigation need at} \\ \text{use for period} & \text{for period} & \text{end of period.} \end{array}$$

Using this form of balance water may be replenished on a daily or weekly basis or when the deficit (irrigation need) has exceeded a set value. Sprinkler irrigation systems can also be used to apply large volumes of water over a short period of time and can also be used for frost protection. Water for frost protection cannot easily be provided by the other types of irrigation system.

C. Effects of Controlled Water Stress

Excessive irrigation will favor vegetative growth and may adversely affect crop production. It has been suggested that the development of an apple can be divided into the following periods:

a. A period of 40–50 days from full bloom.
b. The period until the end of active shoot growth.
c. The period up to harvest.
d. The postharvest period.

Irrigation in period (b) will favor vegetative growth, while that in period (c) will benefit cropping. The risks of excessive irrigation mean that deficit irrigation, i.e., the replacement of only 25% of the water lost by transpiration, economizes on water use, avoids excessive vegetative growth, and may reduce storage disorders. Different fruit species vary in their response to this type of treatment. In Golden Queen peach where little fruit growth occurs in period (b), irrigation can be so reduced without adverse effects on crop and with a useful restriction of vegetative growth. Where there is little separation of (b) and (c) the usual situation in apple (Fig. 1), this type of system is not effective.

VII. Integrated Pest Management (IPM)

The intention to minimize the use of chemical control agents, to maximize the use of natural control agents, and to integrate the various inputs and management practices within the orchard have led to the development of IPM systems. These place emphasis on the management of interactions, rather than solving the consequences of targeted actions. For example, the control of the availability of nitrogen will reduce the amount of vegetative growth produced by both the tree and cover crop, and, as a consequence, reduce the susceptibility of the tree to a range of fungal diseases. IPM systems rely on a mixture of chemical, biological, and cultural controls. Currently there are limits to the extent to which the control of fungal disease and invertebrate pests can be achieved in the absence of chemicals. The use of natural predators seems to be most advanced for the control of insect and mite pests. Although weed control is difficult without herbicides, the management of the orchard floor and the use of mulches can give some weed control and minimize the need for additional water and nutrients. [*See* INTEGRATED PEST MANAGEMENT.]

VIII. Environmental Impact of Orchard Management

All agricultural and horticultural systems have an impact on the wider environment. The major components of the impact of orchard systems are

a. The movement of pesticides from the orchard by drift, or leaching.
b. Effects of pesticides on nontarget organisms, e.g., "weed" species of low competitive effect, species of conservation potential, or natural predators, etc.
c. The leaching of added nutrients especially NO_3.
d. Deterioration of soil structure and soil fertility as a consequence of cultivation, machinery movement, or herbicides use.
e. The loss of biodiversity necessitated by planting an orchard system in an area which might otherwise have been a species rich pasture or mixed woodland.

In all agricultural systems the major impact is due to situation (e). Orchards traditionally receive many fungicide applications and rely on soil-acting herbicides. Both of these crop protection practices result in negative environmental effects. The herbicides, atrazine and simazine, are among the most commonly detected pesticides in water. It is probable that at least some of these residues come from orchard applications. The frequent application of fungicides to orchards can result in their leaching to groundwater and modifications to soil microbial populations and processes. Despite these effects, the perennial nature of orchards, the infrequency of soil disturbance, and the low rates of nitrogen used probably mean that the environmental impact of orchards is less than that

of many other agricultural systems. This will be especially the case for orchards where much of the orchard floor is covered by grass.

IX. Future Developments

Crystal ball gazing is an imprecise art. Factors which seem likely to impact upon orchard soil water and nutrient management are the following:

a. Organic production systems. The demand for the production of fruit without the use of agrochemicals will increase as a consequence of the development of more sustainable biologically based agro ecosystems. The attainment of this will require the development of effective means of nonchemical weed control or an increased availability of water for irrigation. It is likely to be influenced also by the rate of genetic improvement of cultivars and rootstocks.

b. Global Climate Change. Current scenarios suggest an increase in mean temperature which will result in an increased need for irrigation in orchards and will increase the difficulty of growing fruit by using only soil water. More varied climatic conditions will accentuate the variations which already occur in the yields of orchards from year to year. More severe winters and higher and more uneven risks of late spring frosts will increase this variation, as will the impact of heavy rain during the growing season.

c. Novel genotypes. The development of rootstocks giving increased growth control and higher harvest indexes have been responsible for increases in mean yields over the past two decades. These can be expected to continue as can the production of varieties with enhanced resistance to pests and diseases.

d. Energy availability. While like all perennial cropping systems orchards need less energy than annual cropping systems the spray programs currently employed use significant amounts of energy. Modification of spray technology will reduce orchard energy budgets.

e. Transgenic crops. Given the need to increase resistance to pests and diseases the introduction of genes from novel sources seems certain to occur.

f. Genetically modified micro-organisms. These are likely to benefit orchard systems in respect of more flexible AM strains and also in relation to biological control agents, perhaps including the use of modified plant pathogens for weed control.

g. Availability of crop protection compounds. The tightening of chemical approval procedures worldwide is causing real problems in relation to the availability of weed control materials for minor crops, including all horticultural crops. Here the costs of product testing, registration, and liability make it financially unattractive for chemical firms to provide materials for orchard use. As older materials are lost, due to the need for reevaluation, they are not being replaced.

Bibliography

Atkinson, D., Jackson, J. E., Sharples, R. O., and Waller, W. M. (1980). "Mineral Nutrition of Fruit Trees." Butterworths, UK.

Atkinson, D. (1980). The distribution and effectiveness of the roots of tree crops. *Horticult. Rev.* **2.**

Faust M. (1989). "Physiology of Temperate Zone Fruit Trees." Wiley, New York.

Ryugo, K. (1988). "Fruit Culture: Its Science and Art." Wiley, New York.

Sadowski, A. (1990). Diagnosis of nutritional status of deciduous fruit orchards. *Acta Horticult.* **274.**

Tamari, J. (1986). "Root Location of Fruit Trees and its Agrotechnical Consequences." Akedemiai Kiada, Budapest, Hungary.

Orchard Management Systems

DAVID C. FERREE, *Ohio State University*

Glossary

Cultivar Cultivated variety
Heading cut Pruning cut that removes part of a shoot or branch and leaves a portion to regrow
Precocity Characteristic of early development of flowering and fruiting in the life of a tree
Pruning Partial or complete removal of vegetative or fruiting wood from a plant to control its size, remove broken or damaged tissue, alter plant shape, remove unnecessary growth, or balance fruiting and vegetative growth
Rootstock Root system of a fruit tree either grown from seed or asexually propagated upon which the main scion cultivar is budded or grafted
Spur-type Classification of growth and flowering habit of apple in which more lateral buds form spurs and fewer form shoots than occurs in standard growth habit cultivars
Thinning cut Pruning cut that removes a shoot or branch completely back to its base of origin
Tree training Physical techniques that direct the shape, size, or direction of plant growth and establish the orientation of the plant in space. Growth can be repositioned by bending, spreading, or supporting, or removed by pruning to conform to a desired tree shape or form
Whip A 1-year-old nursery tree without branches

An orchard management system defines a management program and plan for the establishment and maintenance of a commercial orchard of fruit trees. A successful orchard management system integrates and coordinates into a cohesive plan many horticultural components; cultivar, rootstock, spacing, tree quality, tree arrangement, tree shape, pruning, training, planting system, tree support, and pest control. Economic components such as management skills, market type, equipment, capital, and land available also influence and must be integrated into the management system. The management of the soil and water environment is covered in another article, but these factors naturally influence the above-soil orchard management system. [*See* ORCHARD MANAGEMENT: SOIL ENVIRONMENT AND RESOURCES.]

I. Components of Orchard Management Systems

The ultimate goal in selecting an orchard management system is to improve profitability and efficiency of the orchard. This goal is achieved by selecting a system with the components best suited to the site, marketing strategy, management skill, and economic situation of the orchard operator. The genetic components of cultivar, rootstock, and tree quality form the basis of the system and greatly influence the management techniques and practices involved in the system. The most important of these cultural practices are tree spacing, tree arrangement, support system utilized, pruning, training, and the overall planting system. Each of these practices interrelate and will be discussed in detail.

Modern orchard management systems are all designed around the principle that orchard efficiency, or increased production per labor input, is improved by increasing planting density. This principle is based on the increase in leaf area and concomitant increase in light interception resulting from increasing the number of trees per unit of land provided. The more rapid establishment of productive potential from the land results in an earlier economic return. However, if tree density is increased too much, fruit size and quality are decreased and the orchard may require an uneconomical investment in labor to return orchard efficiency. Thus, it is critical to understand the interaction of the various components of an orchard management system so that an efficient combination can be made.

A. Orchard Site

Most fruit crops have temperature limitations that determine where they can be grown successfully. Many temperate species have a winter chilling requirement to break rest; for example, apple requires 1000 to 1600 hr and peach, 400 to 1000 hr of temperatures of 7°C or lower to satisfy this requirement. Adequate moisture also limits the areas where fruit can be grown successfully and in arid or semi-arid areas, irrigation becomes necessary. Available sunlight, although not normally a limiting climatic factor, gives some areas a distinct advantage over more cloudy areas. The apple production areas in the arid regions of Washington state have nearly 30% more sunlight in the early part of the growing season than eastern areas of the United States. These more advantageous areas consistently have improved fruit set, fruit size, and yields compared to areas with less available light.

Within a region with acceptable climate, the actual site selected for the plantings is very important, as no other single factor has a greater influence on the profitability of an orchard. The most critical elements of a site are its potential for frost and the soil characteristics. In rolling topography the best orchard sites are normally high on gradual slopes, avoiding lower areas where cold air can accumulate in the spring and create frost problems. [See MICROCLIMATE.]

In some areas it is possible to plant an orchard in close proximity to a large body of water that provides a cooling effect in the spring, which delays development and reduces the risk of frost damage to flowers. If the region has large temperature inversion conditions in the spring, wind machines can be used to mix upper level, warmer air with cooler air close to the soil surface to avoid frost problems. In the past, heating with special orchard oil heaters was often used, but increasing costs of fuel and air pollution problems have reduced the usage of this form of frost control. In areas where sprinkler irrigation is normally used and adequate water is available in the spring, irrigation can be used for frost control.

Strong consistent winds cause another site concern, as windy conditions during bloom can reduce fruit set and cause fruit finish problems. In some areas with strong prevailing winds, windbreaks are established to lessen this problem.

Hail is another site-related problem that must be considered. Growers in some areas with a high frequency of hail install hail nets over the trees. These nets have the disadvantage of lowering light penetration and being very expensive, and thus are normally not economically feasible. Therefore, locations prone to frequent hail should be avoided as orchard sites.

B. Economic Considerations and Marketing

Before choosing an orchard system, careful consideration must be given to the potential market for the fruit and the economic limitations. Normally if the main market is processing, and cosmetic factors of fruit color and finish are less important, a system will be selected that minimizes orchard establishment costs and provides high yields with limited or reduced inputs of such cultural practices as pruning, training, and tree support. If fresh fruit is to be the primary market and large fruit size, fruit color, and finish are critical, an orchard system should be selected that enhances these attributes and makes the necessary cultural practices efficient to perform. In a market outlet designed for the consumer to pick the fruit, reduced tree size becomes critical to avoid the liability of the public using ladders. [See FOOD MARKETING SYSTEMS.]

Economic considerations can also have a strong influence on the orchard system selected. Several studies have shown that if ample establishment capital is available and land is limiting, a more intensive system that produces greater earlier returns is preferred. However, if capital is limiting and ample prime orchard land exists, a less intensive system is preferable. In several European countries which have limited new land available and government subsidies available for new orchard establishment, orchard intensification is much more advanced than in other fruit producing areas with different economic conditions.

II. Cultivars and Rootstocks

A. Cultivars

Apples, pears, and peaches are generally sold with the cultivar name used as a part of the marketing strategy. This is not the case with other produce (e.g., nuts, bananas, corn, or beans). The use of a cultivar identification provides an opportunity to establish a market niche, and because of unique cultivar taste or other characteristics appeals to select segments of consumers. Niche marketing makes it more difficult to introduce a new cultivar that may have advantages in terms of disease resistance or yield efficiency. Cultivars with such advantages have provided major advances in crops without cultivar identification. Because of the special name recognition, fruit from some cultivars bring much higher prices than others. Since the price received for fruit is the most critical economic factor affecting orchard profitability, cultivars that return high prices reach the break-even point earlier. High value cultivars may justify greater initial investment and more detailed cultural management than those that receive only average prices.

Cultivars differ in growth habit, time of bloom, precocity, and other factors that influence and must be taken into account in an orchard management system. Mutations of a number of cultivars have been found that have a spur-type growth habit. Spurs are short, compact fruit bearing shoots and a tree with spur habit has an increase in the number of spurs per linear unit of growth. A cultivar with a spur habit can range from 30 to 40% smaller than the same cultivar without the mutation for spur habit. Most of the currently planted strains of 'Delicious', the most important and widely planted apple cultivar, have spur habit with varying degrees of tree size reduction and upright growth habit. [See Cultivar Development.]

Some apple cultivars such as 'Rome Beauty', 'Fuji' and 'Granny Smith' bear most of their fruit at the end of medium-long terminal shoots. Cultivars with this habit are not easily adapted to training systems such as the slender spindle or trellis forms. Since apples and pears require cross-pollination, the bloom period of cultivars intended to provide pollen for each other must overlap. This is particularly important in more southern- or maritime-producing areas where the bloom period is extended as opposed to the rather short bloom period in more northerly production areas.

B. Rootstocks

The rootstock forms the foundation for the tree and is the most important of all the components of an orchard system to be placed on the chosen site. Rootstocks influence tree size and, therefore, tree density and labor efficiency. They also influence precocity or early-cropping and the need for tree support, disease susceptibility, and soil tolerance of the trees.

Size-controlling apple rootstocks have been known since early Roman times and French gardeners used them in formal palace gardens. However, their use in commercial orchards was not widespread until the early 1900s after the East Malling Research Station (UK) collected and classified them. East Malling and a number of other research stations began rootstock breeding programs and many new rootstocks have been introduced as a result of these programs. Before discussing how rootstocks interact with other components of the orchard, the standard rootstock in each size class will be briefly described in order from largest to smallest.

1. Seedling

Historically nearly all fruit trees were grown on rootstocks produced by planting the appropriate seed and then through vegetative propagation (budding or grafting); the scion cultivar was joined to the young seedling. Most peaches, plums, and cherries are still propagated on seedling rootstocks, but now apple and pear are usually propagated on clonal rootstocks because of their size-controlling ability and precocity. Seedling rootstocks produce full size trees with strong graft unions, but generally have low precocity and vary greatly tree to tree in their disease susceptibility and propensity to sucker. [See Plant Propagation.]

2. MM.111

Produced as part of the Malling Merton (M.M.) series from a 'Northern Spy' × MI.793 cross. It produces a tree 85–90% the size of seedling and is easy to propagate in the nursery. It is tolerant of droughty soil conditions and adaptable to many soil types and climatic areas. It is resistant to woolly aphid and intermediate in susceptibility to crown rot, fireblight, and apple scab and moderately susceptible to powdery mildew. Although trees on MM.111 out-produce seedling in long-term trials, this rootstock lacks precocity. Generally, it is nonsuckering and is often used as a rootstock for dwarfing interstem trees.

3. MM.106

It is from a cross of 'Northern Spy' × M.1 and is easy to propagate in the nursery. It produces a tree 60–75% the size of seedling, is well anchored, nonsuckering, and precocious and productive. Cultivars on MM.106 tend to grow late in the season, defoliate late, and develop dormancy slowly, which can result in increased trunk injury due to sudden drops in autumn temperatures. The tendency to grow late in the year has also caused an increase of fireblight in susceptible cultivars. The rootstock itself is resistant to woolly aphid, moderately susceptible to crown rot, and intermediate in susceptibility to fireblight, apple scab, and powdery mildew. MM.106 is the primary clonal rootstock used in Australia, New Zealand, Africa, and South America and it is widely planted in the United States in areas without a crown rot problem.

4. M.7

This rootstock was one of the original stocks collected and classified by East Malling and has been the most widely planted clonal rootstock in the United States. It produces a tree 55–65% the size of seedling and it is adaptable to a range of soil types. It tends to sucker from the roots and in some soil types some trees may lean due to less than desirable anchorage. M.7 is moderately resistant to crown rot, powdery mildew, and fireblight. It is susceptible to woolly aphids and intermediate in resistance to apple scab. M.7 is the most sensitive of the commonly planted Malling rootstocks to winter cold temperature damage, especially if little or no snow cover exists.

5. M.26

This rootstock is from a cross of M.16 and M.9 and produces a tree 40–50% the size of seedling. It is very precocious and productive and generally is nonsuckering. It has demonstrated partial incompatibility with some apple cultivars (e.g., 'Blaxtayman', 'Granny Smith', 'Holiday', and 'Rome Beauty'). M.26 is the most winter hardy of the Malling rootstocks now used commercially. On some cultivars in deep, well-drained soil trees on M.26 may not need support, but vigorous cultivars on less desirable soils need to be supported. M.26 is intolerant of wet soil conditions. M.26 is very susceptible to fireblight and woolly aphids, moderately susceptible to crown rot and powdery mildew, and intermediate in susceptibility to apple scab. M.26 has been widely planted in many fruit growing areas, particularly where soil conditions are conducive to its growth and fireblight is not a problem.

6. M.9

This rootstock was selected in France in 1869 and called Jaune de Metz. It is the most widely used rootstock in the intensive production areas of Europe. M.9 produces a tree 25–35% the size of seedling and requires support due to the brittle nature of its root system. It tends to produce root suckers and is best adapted to deep soils with ample moisture holding capacity. Trees on M.9 are very precocious and productive. Studies have shown that it will set fruit with fewer seeds than the same cultivar on other rootstocks. M.9 is susceptible to fireblight and woolly aphid, resistant to crown rot, moderately resistant to powdery mildew, and intermediate in susceptibility to apple scab. Currently several extensive European tests are comparing subclones of M.9 that differ in tree size, fruiting, and other characteristics, but the most desirable selections have not yet been identified. These trials may provide a range of M.9 selections allowing for small site adjustments which may be needed.

7. M.27

M.27 resulted from a cross of M.13 and M.9 and is very dwarfing, resulting in a tree 15–20% the size of seedling. Trees on M.27 are precocious and productive but need support. Several studies have shown smaller fruit size on trees growing on M.27, compared with similar crop levels on other rootstocks. M.27 is rated as resistant to crown rot, susceptible to woolly aphids, moderately susceptible to fireblight, moderately resistant to powdery mildew, and intermediate in susceptibility to apple scab. M.27 should only be used with vigorous large fruited cultivars in very intensive plantings.

These brief detailed descriptions of the most widely planted apple rootstocks indicate clearly that each rootstock has advantages and disadvantages. Considerable research is in progress to develop and identify better rootstocks in each general size class that solve local problems or that are better adapted to certain growing areas (Fig. 1). Some of the rootstocks listed below each general size class are currently being planted commercially, while others in this list are being tested in research trials. However, it is clear that a range of rootstocks exists resulting in a series of ultimate tree sizes that can be tailored to fit a number of orchard management systems.

C. Interaction of Rootstocks and Other Orchard System Components

On orchard sites that have consistent spring frost problems the most dwarfing rootstocks should be avoided, since frost injury occurs first in the lower portion of the canopy. Frost control measures such as wind machines and irrigation can only partially overcome the inherent problem of a poor site and taller trees should be considered.

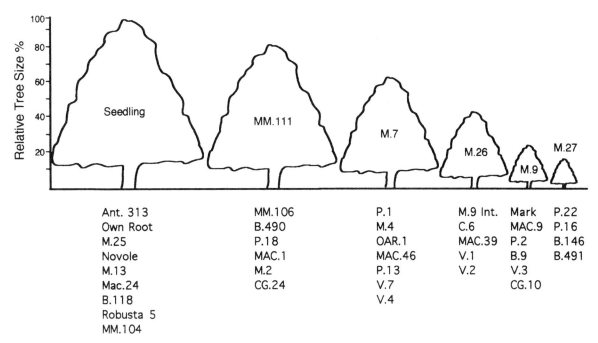

Ant. 313	MM.106	P.1	M.9 Int.	Mark	P.22
Own Root	B.490	M.4	C.6	MAC.9	P.16
M.25	P.18	OAR.1	MAC.39	P.2	B.146
Novole	MAC.1	MAC.46	V.1	B.9	B.491
M.13	M.2	P.13	V.2	V.3	
Mac.24	CG.24	V.7		CG.10	
B.118		V.4			
Robusta 5					
MM.104					

FIGURE 1 Relative tree size class produced by apple rootstocks. The rootstock presented in the drawing is the most widely planted in this size class and the ones listed below it are others producing approximately the same size tree. © David C. Ferree.

Poor soils that are shallow with little moisture holding capacity have a hard pan that restricts root growth or heavy, poorly drained soils should not be planted with the more dwarfing rootstocks. Rootstocks that direct most of the photosynthates to fruit, rather than vegetative growth, such as M.9, M.26, and M.27, will not produce adequate growth to sustain consistent cropping in poor soils. Heavy or poorly drained soils foster the development of the disease organism responsible for crown rot and significant tree losses can occur on these soils.

Cultivars that are normally weak growing, such as 'Spur Delicious', will not make adequate growth and can "runt out" on the most dwarfing rootstocks. These cultivars should be grown on a more vigorous rootstock such as M.7 to insure a balance of fruiting and vegetative shoot growth. Vigorous cultivars such as 'Mutsu' may need a more dwarfing rootstock (e.g., M.27) to produce the same size tree as average cultivar on M.9, if these cultivars are to be combined in an intensive orchard system.

Rootstocks provide the most economical and permanent means of reducing tree size and this in turn allows an increase in planting density to produce earlier yields and more rapid returns on investment. Rootstocks are the primary means of providing the lowest possible labor costs per unit of production through their precocity and ability to maximize photosynthate going to fruit and reduce the amount going into wood. Numerous studies have shown that harvest labor is more efficient when it can operate from the ground thus avoiding the use of ladders. Fruit quality is also improved as tree size is reduced because harvesting from the ground results in less bruising and improved fruit firmness and soluble solids result from more of the canopy receiving adequate light in smaller trees.

One of the advantages of reducing tree size that is often overlooked is the reduction in pesticide needed to adequately control insects and diseases. A 50–75% reduction in pesticide application has been shown to be possible with well-trained dwarf trees. This reduction not only lowers management costs, but also reduces environmental exposure to unnecessary pesticides. Air blast sprayers use considerable energy to deliver spray droplets to the tops of large trees and much of the spray misses the tree foliage and is lost to the environment. A much higher proportion of the spray hits the leaf target of small trees that are much closer to the spray nozzle. Apple scab and other diseases are particularly a concern in humid growing areas requiring 12–15 spray applications per year to produce quality blemish-free fruit. It is anticipated in the near future that spray that misses the tree canopy will need to be recaptured and returned to the spray tank to avoid environmental contamination. Capture will be much easier on small dwarf trees and nearly impossible on very large trees. [See Pest Management, Chemical Control.]

D. Rootstocks for Other Fruit Crops

Although a few size-controlling rootstocks exist for most of the other fruit crops, none has a whole range of stocks similar to those available for apple. Several types of quince have been used commercially as rootstocks in Europe to dwarf pear, particularly Quince A and Quince C. Due to lack of cold hardiness, these have not been used widely in the United States. A number of research trials have compared clonal rootstocks for peach, but none has found wide acceptance due to problems with incompatibility and, generally, a lack of tree survival. Currently, several groups of rootstocks for controlling tree size of cherry look very promising in early results from research trials, but have not yet gained commercial acceptance.

III. Tree Density, Arrangement, and Size

A. General Influence of Increasing Tree Density

As trees are planted closer together, growth of an individual tree is reduced and the branch orientation tends to become more upright. The root system is also generally smaller, sends down more sinker roots, and the density of roots per unit volume of soil is increased. The tree to tree competition that causes these shifts in growth also causes a reduction of fruit size and, if tree density becomes too high, fruit color and quality can also be reduced. These changes to an individual tree must be recognized in relation to the increase in total canopy (vegetative growth) and yield on a per unit land basis. Depending on the tree density compared, the yield advantage per hectare of higher density plantings may extend up to 10 years after planting. When the available space is filled with productive canopy, yields generally plateau. However, if excessive pruning is used to control tree growth at close spacing, the yield advantage of increased tree density can be lost. Thus, a balance must be achieved between the time required to fill the space allotted for canopy and the influence of cultivar, rootstock, and the cultural techniques necessary to balance growth and fruiting.

The general effects of increasing plant density are amplified as rectangularity increases. Rectangularity is the ratio of the wide (between row spacing) to the narrower (in-row spacing). The severity of competition is determined more by the proximity of adjacent trees than by overall tree density. Thus, the effects of a spacing of $2 \text{ m} \times 10 \text{ m}$ will have a much greater effect on growth and fruiting than a spacing of $4 \text{ m} \times 5 \text{ m}$, even though both have a tree density of 500 trees/ha.

B. Tree Arrangement

Trees can be arranged in single rows, double rows, or multi-row beds with a travel aisle between beds. Most research and practical experience indicate that single-row systems are easiest to manage and result in higher fruit quality. Light penetration to the interior of multi-row systems is generally insufficient for optimum red color development. It is also difficult to achieve adequate spray penetration and special herbicide equipment must be developed for multi-row systems. If very high tree density is a requirement for early returns, multi-row systems are necessary. In multi-row systems less land area is devoted to roadways and aisle middles and more to trees.

Single rows are more appropriate for hilly sites. Tree access for picking and other cultural operations are also much easier. The space devoted to aisle ways is often determined by equipment size. One of the problems for growers with large equipment and making their initial planting of an intensive orchard on small trees is the wasted space between rows. Small narrow specialized equipment is necessary for greatest efficiency of intensive dwarf orchards. Several studies have shown that orienting hedgerows north–south (N–S) in temperate zones provides a distinct advantage in canopy light distribution and long-term yields compared to E–W oriented hedgerows. This advantage disappears as you approach the equator and the angle of the sun is higher.

A general rule indicates that tree height should not be more than twice the open space between rows. Tree height that exceeds this relationship results in one row shading the adjacent row during part of the day. Most studies show that generally tree shape should be wider at the bottom than at the top to expose as much of the canopy surface area to direct sunlight as possible. In large trees, as much as 25% of the tree canopy receives less than 30% full sunlight, which is the threshold light level necessary to maximize photosynthesis and flower initiation. The proportion of the canopy receiving inadequate light in dwarf trees is very small and is one of the primary reasons for improved efficiency and fruit quality.

C. Tree Quality

Normally deciduous fruit trees are sold as bare-root nursery trees with a single stem called whips. In widely spaced orchards these newly planted trees would be pruned back hard and vegetative growth promoted for several years. As orchards were planted more intensively, not as much time was needed for trees to fill the allotted space and research showed that branched trees would increase early fruiting. Some cultivars grown very rapidly in the nursery will branch the first year in the nursery and are called feathered maidens, while others need to be cut off and grown a second year. These branched trees filled their allotted space earlier and flower and fruit a year or two earlier than trees planted as whips. Many growers now demand branched, high-quality trees because of the advantage they provide.

IV. Pruning, Training, and Tree Support

A. Pruning

Pruning is the art and science of cutting away part of a plant to improve tree shape, to influence growth, to enhance flowering or fruiting, to repair injury, or to contain the plant. There are two types of pruning cuts which have different physiological influences. The first is the heading cut, which removes a portion of a shoot leaving a portion to regrow. The second is a thinning-out cut, which removes entirely a shoot, branch, or limb. Heading cuts stimulate vegetative growth, decrease flowering, and promote an upright growth habit and often reduce light penetration into the canopy. Thinning-out tends to have the opposite effects and the combination of the two can be used to manipulate the tree toward the desired goal of balanced growth and fruiting.

Some general principles of pruning are: (1) it is a dwarfing process; (2) the greater amount of wood removed in pruning, the greater the dwarfing effect; (3) the plant response is localized to the immediate vicinity of the cut; (4) fruit set and fruit size are normally increased by pruning, but not in direct proportion to pruning severity; (5) pruning reduces flowering. These principles determine the influence of pruning on the balance of vegetative and fruiting wood which is necessary to sustain production of high quality fruit.

B. Training

Training refers to changing the orientation of the plant or its parts in space. Many training techniques such as spreaders, weights, and tying branches to a support system use bending as an important aspect to change the orientation of the plant part. Bending causes the plant to produce the plant growth hormone, ethylene, and also decreases the influence of the hormone auxin in the upright shoot tip. Ethylene causes a reduction in vegetative growth and increases flowering. The change in auxin distribution results in an increase in the number of buds forming lateral shoots. The use of these practical bending techniques allow fruit growers to shift the balance of growth from vegetative to fruiting. The more a vigorous vegetative branch is bent, the greater the reduction in terminal growth and the increase in flowering. Lateral shoots are also induced to grow out along the branch in response to its change in orientation. If a branch is growing too weakly, it can be tied to a more upright orientation and its vegetative growth increased. These bending techniques provide the grower an opportunity to alter the tree or portions of it to create the desired balance.

The training technique of notching can be used to initiate branches on young trees when they do not develop naturally. A latent bud is selected in an area on the main trunk where a branch is needed and a thin notch of bark removed above the bud in the spring just as growth is starting. Normally this notch should be 2–3 mm wide and removed from approximately a third of the trunk. The notch interrupts the auxin signal from the apical tip and the latent bud is induced to grow. The level of success in inducing lateral growth decreases as the age of wood increases. The crotch angle of these induced shoots will be very narrow and the young shoot should be carefully bent outward to widen the angle with the main trunk.

Scoring or girdling provides the grower with another training technique to reduce growth and increase fruiting. Scoring is a single cut through the bark and cambium down to the wood around a trunk or limb with no bark tissue removed, while girdling removes bark tissue. This technique operates on the principle of temporarily interrupting translocation through the phloem and results in a reduction of growth and an accumulation of carbohydrates above the scoring cut that increases flower formation and fruit set. These techniques are very useful on trees that are too vigorous or when frost has eliminated an early fruit crop that is needed to slow growth. Scoring has been successful in causing return bloom in biennial cultivars such as 'Fuji.' Normally, the preferred time is 10–14 days after bloom. If it is desired to increase fruit set on trees with a light bloom, the time can be advanced to full bloom.

C. Tree Support

In less intensive orchard systems of the past, several years were invested in pruning and training newly planted trees to develop a strong branch scaffold system to support the fruit crop. Under recent economic conditions, the time to develop this support structure often cannot be justified and trees are planted intensively and an artificial support system is installed to support the fruit load. Some rootstocks such as M.9 and M.27 have brittle root systems that will break if the tree has a heavy crop and these trees must be supported. A pole or trellis system also provides a structure so that limbs can be evenly spaced resulting in a more uniform distribution of growth in the canopy and creating a better balance of vegetative growth and fruiting by using bending techniques. Often the combination of a precocious rootstock and cultivar will result in excessive fruiting on a young tree and the leader and branches may bend so much that the tree is "runted out." With a support system in place, the leader and primary scaffold branches can be tied up to promote vegetative growth.

D. Canopy–Light Relationships

The ultimate goal of pruning, training, and the support system is to expose as much of the canopy as possible to desirable light conditions. Approximately 30% of full sunlight is required to maximize photosynthesis and initiate flower buds and nearly 50–60% of full sunlight to develop maximum red color. The fruit quality attributes of size, firmness, and soluble solids are also increased under relatively high light conditions and are decreased in deeply shaded areas of the canopy. The selective removal of branches in pruning or changing their orientation with training opens the canopy for light penetration and improves yield and fruit quality. It is not only the total amount of light intercepted by the tree that determines ultimate yield, but also the proportion of the canopy above the minimum threshold to initiate and sustain production. The perennial nature and permanent scaffold systems of tree fruits make them ideal subjects to be manipulated to optimize production.

V. Planting Systems

There are many planting systems that have been named based on the combination of the components of the orchard management system. Each of these systems has many details that are critical to the success of that system, but cannot be covered in the space allotted here. The general advantages and disadvantages proposed by the system originator and the general characteristics of a selected range of systems will be described.

A. Pyramidal-Shaped Systems

These systems are characterized by having the base of the tree wider than the top, giving the whole tree a conic or pyramidal shape (Fig. 2). These systems evolved as an improvement over the vase or open center systems that many fruit trees assume if left to assume their natural shape. Astute growers and researchers recognized if the top of a tree was wider than the bottom, much of the bottom became shaded and unproductive.

1. Central Leader

Although this general shape has been described for many years, the work of Dr. Don Heinicke in the United States and Dr. Don McKenzie in New Zealand, beginning in the early 1960s, was primarily responsible for it being widely adopted. This system is characterized by the use of a free-standing semi-dwarf to semi-standard rootstock such as M.7, MM.106, or MM.111. The trees would be spaced 3 to 5 m in the row and 5–8 m between rows, giving densities of 250 to 666 trees/ha and trees were generally unsupported.

In the United States the system was particularly designed to improve fruit size and color of spur-type 'Delicious' that were being widely planted. These spur-type trees produced few side branches on scaffold limbs with all scaffold limbs being nearly the same size and very upright. Training the trees using limb spreaders to bend scaffold limbs established the dominance of the central leader. Since spur 'Delicious' produced few side branches, a heading pruning cut on the leader and tips of the scaffold was used to induce branching.

Normally three tiers of scaffolds, separated by 0.5–1 m on the leader, would be developed so that each successive tier would be shorter than the others. The New Zealand approach differed slightly by pruning to create ladder bays or gaps so a ladder could be placed and the entire tree height could be picked. Central leader trees in New Zealand were also taller and less heading pruning was used with cultivars commonly grown there. The relatively severe pruning

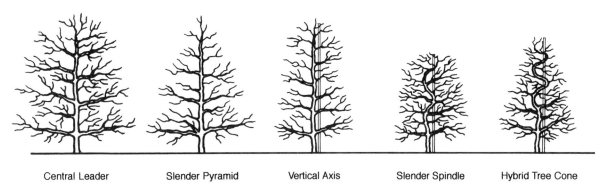

Central Leader   Slender Pyramid   Vertical Axis   Slender Spindle   Hybrid Tree Cone

FIGURE 2 Diagrams of the pyramidal-shaped orchard planting systems showing relative size, shape, and tree form of the central leader, slender pyramid, vertical axis, slender spindle, and hybrid tree cone. © David C. Ferree.

used during the years of tree development delays cropping and is the main disadvantage of this system. Minor modifications of these two approaches to central leader trees have been widely used and several economic evaluations suggest that the relatively low establishment cost and reasonable production resulted in a profitable return on investment. The central leader system predominates in the apple industries in New Zealand, Australia, and North America.

2. Vertical Axis System

This system was introduced in the late 1970s by Mr. J. M. Lespinasse in France and is based on his long-term studies of the natural fruiting habits of apple trees. The vertical axis is designed to begin cropping early not only for economic reasons, but also to use the crop to limit tree growth. Rootstocks for this system can be M.9, M.26, and, with small stature cultivars, even M.7 and MM.106. Trees are normally spaced 1.5–4 m in the row and 3.5–6 m between rows, giving tree densities of 400–1900 trees/ha. Minimal support for the leader is supplied by a one- or two-wire trellis and a small training pole by each tree to keep the leader straight.

The trees are normally not pruned at planting and the leader is never pruned until it reaches the desired tree height and bends over due to the weight of the fruit. A tier of lower permanent scaffolds is developed and all laterals above this bottom tier are considered temporary and will be partially removed or replaced after fruiting. Since the trees are pruned very little and receive only minimal summer pinching to keep the leader dominant in the first few years, the trees tend to flower and fruit early. On fertile soils, the trees can become quite tall (4–6 m) before cropping in the top of the tree slows growth and causes the limbs to bend. This system is very efficient after estab-

lishment because of the reduced labor costs through simplified minimal pruning and training. The increased density and minimal pruning result in early crops, but the disadvantage is the tall tree height developed prior to the desired balance between cropping and growth is achieved on fertile soils.

3. Slender Pyramid

This system, developed in the 1980s in New Zealand by Tustin and others, combines many of the desirable characteristics of the central leader and vertical axis systems. This system seems ideally suited to areas of high light and nearly ideal growing conditions with fertile soils. Rootstocks would normally be M.26 or MM.106. The trees are planted slightly closer than they would be in the central leader system with the trees given minimal training support by one or two wires that support the leader. Trees are generally unpruned at planting and selective pinching out young growing shoots during the season is used to establish a strong lower scaffold system of up to eight limbs (double that of the central leader). Weaker fruiting wood is developed in the leader above the lower tier of heavier wood and removed as needed with renewal pruning. Dormant pruning only removes wood that has fruited and pendant limbs, which developed poor quality spurs and fruit. When the desired canopy volume is filled, scaffold limbs from each of the tiers are removed over several years down to four/tier, similar to the number established in the central leader trees. The slender pyramid system appears to be well suited to areas with ideal soil and environmental conditions.

4. Slender Spindle

This very intensive system was developed in the 1950s in Holland and Germany. The slender spindle is almost always planted on M.9 rootstock or with some vigorous sites, rootstocks the size of M.27 may be used. Spacing range from 1 to 2 m between trees in the row and 2.75 to 3.5 m between rows, giving tree densities of 1400–3500 trees/ha. Densities are even higher if multiple row beds are utilized. Well-feathered trees or 2-year branched trees are planted and unpruned at planting. Each tree is supported by a pole which serves as a base to tie limbs down or up to balance growth and cropping. Training is very detailed and each branch is adjusted to achieve the desired goal of early cropping and growth contained to the allotted space. Some of the lower scaffolds may be permanent, but most others are pruned to renew growth after cropping. This system has resulted in early cropping and sustained cropping at high levels has been achieved in fruit growing areas with lower light levels and poorer growing conditions. Fruit on this system are well exposed to light and develop optimum red fruit color. In regions with high natural light levels sunburn damage on the fruit can be a problem. Slender spindle is widely planted in Europe and is currently being planted in the United States and Canada. This system depends on a dwarf, precocious rootstock and intensive training through bending to achieve success. Several economic studies point out the high cost of establishment and considerable labor necessary to maintain these plantings. However, high early yields have been achieved and in areas where land is limiting, this system has been economically successful.

5. Hybrid Tree Cone (HYTEC)

This system originated in Washington by Dr. Bruce Barritt in the early 1990s and combines many aspects of the vertical axis and slender spindle. This system utilizes the small precocious rootstocks (e.g., M.9 and M.26), is generally supported by a post, as is the slender spindle, but is intermediate in height between the slender spindle and central axis. Fruit are often damaged by sunburn in high light areas when trained to very open systems, such as the slender spindle, and the increase in fruiting branches and added height of the HYTEC system alleviates this condition. Although this system has great promise, commercial experience at this time is limited. Just as with the slender spindle, the considerable bending required must be paid for through improved fruit quality of high value cultivars.

B. Trellis Systems

1. Palmette

The Palmette was developed in the 1950s in northern Italy and southern France. A four to five wire trellis, 4–5 m high is established and the trees are trained by orienting all primary branches along the wires. The branch arrangement on the trellis varies, with the horizontal and oblique palmette arrangement most common. This narrow tree form allowed rows to be closer together and thus tree density and early yields were increased (Fig. 3). This system has been successful with apple, pears, and peaches grown on seedling or semi-dwarf rootstocks. The most significant problem is the development of a nonfruitful area in the lower canopy which can be averted by allowing the lower canopy to be slightly wider than the top and detailed pruning to insure a steady progression of young fruiting wood. Although this system has been tried on a very limited scale in many fruit growing areas, it has only achieved widespread use in Europe. Generally new plantings are not going in using this system.

2. Low Trellis Hedgerow

This is a small version of the palmette, developed by Dr. Loren Tukey in Pennsylvania. This system depends on M.9, M.26, or M.27 rootstocks and generally four wires are used with the top wire 2 m above ground. In several studies, this system was very efficient, having a higher ratio of fruit to either trunk area or leaf area than other systems such as the slender spindle or central leader. Light distribution within the canopy was also high in this system.

Although limited commercial plantings of this system exist in most eastern North American fruit growing areas, it has not been widely used. The limiting factors have been the high cost of establishment and growers dislike for wires impeding movement around the trees.

3. Solan

This is a new system developed by Mr. J. M. Lespinasse in France, designed for production of terminal fruiting cultivars such as 'Granny Smith' or 'Rome Beauty' which are not well adapted to vertical trellis or the slender spindle system. This trellis consists of 1.5-m posts with two wires at the top, with the primary scaffolds trained along the wires and the side branches pendant with the weight of fruit.

4. Lincoln Canopy

This horizontal trellis system was developed by Mr. John Dunn of New Zealand in the 1970s, especially for mechanical harvesting. Past work had shown that most bruising of mechanical harvested fruit occurred when falling fruit hit other fruit or limbs. Producing fruit on a single horizontal plane with the branches

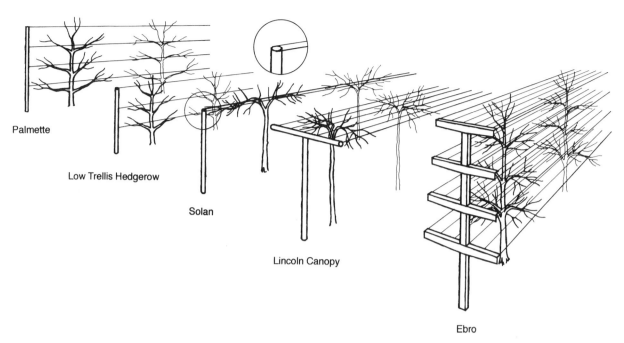

FIGURE 3 Diagram of the trellis forms show relative size, shape, and tree form of the vertical forms: palmette, low trellis hedgerow, solan; and horizontal forms, Lincoln canopy and Ebro. © David C. Ferree.

fastened to a "T"-shaped trellis would alleviate much of this problem. The system greatly reduced fruit bruising, but the trellis system is expensive to establish and in areas with less than optimum growing conditions, it requires several years to cover the trellis with fruiting wood. Research work on light relations and fruit size and quality show that poor light and quality conditions can develop in a single year when the canopy is concentrated in one plane. Thus, great care must be used to insure adequate light conditions so fruit size and quality can be maintained. Some pruning is done mechanically in the summer, but winter thinning-out pruning must still be done by hand. Although tested in many areas, Lincoln canopy is not widely planted commercially.

5. Ebro

This four-tier, horizontal trellis was developed by Roger Evans in New Zealand, specifically to produce high-quality fruit. Under environmental conditions that induce high vigor, light quickly becomes limiting in the lower tier and yield and fruit quality are reduced. Although early yields are high and quality generally desirable, sustained productivity may be a problem, unless management is very detailed. A number of other versions of a multi-tiered, horizontal canopy have been considered. One of the most unique (attempted in West Virginia) was to have a red-fruited

cultivar as the top tier and a yellow- or green-fruited cultivar which would need less light for color development, as the bottom tier.

6. Tatura Trellis

A number of researchers have reported on the influence of a "V"-shaped trellis on productivity of several fruit crops (Fig. 4). Some of the earliest and most detailed studies were conducted by Dr. D. J. Chalmers and B. van den Ende at the Tatura Research Station in Australia. Yields, particularly of stone fruit, have been high and reports from New York indicate that because the system can be designed to intercept more light than several other systems, it is very efficient. The cost to establish the trellis is high, but the training is simple and can be accomplished by unskilled workers. A version of this shape is being investigated and evaluated in many fruit growing regions.

7. MIA

A system developed by the Murrumbidgee Irrigation Area (MIA) is the inverse of the Tatura trellis and is a truncated "A"-frame with the inward sloping arms treated as a unit. The proposed advantages of this shape are that instead of doing all the work inside the canopy as required by the Tatura trellis, work can be done from the outside resulting in less elaborate

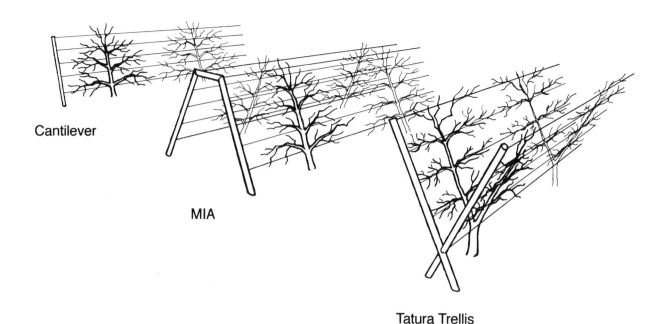

Cantilever

MIA

Tatura Trellis

FIGURE 4 Diagram of other trellis forms showing relative size, shape, and tree form of Tatura, MIA, and cantilever. © David C. Ferree.

equipment. A limited number of commercial plantings have recently been established on this system.

8. Cantilever

Most of the other systems reported here are designed for rows running north–south. This system was designed to run east–west with the sloping wall toward the sun (south in the northern hemisphere). This single slanting trellis has only been tested in a limited way, but demonstrates the potential of changing the trees orientation to influence light absorption.

Most of these training systems have been tried with several different fruit crops. It must be pointed out that only general tree forms and likely combinations are mentioned here and many modifications of each exist. Most modifications are designed to counter particular problems of a particular cultivar, site, or environmental region. To be successful, detailed and comprehensive programs for pruning and training must be developed for each system.

VI. Tailoring an Orchard System to Specific Needs

Due to the cost of capital, cost and availability of capable labor, and the changing market, fruit growers must tailor an orchard system that meets their specific needs. Since orchards are long-term investments it is not easy to change after a system is chosen and planted. Thus, it is important that those considering the establishment of a commercial fruit planting understand the components and interrelationships of an orchard system.

Commercial growers have changed to more intensive systems in the last 10–15 years than ever before in history. Those refusing to change or those who made a serious mistake in orchard system selection will not be in business in the future because of the intense competition worldwide. This article provides only an overview and much more detailed information would be necessary to properly plan a successful new orchard.

Bibliography

Barritt, B. H. (1992). "Intensive Orchard Management" Good Fruit Grower, Washington.

Forshey, C. G., Elfving, D. C., and Stebbins, R. L. (1992). "Training and Pruning Apple and Pear Trees." Amer. Soc. Hort. Sci. IPC, MI.

Jackson, J. E. (ed.) (1981). Symposium on research and development on orchard and plantation systems. *Acta Horticulturae* (114).

Jackson, J. E. (ed.) (1986). Third International Symposium on research and development on orchard and plantation systems. *Acta Horticulturae* (160).

Peterson, A. B. (1989). "Intensive Orcharding." Good Fruit Grower, Washington.

Rom, R. C., and Carlson, R. F. (1987). "Rootstocks for Fruit Crops." Wiley, New York.

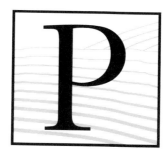

Peanuts

JOHN BALDWIN,
University of Georgia Cooperative Extension Service

Glossary

Botanical types The cultivated peanut consists of three botanical types: Spanish, Virginia, and Valencia; Runner type is a marketing classification for medium seed size varieties of the Virginia botanical type

Market types Four predominant types are marketed in the United States which are Runner, Virginia, Spanish, and Valencia; each is distinctive in size and flavor

Peanut *Arachis hypogaea* L. is a four-foliate legume with a prominent taproot, yellow sessile flowers, and subterranean fruit; the peanut is native to South America, originating in Bolivia

Production area Seven states account for 98% of all peanuts grown in the United States; Georgia grows about 42% of all peanuts followed by Texas (16%), Alabama (14%), North Carolina (9%), Virginia (6%), Oklahoma (5%) and Florida (5%)

The peanut, while grown in tropical and subtropical regions throughout the world, is native to the Western Hemisphere. It originated in South America and spread throughout the New World as Spanish explorers discovered the peanut's versatility. Although there were some commercial farms in the United States during the 1700s and 1800s, peanuts were not extensively grown. After the Civil War, the peanut remained basically a regional food associated with the southern United States.

After the Civil War, the demand for peanuts increased rapidly. By the end of the 19th century the development of equipment for production, harvesting, and shelling of peanuts, as well as processing

techniques, contributed to the expansion of the peanut industry. The new 20th century labor-saving equipment resulted in a rapid demand for peanut oil, roasted and salted peanuts, peanut butter, and confections.

Also associated with the expansion of the peanut industry is the research of George Washington Carver at Tuskegee Institute in Alabama at the turn of the century. The talented botanist recognized the intrinsic value of the peanut as a cash crop. Dr. Carver proposed that peanuts be planted as a rotation crop in the Southeast cotton-growing areas where the boll weevil insect threatened the region's agricultural base. Not only did Dr. Carver contribute to changing the face of southern farming, but he also developed more than 300 uses for peanuts, mostly for industrial purposes. As peanut consumption continued to rise, in 1934 the United States government instituted programs to regulate the acreage, production, and price of this food item. Federal government production controls were lifted during World War II to meet the heavy demand for oils required for the United States war effort. Controls were re-established in 1949 and in 1977; a two-tier price support system was initiated and continues with renewal every 5 years as a part of the overall Farm Bill adopted by Congress.

Peanut production is contained predominantly in seven states which are designated for regulatory and market type purposes as the Southeast, Southwest, and Virginia–Carolinas. Unlike other countries where the end products are peanut oil, cake, and meal, the primary market for United States peanuts is edible consumption. Only 15% of United States production is normally crushed for oil. Peanuts are the 12th most valuable cash crop grown in the United States with a farm value of over one billion dollars.

I. World Production

India, China, and the United States produce about 70% of the world's peanuts (Table I). The major

TABLE I

Peanut Production in Specified Countries, 1986–1991

Country	Year					
	1986	1987	1988	1989	1990	1991
	(1000 metric in-shell tons)					
India	5875	5854	9000	8088	7300	7600
China	5882	6170	5693	5365	6368	6000
United States	1677	1640	1806	1810	1634	2279
Indonesia	750	786	843	875	890	920
Senegal	817	932	690	815	670	695
Burma	544	519	438	459	505	500
Sudan	380	435	450	400	325	400
Argentina	518	450	243	336	475	400
Brazil	195	170	156	137	150	150
South Africa	235	204	163	113	137	135
Other	3705	3818	3797	3661	3660	3782

Source: U.S. Department of Commerce, October 1991.

TABLE II

U.S. Peanut Production by State

State	Area harvested			Yield		
	1990	1991	1992	1990	1991	1992
	1000 acres			Pounds/acre		
Alabama	256	277	236	1510	2305	2505
Florida	94	118	80	2480	2370	2530
Georgia	770	895	673	1750	2490	2705
New Mexico	20	23	21	2500	2250	2760
North Carolina	164	162	153	2900	2850	2660
Oklahoma	106	106	98	2220	2300	2410
South Carolina	13	14	13	2230	2400	2500
Texas	289	325	305	1850	2100	2230
Virginia	97	96	93	3195	3200	2755
United States	1810	2016	1672	1991	2444	2562

Source: Crop Production Report, USDA Agricultural Statistics Board, 1992.

suppliers to the export market are the United States, China, and Argentina. Although United States peanuts represent approximately 10% of the world peanut production, the United States has become one of the leading world exporters, accounting for more than one-third of the world peanut trade. Other origins, such as India and several African countries, periodically enter the export market, depending upon their crop quality and world market demand.

II. U.S. Production Areas: Market Types

There are three major peanut producing areas in the United States (Southeast, Southwest, and Virginia–Carolinas). Although peanut allotments are located in 16 states, 98% of the peanuts produced in the United States are located in the following 7 states listed in ranking order of acreage: Georgia, Texas, Alabama, North Carolina, Oklahoma, Virginia, and Florida. Acreage and average yield for each of the seven major producing states for the 1990–1992 growing seasons are contained in Table II.

United States peanuts are marketed as four basic market types: Runner, Virginia, Spanish, and Valencia. These four market types can be grown in each producing area of the United States but have been predominantly grown by area and state as follows:

Runner. Runners have become the dominant type due to the introduction in the early 1970s of the Florunner variety which showed tremendous yield advantage and quality over previously released Runner varieties.

Runners have rapidly gained wide acceptance because of their kernel size range. About 54% are used for peanut butter, with the rest used in candy and snacks. Runners, grown mainly in Georgia, Alabama, Florida, Texas, and Oklahoma, account for 75% of total United States production.

Virginia. Virginias have the largest kernels and account for most of the peanuts roasted and eaten in shells. When shelled, the larger kernels are sold as salted peanuts. Virginias are grown mainly in southeastern Virginia and northeastern North Carolina. Virginia-type peanuts account for about 21% of total United States production.

Spanish. Spanish-type peanuts have smaller kernels covered with a reddish-brown skin. They are used predominantly in peanut candy, with significant quantities used for salted nuts and peanut butter. They have a higher oil content than the other types of peanuts which is advantageous when crushing for oil. They are primarily grown in Oklahoma and Texas. Spanish-type peanuts account for 4% of United States production and have been steadily replaced in these states over the past decade by Runner types.

Valencia. Valencias usually have three or more small kernels to a pod. They are very sweet peanuts and are usually roasted and sold in the shell. They also are popular throughout the United States growing areas for fresh use as boiled peanuts. Valencias account for less than 1% of United States production and are grown in New Mexico.

III. Peanut Growth and Physiology

Developing an understanding of how a peanut plant grows and develops aids in making sound management decisions which affect yield and quality. Crop growth from seedling emergence through pod maturation must be understood to apply the correct management practices as well as make certain that the practices applied result in a positive effect on yield and quality.

A. Seed Structure and Seedling Development

The peanut seed is made up of two cotyledons or seed leaves and an embryo. The peanut embryo consists of a plumule, hypocotyl, and primary root. (Fig. 1). The plumule becomes the stems and leaves above the cotyledon leaves. The hypocotyl is the stem below the cotyledon leaves and above the primary root. At germination and emergence, the hypocotyl and primary root are known collectively as the radicle. Peanut seed are some of the most delicate seed in commerce today and subject to mechanical injury and invasion by seed-decaying fungi. The seed's radicle protrudes from the end of the seed and is easily damaged. Radicle injury leads to decay, poor germination, and abnormal growth.

The germination process begins when the peanut seed imbibes water. The water intake is uniform around the seed surface, and increases as temperature increases. When the seed moisture level reaches 35%, germination can occur, resulting in metabolic activity, cell division, and elongation. Peanut seed will germinate in soil temperatures of 41° to 104°F, but

germinates best at 68–95°F. As the embryo grows, the seed coat (testa) ruptures and the seedling emerges.

The first visible sign of germination is emergence of the radicle, which takes 1 to 2 days, depending on soil moisture levels and temperature. The seedling uses food reserves in the cotyledons during the first days of growth, and after 5 to 10 days, the newly developed root is able to absorb minerals and the emerged leaves will eventually begin supplying photosynthate for sustained growth. After about 5 days the tap root is 4 to 5 in. long and lateral roots begin to develop. Lateral roots are arranged in four distinct positions off the primary root. Secondary roots then develop from the lateral roots.

Emergence through the soil, or "cracking," begins 7 to 10 days after planting. However, planting depth, soil temperature, seedling vigor, and moisture all affect seedling emergence and germination time. Normally, all seedlings which are going to emerge do so within 30 days after planting. Dry and cool soils can slow germination and emergence for up to 3 weeks. Seed that have not germinated or emerged in 3 to 4 weeks have probably been killed by seedling pathogens.

B. Plant Development: Emergence to Bloom

As the plant grows, the root develops faster than the shoot, helping the plant to survive. The plumule pushes upward, causing the soil surface to crack and form an opening for plant emergence. After emergence, the plumule is called a shoot and consists of a main axis (main stem) and two cotyledonary lateral branches. At emergence, the main stem has at least four immature leaves and the cotyledonary lateral branches have one or two. The seedling develops slowly, showing as few as 8 to 10 fully expanded leaves 20 days after planting, all of which were present in the embryonic state of dormant seed. The hypocotyl and root system develop rapidly.

Leaves are attached to the main stem at joints, or nodes. There is a distinct pattern by which these leaves are attached. There are five leaves for every two rotations around the main stem, with the first and fifth leaves located one above the other. Leaves attached to the cotyledonary laterals and other lateral branches are two-ranked, so there is one leaf at each node, alternately occurring on opposite sides of the stem.

Peanut leaves have four leaflets per leaf, and the leaflets are pinnately compound. Leaflets are elliptical, with a hairy appearance and a prominent midvein. Terminal leaflets are usually longer and wider than

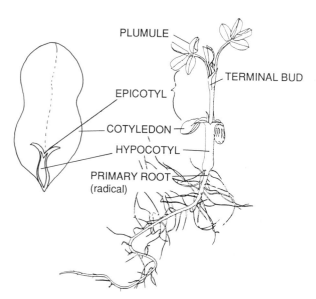

FIGURE 1 Peanut morphology. [From Guerke, W. R. (1993). Unpublished, Dir. Seed Div., Georgia Department of Agriculture, Tifton.]

lateral leaflets. The main stem and cotyledonary laterals determine the basic branching pattern of the shoot. The main stem develops first, and in runner-type plants, the cotyledonary laterals eventually are longer than the main stem. Additional branches arise from nodes (joints from which leaves, stems, and pegs develop) on the main and lateral stems.

The growth habit of peanut foliage is described as bunch, decumbent, or runner. Branches on bunch or erect plants such as the spanish and Valencia market types grow upright. Branches on runner or prostrate plants, such as several virginia varieties, grow flat. Decumbent growth exhibits a combination of bunch and runner characteristics. Runner-type plants have suppressed branching from the main stem and a strong tendency for continued branching of the cotyledonary laterals.

The amount of vegetative growth between emergence and first bloom is largely dependent upon genetics, environmental conditions, and cultural practices. Rainfall, temperature, soil fertility levels, pest control, and seeding practices influence vegetative growth before first bloom. Peanuts are indeterminate in vegetative and reproductive development, so vegetative growth continues after the plant flowers and begins to set fruit (pods).

High and low temperatures can impair peanut growth and development. Peanut, being of subtropical origin, is more sensitive to low temperature damage than many of the crops originating in more temperate climates. However, peanut yields can also be adversely affected by very high temperatures. The influence of temperature on peanut physiology is complex and often compounded with other stresses that can be associated with temperature stress. For example, high temperatures are often associated with drought and high light intensities. As soil water becomes limited, a plant's stomata will close, impairing transpirational cooling of the leaf and causing leaf temperature to rise above ambient temperature.

The optimal temperature for peanut plant photosynthesis and dry matter production is approximately 86°F. The optimal day/night temperature combination for peanut plants is between 86° and 95°F for daytime temperatures, and 68° to 77°F for nighttime temperatures. In general, peanut plants can live and grow at temperatures between 50° and 95°F, although little growth occurs below 60°F. Peanut plants can survive at temperatures greater than 95°F, and are known to withstand temperatures as high as 120°F for short periods.

C. Plant Development: Bloom to Pod Maturity

Approximately 30 days after emergence, peanut plants begin to produce flowers. The number of flowers per day peaks 2 to 4 weeks after floral initiation and declines during late pod fill. Environmental conditions such as drought or high temperatures will reduce the number of flowers produced. Interestingly, peanut flowering is inhibited when day and night temperatures vary by more than 36°F. In addition, flower set, and peg formation, is inhibited when night temperatures exceed 95°F. The best soil temperature for pod growth is about 86°F, similar to that for best shoot growth. However, researchers have found that soil temperatures of about 80°F generally produce larger seed than higher soil temperatures.

Flowers initiate within individual axils at each node. Several flowers can develop at a single node. The peanut flower is a perfect flower, meaning male and female structures are present in the same flower (Fig. 2) allowing self-pollination. The peanut flower has a showy yellow bloom that is papilionaceous (butterfly-shaped). The flower has standard petals, which are the larger petals that spread open, and two wing petals, which are the smaller petals that enclose the reproductive structures. The keel, a thin tissue wrapped around the male and female parts, is between the wing petals.

The female structures are collectively called the pistil and include the stigma, style, and ovary. The stigma is the receptive and germination surface for pollen grains. The style is the tissue column that connects the ovaries to the stigma and through which the pollen tube grows. In peanuts, the style, or hypanthium, is also the flower stem. The ovaries are at the base of the style and are fertilized by the pollen tubes that travel down the style.

The male structures are collectively called the stamen and include the anthers and filaments. The an-

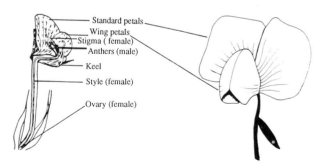

FIGURE 2 The peanut flower. [From Beasley, J. P., Jr. (1990). "Peanut Growth and Development." Univ. of Georgia, Cooperative Extension Service. Bulletin No. SB 23-3, Fig. 9, p. 5.]

thers bear pollen grains and the filament is the stalk that holds the anthers.

When the peanut flower first emerges, the petals are folded together. In the early morning of the following day, the standard petals unfold and the pollen is shed and attaches to the stigma. The first fertile pollen grains that germinate and produce pollen tubes travel down the style and fertilize the ovaries.

The fertilized ovary begins to elongate and, because of geotropism (growth induced by gravity), bends toward the soil surface and extends downward from the reproductive node of branches (Fig. 3). This structure is called the peg, or gynophore, and is first visible about one week after fertilization. The deteriorated flower remains attached to the tip of the peg for several days before falling off. Pegs enter the soil 8 to 12 days after the flower is pollinated.

The tip of a peg is sharp, allowing it to penetrate the soil easily. The developing peanut fruit is in the tip of the peg and begins to enlarge soon after entering the soil if water and calcium are present. Several pegs can develop at a single node. After pegs enlarge, the fruit is referred to as a pod. Because of the indeterminate fruiting habit of peanuts, pods of various maturities and sizes can be present on a single node at harvest.

During the period 30 to 60 days after emergence, the plant's energy is used for vegetative growth and the beginning of flowering and fruit production. From 60 to 110 days after emergence, the peanut plant undergoes flower production, pegging, pod formation, and pod fill. During the early stages of formation, the pod tissue is soft and watery. As the pod develops, the hull and seed begin to differentiate. The cell layer (mesocarp) just below the outer cell layer (exocarp) of the pod changes from white to yellow to orange to brown to black as it matures, providing

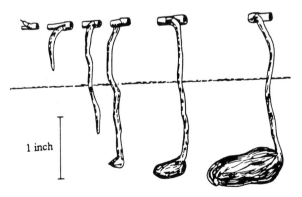

FIGURE 3 Peanut peg growth and development. [From Beasley, J. P., Jr. (1990). Univ. of Georgia, Cooperative Extension Service Bulletin No. 5B 23-3, Fig. 11, p. 5.]

a color indication of optimum harvest time. With favorable temperatures, nutrient, and moisture conditions, mature fruit will develop in 9 to 12 weeks from the time a peg enters the soil. Peanut varieties also will vary in the length of time to develop mature pods.

Water and nutrients are absorbed by the pod and enter the developing seed by diffusion. The seed is attached to the inner hull layer by the funiculus (stalk) as the pod matures. The funiculus functions as an umbilical cord, transporting water and nutrients to the kernel. Physiological maturity or late-season stress may cause the funiculus to shrivel and the seed to separate from the pod, ending water and nutrient transport to the seed and further development of that individual seed ceases. Very high temperatures (above 95°F) will increase a peanut plant's respiration rate, decrease the efficiency of many enzyme systems and delay development of the crop. As with excessively low temperatures, high temperatures are known to cause plant cell membrane damage and electrolyte leakage. Excessively high soil surface temperatures can burn peg tips resulting in reduced pod set. In general, average soil temperatures of 80° to 90°F are considered optimal for pod development.

IV. Production Practices

A. Rotations

A good crop rotation program is necessary for successful peanut production. The peanut plant responds to both the harmful and beneficial effects of other crops grown in the fields. Research shows that long rotations are best for maintaining peanut yields. A 3-year rotation, 1 year with peanuts, followed by 2 years of grass-type crops such as corn, has been effective in improving peanut yield and quality. Grass crops reduce nematode and soil-borne disease problems and permit better control of some weed species. These crops respond to heavy fertilization but leave adequate residual nutrients for healthy peanut growth.

Peanuts following tobacco, soybeans, or other legumes and some vegetable crops, generally have higher disease losses than peanuts following corn, grain sorghum, or small grains. Cotton is a good rotational crop except that the taproot and other plant residue is difficult to manage. If peanuts must be planted directly behind cotton or tobacco, the taproots should be ripped up and shredded early in the fall to allow for maximum decomposition before land preparation. Collar rot is frequently more severe in a rotation with cotton.

A small grain cover crop can be planted to reduce the disease and nematode problems and to lessen the possibility of soil erosion during the winter. Growers who fail to follow an effective rotational program can expect a buildup of diseases such as southern stem rot and sclerotinia blight.

B. Land Preparation

Peanut growers have, historically, used the moldboard plow equipped with trash covers to prepare a smooth, uniform, and residue-free seedbed for planting. The burial of old crop residue and weed seed has been effective in the long-term suppression of soil-borne diseases and short-term suppression of some weed problems.

Land preparation begins with the disposal or management of the previous crop residue. In order to promote decomposition during the winter, crop litter should be chopped or shredded and disked lightly. Seeding a cover crop can help reduce soil erosion during the winter months. Applications of lime, phosphorus, and potassium can be applied at this time if needed.

Most peanut soils should be broken in the spring with a moldboard plow. Chisel plowing may be acceptable in fields with light crop residue or in heavier clay content soils which tend to clod. Harrowing to break up clods and leave a smooth, firm, and clean seedbed usually follows plowing. Many peanut growers bed their peanut fields either in the fall or spring. Beds should be prepared several weeks before planting to allow them to warm up and gain uniform moisture levels. Many growers prefer planting on raised beds rather than planting flat. Beds often give faster germination and early growth, provide drainage, and may reduce pod losses during digging. The top of the seed row should be level with or slightly higher than the middles. This will help prevent "dirting" of the peanuts during cultivation.

The goal in land preparation is to provide soil conditions for rapid and uniform seed germination, good root penetration and growth, and steady plant and pod development. Erratic emergence and irregular growth result in uneven pod set and maturity with some plants having a taproot crop and others a limb crop. This circumstance is sometimes referred to as a "split crop."

C. Seed Selection and Planting

Planting sound, mature, disease-free seed of known pedigree, purity, and performance should be a goal of every peanut producer. Handle seed carefully when loading or unloading to prevent damage. Gently place bags or seed in trucks or storage buildings. Do not drop or throw seed bags as mechanical damage may occur. County offices of the Cooperative Extension Service can provide performance data on varieties that are tested each year. Use planters that provide good seed to soil contact and will not damage the seed. Plant at a uniform depth of 2 to 3 in. in sandy soils and 1.5 to 2 in. on heavier soils. Strive for the proper plant population of three to four plants per foot of row. When planting, remember to consider seed germination and seed count per pound. Most Extension Services in peanut producing states recommend four to six seed per foot of row for Runner and Spanish type and three to four seed per foot of row for Virginia types.

A decision producers face when planting peanuts is what row pattern to utilize. The majority of peanuts are planted in a standard row pattern where the rows are equidistance apart. Most rows are planted two per bed spaced 36 in. apart. A variation of this pattern that is used in some parts of peanut producing states is 32-in. rows on 64-in. beds. Where cotton and peanut rotation is the predominant production system, peanut rows have been adapted to the cotton regime and bed widths are 76 in. and rows are 38 in. apart.

The twin row pattern is becoming popular in some areas. In this pattern, a paired set of rows are planted 7 to 10 in. apart on each side of a bed. It is difficult to plant a set of twin rows closer than 7 in. and digging and inverting becomes more difficult when a set of twin rows are spaced wider than 10 in. apart. High plant populations or twin rows generally result in fewer pods per plant but pod size is larger. The seeding rate for twin rows should equal that for single rows. There are several documented advantages to the twin row pattern, the most dramatic of which is more rapid canopy closure. More rapid canopy closure can improve the plants competition against emerging weeds. Quicker canopy closure maximizes light interception for efficient energy use and biomass production by the plant and cools the soil temperature.

Planting should occur as soon as soil conditions are favorable for rapid germination and emergence. Use a planting schedule that is compatible with harvesting capacity. Late planting generally means a late harvest, reduced yield and quality, and increased risk of freeze damage. Generally the best planting dates for peanuts are between April 20th and May 20th. However, it is best to plant according to field conditions and expected weather patterns rather than calendar dates.

As a general rule, peanuts should be planted as soon as the risk of a killing frost is over. Early plantings usually give higher yields, more mature pods, and permit earlier harvesting.

Peanuts should not be planted until the soil temperature at a 4-in. depth is 65°F or above for 3 consecutive days when measure at mid-morning. Favorable weather for peanut germination should also be forecast for the next 72 hr after planting. The soil should be moist enough for rapid water absorption by the seed.

D. Fertilization

A major benefit of an effective crop rotation program is that peanuts respond better to residual soil fertility than to direct fertilizer applications. For this reason, the fertilization practices for the crop immediately preceding peanuts are extremely important. Grass-type crops generally respond well to direct application of fertilizer. Growers can fertilize these crops for maximum yields and, at the same time, build residual fertility for the following crop of peanuts. Peanuts have a deep root system and are able to utilize soil nutrients that reach below the more shallow root zone of the grass-type crops.

The peanut crop is usually produced without applying any fertilizer materials during the production year. However, soil tests should be taken and the data used to determine the fertilizer and lime needs for peanut production. A balanced fertility program, with emphasis on adequate levels of phosphorus, potassium, calcium, and magnesium, is essential for high yields. If peanut fields need these materials, they should be broadcast before land preparation. Fertilizing peanuts requires that there be an understanding of growth characteristics and nutrient needs of the plant and the ability of the soil to provide these needs.

Calcium (Ca) is one of the most important nutrients needed for the development of high-quality, top-yielding peanuts. Since the early 1940s it has been known that peanuts need an adequate supply of calcium in the pod or pegging zone, which is the top 3 in. of soil. This is especially true for the production of Virginia-type peanuts and peanuts grown for seed. Yields are reduced by a lack of calcium more often than by any other nutrient on many peanut soils, particularly deep sands. This is due to the unusual fruiting habit of the peanut, and the inability of the peanut plant to translocate calcium from the root system to the pod. The calcium deficiency symptom that appears most often is the presence of aborted or shriveled fruit. Other deficiency symptoms include a darkened plumule, or "black heart," poorly germinating seed, seedling abnormalities, abundant green foliage late in the season due to poor seed development, and pod rot.

Calcium is generally applied as lime or calcium sulfate (gypsum) when recommended by soil test. Supplemental calcium is always recommended on Virginia types and peanuts which are to be used for seed.

Two other elements often found to be deficient in peanuts are manganese and boron. Manganese deficiency usually occurs when soil is overlimed. Increasing the soil pH reduces the plant's uptake of manganese. Symptoms of manganese deficiency are interveinal chlorosis. A deficiency can be corrected by foliar application of manganese sulfate.

Boron plays an important role in kernel quality and flavor. Boron deficiency may occur in peanuts produced on deep, sandy soils. Deficient kernels are referred to as having hollow hearts. The inner surfaces of the cotyledons are depressed and darkened and are graded as damaged kernels. A general recommendation is to apply boron to the plant as a foliar spray in the first and/or second fungicide spray. Some producers apply boron in the preplant fertilizer or preplant incorporated herbicides.

Zinc toxicity in peanut has become more of a problem in recent years, particularly when following corn in rotation. Zinc toxicity generally occurs on soils low in pH (below 5.4), where high levels of zinc have been applied in corn fertilizer, or where the calcium to zinc ratio is less than 50:1. Toxicity symptoms begin with stem splitting and, if severe enough, will cause eventual death of the plant. The variety Southern Runner has been identified as having a higher tolerance to zinc than other currently grown varieties.

Peanuts belong to the legume family of plants. Roots of peanuts can be infected by *Rhizobium* bacteria. Nodules form on the roots at infection sites. Within these nodules, the bacteria can convert atmospheric nitrogen into a nitrogen form that can be used by plants. This symbiotic relationship provides sufficient nitrogen for peanut production if the roots are properly nodulated. Therefore, direct applications of nitrogen to peanuts are not generally needed. Commercial peanut inoculants are applied at planting and applied in the seed furrow when recommended. *Rhizobium* survive well in most soils and generally inoculation is not recommended unless peanuts have not been planted in a field in the past 4 to 5 years. Some areas of the country have shown some response when inoculants are applied to fields which contain peanuts every 2 to 3 years. Peanuts are in the same cross-inoculation group with several other plant species and are classed in the cowpea group. The southeast has abundant Florida beggarweed and kudzu which makes it difficult to show response to commercial inoculants.

E. Water Requirements

Irrigation can increase peanut yield and quality during dry weather. Peanuts generally require 20–24 in. of water during the growing season for maximum yield. This is either through rainfall or supplemental irrigation plus rainfall. Adequate moisture is important during all stages of the peanut plant's life cycle. The value of adequate moisture varies according to the three growth stages of the plant. The first stage is from germination through early vegetative growth. Early season moisture is important in establishing a good stand and ensuring herbicide effectiveness. The second stage is from 50 to 110 days after planting. This is the critical flowering, pegging, pod-addition, and pod-filling period that helps to determine yield. The third stage is from 110 days and harvest. Late season droughts increase the possibility of aflatoxin formation. Total water needs are greatest during the second stage followed by the third and first stages, respectively.

The amount of water needed by the peanut plant depends on its stage of growth and the soil type. Weather conditions, including temperature, rainfall, wind speed, and relative humidity are also important in determining total water use by the peanut plant. Figure 4 shows the general water use by peanuts from planting to near harvest.

Water availability affects the yield, uniformity, and germination of peanut seed. Water deficits affect vegetative growth the first 7 weeks after germination and result in a decreased plant growth rate. Slight water deficits during this early growth stage do not affect yield much but can reduce vine growth. Early season droughts delay pegging and pod formation, although pod formation resumes when the plant is watered.

During flowering, water stress can delay or inhibit flower formation. After flowering, peg penetration requires adequate moisture. Once active pegging and pod formation have begun the pegging zone should be kept moist. This improves calcium uptake which is essential for proper pod development. Failure of pegs to penetrate and develop pods results from the following: low relative humidity; high soil temperatures; soil compaction; low turgor pressure in dry soil, insect damage to pegs, pods or roots; herbicide damage; and reduced calcium uptake by the developing pods. A lack of water in the pegging zone during pod addition and development results in more aborted seeds, more one-seeded pods, a less mature crop, and a lower calcium content in the seed, which in turn affects germination and seed quality. Water deprivation can reduce germination by 40%, particularly in Virginia types. Irrigation or rainfall during pod addition and pod fill periods improves germination percentages. Too much water can cause problems by promoting excessive vine growth, disease, peg deterioration, and nonuniform maturity late in the season.

F. Pest Problems

Insects, weeds, and diseases are the primary problems during the growing season affecting peanut yield and quality. Integrated pest management (IPM) programs allow growers to treat pests when they reach economic threshold levels and to apply the appropriate pesticides. [See INTEGRATED PEST MANAGEMENT.]

Part of increased peanut productivity hinges on the control of various pests. Modern control of most insects, weeds, and diseases has been achieved with chemical agents. Our reliance on chemicals has caused much uneasiness and concern about agricultural pest control. The possibility of unexpected, detrimental effects of these agricultural chemicals on man and his environment at present and in the future is a principal concern of many observers today. Because of this, increased emphasis is being placed on pest management practices which will enhance crop production with minimal effect on the environment. [See PEST MANAGEMENT, CHEMICAL CONTROL.]

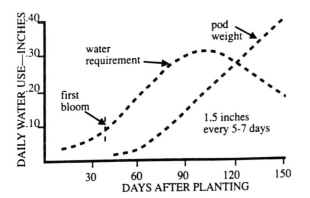

FIGURE 4 Daily water use for peanut plants. [From Stansell, J. R., *et al.* (1976). Peanut responses to soil water variables in the southeast. *Peanut Sci.* **3,** 44–48.]

In addition to chemical use, several other cultural-based pest management practices are used in peanut production such as deep turning of the soil in land preparation. This has effectively reduced some weed and disease problems. Crop rotation has been widely used in peanut production. This practice has allowed the use of herbicides on the rotational crops which reduces weed problems when peanuts are planted. Crop rotation has been an effective measure in reducing disease problems in most fields. For nematode reduction the use of grass crops, such as corn and/or small grains, has been very effective. Cotton also does not support root-knot nematodes. Planting behind grass crops may increase some insect problems such as cutworms. A close examination for cutworms and soil insects should be made as land is being prepared and treatment can be made if excessive numbers are found.

Cultivation has been an important management tool for weed control and can still be used in peanut production if done with precision. Otherwise, soil moved onto plant parts can increase the incidence of several peanut diseases. Protection of beneficial insects by delaying insecticide applications is also an important management tool in an overall insect control program.

Pests that consistently cause problems include leafspots, white mold, Rhizoctonia limb rot, Sclerotinia blight, Cylindrocladium black rot (CBR), tomato spotted wilt virus, foliage-feeding, and soil insects, nematodes, and weeds. Insect stresses on the physiology of the plant result primarily from feeding activities which decrease the total biomass of a plant and trigger hormonal and metabolic changes. Insect pests of peanut are generally classed into defoliating, foliar-sucking, or underground feeders. [See PLANT PATHOLOGY.]

Defoliating insects, such as armyworms, corn earworm, cutworm, green cloverworm, loopers, velvetbean caterpillar, and the rednecked peanutworm, primarily decrease photosynthetic area, although their feeding habits can extend to pegs, pods, and young shoot tissue. Foliar-sucking insects and mites, such as the two-spotted spider mite, leafhoppers, three-cornered alfalfa hopper, aphids, and thrips, damage a plant's vascular system and serve as carriers for several viral diseases. Their feeding habits rob plants of sugars and amino acids produced in the leaves and transported to the actively growing regions of the plant through the phloem.

Soil-borne insects, such as the lesser cornstalk borer, white-fringed beetle, southern corn root-worm, and wireworms, can cause above ground symptoms similar to those of the root-rotting fungi. These insects reduce root area, decreasing the plant's ability to absorb water and nutrients, damage pods, and predispose the roots and pods to attack by bacterial and fungal pathogens.

Numerous pathogens gain entry into plants through associations with insects. Other pathogens enter through mechanical damage to the plant. Many plant physiological processes, including photosynthesis, the movement of water and nutrients, growth and storage of food, metabolic and catabolic rates, and even the plant's hormonal balance, are affected by microbiological pathogens. Pathogens disrupt normal plant functions. Some pathogens are able to move throughout the host in the vascular tissue.

Viruses often produce symptoms described as mosaics, or chlorosis, resulting from damage to the chloroplast. These symptoms tend to be seen on young leaves. Shoot swelling or puckering due to the disruption of the vascular tissue is another sometimes symptom of viral infections. [See PLANT VIROLOGY.]

Like viral infections, fungal and bacterial infections can disrupt translocation of water and nutrients, reduce photosynthetic rates, change metabolic rates, and reduce root, shoot, and fruit growth. The pathogens causing early and late leafspots and rust enter leaf tissue and colonize inside the cells of the leaf surface, and occasionally can be found on stem tissue. Foliar diseases reduce photosynthetic rates of the canopy by defoliation, by decreasing the photosynthetic efficiency of diseased attached leaves, and by decreasing the amount of light reaching other leaves underneath the infected leaves.

Soil-borne diseases generally affect a plant's ability to absorb, translocate, and store nutrients by attacking the root, stem, and pod. While most soil-borne pathogens do not attack the vascular system specifically, vascular tissue will be infected and degraded along with the other cell types. Nematodes reduce the effective root area of a plant, interfere with nutrient transport, and can parasitize newly developing pegs and pods. Foliar tissue generally becomes chlorotic when nematodes infect a plant, probably due in part to reduced nitrogen fixation.

Weeds typically interfere with the peanut's physiological functions indirectly by competing with the crop for water, nutrients, and light. In addition, some weeds, such as giant foxtail, large crabgrass, johnsongrass, quackgrass, and common sunflower, produce allelopathic substances. These toxins enter the soil and slow crop development more than would normally be expected as a result of competition for light, water, and nutrients. [*See* WEED SCIENCE.]

G. Maturity Determination

The best tasting peanut is a mature peanut. Growers must harvest and deliver mature peanuts to the market. Maturity affects flavor, grade, milling quality, and shelf life. The indeterminate fruiting pattern of peanuts makes it difficult to determine when optimum maturity occurs. The fruiting pattern may vary considerable from year to year, mostly because of different weather conditions. Therefore, each field should be checked before digging begins.

Determining when to harvest is one of the most important decisions growers make each year. Peanuts may gain 300 to 500 pounds per acre and 2 to 3% in grade during the 10-day period prior to optimum harvest. Losses greater than 300 to 500 pounds per acre may occur if harvest is delayed past optimum maturity. Immature peanuts have poor flavor. They are more likely to deteriorate in storage and be damaged by insects, and are more susceptible to aflatoxin contamination in storage.

The best time to harvest, of course, is when the crop has the highest percentage of sound, mature kernels and the highest yield. Perfect timing is not easy to determine mainly because the peanut is an indeterminate plant. The plant continues to set new pods and shed old pods until it is harvested or dies.

Determining when to harvest is a decision of risk management. Growers must balance the risk of losing mature pods because of weakened pegs and pod rots with the potential gain from pods still maturing. Runner and Spanish-type peanuts should have 75 to 80% dark inner hulls for ideal maturity. Virginia-type peanuts should have 60–65% dark inner hulls or deep pink seed coats. These percentages give growers their highest profits and provide the industry with flavorful peanuts that have good milling qualities.

A method of determining maturity was implemented in 1984 by USDA and University of Georgia researchers. The "Hull Scrape" method utilizes a maturity profile instead of relying on a percentage method. From the pod-maturity profile, growers are able to predict the date of optimum maturity. They can then schedule the order for harvesting of various fields. This method can also be used to determine from the maturity profile the effects that various production practices and stresses that occurred during the growing season and the effects these had on pod set and maturity.

The Hull Scrape method is based on color changes in the middle layer of the peanut hull as the kernels mature. The color begins with white when the kernels are watery and poorly defined. The color changes to light yellow, dark yellow, orange, brown, and finally to black when the kernels are completely developed. Even at optimum maturity, there will be a range of the above listed colors due to the indeterminate nature of peanut pod maturity.

Heat units or growing degree days are being evaluated as a means of determining maturity, particularly in growing areas like North Carolina and Virginia where cool or freezing temperatures can occur in early fall. One growing degree day (base 56°F) is accumulated when the average daily high and low temperature is 57°F. If the average daily high and low temperature were 76°F, then 20 growing degree days would be accumulated for that day. Research on Virginia-type peanuts has shown that 2500 growing degree days are needed for maturity.

H. Harvesting

Peanut handling starts with digging and ends when the finished peanut product reaches consumers. At the farm level, handling involves digging, shaking, windrowing, combining, curing, and sometimes storing. Growers can maintain or reduce quality during each of these steps, but cannot make up for quality lost during the growing season.

1. Digging: The First Step for Harvesting Peanuts

The modern peanut digger–inverter lifts peanuts from the soil, separates pods from the soil, elevates the vine mass, and inverts and windrows the vines, exposing pods to the air for curing.

Peanut inverters can be classified into two types. One type uses a conventional rattler bar system for moving the peanuts upward from the digger blades to the point where the vines are inverted. The second type uses a tangent chain or chain rod combination to move the peanuts up to the inverter attachment. The first type of machine is the most common and will be discussed in detail.

The peanut plant passes through three stages: digging, shaking or dirt removal, and inversion. Digging is accomplished by cutting the peanut taproot with a horizontal blade just below the pods. This blade has a slight forward pitch to lift the plant onto a shaking conveyer just after the taproot is severed. The shaking conveyer is made up of horizontal bars that ride over small rubber wheels. These wheels cause the bars to vibrate, which aids in removing soil from the plants and pods. As vines exit the shaking conveyer, they engage the inversion wheels and rods that flip and windrow the plants.

2. Combining or Threshing Peanuts

All peanuts produced commercially in Georgia are harvested directly from the windrow with a combine. The peanut combine removes peanut pods from the vines, separates pods from vines and other material, and delivers pods to an overhead basket.

When the windrowed peanuts have reached a moisture content of 18 to 24%, which usually occurs 2 or 3 days after digging, they can be combined with minimum mechanical damage. There is less combine damage when the plant has partially dried than when it is very green or very dry. Most peanut growers prefer this time for combining when all quality and cost factors are considered.

An improperly set combine can result in reduced peanut yield and a product with excessive pod damage, loose shelled kernels (LSKs), and foreign material.

3. How a Combine Works

The combine places the windrow into the feeder auger using a pickup head. The feeder auger delivers the material into the threshing mechanism. The threshing mechanism consists of picking cylinders that operate over concaves. Stripper bars slow the movement of vines.

Most of the peanut pods and small vine material fall through the concaves onto the shaker pan. The material that does not fall through is transferred rearward, either by straw walkers or by some type of agitation device, where further separation takes place. The material that has fallen on the shaker pan is conveyed rearward onto an oscillating chaffer and sieve or other type of separation device. Openings in the chaffer or other mechanism allow the peanuts to fall through but retain the trash. As the material moves rearward, an air blast directed upward through the sieves aids in separating pods from small vine material and other foreign material.

The peanut pods fall onto stemmer saws that remove the stems. Then the air delivery system conveys the pods to a storage basket. Following combining, the harvested peanuts are placed on drying wagons and delivered to buying points.

I. Quality

United States peanuts set the benchmark for premium quality standards in the world. Unlike other coun-tries, all peanuts produced in the United States are inspected by licensed government personnel from USDA/Federal State Inspection Service; certified as safe for consumption; and graded along industry standards. They are then reinspected continuously throughout the manufacturing process by the industry as well as the Food and Drug Administration.

The United States peanut industry maintains a Code of Good Practices for the production and processing of peanut products and actively supports ongoing research at universities, colleges, government, and private facilities. The United States peanut industry established all of these practices and procedures voluntarily to insure its position as the premium quality producer in the world.

Although it is only the third-largest producer of peanuts in the world, the United States is the leading exporter. This is due to America's preeminent quality standards and the United States government's peanut program. Annually, the United States exports an average of $225 million of peanut products: primarily to Europe, Canada, and other nonproducing countries. United States peanuts command a premium price on the world market due to their quality. These quality aspects include good flavor, long shelf-life, and minimal defects from damage, foreign material, aflatoxin, and chemical residues.

Peanuts have been an important food source for American families and people throughout the world. Volumes of information have been published on all aspects of peanuts including history, production, harvesting, marketing, government programs, and nutrition.

Once a producer delivers Farmer Stock peanuts to the sheller or buying point, there is no opportunity to improve quality. "Quality" has different meaning to different areas of the peanut industry. To the producer, high quality means a Total Sound Mature Kernel (TSMK) grade of 75 or higher with no Segregation III (visible *A. flavus* mold) peanuts or with few damaged kernels or loose shelled kernels (LSKs) present. Harvesting at proper maturity, following sound pest control management throughout the season, and developing careful and precise harvest and curing techniques will help provide the quality needed for processors and manufacturers of peanut products to provide to our consumers. To the consumer, quality means a nutritious, flavorful, wholesome product that is safe to eat. The peanut industry has identified five major areas of quality concerns which need to be addressed and continually improved upon. These are aflatoxin, foreign material, chemical residues, flavor, and kernel maturity. Each segment of the industry addresses these problems and it all starts with the grower delivering the best quality possible.

Bibliography

Bader, M. (1992). Peanut Digger and Combine Efficiency. Univ. of Georgia Cooperative Extension Service Bulletin 1087.

Gregory, W. C., Gregory, M. P., Krapovickas, A., Smith, B. W., and Yarbrough, J. A. (1973). Structures and genetic resources of peanuts. *In* "Peanuts: Culture and Uses," p. 47–133. American Peanut Research and Education Association, Stillwater, OK.

Hartzog, D. L., *et al.* (1990). "Peanut Production in Alabama." Alabama Cooperative Extension Service. Circular ANR-207.

Johnson, W. C., *et al.* (1985). "Georgia Peanut Production Guide." Univ. of Georgia Cooperative Extension Service, SB23.

Ketring, D. L., Brown, R. H., Sullivan, G. A., and Johnson, B. B. (1982). Growth physiology. *In* "Peanut Science and Technology," pp. 411–457. Yoakum, TX.

Kvien, C. K. (1994). "Management of Physiological and Environmental Disorders of Peanut in Peanut Health Management." American Phytopathological Society, in press.

Sholar, R. (1993). "Peanut Production Guide for Oklahoma." Oklahoma State Univ. Cooperative Extension Service, Circular E-608.

Sullivan, G. A., *et al.* (1993). "1993 Peanuts." North Carolina Cooperative Extension Service Bulletin AG-331.

Swann, C. W. (1993). "Peanut Production Guide." Tidewater Agric. Experiment Station Information Series No. 311, Virginia Cooperative Extension Service.

Womack, H., *et al.* (1981)." Peanut Pest Management in the Southeast." Univ. of Georgia Cooperative Extension Service, Bulletin 850.

Pest Management, Biological Control

HEIKKI M. T. HOKKANEN, *University of Helsinki, Finland*

Glossary

Biopesticide application Application of a product whose "active ingredient" is a living organism; the application is similar to that of chemical pesticides; synonyms: inundative biological control, repeated release method

Classical biological control Introduction of an exotic natural enemy into a new environment, to permanently control a pest which appears not to have effective local natural enemies; traditionally used against exotic pests to reconstruct the pest–enemy balance of the native habitat, but effective in other contexts as well; synonyms: inoculative biological control, one-time-release method

Parasitoid Insect living and feeding within another insect, finally killing the host; in contrast to a true parasite, which does not directly kill its host

Biological control can be defined as the conscious use of living organisms (biocontrol agents) to restrict the population sizes of unwanted organisms (target pests). Biocontrol agents may be predators, parasites, pathogens, or competitors of the target pests (including animal, weed, and disease pests). Many other valuable pest control tactics—such as the use of phero-mones or host plant resistance—are often included in the concept of biological control. As they do not easily relate to the same basic principles as the main body

of biological control, they are treated here only briefly in connection with the other methods.

I. Principles and Strategies of Biological Control

A. Ecological Principles

Biological control relies on the general capacity of ecosystems to resist great changes in population densities of any of its component species. The complex natural feedback mechanisms involved in keeping population numbers in check are utilized in various ways and to varying degrees by the different approaches to biological control. Typically, one or several factors contributing to natural control of populations are exploited. In a broad sense the natural enemies of pests, which comprise the bulk of the arsenal used in biological control, include parasitic and predatory arthropods, disease-causing organisms such as bacteria, viruses, fungi, nematodes, and protozoans, and competitors such as antagonistic fungi or weakened conspecifics (e.g., hypovirulent disease strains or sterilized individuals).

The different strategies and approaches of biological control rely on different ecological principles. In most cases a very specific, close interaction between a monophagous natural enemy and its host pest species is essential. In some situations the strategy is based on enhancing the general ecosystem buffering against a range of potential pests through generalist predators, parasitoids, and pathogens. Clearly, different ecological principles are important in the various situations.

Generally the typical factors relating to the ecology of the target system, which need to be considered while carrying out biological control, include features of the habitat (e.g., ecological stability, patchiness, system management, presence of off-target, native fauna and flora), of the pest (e.g., population growth

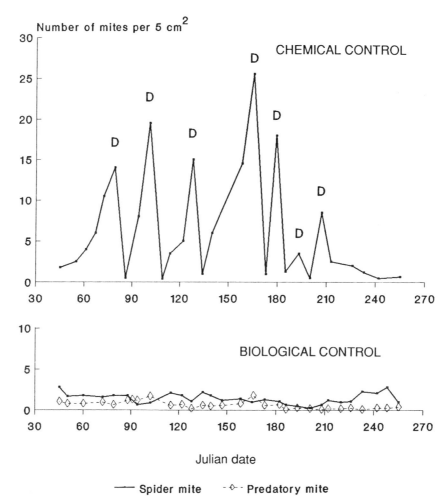

FIGURE 1 Comparison between chemical and biological control of the spider mite *Tetranychus urticae* in a commercial cucumber crop in Finland. D, treatment with the acaricide dicofol; predatory mite, *Phytoseiulus persimilis*. Economic threshold level of spider mite is 20 mites per 5 cm². The predatory mite was introduced in the middle of February, and provided a season-long, effective control of the pest. (Reproduced with permission from Markkula *et al.* (1972). *Ann. Agric. Fenn.* **11**, 74–78.)

and regulation, mobility, resistance to controls, host spectrum), and of the natural enemy (e.g., target host specificity, population dynamics, evolutionary aspects).

One can find in biological control a variance from total dependence on ecological interactions to practical independence—cases where biocontrol agent survival, reproduction, and dispersal are all compensated for and determined by the system operator: man. The choice of method is often related to the types of habitat and pests in question, explained in part by the ecological differences between species and by the system characteristics.

A further basic consideration is that the need for control, or success or failure of the resident natural regulating factors to keep the pest population levels low enough from man's point of view, is an economical assessment by man. This is very variable depending on the type of crop, pest damage, etc., and as such does not easily relate to any general ecological theory or principle.

B. Practical Approaches

According to the way in which beneficial organisms are utilized, four main approaches are possible:

1. New, effective natural enemies are introduced into the ecosystem, typically against a specific target pest. The aim is a long-term establishment of the biocontrol agent and permanent suppression of the pest. This is often called the one-time, or inoculative release method,

or classical biological control. It has mostly been used against exotic pests, which sometimes lack effective natural enemies in their new habitat, but it can equally well be employed against native pests. This method has proven particularly useful in relatively stable forest, orchard, and pasture ecosystems, where a continuous natural enemy–pest interaction can be established.

2. Biocontrol agents are often released repeatedly, in great numbers, very much in the same way as chemical pesticides. A quick remedy of an acute pest problem is the objective of this approach, rather than long-term control. The use of biopesticides is also known as the inundative release method of biological control. It is typically used in places where, and at times when, the natural spread of the biocontrol agents is not possible or sufficient, for example in annual crops and in greenhouses, but also increasingly in forests and orchards.

3. An intermediate possibility between the classical and the biopesticidal methods is that the control agent is reintroduced for each season, after which it remains effective until winter. This seasonal inoculative method is widely used in greenhouses (Fig. 1), but also outdoors for example where an otherwise effective natural enemy cannot survive the winter or other seasonal/environmental conditions.

4. A fourth, potentially very important method of biological control is to make more effective use of the control agents already present in the ecosystem: enhanced natural control. This may be obtained through habitat management, alteration of cropping practices, or by other means, after obtaining a thorough knowledge of the ecological requirements and interactions in the system. This method is also compatible with, and can enhance the effect of, introduced biocontrol agents.

C. Present Overall Significance

The importance of biological control may best be appreciated by considering the fact that only a minute fraction of the potential pest species actually do cause economic losses at a given location and a specific cropping season. Worldwide, for example, only a few hundred species of insects are considered as serious pests, and about 10,000 species out of millions are ever recorded as pests. Natural population controls are therefore extremely important for the protection of crops, plantations, forests, livestock, and health.

It may be estimated that, worldwide, natural or enhanced natural controls currently satisfy at least 95% of all crop protection needs. The remaining pests, diseases, and weeds may occasionally cause, however, 100% crop losses if left unchecked.

It is impossible to quantify the importance of, or the area under, biological control resulting from new organism introductions. Moreover, the results obtained through classical biological control tend to be forgotten because they do not require any further actions. A recent example of the magnitude of importance of classical biological control is the protection of cassava in 34 affected African countries covering most of the continent. In another instance, some 25 million ha of rangeland in Australia was cleared of prickly pear cacti (*Opuntia* spp.) by introduction of two insect species. This control of *Opuntia* has since been repeated in 16 other countries.

The introduction of new organisms has solved probably not more than a percentage of the acute pest control needs. The economic consequences of these introductions have, however, been far greater than this small share would indicate.

The role of using beneficial organisms as biopesticides is easier to estimate. Egg parasitoids (*Trichogramma* and others) are the most widely used arthropods for this purpose. They are used annually on more than 20 million ha in Russia and the other CIS (Commonwealth of Independent States) republics, and on over 2 million ha in China to control caterpillars and other pests on maize, fruits, cotton, sugarbeet, sugarcane, vegetables, etc. (Table I). They are also used on some 350,000 ha in the United States mainly on cotton and sugarcane, and on over 10,000 ha of maize in Europe. In orchards predatory mites and insect parasitoids are used as biopesticides on about 13,000 ha in the United States and 10,000 ha in Europe. Of the European glasshouse area 30% employs biological control, mainly on vegetables. Worldwide the figure is much smaller, about 5%.

Sterile insects are regularly utilized as biopesticides to eradicate the constantly reappearing Mediterranean fruit fly (*Ceratitis capitata*) in California, as well as to control invasions by the screw-worm fly (*Cochliomyia hominivorax*) in the southwestern United States. Recently this technique was successfully applied to eradicate the potentially devastating, first-time invasion of the screw-worm fly in Africa.

The sales of microbial pesticides in the world currently amount to close to 1% of the sales of chemical pesticides. In terms of area treated they still lag far behind *Trichogramma* worldwide, but this is rapidly changing with the increasing use of *Bacillus thuringiensis* (Bt)-based products. At present this bacterium is primarily used against forest pests, but also against pests on vegetables, potato, etc. Bt currently accounts for over 90% of the world use of microbial pesticides. Other bacteria, viruses, and fungi are used as biopesticides against insects, weeds, and plant pathogens to

TABLE I

Examples of the Extent of Biological Control, Worldwide

Beneficial(s)	Pest(s)	Crop	Area (ha)	Region
Trichogramma spp.	*Helicoverpa armigera,*	Cotton, maize,	17,000,000	CIS
	Ostrinia nubilalis,	vegetables, orchards,	2,000,000	PRC
	Mamestra brassicae,	fruits, cereals,	335,000	USA
	Tortricidae, many	sugarbeet, sugarcane,	10,000	Europe
	other Lepidoptera	pulse crops		
Bracon hebetor	*Agrotis segetum, H. armigera*	Cotton	1,800,000	CIS
Phytomyza orobranchia	*Orobranche cumana*	Vegetables, sunflowers, tobacco	170,000	CIS
Cryptolaemus,	Scales and mealybugs	Citrus, vegetables	92,000	CIS
Chrysoperla,			13,000	USA
Aphytis,				
Pseudaphycus				
Predators, parasitoids	Mites, aphids	Orchards	10,000	Europe
Phytoseiulus persimilis	*Tetranychus urticae*	Greenhouse vegetables	5,350	CIS
			5,100	others
Encarsia formosa	*Trialeurodes vaporariorum*	Greenhouse vegetables	1,050	CIS
			2,350	others
Predators, parasitoids	Aphids, thrips, leafminers	Greenhouse vegetables	760	CIS
			600	others
Predators, parasitoids, pathogens	Many	Many crops: enhanced natural control	Large	Worldwide
Predators, parasitoids, pathogens	Many	Many crops: classical biocontrol	Vast	Worldwide
Many	Many	Most crops: natural control	Immense	Worldwide

a small extent, but their use is estimated to rapidly increase in the near future. Biopesticides are expected to increase their share of the total pesticide sales up to 10–20% by the year 2000.

D. Place in Pest Management

Biological control is a cornerstone of any sustainable pest management strategy. It is an essential component of integrated pest management (IPM), or overall, integrated production (IP) of crops. It is also one of the very few possible and practical alternatives to the use of chemical pesticides. As a sole method of pest control in a specific target crop, biological control is seldom sufficient. Therefore, the requirements of biological control must be integrated with the needs and uses of other control tactics in such a way that a synergistic outcome is obtained. For example, chemical pesticides, when used, must be harmless to the key biological control agents in the target ecosystem. [See INTEGRATED PEST MANAGEMENT; PEST MANAGEMENT, CHEMICAL CONTROL.]

II. Introduction of New Organisms

A. Present and Potential Uses

Up to now there have been more than 5000 introductions of beneficial species for classical biological control, worldwide. About 4300 of them are insect parasitoids and predators for the control of insect pests, and some 700 are biocontrol agents (mainly insects) for the control of weeds. A small number of insect pathogens and other microbes have also been introduced with the aim of a permanent establishment and continuous control of the target pest.

Altogether 416 species of insect pests in a total of 1275 different control projects, worldwide, have been the target of classical biological control. Of the target pest species, 75 have been brought under complete control, 74 under substantial, and 15 under partial control in at least one country. Of the individual projects, 156 have been rated as complete successes, 164 as substantial, and 64 as partial successes. Examples of the outstanding successes include (cf. Table II):

TABLE II

Economic Assessments of Some Classical Biological Control Programs

Pest	Region	Annual savings[a]	Costs of control program
Rhodes grass scale (*Antonina graminis*)	Texas (1974–1978)[b]	194	0.2
Alfalfa weevil (*Therioaphis trifolii*)	USA (1954–1986)	77	1.0
Cassava mealybug (*Phenacoccus manihoti*)	Africa (1984–2003)	96	14.8
Skeleton weed (*Chondrilla juncea*)	Australia (1975–2000)	13.9	3.1
Water fern (*Salvinia molesta*)	Sri Lanka (1987–2112)	0.50	0.22
Potato tuber moth (*Phthorimaea operculella*)	Zambia (1974–1980)	0.09	0.04
Winter moth (*Operophtera brumata*)	Canada (1954–1964)	0.20	0.16
Two spotted spider mite (*Tetranychus urticae*)	Australia (1975–2000)	0.9	0.9
White wax scale (*Ceroplastes destructor*)	Australia (1975–2000)	0.09	1.4
Wood wasp (*Sirex noctilio*)	Australia (1975–2000)	0.8	8.2

[a] In U.S. $ millions.

[b] The years in parentheses are those of the period used in calculating the discounted benefits shown in column 3 as "Annual savings."

Cottony cushion scale, *Icerya purchasi:* Threatened to wipe out the California citrus industry at the end of the 19th century. In 1888 the coccinellid beetle *Rodolia cardinalis* was introduced from Australia and New Zealand, and within months the pest virtually disappeared. This complete success was repeated in some 40 countries around the world. *R. cardinalis* is the single most successful biological control agent up to date.

Cassava mealybug, *Phenacoccus manihoti:* Cassava, a staple food crop for over 200 million people in Africa, was devastated by the accidental introduction of the mealybug in the early 1970s. Through the largest biocontrol operation ever, the parasitoid *Epidinocarsis lopezi* has been distributed over very large areas throughout the continent, with outstanding success.

Rhodes grass mealybug, *Antonina graminis:* Native of Asia, this insect is an important pest of many forage grasses. It invaded Texas in the 1940s, infesting an area of 16 million ha. A parasitoid (*Neodusmetia sangwani*) was imported from India in 1959. It was spread aerially from airplanes all over the infested area, and quickly controlled the pest completely.

For the classical biological control of weeds a total of 267 individual projects are currently listed, concerning 125 species of weeds as targets. The success rates are clearly higher than for the control of insects, possibly due to the much more stringent procedures and generally better ecological knowledge of the system. Some of the spectacular successes are:

Prickly pear cacti, *Opuntia spp.:* These cacti were brought to Australia around 1840, and by 1925 had taken over 25 million ha of rangeland. A South American moth, *Cactoblastis cactorum*, and a scale insect, *Dactylopius opuntiae*, from California, cleared the whole area within 10 years of introduction, reducing the density of cacti from about 1250 to around 27 plants/ha.

Water hyacinth, *Eichhornia crassipes:* This floating plant from the Neotropics, claimed to be the world's worst aquatic weed, invades and blocks waterways around the tropical world. In Louisiana, some 445,000 ha of invaded waterways in 1974 were reduced to 122,000 ha by 1980 through the introduction of the weevil *Neochetina eichhorniae* from Argentina. This and other natural enemies are used also elsewhere with success to control the weed.

Salvinia, *Salvinia molesta:* Salvinia, a South American floating fern, is a serious aquatic weed in Australia, Southeast Asia, the Pacific region, and in Africa. The beetle *Cyrtobagous salviniae* from Brazil has quickly controlled the weed in most areas; the control project in Papua New Guinea was awarded the UNESCO Science Prize in 1985 for its success.

Tansy ragwort, *Senecio jacobaeae:* This European weed occupies pasture, range, and forestlands in western United States, Canada, Australia, and New Zealand. A flea beetle (*Longitarsus jacobaeae*) and a moth (*Tyria jacobaeae*) have given 99% control of the weed in California, excellent control in Oregon, and good or some control elsewhere.

Skeleton weed, *Chondrilla juncea:* The plant was a major weed of wheat cropping in Australia, causing annual production losses of AUD 25 million, and control costs of AUD 5 million. In the 1970s several strains of a host specific rust fungus (*Puccinia chondrillina*) were introduced from eastern Europe, resulting in complete control of the dominant form of the weed. [*See* WEED SCIENCE.]

B. Components of Success

Compared to chemical control, the success rates of classical biological control are outstanding. While only about one out of 15,000 tested chemicals ends up as a chemical pesticide meeting the requirements of efficacy and safety, approximately 12% of the individual classical biological control projects against insects have resulted in complete control of the target pest. Further, 18% of projects have resulted in substantial or partial success. The figures for individual pest species are even better: 18% of the target insect species have been completely controlled in at least one country, and a total of 40% of target species have been controlled to some degree. Classical biological control of weeds has been even more successful than insect control: the success rates are approximately twice as high. Overall, considering the permanency of the results of classical biological control, these figures are very impressive.

Research aiming at still improving the success rates of new organism introduction for biological control indicates that significant progress is possible. A better understanding of the critical components, the functioning of the target ecosystem and its processes, and of the most important organisms in question, will be necessary for such an improvement.

Of the key components possibly affecting the success, most are subject to manipulation by man, while some are not. The manageable factors likely to influence the success of biocontrol introductions include (i) the choice of the natural enemy species (e.g., its reproductive and host utilization characteristics, evolutionary relationship to pest, environmental adaptability), (ii) procedures of natural enemy use (e.g., release strategy and techniques, genetic diversity of control agents), and (iii) infrastructural support to the project (e.g., intellectual, material, administrational).

Important factors, which may be manageable to some degree include several habitat characteristics such as its physical structure (patchiness), diversity, stability, the presence of antagonists, competitors, and mutualists, as well as human interference. Nonmanageable features, which nevertheless influence the biocontrol success, include (i) the target pest characteristics (e.g., taxon, reproductive potential, and feeding behavior), and (ii) the climate of the target ecosystem. On the other hand, the success of classical biological control has proven to be quite insensitive to general factors such as geographical location, or whether the pest is of exotic or native origin.

C. Economics of Control

It is clear that many projects in classical biological control yield negligible economic results. On the other hand, those projects which are successful can easily cover the costs of a large research agency over a decade or more.

Earlier data on the economics of classical biological control programs in California suggested a rough overall return to investment ratio of 30:1, which compares favorably to that of about 5:1 for chemical pesticides. Only recently have some rigorous economic

analyses been made, notably in Australia. The data in Table II show that **annual** savings can exceed a **one-time** investment by almost 1000-fold; for the projects included the mean is 13:1. On a longer-term basis, therefore, it seems likely that early economic evaluation has been conservative rather than optimistic, confirming that investment into biological control is highly profitable.

III. Biopesticides

A. Currently Available Organisms

1. Insect Control

For insect and mite control, several species of parasitic and predatory arthropods are being used fundamentally in the same way as chemical pesticides. The extensive use of *Trichogramma* spp., egg parasites of many important lepidopterous pests (e.g., European corn borer *Ostrinia nubilalis* in maize, leafrollers in apple orchards, and the cabbage moth *Mamestra brassicae* on vegetables), was already mentioned. Several other species of parasitoids are available and are used as biological pesticides against many other pests, such as scale insects, mealybugs, and house flies. A few species of lady beetles (Coccinellidae) and lacewings (*Chrysoperla* spp.), which are general predators of small insects, are also on the market. Very important to the greenhouse industry, and outdoors for example in apple orchards, are predatory mites of the genera *Phytoseiulus*, *Amblyseius*, and *Typhlodromus*. These are utilized also for seasonal inoculation and natural enemy conservation programs in these crops.

Microbial pesticides have rapidly penetrated the pesticide market, and their use is growing faster than any other type of pest control. For insect control, products based on bacteria, viruses, fungi, nematodes, and a protozoan are at present on the market (Table III). By far the most extensive is the use of several dozen different products based on the bacterium *B. thuringiensis* (Bt). In the United States, over 450 uses and formulations are registered. Various strains of the bacterium produce some 40–50 different endo- and exotoxins, with activity against insects. The most common types of toxins kill lepidopterous larvae, and they have been used since their discovery in the 1950s for the control of at least 75 pest species. A strain of Bt which kills mosquito larvae was discovered in 1977 and has been extensively used in malaria control programs of the World Health Organization. In 1983 another pathotype was isolated, with activity

TABLE III

Examples of Commercialized Microbial Pesticides

Control organism	Target pest(s)	Remarks
Bioinsecticides		
Bacillus thuringiensis (several strains)	Many Lepidoptera	Annual production about 5000 ton; available since 1950s; dozens of products; main users Canada, United States, CIS, PRC, Europe
B. thuringiensis tenebrionis	*Leptinotarsa decemlineata*	Available since 1990; used on 30,000 ha in CIS (1992); about 10,000 ha elsewhere
B. thuringiensis israelensis	Mosquitoes, blackflies	Available since 1980s; on >10,000 ha/a in Europe; on large areas worldwide
B. popilliae	*Popillia japonica*	A few 1000 ha/a in North America; since 1940s
Insect viruses	About 20 spp. Lepidoptera, Hymenoptera	Nine products for six pest spp. available in Europe; in United States six products for six spp.; in CIS many products for about 15 target species
NPV	*Helicoverpa* spp.	In the United States since 1975 on cotton, soybean, maize
NPV	*Autographa californica*	In the United States since 1981 on vegetables
GV	*Cydia pomonella*	In the United States since 1981 on fruit trees; in Europe since 1987/89 (CH/FRG)
GV	*Agrotis segetum*	In Denmark since 1989
NPV	*Mamestra brassicae*	France
Beauveria bassiana	Many pests, e.g., *L. decemlineata Ostrinia nubilalis*	Registered in the United States in 1962; most widely used in CIS; also in France
Metarhizium anisopliae	Spittle bugs, cockroaches, coffee leafminer	In Brazil on sugar-cane about 50,000 ha/a treated; United States
Aschersonia aleyrodis	Whiteflies	CIS
Hirsutella thompsonii	Mites	United States on citrus (withdrawn from the market)
Verticillium lecanii	Aphids, whiteflies	Europe, North America, in the greenhouse
Paecilomyces lilacinus	Nematodes	Philippines; on potato, etc.
Steinernema carpocapsae, S. feltiae, Heterorhabditis bacteriophora, H. megidis	Many pests in high value crops; sciarid flies, *Otiorrhynchus* spp.	Ornamentals, home gardens, tree nurseries; in PRC hundreds of ha/a orchards treated against *Carposina nipponensis*, >100,000 shade trees in injected/a
Biofungicides		
Trichoderma spp.	*Chondrostereum purpureum*	Reg. in France (1976), U.K., Belgium, Switzerland, Sweden, and Chile (1987) for use on fruit trees
T. lignorum	*Rhizoctonia, Pythium*, etc.	In CIS on about 4000 ha/a on greenhouse crops
Pseudomonas putida	*Fusarium*	In the United States on cotton; reg. in 1988
Streptomyces griseoviridis	*Fusarium, Alternaria, Pythium*	Reg. in Finland (1991), United States (1993), and eight other countries; on ornamentals and vegetables
Agrobacterium radiobacter	*Agrobacterium tumefaciens*	Reg. in Australia (1989) on ornamentals and fruit trees
Gliocladium virens	Damping-off diseases	United States, reg. in 1992, on vegetable and ornamental seedlings
Bioherbicides		
Colletotrichum gloesporioides	*Aeschnomene virgica*	United States, available since 1973; in rice and soybean
Phytophthora palmivora	*Morrenia odorata*	United States, for use in citrus; currently unavailable

against certain species of beetles. Products based on this new isolate are on the market already for the control of the Colorado potato beetle. Product development with Bt has recently produced a conjugant strain, which combines two different kinds of toxins to provide protection against the potato beetle and caterpillars at the same time. Already before Bt another bacterium, *Bacillus popilliae,* has been sold for

the control of the Japanese beetle *Popillia japonica.* This bacterium has a narrow host range, and unlike Bt, it cannot be grown on artificial medium.

Insect viruses have successfully been tested under practical conditions for the control of about 30 different pest species, but only a few have been developed to marketable products. The most important is the nucleopolyhedrosis virus of the pine sawfly *Neodiprion sertifer,* on sale and widely used at least in the United States, U.K., and Finland. A granulosis virus of the codling moth *Cydia pomonella,* a serious pest on apple, is another virus product on the market in North America, as well as in Europe. In Eastern European countries and in the CIS, over a dozen different viral products have been on the market for controlling insect pests, particularly in orchards. Commercial interest in developing viruses as insecticides has lagged behind that in Bt because viruses kill the target insects much more slowly, typically have a narrow host spectrum, and are too costly to produce.

Of the over 750 species of fungi known to be pathogenic to insects, only five have been commercialized, and none is widely used yet. *Beauveria bassiana* and *Metarhizium anisopliae* are the best known of these fungi. They can kill a wide range of insects and are used for example to control the Colorado potato beetle (CIS), spittle bugs on sugarcane (Brazil), and European corn borer (France). Other commercialized fungi are *Aschersonia aleyrodis* for whitefly control (CIS), *Hirsutella thompsonii* for mite control on citrus (United States), and *Verticillium lecanii* for aphid and whitefly control (Europe).

Insect killing nematodes of the genera *Steinernema* and *Heterorhabditis* are produced in many countries. Nematodes have been commercially available for only about a decade, and their use is rapidly growing. At present nematodes are sold at least in the United States, Australia, Europe, and China to control many different pests primarily on high value crops such as greenhouse crops, strawberries, field vegetables, and tree nurseries. New, more cost-effective production technologies may soon make it possible to use them also in other crops.

An intracellular protozoan parasite of grasshoppers and locusts, *Nosema locustae,* is available in the United States for the control of these pests on rangeland.

2. Plant Disease Control

Biopesticides to control plant diseases and weeds are under intensive development. A few products are on the market already, and many more are expected to come soon. At least five different biofungicides are currently available commercially. Fungi of the genus *Trichoderma* inhibit the growth of several important plant pathogens, such as *Pythium, Rhizoctonia,* and *Sclerotium.* Products based on *Trichoderma* are available in several countries; in CIS they are used on about 4000 ha annually. For the biological control of *Fusarium* diseases on cotton a product based on the bacterium *Pseudomonas putida* was commercialized in

the United States in 1988. Another biofungicide entered the market in the United States in 1992, based on the fungus *Gliocladium virens,* for the control of damping-off diseases of vegetable and ornamental seedlings in the glasshouse. In Finland, an actinomycete *Streptomyces griseoviridis* was discovered, developed, and registered (in 1991) for the control of *Fusarium, Alternaria, Rhizoctonia,* and *Pythium* on ornamentals; it has also proven effective on cole crops, cucumber, wheat, and oil palm. Worldwide registration of the product is in progress. The fifth product currently available is for the control of crown gall (*Agrobacterium tumefaciens*) on ornamental plants in Australia; the biocontrol organism is *Agrobacterium radiobacter.* [*See* FUNGICIDES; PLANT PATHOLOGY.]

3. Weed Control

Not many bioherbicides are currently available, although the first one has been on the market for over 20 years: the fungus *Colletotrichum gloesporioides* for the control of Northern joint-vetch (*Aeschynomene virgica*) on rice and soybean in the United States. Another product based on *Phytophthora palmivora,* for the control of milkweed vine (*Morrenia odorata*) in citrus in the United States, proved to be so effective that only one treatment was necessary over many years; therefore, the demand for the product decreased and finally it disappeared from the market.

B. Possibilities and Constraints

Steady progress in research concerning the discovery, production technology, ecological requirements and interactions, and other important aspects of biological control agents promises an accelerating rate of practical uses for most of the different approaches to biological control. For example, the rearing of *Trichogramma* on artificial insect eggs, already practiced in China, will significantly reduce the production costs of this parasitoid. Similar progress is expected for example in the production of insect killing nematodes.

New, effective biocontrol organisms are discovered constantly. New strains of *B. thuringiensis* appear to produce many narrow-range insecticidal toxins that only await their discovery. The most recent example is the isolation of Bt, which effectively kills scarabeid beetle larvae (grass grubs). Only a little earlier another strain was found to carry nematicidal properties.

Insect viruses and fungi are grossly underemployed as biocontrol agents. Many of them are currently difficult or impossible to rear on artificial media, which effectively restricts their practical use. They are also easily inactivated when exposed to sunlight; in addition, fungi require a relatively humid microclimate. Production and formulation technology, however, can be improved (e.g., the use of insect cell cultures, and oil-based formulations), and for example some new strains of fungi appear quite tolerant to environmental extremes. Thus, promising mycoinsecticides based on recently discovered strains of *Metarhizium flavoviride* and other fungi are now intensively developed for the control of the desert locust—an unlikely target for an entomopathogenic fungus *a priori*.

The biological control of plant pathogens has made strong progress during the last decade, as indicated by the several new products being introduced to the market. This trend is likely to continue as product registration procedures for microbial pesticides in general become a routine matter. Lack of understanding of the ecological interactions in soil, or on the phylloplane, as well as constraints posed by the physical environment, appear as the factors most delaying the discovery and practical development of these methods.

Concerns about the safety to target and nontarget crops and other plants have been the main reason for the slow progress in weed control using bioherbicides. Plenty of opportunities for expanding this field exist, however, and many new products are expected to reach the market within the next 10 years—up to a total of 30 products by the year 2000, according to one estimate.

The major constraints limiting the development and/or the use of biopesticides include (i) the speed of effect (often too slow, or "inconspicuous"), (ii) inconsistent performance, (iii) specificity (often too narrow), (iv) production costs, and (v) heavy and expensive registration process. As most of these problems can be solved with innovative research and breakthroughs in technology, it is certain that biopesticides will continue to increase their share very rapidly as a means of crop protection.

IV. Genetic Engineering for Biological Control

A. Control Organisms

Genetic improvement of biological agents is relatively new. Through traditional selection pesticide-resistant strains of predatory mites and a parasitoid have been obtained and are used in integrated control programs in orchards in the northwest United States.

In order to improve a natural enemy, the factors limiting the efficacy must be known. Then the genes carrying the desired trait have to be available. Extended genetic variability can be provided by different species, from which the desired genes must be identified, cloned, and inserted into the genome of the natural enemy. Genetic engineering techniques make improvement much more efficient compared to traditional artificial selection procedures because once a useful gene is cloned, it can be used to transform many beneficial species.

Only very few biocontrol agents have been genetically engineered so far, and almost exclusively for research purposes. The number of experiments, and of the species involved, is increasing exponentially. Field trials have involved over a dozen species of microbes, mostly baculoviruses, bacteria, and fungi. Overall, hundreds of field tests involving genetically engineered organisms have been done all over the world. About 80% of them have involved modified plants, 15% bacteria, 5% viruses, and 1% have been done with modified fungi. This breakdown is representative for biological control organisms as well.

Currently only one commercially available, living, genetically engineered biocontrol agent exists: a modified form of *Agrobacterium radiobacter* (K84), which kills the crown gall disease *A. tumefaciens* on fruit trees and roses. The product was developed in Australia, and came to the Australian market in 1989. Numerous earlier attempts to control crown gall by chemical, physical, or management procedures had limited success. Biological control with unmodified K84 is hampered because *A. tumefaciens* can become resistant to the antibiotic produced by K84. This is caused by a small piece of DNA in the K84 plasmid transferring to the pathogen. The engineering involved removing the piece of DNA in K84 to create a new bacterium which controls the crown gall without the possibility of resistance developing, at least through the same mechanism as before.

Of the insect control agents *B. thuringiensis* has been the favorite target for genetic engineering during the past years. The genes for all the major proteins that account for the insecticidal properties of Bt have been cloned and sequenced. Nucleotide sequences are now known for more than 20 Bt genes that encode proteins active against lepidopterans, 8 genes encoding proteins against dipterans, and 2 genes encoding proteins active against coleopterans.

The ease with which Bt can be mass produced, and its toxin-encoding plasmids and genes manipulated, has also greatly contributed to the rapid advances made with this bacterium. These have led to the development of new types of microbial insecticides and a series of transgenic crop plants. To increase the environmental stability and effectiveness of the various Bt toxins in the field, genes encoding proteins active against beetles and caterpillars have been cloned into the rhizobacterium *Pseudomonas fluorescens*. After fermentation the bacteria are killed and the cell walls hardened chemically. The endotoxins are thereby microcapsulated, resulting in insecticides with greatly enhanced residual activity. The product obtained a full registration in the United States in 1991.

Genetic engineering may greatly enhance the properties of insect viruses. For example, using the recombinant DNA techniques the *Autographa californica* virus (AcMNPV) has been engineered to kill insects more quickly by expressing either enzymes or toxins soon after host invasion. Of particular interest is the possibility of making viruses produce insect neurohormones, which can cause rapid physiological disruptions in extremely narrowly defined target hosts. This strategy is in its early stages of development, but there is little doubt that in the near future we will have viruses with extended or specifically designed host ranges, capable of killing insects within 24–28 hr. These recombinant viruses should have an advantage for use against hosts which are not easily controlled by Bt.

The first successful experiments aiming at the genetic improvement of arthropod natural enemies of insects utilizing the rDNA techniques, have already been reported. Classical genetic improvements have been targeted for improved climatic tolerance, improved host finding ability, changes in host preference, improved synchronization with host, insecticide resistance, nondiapause, and induction of thelytokous (all female) reproduction. The rDNA techniques offer the possibility that genetic manipulation becomes more efficient and more creative: beneficial genes isolated from one species could be introduced to many if efficient transformation systems can be found.

To enhance bioherbicide performance of plant pathogens, many aspects of the pathogen such as increased virulence, improved toxin production, altered host range, resistance to crop protection chemicals, altered survival or persistence in soil, broader environmental tolerance, increased propagule production in fermentation systems, and enhanced tolerance to formulation processes can be the targets for genetic improvement. No engineered bioherbicides are expected to become commercially available in the near future.

Beside the control of crown gall, genetic engineering may eventually help in controlling other important plant diseases. Antagonists such as *Trichoderma* spp. are being studied for their possible genetic manipulation, or as a source of valuable genes for disease resistance to be engineered directly into plants. Major successes include the incorporation of virus resistance into crop plants such as potato and cucumber.

B. Crop Plants

The first published reports of successful engineering of crop plants to produce insecticidal or antifeedant proteins are from 1987. The crop plants were tobacco and tomato, producing the δ-endotoxin of *B. thuringiensis* to make them resistant against caterpillars. By 1992 transgenic crop plants expressing the Bt toxin had already been produced for about 50 different species, including potato, cabbage, sugarbeet, rice, soybeans, maize, rapeseed, sunflower, clover, apple, kiwi, walnut, and poplar. About a dozen crop species expressing at least 11 different genes encoding pesticidal properties have been field tested in the United States. The first commercially available transgenic cultivars are expected by 1995. North America, Western Europe, Australia, New Zealand, and Japan have all been very active in developing these techniques.

Instead of inserting the protective insecticidal genes directly into the crop plant genome, they can be engineered into associated organisms. Two bacteria have been successfully tested for this purpose. *Pseudomonas fluorescens,* which colonizes the root systems of crops, has been engineered to express Bt toxin to provide continuous protection against such pests as the corn rootworm. Another bacterium, *Clavibacter xyli* spp. *cynodontis,* colonizes the vascular system of many grasses, including maize. A transgenic strain of the bacterium expressing the Bt insecticidal toxin has been successfully tested on maize for protection against the European corn borer.

The new rDNA techniques make the conventional breeding for resistance much more efficient than before and make it possible to combine resistance factors from completely unrelated organisms (e.g., from the neem-tree or from insects or viruses). First efforts in this direction include the incorporation of the cowpea trypsin inhibitor (CpTI) gene into elite, high-yielding sweet potato varieties for insect resistance. [*See* PLANT BIOTECHNOLOGY: FOOD SAFETY AND ENVIRONMENTAL ISSUES; PLANT GENETIC ENHANCEMENT.]

V. Enhanced Natural Control

When the naturally occurring biological control is inadequate, the first action should be to investigate why. Often the efficiency of the indigenous natural enemies is hampered by factors which are relatively easy to change. For example, the timing of pesticide applications might be modified to coincide with maximum kill of the pest and minimum kill of the most important natural enemies. Similarly, the active ingredients, rates, formulations, and location of pesticide application could be altered to maximally conserve the natural enemies. Other means of conservation include the maintenance of refuges, crop rotation planning, management of crop residues, avoidance of mechanical destruction of natural enemies through minimum tillage, etc., always depending on the ecology of each particular situation. [See PEST MANAGEMENT, CULTURAL CONTROL.]

Management may be particularly important in ecosystems such as in annual crops, where the natural feedback mechanisms have difficulties in operating effectively. The immense diversity of resident antagonists is easily overlooked, and their potential underestimated. For example, over 600 species of predaceous arthropods are known to be present in Arkansas cotton fields, more than 750 predator species in alfalfa in New York, and some 100 species in Florida soybean fields. Similar, if not greater, diversity is known for insect parasitoids.

Stable equilibrium population densities for pest–enemy interactions in annual crops may in most cases be unattainable. However, in virtually all annual cropping systems large numbers of predatory beetles, true bugs, and spiders can be found, and in the tropics, also ants. Generalist predators are not coupled with the pest species in the same way as the specialists are. Therefore, in annual crops enhancing predators should aim at increasing the predator buffer in the crop. Ideally, a stable, high-density guild of predators would be formed, capable of rapid aggregation to high-density "hot spots" of colonizing pests and capable of switching to pest species from alternative food.

Enhancement and skillful exploitation of the indigenous biological control offer possibly the greatest potential for reducing significantly the use of chemical pesticides in agriculture. Such manipulation may be used to improve the control of all pests, and it may be utilized in synchrony with other methods of biological control, such as classical introductions or inoculative or inundative releases. Examples of the use of such methods include the conservation of epigeial predators for the control of cereal aphids, parasitoids of the rape blossom beetle *Meligethes aeneus*, enhancement of *Entomophthora muscae* for the control of the onion fly *Delia antiqua*, *Zoophthora* sp. mycosis of the alfalfa weevil *Hypera postica*, and the enhancement of *Trichoderma* spp. and other antagonists of plant pathogens. Fundamental to all these activities is a thorough understanding of the functioning of the particular agroecosystem to be manipulated. Learning how to conserve and maximally utilize indigenous natural enemies is the most effective way to increase the use of biological control in agriculture.

VI. Benefits and Risks

The various forms of biological control offer significant benefits with respect to questions of pest control, the environment, micro- and macroeconomics, and the society. Some risks, however, can also be identified.

Concerning pest control, biological control offers help in situations where other forms of control have failed, or cannot be used. Unfortunately, this—rather than the other positive features—has been a major reason to develop biological control in the first place: sheer despair with pests that have become resistant to all known pesticides (e.g., pests on cotton, in the greenhouse, or the sweet potato whitefly *Bemisia tabaci* on vegetables and ornamentals in the southwestern United States).

Biological control can be used to control pests which are out of reach with chemicals due to either their location or the timing of control (e.g., many root feeders, or pests that attack at times when chemicals cannot be used for residue safety or other reasons). In addition, most forms of biological control have at least a season-long control effect, while some solve the pest problem forever.

Another major benefit of biological control is its virtually absolute safety to the user, in contrast to the approximately 500,000 human pesticide poisonings annually.

As a method of pest control, however, biological control has been grossly underemployed, and therefore its practical applications as yet are very limited. Therefore, even when a farmer would want to use biological control, it is likely that for his particular case there is no method available at the moment. Also, biological control currently is often clearly more expensive to the farmer than chemical. In terms of reliability biological control is generally viewed as more prone to poor control effect than chemical pesticides. This may be due to the usually more dramatic visual killing effect of chemicals, and to the fact that biological control normally is a preventive rather than corrective method of control. Therefore, when an acute, visible pest problem is there, it usually is too late to apply biocontrol measures to save that crop. Biological control methods also often require more skills, training, and labor than simple pesticide sprays.

A further risk to be considered is the risk of resistance development to the applied control measures. Evolution of resistance in the target pest to such a degree as to diminish the control effect is exceedingly rare in biological control using the classical approach. Pests do have a great capacity to evolve resistance to any single controlling factor, and this should not be overlooked. The first cases of significant resistance in the field to Bt bioinsecticides are already a reality at several locations and should warn scientists and practitioners against wasting valuable biocontrol organisms, or their genes, in an ecologically naive, single-approach control scheme. The production of transgenic, "insect resistant" crops requires particular attention in this respect.

Any kind of introduction of a species with the capacity of permanent establishment into a new ecosystem, be it for classical biological control, use of a biopesticide where the control agent is of exotic origin, or a genetically modified organism even if of local origin, poses an unpredictable risk to the target environment. The organism may not behave as expected and can become a pest of crop plants or a threat to nontarget organisms. There are known cases (unrelated to biological control) of a very small genetic change turning a previously harmless commensal into a pathogenic or virulent pest that demonstrate this possibility.

So far the modern methods of biological control, including those of "classical" biological control, have a superb record of environmental safety. Including the early, carefree introductions up to 100 years ago, extremely few problems from using these methods have arisen, and no serious cases can be pointed out. One, or at most a few, disputed extinctions of the target or nontarget species as a result of early classical introductions into tropical islands may have taken place. It is difficult, however, to distinguish these from the general pattern and dynamics of species extinctions and recruitments in island ecosystems. None

of the thousands of introduced biocontrol organisms have so far become a major pest on crop plants.

There are legitimate concerns about the safety of all human activities, including pest control and the use of biological control, to nontarget species. Particular attention has been given to rare and endangered, native species. Any current form of biological control appears as very safe to these nontarget organisms, compared to alternative control methods, and in particular compared to other kinds of threats such as urban development, forestry practices, etc.

Enormous economic benefits at the macrolevel can be gained through biological control (cf. Table II). Besides the direct value of the protected crops, the improved quality of the environment, human health, and other benefits should be taken into account. At the microlevel, however, the sometimes unreliable performance of biological control may force the farmer to use chemical control for reasons of income security. Currently farmers must use whatever means there are to maximize their economic return and its security. Biological control is under these circumstances used much less than it could be if income security were ensured externally as in cases of, for example, severe frost, drought, or hurricane damage in many countries.

It must be stressed that much of the responsibility for researching and developing methods of biological control rests in the public sector, because except for bioinsecticides and transgenic crops, these methods are unattrative to private companies. The benefits of biological control also often extend far beyond a single farmer's field, making it difficult to collect payments for the treatments. Alternatively, extraordinary cooperation from the farmers' side is necessary. Classical biological control also suits very well the crop protection needs of many of the developing countries. In this case the research and development is often funded by international donor agencies. Two excellent recent examples of such work include the Africa-wide biocontrol program on cassava pests and the development of biopesticides to control the desert locust, also in Africa.

This "division of labor" between the public and private sector also has implications on policy issues regarding what kind of biological control is favored. The tools of regulatory policy, international cooperation, and public and private research funding can be used to determine how pests are to be controlled in the future. The public sector must ensure sustainability as the prime goal of pest control, including biological control, as well as of agriculture in general.

Bibliography

Baker, R., and Dunn, P. H. (eds.) (1990). "New Directions in Biological Control: Alternatives for Suppressing Agricultural Pests and Diseases." A. R. Liss, New York.

DeBach, P., and Rosen, D. (1991). "Biological Control by Natural Enemies," 2nd ed. Cambridge Univ. Press, Cambridge.

Hokkanen, H. M. T., and Lynch, J. M. (eds.) (in press). "Biological Control: Benefits and Risks." Cambridge Univ. Press, Cambridge.

Hokkanen, H. M. T. (1985). Success in classical biological control. *CRC Crit. Rev. Plant Sci.* **3,** 35–72.

Hoy, M. A., and Herzog, D. C. (eds.) (1985). "Biological Control in Agricultural IPM Systems." Academic Press, Orlando.

Hussey, N. W., and Scopes, N. (1985). "Biological Pest Control. The Glasshouse Experience." Cornell Univ. Press, New York.

Krieg, A., and Franz, J. M. (1989). "Lehrbuch der biologischen Schädlingsbekämpfung." Paul Parey, Berlin/Hamburg.

Pickett, C. H., and Bugg, R. L. (eds.) (in press). "Enhancing Natural Control of Arthropod Pests through habitat Management." Ag Access Corp., Davis, CA.

Pimentel, D. (ed.) (1991). "CRC Handbook of Pest Management in Agriculture," 2nd ed. CRC Press, Boca Raton, FL.

Pest Management, Chemical Control

CHESTER L. FOY, *Virginia Polytechnic Institute and State University*

Glossary

Adjuvant Ingredient in a pesticidal or other agricultural chemical prescription that aids or modifies the action of the primary ingredient(s); adjuvants include activators, antidrift agents, antifoam agents, chelates, conditioners, penetrants, safeners, stickers, spreaders, surfactants, wetting agents, etc.
Integrated pest management (IPM) Integration of cultural, mechanical, genetic, biological, and chemical methods to obtain effective economical pest control with minimum effect on nontarget organisms and the environment
Pest Unwanted organism (animal, plant, bacteria, fungus, virus, etc.); agricultural pests include insects, plant diseases, weeds, nematodes, snails and slugs, rodents, birds, and other wildlife
Pesticide Chemical or other agent that will destroy a pest or protect something from a pest
Sustainable agriculture Integrated system of plant and animal production practices having a site-specific application that will, over the long term, satisfy human food and fiber needs; enhance environmental quality and the natural resource base upon which the agriculture economy depends; make the most of non-renewable resources and on-farm resources, and integrate, when appropriate, natural biological cycles and controls; sustain the economic viability of farm operations; and enhance the quality of life for farmers and ranchers, and society as a whole

Chemical control may be described as the judicious use of chemicals rather than other methods, such as physical, mechanical, cultural, biological, or genetic, for the practical manipulation (control, suppression) of agricultural pest populations. Much progress in pest management has been made possible by the judicious and timely use of pesticides and other agricultural chemicals. Practical benefits that can be and have been obtained through their use are lower production costs, increased yields, improved quality of farm products and recreational use areas, reduced sanitation costs on nonuse areas that serve as potential reservoirs for reinfestation (or reinfection), and/or prevention of undesirable chemical residues in the environment.

I. U.S. Agriculture at a Glance

A. Current Status and Worldwide Demands

The United States is the most efficient food-producing country in the world. According to Dr. C. Everett Koop, M.D., former U.S. Surgeon General, the U.S. food supply is the safest and most abundant in the world and pesticides are one of the most important tools that have made that abundance possible. U.S. farmers spend approximately $5 billion annually for pesticides which accounts for less than 3.5% of total production expenses. Pesticides contribute about 30% to crop yields. Currently, 2% or less of the U.S. population is engaged in production agriculture, but one American farmer produces enough food for approximately 129 people. The United States produces about one-half of the world's corn and two-thirds of the world's soybeans. Food costs today account for less than 11% of a U.S. family's budget. If the use of agricultural chemicals were banned, food prices would rise by at least 50% according to the U.S. Office of Technology Assessment.

The requirement for efficient worldwide food production has never been greater, for several reasons: (1) five billion people now inhabit the earth, with an additional one billion to be added each decade well into the next century; (2) for a variety of reasons, there is an ongoing loss of land devoted to food production (approximately one-half acre of cropland was available to support food and fiber needs of one person in 1960 compared to one-third acre today; (3) there is an increased emphasis on alternative, sustainable, or low-input agriculture, which many believe will lead to reduced crop yields on a per-unit-of-land basis; and (4) there is the present and continuing threat of loss of registration of agrichemicals important to agriculture.

B. Sustainability: An Essential Goal

The emphasis of sustainable agriculture is not to eliminate the use of pesticides and other important agricultural chemicals. In many instances these chemicals are necessary. Rather, the emphasis is to seek ways to reduce their use and increase their effectiveness to improve and maintain environmental and economic sustainability. Sustainable agriculture systems of the future will rely as much on information as on products. [See SUSTAINABLE AGRICULTURE.]

C. Major Pests and Their Toll

Crop losses are severe in many areas of the world, especially in developing countries and where irrigation allows pests to survive throughout the year. From planting to consumption, an estimated 50% or more of the world's production of food is lost to weeds, insects, diseases, and other factors. Insect-borne diseases kill or disable millions of humans and world losses from insects, diseases, weeds, and rodents are estimated at $100 billion annually. Control of harmful pests is vital for the future of the agriculture industry and human health.

The world's major source of food is plants. Wheat, rice, corn, potatoes, sweet potatoes, soybeans, millet/sorghum, beans, peas, sugarcane, and cassava are the crops from which most of the world's food is derived. Pesticides help farmers compete with 10,000 species of insects, 1500 plant diseases, and 1800 kinds of weeds to produce the world's food. Weeds are the principal agricultural pests on most farms and it has been estimated that more energy is expended on weeding of crops than on any other single human task. Approximately 250 million of the world's 350 million farmers still rely on hoes and wooden plows to weed and cultivate their crops. Weeds compete with crop plants for water, nutrients, and sunlight. They can complicate harvest and contaminate crops with weed seed, and some release toxins that reduce crop growth or attract insects and viruses. Poisonous weeds adversely affect humans and all classes of livestock through reduced productivity and even death. [See WEED SCIENCE.]

The ten most troublesome weeds for 1993 in the United States according to *Farm Chemicals,* a magazine published by Meister Publishing Compnay (Willoughby, OH), are listed below. The list was compiled by consulting weed control specialists in each state.

1. Nutsedge species (yellow, purple)—purple nutsedge ranks first among the world's worst weeds; yellow nutsedge ranks sixteenth
2. Pigweed species (smooth, redroot)—smooth pigweed ranks fourteenth among the world's worst weeds
3. Foxtail species (giant, green, yellow)
4. Morning glory species (annuals)
5. Field bindweed—ranks twelfth among the world's worst weeds
6. Velvetleaf
7. Common lambsquarters—ranks tenth among the world's worst weeds
8. Canada thistle
9. Johnsongrass—ranks sixth among the world's worst weeds
10. Common cocklebur

Insects cause losses in yield and quality by feeding or as vectors of disease. Losses can be serious and result in total crop losses in some fields. All the insects that are harmful to plants and may cause agricultural losses are too numerous to mention. The following insects are included on several insecticide product labels representing 15 crop protection chemical companies; the list was compiled from 125 product labels. Several species of the following are mentioned:

1. Aphids	9. Loopers
2. Caterpillars	10. Maggots
3. Cutworms	11. Moths
4. Flea beetles	12. Stinkbugs
5. Grasshoppers	13. Thrips
6. Leafhoppers	14. Webworms
7. Leaf miners	15. Weevils
8. Leaf rollers	16. Wireworms

[See ENTOMOLOGY, HORTICULTURAL.]

Individual species listed on 25 or more of the product labels include:

1. Beet armyworm	13. Fruittree leafroller
2. Boll weevil	14. Green cloverworm
3. Bollworm	15. Imported cabbage
4. Cabbage looper	worm
5. Codling moth	16. Lygus bug
6. Colorado	17. Mexican bean beetle
potato beetle	18. Peach twig borer
7. Corn earworm	19. Pink bollworm
8. Corn rootworm	20. Tarnished plant bug
9. Cotton leafperforator	21. Tobacco budworm
10. Diamondback moth	22. White fly
11. European corn borer	23. Yellowstriped
12. Fall armyworm	armyworm

More than 50 types of mites are also mentioned on the insecticide product labels.

Plant diseases reduce crop quality and yield. Losses from diseases in the United States are estimated to be 15% of the total agricultural production. A major loss resulting from disease management expenses. Farmers continue to grow susceptible varieties because of the quality of taste and yield. Some diseases can be kept in check only by repeated applications of pesticides. [See PLANT PATHOLOGY.]

Soilborne fungi cause some of the most widespread and serious plant diseases. Some of these diseases are:

1. Black rot	5. Root rot
2. Crown and stem rot	6. Stem rot
3. Dampingoff	7. Southern blight
4. Leaf blight	8. Watery soft rot

Airborne fungi can cause devastating losses because they can be spread long distances by the wind. Some important diseases caused by airborne fungi are:

1. Anthracnose	9. Leaf blight
2. Blast	10. Leaf rust
3. Brown spot	11. Leaf spot
4. Cedar-apple rust	12. Loose smut
5. Corn smut	13. Powdery mildew
6. Covered smut	14. Scab
7. Downy mildew	15. Soft rot
8. Gray mold	16. Stem rot

Important fungal diseases of trees include:

1. Annosus root and butt	4. Fusiform rust
rot	5. Little leaf
2. Chestnut blight	6. Oak wilt
3. Dutch elm disease	

Several important diseases are caused by bacteria and include:

1. Barn rot	5. Soft rot
2. Canker	6. Storage rot
3. Crown gall	7. Wilt
4. Fire blight	

Viruses cause the following diseases, among others:

1. Barley yellow dwarf	5. Tobacco etch
2. Cucumber mosaic	6. Tobacco mosaic
3. Maize chlorotic dwarf	7. Tobacco vein mottle
4. Maize dwarf mosaic	8. Vein banding

[See PLANT VIROLOGY.]

Some important plant diseases are caused by nematodes, microscopic roundworms that live in water or soil, or as parasites of plants and animals. Root knot is one of the most important plant diseases in the world and the root-knot nematode parasitizes more than 3000 plant species. Root-knot and lesion nematodes enter the roots of plants and feed on the inside. Lesion nematodes occur all over the world and feed on many crops. Damage by lesion nematodes leads to increases in other root diseases. Stem and bulb nematodes are hosted by more than 400 species of plants and are serious pests of many crops. Stubby-root nematodes live in the soil and feed on roots and other underground parts from the outside.

Birds destroy crops, contaminate food products with feces, and transmit diseases to humans, poultry, and dairy animals. Rodents (rats, mice, squirrels, rabbits, woodchucks, gophers) damage cultivated crops and stored products. Snails feed in the field and greenhouse and host parasites that are harmful to humans and animals. Algae, aquatic nonvascular plants with chlorophyll (seaweeds, pond scums, stoneworts), create problems in public water systems, swimming pools, irrigation water, farm ponds, fish ponds, lakes, fish hatcheries, greenhouses, etc.

D. Integrated Pest Management

Integrated pest management (IPM) has gained increased attention in recent years as a potential means of reducing commodity losses to pests and reducing reliance on chemical pest control. The focus of IPM is suppression, that is, holding pest populations below threshold damage levels, as opposed to eradication of an entire pest population. IPM involves a combination of pesticides with the following practices: the use of predators, parasites, or pathogens to reduce pest populations (biological control); the use of cultural practices associated with crop production, such as crop rotations, tillage practices, cover crops, intercropping, fertilizer management, altering planting dates, field sanitation, etc. (cultural control); the use of plants resistant to certain pests (host-plant resistance); the use of measures that kill a pest, disrupt its physiology other than by chemical means, exclude it from an area, or adversely alter its environment (physical and mechanical control); and the use of biochemical and microbial tools developed from biotechnological research (biorational control). [See INTEGRATED PEST MANAGEMENT; PEST MANAGEMENT, BIOLOGICAL CONTROL; PEST MANAGEMENT, CULTURAL CONTROL.]

Pesticides are an indispensable part of IPM. They are the first line of defense in pest control when crop injuries and losses become economic, and they are the only answer to a severe pest outbreak or emergency.

II. Classification of Pesticides

A. Herbicides, Desiccants, and Defoliants

Herbicides are chemicals or cultured biological organisms (bioherbicides, mycoherbicides) used to control unwanted plants. Herbicides and cultivation are the most widely used weed management tools. To date, almost 400 herbicides have been registered or are in the registration process, and these form the active ingredients of thousands of commercial products. About 100 herbicides inhibit photosystem II electron transport, 37 inhibit branched amino acid synthesis, 32 are active auxins, and 28 interfere with microtubular synthesis or function. [See HERBICIDES AND HERBICIDE RESISTANCE.]

Herbicides are used on almost all the corn and soybean acreage in the United States. For example, in 1992, 90% of corn acreage and 97% of soybean acreage was treated with herbicides. Most cotton and wheat acreage is also treated with herbicides at least once per crop. A trend toward the use of postemergence herbicides is likely because some of these products are considered relatively safe environmentally as they are not applied directly to the soil and are broken down rapidly; also, they are applied after a weed problem develops, fitting the "only as needed" approach that many growers are adopting with pesticides. Low-dosage herbicides (sulfonylureas and imidazolinones) are applied at a few grams per hectare rather than kilograms per hectare, and they score favorably on most environmental tests, especially with regard to their effects on humans and wildlife. Sulfonylurea products have been or soon will be commercialized for most of the world's major crops, including wheat,

barley, oat, corn, rice, soybean, oilseed rape, sugarbeet, and cotton. Oxyphenoxy acid esters are also applied at low dosages. Herbicide mixtures are often the best method to improve weed control and reduce herbicide rates. There is more interest in application of dry bulk fertilizers impregnated with herbicides. Microencapsulation, encapsulating dry herbicides into microscopic polymer shells, offers timed-release activity and less soil mobility in heavy rains.

Desiccants are types of pesticides that draw moisture (liquids) from a plant or plant part, causing it to wither and die. They are used to speed the drying of crop plant parts such as cotton leaves and potato vines.

Defoliants are chemicals applied to plants to cause leaves to drop prematurely. They facilitate harvest operations by accelerating leaf fall from crop plants such as cotton, soybean, and tomato.

B. Insecticides, Miticides, Larvicides, and Ovicides

Insecticides are pesticides used to control or prevent damage caused by insects. Organophosphate and carbamate insecticides have been successful for many years and are still used to control many different insects. Pyrethroids are dominant for many insecticide uses at present, but many of the most difficult insects to control are highly resistant to a variety of pyrethroids. Other classes of insecticides are organosulfur, formamidines, thiocyanates, dinitrophenols, and botanicals. Newer compounds, such as avermectin, hydramethylnon, insect growth regulators, and others, have recently arrived on the market. The development of IPM strategies has resulted in a more judicious use of insecticides and a corresponding decline in the total amount of insecticide used. However, the implementation of the true IPM concept of relegating insecticides to a minor position in insect control has not occurred in most cases. Some experts believe that the amount of insecticide used will continue to decline because of further implementation of IPM and because many of the newer insecticides are effective at extremely low doses.

Miticides are pesticides used to control mites and ticks. They are similar in action to insecticides and often the same chemicals kill both mites and insects. Larvicides are agents used for killing larval pests, and ovicides are agents that kill eggs (insecticides effective against the egg stage).

C. Fungicides

Fungicides are pesticides used to control organisms that cause rots, molds, and plant disease (fungi).

Chemical control and selection of varieties immune to fungi that cause plant diseases are the only alternatives to indirect practices of good husbandry (crop rotation, general hygienic measures, and evasive tactics). Fungal diseases are less obvious (visible) than weeds and insects and are basically more difficult to control with chemicals than are insects because the fungus is a plantlike organism living in close quarters with its host. There are now approximately 200 fungicides and most act as protectants, preventing spore germination and subsequent fungal penetration of plant tissues. [See FUNGICIDES.]

The modes of action for fungicides are classified as follows: inhibitors of the electron transport chain, inhibitors of enzymes, inhibitors of nucleic acid metabolism and protein synthesis, or inhibitors of sterol synthesis.

Inorganic fungicides currently used in agriculture are copper and sulfur, which are very effective against many fungal diseases. However, they can retard growth in sensitive plants. A variety of organic fungicides have been developed and include dithiocarbamates (ethylene bisdithiocarbamates or EBDCs), thiazoles, triazines, substituted aromatics, dicarboximides, systemic fungicides (oxathiins, benzimidazoles, pyrimidines, phenylamides, triazoles, piperazines, organophosphates, imides), dinitrophenols, quinones, and aliphatic nitrogens. Other fungicides include fumigants and antibiotics. EBDCs are the oldest and most widely used fungicides in the world. They have been used to control some 400 pathogens on more than 70 crops worldwide. Mancozeb, maneb, and metiram are the EBDCs registered for food use in the United States.

D. Nematicides

Nematicides are pesticides used to control nematodes. Nematicides available commercially are either halogenated hydrocarbons, isothiocyanates, organophosphate insecticides, or carbamate or oxime insecticides. Halogenated hydrocarbons lodge in the primitive nervous system of nematodes and kill primarily through physical rather than chemical action. Isothiocyanates inactivate sulfhydryl groups in amino acids; organophosphates and carbamates inhibit cholinesterase. [See NEMATICIDES.]

E. Other Pesticides (Rodenticides, Avicides, Molluscicides, and Algicides)

Rodenticides are pesticides used to control rats, mice, rabbits, and their relatives. The most successful ro-

denticides are coumarins (anticoagulants) that inhibit prothrombin formation, the material in blood responsible for clotting, and they also cause capillary damage, resulting in internal bleeding.

Avicides are agents used to control birds. They include repellents, perch treatments, toxic baits, soporific agents, chemosterilants, and stressing agents. Molluscicides are pesticides used to control snails and slugs. Baits are often used to attract and kill these pests. Algicides are agents used to kill algae. They include inorganic chlorines, copper compounds, quaternary ammonium halides, and miscellaneous organic compounds (herbicides, fungicides, etc.).

F. Plant Growth Regulators

Plant growth regulators are chemicals used to increase, decrease, or change the normal growth or reproduction of a plant. They include auxins, gibberellins, cytokinins, ethylene generators, and inhibitors and retardants. Auxins are compounds that induce elongation in shoot cells. They are used to thin fruit crops, increase yields of certain crops, assist in rooting of cuttings, and increase flower formation. Gibberellins stimulate cell division, elongation, or both and are used to stimulate growth. Cytokinins induce cell division in plants. They prolong the storage life of green vegetables, cut flowers, and mushrooms, and also stimulate lateral bud development in nursery and greenhouse production of ornamental plants. Ethylene produces numerous physiological effects and can be used to regulate different phases of plant metabolism, growth, and development. Inhibitors and retardants inhibit or retard certain physiological processes in plants. [See PLANT PHYSIOLOGY.]

III. Trends in Pesticide Manufacturing and Sales

Trends in the pesticide industry are toward fewer and larger companies that produce pesticides that are more potent, more selective, more site specific in their modes of action, and safer to humans and the environment.

Pesticides represent a $25 billion industry worldwide and sales total more than $7 billion annually ($7.9 billion in 1992) in the United States. Domestic sales of herbicides and insecticides increased 5.5% each and fungicides increased by nearly 8% in 1992 compared to 1991. Export sales of U.S.-produced pesticides dropped 1% in the same period. According to the National Agricultural Chemicals Association, herbicides represented 63.1% of total U.S. pesticide sales in 1992. Insecticides accounted for 24.4%, fungicides for 8.2% and other pesticides for 4.5%. More than 1400 trade names of pesticides are provided in the *Farm Chemicals Handbook 1993* (Buyers Guide Section) published by Meister Publishing Company, (Willoughby, OH). Herbicides, insecticides/acaricides (miticides)/nematicides, and fungicides represent over 1150 of these trade names. Growth regulators, biocontrol agents, rodenticides, algicides, defoliants/desiccants, bactericides, and molluscicides account for the remainder.

The costs to comply with Environmental Protection Agency (EPA) regulations continue to be a critical issue for the crop protection industry. Herbicide and insecticide prices have risen over the past three years according to the U.S. Department of Agriculture. Pesticide manufacturers increased expenditures to research and develop additional data to reregister older products. Also, many manufacturers have embarked on expensive biotechnology research. Almost all publicly traded agriculture biotechnology companies continue to show net losses as they struggle to gain market share and refine seed and crop protection products. Sales of biotechnology-derived insecticides are expected to increase rapidly, however. Dealer costs have risen, particularly for liability insurance. Growth in the pesticide market is expected to reach a near standstill through 1996 as a result of (1) market saturation—a manufacturer must displace a product or demonstrate new benefits for additional treatment; (2) regulatory problems—responses to environmental groups and public pressure are likely to cause more pesticide products to be restricted or banned in the United States; and (3) pirated products.

The number of U.S. pesticide suppliers has dropped from over 25 to 18 in recent years. Six companies—Ciba, DuPont, American Cyanamid, Zeneca, Monsanto, and Dow Elanco—are expected to increase their share of total pesticide sales from 68% in 1990 to at least 70% by the year 2000. Many believe that these companies will have long-term success in the pesticide market.

IV. Adjuvants

Adjuvants are important to the production, marketing, application, and effective use of pesticide products. Their importance to agriculture parallels that of pesticides themselves. Adjuvants that benefit pesticides are of two general types: formulation adjuvants—additives already present in the container when purchased by the dealer or grower—and spray adjuvants—substances added along with the formulated or proprietary product to the diluent (carrier), which is most commonly water, just before spray application in the field. Spray adjuvants are used more extensively with herbicides than with other classes of pesticides.

As pesticide formulations change and environmental concerns become more of an issue, the need for different adjuvants increases. Agrichemical manufacturers are moving toward recommending specific adjuvants that their research has identified as maximizing the efficacy of their products. Some researchers claim that up to 70% of the effectiveness of a pesticide can be dependent on the spray application. The effectiveness of the application often depends on adjuvants.

There is an increasing need to reduce pesticide dose rates as a consequence of the public's growing demand for a contaminant-free environment. Better manipulation and control of the distribution and fate of pesticides in the biosphere are desired. Adjuvants will play a critical role in pursuing the strategy of using less pesticide. In many cases, pesticide dose rates can be reduced by addition of adjuvants to commercially formulated pesticides.

There are now over 200 U.S. EPA-registered pesticides that have specific recommendations for the use of spray adjuvants. Thousands of adjuvants are now available and sales of emulsifiers and spray adjuvants represent about 6% of the U.S. pesticide market.

V. Risk (Cost) vs Benefit Analysis

Constant monitoring and assessment of risks (costs) vs benefits are essential when applying advances in agricultural science and technology. The objective is to provide more high-quality food, feed, fiber, and shelter more efficiently and economically, while at the same time preserving and improving the quality of life and the environment. If potential total losses from not controlling pests exceed the treatment cost, then treatment is justified.

VI. Society's Concerns Regarding Pesticides

A. Environmental (Soil, Air, Water, and Food)

Fertile and healthy soil is vitally important if we are to meet the increasing demand for food. Poor soil practices and misuse result in poor crop yields and quality. Pesticides that remain in the soil for long periods may limit planting to only certain crops. [*See* SOIL POLLUTION.]

Plants and animals require air to live. It is their source of oxygen and it receives their carbon dioxide waste. Air can move particles a great distance, and airborne pesticides may be harmful to both human and wildlife health and safety.

Water, one of our most plentiful and vulnerable resources, is necessary for all life. Humans and wildlife need nonpolluted water for drinking and bathing. Most fish and other freshwater and saltwater life can survive only slight changes in their water environment. Responsible pesticide use is essential to prevent plant or animal poisoning or the accumulation of dangerous residues.

Misuse of pesticides can result in illegal residues in meat and food crops. According to a 1992 survey by the Food Marketing Institute, pesticide residues or chemicals were cited by 31% of all consumers as being the greatest threats to the safety of their food. However, the Food and Drug Administration (FDA) reports that the U.S. food supply remains virtually pesticide free. Test results indicate that 99% of our food has pesticide residue levels within EPA tolerance (the maximum amount of residue that may remain on a harvested crop is called a tolerance) and over one-half is residue free.

Farmers are as concerned as consumers in metropolitan areas about environmental problems and food quality. A national Gallup poll sponsored by Sandoz Agro, Inc. indicated that farmers view the contamination of soil and ground and surface waters by pesticides as the most serious agricultural environmental problem. Ninety-two percent of those surveyed indicated that they are likely to use fewer pesticides in the future.

1. Residues and Carryover to Crops

Some pesticides are broken down quickly by heat, light, moisture, soil organisms, and other chemical reactions in the environment, resulting in little or no residues. Others may leave residues for weeks, months, or years. Persistent pesticides can be very useful for long-term control of insects, diseases, and weeds; however, their residues may remain on food or feed and be hazardous when consumed. Federal law requires that a tolerance be set for every use of each pesticide. If the residue exceeds the set tolerance, the crop may not be marketed or sold.

Pesticides that remain in the soil may interfere with crops that are planted at a later date. Product chemis-

try data showing a pesticide's fate in the environment, data regarding potential ecological effects, and residue chemistry data are required for support registration.

2. Groundwater Contamination (Water Purity)

Widespread use of pesticides has resulted in their appearance in surface water and groundwater used as primary sources for drinking. Their source of origin can be difficult to determine because pesticides can migrate considerable distances in water and air. The problem appears to be much less than public perception has indicated, especially in relation to herbicide contamination. The National Well Water Survey conducted by the Monsanto Agricultural Company involved 1430 wells from 89 counties in 26 states that were sampled from June 1988 to May 1989. Fewer than 0.02% had traces of alachlor above EPA's level of 2 parts per billion. Similar results were obtained for metolachlor, atrazine, cyanazine, and simazine. Many pesticide products include groundwater warning statements in their labels. [*See* GROUND WATER.]

B. Human Health and Safety (Consumers vs Applicators and Handlers)

Increased attention has been given to the safety of the American food supply, particularly that of infants and children. A recent report by the National Academy of Sciences concluded that there is "a potential concern" that some children may be ingesting unsafe amounts of pesticides; however, the report did not indicate that pesticide residues in the diet of infants and children pose significant risks. The U.S. Department of Agriculture, EPA, and FDA announced that they will launch a new cooperative program geared toward reducing pesticide use and promoting sustainable agriculture.

New packaging (water-soluble packets, tablets, closed transfer systems, shipping in bulk and/or refillable containers, etc.) and production formulations of pesticides are helping to reduce container disposal and worker exposure concerns.

EPA requires waiting intervals between application of pesticides and worker reentry into treated fields. Pesticide labels contain signal words that give an indication of the potential hazard. EPA also proposes clothing restrictions for mixers, loaders, and applicators of pesticides.

VII. Regulatory Climate

The use of pesticides on both food and nonfood crops is strictly controlled in the United States. The registration and regulation of pesticides is carried out by the EPA under two statutes: the Federal Insecticide, Fungicide, and Rodenticide Act (FIFRA) and the Federal Food, Drug, and Cosmetic Act (FFDCA). In addition, individual states have laws controlling the sale and use of pesticides.

The number of registered pesticides is being reduced by state and federal regulatory actions and by private sector decisions to withdraw individual products. The registration process is so costly that manufacturers sometimes choose to cancel a product rather than meet the new requirements unless the product can compete in a major field crop market.

According to the National Agricultural Chemicals Association, the development and testing of a pesticide by the manufacturer takes 8 to 10 years, costs $35 to $50 million, and requires more than 120 separate tests before it is registered by the EPA. Crop protection companies spend an average of 10% or more of their revenue on research and development; $1.88 billion was spent in 1992. Only one in 20,000 pesticides makes it from the laboratory to market.

Pesticides remain the most powerful tool in pest management. They must be regarded as a valuable resource and be used wisely. The newest pesticides to be developed are generally more selective and less toxic to humans. They are more active and require precise methods of application.

Nobel Peace Prize winner Dr. Norman Borlaug states: "Environmental extremists have been free to spread falsehoods which have led to unfair and restrictive public policy decisions on agriculture. There are no technical or agronomic barriers to prevent meeting food demand well into the next century, only political barriers".

Bibliography

"Crop Protection Chemicals Reference." (1993). 9th ed. Chemical and Pharmaceutical Press, Paris, and John Wiley & Sons, New York.

Devine, M. D., Duke, S. O., and Fedtke, C. (1993). "Physiology of Herbicide Action." Prentice–Hall, Englewood Cliffs, NJ.

Foy, C. L. (ed.). (1992). "Adjuvants for Agrichemicals." CRC Press, Boca Raton, FL.

Lucas, G. B., Campbell, C. L., and Lucas, L. T. (1985). "Introduction to Plant Diseases: Identification and Management." AVI Publishing, Westport, CT.

Matthews, G. A. (1992). "Pesticide Application Methods," 2nd ed. Longman Scientific and Technical. Essex, and John Wiley & Sons, New York.

Meister Publishing Company. (1992). "Insect Control Guide." Meister Publishing Co., Willoughby, OH.

Meister Publishing Company. (1994). "Weed Control Manual." Meister Publishing Co., Willoughby, OH.

Pimentel, D. (ed.). (1991). "Handbook of Pest Management in Agriculture," 2nd ed. CRC Press, Boca Raton, FL.

Ware, G. W. (1989). "The Pesticide Book." Thomson Publications, Fresno, CA.

Weed Science Society of America. (1989). "Herbicide Handbook," 6th ed. Weed Science Society of America, Champaign, IL.

Pest Management, Cultural Control

VERNON M. STERN, *University of California*

PERRY L. ADKISSON, *Texas A&M University*

Glossary

Economic injury level Lowest population density that will cause economic damage; economic damage is the amount of injury that will justify the cost of artificial control measures; consequently, the economic injury level may vary from area to area or season to season, or with a changing scale of economic values

Economic threshold Density at which control measures should be determined to prevent an increasing pest population from reaching the economic injury level

Key pest Pest that is a perennial, persistent threat dominating chemical control practices; in the absence of deliberate control by humans, its population density often exceeds the economic threshold one or more times during the growing season

Agroecosystems vary widely in stability, complexity, and the area they occupy. The kinds of crops, agronomic practices, changes in land use, and weather are important elements affecting the degree of stability of an agroecosystem. All of these, except perhaps weather, are subject to manipulations that influence pest or natural enemy populations. Often the agroecosystem may lack only a minor key factor or feature to adversely affect a pest or favorably modify the environment to increase the effectiveness of its natural enemies. These needs can often be met by proper use of cultural controls.

I. Introduction

The two basic principles in the cultural control of arthropod pests are (1) manipulation of the environment to make it less favorable to the pest and (2) manipulation to make it more favorable to their natural enemies. Both may be used together to prohibit, reduce, or delay pest population increase.

Classical use of cultural practices has involved such measures as stalk destruction, plowing and tillage, crop rotation, timely harvesting, selected planting dates, use of trap crops, barriers, flooding, and the planting of special plants to increase natural enemies. Recently, such techniques as selective use of preharvest chemicals, modification of crop variety, manipulation of pest populations using pheromones, and selective employment of insecticides also have helped suppress the pests and preserve their natural enemies. [*See* PEST MANAGEMENT, BIOLOGICAL CONTROL; PEST MANAGEMENT, CHEMICAL CONTROL.]

Cultural methods require a thorough knowledge of production of the crop as well as of the biology and ecology of the pest and its natural enemies to integrate the techniques for pest control into proven agronomic procedures for crop production. This knowledge permits assessment of the agronomic procedures that favor a particular pest and enables agronomic changes to be made to reduce pest population numbers and damage.

However, pest control is only one factor in crop production. Changes in agronomic procedures should not lead to other problems. If this occurs, the most valuable practice should be followed. For example, the last sustained and widespread grasshopper outbreak in the United States occurred in the 1930s, a period of severe drought combined with high winds and temperatures. Wind erosion was also a serious problem, and the United States Department of Agriculture (USDA) and state agencies initiated strip cropping to reduce soil erosion on the western Great Plains and elsewhere. However, in certain years, the strip cropping favored buildup and damage from grasshoppers, mainly *Melanoplus mexicanus* Saus. In this case, wind erosion was by far the more serious long-term problem and strip cropping prevailed as an agronomic procedure because, in most cases, grasshoppers could be controlled with baits and their outbreaks were sporadic.

Many cultural measures for pest control are closely associated with ordinary farm, forest, or water management practices. They can be simple and inexpensive because they can often be carried out with only slight modification of routine management operations. In addition, they do not contribute to pest resurgence, resistance to pesticides, undesirable residues, and contamination of the environment. Moreover, some insects are easier to control by cultural methods at a stage when they are doing no damage or at the end of a season than at a stage when they cause serious damage. This is especially true when the pest has no important natural enemies and is very difficult to control with available chemical methods, for example, the pink bollworm *Pectinophora gossypiella* (Saunders), or when the pest has developed resistance to pesticides, for example, the cotton leaf perforator *Bucculatrix thurberiella* Busck.

For various reasons, growers may not use or have abandoned cultural controls to reduce pests populations, including domestic use of the infested crop residues combined with primitive agriculture, as in the case of lepidopterous stem borers attacking cereal crops in Africa, south of the Sahara; education problems in the case of the sorghum midge *Contarinia sorghicola* (Coq.), in Nigeria and in Ghana; similar reasons for weevils and soil pests attacking a variety of crops in Rhodesia; advancements in farm technology in the case of corn rootworms in midwestern United States (i.e., mechanized corn pickers, readily available fertilizers, insecticides, and herbicides); lack of enforcement of laws requiring cotton stalk destruction for the pink bollworm in parts of Turkey and for spiny bollworm, *Earias insulana* Boisduval, in Iraq.

Other reasons for grower disinterest in cultural control methods are that these preventive measures are usually applied long in advance of the actual pest outbreak. Growers may not accept them for this reason or, if they are accepted, they may use them too late or improperly. Moreover, a reluctance to use them often arises because they may not provide the complete economic control that is often attainable using an insecticide, even though a population reduction can delay buildup to damaging levels and can reduce the number of insecticide applications necessary for crop protection.

Cultural practices can be used by individual growers for controlling pests such as the meadow spittlebug *Philaenus spumarius* (L.) or corn rootworms. They may also be used to achieve partial suppression of pests in conjunction with other control measures, especially chemicals. In other cases, cultural practices require community or area-wide adaptation to achieve population suppression over a relatively large geographical area. This option is especially important for insects that migrate or disperse a considerable distance, such as the pink bollworm.

When direct suppression of the pests is involved, a vulnerable stage in its life cycle, or one of its special behavioral patterns in relation to the habitat, is often singled out for manipulation. For example, corn rootworms have one generation per year and lay their eggs in corn fields in late summer and fall; the larvae hatch the next spring. The life cycle can be broken by crop rotation. Pink bollworm larvae enter diapause in October and November. Early defoliation or desiccation of cotton in late August and early September causes high larval mortality because most of their food is destroyed. Shredding the stalks and plowing them under, and winter weather, kills most of the remaining larvae. The Hessian fly *Mayetiola destructor* (Say) emerges from the summer generation in the fall and must oviposit within 4–5 days because it soon dies. Oviposition can be essentially eliminated by delaying the planting of the winter wheat until after the fall emergence. Long-term ecological research may be necessary before such cultural methods can be implemented.

The first consideration in any insect pest management program should be given to the "key" pest species, including their biological and behavioral characteristics, their natural enemies, their main and alternative food supplies, and the direct and indirect influence of other environmental factors. Key pests are serious, perennially persistent species that dominate control practices and, in the absence of deliberate control by humans, reach or exceed their economic thresholds one or more times each year.

Although the total number of potential pests species in a crop may be high, the number of key pests involved is low, usually only one or two in an area. For example, in the San Joaquin Valley of California, the key pests of cotton are lygus bugs: *Lygus hesperus* Knight and *L. elisus* van Duzee. On grapes in the same area, the key pests is the grape leafhopper *Erythroneura elegantula* Osborn.

In both cases, untimely, too frequent, and often unnecessary chemical treatments have created secondary pest problems. Cultural controls have been developed for lygus bugs and are being studied for the grape leafhopper.

II. Cultural Practices to Reduce Overwintering Pest Populations

Many pest species remain in or on the stalks, stems, and other parts of their host plant during the winter. Destruction of these parts by shredding, burning, and plowing can greatly reduce the overwintering populations. A stalk destruction program followed by plowing was developed in Texas for control of the pink bollworm *P. gossypiella*. This pest lays its eggs on the fruiting forms of cotton; immediately after hatching, the larvae burrow into the flower buds and bolls. This behavior makes chemical control expensive and unsatisfactory. In addition, no effective natural enemies have been found. [*See* COTTON CROP PRODUCTION.]

Early research showed that the pink bollworm diapauses in the last larval instar, mainly in the seeds of bolls that remain in the field after harvest. Later studies showed that diapause is controlled by photoperiod and that induction occurs in early September when day length becomes less than 13 hr. Diapause incidence then increases rapidly and attains a maximum in mid-October and early November, the seasonal onset of diapause can be predicted with precision at any location.

This discovery of the place and timing of diapause in the pink bollworm was used to achieve heavy suppression of overwintering populations by modification of certain cotton cultural practices. Until this time, cotton growers usually allowed the plants to grow in a way that resulted in a lengthy period of boll opening and harvesting; plants often continued to grow and bolls to open after harvest was completed; plants were left undisturbed in the fields through the winter; next year's cotton planting times were varied at the will of the growers. Such practices provided excellent overwintering conditions for the pest. Anti-pink bollworm cultural control eliminated or modified all these conventional practices. The mature cotton plants were managed using defoliants and desiccants so that bolls opened at nearly the same time and were promptly harvested. Soon after harvest, the plants were shredded mechanically and plowed deeply into the soil to prevent growth of new fruiting forms that provided food and diapause sites for the overwintering pest generation. New cotton crops were not allowed to be planted in the spring until after a designated time, which was set well after adult moths from the overwintering generation emerged and died.

Early defoliation provides the initial step for pink bollworm suppression, followed by the remaining cultural procedures because most of the overwintering generation comes from eggs laid after mid-September. If a preharvest defoliant or desiccant is applied in late August or early September (before days are short enough to induce diapause) the mature bolls open and immature fruiting forms either shed or dry up and little suitable food is left for larval development. However, if the application of the desiccant or defoliant is delayed until early October, it is relatively ineffective in reducing potential overwintering larval numbers because most of the larvae have already reached the diapause stage by this time.

When cotton is mechanically stripped, virtually all pink bollworm larvae are carried to the gin since the stripper leaves almost no bolls in the field. Almost 100% of the larvae are killed by the ginning process. When cotton is harvested by a spindle picker, some immature bolls are left in the field; these may constitute a source of infestation the next season. However, a rotary stalk cutter may kill 50–85% of the larvae that remain on standing stalks after harvest. When the stalks are plowed under immediately after shredding, the combined mortality from stalk shredding, plowing, and winter weather may exceed 90%.

These cultural procedures have been so successful in Texas that insecticides are seldom needed for the control of this pest. The effectiveness of this program has largely depended on legislation that prohibits planting before an established date in the spring and requires plowing up of the crop by another date in the fall. Growers who do not comply with these dates are subject to a fine. The success of this program also has a "social" aspect because few if any growers care to be accused of contaminating their neighbors' fields.

A similar type of cultural program utilizing stalk destruction and black-light trapping of the tobacco hornworm *Manduca sexta* (Johannson) and tomato hornworm *M. quinquemaculata* (Haworth) attacking tobacco has been developed in North Carolina. Insecticidal treatments on tobacco for these two pests in a 113-square-mile study were reduced more than 90% following wide-scale implementation of the program. A reduction by 60% occurred in applications for all tobacco pests.

III. Use of a Host-Free Season

Where the winters are not cold enough to kill cotton roots, the land may be left unplowed after the stalks are shredded and the old plant stubs will grow new sprouts for producing the next crop. This practice, referred to as "ratooning" or "stub cotton," has been discouraged and prohibited by law in many areas of the world because it can provide a continuous source of food for pest species and results in higher winter survival of pests, such as the pink bollworm, when the infested crop residue is not plowed under. Rigid enforcement can be of tremendous benefit even when the pest can survive on native host plants, which in some areas may not be common enough to pose a problem. This occurs with the cotton leaf perforator in southern California.

The elimination of ratooning cotton or its restriction to 1 yr in dry areas has reduced a number of pests successfully in Peru. The scale *Pinnaspis minor* Mask., established in Peru in 1905, had, by 1920, spread into the entire coastal area where cotton was grown. Several parasites—*Aspidiotiphagus citrinus* Crawf., *Prospaltella aurantii* (How.), *P. berlesei* (How.), *Aphelinus fuscipennis* How., and *Signiphora* spp.—and predators—*Microweisia* spp. and *Scymnus* spp.—attacked this scale. However, the practice of ratooning cotton favored the pest over its natural enemies, particularly in the hotter and drier areas. The scale also has a number of other hosts (*Gossypium raimondii, Ricinus communis, Manihot utilissima, Tessaria integrifolia, Cassia* spp., *Sida panniculata, Malachra* spp., *Malvastrum* spp.) that favor its persistence. Satisfactory control of *P. minor* was achieved by prohibiting ratooning cotton in the more northern valleys and by reducing the number of ratooning seasons, and/or reducing the height of the ratoon stumps, and eliminating wild hosts within and around the fields in the middle coastal area.

The weevil *Eutinobrothrus gossypii* Pierce girdles cotton at the base of the stem and root. This wilts or kills the plants if girdling is complete. Laws prohibiting ratooning decreased populations below the economic-injury level when combined with complete clearing of the fields of other host plants. This included uprooting the plants within the row, then plowing as many as three times crosswise, and hand removing any remnants of roots. Attempts to simplify the operation usually resulted in increased weevil damage.

Experience has also shown that, when ratooning is suppressed, both the square weevil, *Anthonomus vestitus* Boh., and the lesser bollworm, *Mescinia peruella* Schaus., are reduced to negligible numbers. In the Cañete Valley, the incidence of damage from the square weevil dropped to 1% or less and the lesser bollworm, which ordinarily destroys more than 10% of the late-formed bolls, essentially disappeared, because cotton is the only known host of this latter pest.

In Peruvian valleys in which irrigation water is insufficient to germinate and sustain new seed plantings, ratooning cannot be suppressed as a regular practice. Several devices have been used to keep *E. gossypii, A. vestitus,* and *M. peruella* at low levels. In the case of *E. gossypii,* ratooning is limited to 1 yr; the plants selected for ratooning are those that show little or no damage. The overwintering populations of *A. vestitus* that can be found as adults and larvae, and of *M. peruella* as larvae on terminal cotton buds, can be drastically reduced by goat or sheep browsing on the sprouts. A measure to enhance natural enemy effectiveness is to hand pick the early-attacked squares and place them in rearing chambers. As natural enemies emerge, they are collected and returned to the fields whereas unparasitized larvae are destroyed.

Populations of the red stainer *Dysdercus peruviansus* Guerin in Peruvian cotton come from two sources: (1) remnant populations favored by ratooning, abandoned fields, or lack of clean fallowed fields, and (2) mass immigrant populations from wild hosts. In valleys with adequate water, the problem has been solved through suppression of ratooning and the destruction of the more important host plants (*Sida panniculata, Malachra* spp.) in and near cotton fields. Where water is scarce and ratooning is permitted for 1 yr, alternative host destruction is combined with control measures using trap plants (*Urocarpidium* spp.) which are much more attractive to the red stainer than cotton.

Urocarpidium is planted in strips that can be treated to eliminate the red stainer without disturbing the natural enemies of other pests in the cotton. In addition, cotton seed is soaked in a pesticide solution and distributed in small heaps spaced 25 m along the cotton row in every fifth row. This apparently attracts the red stainers and they are killed by the poisoned seed.

IV. Use of Crop Rotation

The cultural practice of rotating a crop that is attacked by an insect with another crop that is not a host is most effective against pest species with a restricted host range, limited power of dispersal, or certain special behavior.

The larvae of the northern corn rootworm *Diabrotica longicornis* (Say) and the western corn rootworm *D. vigifera* LeConte feed mainly and preferably, but not exclusively, on corn roots and can be serious pests of corn in the midwestern United States. These two rootworms have one generation each year and their eggs are laid in corn fields in late summer and early fall. The adults die soon after oviposition and the diapausing eggs usually hatch in the following late spring, although some eggs of the northern corn rootworm will go through two winters before hatching.

Using knowledge of the single generation per year and the preference for ovipositing in corn fields, Forbes, a professor at the University of Illinois, recommended crop rotation to curtail the life cycle of the northern corn rootworm. A 4-yr crop rotation of corn, corn oats, and clover was generally adopted for the level lands of northern Illinois and Iowa. In other areas, sorghum, alfalfa, soybean, and other crops were used in rotation with corn. Although few growers maintained a systematic consistent rotation pattern, corn was seldom planted consecutively for more than 2 or 3 yr. The rotation program maintained the northern corn rootworm as a minor pest until the late 1940s. The same situation applied to both rootworm species in Nebraska.

This corn rootworm situation changed after World War II. Before the development of the organochlorine insecticides, soil insect control was virtually impossible except in limited cases, where cultural methods and a biological method (for Japanese beetle, *Popillia japonica* Newman) were effective. When DDT became available in the late 1940s, followed by the cyclodiene insecticides a few years later, general soil insect control became practical. Aldrin and heptachlor were particularly effective. These chemicals were quickly and extensively adopted for control of soil insects. They were often misused, resulting in more rapid development of resistance to them in many species and needless contamination of large acreages of agricultural land. In addition, these insecticides, ab-

sorbed by soil particles, have been transported by erosion into streams and other water systems, causing significant pollution.

With effective soil insecticides, herbicides, cheap commercial fertilizers, new high-yielding hybrids, and the development of efficient farm machinery, farmers in the corn belt began to grow corn continuously in the same field and usually on their best non-erodible land. The three classes of chemicals mentioned required a small outlay of money, and profits were greater from continuous planting of corn than from rotating with other crops. Thus, changes in farm technological, economic, and social factors contributed to general abandonment of a once successful cultural control program for corn rootworms. However, continuous corn cropping has now led to increased rootworm populations in all corn growing areas of the midwest. The use of insecticides has also increased markedly. In 1966, 33% of the United States corn acreage was treated, an increase of 50% from 1964. Moreover, whereas these pests were of minor importance under cultural control up to the late 1940s and early 1950s, they are now major and disturbing corn pests because in many areas they now have resistance to the formerly effective materials used against them.

V. Use of Harvesting Procedures

Harvesting procedures involving crop maturity, time of harvest or cutting practices, selective harvesting, and strip harvesting can be of considerable assistance in suppressing a variety of insect pests and increasing their natural enemies, thus affecting yields. A delay in harvesting cotton can reduce both its quality and quantity because of adverse weather factors; a subsequent delay in the shredding and plowing under of the cotton stalks favors overwintering populations of the pink bollworm.

A. Time of Harvest

Harvesting can have a marked effect on the insects in certain crops. In contrast to cereals and other seed crops which are usually dry at harvest time, forage crops such as alfalfa are harvested green and may be cut several times a year. Harvesting suddenly changes the insects' physical environment, which generally becomes much hotter and drier. Insects such as lygus bug adults and various parasites are caused to leave, seek shelter, or die. With regrowth of the alfalfa, many insects such as the alfalfa caterpillar, *Colias eurytheme* Boisduval, return to the field. This species prefers to oviposit on short regrowth alfalfa stems. Following heavy butterfly flights, larval populations often reach damaging numbers in these fields.

Another example of harvesting that affects future pest numbers concerns the meadow spittlebug *P. spumarius* (L.). Damaging infestations of nymphs vary from field to field. In 1953, entomologists were able to explain this in Ohio by (1) the time of cutting alfalfa field, 92) the time of first egg development and maximum oviposition, and (3) the fall behavior of the adult spittlebugs. When adults emerge from the spittle masses, their dispersal is brought about gradually by a "hardening off" of the maturing plants where fields are left uncut or abruptly by the first cutting in early June. Some adults return to these fields as alfalfa regrowth occurs, but they will leave when the plants mature or the field is cut a second time. The adult dispersals proceed through the season, but decrease when the first fully developed eggs appear in the females during the latter part of August. A preoviposition period of 3–4 wk occurs between the time of full development of the first eggs and maximum egg deposition in September and October.

When comparing adult spittlebug numbers with alfalfa conditions during the preoviposition period, the number is greatest in fields with the most succulent foliage. When the last cutting is removed in early August, plant regrowth is rapid. Large numbers of adults accumulate in these fields during the preoviposition period and large nymphal infestations occur the next spring. On the other hand, when the last cutting occurs in July, the plants attain advanced maturity by late August and these fields are unattractive to the adults. Likewise, when the last alfalfa cutting occurs in early September, the majority of adults leaves the field before ovipositing. Thus, a harvesting schedule that produces succulent foliage in September generally promotes large spittlebug infestations the following spring.

B. Selective Harvesting

Each year bark beetles, mainly the western pine beetle *Dendroctonus brevicomis* LeConte, destroy millions of board feet of lumber in the western United States. In contrast to many pests that feed on their host plant but do not kill it, the western pine beetle, when successfully established in a tree, changes it from a living to a dying and then a dead organism; decay follows.

Early forest entomologists and silviculturists observed that the western pine beetle has a tendency to select mature slow-growing or decadent trees on poor sites, seemingly avoiding younger, vigorously growing trees. Denning used this idea and described 7 categories for ponderosa pine based on their silvic characteristics. Keen expanded the categories to 16, based on four age groups and four degrees of crown vigor. Records over a 7-yr period involving approximately 39,000 beetle-killed trees showed that they were largely mature and overmature trees with poor to very poor crown vigor.

The value of the tree susceptibility classifications was that, if the type of tree most likely to be attacked and killed could be recognized, it would be possible to make a light-cut of beetle-susceptible trees to improve sanitation, reduce subsequent pine beetle infestations, and salvage valuable trees. However, other foresters commented that the application of Keen's system would have entailed removal of a considerable percentage of the standing virgin timber along the eastern slopes of the Sierra and Cascade mountains.

In 1942, California foresters designed a more practical risk-rating system for selecting individual trees, based on four categories: low risk, moderate risk, high risk, and very high risk. Trees in the high and very high risk categories were those most likely to be attacked and die in the near future. This system proved to be very successful in terms of timber management objectives and is still used by foresters in sanitation and salvage logging on the drier eastern side of California, Oregon, and Washington. For unknown reasons, in mixed tree stands on the western slopes of the Sierra and Cascade mountain ranges, which receive more rainfall, the risk-rating system for selective timber harvesting has not been too reliable.

C. Strip Harvesting

As noted earlier, lygus bugs, primarily *L. hesperus,* are key pests of cotton in the San Joaquin Valley of California. During May, most cotton fields are still in the seedling stage and unattractive to *Lygus*. By mid-June the situation changes and often large numbers of adults can be found in the fields. The adults come from two sources: alfalfa and safflower. Thousands of acres of alfalfa are grown and adjoin cotton fields through the middle part of the valley (N–S) and across the southern part of the valley (E–W). When an entire alfalfa field is suddenly mowed, *Lygus* is drastically affected. The humidity drops and the temperature quickly rises. This sudden change destroys the lygus bug's habitat, food, shelter, and oviposition sites. During hot weather, nearly all the adults leave solid cut field within 24 hr of cutting.

The problem then is to stabilize the alfalfa hay environment to keep the bug adults in the alfalfa habitat where they do little or no damage. This can be done by strip cutting the alfalfa fields. Under the strip-cutting technique, the alfalfa is harvested in alternate strips so that two different aged hay growths occur in a field simultaneously. When one set of strips is cut, the alternate strips are about half-grown. The field becomes a rather stable habitat because the lygus bugs move from the cut strips to the half-grown strips instead of flying to adjoining crops, such as cotton, as they do when the entire field is cut at one time.

Since natural enemies (predators) of the *Lygus* also move from strip to strip, no increase in the lygus bug population in the alfalfa occurs. Moreover, when the *Lygus* adults move into the uncut strips, they deposit eggs in the half-grown hay. However, these strips mature and are cut in about 2 wk. The time required for *Lygus* development (egg to adult) is much longer and most of the nymphs and the unhatched eggs are destroyed by high temperatures and the drying of the mowed alfalfa.

Strip harvesting also reduces other pest problems in alfalfa itself, particularly as a result of more effective biological control. In strip-cut fields, the pea aphid *Acyrthosiphon pisum* (Harris) and its parasite *Aphidius smithi* Sharma & Rao persist during mid-summer, a time when populations of both species are disrupted in solid-cut alfalfa. In the latter fields, *A. smithi,* being host-density dependent, is particularly hard hit, not only by the adverse physical conditions but also by the protracted scarcity of its host. It is virtually eradicated from the fields and does not show vigorous activity again until autumn. In contrast, in strip-cut alfalfa the parasite remains abundant, in continuous interaction with its host, and quickly responds to the aphid upsurge in late summer.

Although intensive analyses have not been made of other parasite–host relationships, some indications exist that parasites of lepidopterous pests are also favored by strip cutting. This appears to be true for parasites of the alfalfa caterpillar *C. eurytheme,* the beet armyworm *Spodoptera exigua* (Hübner), and the western yellow-striped armyworm *S. praefica* (Grote). Evidence also suggests that spotted alfalfa aphid *Therioaphis trifolii* (Monell) populations are reduced because of more effective biological control in strip-cut fields.

Bashir and Venkatraman found that the efficiency of *Zelomorpha sudanensis* Gahan (an important parasite of the beet armyworm *S. exigua* attacking alfalfa near Khartoum, Sudan) is high only during the winter. The frequent early cutting of alfalfa and the summer "dead season" cause unfavorable changes in the microclimate and deprive the adult parasites of nectar-bearing plants. These researchers suggested that parasite efficiency could be increased by strip cutting alfalfa throughout the growing season. The alternating uncut strips would serve as favorable habitats for the natural enemies and would increase their efficiency.

VI. Use of Habitat Diversification

The replacement of natural communities with monocultures of agricultural crops has caused general faunal impoverishment, whereas certain species of phytophagous arthropods have become extremely abundant. Many of these are pests and many have a high degree of vagility, often colonizing the disrupted agroecosystems ahead of their natural enemies.

Burnett, Odum, and many others comment that, as biotic complexity increases, particularly with reference to the number and kinds of trophic interactions, the stability of the agroecosystem will increase. An opposing viewpoint of this ecological "dogma" is taken by van Emden and Williams, who argue that species diversity does not necessarily cause greater stability. However, some examples of diversification of the crop environment show that it is more favorable to natural enemies and less favorable to pests.

An example is the sunn pest, *Eurygaster integriceps* Puton, and its scelionid egg parasites, mainly *Trissolcus* spp. (= *Asolcus*). Kamenkova, Viktorov, and others report that sunn pest egg parasites are very efficient in areas with small wheat fields surrounded by diverse vegetation. Under these conditions (e.g., Armenia) the polyvoltine egg parasites have a number of other pentatomid hosts and favorable hibernating places. On the other hand, in areas with extensive wheat monocultures (North Caucasus, lower Volga region), the density of scelionids is usually low. However, their numbers and efficiency increase near forests and in winter wheat fields following corn and sunflower, which are inhabited by other pentatomid hosts. Thus, the efficiency of sunn pest parasites can be increased by changes in the rotation practice and the spatial distribution of crops.

Doutt and Nakata studied vineyard ecosystems in California to determine the reason for different population levels of the grape leafhopper *E. elegantula*. *Anagrus epos* Girault was found to be an effective egg parasite of this leafhopper in the north coastal region, but the parasite was not effective in most of the San Joaquin Valley, where the grape leafhopper is a chronic pest. During the winter, the grape leafhoppers are all adults in reproductive diapause whereas *A. epos* has no diapause. In the north coastal region and similar areas, *A. epos* can breed through the winter on eggs of another leafhopper, *Dikrella cruentata* (Gillette), that occurs on native and introduced blackberries (*Rubus* spp.). In these wooded areas, sufficient water exists for trees to grow along the stream beds to provide shade for these host plants.

On the other hand, the southern San Joaquin Valley was virtually reclaimed from a desert. It lacks the trees to provide shade for the wild grape (original host of the grape leafhopper) and blackberries except along the banks of major, continuously flowing streams, where populations of *D. cruentata* exist in very isolated spots. R. L. Doutt (personal communication) has suggested planting trees around water-holding irrigation ponds to provide shade for the blackberries to increase populations of *D. cruentata* as a host for *A. epos* during the winter months. The growers will not do this because they fear these trees will provide nesting sites for birds that feed on the grapes. Nevertheless, research is continuing to find possible ways to diversify the vineyard areas to provide a suitable habitat for *D. cruentata*.

Many experiments have been conducted using trap crops to attract the pest species or to provide a more favorable habitat to increase natural enemies. An example is the interplanting of alfalfa strips in cotton (20-ft strips every 250–500 ft of cotton row). Stern and Sevacherian showed that *L. hesperus* prefers alfalfa over cotton as long as the alfalfa remains in a lush growing condition.

In Peru, the planting of corn in neighboring fields or rows of corn interplanted in cotton fields (one row to every fifth or seventh row of cotton) favors reproduction of many natural enemies of cotton pests, which also build up earlier in corn than in cotton. Beingolea commented that the benefits are real, although not easy to establish.

As a substitute for, or supplement to, interplanting special plants within a main crop to increase natural enemies, the transfer of plants densely populated with natural enemies can often be used to advantage. The introduced parasites *Praon palitans* Mues. and *Trioxys*

utilis Mues. of the spotted alfalfa aphid *T. trifolii* were rapidly spread throughout California by cutting and spreading infested hay. This technique, along with (1) the use of resistant alfalfa varieties, (2) the use of a selective insecticide (Demeton) to permit maximum survival of the parasites and lady beetles, and (3) the aid of a fungus disease reduced the annual cost of damage and chemical control from about $13 million in 1955 to about $1 million by 1958. Thereafter, this insect became a very minor alfalfa pest.

Similar parasite movements have been used in the former Soviet Union. Kolobova reported that *Bruchophagus roddi* Gussako., an important pest of seed alfalfa in the Ukraine, has several species of parasites, especially *Habrocytus medicaginis* Gahan, *Liodontomerus perplexus* Gahan, and *Tetrastichus bruchophagi* Gahan. When the alfalfa seed crop is harvested in early August, up to 50% of the pest population is in diapause. However, the majority of the parasites (99% of *H. medicaginis* and 82% of *T. bruchophagi*) completes its development and many die without parasitizing new hosts because the early harvesting precludes oviposition by any *B. roddi* that have not entered diapause.

The situation is quite different when the alfalfa seed crop is harvested at the end of August. At this time, parasitization of *B. roddi* is about 75% and most of the parasites are then in diapause. Thus, the chaff from alfalfa seed harvested in late August contains a great many hibernating parasites. By saving the chaff, these hibernating parasites can be easily moved to other fields where their benefits can be gained the following year.

Beingolea reported similar movements of plants containing natural enemies in the Cañete Valley of Peru. Corn plants carrying great numbers of cotton green leaf roller egg masses per plant, heavily parasitized by *Trichogramma* spp. and *Prospaltella* spp., were moved by truck loads. Corn tassels carrying anthocorids and mirids were also moved into crops to increase the local populations of these parasites.

Studies on the population dynamics of the weevils *Apion apricans* Hbst. and *A. aestivum* Germ. on wild and cultivated clovers in the former Soviet Union revealed a low pest infestation and high parasite activity in the wild habitats, as contrasted to the cultivated clover fields. *Apion* spp. hibernate as adults under the litter in forests and forest "windbreak" belts. *Spintherus dubius,* which has two generations per year, hibernates as full-grown larvae in the clover heads. Thus, the parasite population is substantially reduced by seed harvesting. However, a large number of hibernating parasites remains in the chaff. Therefore, the parasite abundance can be increased substantially by leaving the chaff in piles in the clover beds to permit parasite emergence the following spring.

VII. Use of Planting Time

Control of some insect pests can be achieved by manipulating the planting date so that the most susceptible stage of the crop coincides with the time of the year when the pest is least abundant.

The Hessian fly *M. destructor* (Say) has been a severe pest of wheat, its most important host. It normally has two full generations per year but, under certain weather conditions, there may be more. When there are two generations, the eggs of one generation are laid in the fall, the winter is passed in the maggot stage in a brown puparium, the "flaxseed" stage. In the spring, the maggot pupates and most of the adults emerge in April. These adults give rise to the second generation; much of the summer is also passed in the "flaxseed" stage. Adults from the summer generation usually begin to emerge in late August.

Wheat varieties with variable degrees of resistance to the Hessian fly are not available. However, for many years, delayed planting coupled with measures to destroy volunteer wheat on which the fall generation might deposit their eggs and the plowing under of infested wheat stubble after harvest were the only effective controls.

The adult Hessian flies from either generation emerge over a 30-day period but the individuals live only 3–4 days. Thus, the planting date for winter wheat in a given climatic zone can be delayed so most of the adults have emerged and died before the wheat has grown to a stage that can be attacked.

Some variation exists in time of adult emergence in different years because development is largely dependent on the weather. However, tests have shown that it is nearly always possible to plant late enough for fly control yet early enough to secure sufficient plant growth for the wheat to withstand winter cold. In Illinois, 8 yr of experiments on planting before and after the "safe" planting date gave an average gain of 5.8 bushels per acre when the wheat was planted after the "safe" date of planting.

VIII. Use of Plowing and Tillage

Soil tillage is used in nearly all commercial crop production and is probably the most widely used method of suppressing pest species, excluding chemical control. Several examples of effects of plowing on insect pests have already been discussed. However, plowing can be used to affect the pest species and their natural enemies differentially. Telenga and Zhigaev found that deep plowing may destroy up to 95% of the egg parasite *Xenocrepis* (= *Caenocrepis*) *bothynoderes* Gromakov, which hibernates in surface soil layers in the eggs of its weevil host *Bothynoderes punctiventris* Germer. The replacement of deep plowing by surface disc tillage greatly increased parasite survival and was twice as effective as deep plowing in reducing the pest population. [*See* TILLAGE SYSTEMS.]

Bobinskaya, Aleynikova, and Utrobina showed that such surface cultivation greatly reduces the numbers of elaterid larvae in field crops. Studies in the Kourgan District showed that under surface tillage, soil-inhabiting insects build up mainly in the upper soil layer (0–10 cm) and are easily accessible to carabids. Under deep plowing, they are turned into the soil (up to 30 cm) where the activity of such predators is greatly reduced.

IX. Management of Drift of Chemicals and Road Dust

The problems associated with drift of agricultural chemicals and of road dust are not clearly within the scope of cultural controls, yet both deserve mention because they are serious problems.

The problems arising from drift of agricultural chemicals have been reviewed by Adkisson and Yates. In addition to direct kill of predators and parasites in the treated area, other side effects may occur. Good evidence suggests that drift of certain pesticides can cause pest outbreaks on property adjacent to the treated area. In certain areas of Israel, citrus groves lie adjacent to cotton fields. The integrated control program on citrus is very restricted to the use of certain chemicals. However, aerial sprays of organophosphorus and chlorinated hydrocarbon pesticides may be repeatedly applied to the adjacent cotton. Drift of these materials into the citrus groves often upset the biological equilibrium, resulting is serious outbreaks of California red scale *Aonidiella aurantii* (Maskell). Legislation was enacted forbidding aerial spraying of cotton with nonselective pesticides within a distance of some 200 yards from a citrus grove.

Adkisson reported that methyl parathion applied at 1–2 lb/acre for control of bollworm, *Heliothis zea*, and tobacco budworm, *H. virescens,* in cotton in Texas has caused upsets in adjacent citrus, mainly of brown soft scale *Coccus hesperidum* L.

Another hazard due to improper timing and/or drift of insecticides is the effect on honeybees and other pollinating insects. The effects on honeybees are well known as are possible honeybee kills at various distances from a treated area.

Road dust is an inevitable companion of cultivation and of unpaved access roads for farm machinery. De-Bach reported the inhibition of effective parasitism and predation of the California red scale *A. aurantii* by dust; he has artificially increased California red scale populations by purposeful road dust applications. As a means of reducing dust in citrus groves, DeBach suggested overhead sprinkling, cover crops, and dust-reducing road surfaces.

Das also commented that tea bushes (*Camellia sinensis*) along dusty roads are often severely damaged by the red spider *Oligonychus coffeae* (Nietner). He noted a low incidence of predators in dusty leaves. Similar situations have been noted on cotton and other crops in many countries. A common practice among growers in California is to sprinkle daily all dirt roads commonly used by automobiles, trucks, and other farm equipment.

Bibliography

Stern, V. M. (1973). Economic thresholds. *Annu. Rev. Entomol.* **18,** 259–280.

Stern, V. M., Smith, R. F., van den Bosch, R., and Hagen, K. S. (1959). The integration of chemical and biological control of the spotted alfalfa aphid. The integrated control concept. *Hilgardia* **29(2),** 81–154.

Photosynthesis

DONALD R. ORT, *USDA-Agricultural Research Service and University of Illinois*

Glossary

Chloroplast Subcellular organelle in higher plants in which the fundamental energy transformation processes of photosynthesis occur

Photorespiration Complicated biochemical pathway which begins with the energetically wasteful fixation of O$_2$, rather than CO$_2$, by rubisco in the chloroplast stroma; remainder of the pathway, involving two other cellular organelles, is necessary to recover the photosynthetic intermediates diverted from photosynthetic carbon reduction by O$_2$ fixation

Photosynthetic carbon reduction cycle (PCR cycle) An intricate biochemical pathway of the chloroplast stroma which utilizes light energy stored as ATP and NADPH to drive the synthesis of carbohydrates from carbon dioxide and water

Photosystem Higher plant chloroplasts have two separate photosystems designated I & II; these are multisubunit, chlorophyll-containing enzyme complexes that span the thylakoid membrane; photosystems function to convert the energy of absorbed light into chemical forms

Rubisco Ribulose 1,5-bisphosphate carboxylase/oxygenase is responsible for the fixation of the competitive substrates CO$_2$ and O$_2$ to ribulose 1,5-bisphosphate; carboxylation reaction is the first step in the photosynthetic carbon reduction cycle whereas the oxygenation reaction is the first step in photorespiration

Stroma Liquid phase of the chloroplast, located outside the thylakoid membrane, where photosynthetic carbon reduction, starch synthesis, and other important biosynthetic pathways take place

Thylakoid membrane Structurally intricate membrane system within chloroplasts which contains chlorophyll and all of the enzymes necessary for light-driven electron transport, proton accumulation, and ATP formation

The term *photosynthesis* derives from Greek and has the meaning "formation by light". It is the process by which higher plants, algae, and certain species of bacteria transform and store solar energy in the form of energy-rich organic molecules. These compounds are in turn used as the energy source of all growth and reproduction in plants. As such, virtually all life on the planet ultimately depends on photosynthetic energy conversion. This brief article focuses on the chloroplast where the basic biochemistry and biophysics of photosynthesis occur within the cells of higher plants.

I. The Chloroplast

A. Structure and Morphology

All of the fundamental energy conversion processes of photosynthesis in higher plants take place within a subcellular organelle known as the chloroplast. Chloroplasts occur most abundantly in leaf mesophyll cells where they typically number between 50 and 200 per cell depending upon species, growth conditions, and developmental stage of the leaf. Mature chloroplasts are generally lens shaped and range in size from ~1 to 3 μm across by ~5 to 7 μm in their longest dimension. Chloroplasts are bounded by two distinct envelope membranes of which the inner envelope membrane contains specific transporters that mediate the flow of photosynthate from the chloroplast to the cytoplasm where sucrose synthesis takes place.

The photosynthetically active, chlorophyll-containing membranes of higher plant chloroplasts are flat lamellar vesicles called thylakoids. These are frequently tightly appressed to the outer surfaces of other lamellae forming structures known as grana stacks. Although the extent of this "stacking" is variable, and altogether absent in the specialized bundle sheath cell chloroplasts of C_4 plants discussed later, thin-section electron micrographs show the high incidence of surface contacts in mesophyll cell chloroplasts of both C_3 and C_4 plants. Figure 1 depicts a cut-away view of a chloroplast and shows the appearance of thylakoid membranes in both two and three dimensions. The simple and relatively familiar two-dimensional appearance of thylakoid organization belies the highly intricate three-dimensional structure of these membranes.

The central feature of thylakoid membrane three-dimensional organization is that of multiple tilted membrane planes arranged helically around the surface of a cylindrical core of stacked flat lamellar vesicles or discs (Fig. 1). These tilted membranes, the

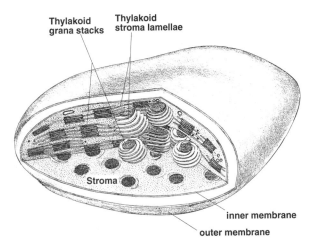

FIGURE 1 Chloroplast thylakoid membrane architecture. This drawing of a sectioned chloroplast shows the relationship of thylakoid membranes viewed in two dimensions compared with the appearance viewed in three dimensions. The drawing shows that the appressed membranes, often called grana stacks, seen in two dimensions arise from viewing in cross-section a cylindrical core of grana discs and that the helically arranged membrane sheets are the origin of the unappressed or stroma membranes visible in two dimensions. The thylakoid membranes are surrounded by the stromal phase where the reactions of CO_2 reduction, starch synthesis, fatty acid synthesis, chlorophyll synthesis, and numerous other biosynthetic pathways critical to plants occur. The chloroplast is bounded by two envelope membranes. [Modified from Ort, D. R. (1986). "Photosynthesis III. Photosynthetic Membranes and Light Harvesting Systems," Fig. 4.2, p. 147. Copyright © 1986 by Springer-Verlag, New York.]

so-called stroma lamellae, form attachments to the individual discs such that there is a continuous internal lumen between the two membrane types. Neighboring grana stacks are connected by narrow bridges between the tilted membranes thereby further extending the interconnection of lumen spaces.

The glycerol lipid composition of the thylakoid membrane is markedly different from that of other membranes in plants. About 70% of the lipid forming the thylakoid bilayer is contributed by two unusual and highly unsaturated glycolipids, monogalactosyldiacylglycerol and digalactosyldiacylglycerol. Although these glycolipids are found in certain photosynthetic bacteria and cyanobacteria, they are exceedingly rare in nonphotosynthetic membranes or organisms. Nevertheless, by virtue of being the major glycerol lipids of thylakoid membranes, these two galactolipids are the most abundant bilayer lipids in the biota. Thylakoid membranes typically contain only three other glycerol lipids. Phosphatidylglycerol and phosphatidylcholine are frequently encountered in a wide range of membrane types but diacylsulfoquinovosylglycerol is quite rare elsewhere in nature. In addition to the glycerol lipids of the bilayer, thylakoid membranes contain an abundance of the lipid chlorophyll in amounts equivalent to nearly $0.1M$ (i.e., on the order of 10^9 chlorophyll molecules per chloroplast). Unlike glycerolipids, chlorophyll is strongly associated with a group of specialized membrane proteins. [*See* PLANT BIOCHEMISTRY; PLANT PHYSIOLOGY.]

B. Evolutionary Origin

An endosymbiotic origin of several eukaryotic organelles, including the higher plant chloroplast, is very well supported by morphological, biochemical, and molecular biological evidence. The chloroplast genome is circular, typical of bacteria, and transformed bacterial cells generally express chloroplast genes without problems. Protein synthesis in chloroplasts also has strong bacterial character including 70 S ribosomes, N-formylmethionyl-tRNA initiation, as well as the compatibility of tRNAs, amino acyl-tRNA syntheses, and elongation factors between chloroplasts and bacteria. In addition, fatty acid synthesis in chloroplasts is distinctly prokaryotic in nature and, perhaps most convincing of all, is the exceptionally close functional and structural parallel between the photosynthetic apparatus of chloroplasts and bacteria.

The evolutionary progenitor of the higher plant chloroplast is generally agreed to have been an oxygen-evolving photosynthetic prokaryote similar to cyanobacteria currently in existence. In fact, there are examples in which the invading endosymbiont retains a vestigial cyanobacterial cell wall. These differ sufficiently from fully "evolved" chloroplasts that they are termed cyanelles rather than chloroplasts. It is believed that the ancestry of the chloroplast in higher plants traces to an organism closely related to *Prochloron didemni*, an apparently obligate ectosymbiont living in association with a marine sea squirt (*Didemnum*). *Prochloron* contain both chlorophyll a and b and also display appression of thylakoid membranes and the formation of three-dimensional membrane structures resembling simple grana stacks.

II. Light Energy Conservation

A. Light Absorption

It is axiomatic that only light which is absorbed can participate in photochemical reactions. The action spectrum of higher plant photosynthesis includes light at wavelengths between about 350 and 700 nm. About 55 to 60% of the sunlight incident on the earth's surface falls within this photosynthetically active wavelength interval. All of the pigments that participate in the absorption of light for photosynthesis in higher plants are located within the thylakoid membrane and fall into two general classes of compounds, chlorophylls and carotenoids. Chlorophylls, of which higher plants have two chemically distinct forms designated a and b, have complex ring structures which place them in a group of related compounds that include the pigments of hemoglobin and cytochromes. Chlorophyll appears green by virtue of the fact that this pigment intensely absorbs red and blue light leaving only the center of the visible light spectrum (i.e., green) to be reflected. Carotenoids, which are in much lower amounts in the thylakoid than chlorophyll, are linear polyenes of which β-carotene is a familiar example.

The first step in photosynthesis is the redistribution of electrons in a chlorophyll (or carotenoid) molecule as the result of the absorption of a photon of light. The formation of this excited state of chlorophyll occurs on an exceptionally rapid time scale ($\sim 10^{-15}$ sec) and represents the first step in photosynthetic energy storage. A basic feature of the photosynthetic apparatus is that the vast majority of light-absorbing pigments within the thylakoid membrane act as a large antenna to intercept light and rapidly

transfer the energy to specialized reaction centers from which subsequent steps of photosynthesis proceed. This array of antenna pigments functions to increase the absorption cross-section in two basic ways. First, the absorption cross-section increases in proportion to the number of molecules in the antenna associated with each reaction center. For example, in crop plant species, there are typically 250 to 300 antenna chlorophyll molecules associated with each reaction center. Second, the antenna contains a number of different chemical species of pigments which broaden the absorption spectrum of the antenna array and thereby increase the likelihood of efficiently capturing photons over a broader range of wavelengths.

An essential feature of an antenna system is that it must efficiently transfer energy to the photosynthetic reaction center with which it is associated. Excitation energy migrates through the pigments within an antenna array by resonance energy transfer. In this process the excitation energy, that is the excited state of a pigment molecule, is transferred from one molecule in the antenna to another. An important conceptual feature about the organization of antenna pigments is that excitation energy is "funneled" to the reaction center through a sequence of pigments with progressively lower energy excited states. Thus, as depicted in Fig. 2, excitation energy transfer toward the reac-

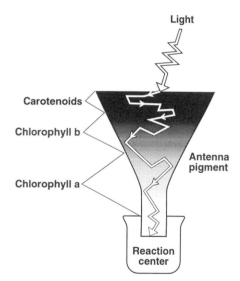

FIGURE 2 The funneling of excitation energy through the antenna pigment array to the reaction center. This figure depicts that resonance energy transfer in the antenna array is energetically downhill since a small amount of heat is released upon excitation transfer to a pigment with a lower energy excited state. Thus, the incremental loss of energy when the excited state is transferred from a chlorophyll b to a chlorophyll a molecule makes back transfer within the antenna array much less likely giving an otherwise random resonance energy transfer process, directionality.

tion center is energetically downhill and a small amount of heat is released on each successive transfer. This incremental loss of energy makes back transfer within the antenna array much less likely giving directionality to the otherwise random process of resonance energy transfer. In photosynthetic systems up to 99% of the photons absorbed by antenna pigments have a portion of their energy transferred to the reaction center. The exceptionally high efficiency of resonance energy transfer in photosynthetic antenna arrays arises from the precise orientation of the pigments within the antenna that is created by the organization of the pigment-binding proteins. All photosynthetically active pigments in chloroplasts are bound to proteins. There are a number of different membrane proteins that bind chlorophyll and the various carotenoids. This association with proteins is responsible for conferring a variety of specific physical and functional characteristics to the pigments.

B. Electron and Proton Transfer Reactions

The photosynthetic membranes of plants perform a remarkable feat: they convert a portion of the energy available in light into the chemical energy of ATP and NADPH. In this way photosynthetic membranes convert a very transient form of energy into forms which are stable over long periods of time, stable enough to be used at a later time for energy-requiring biochemical processes such as the reduction of CO_2 to carbohydrate. We have already discussed the first step in the energy transformation, the absorption of light in the antenna array. This energy, stored in the excited state of pigment molecules, is used to drive a series of oxidation-reduction reactions which take place within the thylakoid membrane. Figure 3 depicts the four major protein complexes within the thylakoid membrane that are responsible for the conversion of energy stored as the excited state of a pigment molecule ultimately into ATP and NADPH. Three of these complexes, photosystems I & II (PS I & PS II) and the cytochrome b_6f complex (cyt b_6f) are involved in light-driven electron and proton transfer.

The PS II complex from a typical crop plant contains at least 16 different polypeptides (currently of uncertain stoichiometry) which, as with all of the intra-thylakoid complexes, are a mixture of chloroplastic and nuclear gene products. These polypeptides bind and properly orient on the order of 200 chlorophyll a molecules, 100 chlorophyll b molecules, 50 carotenoid molecules, 2 plastoquinones (Q_A, Q_B), 1

iron, 2 pheophytin a molecules, 1 or 2 cyt b_{559} molecules, 4 atoms of manganese, and an underdetermined amount of chloride and calcium. PS II mediates the one-electron charge transfer events of the reaction center with the four-electron oxidation of two water molecules and the two-electron reduction of plastoquinone. There is a great deal that is known about the mechanisms of these concerted processes which is beyond the scope of this short overview. For our purposes it is most important to understand that energy is transformed and stored during the operation of PS II because the component reactions of PS II occur in an ordered fashion and asymmetrically across the thylakoid membrane. As depicted in Fig. 3, light energy is used to create, on the lumenal side of the membrane, an extremely powerful oxidant (by the oxidation of a pair of specialized chlorophyll a molecules, $P_{680} \rightarrow P_{680}^+$) capable of removing electrons and dissociating protons from water. Almost simultaneously, an electron is transferred toward the opposite side of the membrane reducing first a pheophytin a molecule and, within a few hundred picoseconds (10^{-12} sec), a plastoquinone molecule (Q_A) which is bound to PSII on the stromal side of the membrane. Soon thereafter electrons are transferred to a second plastoquinone (Q_B) which in turn results in the uptake of protons from the stromal phase. It is perhaps worthwhile to digress and mention that triazine herbicides such as atrazine function by inhibiting the binding of plastoquinone to the Q_B site on PS II. In terms of energy storage, the net result of these extremely rapid electron transfer reactions of PS II is the utilization of light energy to separate charge and the chemical activity of protons on opposite sides of the membrane (i.e., electrochemical potential energy) as well as the creation of a reductant (redox energy), reduced plastoquinone, that is far more energy rich than water which was the original source of the electrons.

The input of light energy occurs twice in higher plant photosynthesis through the participation of a second photosystem, PS I. Although PS I and PS II differ markedly in both structure and detailed mechanism, both photosystems bring about the same sorts of energetic transformations. In the case of PS I, the primary light-induced oxidant P_{700}^+ is a less powerful oxidant than its PS II counterpart, but is nevertheless fully competent to oxidize the small copper-containing protein, plastocyanin, located in the thylakoid lumen. As with PS II, the reactions of PS I produce an electric potential across the thylakoid and generate a very potent reductant in the form of

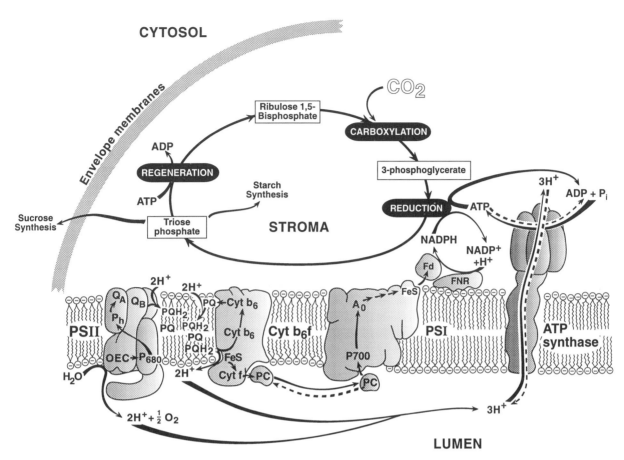

FIGURE 3 This schematic drawing of a chloroplast depicts the functions of the major protein complexes of the thylakoid membrane and the coupling of their activities to the photosynthetic carbon reduction cycle which takes place in the chloroplast stroma. Illustrated is the concept of light-driven linear electron flow coupled to the net accumulation of protons in the thylakoid lumen which is in turn used to drive the reversible ATP synthase in the direction of net ATP formation. In addition to the energy stored in ATP formation, energy derived from absorbed light is also stored by the reactions of the thylakoid membrane in the formation of NADPH. Photosynthetic carbon reduction is shown as a three-stage cycle. Carboxylation: a molecule of CO_2 is covalently linked to a carbon skeleton. Reduction: energy in the form of ATP and NADPH is used to form simple carbohydrate. Regeneration: energy in the form of ATP is used to regenerate the carbon skeleton for carboxylation. OEC, oxygen-evolving complex; P_{680}, reaction center chlorophyll of PS II; Ph, pheophytin acceptor of PS II; Q_A & Q_B, quinone acceptors of PS II; PQ & PQH_2, plastoquinone and reduced plastoquinone; cyt, cytochrome; FeS, Rieske iron sulfur protein; PC, plastocyanin; P_{700}, reaction center chlorophyll of PS I; A_0, primary acceptor of PS I; FeS, bound iron sulfur acceptors of PS I; Fd, soluble ferredoxin; FNR, ferredoxin-NADP reductase.

reduced ferredoxin which is in turn the direct source of electrons for NADPH formation.

The light-driven oxidation/reduction reactions of the two photosystems are interconnected through the catalytic activity of the cytochrome b_6f complex. Both photosynthetic and respiratory membranes of prokaryotes and eukaryotes contain closely related protein complexes that function to oxidize a low potential quinol and reduce a high potential metalloprotein. In the case of higher plant chloroplasts, the cytochrome b_6f complex catalyzes the energetically downhill reaction of oxidizing plastoquinol produced by PS II and reducing plastocyanin. In comparison to the pho-

tosystems, the cytochrome b_6f complex is structurally uncomplicated consisting of single copies of just five polypeptides. The redox-active components consist of two b-type cytochrome hemes (cytochrome b_6), a c-type cytochrome heme (cytochrome f), and a Fe_2S_2 center (Rieske iron-sulfur cluster). In addition to its role in connecting the electron transfer reactions of PS II with those of PS I, the cytochrome b_6f complex plays a central role in storage of energy that is subsequently used to synthesize ATP. A portion of the redox free energy that is available from the energetically downhill reactions catalyzed by the cytochrome b_6f complex is captured in transmembrane pH difference that is created because the reduction and protonation of plastoquinone occurs on the stromal side of the membrane whereas the oxidation and proton release occurs on the lumenal side of the thylakoid (see Fig. 3).

C. ATP Formation

As depicted in Fig. 3, there are two redox couples within the photosynthetic electron transfer sequence that result in the accumulation of protons within the thylakoid lumen and thereby result in the formation of a pH difference across the membrane. Water and plastoquinol are alike in the respect that both are hydrogen donors and oxidations thereby yield hydrogen ions. Since the oxidations take place across a membrane, a transmembrane pH difference is generated, and this potential can be used in turn for ATP synthesis. Figure 3 also illustrates an important concept in photosynthetic energy coupling, that the transmembrane pH difference intervenes between two otherwise independent processes of electron transfer and ATP formation.

The enzymatic coupling of the transmembrane electrochemical potential with the energy-requiring reaction of ADP phosphorylation is performed by a reversible ATP synthase or coupling factor which is located in the thylakoid membrane. This is a structurally complex multisubunit enzyme complex for which the catalytic mechanism is poorly understood despite more than two decades of active research attention. Structurally the enzyme complex has to distinct domains. Five subunits make up the hydrophilic domain that protrudes into the stroma and contains the catalytic sites which are involved in ADP binding and phosphorylation. The integral membrane portion of the chloroplast coupling factor complex contains four different polypeptides. While it is clear that this portion of the enzyme is involved in conducting protons across the thylakoid membrane to the active site located in the hydrophilic domain, it is undoubtedly an oversimplification to consider the membrane-localized domain of the enzyme simply as a noncatalytic proton channel.

The nonuniform and unpredictable light environment that plants encounter necessitates that the catalytic activity of the chloroplast ATP synthase be carefully regulated. Otherwise, the thermodynamically favored hydrolysis of ATP could quickly drain chloroplast and possibly cellular energy pools when light is limiting or absent. On the other hand, these regulatory controls must respond if energy is to be stored efficiently as light levels increase. Current evidence suggests that there is a regulatory hierarchy consisting of three principle components: the electrochemical potential across the thylakoid membrane, the oxidation state of a cysteine bridge within the hydrophilic domain of the ATP synthase, and the interactive binding of ATP, ADP, and P_i to the enzyme complex. These controls act together to allow catalytic inactivation of the ATP synthase at night and dynamic modulation of activity as light levels change during the day.

III. CO_2 Reduction

A. The C_3 Photosynthetic Carbon Reduction Cycle

Although ATP and NADPH are chemically stable photosynthetic energy storage forms, plants do not accumulate high levels of these compounds. Instead, ATP and NADPH are rapidly utilized in the biosynthesis of carbohydrate from atmospheric CO_2 and water. This complex biosynthetic pathway, known as the C_3 photosynthetic carbon reduction cycle (PCR cycle), takes place in the stroma of the chloroplast and involves more than a dozen different enzymatic conversions. The overall net reaction of the PCR cycle is

$$CO_2 + 3 \text{ ATP} + 2 \text{ NADPH} + 2 \text{ H}^+ + 2 \text{ H}_2O \rightarrow$$
$$\text{}^1/_6(C_6H_{12}O_6) + 3 P_i + 3 \text{ ADP} + 2 \text{ NADP}^+.$$

Thus, the PCR cycle consumes 12 molecules of NADPH and 18 molecules of ATP for each molecule of glucose formed. The subsequent metabolism of glucose provides transport (e.g., sucrose) and storage (i.e., starch) carbohydrate forms and the oxidation of glucose drives the energy-requiring reactions elsewhere in plant cells.

Figure 3 depicts the three central subprocesses within the PCR cycle. Atmospheric CO_2 and water are covalently attached to a five-carbon acceptor molecule in the carboxylation step of the cycle to produce two, three-carbon intermediates. It is these three-carbon compounds from which the C_3 designation of the PCR cycle comes. The enzymatic attachment of CO_2 to ribulose 1,5-bisphosphate to form two molecules of 3-phosphoglycerate is catalyzed by rubisco (ribulose bisphosphate carboxylase/oxygenase). This is an exceptionally abundant enzyme in leaves often accounting for 40% or more of the soluble leaf protein in C_3 plants. In fact, rubisco is the most abundant enzyme in the biota estimated at 10^{10} kg globally.

Rubisco is also a rather large enzyme made up of eight large and eight small subunits with a total molecular mass of about 560 kDa. As mentioned earlier for the thylakoid membrane protein complexes, the rubisco enzyme is composed of both chloroplastic (large subunit) and nuclear (small subunit) gene products.

The reduction process of the PCR cycle involves a substantial input of energy in order to form carbohydrate from a carboxylic acid. ATP is utilized to phosphorylate 3-phosphoglycerate to form 1,3-phosphoglycerate. Thereafter, triose phosphate (glyceraldhyde-3-phosphate) is formed by reduction utilizing redox energy stored in NADPH. Triose phosphate is the principle branch point within the PCR cycle (see Fig. 3). A portion of the triose phosphate is translocated across the chloroplast envelope membranes enroute to the synthesis of sucrose which occurs in the cell's cytosol. Triose phosphate is also basic building block for starch which is synthesized in the chloroplast stroma. Finally, triose phosphate must be reinvested in the PCR cycle.

Regeneration of the CO_2 acceptor, ribulose 1,5-bisphosphate, is clearly an essential feature for the cyclic operation of photosynthetic carbon reduction. In fact, 75% of the enzymatic conversions which comprise the PCR cycle are involved in regenerating ribulose 1,5-bisphosphate from glyceraldhyde-3-phosphate. The regeneration process requires energy in the form of photosynthetically-produced ATP and involves 10 intermediate compounds.

An essential feature in photosynthesis is the coordination of the light-driven reactions of the thylakoid membrane with the operation of the PCR cycle. This coordination is accomplished through a variety of inter-related processes. For example, the catalytic activity of rubisco is dependent upon several factors which coordinate its activity with light reactions. The stromal protein rubisco activase plays a central role in the activation of rubisco. Although the detailed mechanism of activase interaction with rubisco remains to be worked out, it is known that activase function is dependent on stromal ATP levels generated by thylakoid membrane reactions. In addition rubisco activity is modulated by the pH and Mg^{2+} concentration of the stroma which change in response to light-driven proton transport mentioned earlier. Light also indirectly controls the activity of several other enzymes in the PCR cycle through reversible redox modulation of regulatory thiol/disulfide groups on the enzyme. In the light, specific disulfide groups on these enzymes are reduced to the corresponding thiols by the small stromal protein thiore-

doxin. Since thioredoxin is reduced by light-driven electron transport through PSI, the catalytic activation of these PCR cycle enzymes is regulated by and coordinated with thylakoid membrane activity.

B. Photorespiration

Oxygen-evolving photosynthesis is considered to be responsible for the oxygenation of the earth's atmosphere approximately 2 billion years ago. Thus, the molecular evolution of the photosynthetic apparatus occurred in an anoxygenic atmosphere. These facts probably explain a curious feature about rubisco which is the cause of a major inefficiency in photosynthesis. In addition to the carboxylation of ribulose 1,5-bisphosphate with atmospheric CO_2, rubisco will also catalyze its oxidation to yield a single molecule of 3-phosphoglycerate and 2-phosphoglycolate, a two-carbon compound which is rapidly dephosphorylated to form glycolate (Fig. 4). This processes represents an inefficiency in photosynthesis because a molecule of CO_2 acceptor is consumed without any net carbon uptake. In a typical C_3 crop plant, such as soybean, the rate of rubisco-catalyzed oxygenation is about 20% of the rate of CO_2 fixation. Molecular oxygen and CO_2 are competitive substrates and the competition appears to an unavoidable consequence of the rubisco reaction mechanism which evolved in an oxygen free atmosphere.

An intricate mechanism has evolved, involving three different organelles in the plant cell, in order to recover the C_2 carbon skeletons that were diverted from the PCR cycle by the oxygenation reaction. This pathway, sometimes referred to as the C_2 photorespiratory carbon oxidation cycle (PCO cycle), is outlined in Fig. 4. This scavenging operation is energetically expensive. Two molecules of phosphoglycolate (i.e., total of four carbons) are converted by the PCO cycle into one molecule of 3-phosphoglycerate at the expense of an ATP molecule. Thus, the PCO cycle returns 75% of photorespiratory carbon to the PCR cycle with the remainder released as CO_2 when two molecules of glycine are converted into serine in the mitochondria.

C. C_4 Photosynthetic Metabolism

The PCO cycle discussed above makes the best of a bad situation created by the oxygenase activity of rubisco. An alternative "strategy" taken by some plant species is to prevent or greatly reduce the rubisco oxygenase activity by exploiting the competition be-

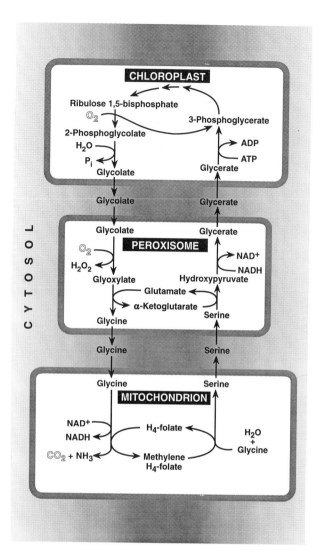

tween CO_2 and O_2 as alternative substrates. Plants such as maize, sorghum, and sugar cane, referred to as C_4 plants, greatly suppress or even eliminate the oxygenation reaction of rubisco by concentrating CO_2 in specialized leaf cells that contain rubisco. Plants such as maize, sorghum, and sugar cane have a distinctly different leaf anatomy than the typical C_3 leaf for which soybean is a familiar example. C_4 plant species have two distinct photosynthetic cell types in which chloroplast-containing mesophyll cells surround chloroplast-containing bundle sheath cells which in turn encircle the vascular bundles of leaf. The cytoplasms of mesophyll cells and bundle sheath cells are extensively interconnected by plasmodesmata.

The basic pathway of C_4 photosynthesis and the interplay of the two photosynthetic cell types are show in Fig. 5. In C_4 plants the assimilation of atmospheric CO_2 takes place exclusively in the mesophyll cell and involves the carboxylation of phosphoenolpyruvate by PEP carboxylase to form oxaloacetate. Unlike rubisco, O_2 is not a competitive substrate for

FIGURE 4 Photorespiratory carbon oxidation cycle of C_3 plants. Operation of the PCO cycle involves three separate plant cell organelles. The production of phosphoglycolate by the rubisco-catalyzed oxygenation of ribulose 1,5-bisphosphate diverts photosynthetic intermediates from the PCR cycle by competing with carboxylation. A phosphatase produces glycolate which leaves the chloroplast via a specific envelope transporter enroute to the peroxisome where it is metabolized to form glycine. The subsequent oxidation of glycine in the mitochondria results in the release of photorespiratory CO_2. Tetrahydrofolate serves as the carrier of the remaining one-carbon fragment which combines with a second molecule of glycine to form serine. A transamination reaction in the peroxisome converts serine to hydroxypyruvate which is in turn reduced to glycerate. Seventy-five percent of the photorespiratory carbon reenters the PCR cycle when photosynthetically produced ATP is used to phosphorylate glycerate in the chloroplast stroma. [Adapted from *Plant Physiology* by L. Taiz and E. Zeiger. Copyright © 1991 by The Benjamin/Cummings Publishing Company. Reprinted by permission.]

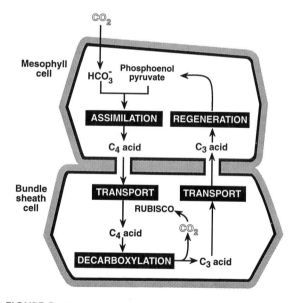

FIGURE 5 The C_4 photosynthetic carbon metabolism pathway suppresses photorespiration by concentrating CO_2 at the sites of carboxylation by rubisco. The C_4 pathway involves two different leaf cell types, mesophyll cells and bundle sheath cells. C_4 photosynthetic carbon reduction is shown as a four-stage cycle involving: (1) assimilation of CO_2 into a four-carbon acid in the mesophyll cell, (2) decarboxylation of the four-carbon acid producing a high concentration of CO_2 within the bundle sheath cell where the C_3 PCR cycle takes place, (3) transport of the resulting three-carbon acid, through plasmodesmata, back to the mesophyll cell, (4) regeneration of the CO_2 acceptor, phosphoenol pyruvate. [Adapted from *Plant Physiology* by L. Taiz and E. Zeiger. Copyright © 1991 by The Benjamin/Cummings Publishing Company. Reprinted by permission.]

PEP carboxylase. Oxaloacetate is converted to a four-carbon acid, either aspartate or malate, transported to the bundle sheath cell, and decarboxylated to release CO_2. The transport and decarboxylation process results in a significant elevation of the CO_2 concentration in the bundle sheath cell where rubisco and the other enzymes of the C_3 PCR cycle are localized. The elevated concentration of CO_2 effectively competes with oxygen virtually eliminating the first step in photorespiration, phosphoglycolate formation. The C_4 cycle is completed by the transport of a C_3 acid back to the mesophyll cell where regeneration of the original CO_2 acceptor, phospholenolpyruvate, takes place. Three metabolic variants of the C_4 pathway are known exist. These differ in both the C_4 (malate versus aspartate) and C_3 (alanine versus pyruvate) acids that are exchanged between the mesophyll and bundle sheath cells as well as in the C_4 acid decarboxylating enzyme in the bundle sheath cells.

Although the C_4 photosynthetic pathway effectively suppresses photorespiration, it is important to recognize that there are substantial energetic costs tied up in achieving elevated CO_2 levels within bundle sheath cells. There is a net cost of 2 ATP molecules for each CO_2 transported to the bundle sheath increasing by 40% the ATP cost of overall CO_2 reduction compared to using the C_3 PCR cycle alone. It is estimated that only about 5% of all higher plant terrestrial species are C_4 suggesting that the pathway with its higher energetic costs has an advantage in relatively few habitats. However, the much higher representation of C_4 species in hot/dry climates is well documented and probably reflects the higher water use efficiency that is associated with C_4 metabolism as well as the increasing affinity of rubisco for O_2 with increasing temperature.

It is worth noting that, with a future which portends a continued rapid increase in atmospheric CO_2 levels, we may expect this to suppress photorespiration and be reflected in increased biomass accumulation and agricultural production. Indeed, there are reports such production enhancements can already be seen and attributed to increases in atmospheric CO_2 which have already occurred. However, as the sophistication of the research into this issues grows, the complexity of the overall interaction of agricultural plants with the changing climate associated with increasing atmospheric CO_2 is becoming increasing evident.

Bibliography

Baker, N. R., and Ort, D. R. (1992). Light and crop photosynthetic performance. *In* "Crop Photosynthesis: Spatial and Temporal Determinants" (N. R. Baker and H. Thomas, eds.), pp. 289–312. Elsevier Science, New York.

Gregory, R. P. F. (1989). "Photosynthesis." Blackie, London.

Lawlor, D. W. (1987). "Photosynthesis: Metabolism, Control and Physiology." Wiley, New York.

Ort, D. R., and Oxborough, K. (1992). In situ regulation of chloroplast coupling factor. *Annu. Rev. Physiol. Plant Mol. Biol.* **43**, 269–291.

Taiz, L., and Zeiger, E. (1991). "Plant Physiology." pp. 179–218 and 219–248. Benjamin/Cummings, Redwood City, CA.

Walker, D. (1992). "Energy, Plants and Man." Oxygraphics, Brighton, U.K.

Plant Biochemistry

PRAKASH M. DEY, *University of London*

Glossary

Biosynthesis Traditionally this term is used to describe reaction sequences which occur in living systems leading to products. It also describes the production of chemicals using growing plants or microbial/plant cell cultures.

Cell wall Plant cell walls are distinct from microbial cell walls and can be termed primary and secondary walls, the former is laid down while the cell is growing. On cessation of growth, a secondary cell wall is deposited on the primary wall. The primary wall consists of various types of carbohydrate polymers and some proteins whereas secondary walls also have phenolic polymers.

Photosynthesis This process occurs in green plants within the chloroplast which contains photosynthetic pigments. Here light energy is used to transform CO_2 and H_2O into organic compounds.

Phytoalexins These compounds are also referred to as "plant antibiotics" and are toxic to pathogens. They can be produced by host plants in response to pathogen attack or other stimuli. Biosynthesis of such plant defence compounds is triggered by elicitors which are low-molecular-weight compounds originating from the pathogen or the host.

Plant biochemistry is the area of study of synthesis and utilization of chemical components of plant mate-

rials that are important for the survival and functioning of plants. Many aspects of plant biochemistry can be regarded as common also to microorganisms and animals; however, there are processes unique to plants only, for example, photosynthesis, cell wall synthesis, nitrogen fixation, and plant defence mechanisms.

I. Introduction

The bulk of literature and knowledge in the field of plant biochemistry is relatively limited as compared to microbial and animal biochemistry. They all are constituted with the same basic elements (C, H, O, N, S, P) and chemical components (carbohydrates, proteins, lipids, and nucleic acids); however, there exist major or minor differences in their metabolic processes. These differences are adapted to suit the requirements of the living organism. During the course of evolution, plants have developed unique processes.

Nutritionally, all green plants (higher plants) are autotrophic, requiring for their existence the six elements supplied by CO_2, H_2O, NO_3^-, SO_4^{2-}, PO_4^{3-} plus small amounts of other elements. However, they can be regarded as heterotrophic at the early stages of seed germination when they are unable to photosynthesize and are dependent upon the stored complex organic compounds. On the other hand, a wholly parasitic plant which is unable to photosynthesize depends entirely upon complex nutrients derived from the host. The energy required for the synthesis of constituent chemical components of the cell is supplied by the sunlight and is trapped by the pigments of green plants. Thus, through photosynthesis, energy is trapped in the form of carbohydrates from atmospheric CO_2 and H_2O as the initial raw materials. Carbon dioxide and water supply carbon, oxygen, and hydrogen. The oxygen of phosphate, however, remains bound to this element during metabolism. The chemical bond energy can be further utilized or released by a process of chemical bond breakage. The bond breakage of carbohydrates requiring oxygen can also lead to the production of CO_2 and H_2O: this is referred to as respiration. A substantial amount of energy released during the process is chemically conserved in the form of adenosine 5'-triphosphate (ATP). This "high-energy" compound is recognized as the universal biological source of energy. It is further utilized in many different ways for the synthesis of a variety of cell components and also to carry out different physiological "work." Thus, green plants have the unique ability to convert light energy into chemical bond energy which is required to drive numerous biochemical processes in the cell. This chemical energy is required for sustenance, growth, and reproduction of the whole plant.

II. Plant Cell

A. Cell Constituents

Four main classes of organic compounds make up nearly 90% of the biological material in the cell. These are carbohydrates, lipids, proteins, and nucleic acids which are present mainly in macromolecular forms. Smaller compounds of varying molecular weight also occur which may serve as precursors to or are degradation products of macromolecules. These could function as source of energy for the living cell.

Monosaccharides are the building blocks of oligo- and polysaccharides. They could consist of three to seven carbon structures and exist as cyclic or acyclic forms; members with five or higher carbon atoms are generally cyclic. Because of the presence of asymmetric carbon atoms in them, they display optical activity by rotating the plane of polarized light. It gives rise to levo (e.g., L-sugars) or dextro (e.g., D-sugars) rotating molecules. Most monosaccharides of plant origin belong to the D-configuration. Occurrence of free monosaccharides is rare; they exist in derivative forms or as constituents of di-, oligo-, or polysaccharides and serve as cell reserves or as part of structural molecules. Sucrose, the raffinose family of oligosaccharides and starch, is regarded as the main carbohydrate reserve whereas cullulose and pectins are structural components.

Lipids are a heterogenous group of compounds, generally hydrophobic in nature and soluble in organic solvents. Phospholipids are hydrophilic and soluble in aqueous media. Lipids occur as storage components in many seeds and also in association with cellular membranes. Many sterols and carotenoids can be classified as lipids; however, a large class consists of acyl lipids. Phospho- and glycolipids are recognized as polar acyl lipids whereas glycerides and waxes are neutral lipids. Glycerides are fatty acid esters of glycerol and accumulate as oil reserves in seeds and fruits.

Proteins are polymeric macromolecules consisting of L-amino acids linked together via peptide bonds. They may be involved with a multitude of functions in the cell, such as storage, catalysis, transport of molecules across membranes, metabolic regulation, energy-generating phenomena, cellular structure, and many other functions. They are generally synthesized from the 20 well-known amino acids. The protein content of the vegetative parts of plants is low as compared with seeds. Some legume seed protein can amount to 30–35% of its total weight; therefore, they make up a large proportion of human dietary proteins. Unlike humans, plants can synthesize all 20 amino acids found in proteins; however, cereal and legume proteins are low in certain essential amino acids, e.g., Lys, Met, Thr, and Tyr. In leaves the major proportion of proteins are found in the chloroplast. Enzymatic proteins are present in all parts of the cell.

The biological activity of proteins is destroyed by disrupting the different levels of its organization in the molecule. The primary structure formed via the peptide bonds constructs polypeptide ribbons which could be interlinked with the aid of disulfide bonds originating from the thiol group of the amino acid cysteine. Intrapeptide disulfide bond formation brings together distant parts of the polypeptide ribbon. When the ribbons are coiled up or associate themselves in the form of layers of sheets, they take up a secondary structure stabilized by inter- or intrapeptide hydrogen bonds, respectively, between the peptide groups. Further folding and twisting of the secondary structured entities give rise to the tertiary structure stabilized by a number of noncovalent forces, e.g., hydrogen bonds and ionic and hydrophobic interactions. Such molecules display a full three-dimensional structure incorporating specific sites and shapes required for their biological activity. Protein units can oligomerize by noncovalent aggregation which may also be an essential feature for the whole protein to be biologically active. Therefore, disruption of noncovalent bonds by physical or chemical means denatures the protein destroying its organizational pattern and this results in a loss of biological activity.

Nucleic acids, deoxyribonucleic acid (DNA), and ribonucleic acid (RNA) are polymeric molecules consisting of pentose sugar (2-deoxyribose in DNA and ribose in RNA), a nitrogen-containing base and orthophosphate. The mode of specific linkages between these constituents is shown in Fig. 1.

The five nitrogen bases of nucleic acids belong to either purine (adenine and guanine) or pyrimidine (cytosine, uracil, and thymine) ring structures. DNA is the source of genetic information in a cell whereas RNA is responsible for the expression of genetic information in the form of the final product protein. The composition and sequence of bases in the nucleic acids are responsible for their function and the nature of the protein they eventually produce.

The purines adenine and guanine and the pyrimidine cytosine are present in both DNA and RNA; however, in addition, DNA contains thymine but not uracil. RNA contains uracil but not thymine. The polymeric DNA ribbon acquires a secondary structure by pairing up with another strand of DNA in the form of a double helix. Intermolecular hydrogen bonds are formed between the bases of each strand involving adenine–thymine and guanine–cytosine. This stabilizes the secondary structure in an antiparallel arrangement in which the 5′-ends and the 3′-ends of the strands are aligned together. Since all the bases are linked by hydrogen bonds, the DNA molecule will have equimolar proportions of the bases. The major site of plant DNA is the nucleus. A small fraction (ca. 2%) occurs in chloroplasts and mitochondria; the former is termed satellite DNA and is different from the nuclear DNA in density, molecular weight, and nucleotide composition.

The natural RNAs occur in single strands. They can acquire secondary structure by forming intramolecular hydrogen bonds between guanine–cytosine and adenine–uracil. RNA is located in the ribosomes, mitochondria, and chloroplasts which occur in the cytosolic part of the cell. RNAs are classified into three major types, ribosomal RNA (rRNA), messenger RNA (mRNA), and transfer RNA (tRNA). The rRNA constitutes approximately 70% of the total RNA and has two forms, one with an M_r of just above one million associated with proteins in the large subunit of ribosome, and the other with an M_r of 0.7 million associated with proteins in the small subunit. The RNAs are complementary in their base sequence with respective sections of DNA strand. mRNAs (ca. 10% of the total) act as the intermediary between the DNA and the corresponding final product, the protein which has a specific amino acid sequence. The tRNA (ca. 20% of the total) occurs as a mixture of a large population of molecules each containing 60–80 nucleotides. One nucleotide is responsible for each amino acid of the synthesized protein. They bind to individual amino acids and function as the last step in the synthesis of a protein. The single-stranded tRNA acquires a secondary structure by complementary base-pairing in four distinct regions of the molecule and forming a tri-lobed clover leaf-like outline. Each lobe has unpaired regions. The region of the central lobe contains a sequence of three nucleotides (termed anticodon) which is recognized by a complementary sequence (termed codon) of mRNA. This is a vital step in protein synthesis and is responsible for a specific sequence of amino acids in the protein. The tRNA may contain other nucleotides and derivatives in addition to the four known bases.

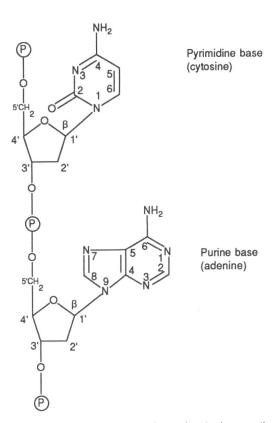

FIGURE 1 Linkage of pyrimidine and purine bases to ribose forming respective nucleosides. Ribose residues are linked together via phosphate diester bonds forming the backbone of an extended polymeric structure.

B. Cell Structure

It is difficult to present a general picture of plant cells as their structure and function vary widely. There are dead cells, such as xylem cells for fluid transport, cells of sieve tubes and phloem which lack nuclei, and parenchymatous cells which constitute nearly 80% of all cells. The latter are metabolically active cells and can be subdivided according to their diverse functions.

The metabolic processes within the cell are mostly compartmentalized as one process in the cell may negate another process. For example, carbohydrates can be synthesized from atmospheric CO_2 and then can be utilized within the same cell. Membranes play a key role in the maintenance and function of the compartments. They are permeable barriers and are crucial in a controlled transport of various molecules across them, thus allowing selective flow of metabolites amidst complex processes of cell metabolism. They can also play important structural and functional roles, e.g., in photosynthesis. An illustrative structure of a metabolically active plant cell is depicted in Fig. 2.

There are two main parts of a parenchymatous cell: the cell wall and the protoplast. The cell wall is the boundary of the cell; it confers strength and maintains the integrity by surrounding the protoplast. The wall is a layered structure, the primary wall and the inner secondary wall. In adjoining cells, two primary walls do not meet directly and are separated by a layer called middle lamella. This consists mainly of pectic polysaccharides rich in galacturonic acid. The cell wall consists mainly of polysaccharides (cellulose, pectin, and hemicellulose) and only minor amounts of proteins (structural proteins and enzymes). Phenolics are also present.

The protoplast is the metabolically active part of the cell and it includes the cytoplasm and the vacuole. The latter has a membranous wall, termed tonoplast, and contains a sap. The cytoplasm is also surrounded by a membrane (plasmalemma) and contains a fluid material, cytosol. The cytosol has various functional particles and bodies, the largest is the nucleus surrounded with a perforated double membrane which has continuity with the membrane of another organelle, the endoplasmic reticulum. This membrane is, however, not connected with either plasmalemma or tonoplast. Nuclear membrane surrounds the nucleoplasm which has the chromosomes. Nucleolus is also embedded in nucleoplasm where synthesis of ribosomes occurs and the latter is the site of rRNA synthesis. Embedded in the cytosol are also chloroplasts (particles classified under the termed plastids), mitochondria, Golgi apparatus, ribosomes, and microbodies, such as peroxisomes, glyoxysomes, and spherosomes. Chloroplasts contain green chlorophyll and yellow carotenoid pigments and are the sites of photo-

synthesis. Mitochondria are the powerhouse of the cell and aerobic oxidation of pyruvate to CO_2 and the phosphorylation of ADP to ATP takes place at this site. Golgi apparatus consists of Golgi bodies (dictyosomes), a stack of nearly five flattened sacs, termed cisternae. The upper flattened face of the outermost cisternae acquires membrane from the endoplasmic reticulum whereas the other outermost cisternae membrane has continuity into cytoplasm. The synthesis of the complex polysaccharides of the cell wall takes place in Golgi bodies and the vesicles transport these polysaccharides for incorporation into the cell wall. Golgi apparatus is also the site of glycosylation of proteins and subsequent processing. Ribosomes occur as particles that either are free entities or are attached to endoplasmic reticulum (termed microsomes). Ribosomes also form polysomes and are concerned with protein synthesis. Peroxisomes and glyoxisomes are also particles which are surrounded with membranes, the former being involved with photorespiration in photosynthetic cells and the latter with the lipid-storing tissues of endosperm and cotyledons of seeds. Glyoxysomes contain enzyme systems for the conversion of long-chain fatty acids to succinate. Spherosomes are also associated with lipid-storing tissues. [See PLANT CYTOSKELETON; PLANT PHYSIOLOGY.]

III. Photosynthesis

Approximately 3.5 billion years ago all living organisms probably resembled the primitive bacteria of today. In the absence of the ozone layer most of the organic materials of the organisms must have been degraded by direct radiation from the sun. Natural selection then must have favored the pigmented organisms to absorb the radiations. Through evolution, pigmented organisms developed an ability to split then abundantly available hydrogen sulfide via their pigments for storage of the radiation energy in the form of accessible chemical energy. Some of the organisms with this ability can still be found in sulfur-rich springs and marshes. Further evolution led to water-splitting organisms trapping solar energy in green pigments. These are today's cyanobacteria (formerly known as blue-green algae). Plants inherited this ability from photosynthesizing bacteria. The oxygen atom evolved from the water molecule supports all animal life on Earth. The hydrogen is joined to atmospheric CO_2 via biochemical processes to build organic molecules with general formula $(CH_2O)_n$.

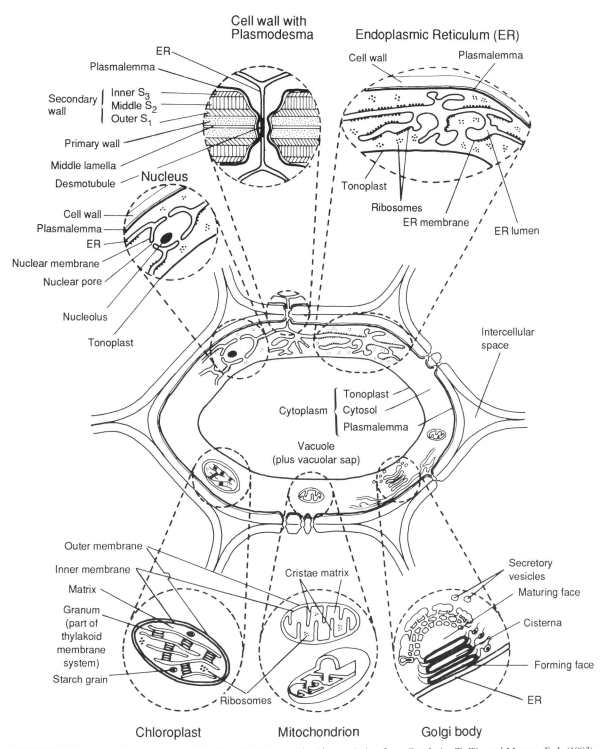

Cell wall with Plasmodesma

ER
Plasmalemma
Secondary wall { Inner S₃ / Middle S₂ / Outer S₁ }
Primary wall
Middle lamella
Desmotubule

Endoplasmic Reticulum (ER)

Cell wall
Plasmalemma
Tonoplast
Ribosomes
ER membrane
ER lumen

Nucleus

Cell wall
Plasmalemma
ER
Nuclear membrane
Nuclear pore
Nucleolus
Tonoplast

Intercellular space

Cytoplasm { Tonoplast / Cytosol / Plasmalemma }

Vacuole (plus vacuolar sap)

Outer membrane
Inner membrane
Matrix
Granum (part of thylakoid membrane system)
Starch grain

Cristae matrix

Ribosomes

Secretory vesicles
Maturing face
Cisterna
Forming face
ER

Chloroplast　　　　**Mitochondrion**　　　　**Golgi body**

FIGURE 2 Diagrammatic presentation of a plant cell. [Reprinted with permission from Goodwin, T. W., and Mercer, E. I. (1983). "Introduction to Plant Biochemistry," Fig. 3.1, p. 19. Copyright © 1983 by Pergamon Press Ltd. Headington Hill Hall, Oxford, Ox3 OBW, UK.]

These molecules give rise to carbohydrates and fats, and by incorporation of nitrogen and sulfur nucleic acids and proteins are formed. [See EVOLUTION OF DOMESTICATED PLANTS.]

The flat leaves are well designed to carry out efficient photosynthesis; the top and the bottom faces consist of single layers of protective cells which are transparent to light. Loosely fitted and spongy mesodermal cells are sandwiched between the two layers. The pigmented chloroplasts which absorb the sunlight are contained in these cells. On the epidermis there are small pores known as stomata, surrounded by guard cells; pores allow CO_2 to diffuse in and O_2 to diffuse out. Water reaches the leaves through the root system and the xylem vessels. Stomata open up during the day for photosynthesis to take place; however, some inevitable water loss also occurs through these openings. Many desert plants have adapted themselves to overcome this disadvantage by modifying the photosynthetic process. This is known as Crassulacean acid metabolism (CAM) and the stomata in these plants open only at night to minimize water loss. Here the entry of CO_2 builds up a store of a transient chemical which is then converted during the day time into the organic compounds of normal plants while the stomata are firmly shut. Photosynthesis has been shown to occur in two phases: (a) the light reaction, conversion of light by the pigmented chloroplasts into chemical energy in the form of ATP and NADPH via splitting of water and CO_2 evolution, and (b) the dark reaction, utilization of ATP and NADPH for the reduction of CO_2 to produce carbohydrates which can occur to some extent in the absence of light. The photosynthetic pigments in the chloroplasts are the receptors of light energy. There are two functional assemblies of the pigments, photosystem I and photosystem II, which are connected to each other via characteristic electron transport chains. Hill's discovery of the electron transport chain and evolution of oxygen can be summarized in the following equation, known as the Hill reaction:

$$H_2O + A \xrightarrow{\text{light}} AH_2 + O_2,$$

where A is the hydrogen (electron) acceptor and AH_2 is its reduced form. Thus, the direction of electron flow is the reverse of respiration. $NADP^+$, a common component of chloroplast, works well as electron acceptor and yields NADPH. This coenzyme is then available to reduce NADP-linked conversion of metabolites via specific dehydrogenases. $NADP^+$ is therefore a common carrier of electrons from water and CO_2 is the terminal electron acceptor in green plants. The process is called photosynthetic electron transport. Chloroplasts also produce ATP when illuminated in presence of ADP and phosphate. This major mechanism of conversion of light energy into chemical energy is called photophosphorylation. This process is coupled to photo-induced electron transport between photosystems I and II. One molecule of ATP is generated per pair of electrons transported from H_2O to $NADP^+$. [See PHOTOSYNTHESIS.]

In the dark phase of photosynthesis, CO_2 is reduced and converted to organic matters of the cell. The dark phase is confined to the utilization of ATP and NADPH. Light is, however, required for CO_2 fixation. In C_3 plants guard cells open only in the light to let CO_2 into the leaf. There are three possible methods of CO_2 assimilation, the C_3 pathway or the Calvin cycle, C_4, and CAM (Crassulacean acid metabolism) pathways. The C_3 plants generally grow in temperate climate, C_4 plants are tropical, and CAM plants grow in arid environment. CO_2 is converted into carbohydrate via the Calvin cycle (Fig. 3) in C_3 plants, but this pathway also participates in C_4 and CAM plants. As shown in Fig. 3 the key reaction of the C_3 pathway is the carboxylation of ribulose di-P which is the rate-limiting step in the Calvin cycle. The carboxylase is allosterically activated by fructose 6-P which turns on the Calvin cycle and is inhibited by fructose 1,6-di-P which prevents the carboxylation reaction. Several crucial enzymes of the Calvin cycle are activated by light.

Figure 3 also indicates various points where the carbon can leave the Calvin cycle to form molecules other than carbohydrate. C_3 plants suffer from some major disadvantages: first, water loss during dry hot days (through open stomata) and second, they are sensitive to oxygen. The latter is because ribulose bisphosphate carboxylase (Rubisco) can also act as an oxygenase and therefore transforms some of the ribulose 1,5-di-P to phosphoglycolate and decreases the yield of 3-phosphoglycerate and ultimately of sugars. Phosphoglycolate is diverted to the photorespiratory pathway and one of the products, glycine, is metabolized in the mitochondria to produce serine and CO_2 (Fig. 4).

Figure 5 shows the pathway (Hatch-Slack or C_4 pathway) by which C_4 plants assimilate CO_2. Two different types of photosynthetic cells occur in these plants: mesophyll cells where phosphoenolpyruvate (PEP) is carboxylated to oxaloacetate which is then reduced to malate and transported to the second type of cells called bundle-sheath cells. In some C_4 plants aspartate is formed instead of malate. Both can be then decarboxylated to pyruvate which is then returned to the mesophyll cells and converted back to PEP. CO_2

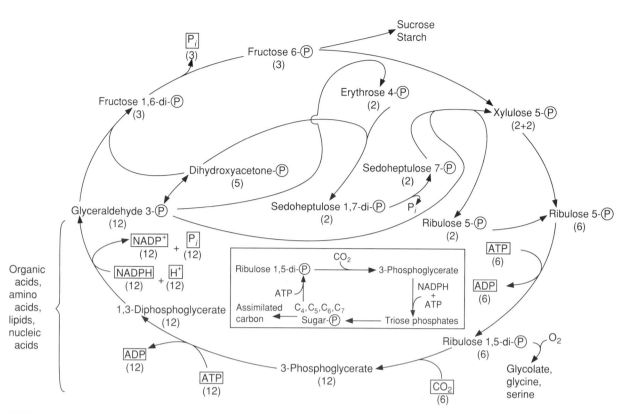

FIGURE 3 The photosynthetic carbon reduction cycle: Calvin cycle or C₃ pathway. The insert shows a summarized version of the pathway.

formed by decarboxylation is therefore concentrated in the bundle-sheath cells and is fixed by the Calvin cycle. The bundle sheath cells are the major sites of photosynthesis in C₄ plants.

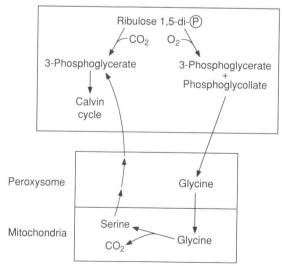

FIGURE 4 The photorespiratory pathway (summary).

CAM plants take up CO_2 in the dark, convert it in the presence of PEP to oxaloacetate, which is stored as malate in the vacuoles (as in C₅ plants, Fig. 5) in contrast with C₄ plants where it is synthesized in light. In the daylight the stomata close and the internally generated CO_2 from malate is consumed by the Calvin cycle. The main difference between C₄- and CAM-CO_2 fixation is that the carboxylation step is specially separated from Calvin cycle in C₄ plants

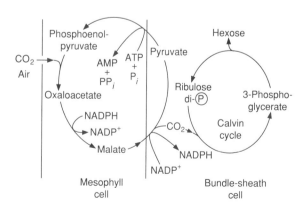

FIGURE 5 The C₄ pathway (Hatch-Slack pathway) of photosynthetic CO_2 fixation.

(different cell types, Fig. 5) whereas in CAM they are separated by time (i.e., light or dark periods). In conclusion, the CAM plants minimize water loss in the arid environment during the day time. This is, however, achieved at the expense of energy to store and reuse malate; therefore, most CAM plants are slow growers.

IV. Primary Metabolism

The primary metabolism involves the formation of basic molecules of the cell, such as, carbohydrates, lipids, amino acids, and nucleic acids. Generation of energy is required for growth and survival of the cell. The metabolic processes are inter-related with each other and a fine balance exists between the biosynthesis and breakdown of the metabolites. Thus, the cell grows in an orderly and highly controlled manner.

A. Carbohydrates

In the previous section the two phases of photosynthesis have been discussed. The sugars produced by photosynthesis are further transformed into various metabolites within the cell. The energy currency from the carbohydrates is derived principally via glycolysis and tricarboxylic acid cycle (TCA or Krebs cycle).

In glycolysis the starting metabolites are the hexose sugars which are produced by the storage carbohydrates, starch and sucrose. Basically, the glycolytic degradation of hexoses is via (a) formation of a number of hexose phosphates, (b) their interconversions, (c) formation of a series of triose phosphates, and finally, (d) conversion to pyruvate. The process does not require oxygen and can occur in both aerobic and anaerobic conditions; the overall reaction being

$$C_6H_{12}O_6 + 2\,P_i + 2ADP \rightarrow 2\,\text{Pyruvate}$$
$$\text{Hexose} \qquad\qquad + 2ATP + H_2O.$$

Under aerobic conditions NADH formed is converted to NAD^+ by the mitochondrial electron transport chain to ensure continuation of glycolysis. But under anaerobic conditions (rare in plants) reoxidation of NADH is prevented. The reactions of glycolysis occur in both the cytosol and plastids. The regulation of the pathway is affected at two points, first in conversion of fructose 6-P to fructose 1,6-di-P catalyzed by phosphofructokinase, and second, in conversion of phosphoenolpyruvate to pyruvate which is catalyzed by pyruvate kinase.

Under aerobic conditions the tricarboxylic acid cycle (TCA cycle; Fig. 6) oxidatively decarboxylates pyruvate in the mitochondria to acetyl-CoA and further to two molecules of CO_2 with an overall generation of four NADH, one $FADH_2$, and one ATP in a single turn of the cycle. A number of metabolites of the cycle provide precursors which could be utilized for the synthesis of other cell components, for example, in the synthesis of amino acids, pyrimidines, and lipids. This cycle is particularly important in nonphotosynthetic cells for generating ATP or reductant. However, excessive diversion of the metabolites to other outlets may potentially lead to cessation of the cycle. The regulation of the cycle takes place at several points. First, a decreased input of acetyl-CoA (mainly from pyruvate or fatty acids) into the mitochondrion could limit the initial step. In addition, a number of enzymes of the cycle are also regulated by various metabolites, e.g., acetyl-CoA, succinyl-CoA, NADH, AMP, ADP, ATP, and oxaloacetate. The overall reaction of the TCA cycle is as follows:

$$\text{Acetyl unit} + 3NAD^+ + FAD + ADP +$$
$$P_i \rightarrow 2CO_2 + 3NADH + FADH_2 + ATP.$$

In addition to the above pathways of carbohydrate catabolism, oxidative pentosephosphate pathway (OPP or hexose monophosphate shunt) can also operate as an alternative or a supplement. The enzymes of the pathway have been found in the cytosol of both photosynthetic and nonphotosynthetic cells. Some of

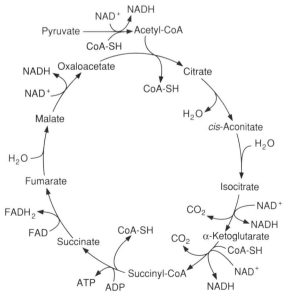

FIGURE 6 The reactions of citric acid cycle (also called tricarboxylic acid cycle or Krebs cycle).

the enzymes are also found in the chloroplasts. Several of the intermediates are shunted around the glycolytic pathway. This provides versatility to the organism for conversion of the carbon source to form aromatic amino acids and phenolic substances. The pathway begins with the phosphorylation of glucose to glucose 6-P by hexokinase which is then oxidized to yield various metabolites and reducing potential (NADPH) necessary for biosynthetic processes. The important intermediates are: fructose 6-P, ribulose 5-P, ribose 5-P, xylulose 5-P, erythrose 4-P, sedoheptulose 7-P, and glyceraldehyde 3-P. The generation of NADPH in the cytosol is considered to be very important.

The major carbohydrates in plants are sucrose and starch which are synthesized as a result of photosynthesis. Sucrose is synthesized via UDP-glucose (a high-energy glucose donor) in the cytosol. There are two possible routes of synthesis catalyzed by sucrose phosphate synthase and sucrose synthase, respectively:

1. UDP-Glucose + fructose 6-P → UDP + sucrose 6-P → sucrose + P_i
2. UDP-Glucose + fructose ⇌ UDP + sucrose.

The sucrose phosphate synthase route is widely accepted and the second route is considered to be involved mainly in the sucrose-breakdown process. Cleavage of sucrose can also be achieved by invertase yielding glucose and fructose. Most of the sucrose synthesized is translocated via phloem to other parts of the plant either for further utilization or for storage as reserve in the vacuole of parenchymatous tissue. The polysaccharide starch (amylose) is synthesized in the chloroplast by starch synthase:

ADP-Glucose + (glucose)$_n$ → ADP + (glucose)$_{n+1}$. Starch primer

Glucose phosphorylase provides the glucose donor ADP-glucose:

Glucose 1-P + ATP → ADP-glucose + PP_i.

Branched-chain starch, amylopectin, is formed by a branching enzyme (a transferase) which transfers a preformed section of the straight-chain polymer to the C-6 of the glucose residue of another part of the starch molecule. Amylopectin, therefore, comprises α-(1 → 4)-linkages with α-(→ 6)-linked branches. Starch breakdown is mediated by the action of hydrolases (maltase and α-amylase) and phosphorylases: the former either removing one glucose at a time from the nonreducing end of the starch chain or cleaving the internal glycosidic linkages. The phosphorylase

catalyzes cleavage of the nonreducing glycosyl residue producing glucose 1-P. In storage tissues such as in seeds and tubers, starch is deposited over a longer period of time and is mobilized during germination to support new growth. In some plant storage tissues (e.g., tubes of Jerusalem artichoke) the fructose polymer fructan accumulates on a long-term basis. Sucrose is the precursor for the synthesis of this polysaccharide and two specific ructosyltransferases are involved. Depolymerization is catalyzed by fructan exo-hydrolase.

The structural carbohydrates in plants are those which constitute the cell wall. The main polysaccharides are: cellulose, hemicelluloses (xyloglucan, xylan, mannan, glucomannan, callose, arabinogalactan), and pectin (homogalacturonan, rhamnogalacturonan, arabinan, galactan). The common precursors for biosynthesis of these molecules are the nucleoside diphosphate sugars: in most cases UDP-sugars. The sugar moiety is enzymatically added to the growing chain of polysaccharide molecules. The synthesis takes place in the endomembrane system. The nucleotide sugars are formed in the cytosol which are transferred to the Golgi for assembly of polysaccharides. Specific glycosyl transferases catalyze the biosynthesis. Degradation of the polysaccharides is catalyzed by specific hydrolases.

B. Lipids

Plant lipids include neutral lipids, waxes, phospholipids, and glycolipids. The main neutral lipids are triacylglycerols and are abundant in oil-containing seeds. The acyl part consists of either the saturated fatty acids (e.g., palmitic, lauric, stearic, myristic, arachidonic) or unsaturated fatty acids (oleic, linolenic, linoleic). Those containing saturated fatty acids (solid at room temperature) are termed fats while those containing unsaturated fatty acids (liquid) are termed oils. The neutral lipids act as energy reserve in many plants. They are also commercially important as fats and oils. The seed storage oils are synthesized in the endomembrane system from glycerol 3-P and acyl-CoA, catalyzed by specific acyl transferases. The mobilization process involves lipases which hydrolyze the fatty acids from glycerol molecule. The fatty acids are converted to acyl-CoAs and undergo β-oxidation followed by entry into glyoxylate cycle in glyoxisomes and probably form sucrose in developing seedlings.

Waxes are generally esters formed between highly saturated fatty acids and aliphatic alcohols. These compounds make thin coats on the epidermal layers of leaves, fruits, flowers, and stems. Their function is probably protective and to restrict water evaporation from the surface. Their synthetic pathway is via acyl-CoA and transfer of acyl moiety to a primary alcoholic group. The process of synthesis and many of the enzymes involved are poorly understood. Among the phospholipids, phosphatidyl choline, phosphatidyl ethanolamine, and phosphatidyl inositides are mainly found in seed tissues while phosphatidyl glycerol occurs in leaf tissues. Phospholipids are components of nonchloroplast membranes. The fatty acid groups are attached via acyl-CoA to C-1 and C-2 of glycerol 3-P. The main glycolipids in plants are mono- and digalactosyldiglycerides and are the major lipid constituents in chloroplasts.

C. Nitrogen Metabolism

Most important nitrogen-containing molecules in plants are amino acids, proteins, and nucleic acids. In some plants atmospheric nitrogen can be fixed to synthesize ammonia via nitrogenase-catalyzed reactions carried out by bacteria in root nodules of legumes. In most plants nitrate is the main source of nitrogen and is absorbed by the root system. Nitrate is reduced to ammonia which is catalyzed by nitrate reductase (NADH is required). The ammonia is subsequently converted to organic nitrogen in the form of glutamine, a reaction catalyzed by glutamine synthase:

$$Glutamate + ammonia + ATP \xrightarrow{Mg^{2+}} glutamine + ADP + P_i$$

Previously it was thought that glutamate dehydrogenase was the key enzyme for reductive amination of 2-oxoglutarate to glutamate, but this enzyme is now believed to take part in the breakdown of glutamate. Glutamate synthase is another important enzyme which can transfer the amide group of glutamine to 2-oxoglutarate to yield glutamate. Thus, ammonia is assimilated with the participation of both, glutamine synthase and glutamate synthase. The most ubiquitous enzymes, transaminases, then catalyze the transfer of an α-amino group from an α-amino acid to 2-oxoacid to synthesize a new amino acid and a new oxoacid. A range of amino acids is therefore produced (except proline). The reversibility of the reaction gives flexibility to the supply/demand of particular amino acids. The main pathways that supply the carbon skeleton for the synthesis of amino acids are glycolysis and the tricarboxylic acid cycle.

Amino acids are the main source for the synthesis of purines and pyrimidines which are the essential components of nucleotides (nucleotide sugars are essential for the biosynthesis of complex carbohydrates) and nucleic acids. Chlorophyll is also derived from amino acids. In the *de novo* synthesis of pyrimidine nucleotides, CTP and UTP, glutamine is enzymatically converted to carbamoyl-P with the uptake of CO_2. Carbamoyl-P then combines with aspartate to form *N*-carbamoylaspartate followed by a series of reactions. The pyrimidine ring is formed first, with the subsequent attachment of a ribose 5-P group via phosphoribosyl pyrophosphate. Although little is known about purine nucleotide synthesis in plants, it is suggested that purine synthesis involves, first, ribosylamide 5-P synthesis from glutamine and then phosphoribosyl pyrophosphate. Other structures are then added on, and AMP and GMP are the final products. Deoxynucleotides are formed by reduction of ribonucleotides. Nucleotides are also formed from free pyrimidines and purines which in turn result from the metabolic turnover of RNA.

Incorporation of radioactively labeled deoxyribonucleoside triphosphate into DNA-like product was initially demonstrated in the presence of DNA, a divalent cation, four deoxyribonucleoside triphosphates, and an enzyme preparation from mungbean seedlings. As in bacterial systems, various DNA polymerases are involved in the above synthesis. DNA ligase catalyzes the synthesis of phosphodiester bonds and requires ADP. DNA replicates autonomously in addition to forming template for the production of RNA. The enzyme RNA polymerase binds to the DNA molecule at a specific site and catalyzes initiation of RNA synthesis and elongation of the RNA strand. There are three classes of this enzyme producing three major RNAs. The RNA transcripts are processed in the nucleus and then transported to the cytoplasm. The translation of mRNA into protein occurs in the cytoplasm. tRNA provides the recognition required for translation of mRNA into polypeptides. A codon in the mRNA transcript is responsible for each amino acid in a polypeptide. Protein synthesis involves (a) initiation of a polypeptide chain, (b) elongation of the chain, and (c) termination of the chain synthesis. Many proteins are post-translationally modified. Phosphorylation, glycosylation, conversion of proline to hydroxyproline, etc., are examples of post-translational modification of newly synthesized proteins.

V. Secondary Metabolism

There are numerous naturally occurring compounds in plants which are regarded to be nonessential metabolically; nonetheless, they are important in ensuring biological viability. Some compounds determine color of flowers which aids in insect pollination, some are toxic molecules providing protection against predators, some act as antibiotics preventing infection by fungal pathogens, and some of the secondary metabolites are pharmacologically and commercially important (e.g., alkaloids, rubber, essential oils, pigments).

A. Phenolics

There are many thousands of plant-derived compounds which contain phenolic residues. They can be categorized into groups based upon the number of carbon atoms in their skeleton. Except for flavonoids, most of them arise from phenylalanine or shikimic acid. Derivatives of catechol (urushiol from poison ivy), resorcinol (tetrahydrocannabinol from *Cannabis*), p-hydroxybenzoic acid, protocatechuic acid, vanillic acid, syringic acid, p-coumaric acid, caffeic acid, ferulic acid, cinapic acid, and gallic acid are present in many plants. The main shikimate pathway provides a number of precursors.

The derivatives of benzoquinone are widely distributed in plants; for example, plastoquinone and ubiquinone play important roles in photosynthesis and mitochondrial electron transport, respectively. Naphthoquinone and anthraquinone derivatives are also of common occurrence (hylloquinone, vitamin K_1 functions in photosynthesis). They also originate from the shikimate pathway.

Flavonoids are water-soluble phenolic derivatives (glycosides), generally occur in vacuoles, and are brightly colored compounds. Their basic structure consists of two aromatic rings joined together with a three carbon unit. They are responsible for the different shades of color in flowers and other parts of plants. Shikimate pathway plays an important role in providing the precursors for their biosynthesis. Participation of acyl-CoAs, cyclization, isomerization, aryl migration, hydroxylation, O-methylation, O- and C-glycosylation, C-alkylation, and polymerization, all enzyme-catalyzed reactions, finally yields the various products.

Lignins, tannins, and melanins are all polymeric molecules containing phenolics. Lignins are very stable and resistant molecules and are important cell wall constituents of supporting and conducting tissues (xylem) of vascular plants. They confer strength to the cell wall and protect it from chemical, physical, and biological attacks. The building blocks of the macromolecules are phenyl propane (C6–C3) residues. Different aromatic aldehydes are produced when they are subjected to mild oxidation (mostly vanillin and syringaldehyde in dicots). Tannins are of two types: hydrolyzable and condensed tannins. The former usually contains glucose at the core of the molecule and all its hydroxylic groups are esterified with either gallic acid or hexahydroxydiphenic acid. Condensed tannins are made up from only phenols of the flavone type and do not contain any sugar in the structure. Melanins are deep brown or black pigments found in seeds and spores.

B. Terpenes and Terpenoids

The key process in plant life, photosynthesis, depends upon the existence of specific derivatives of terpenes and terpenoids, e.g., carotenoids and chlorophylls. Terpenes and terpenoids can be derived from basic branched unit of carbon skeleton

and can be classified as: hemiterpene (5 carbon isoprene), monoterpene (10 carbon, geraniol), sesquiterpene (15 carbon, farnesol), diterpene (20 carbon, gerenylgereneol), sesterterpene (25 carbon, ophiobolin A), triterpene (30 carbon, squalene) tetraterpene (40 carbon, phytoene), and polyterpene (rubber), most of them take up cyclic forms. Monoterpenes and sesquiterpenes occur widely in higher plants as components of essential oils and can be recognized by their strong smell. Most sterols are derived from cyclic triterpenes; steroid glycosides, steroid hormones, and steroid alkaloids are some of the pharmacologically important compounds. Carotenoids are tetraterpene pigments, are localized in the chloroplasts, and contribute to light harvesting within the photosynthetic assembly in photosynthesis and also play a protective role against harmful radiations. They are also responsible for fruit pigmentation, e.g., in ripe tomato and in red pepper. Carrot is the well-known source of β-carotene. The polyterpene rubber which consists of 1500–60,000 isoprene residues is produced in the latex of a number of genera of angiosperms. However, *Hevea brasiliensis* is the most commercially successful plant for the production of rubber.

The basic concept for the biosynthesis of terpenoids involves polymerization of isoprene units via its active derivative, isopentenyl pyrophosphate (IPP). This compound is derived from mevalonic acid for which acetyl-CoA is the precursor. A number of steps catalyzed by specific enzymes are involved in the whole biosynthetic process.

C. Nitrogenous Compounds

Purines are oxidized to ureids which are substituted ureas (NH_2-CO-NHR and RNH-CO-NHR) such as, allantoin, allantoic acid, and citrulline. In a variety of plants these compounds act as nitrogen transporters. Several legumes, referred to as tropical legumes (soybean, cowpea, mungbean), switch to ureid synthesis from glutamine and asparagine when their root systems are nodulated. The temperate legumes (pea, lupin, alfalfa) continue to synthesize glutamine and asparagine. The bacteroids in root nodules reduce N_2 to NH_3 which is assimilated into glutamine and then transformed to purine. The purine, xanthin, is exported to the cytosol to be oxidized to uric acid by xanthin dehydrogenase. The action of uricase and allantoinase yield, respectively, allantoin and allantoic acid.

The heterocyclic nitrogen-containing compound pyrrole is the basic constituent of tetrapyrroles providing the metalloporphypin derivatives formed by chelation with iron or manganese. Iron porphyrins are present as the prosthetic groups of cytochromes, leghemoglobins in root nodules, and enzymes such as catalase and peroxidase. Magnesium porphyrins are the chlorophylls. The preferred biosynthetic pathway outlines the following conversions: Glutamate → glutamyl 1-semialdehyde → 5-aminolevulinate → porphobilinogen → uroporpyrinogen III → protoporphyrin IX. This last compound (a tetrapyrrole) is the branch point for the formation of either magnesium or iron porphypins. Magnesium is inserted into the protoporphyrin IX ring by Mg-chelatase. This is followed by further enzyme-catalyzed reactions to yield chlorophyll-*a*.

Alkaloids constitute a very large group of nitrogen-containing plant secondary products, many of which have pharmacological importance. For most of them the biosynthetic precursors are amino acids. They accumulate in actively growing tissues, epidermal and hypodermal cells, vascular sheaths, and latex vessels. They are frequently translocated to other tissues from their sites of synthesis. Their biochemical and physiological role is not clear, however, in several cases they offer protection against predators.

VI. Plant Hormones

The term hormones was first coined in the context of higher animals to depict chemical substances synthesized in minute quantities which are secreted into the blood stream and transported to distant organs and tissues, thereby influencing their function. The more appropriate term in the plant context will be plant growth substances or plant growth regulators. Unlike in animals, they are not necessarily transported away from their sites of synthesis. The major classes of the compounds are: ethylene, auxins, cytokinins, gibberellins, and abscisic acid.

A. Ethylene

Ethylene gas ($CH_2 = CH_2$) as a plant hormone has been known since the last century. In angiosperms all tissues are capable of synthesizing this hormone; its site of action is close to its site of synthesis although it is sufficiently water soluble and can be transported across all membranes. The precursor of ethylene synthesis is L-methionine which is transformed to amino cyclo-propyl carboxylic acid (ACC) by ACC synthase. This is then oxidized by ethylene-forming enzyme (ACC oxidase) to yield ethylene. The physiological role of ethylene is diverse. For example, it stimulates the ripening of fleshy fruits, abscission of leaves, radial swelling of stems, horizontal growth of stems, adventitious root formation, flowering and fading of some flowers, and epinasty (downward curve of leaves), and inhibits root and stem elongation. The molecular mechanism of ethylene action is not sufficiently clear; however, it can rapidly stimulate the synthesis of a host of cell wall degrading enzymes in fleshly fruits. It is not certain whether a ethylene-specific receptor/binding protein exists, nor is it known how are the signals triggered or conveyed for the synthesis of specific proteins.

B. Auxins

Although the growth-promoting effect of diffusible chemical auxin was known at the turn of the century, it was not until 1972 that the compound was identified as indole-3-acetic acid (IAA). Several derivatives of IAA have since been isolated and characterized. L-Tryptophan is the primary precursor of IAA biosynthesis which involves tryptophan transaminase, indole pyruvate decarboxylase, and indoleacetaldehyde dehydrogenase. IAA can be oxidized *in vivo* by several of the peroxidase isoenzymes and also by IAA oxidases lacking peroxidase activity. These reactions are significant in the light of physiological control of IAA level in plants.

The physiological role of auxins is varied and complex. They may have both primary and secondary effects and the entire process is poorly understood. The well-known effect of auxin is to promote cell enlargement. It is produced in shoots and transported by polar transport system; lateral transport accounts for phototropism and geotropism. Auxins induce root formation, vascular differentiation, tropic responses, elongation of internodes, growth of ovary into fruit, ethylene synthesis, cell division and enlargement in callus tissue, cell wall elasticity, and hydrogen ion extrusion. It inhibits root elongation and senescence in excised bean endocarp, and controls abscission. Auxin must enter the target cells which are in the process of differentiating in meristematic tissues. This leads to the prediction of the presence of its receptor, followed by signal transduction and occurrence of subsequent biochemical events. However, the presence of receptor is not universally evident and the IAA response occurs within 10–15 min, thus, excluding the possibility of *de novo* protein synthesis. Many of the slower responses that follow are, on the other hand, the result of protein synthesis.

C. Cytokinins

These compounds promote cytokinesis (cell proliferation/division) in cultured plant cells. Zeatin was the first naturally occurring cytokinin isolated and characterized from the immature corn kernels. Kinetin [6-(furfurylamino) purine] is the first chemically derived cytokinin and a large number of synthetic cytokinins are to date available. Some of the derivatives are extremely effective even in picomole concentrations. The physiological effects of cytokinins are diverse: to mention a few, they can induce elongation in hypocotyls, DNA synthesis, tuberization in excised potato stolons, formation of enzymes (in roots, cotyledons, calli, and seedlings), bud development, and germination of seeds. Translocation of cytokinin in plants is not clearly understood: synthesized molecules in the root tips are transported to other parts probably via the xylem sap.

It is not clear whether cytokinins occur in free form or as a part of tRNA as demonstrated in 1966. However, exogenously added cytokinins in tissues are not incorporated in significant amounts in tRNAs. *N*-(3-*methyl*-2-butenyl)Adenosine is synthesized by tobacco callus cells by attachment of a side chain to an adenosine residue of RNA. Thus, tRNA turnover may be a possible route of cytokinin synthesis. On the other hand in pea root tips the level of free cytokinin was 27 times more than could be obtained from tRNA degradation. The *in vivo* mode of action of cytokinins to produce the physiological effects is still obscure.

D. Gibberellins

Gibberellins were first discovered in the fungus *Gibberella fujikuroi* as tetracyclic diterpenoid compounds and to date a large number of its family members are known. They exist as gibberellic acids and are abbreviated GA_1, GA_2, etc. They have been detected in all parts of higher plants; however, they are less in vegetative than reproductive parts. Immature seeds are rich source of GAs. Biosynthetically they are derived from the tetracyclic diterpene *ent*-kaurene; thus, the starting compound is acetyl-CoA which is first converted via a multistep process to geranyl-geranyl pyrophosphate and then to *ent*-kaurene. GA_{12} aldehyde is then formed and this is the point where the pathway leads to the synthesis of all other GAs. The sites of synthesis are probably the plastids. Structural modifications of GAs, such as 2β-hydroxylation, formation of β-glucosyl ethers, or esters, and formation of peptide-linked GAs render them inactive. This has a regulatory effect on the levels of biologically active GAs. In addition, the derivatives can also be transported, stored, and further regenerated to the active form.

Gibberellins have been shown to induce normal growth in several dwarf mutants of plants and they enhance the elongation growth in normal plants. Cell elongation rather than increased growth was the main cause and this was preceded by the proliferation of endoplasmic reticulum and polysomes. Thus, GAs can control the formation and secretion of a number of biochemically important enzymes, e.g., hydrolases in germinating cereal grains. Application of GA can stimulate flower formation in certain plants and, in some, they can cause morphological reversion to the juvenile form of growth.

E. Abscisic Acid

The sesquiterpenoid abscisic acid (ABA) is a diffusable abscission-accelerating substance in a large number of plants. The chemical synthesis of ABA [3-methyl-5-(1'-hyaroxy-4'oxo-2'-cyclohexene-1'-yl)-cis-2,4-pentadienoic acid] was accomplished and several analogs and ^{14}C-ABA were prepared. Biosynthetically, ABA can be formed directly from farnesyl pyrophosphate; therefore, mevalonate can be incorporated into the molecule. In the alternative pathway which appears to operate in higher plants, ABA can be derived from an oxygenated carotenoid such as violaxanthin via an intermediate known as xanthoxin. The carotenoid synthesis inhibitor fluoridone inhibited $^{14}CO_2$ incorporation into violaxanthin and ABA to the same extent.

In addition to promoting abscission, ABA is known to play other physiological roles, such as geotropism in roots, stomatal closure, induction of bud dormancy, seed dormancy, tuberization, senescence in detached leaves, and ripening of nonclimacteric fruits.

VII. Plant Genetic Engineering

Plants are the most renewable source of energy which harvest sun light; their usefulness at the present time and in the future is unlimited. Plants have always been important for mankind and many of the cultivated crops used by us have undergone improvement over time. Traditional methods of breeding and screening have largely been employed and still contribute extensively to species improvement. This of course did not require much biochemical knowledge of plant metabolism. However, we are now genetically manipulating plant characteristics directly associated with primary and secondary metabolism to suit our needs by introducing appropriately cloned genes into plants. In this way the advantageous characteristics of one species can be transferred to another species. Indeed, bacterial genes encoding proteins with insecticidal activity have been introduced into higher plants conferring resistance against certain insects. Plants are being genetically engineered for various purposes, for example, to improve quality, to achieve novel usages, enhanced productivity, increased protection against pathogens, etc. [See PLANT GENETIC ENHANCEMENT.]

The gene transfer technology takes advantage of the natural tendency of the soil bacteria of the genus Agrobacterium to insert a part of its genome into the genome of the infected plant. This bacterium is the cause of crown gall disease and has evolved a natural system for genetically engineering plant cells for its own survival. Viral-mediated DNA transfer can also be achieved. DNA transfer without biological vector is also possible, for example, direct uptake of DNA by protoplasts, by microinjection of DNA, and liposome-mediated DNA transfer. The microinjection technology offers a more versatile and unique approach by which specific DNA molecules can be targeted in specific intracellular compartments. It overcomes the limitations of the using Agrobacterium for transforming species within its host range. The Agrobacterium infection in susceptible plants is manifested as tumor induction at the infected site. The DNA segment transferred (T-DNA) to the plant is present in the bacterial plasmid, Ti plasmid (tumor inducing). A specific region of the T-DNA has been found to integrate with the genomic DNA of the plant. Plasmid constructs carrying the engineered T-DNA for inserting a desired gene into host are used in genetic engineering for the production of transgenic plants. [See PLANT GENE MAPPING.]

One area of plant species improvement is the enrichment of certain amino acids in seed proteins as food for humans and animals and also in forage crops. Globally almost 80% of proteins for human consumption is derived from plant seeds. Thus, the levels of essential amino acids that are not synthesized by animals can be increased by genetic engineering. Similarly, the content of sulfur-containing amino acids can be increased in pasture grass to improve wool production by sheep. Many toxic contents of agriculturally important plants can be eliminated rendering the products suitable for consumption, for example, removal of the toxic lectin ricin from castor beans, removal of the neurotoxin (β-N-oxalyl-α, β-diaminopropionic acid) from the pulse crop Lathypus sativus, and improving oil seed plants to produce the desired quality and quantity of oil. Altering existing enzymes or expressing new enzymes in plants has produced petunias with a new range of colors, improved the quality of timber with an altered vascular system, and improved the shelf life of some fruits (e.g., in tomato by restricting enzymatic cell wall degradation). Novel uses of plants have been achieved in the field of industrial oils and secondary metabolites. Short-chain fatty acid-containing oils are important in soap making and these are obtained from labor-intensive palm and coconut plantations. Here is the case for introducing appropriate genes into mechanically harvestable crops for the production of the required quality of oils. Similarly, oils with hydroxylated fatty acids could be produced for the manufacture of lubricants, solvents for the synthesis of plastics, and many other industrial products. Secondary metabolites from plants support the multibillion dollar drug industry. Appropriate transgenic plants either in crop form or in cell culture form will help to reduce the cost of production and offer opportunities of growing and producing the raw material in nonconventional environments. High yields of crops have been achieved by increasing the concentration of the carbon source, i.e., CO_2 in photosynthesizing plants. As this is not possible to implement directly, efforts are being made to modify the key regulatory enzyme ribulose bisphosphate carboxylase (Rubisco) for enhanced CO_2 fixation. However, Rubisco also displays oxygenase activity which takes part in photorespiration and competes directly with the carboxylase that is involved in CO_2 fixation. Research is also being carried out in order to produce transformed plants which would contain pest-specific toxins such as protease inhibitor, Bt-toxin (against insect larvae) and herbicide-resistant plant varieties. In the latter case weeds belonging to the same family can be eliminated without affecting the transformed plants. [See PLANT BIOTECHNOLOGY: FOOD SAFETY AND ENVIRONMENTAL ISSUES.]

Bibliography

Anderson, J. W., and Beardall, J. (1991). "Molecular Activities of Plant Cells: An Introduction to Plant Biochemistry." Blackwell Scientific, London.

Dennis, D. T., and Turpin, D. H. (1990). "Plant Physiology, Biochemistry and Molecular Biology." Longman, UK.

Dey, P. M., and Dixon, R. A. (1985). "Biochemistry of Storage Carbohydrates in Green Plants." Academic Press, New York.

Dey, P. M., and Harborne, J. B. (1989–1993). "Methods in Plant Biochemistry," Vols. 1–10. Academic Press, New York.

Goodwin, T. W., and Mercer, E. I. (1983). "Introduction to Plant Biochemistry." Pergamon Press, London.

Harwood, J. L., and Russell, N. J. (1984). "Lipids in Plant and Microbes." George Allen and Unwin, London.

Lea, P. J., and Leegood, R. C. (1993). "Plant Biochemistry and Molecular Biology." Wiley, London.

Lloyd, C. W. (1991). "Cytoskeletal Basis of Plant Growth and Form." Academic Press, New York.

Simmonds, R. J. (1992). "Chemistry of Biomolecules." Royal Society of Chemistry, London.

Plant Biotechnology: Food Safety and Environmental Issues

ERIC L. FLAMM, *Food and Drug Administration*[1]

Glossary

Chromosome Double-stranded linear coil of DNA and proteins, a complete set of which composes an organism's genome

DNA, RNA (deoxyribonucleic acid, ribonucleic acid) Genetic material of all life on earth

Gene Segment of DNA (or RNA in some viruses) containing the information necessary to assemble a protein (or, in some cases, an RNA molecule)

Gene expression Reading of a gene by cell machinery to produce the gene product

Genome Totality of an organism's genetic material

Molecular biological techniques Techniques by which scientists can purify DNA, cut out and isolate a discrete segment, splice it into the DNA of another organism, and have the new DNA segment become an integral and indistinguishable segment of that organism's genome; these techniques also are referred to as gene splicing, cloning, genetic engineering, and recombinant DNA (rDNA) technology

Plant biotechnology Use of molecular biological techniques in plant breeding

Recombination Splicing together of DNA molecules or segments of molecules

Transgenic Organisms containing genetic material introduced by molecular biology techniques

Plant biotechnology enables biologists to develop plants and foods with a greater range of new attributes

[1] The opinions in this paper are the author's own and do not necessarily reflect those of the FDA.

than previously possible. The use of biotechnology per se raises no unique food safety or environmental issues and, in fact, can often minimize the likelihood of certain unwanted events occurring. However, the new substances or attributes may present a different set of safety issues than had been associated previously with that particular plant or food.

I. General Attributes of Plant Biotechnology

A. Plant Breeding

All plant breeding entails, first, the introduction of new traits into a plant and, second, the selection of progeny with the desired traits. With nonmolecular methods, breeders work at the whole-organism level. Breeders can cross (mate) plants with different desirable characteristics and look for those that have the desired combination of traits, or they can introduce random mutations by various mutagenesis techniques and look for plants with desirable characteristics. Once the breeder obtains plants that appear to have the desired characteristics, he must subject them to progressively larger evaluations, from greenhouse to small plots, to larger plots, to multisite and multienvironment trials to determine whether the new plant line has any unforeseen undesirable attributes and whether it performs satisfactorily in expressing the desirable traits. [*See* PLANT GENETIC ENHANCEMENT.]

With the tools of molecular biology, breeders can introduce into plant cells discrete segments of DNA containing particular genes known to encode the desired traits. Thus, the molecular techniques give breeders the ability to target their modifications to an extent never before possible. Also, because all known organisms use the same genetic material and the same genetic language, genes from one organism will still encode the same proteins when expressed by another organism. Therefore, the molecular techniques give breeders the ability to introduce a greater range of traits than previously possible.

However, once the breeder has selected plants with the newly introduced traits, he must subject them to the same kinds of evaluations necessary for plants derived with any other breeding techniques. Thus, the molecular techniques are an advance on previous breeding techniques only in the first step of breeding, that of introducing a new trait to a plant.

Scientists are using the techniques of molecular biology to study the processes involved in plant growth and development, with the goal of developing plants with increased resistance to insects, plant viruses, and herbicides; with longer times to ripen or to rot; with increased starch content; with more nutritionally balanced proteins; and with other improvements.

Molecular biology techniques have been incorporated into all aspects of biology, in fields as diverse as parasitology, entomology, immunology, and anthropology. Its use in agriculture has engendered controversy, because it entails introductions into the environment, because it entails modifying food, and because few people see the need to improve food production.

Many of the concerns raised about biotechnology focus on specific products, for example, that certain foods may be unsafe or that certain traits in plants may cause environmental problems. Others focus on biotechnology itself, for example, that it is unnatural or that transcending breeding barriers may yield unpredictable results. Although this article primarily addresses food safety and environmental questions, some of the concerns raised about biotechnology itself are addressed briefly.

B. Is Biotechnology Unnatural?

All breeding may be considered unnatural, because breeding substitutes human selection for natural selection. However, biotechnology might be considered more unnatural, primarily because it entails the use of laboratory techniques and provides the ability to transcend natural mating barriers. Whether "unnaturalness" can have gradations seems more an aesthetic question than a scientific one. Although some individuals might argue that a tomato containing a frost-resistant gene copied from a fish is the pinnacle of "unnaturalness," because such a tomato could never arise naturally, others might argue that a tomato that cannot transfer pollen (as is the case for many commercial tomato varieties) or a seedless watermelon,

is more unnatural because it cannot propagate itself. Interestingly, other "unnatural" breeding techniques, such as laboratory-assisted crosses between plants of different species and genera, and mutagenesis and regeneration from tissue culture, are not usually tarred with the "unnatural" label, nor is a child born from artificial insemination or *in vitro* fertilization considered unnatural.

C. Is Interspecies Gene Transfer Intrinsically Risky?

All known organisms are related, sharing a common evolutionary history and, therefore, a common genetic code. For this reason, genes from one kingdom can function when transferred to another kingdom. [See Transgenic Animals.]

Because only a single genetic code exists, taxonomic differences have little meaning at the level of the gene. A gene is a gene, and its product is the same protein, regardless of the organism in which it resides. In fact, plants, microbes, and animals contain many genes in common. A gene from a cow is not "cow-like", and does not confer "cow characteristics" to the organism into which it is introduced. Nor would a "cow gene" in a plant represent physical cow material. Rather, it would be the plant's copy of a gene found in and originally copied from the cow genome. For this reason, rabbinical koshering authorities have ruled that the bovine enzyme chymosin, used to make cheese, is not considered "meat" when it is derived from recombinant microbes instead of from cows (as rennet).

The ability of newly introduced genes to significantly alter their new host is tempered by the fact that the product of the genes must be integrated into the developmental stages of the organism, and into the overall physiology and morphology of the cell, tissue, and organism. If the newly introduced genes disrupt that system, the organism will not develop or will die. Thus, the power of genetic engineering, although great compared with other techniques, cannot transform an organism into something completely different from what it was.

Some observers have argued that there may be risks that have yet to be identified, as was the case with nuclear power and pesticides. For example, might there be long-term effects, or might there be some unexpected interactions between the newly introduced substance and other substances in the diet? However, nuclear power and pesticides, by their very nature, raise health and environmental concerns. Nuclear power makes use of extremely hazardous materials; pesticides are designed to be toxic and to change the ecology of the environment in which they are used. A more appropriate model for the use of molecular biology to modify organisms is domestication, the genetic manipulation of plants, animals, and microbes for human purposes. This is not to say that domestication is risk-free, but that it is not intrinsically hazardous, as are the other models.

Regarding the question of long-term risks or risks from unexpected chemical interactions, these apply primarily to novel substances in the diet. With genetic engineering, the new substance introduced is usually a protein, or a modified fat or carbohydrate. In most cases, these are substances commonly found in the diet, are generally digested and metabolized, do not bioaccumulate, and do not raise the kinds of questions of chronic exposure and synergistic effects associated with novel chemicals that do not share those characteristics.

II. Food Safety Concerns

Plant biotechnology raises two principal food safety concerns. First, it allows scientists to introduce into a plant a wide variety of new proteins. These proteins themselves may raise safety issues, or the effects of the proteins on metabolic pathways in the plant may raise safety issues. Second, the introduced DNA segment, depending on where in the plant chromosome it integrates, may physically disrupt functions of the host chromosome, thereby causing unanticipated changes in the plant that may raise safety issues.

A. New Substances in Food

All breeding methods have the potential to alter the composition of food. Most often, the changes involve increasing or decreasing the concentration or modifying the structure of substances already in food. However, molecular techniques and matings between cultivated varieties and wild or weedy relatives also can introduce new substance into food. The new substances that are introduced are proteins, which may be introduced for their own properties or to alter the structure of fats, carbohydrates, or other proteins. [See FOOD CHEMISTRY; FOOD COMPOSITION.]

1. Proteins

Tens of thousands of different proteins are found in the diet. In general, individual proteins do not pose

health risks when eaten. Even rattlesnake venom is rendered harmless by digestion (explaining the ability of rattlesnakes to eat their kill). Nevertheless, a number of classes of proteins (including rattlesnake venom) would raise significant safety concerns if introduced into food. These proteins include toxins, hemagglutinins, enzyme inhibitors, vitamin-binding proteins, vitamin-destroying proteins, enzymes that release or activate toxic compounds (e.g., cyanogenic glucosidases), and selenium-containing proteins. [See FOOD BIOCHEMISTRY: PROTEINS, ENZYMES, AND ENZYME INHIBITORS.]

Although many of those kinds of proteins are found in foods, the method of preparation of those foods frequently entails steps (such as cooking or soaking) that inactivate or remove the offending proteins. Transferring such proteins into new foods would be inadvisable in principle. They could cause safety problems if these foods are not prepared in the same manner as the traditional foods containing them, or if the proteins are not similarly susceptible to the effects of such treatment when in the new food. A conservative approach for minimizing the risk attached to the introduction of new proteins in food would require that such proteins be assumed to be dangerous unless suitable information contradicts that assumption.

Several kinds of information can establish the safety of a protein, most notably whether the protein is already safely consumed as part of other foods or whether it belongs to a class of proteins that is safely consumed. Thus, if a protein is moved from corn to wheat, or from fish to tomatoes, there would be no reason to suspect the food safety of that protein (apart from possible allergenicity, which is discussed in the next section).

Similarly, if the biological or biochemical function of a protein is known to be the same as that of proteins safely consumed in food, and if that function raises no food safety concerns, there would be little reason to suspect the food safety of that protein even if it were not derived from a food-use organism. An example of such a protein would be a bacterially derived herbicide-insensitive enzyme that can substitute for an herbicide-sensitive plant enzyme. Of course, the safety of using a new herbicide on that plant would have to be assessed.

If, however, the biological or biochemical function of the protein is not known, the function raises a safety concern, or the protein is derived from a food whose safe consumption requires cooking or soaking (for example, lima beans or cassava), further studies or safety testing would appear warranted. For example, a protein known to be toxic to insects, such as the pesticidal delta endotoxin of *Bacillus thuringiensis*, clearly raises more safety questions than does a seed storage protein. Also, nutritional issues would exist for proteins that would be major new constituents of a food, for example, a seed storage protein introduced to improve the amino acid balance of a food.

2. Allergenicity

All known food allergens (save one particular RNA molecule in shrimp) are proteins. The converse does not appear to be true, however; relatively few of the proteins in any particular food are likely to be allergens. Although many breeding techniques can alter the levels of particular proteins in a food, and can even introduce new proteins into food (either by crosses with wild or weedy relatives or by causing the expression of quiescent genes or metabolic pathways), genetic engineering can transfer proteins between unrelated food organisms. Only with genetic engineering can one move a peanut protein into a tomato. The obvious concern is that developers not create foods with unexpected allergenic properties.

Although we think of a food as triggering an allergic reaction, the allergenic response is really to particular proteins in that food. Most food allergies are to proteins in plants, marine organisms, milk, and eggs. Despite the high intake of meat and poultry, and their high protein content, they rarely provoke food allergies. In the United States, foods that are commonly allergenic include fish, shellfish, legumes, nuts, milk, and cereals. The foods responsible for most allergies will vary somewhat with the eating habits of the population; as a food is eaten more widely, more people will be exposed to it, and will be exposed to it more often. The development of allergies in the United States to kiwi fruit is an example of this effect.

Proteins that are food allergens tend to have certain physical characteristics in common. They range in size, generally, between 10,000 and 70,000 daltons. They also are resistant to digestion, to acid, and to heat, all properties enabling them to survive processing, cooking, and passage through the digestive system intact. They are glycosylated (have sugar molecules attached to them), which may itself trigger sensitization or may simply reflect the fact that glycosylated proteins are resistant to digestion.

To determine whether a protein from a particular food is responsible for that food's allergenicity, one can test sera from sensitive people. Because not all people allergic to a particular food will always be allergic to the same protein in that food, it is necessary to test sera from a representative sample of the sensitive population. When a representative sample of people allergic to a particular food cannot be identified, or when a protein is derived from a nonfood source (for which no allergic population yet exists), predictions of potential allergenicity must rely on information about the source of the protein, its physical and chemical characteristics, and the level at which it will be present in the food. No generally accepted animal or *in vitro* tests are presently available to predict allergenicity.

3. Fats and Oils

Researchers currently are developing plants with altered oils. For example, breeders are developing rapeseed plants that produce oil of increased saturated fat content for use in margarines and candy. Such changes can have nutritional implications, especially were the oils to be mistaken for their low saturated fat namesakes, or were they to significantly alter the overall pattern of dietary fat intake.

Safety issues could arise for oils modified to contain fatty acids of known toxicity, such as erucic acid, and oils with characteristics significantly different from those of dietary fats and oils (and, therefore, lacking empirical evidence of safety), for example, having chain lengths longer than 22 carbons, cyclic substituents, or functional groups not normally present in dietary fats and oils.

4. Carbohydrates

Breeders are modifying plants to produce higher concentrations of starch or to alter the starch's amylose and amylopectin content or structure. In general, such modifications would not raise safety concerns, because the starches would be physiologically and functionally equivalent to starches commonly found in food. Nutritional questions might arise if a food were modified to contain high concentrations of indigestible carbohydrates (fiber), however. [*See* FOOD BIOCHEMISTRY: LIPIDS, CARBOHYDRATES, AND NUCLEIC ACIDS.]

B. Unintended Effects

When introducing new traits, plant breeders often find that they get more than they bargained for, that is, that a desired modification has unexpected and undesired attributes (for example, lowering the level of a bitter substance may cause the plant to be more susceptible to insect infestation, or altering a metabolic pathway may cause alterations in other interrelated pathways) or that other traits were inadvertently introduced or modified by the breeding technique.

From the perspective of food safety, an elevation in levels of natural toxicants is the unintended effect of most concern. Virtually all plants produce toxic substances, presumably as protection against pests. Plants often have several different metabolic pathways that yield toxins. However, most domesticated food plant species produce relatively low levels of toxicants in the edible part of the plant, although toxicant levels may be at unsafe levels in other parts of the plant or in nondomesticated relatives. [See Food Toxicology.]

Breeders rely on a number of practices to insure that they do not develop varieties with elevated levels of toxins, including taste tests (many toxicants are extremely bitter) and chemical analyses as indicated by the characteristics of the plant and its pedigree. Breeders have quite a successful record in producing safe new varieties. Is there reason to believe that biotechnology will change that? To examine that question, it is worth looking at the mechanisms by which new or elevated toxicant levels may be produced.

Breeding can inadvertently induce the production of new or elevated levels of toxicants by three general means. When intentionally altering a metabolic pathway, the breeder may get unanticipated results, such as raising the level of toxic intermediates or affecting other metabolic pathways that interact with the intentionally altered one. The breeder may unintentionally cause a mutation that affects a toxicant-producing metabolic pathway, or the breeder may unintentionally introduce DNA that encodes proteins involved directly or indirectly in controlling a toxicant-producing metabolic pathway. Intentionally altering metabolic pathways can lead to unanticipated results, regardless of method used. Biotechnology makes it easier to modify pathways, but also easier to target the modifications and, because the modified pathways are identified, easier to know what unexpected effects to monitor for.

Molecular techniques can introduce secondary mutations, and are likely to do so more frequently than are crosses between plants of the same species, because the molecular techniques are unable to target the site on the chromosome into which the newly introduced DNA will insert. Depending on where in the chromosome the DNA segment inserts, it may physically disrupt a gene or gene-control region, with the concomitant loss of gene function or gene control. If the insertion disrupts a region that controls or is part of a metabolic pathway, it may cause the production of new or elevated levels of toxicants. Thus, breeders using molecular techniques should not decrease their vigilance in looking out for unintended effects, especially when working within a genus known to contain member species that produce unsafe levels of toxicants. However, biotechnology is not unique in being subject to these kinds of unintended effects. Crosses between plants of different species or genera, for example, frequently cause chromosomal rearrangements in progeny, and can activate mobile genetic elements (e.g., transposons) that can move from one site in the chromosome to another and may disrupt gene function or gene control, depending on where they insert on the chromosome. Thus, wide crosses, like molecular techniques, are also subject to the second mechanism by which toxicant levels may be elevated. [See Transposable Elements in Plants.]

Plants developed through somaclonal variation are also subject to chromosomal rearrangements and other kinds of mutations. In this technique, plants are regenerated from cultured plant cells or tissues. The tissue culturing and regeneration, by permitting preexisting mutations to survive or by inducing new mutations, gives rise to plants with new characteristics, sometimes as extreme as miniaturization or seedlessness. Multiple mutations are produced, are random, frequently include chromosomal rearrangements, and theoretically could cause the production of new or more toxicants.

Biotechnology practically eliminates the likelihood of unintentionally introducing new proteins because it enables the researcher to target precisely the DNA to be introduced or mutated. Cross hybridization, however, is quite prone to this; breeders have no control over what DNA is brought in from a cross, and must perform extensive back-crosses between progeny and one of the original mating parents to eliminate as much undesired genetic material as possible. In cross-hybridizations, the more closely related are the parents, the less of a problem will be the introduction of extraneous genes, since most of the traits in the two parents will be the same. However, the more closely related are the parents, the less new genetic information is available to encode new traits. Breeders therefore sometimes turn to wild or weedy relatives from other species or genera for traits of interest, for example, for resistance to particular diseases or insects. In these wide crosses, much of the introduced DNA will encode undesirable traits, possibly including the production of toxicants different from or at higher levels than those found in domesticated species.

Historically, only a few cases have been documented in which foods with high toxicant levels have reached or almost reached the market. In the two cases in which the cause of the elevation is known, the cause was a cross with a high-toxicant-producing relative. In one case, the high-solanine potato Lenape (which can cause stomachaches when enough potato peel is eaten) resulted from a cross between high and low solanine-producing potatoes. In the other case, high-psoralen celery (which causes dermatitis in workers handling it) resulted from a cross between high and low psoralen-producing celery. The toxicant levels appear not to be related to any mutations in the plants. Several isolated cases of squash, both commercially canned and home-grown, with elevated toxicant levels have been reported. These are presumed to have resulted from mutations or chance outcrosses with a wild species, and do not appear to have resulted from breeding.

III. Environmental Issues

Biotechnology has great potential to provide alternatives to many of the agricultural practices that contribute to environmental problems. These alternatives, however, may have their own adverse environmental consequences. This section will discuss two environmental concerns often raised about biotechnology itself, and subsequently discusses environmental issues associated with several applications of biotechnology.

A. General Issues

1. Genetically Engineered Organisms and Exotic Organisms

It is sometimes argued that the introduction of transgenic organisms into the environment raises the same kinds of risks as does the introduction of exotic organisms, raising the specter that a transgenic potato may be turn out to cause problems similar to those caused by kudzu vine or zebra mussel. However, modifying a few genes in an organism does not render that organism completely new. Thus, reintroducing such a modified organism into an environment where its progenitor's behavior is well known is quite different from introducing a new organism into an environment where its behavior is completely unknown.

An organism has a complex web of survival characteristics that it brings with it to a new environment. If that new environment provides the natural resources necessary for the organism's survival, and lacks adequate competitive forces to keep the organism's population under control, the organism may become so successful that it becomes a significant pest. In contrast, a genetically modified organism introduced back into an environment where its progenitor's behavior is known will still encounter all or most of the previously existing constraints on its survival. It is certainly possible to envision organisms for which a particular kind of new characteristic could significantly alter its ability to succeed. However, the likelihood of this occurring in an unpredicted fashion is low for highly domesticated organisms lacking the ability to thrive without human assistance, such as annual row crops, reintroduced into areas where they have long been grown. The recent British study (Crawley *et.al.*, 1993), showing that engineered herbicide-resistant and kanamycin-resistant rapeseed are no more invasive than their nonengineered counterparts, was as notable for its goal of demonstrating the completely obvious as for its scientific rigor in achieving that goal.

2. Spread of a Molecularly Introduced Trait

Another oft-raised concern is that a trait introduced into a plant or other organism may spread to other organisms. Clearly, environmental safety concerns must be considered not only for the organisms into which a new trait was intentionally introduced, but also for those organisms that can exchange genetic information with the modified organism. For plants, these would be cross-hybridizable relatives within pollen range.

However, a popular perception appears to be that genes introduced into plants or microbes by molecular techniques are more prone to spread to other organisms than are the genes native to those organisms. This perception perhaps arises from a belief that the relative ease with which a DNA segment is introduced into an organism via molecular means somehow allows that organism to transfer it just as easily to other organisms. In fact, once a gene is stably incorporated into the plant or microbe genome, it is no more foreign, and no more susceptible to transfer, than any other segment of the genome.

B. Specific Applications

1. *Bacillus thuringiensis* δ-Endotoxin Pesticide

Bacillus thuringiensis (BT) is a common soil bacterium that has found wide use for its pesticidal properties. More than 4,000 strains have been isolated throughout the world. These produce several different types of toxins, of which the δ-endotoxins have proven to have great utility as insecticides. [*See* Ento-mology, Horticultural.]

BT δ-endotoxin has a number of attributes that make it a relatively environmentally benign pesticide. It is biodegradable and does not bioaccumulate. None of the endotoxins have any known toxicity to vertebrates, and it is very selective in its toxicity. Different BTs are toxic to different insects. For example, some δ-endotoxins are toxic only to certain species of caterpillars of the order Lepidoptera, others to beetles from the order Coleoptera, and others to mosquitoes and blackflies of the order Diptera.

Traditionally, BT spores have been sprayed onto crops. For some applications, this has been very efficacious and for some crops/insects, BT is the only available insecticide. Because BT is toxic to a very narrow range of insects, it causes few of the problems associated with many chemical insecticides, such as simultaneously killing off beneficial insects that help control insect pests.

The traditional uses of BT are limited by the endotoxin's rapid degradation, and because spray applications do not reach insects infesting roots or the inside of a plant. Molecular biologists have used a number of approaches to overcome these liabilities, including cloning the endotoxin gene into plants so the pesticide is continuously produced by the plant, and cloning it into bacteria that can persist longer when killed and fixed, or that can colonize the plant (either the root and leaf surfaces or inside the plant) and continuously provide protection.

The principal environmental issues associated with the new uses of BT δ-endotoxin is whether they will increase the speed and frequency of development of insect resistance to BT, and whether the δ-endotoxin gene or the microbe carrying the gene will spread to other plants and, if so, whether it will provide a significant competitive advantage to those plants, resulting in weediness. As new BT endotoxins are isolated, effects on nontarget insects could also be an issue.

Until recently, traditional applications of BT had not engendered significant insect resistance. However, in Hawaii, the Philippines, and certain regions of the continental United States, extensive and frequent BT applications have led to the evolution of diamondback moth populations with high levels of BT resistance. Thus, the continuous presence of the δ-endotoxin, as occurs in engineered plants that express the toxin and in plants colonized with δ-endotoxin-producing microbes, and the increased

persistence of the δ- endotoxin in one of the killed-bacteria applications, could lead to an increase in insect resistance.

Several theoretical strategies are available for preventing or delaying the development of insect resistance to pesticides. One such approach is to provide refuges for sensitive insects. A sufficient population of sensitive insects might, by mating with resistant insects, slow the development of insect populations homozygous for resistance. If the toxin, where present, was in doses high enough to kill the heterozygotes, evolution of homozygous resistant insects could be hampered. Refuges could be provided by planting mixtures of resistant and sensitive seed, or by having the endotoxin expressed only in certain tissues of the plant, or at specific times, for example, after stress induced by significant insect infestation. Other approaches for managing insect resistance include using two or more different toxins, so that insects resistant to one toxin will still be killed by the other, and providing only low doses of toxin, high enough to minimize economic damage to the crop plant but low enough to minimize selective pressure on the insects. How well these and other approaches will work, and how well industry will be able to implement them in light of potential competitive pressures to sell products with high kill rates, remains to be seen.

The potential for spread of the trait to other plants depends on the mode of delivery of the endotoxin. Plants that express the gene can transfer it to other plants with which they can cross-hybridize. The ability of engineered microbes to spread will depend on the characteristics of the particular microbes, including the plants they are able to colonize and how they might be spread to them. The ecological significance of the transfer of the trait to other plants will depend on how important the endotoxin-sensitive insects are in controlling the population of those plants.

2. Viral Resistance

Farmers use a variety of methods to combat viral disease in crop plants. They may modify planting and harvesting procedures in an attempt to reduce the spread of infection. They may plant virus-free seed or vegetative stock when they are available. They may pre-inoculate their crop with mild strains of a particular virus, thereby providing cross-protection, by an unknown mechanism, against subsequent infection by more virulent strains. They may spray their crops with oils and insecticides to minimize transmission of virus by insects. All these methods are expensive and labor-intensive, and are not available or effective for many crops. [*See* PLANT VIROLOGY.]

Breeders have also developed plant varieties that resist viral disease. Although breeders have had some notable successes using traditional cross-hybridization methods, they have been limited by their ability to find genetically compatible virus-resistant wild relatives of susceptible crops. Molecular techniques have greatly expanded breeders' ability to develop virus-resistant varieties. In a refinement of the pre-inoculation/cross-protection technique, breeders have developed virus-resistant plants by introducing into the plant genome a copy of a coat protein or replicase gene of a particular virus. These techniques hold particular promise for developing countries in which viral disease threatens the viability of staple crops such as rice and cassava.

A number of environmental issues pertain to the use of virus-resistant plants, including the extension of virus host ranges by encapsidating the nucleic acid of one virus with the coat protein of another (transcapsidation), the creation of more virulent viruses by recombination between infecting virus and the viral DNA segment in the plant chromosome; synergistic effects between infecting virus and proteins encoded by the viral sequences contained in the plant genome, and the transfer of resistance to wild or weedy relatives of the engineered plant. Note that these issues are pertinent also for many of the traditional plant protection strategies. For example, plants that are naturally virus resistant often are resistant to the disease effects of infection, rather than to infection and virus replication. Therefore, such plants often are infected with large numbers of virus, orders of magnitude higher than in transgenic plants protected by introduced viral coat protein, and would be subject to the same issues of transcapsidation, recombination, and synergism. Similarly, plants protected by pre-inoculation with mild virus strains also often are infected with large numbers of virus.

a. Transcapsidation

Transcapsidation commonly occurs in plants simultaneously infected with closely related viruses, and also occurs in viral-coat-protein-producing transgenic plants. The frequency of transcapsidation in viral-coat-protein-protected transgenic plants relative to that in nonprotected plants co-infected with two viruses is hard to predict. Although the amount of coat protein in transgenic plants is usually 100- to 1000-fold lower than in virally infected plants in which transcapsidation has been observed, the coat protein will be continuously available in the transgenic plant.

The principal environmental risk posed by transcapsidation is that the virus may be spread more quickly or more widely than usual if, as a result of its new coat, it can now be spread by a different vector. The likelihood of significantly expanding host range is minimized by the fact that, on replication in the new host, the virus would be encapsidated with its own coat protein and would no longer be transmitted by the new vector.

Transcapsidation in transgenic virus-resistant plants may be minimized through the use of viral sequences other than coat protein that prevent viral replication, or through the use of mutations in the coat protein sequence that prevent transmission. Current research indicates that both approaches are promising.

b. Recombination and Template Switching

Recombination and template switching, in which replicated nucleic acid (as in progeny virus) will contain segments of nucleic acid from an exogenous source, occur between the nucleic acids of co-infecting viruses and between the nucleic acid of infecting virus and plant DNA. Presumably, recombination and template switching will also occur between infecting viral nucleic acid and viral sequences present in the genome of transgenic plants. Evolutionary principles and experimental evidence argue that such viruses most often will be less fit and less virulent than the wild-type. However, the fact that influenza virus strains sometimes generate, by recombination, altered coat proteins that escape the immune system demonstrates that, at least in some systems, the results of recombination may not always be benign.

Greenhouse and field studies can provide information on the likelihood of recombination and the potential for the particular viral gene contained in the plant genome to increase the virulence of different infecting viruses.

c. Synergism

Synergism, in which the disease caused by co-infecting viruses will be worse than that caused by either alone, occurs between certain types of co-infecting viruses. Greenhouse and field studies can be used to determine whether the viral gene contained in the genome of a transgenic plant has a synergistic effect on various infecting viruses.

d. Transfer of Viral Resistance

Transfer of viral resistance to wild or weedy relatives of the viral-resistant crop can occur in those areas harboring relatives with which the crop can cross-hybridize and produce fertile progeny. The consequences of transferring viral resistance will depend on the degree to which the resisted virus is critical in controlling the growth or spread of the particular plant.

Viral infection generally has a greater effect on agricultural plants than on those in a wild or weedy environment. Agriculture's large tracts of a single crop present more favorable conditions for viral infection than do wild or weedy environments. Additionally, effects on quantity and quality of fruit, although having great economic impact, do not necessarily have corollary ecological impacts.

Greenhouse and field studies can provide information on the potential frequencies of transfer of viral resistance in particular environments, and on the potential ecological effects of such transfers.

3. Herbicide Resistance

One of the more controversial areas of plant biotechnology is its use in developing crops that are resistant to particular herbicides, primarily because the use of herbicides is controversial. However, herbicides have played an important role in increasing food production over the last 50 yr. Especially important are the selective herbicides, those that kill weeds but not the crop on which they are sprayed. Additionally, they can provide certain environmental benefits by allowing low-till agriculture, which limits soil compaction, erosion, and evaporative water loss. [See HERBICIDES AND HERBICIDE RESISTANCE.]

The problems with herbicides are that some are toxic, some accumulate in ground water or elsewhere in the environment, and some are losing their effectiveness because of the development of resistant weeds. Biotechnology can be useful in combatting all these problems by allowing the use of herbicides (1) for which weed resistance is not a problem, (2) that are not toxic (except to weeds), (3) that rapidly degrade in the environment, and (4) that can be applied only when weeds are a problem (thereby minimizing exposure or environmental contamination that arises from spraying fields in the spring prior to planting, when heavy rains wash the herbicides from the bare soil into streams and groundwater).

Also, biotechnology can expand the use of herbicides to treat additional agricultural problems. For example, parasitic weeds, such as broomrapes and witchweeds, attach themselves to crops, often underground, and cannot be removed by mechanical weeding. These plants are endemic in much of Africa and Asia and, in sub-Saharan Africa, more severely de-

press agricultural productivity than do insects or disease. Researchers have estimated that crop yields could be doubled in those areas, without added fertilizer or irrigation, if the parasitic weeds could be effectively controlled. One means of achieving that goal would be to engineer resistance to appropriate herbicides, such as glyphosate and sulfonylureas, into the crop plants. [See WEED SCIENCE.]

One of the reasons that herbicide-resistant plants may not be developed or used to their full potential is that almost all the research in this area is done by the private sector, particularly, by companies manufacturing herbicides. For obvious reasons, companies focus their research in areas in which they can make the most profit, which means focusing on hybrid crops (whose seed cannot be saved from harvest, but must be bought every year from the seed company), grown in North America and Europe (where the largest markets exist), and resistant to herbicides on which the company holds a patent (rather than to the herbicides that are best for a particular purpose).

The principal environmental issues associated with these plants are whether their marketing would increase the use of herbicides, whether the resistance trait could spread to wild or weedy relatives and impede their control, whether the widespread application of the tolerated herbicides could spur the evolution of resistance in weeds exposed to the herbicide, and whether the focus on herbicide resistance diverts resources from research into nonchemical alternatives for weed control.

It is difficult to envision how herbicide-resistant crops could lead to significantly increased use of herbicides in the United States, considering the fact that herbicides are already applied to almost all fields of all major crops. Rather, herbicide-resistant crops should lead to the substitution of one herbicide for another (frequently for several others). Thus, the real issue is the comparative safety of the new herbicide relative to that of the replaced herbicide, requiring a case-by-case evaluation.

Spread of the resistance trait to wild or weedy relatives would limit the utility of the herbicide for managing those plants. Careful consideration should be given before engineering herbicide resistance into crops that are grown in areas where cross-hybridizable relatives are problem weeds, for example, into sorghum which can cross with Johnson grass, a problem weed in the United States. Spread of herbicide-resistance is no more likely from engineered crops than from crops intrinsically resistant to particular herbicides. To date, gene transfer has not been found to be responsible for any herbicide resistances developed by weeds. All have developed from within their own gene pool.

The widespread use of any particular herbicide may lead to the evolution of weeds resistant to it, so widespread use of one resistance gene in crop plants may foster the development of resistance in wild or weedy plants. So too would the engineering of resistance to those herbicides to which weeds have shown an increased ability to develop resistance. Whether resistant weeds will be a problem will depend to a great extent on the intelligence of the individuals developing resistant plants, and on their abilities to overcome competitive pressures pushing them toward achieving short-term gains at the expense of long-term benefits.

To the extent that herbicide-resistance is successful in combatting both weed problems and environmental problems associated with herbicide usage, it may reduce interest in finding nonchemical weed-control strategies. However, because little government money is spent on developing herbicide-resistant plants, and because the companies that are developing herbicide-resistant plants are chemical companies unlikely to expend resources on nonchemical control mechanisms, it is not clear whose resources are being diverted.

4. Kanamycin Resistance

Many of the crops developed with molecular methods contain a gene called the kanamycin resistance or KanR gene. This gene does not cause the plant to produce an antibiotic, but encodes an enzyme that inactivates the antibiotics kanamycin and neomycin. The DNA segment introduced into a plant cell will contain both the gene(s) of interest and the KanR gene. By growing the plant cells in the presence of kanamycin, researchers select cells that have incorporated the KanR gene and, in most cases, the gene(s) of interest as well.

The environmental issues posed by the use of KanR pertain to the potential for the gene to spread to other organisms, and the potential consequences of such spread. The principal organisms to which the gene could spread, as with any genetic trait, would be plants that can cross-hybridize with the KanR plant. It is hard to envision how the presence of the KanR gene could confer a competitive advantage to a plant in any way. Plants are rarely, if ever, exposed to appreciable levels of kanamycin or neomycin; the levels to which a plant conceivably could be exposed from antibiotic-producing soil microbes are ex-

tremely low and would have no appreciable affect on its growth or survival. Additionally, plants (as opposed to plant cells in tissue culture) are relatively insensitive to these antibiotics.

A potentially more significant issue would be the transfer of the KanR gene to bacteria associated with the plant or in the gastrointestinal tract of a consumer of the plant. The concern would be that these bacteria could then become reservoirs for the resistance trait, potentially spreading it to pathogens that otherwise might be susceptible to treatment with kanamycin or neomycin. The degree to which transfer of antibiotic resistance genes from plants to bacteria could be a problem depends on the extent that it would add to the number of antibiotic-resistant microbes already in the environment or in people. Bacteria resistant to the antibiotics kanamycin and neomycin are found in soil, on fresh fruits and vegetables, and in the gastrointestinal tract of humans. Worst-case estimates of the frequency of gene transfer from plants to bacteria indicate that KanR plants would contribute a negligible and undetectable increase in kanamycin-resistant bacteria.

IV. Conclusion

Biotechnology, although unlikely to offer a panacea for solving world hunger or environmental problems, offers great promise for improving agriculture, both in providing crops with improved agronomic and food-use characteristics and in providing alternatives to some environmentally destructive practices. Of particular importance is its potential to alleviate specific agricultural problems in developing countries. Food safety and environmental issues are associated with its use, but these are of the same kind as those associated with other breeding techniques. Whether this technique will be allowed to fulfill its promise, and the extent that its practitioners will use it to address social and environmental ills, remains to be seen.

Acknowledgments

The author thanks David Berkowitz, Bonnie Liebman, James Maryanski, Michelle Mital, James White, and Larry Zeph for their comments and suggestions.

Bibliography

Crawley, M. J., Hails, R. S., Rees, M., Kohn, D., and Buxton, J. (1993). *Nature* **363,** 620–623.

DeWet, J. M. J., and Harlan, J. R., (1974). Ecology of Transgene oilseed rape in natural habitats. Weeds and domesticates: Evolution in the man-made habitat. *Econ. Bot.* **29,** 99–107.

Gressel, J. (1993). Advances in achieving the needs for biotechnologically-derived herbicide resistant crops. *Plant Breeding Rev.* **11,** 155–198.

International Food Biotechnology Council (1990). Biotechnologies and food: Assuring the safety of foods produced by genetic modification. *Reg. Toxicol. Pharmacol.* **12 (3),** 51–5196.

Keeler, K. H., and Turner, C. E. (1991). Management of transgenic plants in the environment. *In* "Risk Assessment in Genetic Engineering" (M. Levin and H. Strauss (eds.), pp. 189–218. McGraw-Hill, New York.

Kessler, D. A., Taylor, M. R., Maryanski, J. H., Flamm, E. L., and Kahl, L. S. (1992). The safety of foods developed by biotechnology. *Science* **256,** 1747–1749, 1832.

McGaughey, W. H., and Whalon, M. E. (1992). Managing insect resistance to *Bacillus thuringiensis* toxins. *Science* **258,** 1451–1455.

National Research Council (1989). "Field Testing Genetically Modified Organisms," Framework for Decisions. National Research Council, Washington, D.C.

Palukaitis, P. (1991). Virus-mediated genetic transfer in plants. *In* "Risk Assessment in Genetic Engineering" (M. Levin and H. Strauss, eds.), pp. 140–162. McGraw-Hill, New York.

Plant Cytology, Modern Techniques

ROBERT W. SEAGULL, *Hofstra University*

DAVID C. DIXON, *USDA-Agricultural Research Service, Louisiana*

Glossary

Classical staining procedures Use of chemical agents to increase optical contrast in biological specimens; includes the use of various pigments, natural dyes, and artificial stains that specifically bind certain cellular components, thus providing increased detection of specific chemical components of cells
Cytology The science of dealing with the cell; the examination of cellular structure using the techniques of light and electron microscopy
Fixation Stabilization and preservation of biological material in as close to a life-like state as possible, through the use of chemical crosslinking agents that form three-dimensional meshwork of cellular components or physical freezing agents that preserve ultrastructure rapidly and maintain it for ultimate chemical preservation
Hybridization Specific and complimentary binding of an oligonucleotide probe to a region of DNA or RNA in the cell; this provides the detection and localization of specific genes or gene products in cells
Immunocytochemistry Use of antibodies as specific binding agents (stains) for the detection of cellular components; primary antibodies specifically bind the component of interest and their location is detected via a secondary binding of an antibody that can be labeled with a variety of easily detected molecules (i.e., radioactive, fluorescent, heavy metal, etc.), and antibodies provide great selectivity for a wide array of macromolecules in the cell
Reporter gene Foreign gene that is inserted into a cell and can be easily detected by some type of chemical processing or staining

Cytology, by definition, is the study of cell structure, function, pathology, and life history. Techniques for the examination of plant cytology (henceforth referred to as cytological techniques) have provided botanists with invaluable insight into the development and function of plants and have facilitated attempts to understand and manipulate plant growth and development in economically relevant ways.

I. Introduction

By using cytological techniques and microscopy, one can examine individual cells, whereas traditional "grind and find" biochemical approaches examine populations of cells. This is a major advantage of microscopic versus biochemical approaches to the study of plants. Because biochemical investigations look at populations of cells, the result of such an investigation represents an average of cellular processes within an organ or tissue that may contain a heterogeneous array of cell types. Averaging of cellular processes can result in a loss of resolution, with underestimates of cellular processes, the synthesis of cellular products, or other differences occurring in a subpopulation of the cells. Provided that the appropriate staining technology is available, cytological techniques can be used to distinguish even subtle differences occurring within and between individual cells in a diverse population. As such, cytological techniques can be used to detect cellular or metabolic changes occurring within a plant tissue or organ at earlier times and at lower levels. Interpreting cytological observations may be difficult. One must determine if observed differences between cells are the result of real biochemical or functional differences or simply a technique-induced artifact. Similar to biochemical studies, major cytological changes are fairly easy to distinguish and interpret; however, minor differences need careful examination and interpretation. The potential for examining many individual cells permits the quantitative, statistical analysis of staining patterns and thus provides a certain amount of confidence in the observations. [*See* PLANT CYTOSKELETON.]

In this article we focus attention on some of the more recent advances in plant cytological techniques. Space does not allow for comprehensive treatments of any of the topics in this article; thus, we highlight those areas that promise to have profound impacts on future plant studies.

II. Light Microscopy

Central to any cytological examination is the equipment used. Light microscopes provide the magnification and resolution needed for detailed study of cell morphology and composition. While light microscopy has been used for hundreds of years for the examination of cells, it is far from obsolete, and new applications are still being invented.

Bright field microscopy remains the technique of choice when one is able to satisfactorily preserve and stain biological samples. As live specimens often have little inherent contrast, bright field microscopy is often used with preserved and treated (stained) tissues. Light is transmitted through the specimen and lenses to the observer. Contrast is obtained through the absorption of light by the specimen or pigments inherent within (i.e., chlorophyll, carotinoids) or added to (various staining agents, see below) the specimen.

Phase-contrast and the newer interference-contrast microscopy remain unrivalled for observing living cells and the dynamics of cellular processes. When light passes through most biological materials there is often no change in amplitude of the light and thus the specimen appears transparent. However, due to the physical nature of biological material, light rays passing through different regions of the specimen (i.e., nucleus, cytoplasm, organelles) may be slowed down relative to rays passing through the background. A phase plate brings together diffracted (slowed) and undiffracted light rays to produce an image with bright and dark areas against a neutral gray background.

Polarization microscopy depends upon modifying the form of illumination to gain specific information about the specimen. Polarizer and analyzer filters allow for the passage of light waves in only one direction. If the two filters are parallel then all the light

passes; however, if they are crossed then no light passes. Biological materials which have ordered arrays of elements can cause a rotation of the polarized light, thus allowing some of the light to pass through the crossed filters. This property of rotating plane polarized light is called birefringence. This form of microscopy finds many applications in plant cytology, especially for the examination of crystalline arrays of wall components.

Fluorescence microscopy uses the physical phenomenon of fluorescence. A chemical compound is fluorescent if it is capable of absorbing ultraviolet light and re-emitting the energy as visible light. An exciter filter, between the light source and the condenser, transmits only ultraviolet light. The barrier filter eliminates ultraviolet light; thus, the image is formed by the re-emitted visible light on a dark background (see Fig. 1). Fluorescence microscopy has grown in application to plant biology due to development of selective fluorescent stains. Linking a fluorescent dye to a specific cell probe (i.e., an antibody raised against a specific protein) or using site-specific fluorescent dyes (i.e., Hoechst 33258 for DNA or Tinopal LPW for β-1,3 and β-1,4 linkage polysaccharide) permits plant material to be stained with far more sensitivity than conventional staining methods.

A new modification to standard fluorescence microscopy is the development of *scanning confocal laser microscopy*. Conventional light microscopes create images with a depth of field of 2–3 μm. As most specimens are thicker than this, superimposed on the image in focus are out-of-focus images of material above and below the focal plane of the microscope. The presence of this material can obscure the focused image and decrease detectability within the specimen. This problem is of particular note with fluorescence microscopy where out-of-focus fluorescence creates a diffuse halo around objects in focus. One approach to minimize this problem is to cut thin sections of biological material; however, this is not a suitable solution for examining living cells or tissues.

The advent of scanning confocal laser microscopy (SCLM) solves this problem by creating images whose depth of field is less than 1 μm. The confocal aspect of the microscope (i.e., narrow focal plane) is generated by pin hole apertures just after the light source and just before the detector. Only light from the focal plane is transmitted through both pin holes, while light from the out-of-focus planes strikes the aperture walls and is not transmitted. The result is an image with remarkable clarity and detail. The increase in signal (focused) to noise (out-of-focus) ratio im-

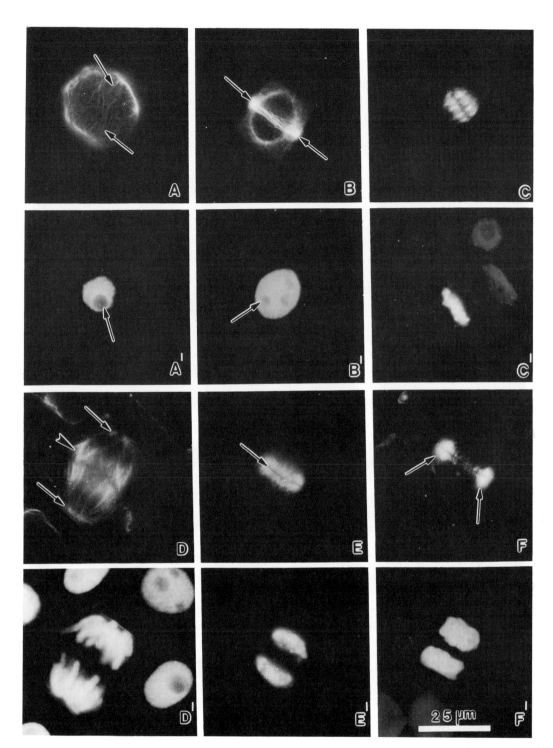

FIGURE 1 Immunofluorescence microscopy of meristematic clover root cells, treated with a primary antibody to the microtubule protein, tubulin (generated in rat), followed by a goat anti-rat secondary antibody tagged with the fluorescent dye fluorescein (A–F). Cells are also stained with Hoechst 33258, a fluorescent DNA intercalating dye, to reveal the organization of DNA (A'–F'). At interphase (A), microtubules form anastomosing arrays (arrow) in the cortical cytoplasm while the nucleus (A') stains intensely for DNA, except for the RNA-containing nucleolus (arrow). During pre-prophase (B), microtubules form a pre-prophase band (arrow) that predicts the future site of cell plate fusion to the parent wall. The nucleus (B') still reflects an interphase condition with prominent nucleoli (arrow). During metaphase microtubules form the spindle (C) and condensed chromosomes (C') form a plate at the spindle mid-plane. By late anaphase microtubules (D) of the spindle have retracted to the spindle poles (arrow) and the phragmoplast array (arrowhead) begins to form at the mid-region. Chromosomes (D') have been separated into two masses. In early telophase microtubules (E) form a compact phragmoplast involved in forming the cell plate (arrow). Chromosomes (E') begin to de-condense to form daughter nuclei. By late telophase microtubules (F) of the phragmoplast are concentrated at the formed edge of the cell plate (arrow). Daughter nuclei (F') have formed.

proves "effective" resolution in the specimen. Scanning of the specimen with the laser is done by a set of rotating or vibrating mirrors. Electronically generating an image from a number of scans increases the signal to noise ratio of the image, thus improving the quality of the fluorescence image.

By eliminating out-of-focus images, SCLM reduces the amount of light available and thus requires greater illumination than non-SCLM techniques. This greater illumination results in increased photodamage and bleaching. Loss of image due to photodamage and bleaching can be reduced by lowering the laser intensity with neutral density filters (without reducing image quality), avoiding "casual viewing" of specimens before image collecting, selecting fluorophores that are resistant to photo-bleaching (i.e., Texas red or rhodamine rather than acridine orange or fluorescein), or altering the size of the pin hole aperture to increase incoming signal.

Images from SCLM are routinely subjected to computer-assisted image enhancement. Due to the limited amount of fluorescence emitted from specimens in SCLM, images are often of extremely low intensity. To enhance and improve detail one can superimpose multiple scans of the same image, using computer imaging, thereby improving the "effective" resolution. Low light video cameras, such as ISIT (intensified silicon target) and CCD (intensified charge coupled devices) allow for direct input of SCLM images into computers for image intensification processing. Computer-assisted changes in focal plane, coupled with image acquisition, facilitates the generation of three-dimensional images of cells, tissue, and organs. Many image processing programs are available for data storage and manipulation.

A. Classical Staining Procedures

Early studies of plant anatomy using light microscopy and simple staining procedures provided not only morphological details but also insight into the chemical composition of cells, tissues, and organs of the plant. Techniques and stains developed over a hundred years ago are still in use providing detailed anatomical and compositional information regarding plant development. Plant tissues are commonly preserved with a fixative, and then embedded into some type of mounting medium (i.e., wax or plastic), sectioned to provide samples thin enough for the transmission of light, and then stained to provide contrast.

Stains or dyes fall into fundamental categories: natural dyes and synthetic ones. Specific formulations and uses of these dyes can be found in more comprehensive treatments of this topic. Many of these dyes by themselves do not bind to biological material, but through the interaction of mordants, these dyes specifically bind to certain cellular components. Most dyes will not penetrate living cells, and thus are only useful for preserved material.

The number of natural dyes in active use is becoming increasingly small, as the active components of these dyes are identified and chemically synthesized. The chemistry of natural dyes is less definitely known than that of the artificial dyes and thus their use often leads to variable results. The most important of the natural dyes are hematoxylin, indigo, carmine (cochineal), and orcein. These dyes are extracted from various natural sources; for example hematoxylin is from logwood of *Hematoxylin campechianum* L. and carmine from the dried bodies of tropical insects. Synthetic dyes are made as derivatives of the hydrocarbon, benzene. The older literature often refers to these stains as "coal-tar" dyes since all of them are made by chemical transformations from one or more substances found in coal-tar. Attached to the fundamental benzene ring structure are certain groups of elements known as chromophores (that impart the property of color to the compound) and auxochrome groups (that produce salt bridges between the dye and the tissue). Auxochromes can be basic (i.e., with amino groups or acidic (with hydroxyl groups), thus imparting to the dye the ability to interact with acidic or basic components of the cell. Differential binding capability provides selective binding properties to dyes. The intensity of binding can be modulated by the pH of staining or destaining solutions. Stains are commercially available as salts, with acid dyes usually made into sodium, potassium, or calcium salts and basic dyes in a chloride, sulfate, or acetate salt. Thus, a dye can be defined as an organic salt which contains chromophoric and auxochromic groups attached to benzene rings. Deciding on which stain to use for a particular purpose requires knowledge of the chemical properties of the structure of interest and of the stain being used. Many published recipes are available for staining specific subcellular structures (i.e., nuclei, mitochondria, starch, etc.) or tissue types (lignified tissue, meristematic zones, etc.). Generally, a specific stain is used to highlight the structures of interest, followed by a "counterstain," to provide contrast and points of reference of general organ or tissue morphology. Counterstains are generally rather nonspecific in their binding to cells and tissues. Combinations of specific stains can be used, resulting in double and triple staining of tissues. A common triple stain for plant material includes safranin, crystal violet, and orange G. Safranin is intended to stain chromatin, lignin, cutin, and perhaps chloroplasts. Crystal violet should stain spindle fibers, nucleoli, and cellulose cell walls. The orange G acts as general background stain for the cytoplasm. In skilled hands, this combination of stains will reveal excellent tissue details and cellular differentiation.

Polychromatic dyes, such as toluidine Blue O, can be particularly useful in revealing tissue organization and chemical composition. The colors which result from the binding of this single dye are dependant upon the chemical nature of the cellular components: carboxylated polysaccharides give a pink color, compounds with benzine rings give a bright green to royal blue color (depending upon the particular chemical structure), and compounds bearing ionized phosphate groups (such as DNA or RNA) stain bluish to purple. These chromatic characteristics result in primary cell walls staining pink, while secondary walls of the xylem stain green and phloem stains bright blue.

B. Immunofluorescence Microscopy

Since well-characterized antisera provide highly specific probes (stains) with which to quantify and locate antigens, immunological techniques have become indispensable in many molecular, cellular, developmental, and physiological analyses. Problems that can be studied with these techniques range from molecular analysis to examinations of evolutionary relatedness of organisms.

As with other stains (see above) used in cytological studies, antibodies offer a degree of versatility unmatched by chemical stains and probes. The mammalian immune system is capable of responding to a virtually unlimited array of foreign substances or antigens; as a result, antibodies with specificity toward almost any molecule found in a biological system can be generated. Antibodies can be produced that specifically recognize and bind proteins, nucleic acids, polysaccharide, lipids, drugs, hormones, etc. However, of the classes of molecules mentioned, proteins contain a greater diversity of unique sites suitable for antibody binding and in general elicit a stronger immune response. For these reasons, antibodies are most commonly used in cytological studies for the immunolocalization of specific proteins.

The first step in a cytological study that uses an antibody probe often involves stabilization of the target molecule within the plant tissue. This is usually accomplished with chemical fixatives such as ethanol/acetic acid, formaldehyde, or glutaraldehyde. Following fixation the plant tissue is often embedded in wax or a plastic resin to facilitate sectioning. Alternatively, the plant tissues may be quick frozen and sectioned directly without prior fixation and embedding. After the tissue has been suitably prepared, it is incubated in a buffered solution containing a primary antibody specific for the target molecule being studied. Following this incubation, the plant tissue is washed to remove excess antibodies that have not bound to the target molecule. Antibodies that have bound to target molecules within the tissue are then detected by incubating the tissue in a solution containing a labeled conjugate. The labeled conjugate is often a secondary antibody that will recognize and bind to the primary antibody. To permit detection of the resulting complex comprising the target molecule, primary antibody, and secondary antibody, the secondary antibody may be labeled with colloidal gold, an enzyme (horseradish peroxidase or alkaline phosphatase are commonly used), or any one of a variety of fluorescent molecules (fluorescein, rhodamine, phycoerythrin, etc.).

The successful use of antibodies in any cytological study is dependent upon at least the following four factors.

1. Antibody Specificity

In most cases antibody specificity depends upon the purity of the target molecule injected into the experimental animal to elicit antibody production. Other factors such as the health of the experimental animal, its pre-immune status, the method of injection, and the amount of target molecule injected can also affect the quality and quantity of antibody produced. Both polyclonal antisera and monoclonal antibodies may be used in cytological studies and the advantages and disadvantages of each should be considered when designing immunocytochemical experiments. Because polyclonal antisera may contain an array of different antibodies, each capable of binding the target molecule at a unique site or epitope, polyclonal antisera generally provide the strongest signal for immunolocalization studies. Another advantage of polyclonal antisera is in studies of evolutionarily conserved proteins in a wide range of species. In this type of study a polyclonal antisera is more likely to contain antibodies that will recognize and bind the conserved protein in all the species in the study. However, antisera may also contain antibodies which are not specific for the molecule of interest. These other nonspecific antibodies will occasionally cause a signal that obscures that of the specific antibody.

In contrast to polyclonal antisera, monoclonal antibodies are very specific and only recognize and bind a single epitope on the molecule of interest. However, because monoclonal antibodies only bind at one site on the target molecule, immunolocalization with monoclonal antibodies may give a weaker signal. These weak signals may be boosted by pooling two or more monoclonal antibodies. Because they are monospecific, monoclonal antibodies may not be appropriate for studying conserved proteins in other species unless the epitope that the antibody binds is highly conserved. For example, tubulin proteins found in plants and animals are very similar. However, because there are differences between plant and animal tubulins, a monoclonal antibody developed against animal tubulin may not bind to a plant tubulin unless the epitope that the antibody binds is in a conserved region found in both plant and animal tubulins. Although monoclonal antibodies are very specific, they will occasionally cross-react with unrelated molecules. For example, a monoclonal antibody that is specific for a particular protein may recognize and bind an epitope comprising only four to seven amino acids. By chance, an unrelated protein may contain the same combination of amino acids and will be recognized and bound by the monoclonal antibody. Because of the potential of both polyclonal antisera and monoclonal antibodies to bind molecules unrelated to the intended target molecule, the specificity of an antibody to its target should be verified by immunoblot, imunoprecipitation, or another appropriate technique before attempting to use the antibody for cytological studies.

2. Controls

When using antibodies as cytological stains, appropriate controls are absolutely essential for distinguishing between a specific signal and nonspecific signals. In addition to cross-reactions with unrelated molecules as mentioned above, nonspecific signals may arise from other sources. For example, autofluorescence of pigments or lignified tissues in plants may cause nonspecific background fluorescence when using fluorescently labeled secondary antibodies. Nonspecific fluorescence may also be caused by glutaraldehyde fixation. Tissues fixed with glutaraldehyde will fluoresce brightly unless they are treated after fixation with sodium borohydride to reduce free aldehydes. Nonspecific background is not limited to fluorescence studies. For example, endogenous peroxidase activity

may cause false signals in cytological studies where secondary antibodies are labeled with horseradish peroxidase. Appropriate controls for immunocytochemical studies may include: pre-immune sera controls, conjugate controls, and if working with transformed or infected plant tissues, nontransformed or uninfected controls. In some instances it may also be appropriate to verify the results obtained with one antibody by using another antibody that is also specific for the target molecule being studied.

3. Effects of Tissue Processing on the Target Molecule

In nearly all cases, some degree of tissue processing is required when doing cytological studies. Processing may include fixation, washing, embedding, sectioning, etc., and can affect the ability of an antibody to recognize and bind its target molecule. For example, inappropriate fixation of plant tissues can decrease the likelihood of antibody binding: Overfixation can mask epitopes recognized by the antibody, while underfixation may allow for extraction or redistribution of the target molecule. Additionally the temperatures and or chemicals required for some procedures may denature target molecules making them unrecognizable to specific antibodies. In contrast to tissue manipulations that decrease the likelihood of antibody binding, some manipulations are necessary to allow access of the antibody into the plant tissues. For example, in order to view the subcellular structure of intact plant cells, the cell wall is often partially digested with wall degrading enzymes to facilitate antibody access to the target molecule.

4. Abundance of the Target Molecule

The abundance of the target molecule in the plant tissue is perhaps the most critical factor influencing the success or failure of any immunolocalization experiment. Unfortunately it is also the single factor over which the scientist has the least control. In general, a protein or other molecule that accumulates to high levels and is concentrated at specific points in the plant tissue is easily detected, while a molecule that is only present at low levels and/or is widely dispersed within the tissue may be difficult to detect. The latter situation is sometimes encountered with proteins that accumulate within the vacuole of a plant cell.

Many proteins and other molecules have been successfully localized within plant tissues with antibody probes. For example, Fig. 1 illustrates the use of antibodies to study changes in the structure of microtubule arrays during the cell cycle. Some of the other uses of antibody probes in plants cytology include the study of developmental accumulation and distribution of enzymes involved in photosynthesis, characterization of the light-induced redistribution of the photosensory protein phytochrome, and characterization the extracellular accumulation of many cell wall proteins.

C. *In Situ* Hybridization

In situ hybridization has been used to study gene expression in animal systems for over 30 years and has more recently been adapted for use in plant systems. This technique is based on the principles of nucleic acid hybridization. Under appropriate conditions, complimentary nucleic acid strands will anneal to each other and form stable hybrids. This hybridization of two complimentary nucleic acid strands is very specific and by using a complimentary strand as a probe, specific sequences of DNA or RNA can be detected. At a cellular level, *in situ* hybridization utilizes these nucleic acid probes to detect specific gene sequences (DNA) or the transcription products of a gene (RNA) within individual cells. As such, *in situ* hybridization can be used to determine the chromosomal location of genes and it can be used to detect nucleic acid sequences of viruses or other pathogens in cells or tissues. Most often, however, *in situ* hybridization is used to detect the RNA transcription products of a specific gene of interest. Because it can detect RNAs within a single cell, *in situ* hybridization has an advantage over more conventional means of measuring RNA accumulation in tissues such as northern blot or RNase protection since it can be used to determine the precise spatial pattern of a gene's expression in a heterogeneous population of cells. An example of *in situ* hybridization's use in determining gene expression in plant tissues is shown in Fig. 2.

The nucleic acid probes used for *in situ* hybridization may be either single-stranded DNA or RNA. However, because they have advantages over DNA probes, RNA probes are more commonly used. For example single-stranded RNA probes are easily generated by *in vitro* transcription; double-stranded RNA/RNA hybrids are more stable, allowing higher wash temperatures to reduce nonspecific binding of probe; RNases can be used to digest probe that has not specifically hybridized to target sequences in the tissue thereby improving the signal to noise ratio. Traditionally the single-stranded probes used for *in situ* hybridization are radiolabeled; however, nonradi-

oactive methods are becoming available and appear to offer a good alternative to radiolabeling.

As with other cytological techniques plant tissues used for *in situ* hybridization require advanced preparation. Nearly all *in situ* hybridization experiments are done with tissue sections. Sections may be cut with a conventional microtome from fixed and embedded pieces of plant tissue or they may be cut directly from frozen tissue specimens with a cryostat. Although not as rapidly prepared, sections cut from fixed and embedded plant tissues typically have superior cellular morphology. After the tissue sections have been cut and mounted on microscope slides they generally require prehybridization treatments to minimize nonspecific background and to facilitate access of the probe to the target nucleic acid sequences. Prehybridization treatments may include treatment with heat, dilute HCl, proteinases, bovine serum albumin, and acetic anhydride.

The experimental conditions for the actual hybridization and the posthybridization washes vary depending on the protocol used. However, in all cases the hybridization temperatures and the composition of the hybridization buffer and the posthybridization washes are established to maximize hybridization of the probe to its target sequence within the tissue and minimize nonspecific binding.

Because radiolabeled probes have traditionally been used for *in situ* hybridization, autoradiography has been the standard method used for data analysis. After hybridization and washing to remove excess probe, the tissue is coated with a thin photographic emulsion. This emulsion is exposed by the bound radioactive probe and then developed and fixed similar to standard photographic film. The tissue is then counterstained and examined with a microscope for silver grains in the emulsion overlaying the tissue. The presence of silver grains in the emulsion indicates the position in the tissue where the radiolabeled probe has bound to the gene of interest or the transcription products of that gene. As mentioned earlier, nonradioactive methods for labeling nucleic acid probes are becoming available. For example, biotin-labeled probes may be employed for *in situ* hybridization. Following hybridization and posthybridization washes, bound probe may be detected with avidin conjugates labeled with fluorescein, alkaline phosphatase, horseradish peroxidase, etc. After a wash to remove the excess conjugate and incubation with an appropriate enzyme substrate (if an enzyme conjugate has been used), the tissue is essentially ready for microscopic examination. These nonradioactive meth-

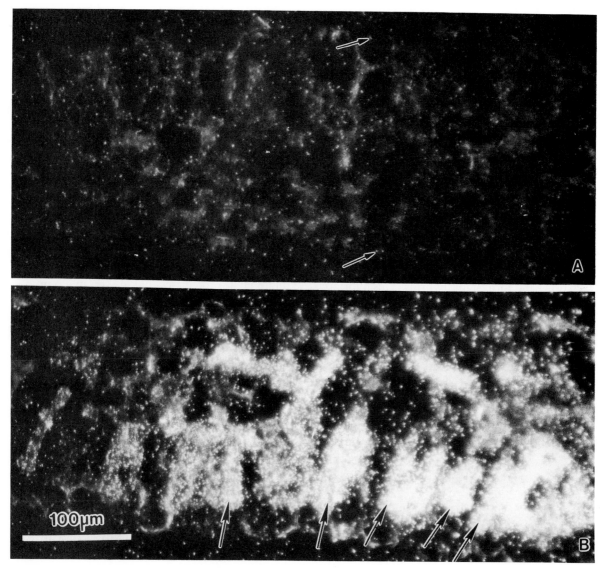

FIGURE 2 *In situ* hybridization. Sections cut from a tobacco leaf infected with tobacco mosaic virus were incubated with radiolabeled(^{35}S)probes to determine the expression pattern of a plant gene induced in response to the viral infection. Exposed silver grains (arrows) in the emulsion overlaying the leaf sections indicate the position of bound probe. (A) This negative control leaf section was incubated with a sense or noncomplimentary probe that will not hybridize to RNAs transcribed from the induced gene. Randomly distributed silver grains represent nonspecific binding of radiolabeled probe. (B) This section was incubated with a complimentary probe that will hybridize with RNAs transcribed from the gene of interest. The location of exposed silver grains (arrows) indicates which cells are expressing this gene.

ods appear to offer a good alternative to radiolabeling since they offer good sensitivity without requiring a photographic emulsion and the associated tedious darkroom work.

D. Gene-Fusion Markers

By using gene-fusion markers it is possible to determine where and when a specific gene of interest is expressed in a plant. Although the technique is widely used to study gene expression, this approach requires considerable preliminary work and manipulation of genetic sequences. First, the gene of interest must be cloned and analyzed to determine which region of the gene acts as the promotor. This promotor region is then fused to the coding region of another gene that codes for a "marker" protein that can be easily detected within plant tissues. The resulting chimeric gene fusion is then used to transform plant tissues to produce transgenic plants. Since the chimeric gene is controlled by the promotor that is identical to the promotor of the gene of interest, the expression of the chimeric gene and the endogenous gene should be identical. However, the promotor of the chimeric gene now regulates the synthesis of a marker proteins rather than the original gene produce. As a result, the accumulation of the marker proteins in plant cells and tissues marks the specific location where the gene of interest is being expressed.

The β-glucuronidase (GUS) gene from the bacterium *Escherichia coli* is widely used as a marker for studies of gene expression and is particularly suitable for use in plants since very little or no endogenous GUS activity has been detected in higher plants. The GUS gene product is an acid hydrolase that catalyzes the cleavage of a wide variety of β-glucuronides. In histochemical studies of gene expression, the GUS gene product is easily detected by infiltrating tissues with the synthetic substrate 5-bromo-4-chloro-3-indolyl β-glucuronide. When this substrate is cleaved by GUS an insoluble blue product precipitates in the tissue and marks the location where the gene in interest is being expressed (Fig. 3). To facilitate detection of the blue stain, chlorophyll may be removed from the plant tissues by washing with ethanol.

In addition to being able to localize the sites of a specific gene's expression, gene-fusion markers may also be used for analysis of promoter sequences of the gene of interest. By making deletions in the promoter of the fusion gene one can determine which regions of the promoter are essential for proper expression of the gene.

As a newly developing technique in the study of plant gene expression, gene-fusion markers have great potential for determining the sites of gene expression in plants and in determining essential elements of promoter sequences. As future improvements are made in the cloning and analysis of genes and in the ability to transform plants this technique will become more widespread.

E. Tissue Printing

Tissue printing refers to the technique of obtaining a physical impression or image of tissue organization on a suitable recording medium. Initially, films of gelatin or starch were used but most recently various membranes (e.g., nitrocellulose, Nytran, Genescreen, and Imobilon) that are developed to specifically bind proteins and nucleic acids have been successfully used for cytological studies. This approach requires minimal tissue preparation since samples are cut with a sharp blade and pressed onto the recording medium. Techniques such as fixation, embedding, and sectioning are not required. Physical prints can be viewed with low power light microscopy and transmitted light. If the light source is placed oblique to the microscope stage, the prints appear as a three-dimensional image, revealing additional detail. The resolution of such prints is inferior to that of fixed and stained sections; however, it is adequate for many purposes, such as immunocytochemical and *in situ* hybridization techniques (Fig. 4).

The ability to obtain a detailed image of tissue structure depends upon tissue firmness. Cells with lignified, cutinized, or silicified walls make pronounced impressions on nitrocellulose and thus reveal large amounts of anatomical information. Tissue prints with softer tissues, such as parenchyma or mesophyll, do not contain as much anatomical detail with nitrocellulose. However, softer media, such as agarose (6% agarose with 15% sorbitol), can be used with soft tissue to reveal good structural details. Use of this softer medium also results in improved resolution of harder tissues.

When the cut surface of a tissue is placed on one of these membranes, the contents of cut cells are transferred with little lateral movement or diffusion on the membrane. The principal steps for cytochemical analysis of tissue prints are (1) the release of the substance (protein or nucleic acid) of interest from the plant tissue, (2) the contact-diffusion transfer of the substance to the membrane, and (3) the retention and binding of the substance to the membrane matrix. The membrane can then be probed with antibodies for specific proteins or nucleotide probes for nucleic acids. With an appropriate secondary color reaction, the location of these productions (proteins or nucleic acids) can be visualized with respect to specific cell or tissue types. Unfortunately processing the prints for histochemical labeling results in a loss in resolution of anatomical detail. This loss of resolution can be diminished if a softer agarose medium is used for printing.

Tissue printing, followed by immunocytochemical staining, has been useful for studying time-dependant and tissue-specific accumulation of specific proteins. Nitrocellulose membranes have a high affinity for proteins and thus should be tried first for printing a new type of protein. For example, various cell wall proteins have been examined for changes in distribution during development or as a result of wounding.

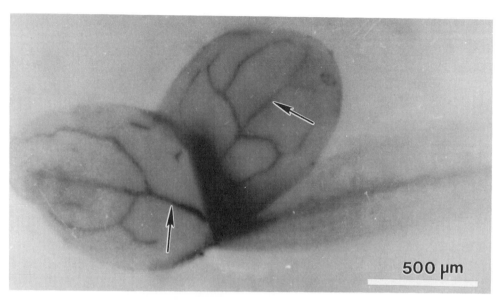

FIGURE 3 A β-glucuronidase (GUS) gene fusion acts as a marker for endogenous gene expression in transgenic plants. The gene coding sequence for the GUS protein, fused to the promoter of a β-1,3-glucanase gene, allows for the detection of sites where the β-glucanase gene is expressed. Once inserted into the genome of a plant, GUS is expressed wherever the β-1,3-glucanase gene is transcribed. The presence of GUS is indicated by the generation of a dark staining pattern in the vascular tissue (arrows) of cotyledons in transgenic tobacco seedlings. (Photograph is courtesy of J. Hennig and D. F. Klessig)

As many wall proteins are ionically bound to the wall, treatment of membranes with various ionic strength solutions helps release specific components to the membrane. Adding methanol to the transfer buffer will increase the capacity and affinity of the nitrocellulose for proteins, thus improving the detection of the protein of interest. Some proteins do not bind nitrocellulose (or bind it poorly). In these cases, other membranes (such as Immunodyne Immunoaffinity membrane) that bind proteins covalently may be used. The type of buffer, salt concentration, and pH of the transfer medium are important in protein binding and must be empirically determined for optimum detection.

Although tissue printing has a lower resolution than *in situ* hybridization for localization of mRNAs, it is a quick and convenient way to screen a large number of tissues from different developmental stages or from different plants at the same time. Both nitrocellulose and nylon membranes can be used for *in situ* hybridization, without pretreatment. However, nylon membranes are generally preferred because they are easier to handle. The general concept is the same as for immunolocalization of proteins, with mRNAs from cut cells being immobilized on membranes with little or no lateral diffusion. Localization is achieved by comparing the site of RNA deposition with the physical impression of the tissue.

This technique facilitates the study of large organs, such as fruit and whole seeds, and allows tissue-level localization of mRNAs. For example, the location of pectin methylesterase mRNA in tomato fruit or the distribution of cellulase mRNA in bean tissue has been used to examine the genetic regulation of fruit ripening. As information from sections of different developmental stages or different plant regions can be printed on the same piece of membrane, tissue printing is well suited for quantitative analysis of developmental or tissue-specific changes in cellular components. A single membrane with numerous tissue prints can be treated with the same hybridization and washing solutions and identical color development procedure, reducing differences between individual prints caused by differences in technique.

Tissue printing is also a rapid and simple technique for the detection and localization of plant pathogens. Suitable probes can be used for either immunolocalization or *in situ* hybridization. Advantages of this approach include: (1) simplicity of techniques and equipment; (2) ease of examining large sample sizes; (3) short processing times allowing rapid determination and identification of pathogens; (4) detection of subliminal and early-stage infections lacking external plant symptoms; and (5) the specificity of the probe can identify the pathogen involved. The greatest strength of this technique is perhaps the ability to

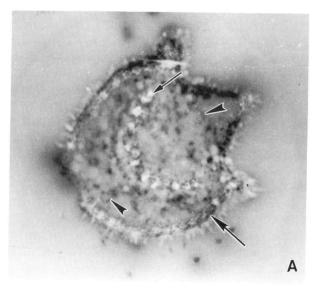
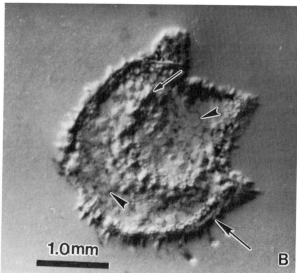

FIGURE 4 A tissue print of a tomato leaf petiole on nitrocellulose paper, viewed with transmitted light (A) and side illumination (B). Thickened or lignified cells such as the epidermal layers (large arrow) and vascular tissue (small arrow) are clearly evident. Softer tissue, such as the cortical cells, are less visible (arrowhead).

precisely localize the tissue targeted by the pathogen and the extent of infection.

III. Electron Microscopy

A. Classical Techniques

Transmission electron microscopy utilizes the higher resolution provided by electrons and the electron scattering ability of heavy metals used as stains. The most significant advantage of electron microscopy resides in the increased resolution which can reveal ultrastructural detail within cells. The basis of using electron microscopy for ultrastructural analysis resides in the fact that if one understands the structure of something (i.e., a cell), one will be better able to determine how it works and how to manipulate its function. Electron microscope technology underwent a rapid growth in the 1960s and 1970s and has now plateaued; however, recent improvements in "user friendliness" have greatly improved the use of this technology.

In electron microscopy, cell components can only be identified by their structure and electron opacity. For this reason, preservation of ultrastructure and the ability to induce contrast are key steps in the electron microscopic examination of biological material. For electrons to penetrate biological tissue, samples must be extremely thin. This is accomplished by embedding preserved material in a suitably hard medium which can be cut into thin (60–80 nm) sections. As most biological material needs to be sectioned, it is essential that cell ultrastructure is adequately preserved.

The current limiting factor for the use of electron microscopy appears to be specimen preparation. Tissue preparation techniques are fraught with potential sources of artifact and the high resolution of electron microscopy (over light microscopy) accentuates the significance of artifacts and the need to attain the best possible preservation of cell ultrastructure. At present tissue preservation is done by two basic methods: chemical fixation and rapid freezing.

Chemical fixation is most widely used to preserve cell structure. A good fixative can be defined as one that rapidly penetrates tissues and extensively crosslinks cell components into three-dimensional configurations that reflect the living condition. Various aldehyde fixatives, such as glutaraldehyde and paraformaldehyde, are most widely used. Paraformaldehyde penetrates tissue more rapidly than glutaraldehyde but does not crosslink as extensively. For this reason combinations of the two are used to achieve both rapid penetration and extensive crosslinking. Care must be taken to maintain ionic and osmotic balances during fixations to reduce cell shrinkage and the redistribution of cellular components.

The major limitation of chemical fixation is the time required to adequately crosslink cellular components. Because of the normal delay in diffusion of the fixative, it is thought that fixation only fixes the cell's response to the fixative and not the living state. Many dynamic, electrophysiological processes occur too fast to be captured by these slow-acting preservatives. Movement within cells may continue for seconds or minutes after the addition of chemical fixatives, indicating that cellular contents undergo a gradual cross linking and may assume a distribution that does not reflect the *in vivo* situation.

As biological material consists largely of molecules containing carbon, oxygen, and nitrogen, with a few atoms of high atomic weight, it is largely transparent in the electron microscope. In order to determine chemical composition, location, shape, and dimensions of cell components, the application of selective staining is necessary to improve contrast. Image contrast is a function of electron scattering when the electron beam passes through the biological material. Contrast is increased through the application of heavy metal stains. The use of osmium, uranium, and lead atoms to bind to specific cellular components increases the detectability of lipid, protein, and carbohydrate components.

In general, biological specimens are stained more than once in order to achieve adequate differential electron opacity. For example, osmium tetroxide is applied after aldehyde fixation and before dehydration. Osmium binds to unsaturated lipids and enhances membrane contrast. Once sections are cut, samples are poststained with uranyl acetate and lead citrate to enhance osmium-induced contrast and also to enhance contrast in structures not readily stained with osmium. Lead and uranium salts bind to nucleic acids and proteins to increase contrast of structures such as nuclei, ribosomes, cytoskeleton, various organelles, etc.

Numerous factors can affect the staining effectiveness of these heavy metal salts. Section thickness determines how much biological material is available for binding to stains, with thicker sections having greater contrast due to the presence of more heavy metal atoms. The pH of the staining solution affects the aggregation state of the heavy metal atoms and also the charge (i.e., binding) characteristics of the biological components. The duration of staining affects how much stain will bind to the specimen, with overstaining resulting in decreased detection due to large aggregates of heavy metal atoms. Temperature can influence staining, with most stains penetrating sections and binding more quickly at elevated temperatures.

B. Rapid Freezing–Freeze Substitution

A better alternative to chemical fixation is cryofixation (physical fixation), a physical procedure used to preserve the distribution and structure of all cell components. Preservation of cell ultrastructure occurs more rapidly if tissues are quickly frozen. The principal aim and advantage of cryofixation is the near-instantaneous arrest of cellular metabolism. Cell morphology is captured in its living state through stabilizing and retaining soluble cell constituents.

The major limitation of this technique is the artificial formation of segregation compartments ("ice crystals") that destroy cell ultrastructure. The formation of true vitrification (amorphous ice) is impossible with most plant specimens, because once the water in the superficial layer of cells has become ice, the deeper layers are unable to dissipate heat fast enough, due to the low thermal conductivity of water and ice. Ice crystals form within cells if freezing rates are less than 10^6 °K sec^{-1}. These fast freezing rates are achieved by several methods, which include rapid plunging into an appropriate cryogen, slamming the specimen on to a suitably cooled copper surface, or using a jet spray of cryogen. Appropriate cryogens include ethane, propane, freon, and isopentane, held at their melting temperature. Liquified nitrogen at its boiling temperature is not suitable as a primary cryogen because of its tendency to form a thermally insulating layer of gas around any warm object, thus reducing the freezing rate of that object. Cooling liquid nitrogen to its melting point is very difficult.

Due to the requirement for extremely fast freezing, tissue samples must be relatively small. As tissues freeze from the outside-in, cells within the tissue are "protected" from the cryogen and freeze more slowly. Under the best circumstances, only about 30 μm of the outer layer of the tissue is adequately frozen to give good ultrastructure. Cryoprotection agents such as glucose or glycerol act as anti-freeze to reduce ice crystal formation and improve the depth of cells in a tissue that are suitably preserved.

In addition to methods at atmospheric pressure, cryofixation may also be carried out at high pressure ($\approx$ 2100 bar). High-pressure cryofixation enables considerably thicker specimens to be frozen (maximum thickness of 600 μm with double sided cooling) without formation of ice crystals. High-pressure cryofreezing opens new perspectives for the cryopreservation of large specimens, such as whole seeds, stems, root, etc. Pressurizing tissues reduces ice crystal formation and growth. Thus, the slower rates of freezing at the center of large tissues (i.e., $500°K\ sec^{-1}$) do not result in artifacts.

Several clear advantages of cryofixation are evident. The physical processes of cryofixation are better understood than the innumerable biochemical reactions possible with chemical fixation, facilitating interpretation of images. Artifacts produced by cryofixation are relatively easily interpreted because only physical processes are involved. Little chemical alteration of cell constituents occurs with cryofixation; thus, there is minimal loss of enzyme activity and antigenicity. The extraction and translocation of solutes in the frozen and freeze-substituted tissue are thought to be less than those involved in conventional chemically fixed preparations. The loss of cell components in chemical fixation is not found in cryopreservation.

Once tissue is frozen in a "life-like" state, that state must be preserved. This is accomplished by freeze substitution of the water in the cell with an organic solvent containing a protein crosslinking fixative. Various chemical fixatives that function at low temperatures are used. Osmium tetroxide is most commonly used; however, if enzyme activity or antigenicity is studied then glutaraldehyde or acrolein may be used since osmium tetroxide adversely affects the preservation of these traits. Uranyl acetate may also be added to further stabilize cell structure. Organic solvents that have been used include acetone (most common), methanol, ethanol, heptane, and diethyl ether. Substitution is carried out at $-80°$ to $-95°C$ and the organic solvent dissolves and substitutes for the cellular ice. Substitution usually takes several days at $-80°C$, after which the temperature of the tissue is gradually raised. To minimize the loss of soluble pools of cell constituents, such as carbohydrates, proteins, and lipids, low temperature ($-70°C$) embedding of tissue is done.

C. Immuno-Electron Microscopy

Combining the high resolution of electron microscopy with antibody–antigen specificity provides plant researchers with a highly versatile technique for studying cell function. If antigenicity is maintained through the tissue preparation procedures then one can localize specific proteins, nucleic acids, or carbohydrates with great precision. Using immuno-electron microscopy also eliminates a common immunofluorescence problem in plant cells, namely autofluorescence. Many plant cells with suberized or lignified walls, chloroplasts, or phenolic or other autofluorescent components are not amenable to immu-

nofluorescence, but can be examined with antibodies and electron microscopy.

As with other electron microscopic techniques, sample preparation is crucial for reliable data collection. Because of the potential problems with loss of antigenicity (discussed above), the tissue preparation method of choice for immunological studies should be rapid freezing–freeze substitution. This preparation appears to be the most likely to maintain the original distribution of molecules, with minimal losses due to diffusion or extraction. The choice of tissue fixation method depends upon the specific antigen–antibody being used. Some antigenic sites are destroyed or masked by exposure to osmium while others are not. The same can be said for aldehyde fixations or embedding techniques. Trial experiments with a variety of procedures need to be done to determine the optimum set of conditions for the specific antibody–antigen complex under examination.

Immuno-electron microscopy facilitates the subcellular and suborganelle dissection of function. Using antibodies to specific proteins, subcellular localization of proteins within specific organelles can be determined. The combination of localization of specific proteins with high-resolution ultrastructural characterization facilitates the examination of cell compartmentalization. For example, subcellular distribution of the photosynthetic enzyme ribulose bisphosphate carboxylase (rubisco) can be achieved using antibodies raised against this enzyme (Fig. 5). As mentioned earlier the use of antibodies as cytological probes is not limited to the study of proteins. Because carbohydrates, in the form of complex polysaccharide, are a major component in plant tissues, there is a great interest in understanding the cellular synthesis of polysaccharide and the organization of these molecules within the cell walls of plants. Many of the recent insights in this area of study have resulted from the use of antibodies as cytological probes. For example, the assembly of the complex polysaccharide, polygalacturonic/ rhamnoglacturonan-1, has been elucidated using antibodies with specificity to different regions of this polysaccharide (Fig. 6).

IV. Conclusions

Plant cytology continues to evolve due to improvements in cellular preservation, development of more specific stains, and the invention of new instrumentation. While improvements have been made in tissue preservation (i.e., rapid freezing), the significance of

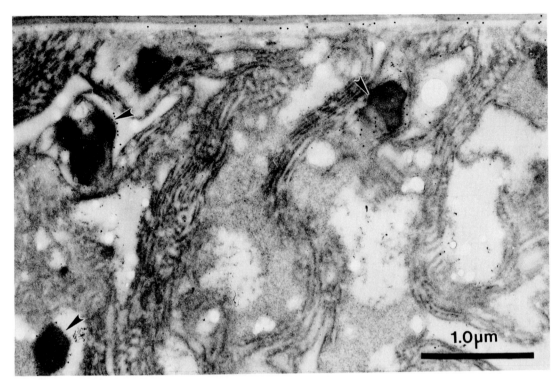

FIGURE 5 Immuno-electron microscopy of the enzyme rubisco (ribulose bisphosphate carboxylase) in the blue-green algae, *Oscilatoria limosa*. Antibodies raised in rabbit to tobacco rubisco were applied to sections of glutaraldehyde/osmium-fixed cells. A secondary antibody (goat anti-rabbit IgG), labeled with 15-nm gold particles, binds to the anti-rubisco and localizes its presence (by an accumulation of gold particles) in carboxysomes (arrowheads).

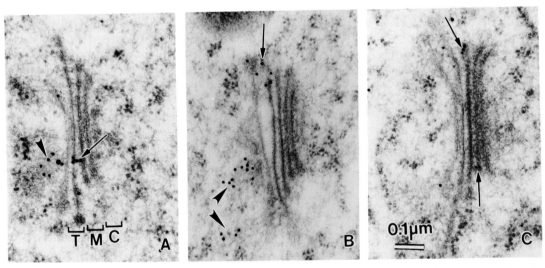

FIGURE 6 Immuno-electron microscopy of glycoprotein synthesis in the Golgi stacks of cultured sycamore maple cells. Electron microscopic observation allows for the discrimination of various parts of the Golgi, i.e., *cis* (C), median (M), *trans* (T), and the *trans* Golgi network (arrowheads) regions. (A,B) Antibodies with specificity for side chains attached to rhamnose residues label the *trans*-cisternae and the *trans*-Golgi network (arrows), indicating that side chain addition is a later event in the assembly of polygalacturonic/rhamnoglacturonan-1. (C) Antibodies specific for methyl-esterified galacturonic acid residues bind to the *cis* and medial compartments of the Golgi (arrow) indicating that this process is compartmentalized in this region. Reproduced with permission from Zhang, G. F., and Staehelin, L. A. (1992). *Plant Physiol.* **99,** 1077.

technique-induced artifacts on interpretations of cell, tissue, organ, and whole plant composition remains a major concern for plant cytologists. New and highly specific probes (i.e., antibodies and specific nucleotides) provide greater resolution and details of subcellular structure and are being used to determine the developmental regulation of plant morphogenesis at both the light and electron microscopic levels.

Most techniques for examining the internal components of plants require some type of tissue preparation and the application of exogenous stains. Recently, advances have been made in the use of noninvasive techniques for the examination of plant anatomy. Nuclear magnetic resonance (NMR) microscopy is being used as a noninvasive technique to observe developmental changes in plant tissues and organs. The theory of NMR microscopy is complex and relates to changes in the magnetic moments of atomic nuclei in response to being bombarded with radio frequency radiation. For complex biological materials with many different atoms, interpreting observations can be formidable. Contrast within a specimen results from the differential distribution of elements with specific magnetic moments. NMR images have remarkable ultrastructural detail of internal tissue distribution and thus provide a way to examine living plant anatomy in a dynamic manner. NMR microscopy has been used to examine changes in oil content in olive seeds and water in maize stems, the effects of environmental stress on spruce leaves, and the distribution of aromatic oils in orange peel and grape berries. Interpretations of images are subject to some controversy; however, the ability to use noninvasive techniques to examine a single plant, organ, or tissue through a specific developmental process may find great application in agricultural science.

Although cytological techniques may be one of the most ancient methods for examining plant anatomy and development, continuous improvement in techniques and equipment have kept cytology on the forefront of investigations in plant morphogenesis. Before one can manipulate plant growth and development in defined and controlled ways, a firm foundation of form and function is required. Plant cytology provides researchers with descriptive details on plant, organ, and tissue structure, sub-cellular compartmentalization, and chemical composition at cellular and subcellular levels.

Bibliography

Berlyn, G. P., and Miksche, J. P., (1976). "Botanical Microtechnique and Cytochemistry." Iowa State Univ. Press, Ames, IA.

Chesselet, M-F. (1990). "*In Situ* Hybridization Histochemistry." CRC Press, Boca Raton, FL.

Clark, G. (1973). "Staining Procedures," 3rd ed. Williams and Wilkins Co., Baltimore.

Conn, H. J. (1940). "Biological Stains." Biotech, Geneva, NY.

Hayat, M. A. (1989). "Principles and Techniques of Electron Microscopy, Biological Applications." 3rd ed. CRC Press, Boca Raton, FL.

Herman, B. and Lemasters, J. J. (1993). "Optical Microscopy: Emerging Methods and Applications." Academic Press, San Diego.

Jensen, W. A. (1962). "Botanical Histochemistry: Principles and Practice." Freeman, San Francisco.

Pawley, J. (1989). "The Handbook of Biological Confocal Microscopy." IMR Press, Madison.

Reid, P. D., Pont-Leziec, R. F., del Campillo, E., and Taylor, R. (1992). "Tissue Printing." Academic Press, San Diego.

Spencer, M. (1982). "Fundamentals of Light Microscopy." Cambridge Univ. Press, Cambridge, UK.

Steinbrecht, R. A., and Zierold, K. (1987). "Cryotechniques in Biological Electron Microscopy." Springer-Verlag, Berlin.

Vaughn, K. C. (1987). "Handbook of Plant Cytology," Vols. I and II. CRC Press, Boca Raton, FL.

Wang, T. L. (1986). "Immunology in Plant Science." Cambridge Univ. Press, Cambridge, UK.

Plant Cytoskeleton

ROBERT W. SEAGULL, *Hofstra University*

Glossary

Cytoplasmic streaming The directed, bulk movement of subcellular components in a cell; this usually appears as strands of cytoplasm along which particles are seen to move in a directed manner

Cytoskeleton A dynamic three-dimensional array of protein filaments found in all eukaryotic cells

Immunocytochemical techniques The use of antibodies as cytological probes to identify and localize specific cellular components

Microfilaments A class of cytoskeletal elements that, when examined with the electron microscope, appear as 8- to 10-nm filaments of indefinite length

Microtubules A class of cytoskeletal elements that, when examined with the electron microscope, appear in cross section as hollow cylinders 25 nm in diameter with 5-nm thick walls; in longitudinal section, microtubules appear as linear elements with relatively straight walls and of indefinite length

Morphogenesis The development of specific shape and function; this can occur at cellular, tissue, organ, or organism levels

Posttranslational modification The modification of amino acid sequence after proteins are synthesized (translated) from RNA

The cytoskeleton can be defined as a dynamic, three-dimensional array of filamentous proteins. Elements of the cytoskeleton are involved with all aspects of plant growth and development, controlling and de-

fining plant cell morphogenesis. Two major structural elements are observed: microtubules and microfilaments. Microtubules appear as hollow cylinders with a diameter of about 25 nm and an indefinite length. They are found in variously organized arrays during the cell cycle and cell morphogenesis. Microtubules function in mitosis and cytokinesis to separate genetic material, define and control the location and synthesis of the cell plate, and somehow regulating the deposition of cellulose microfibrils in the cell wall. Microfilaments appear as 8- to 10-nm filaments, organized into bundles in the cortical cytoplasm or sometimes found singly, associated with microtubules. Bundles of microfilaments function in cytoplasmic streaming by providing linear elements along which various cytoplasmic components move. Both microtubules and microfilaments are composed of a structural protein (tubulin for microtubules and actin for microfilaments) and arrays of regulatory proteins that modify the dynamics and function of the polymer population. The structural proteins of the plant cytoskeleton are being defined and characterized. However, research on the existence and characterization of cytoskeletal regulatory proteins in higher plant systems has just begun.

I. Introduction

The development of plant form and function (morphogenesis) is defined by both genetic and environmental factors. Improvements in sustainable plant agriculture will depend on modifying how these factors impact plant morphogenesis. A key to this approach is the understanding of fundamental mechanisms controlling plant morphogenesis. The ultimate target of this understanding is the development of strategies for the manipulation of plant growth and function to meet current and projected agricultural needs, thus leading to direct and long-term improvements in plant viability and productivity.

The understanding of fundamental mechanisms of plant morphogenesis will play a central role in future improvements of plant agriculture. The directed manipulation of plant development through molecular genetic and classical plant breeding techniques requires knowing potential pivotal morphogenic events and the factors (proteins and genes) that regulate them.

The development of specific function is often preceded by the establishment of specific form and organization. Thus, understanding function must be based on a firm knowledge concerning the establishment of specific form. Specialized tissues and organs (often the product for agricultural exploitation) are the culmination of morphogenic processes at the cellular level. The establishment of specific cell form and the compartmentalization or polarization of subcellular components are key steps in cell development. Understanding the mechanisms and regulation of cellular morphogenesis will provide the foundation for understanding whole-plant morphogenesis.

Cellular morphogenesis uses two fundamental mechanisms to alter cell shape: cell division and cell expansion. After cytokinesis, daughter cells invariably have different sizes and morphologies than the parent cell. Cell division can occur equally, producing identical daughter cells, or unequally, producing distinctly different daughters destined for different developmental pathways. Cell expansion characteristics are regulated by the organization (parallel or random) of reinforcing cellulosic microfibrils in the cell wall. Randomly organized wall microfibrils do not provide sufficient reinforcement to restrict cell expansion; thus, expansion produces isodiametric cells. The deposition of parallel arrays of microfibrils restricts expansion to perpendicular to the axis of microfibril orientation; thus, growth is anisotropic and cells elongate.

Intense research over the past 30 yr has indicated that many of the processes of morphogenesis are controlled through the function of arrays of cytoplasmic filaments, collectively called the cytoskeleton. The cytoskeleton is a dynamic, three-dimensional array of proteinaceous filaments found in all eukaryotic cells (plants, animals, fungi, protists, etc.). Using electron microscopy, three major classes of cytoskeletal fibers have been identified: 7-nm diameter actin microfilaments (MFs), 24-nm diameter microtubules (MTs), and 10-nm diameter intermediate filaments (IFs). These elements differ not only in morphology but

also in protein composition. MFs are composed of a major structural protein, actin, that polymerizes into long double-helical chains and a variable number of actin binding proteins (ABPs) that modify the organization, stability, and function of the actin polymers. MTs are composed of a major structural protein, tubulin, that occurs as a heterodimer composed of nonidentical subunits (alpha and beta tubulin) and an array of MT-associated proteins (MAPs) that modify the physical and functional characteristics of the microtubules. IFs, so named because their diameter is intermediate between MTs and MFs, are broken down into five classes, based on their protein composition and cell-type distribution. IFs are best characterized in animal systems and, until recently, were thought not to occur in plant cells.

This article presents current knowledge of plant cytoskeletal elements, emphasizing where possible its relevance to plant morphogenesis. Much of what is known about cytoskeletal elements (particularly with respect to protein components and genetic regulation) comes from animal systems. Although this literature provides a valuable guide for plant cytoskeletal research, significant differences exist between cytoskeletal components of plant and animal systems. Thus, caution should be used when drawing analogies between the two. Many excellent books and review articles are available on cytoskeletal elements and their role in cell morphogenesis; these should be consulted for a more in-depth review.

II. Microtubules

A. Morphology and Distribution

MTs were first observed in plant cells using electron microscopy during the early 1960s. In cross section, MTs appear as hollow cylinders with a 10- 15-nm electron translucent core and in longitudinal section as variously long, relatively straight structures with parallel edges (Fig. 1). MTs are found at all stages of the plant cell cycle (Fig. 2). During interphase, microtubules are located in the cortical cytoplasm, close to and often appearing connected with the plasmalemma. This early observation provided the foundation for belief that MTs somehow control wall microfibril patterns. Preprophase is marked by the formation of a band of MTs predicting the future site of cell plate connection with the parent wall. During prophase through anaphase, MTs—as part of the spindle—polymerize and depolymerize as the spindle

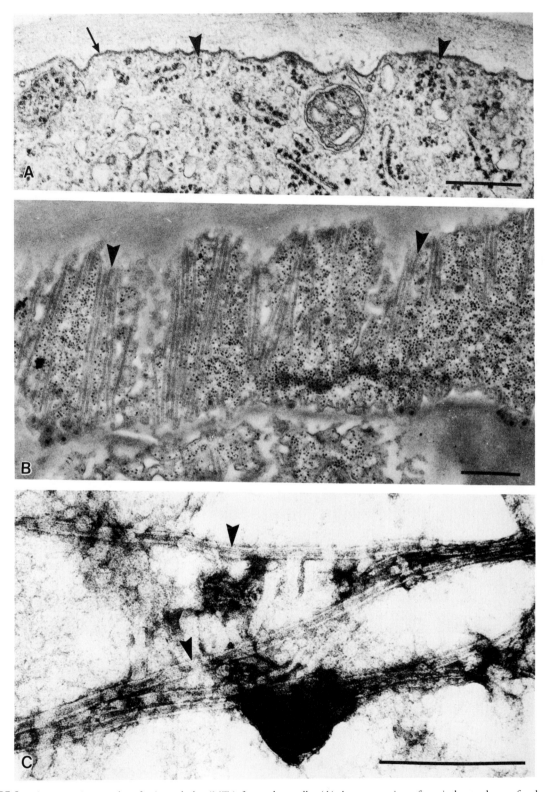

FIGURE 1 Electron micrographs of microtubules (MTs) from plant cells. (A) A cross section of cortical cytoplasm of a developing radish root hair. MTs (arrowheads) appear as small circles near the plasmalemma (arrow). Bar: 0.5 μm. (B) In this section through the cortical cytoplasm of a wheat root cell, MTs (arrowheads) appear in longitudinal view as variously long elements with parallel sides. Bar: 0.5 μm. (C) *In vitro*, negatively stained MTs isolated from cotton cells exhibit their linear morphology and constant diameter. In certain regions, evidence of protofilaments is evident (arrowheads) Bar: 0.5 μm.

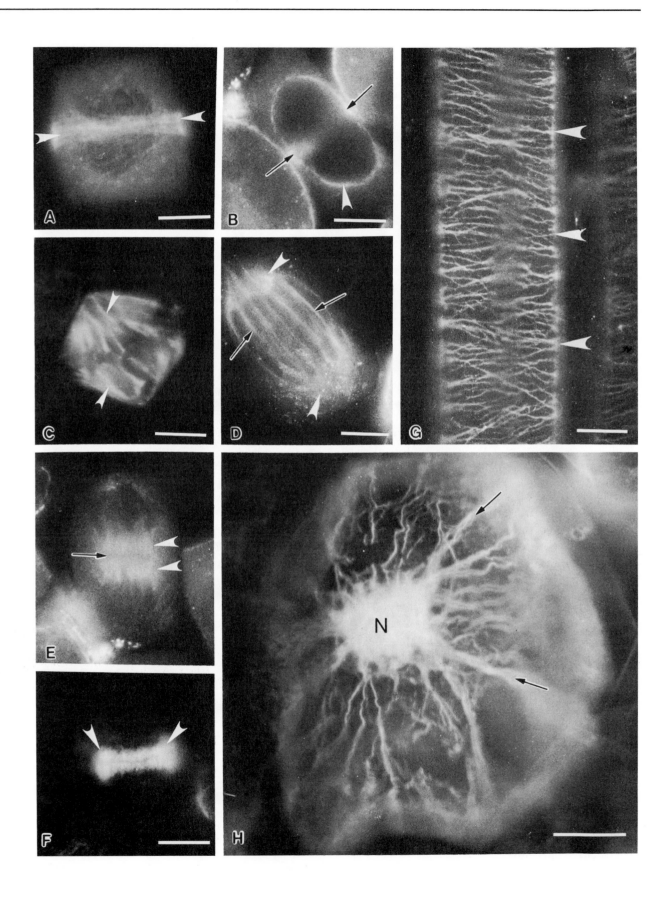

separates chromosomes. Formation of the cell plate during telophase is accomplished through the function of the phragmoplast which contains arrays of MTs oriented perpendicular to the plane of the forming cell wall.

B. Protein Chemistry

1. Homology of Plant and Animal Microtubule Proteins

The study of MT proteins from animal systems (particularly neuronal tissue) has greatly facilitated our understanding of MT function in plants. Brain tissue provides a readily available source of MT proteins. Studies on this system have provided the foundation for much of the current work on plant tubulins and MAPs.

MT proteins in neuronal cell extracts undergo temperature- and concentration-dependent polymerization into MTs. This inherent ability to cycle between pools of protein monomers and polymeric MTs facilitates the purification and identification of the major proteins involved in the formation and function of MT proteins.

Although MTs were first observed in plants in the early 1960s, research specific to plant MT proteins began only in the past 10 yr, partially because of the assumption that plant and animal MTs and their associated proteins were similar. This assumption was based on observations that MTs from both sources have similar morphology (diameter and protofilament patterns), organization (spindle and interphase arrays), and biochemistry (polymerization characteristics and antigenicity). However, other observed characteristics of plant MTs indicate that this assumption may be unwarranted. Plant MTs appear to have different sensitivities and binding constants to certain disrupting agents (i.e., colchicine). They are insensitive to some agents (nocodazole) that are very potent in disrupting animal MTs and are extremely sensitive to other agents (amiprophos-methyl) that have no effect on MTs from animal sources.

Another reason for the lack of study of plant MTs has been the difficulty in purifying MT proteins from plant tissues. Plant cell extracts yield low total protein amounts and concomitantly low MT protein concentrations because of the presence of the central vacuole in nearly all plant cells. As a result, temperature-dependent cycling, which was a critical breakthrough in the study of animal MTs, is not readily adaptable to the study of MTs in plants. In addition, various proteases, oxidases, phenolics, and quinones are present in plant tissues that degrade or bind to MT proteins, thus preventing polymerization of MTs. Alternative procedures, using column chromatography as an initial purification step or taxol (an alkaloid that reduces the critical concentration for MT polymerization) to induce polymerization, have been successfully used to isolate intact plant MTs for protein characterization.

2. Tubulins

Tubulin, the major structural protein of microtubules exists as a heterodimer of about 120,000 Da. These dimers consist of nonidentical α and β polypeptides of about 450 amino acids. The interaction between heterodimers in the polymer (i.e., the β subunit of one heterodimer interacts with the α subunit of the next) results in a polarity within the MT, which is reflected in the relative polymerization rates at the ends of the MT. A "plus" connotation designates the end with the higher polymerization rate, whereas a "minus" designates the slow growing and more rapidly depolymerizing end. Since rates of polymerization and depolymerization are dependent on many factors (see subsequent discussion), these rates are relative, with possible subunit addition and removal occurring at either end.

The amino acid sequence of many tubulins has been determined by direct protein sequencing or by deduc-

FIGURE 2 Immunofluorescence light micrographs of microtubule (MT) arrays at various cell cycle stages in developing clover root cells (A–F), cotton fiber (G), and cultured alfalfa cells (H). (A) The preprophase band (PPB) of MTs (arrowhead) predicts the future site where the developing cell plate will fuse to parent cell walls. (B) During prophase, the nuclear envelope exhibits tubulin fluorescence (arrowhead). Areas of brighter fluorescence (arrows) are the PPB in median optical section as it circumnavigates the nucleus. (C) Metaphase spindle shows kinetochore fibers (arrowhead) containing numerous MTs. (D) In anaphase, the kinetochore fibers have retracted to the spindle poles (arrowheads) and only continuous spindle fibers (arrows) remain in the central region. (E) The phragmoplast consists of two equal arrays of MTs (arrowheads) separated by the developing cell plate (arrow). (F) The phragmoplast grows centripetally toward the cell periphery; the most intense arrays of MTs are at the developing edge (arrowheads). (G) At interphase, MTs are organized into variously parallel arrays (arrowhead) in the cortical cytoplasm. (H) Highly vacuolated cells form a phragmosome, containing arrays of MTs (arrow) that suspend the nucleus (N) at the cell center. Bar: 20 μm.

tion from genomic or cDNA nucleotide sequences of tubulin genes. Tubulins from a wide variety of sources (animal, plant, fungal, and protist) exhibit a high degree of amino acid conservation. Some regions of the protein contain long sequences of nearly 100% homology. Other regions may have different amino acids, but similar charge distributions. The conserved regions are involved in protein secondary structure, formation of the heterodimer, and the binding of various regulatory factors (e.g., GTP, Ca^{2+}). Much of the divergence in amino acid sequence is clustered into specific regions of the protein, such as the carboxyl terminus. Some of these divergent regions are responsible for binding various MAPs that may modify MT function. Plant β-tubulin has a hydrophobic pocket around amino acid 200 that is not present in animal β-tubulin. This difference may account for some of the unique pharmacological properties of plant tubulins. Among plant tubulins are also regions of hypervariability in which no consensus with animal tubulins is observed. These regions presumably are not directly involved in key functions and, thus, can exhibit more variation.

Plant β-tubulins exhibit an array of predicted secondary structures, some of which are similar to animal tubulins whereas others are quite different. Differences in secondary structure may modify the binding domains present in tubulins, thus altering the binding kinetics of self-assembly or various regulatory factors. Mild proteolytic cleavage indicates that a high affinity Ca^{2+}-binding site is located at the amino terminus of both α- and β-tubulin. Binding of the MT regulatory proteins MAP_2 and Tau (see subsequent discussion for descriptions) are located at the C-terminal ends of both α- and β-tubulin, as well as the N-terminal end of α-tubulin. Both α- and β-tubulins are organized into compact globular structures from which a short tail projects. The carboxyl terminus of tubulin is located in this projection and appears on the surface of polymerized MTs, thus facilitating binding of regulatory proteins *in vivo*. Thus, the C-terminal domain of tubulin represents a regulatory component, to which other molecules can bind to regulate subunit assembly or interactions between MTs and other cellular components.

Multiple α- and β-tubulin genes have been identified in all plant species examined to date. For example, in *Arabidopsis thaliana* (the only plant to be extensively examined for the presence of genes and pseudogenes), six α- and nine β-tubulin genes have been identified and all have been shown to be expressed in the plant. Pea, oat, and maize all have multiple functional tu-

bulin genes as indicated using cDNA technology to examine expression. Analysis of tubulin expression in different organs of the same plant shows that certain isotypes may be preferentially expressed in a single tissue, whereas other isotypes are present in all tissues of the plant.

Changes in tubulin isotype patterns occur in response to changing environmental conditions. These observations are consistent with the possibility that part of a plant's response to environmental signals may be made through changes in cytoskeletal components. The number of expressed genes does not necessarily reflect the number of isotypes. For example, in *Arabidopsis*, six different α-tubulin genes have been identified that encode four α-tubulin isotypes. Similarly, nine β-tubulin genes appear to encode eight specific isotypes.

Posttranslational modification of tubulins also generates different isotypes. This area of MT chemistry is very interesting because posttranslational modifications could potentially confer rapid and reversible changes on the properties of microtubules *in vivo*. The types of modifications observed include phosphorylation of various amino acids on the β- subunit, which may be involved in binding various MAP proteins; detyrosinolation of the carboxyl terminus of α-tubulin, thereby exposing the penultimate glutamic acid residue and increasing the affinity between the tubulin and various MAPs; acetylation of a lysine ε-amino group of α-tubulin, thereby inducing increased stability of the MT array; and glycosylation of tubulin for possible interaction with the plasmalemma.

The role of multiple tubulin genes or posttranslational modifications of tubulin in the organization and function of MTs in higher plants remains to be determined. The "multitubulin hypothesis" proposes that specific sets of tubulin (and, thus, MTs) perform different functions at various developmental stages or in different organs. Support for the hypothesis is contradictory. A variety of genetic, biochemical, and cell biological approaches indicates that most tubulin isotypes are functionally interchangeable and, thus, equivalent. The observed variety of tubulins in most eukaryotic cells may have evolved to facilitate the complex patterns of regulatory fine tuning necessary to accomplish the various functions and arrays of MTs that occur during normal development and in response to environmental signals. However, recent experiments (with *Drosophila*) demonstrate a subcellular sorting of tubulin isotypes. Gene complementation studies have shown that one tubulin gene could not substitute for a null mutation of another tubulin gene; thus, the genes are not functionally equivalent.

3. Microtubule Associated Proteins

MAPs refer to a class of proteins that, on binding to MTs, affect their stability, organization, or function. In general, MAPs function to extend the chemical and functional characteristics of MTs. However, unlike tubulin, which has remarkable similarities among organisms and even kingdoms, MAPs exhibit considerable variability in composition and function. MAPs comprise a diverse array of proteins with different molecular weights and amino acid compositions. Various functions of animal MAPS have been demonstrated to include enhancement of MT stability, cross-bridging of MTs to each other or to other organelles, transport of particles, and enhancing MT formation by acting as nucleating sites for polymerization.

MAP proteins have been most extensively characterized in neuronal tissues and are broadly categorized into structural and motor proteins. Structural proteins include high molecular weight MAPs and Tau (τ) proteins. High molecular weight MAPs include the MAP_1 group (consisting of at least three different proteins of 300–350 kDa), the MAP_2 group (two proteins of 270–300 kDa), MAP_3 (180 kDa), and MAP_4, a heat-stable protein (≈ 200 kDa). τ Protein consists of at least four phosphoproteins of between 55 and 68 kDa. τ Proteins have an 18-amino-acid sequence that repeats three or four times in the carboxyl terminus of the protein and is thought to represent the MT binding region. The presence of a similar (67% homologous) repeat motif in MAP_2 indicates that this repeat may represent a widespread common MT binding motif. After binding to the MTs, high molecular weight MAPs and τ protein appear as linear, regularly spaced projections on the polymer surface. MAP_2 and τ protein also share similar sequences in this "projection" region of the protein. These regions have sequences that form a "zipper" motif, capable of intramolecular bonding between adjacent proteins. Interaction between MAPs or τ proteins on adjacent MTs may be involved in bundling MTs.

Motor proteins are those that contain an ATPase (dynein, kinesin) or GTPase (dynamin) capable of generating mechanochemical force in the presence of appropriate nucleotides. These motor proteins function in generating movement between adjacent MTs (cilia and flagella) or MTs and subcellular components (e.g., vesicles, chromosomes, RNA, protein) or may function to regulate MT turnover or bundle length.

Proteins related to dynein, kinesin, or dynamin are found in many animal and fungal cells and contain similar force-generating heads, linked to different cell-specific attachment domains. Motor proteins function in a directional manner, moving toward the plus end (kinesin) or minus end (dynein) of the MT. The plus and minus designation refers to the inherent polarity of the MT (see Section II,B,2).

Although most of our knowledge regarding MAPs comes from animal systems, recent advances in plant MT isolation and characterization are providing some insight into the existence and diversity of plant MAPs. Several reports have been made of MAPs isolated from higher plants. Using taxol to induce polymerization, MTs containing tubulin and various associated proteins have been isolated. From limited information, plant MAPs appear to be different from their animal cell counterparts. However, some immunological cross-reactivity appears to exist with antibodies directed against bovine τ protein and certain of the proteins that co-isolate with plant MTs. MAPs detected to date in higher plants have molecular masses ranging between 40 and 120 kDa and are thus similar to the lower molecular weight MAPs found in animal systems. Some of these proteins appear to induce bundling of MTs. The significance of these observations remains unclear. However, postulating that, given the conserved nature of tubulins, there may be some homology in MAP proteins among plant and animal systems, particularly in the region of the protein that binds tubulin polymer, seems reasonable.

4. Microtubule Heterogeneity in Plants

Plant MTs are not homogeneous among the various arrays formed in plant cells or within the population of MTs in a single array. For example, MTs do not exhibit an "all or none" response to various disrupting agents. Treatment of plant cells with these agents results in disruption of most MTs within the cell; however, in many instances, a subpopulation of MTs may remain intact and functional. This heterogeneity within MT arrays is also apparent after plant cells are subjected to cold treatment. The stability of MTs at low temperatures is of particular interest, since the cold stability of MTs is related to cold hardiness of the plant. Hormone treatments that increase cold hardiness also increase the cold stability of MTs.

The various MT arrays that form within plant cells are also not identical. Entire arrays of MTs may show differential stability in response to developmental cues. For example, MTs in cells making primary cell wall are less stable than MTs in cells making secondary wall. Changes in MT stability may impact many other aspects of plant growth and development. Unfortunately, very little is known about the factors that regulate MT stability in plants. Incorporation of newly synthesized tubulin isotypes, posttranslational modification of existing subunits, or the binding of MAPs may act to alter the stability of MTs; however, the exact influence these events may have on MT stability within living plant cells remains unknown.

5. Microtubule Polymerization and Organization

Much of what is known about the regulation of MT assembly comes from the study of animal cell systems. MTs can form spontaneously *in vitro* if the concentration of monomers exceeds a specific critical concentration. No energy requirement exists for assembly of MTs although GTP is hydrolyzed in the process. Tubulin binds GTP and assembles rapidly to form MTs. Once assembled, GTP may be hydrolyzed, causing a conformational change in the tubulins, destabilizing the polymer, and inducing depolymerization. Polymerization–depolymerization rates are governed by the rate of GTP hydrolysis. GTP–tubulin forms stable polymers, whereas GDP–tubulin polymers are less stable and undergo rapid disassembly. Under conditions of high tubulin concentration, MTs are chimeric with a GTP–tubulin stable cap at the polymerizing end of the GDP–tubulin-containing MT. If polymerization rates fall below the rate of GTP hydrolysis, the stable GTP–tubulin cap will be converted to GDP–tubulin and the MT will undergo rapid depolymerization. Separate growing and shrinking phases of MTs in the same population have been called "dynamic instability."

In vivo experiments with MT polymerization have indicated that MTs in various arrays have very short half-lives, ranging from seconds in spindle fibers to minutes in interphase arrays. However, not all MTs in the cell are homogeneous with respect to dynamic instability, since at interphase a few MTs may have half-lives of 1–2 hr. Rapid turnover may allow the cell to adapt MT arrays to changing environmental conditions. By reducing the rate of GTP hydrolysis, cells can maintain MT arrays at the low tubulin concentrations found in most cells.

In animal systems, most if not all MTs arise from sites of polymerization and organization known as MT organizing centers (MTOCs). Centrosomes, MTOCs found near the cell nucleus, induce MTs to form and radiate out toward the cell periphery with uniform polarity. This organization facilitates intercellular transport. MTOCs function to induce polymerization at low tubulin concentrations and regulate not only the rate of formation but also the pattern. Different arrays of MTs are established in animal cells through the migration and duplication (in spindle formation) of the centrosome.

In higher plant systems, the nuclear envelope (or cytoplasm closely associated with it) may participate in the formation of specific arrays of MTs. The reestablishment of interphase arrays of MTs appears to start at the nuclear envelope, as do the early stages of spindle formation and the establishment of the division plane in vacuolated cells. Work on isolated MT arrays from *Haemanthus* endosperm indicates that organization of the spindle is determined primarily by intrinsic properties of the MTs; the nuclear envelope and kinetochores modify these properties to establish final spindle form and function.

No compelling evidence exists that higher plant cells have a major governing structure comparable to the centrosomes of animal systems. MT arrays appear to generate from various locations in a cell-cycle-dependent manner. Although discrete structures are not evident, protein components characteristic of MTOCs in animal systems may be found in plant cells. A recently discovered member of the tubulin superfamily, γ-tubulin, has been implicated in the spatial control of MT organization in animal and fungal systems. Preliminary evidence indicates that this protein is present in some higher plant systems and is localized in regions of suspected MT polymerization (e.g., spindles, phragmoplast, and interphase arrays). Whether higher plant MTOCs are mobile, changing location to induce various arrays, or whether different MTOCs are present and developmentally regulated is not known. Improvements in cell preservation and detailed serial section reconstruction analysis reveals, in a few systems, the presence of amorphous material with various-sized vesicles embedded in it, specifically located at the terminations of MTs. Interphase arrays, preprophase bands, and bands of MTs associated with the development of localized wall thickenings are all seen to originate from dense aggregates of small, irregularly shaped vesicles and amorphous material located in the cortical cytoplasm. These sites were proposed to function in the nucleation of MTs with other mechanisms (as yet unknown) involved in determining organization. The universality of these regions has come under question. In numerous plant systems, such regions have not been visualized. Plant cells that elongate maintain the density of MTs along the plasmalemma; thus, interpolation of MTs into pre-existing arrays must occur. The polymerization of these MTs has not been correlated with identifiable MTOCs. These elements could form from pre-existing shorter MTs, with the rate of formation governed by tubulin levels or availability of other regulatory factors.

Increasing evidence suggests that MT formation and organization may be separate events. After cytokinesis, interphase arrays are formed and then develop a parallel order. During mitosis, MTs are first formed in a random pattern associated with the nuclear envelope. Subsequently, MTs form a spindle shape in which the poles become more focused during spindle function. Reorganization of MTs after polymerization may be regulated by interactions between adjacent MTs. Interpolation of MTs into existing arrays as the cell elongates and the formation of localized bundles of MTs during xylogenesis may be controlled by intra-MT bridging.

C. Function

Tremendous insight into the organization and formation of cytoskeletal arrays has come with the use of immunocytochemical techniques. These procedures allow the examination of cytoskeletal arrays in hundreds of cells in a relatively short time, facilitating the establishment of trends and norms in cytoskeletal development. MTs are involved in all aspects of plant cell development. Their formation and organization directly regulate two fundamental mechanisms of plant morphogenesis: cell division and cell shape. [*See* PLANT CYTOLOGY, MODERN TECHNIQUES.]

1. Cell Division

Plant cell division is accompanied by the formation of four specific arrays of MTs: preprophase band (PPB), phragmosome, spindle, and phragmoplast (Fig. 2A–F). PPBs function in the establishment of the division plane. The location of the PPB *predicts* where the developing cell plate will fuse to the parent cell walls. The phragmosome (Fig. 2H) functions during division of highly vacuolated plant cells and is responsible for positioning the nucleus and stabilizing it in the division plane. The spindle is involved with the equal separation of chromosomes. The phragmoplast functions in the formation of the cell plate that divides the two daughter cells.

The PPB appears to be specific to vegetative cells and does not occur in any meiotic cells. PPBs are found in all land plants but not in all cells of each species. Certain mosses and bryophytes lack PPBs at certain stages of development. The formation of PPBs appears to be an essential component of all cells that divide by producing a cell plate that grows to meet and fuse with parent walls. PPBs function regardless of the orientation of division, of whether the division is symmetrical or asymmetrical, of whether one or both the daughter cells will proceed to differentiate, or of whether the division is formative or proliferative. Normally the PPB is connected with the nucleus by arrays of endoplasmic MTs.

Since the PPB disassembles long before the formation of the cell plate, a central unresolved question remains concerning the function of PPBs as recognition sites for the developing cell plate. The site of the PPB apparently contains factors (of unknown composition) that assist in final formation and stabilization of the cell plate. This site may function to nucleate, attract, or stabilize phragmoplast MTs, thus facilitating proper cell plate formation. The site may also exert an attractive force on the edges of the developing cell plate to help in alignment and fusion.

When highly vacuolate cells divide, the nucleus is repositioned by a set of transvacuolar strands that suspend the nucleus in the cell center. Within these strands are arrays of MTs, extending from nucleus to cell periphery. Several hours after nuclear repositioning, these strands aggregate to form a diaphragm, termed the phragmosome, that also predicts the position of the future cell plate. Phragmosome formation occurs concomitantly with PPB formation and the two function together to establish the division plane and the location of future cell plate fusion with the parent wall. Although most easily observed in highly vacuolate cells, phragmosomes may be a common feature of all dividing cells.

The perinuclear arrays of MTs just described appear to develop into the plant spindle (Fig. 2C,D). Multiple MT nucleation sites are observed on the nuclear envelope. Convergence of the MTs results in the formation of spindle fibers that, on breakdown of the nuclear envelope, are captured by kinetochores on chromosomes. Specific proteins in the kinetochore (very similar to those found in animal systems) act as specific MT binding sites. When formed, spindle fibers (bundles of MTs connecting the chromosome to the spindle pole) contain two identifiable populations of MTs. Stable MTs, resistant to cold- and drug-induced depolymerization, form the core of the fiber. Numerous skewed and labile MTs diverge from the core, connecting various spindle fibers. Labile MTs connect the spindle as a unit and may function in spindle morphology and chromosome movement.

In telophase, as chromosome masses separate, the phragmoplast forms (Fig. 2E,F). This structure functions to form the new cell wall separating the daughter cells. The phragmoplast consists of two sets of overlapping MTs, both oriented at right angles to the developing cell plate. The origin of the phragmoplast is unclear; MTs possibly come from remnants of the spindle, arise *de novo* from new polymerization at the spindle equator, or come from the reforming nuclei. Recent evidence indicates that intense MT polymerization occurs at the phragmoplast; however, the source of this polymerization remains unclear. Developing edges of the phragmoplast contain aggregates of amorphous material and small vesicles, similar in morphology to previously described plant MTOCs (see preceding discussion). Polarity of MTs in the phragmoplast is consistent with polymerization occurring in the center of the structure, at the developing cell plate. The phragmoplast functions to transport secretory vesicles to the developing cell plate, thereby expanding the plate to the cell periphery.

2. Wall Deposition

Since their initial discovery in the cortical cytoplasm, MTs have been proposed to play a role in organizing cellulose microfibrils in the wall of most higher plant cells. Cellulose microfibrils are made at the cell surface, associated with membrane-bound enzyme complexes. The inner-most layer of wall microfibrils is the load-bearing component of the wall and thus defines the expansion characteristics of the cell wall. Despite extensive investigation over the past 30 yr, evidence for MT involvement in wall deposition remains largely circumstantial. Principle observations that are consistent with MT involvement in wall patterning include the following. (1) When deposited in parallel arrays, the orientation of wall microfibrils mimics the orientation of cortical MTs. (2) MTs are close (possibly physically connected) to the plasmalemma, the site of cellulose synthesis and microfibril assembly. (3) When localized wall deposition occurs, for example, guard and xylem cells, MTs are organized in bundles that parallel the orientation and location of thickenings (Fig. 3). (4) Disruption of the MT patterns results in alterations in subsequent wall deposition. MTs are not involved in the synthesis or polymerization of microfibrils since, in the absence of MTs, cellulose synthesis remains unaltered and wall microfibrils are made with normal morphology. Only the organization of wall microfibrils is altered.

In some plant cells, MTs appear not to be involved in wall patterning. The best examples of this situation are found in various "tip-growing" cells (root hairs and pollen tubes) and/or cells producing helicoidal walls. Helicoidal walls are produced when the orientation of microfibrils within each lamella is helical and the pitch of the helices varies by some angle from one layer to the next. In these examples the inner-most layer of wall microfibrils does not parallel the orientation of MTs. This difference indicates that some event other than the mere presence of cortical MTs is required to control microfibril deposition. The involvement of other factors is also indicated by detailed quantitative analysis of MT/microfibril patterns that show small but statistically significant differences in organization.

Numerous models have been proposed to explain how MTs influence wall patterns. However, no definitive experiments have been done to distinguish between the models. Currently, the most accepted model proposes that MTs function as steering devices, guiding the cellulose synthetic machinery along the plasmalemma. The influence of MTs is indirect, with MTs restricting lateral movement of enzyme complexes by creating areas of reduced membrane fluidity through MT–plasmalemma interaction, thus inducing synthetic complexes to move parallel to the MTs. Improved preservation techniques have revealed the presence of numerous connections between MTs and the plasmalemma. Biochemical isolation of MTs should lead to the identification of bridging proteins.

D. Response to Hormones

Plants respond to their environment by changing their development and morphology. Many of these changes are mediated through plant hormones. Among the various effects of hormones, the regulation of directed cell expansion is especially important. Given the involvement of MTs in controlling cellular morphology, MTs would seem to be involved in the hormone-induced regulation of cell expansion and, thus, in the regulation of plant development by the environment.

Gibberellins are known to influence cell elongation in plants. Hormone-induced increases in elongation appear to involve the reorientation of MTs into arrays that are perpendicular to the axis of cell elongation. This hormone effect can be reduced or eliminated through the application of MT-disrupting agents. Dwarfing in mutants of maize and pea was shown to be the result of longitudinally oriented MTs that

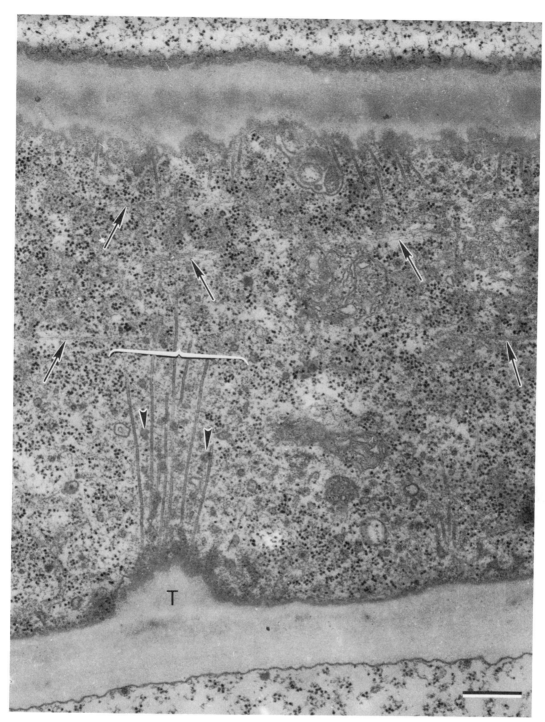

FIGURE 3 Electron micrograph of microtubule (MT) patterns in a developing tracheary element from wheat root. MTs are clustered (bracket) in association with the developing wall thickening (T). Numerous secretory vesicles (arrowheads) are associated with the MTs and may be directed by them to their site of membrane fusion. Bundles of microfilaments (arrows) in the cytoplasm are oriented transversely to the MTs. Bar: 0.5 μm.

become transversely oriented on application of gibberellins.

Ethylene is responsible for inducing stem thickening by inducing lateral cell expansion, often at the expense of cell elongation. Changes in cell expansion are brought about via changes in the orientation of MTs. On ethylene treatment, MTs reorient within 30 min, well in advance of any observed changes in cell expansion characteristics. Each entire array of MTs seems to reorient without any detectable gross disassembly and reassembly.

The role of auxin in phototropic and geotropic responses appears to involve, at least in part, MTs in epidermal cells on the concave and convex sides of phototropically or geotropically curved plant organs. Coleoptiles of various plant species exhibit changes in MT orientation in epidermal cells exposed to appropriate stimuli. Auxin treatment to mimic these responses results in appreciable reorientation of MTs within 15 min and complete transition within 1 hr. Tissue-specific responses (i.e., epidermal cells only) to these hormones are probably related to the availability of appropriate receptors on the cell surface.

Although the ability to regulate orientation of cortical MTs has been associated with all major plant hormones, the mechanism by which this is accomplished has not been clarified. Before this process can be understood, we must know how MT patterns are established and maintained in the cortical cytoplasm (see Section II,B,5). Information is needed on the dynamic instability of MTs in plant cells as well as on the mechanisms for MT stabilization in plant cells. These factors are important in establishing and maintaining cortical arrays of MTs and may be one vehicle through which hormones affect plant morphogenesis.

III. Microfilaments

A. Morphology and Distribution

MFs are best visualized in plant cells using electron microscopy. In longitudinal section, they appear as bundles of 8- to 10-nm filaments in the peripheral cytoplasm (Fig. 4B). Isolated MFs appear as a double helix of monomers, forming long chains (Fig. 4A). Initially, only bundles of MFs were observed in plant cells. However, with improvements in cell preservation and the use of specific fluorescent cytochemical staining, MFs have been visualized in many cell types, organized into bundles or occurring as single filaments associated with various organelles. It is now clear that MFs form three-dimensional arrays extending from the nucleus to the plasmalemma, organized into complex anastomosing networks of variously sized filament bundles (Fig. 4C). Chemically, plant MFs are composed of a major structural protein, actin, and presumably an array of regulatory proteins. Early research indicated that these bundles of filaments were involved in cytoplasmic streaming, with subcellular particles traveling long distances along MF cables. Recent studies have shown MFs located in areas not involved in streaming; thus, they are perhaps involved in other cellular processes.

B. Protein Chemistry

1. Actin

Much of our current knowledge of the protein chemistry of plant MFs is rooted in their similarity to MFs and actin proteins from animal systems. Initially identified from muscle, actin exists in two forms: G-actin, which is a globular monomer with a molecular weight of about 43,000, and F-actin, which is the G-actin monomer polymerized to form a polar helical filament in which each monomer interacts with its four adjacent subunits. Animal actins from a wide variety of sources have been identified and shown to have various isotypes, classified as α, β, and σ actins. In animals, some isotypes are found only in specific tissues, indicating specific functions. However, the significance of these distributions remains unknown.

Animal actin monomers bind to a single adenine nucleotide (ATP or ADP) and several mono- and divalent cations (Ca^{2+}, Mg^{2+}, or K^+). The binding of cations to G-actin increases its ability to associate into polymers. If F-actin or actin binding proteins that mimic filament ends are present, polymerization proceeds rapidly. A critical concentration of G-actin must be achieved before polymerization will occur. Various factors—such as the relative concentrations of actin-associated proteins, divalent cations, and nucleotides—influence the kinetics of actin polymerization. Since actin subunits within filament hydrolyze bound ATP, ATP–actin "caps" can form on filaments, stabilizing them relative to filaments with ADP–actin on their ends. This differential stabilization of filaments is similar to that observed with MTs in the presence of GTP.

The isolation of plant actin was hampered by the lack of high concentrations of actin in plant tissues. Once isolated, however, plant actin was found to have many of the same characteristics as actins from muscle tissue. Based primarily on gene sequence analysis, plant actin exhibits numerous isotypes. In general, good homology exists between the actins of plants, animals, fungi, and protists. However, several plant actins reveal unique, short amino acid sequences that are plant specific. Significant differences exist in the numbers of charged amino acids between plant actins and other actins, resulting in complex arrays of actin isotypes. At least eight isotypes are found in leaves, roots, and hypocotyles. This variability is seen in a number of plant systems. Isotypes are expressed in a tissue-specific manner, but whether they correlate with specific properties of G- or F-actin is not known.

2. Actin Binding Proteins

Muscle actin has an array of binding proteins that function to enact the process of muscle contraction. Myosin is the major protein, which itself forms thick filaments that, on interaction with the thin actin filaments, slide to produce a shortening of the muscle fibers. This process is enzymatically driven by an actin-activated ATPase in the flexible head region of the myosin molecule. Animal myosins are broadly divided into two categories: myosin I has a catalytically active heavy chain of 100–140 kDa and a variable tail region that modifies the binding of myosin to other cellular components; myosin II, with a catalytic subunit of approximately 200 kDa, has a long conserved tail that function to form the large bipolar myosin filaments of muscle. Control of the actin–myosin interaction, in response to Ca^{2+} concentration, occurs via a complex of troponin and tropomyosin molecules.

Other actin binding proteins are found in animal nonmuscle cells. Although this group of proteins is complex and still expanding, several broad categories of proteins are emerging. G-actin sequestering proteins such as profilin promote the disassembly of F-actin by stabilizing the actin in its G form. Severing and capping proteins such as fragmin, β-actinin, and gelsolin induce severing of long filaments and cap the ends of short fragments, thus shortening actin arrays. Cross-linking proteins such as filamin, spectrin, and α-actinin establish links between adjacent microfilaments, inducing three-dimensional networks or bundles of MFs (depending on the concentration of cross-linking agents). Various nonmuscle, myosin-like molecules have been identified that confer motility to organelles along microfilaments or give cells ameboid movement.

Several actin binding proteins have been identified in plant cells. A myosin-like protein that has an actin-activated ATPase has been isolated from several plant species. Recently, several antibodies against animal myosin have been found to react with plant polypeptides. By molecular weight, putative plant myosins fall within the myosin II range. However, several isolated myosins have lower molecular weights. Plant myosin can form bipolar filaments (similar to muscle myosin) in vitro. However, no such filaments have been seen in vivo. Plant myosin is thought to function by associating with various organelles and generating movement of those organelles along actin filament cables. A Ca^{2+}-dependent protein kinase (CDPK) has been identified in plant tissues. This protein is localized along actin cables and may be involved in regulating actin–myosin interaction. Recently, profilin-like proteins (known to function in actin polymerization and organization) have been identified using cDNA cloning in various pollens.

C. Function

1. Cytoplasmic Streaming

Much of the work on MFs and streaming has been done on the giant algal cells of Characean algae (Nitella and Chara). Cell perfusion studies show that the large subcortical MFs of these algae are responsible for moving cellular components and that these MFs contain actin filaments. The direction of movement, along the actin cables, is defined by the polarity of the actin filaments. In some systems, in which movement along a cable is unidirectional, actin filaments within the cable all have the same polarity. Other systems, such as developing root hairs in radish, exhibit bidirectional streaming along cables that contain actin filaments of opposite polarities. Myosin-like molecules were identified in Nitella and are localized with various organelles.

Many plant cells exhibit streaming. The general concepts and regulation of this process are thought to be the same as that found in the Characean algae. The size of plant cells necessitates the specific redistribution of subcellular components. Although general mechanisms for streaming are understood, specific regulation of the process and the mechanisms that define specific destinations remain unknown.

2. Microfilament–Microtubule Interactions

Improvements in the preservation and identification of actin have resulted in the detection of actin in many other areas of the cell that are not involved in cytoplasmic streaming. As a result of difficulties in properly preserving MF arrays, the data on actin distribution and function are far from complete. Many

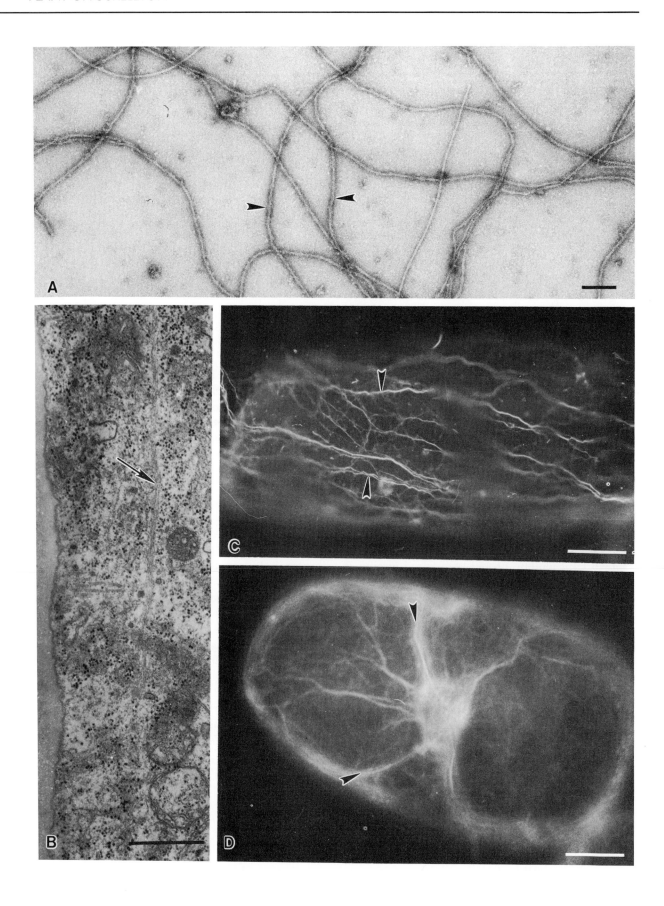

arrays of MTs appear to contain MFs. During interphase, in addition to the large actin cables involved in streaming, a fine anastomosing array of filaments is associated with cortical MTs (Fig. 5). The role of these filaments in the cortical cytoplasm remains unclear. Disruption of these filaments in some plant systems results in an alteration of MT distribution, indicating a possible role in establishing or maintaining the organization of interphase MT patterns.

Throughout the cell cycle, MFs appear to be associated with various MT arrays. Actin patterns appear to be dependent on the degree of vacuolization. In vacuolated cells, actin filaments are involved in the premitotic migration and positioning of the nucleus in the phragmosome, as well as in maintaining the structural integrity of the phragmoplast (Fig. 4d). In meristematic cells, however, nuclear migration and phragmosome formation are typically not observed. Thus, actin filaments may not be required. In many plant systems, a band of MFs coaligns with the PPB of MTs. Actin may be involved in clustering MTs into a narrow band, but disruption experiments indicate that the integrity of actin is not required for the establishment of the PPB. MFs are localized with the mitotic spindle. An actin network forms an elastic cage around the microtubular spindle and may be involved in establishing or maintaining spindle morphology. Actin has also been identified within the spindle, associated with spindle fibers. Evidence is available to both support and refute a role for actin in spindle function (chromosome movement). Thus, more work must be done before a general consensus can be reached.

IV. Intermediate Filaments

Only recently have IF proteins been identified in plant systems. Relative to the literature on MTs and MFs in plants, information on IFs from plants is minimal because of the inability to identify such filaments using electron microscopy and a lack of appropriate probes (disrupting agents, stains, or antibodies) to examine distribution and function.

In many animal cells, IFs constitute a major component of the cytoskeleton. As a result of their extreme stability in cells, they are thought to be functionally stable, not exhibiting the turnover and diversity of organization typical of MTs and MFs. Although diversity exists in protein composition within IFs of animal systems, they can be divided into five groups according to protein primary sequence. All IF proteins contain a common epitope in the carboxyl terminus of the protein, capable of inducing filament formation. The conserved region has been identified by a monoclonal antibody designated "anti-IFA."

Several IF proteins have been identified in plant tissues using the IFA antibody. Immunoblots show that two proteins of 50 and 68 kDa specifically bind the antibody in a variety of plant systems. Immunofluorescence shows the antigen to be associated with microtubules of the PPB, spindle, phragmoplast, and cortical cytoplasm. The anti-IFA antigen appears to be widely distributed in higher plants.

The function of IF proteins remains unclear. Although isolated plant IF proteins can form 10-nm filaments *in vitro*, no such filaments have been seen in the cell. Their association with microtubule arrays, although interesting, has yet to be explained.

V. Conclusions and Future

The cytoskeleton of higher plant cells forms a dynamic interconnected array of filamentous proteins that are involved in most, if not all, aspects of cell growth and development. Ultrastructural and pharmacological evidence is clear that MTs, MFs, and most likely IFs change distribution and function in a developmentally regulated manner. How environmental and developmental signals are transmitted through the cytoskeleton to achieve appropriate morphogenic responses remains an unresolved question.

A. Application to Agriculture

From the preceding discussion it is readily apparent that the cytoskeleton is intimately involved in all as-

FIGURE 4 Electron micrographs (A,B) and fluorescent light micrographs (C,D) of plant microfilaments (MFs). (A) *In vitro*, actin from pea roots shows a linear structure with some evidence of helical patterns of monomers (arrowheads). Bar: 0.1 μm. (B) In thin sections of elongated cells from wheat root, MFs are seen as bundles of filaments (arrow) in the peripheral cytoplasm. Bar: 0.5 μm. (C) Using fluorescent staining with light microscopy, a complex network of variously sized MF bundles (arrowheads) are evident in elongating cells of onion root. Bar: 20 μm. (D) The developing phragmosome in cotton suspension culture cells contain arrays of MFs (arrowhead) with a distribution similar to that of the MTs (compare with Fig. 2H). Bar: 20 μm.

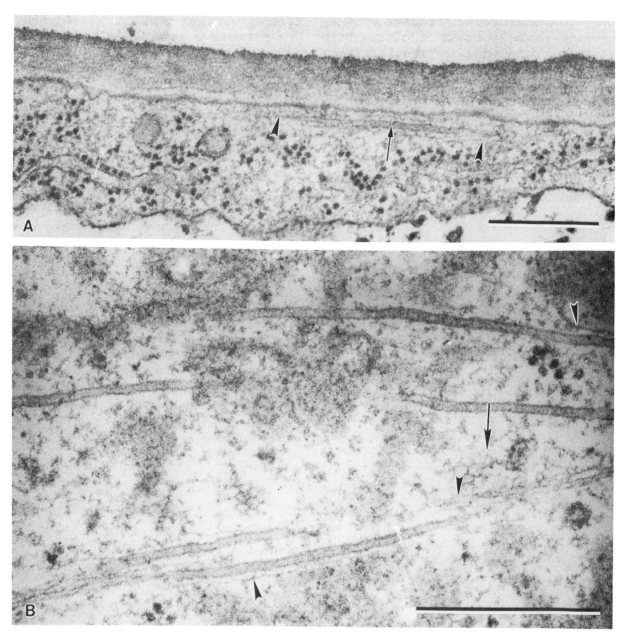

FIGURE 5 Electron micrographs of possible microtubule–microfilament (MT–MF) interaction in developing radish root hair (A) and cotton fiber (B). (A) Single MFs (arrowheads) associate with cortical MTs (arrow), possibly connecting them to the plasmalemma. Bar: 0.5 μm. (B) MFs (arrowheads) connect adjacent MTs and are possibly involved in coordinating MT patterns or function. Bar: 0.5 μm.

pects of plant growth and development. In agriculture we try to optimize specific plant responses to a specific set of growth conditions. Understanding how the plant detects and responds to growth conditions will assist us in manipulating plant growth in economically relevant ways.

In plant species in which wall architecture influences crop value, for example, the production of wood and cotton fiber, manipulation of wall patterns through modifications in cytoskeletal pattern or function may improve crop utility. Specific wall patterns in both wood and cotton fiber are controlled via the organization and location of MTs. Changing the timing or extent of specific changes in MT arrays would have a dramatic impact on final cell wall organization and, thus, would affect the economic utility of the crop.

Cold hardiness in plants may involve the stabilization of MTs. Cold hardiness can be induced by hormone application, which in turn has an affect on MT stability. The mechanism by which MTs are stabilized remains unknown. During cold acclimation, specific proteins are synthesized. Determining if any of these proteins influences MT stability would be important. Cold hardiness also involves changes in the stability of the plasmalemma. Given the observed interaction between cytoskeletal elements and the cell membrane, it is reasonable to propose that cytoskeletal elements may also be involved in membrane stability. This aspect of cytoskeletal function is not well studied in plants, yet may have important applications to agricultural crop species.

B. Major Unresolved Questions

The role of MTs in regulating cell expansion characteristics remains unexplained. If one accepts that MT patterns control wall microfibril pattern, then how are MT patterns established, maintained, or modified? The process of *in vivo* MT polymerization and organization needs to be examined. An examination of MT polarity needs to be done since this information may indicate where MTs are synthesized and how various MT nucleating sites interact to establish interphase patterns. The proteins that connect cortical microtubules to the plasmalemma during the deposition of organized arrays of cellulose microfibrils must be identified. Once identified, a number of pivotal questions concerning the mechanism of MT involvement in wall deposition can be addressed. An increasing number of systems has been identified in which oriented microfibril deposition occurs without apparent involvement of MTs. Research on these systems to determine the mechanisms that operate under these circumstances is needed.

We have detailed descriptions of cytoskeletal organization in many developmental processes. However, we have little information on the control mechanisms that regulate cytoskeletal function. Based on the animal literature, clearly extensive knowledge of the protein composition of plant cytoskeleton is required.

Although complex arrays of MT and MF associated proteins are known in animal systems, in plants only a few associated proteins have been tentatively identified and little or no biochemical information has been generated. Intensive efforts must be made to isolate and characterize not only the major structural proteins of MTs and MFs (i.e., tubulin and actin, respectively), but also plant MAPs and AAPs. Central to our understanding of cytoskeletal function is detailed characterization of the specific regulatory proteins that contribute to the behavior of MTs and MFs in various arrays within the developing plant.

Bibliography

Dowben, R. M., and Shay, J. W. (1981). "Cell and Muscle Motility," Vol. 1. Plenum Press, New York.

Fosket, D. E. (1989). Cytoskeletal proteins and their genes in higher plants. *Biochem. Plants* **15**, 393–454.

Holmes, K. D., Popp, D., Gebhard, W., and Kabsch, W. (1990). Atomic model of the actin filament. *Nature* **347**, 44–49.

Kirschner, M., and Mitchison, T. (1986). Beyond self-assembly: From microtubules to morphogenesis. *Cell* **45**, 329–342.

Lloyd, C. W. (1982). "The Cytoskeleton in Plant Growth and Development." Academic Press, London.

Lloyd, C. W. (1988). Actin in plants. *J. Cell Sci.* **90**, 185–188.

Lloyd, C. W. (1991). "The Cytoskeletal Basis of Plant Growth and Form." Academic Press, London.

Meagher, R. B. (1991). Divergence and differential expression of actin gene families in higher plants. *Int. Rev. Cytol.* **125**, 139–161.

Morejohn, L. C., and Fosket, D. E. (1986). Tubulins from plants, fungi, and protists. *In* "Cell and Molecular Biology of the Cytoskeleton" (J. W. Shay, ed.) pp. 257–329. Plenum Press, New York.

Seagull, R. W. (1989). The plant cytoskeleton. *CRC Crit. Rev. Plant Sci.* **8**, 131–167.

Smirnova, E. A., and Bajer, A. S. (1992). Spindle poles in higher plant mitosis. *Cell Motil. Cytoskel.* **23**, 1–7.

Vallee, R. B. (ed.) (1991). "Methods in Enzymology," Vol. 196. Academic Press, New York.

Plant Ecology

DAVID M. ALM, JOHN D. HESKETH, *USDA-Agricultural Research Service, Illinois*

WILLIAM C. CAPMAN, *University of Illinois*

GREGORIO B. BEGONIA, *Jackson State University*

Glossary

Biome Largest geographical area that can be recognized biologically as a unit; perceived as relatively uniform at the global scale, classified based on climatic factors and subdivided based on edaphic factors or elevation; nearly synonymous with "plant formation," but includes the associated animals and physical geographical factors
Community Recognizable group of interacting populations
Ecological modeling Application of physical principles or systems analysis to derive a set of rules or equations representing the structure or behavior of an organism in its environment; includes modeling of populations, communities, ecosystems, or biomes
Ecosystem Community (plant and animal) plus the associated abiotic factors
Edaphic Resulting from the earth in which plants are rooted; includes slope, aspect, fertility, pH, water availability, soil type, etc.
Microclimate Climate of a small area, such as surrounding an individual plant; sometimes used in reference to azonal (out-of-place) communities, such as on steep, north-facing slopes in otherwise temperate climates
Natural history Study of natural objects, especially involving descriptions of organism life histories as they relate to the biotic and abiotic environments
Population Local group of a single species

Plant ecology is the study of the interactions among individual plants, populations, and communities with their abiotic environment. Although sometimes perceived as a catchall discipline, plant ecology, in the classical sense, involves the study of how biotic–abiotic interactions affect distribution and abundance patterns.

I. Naturalists: The First Plant Ecologists

Plant ecologists study relationships between plants and their environment, which includes other plants, animals, microorganisms, and the soil–air environment. Ecologists working at or above the population scale (Fig. 1) generally seek explanations for the factors contributing to plant species distribution and abundance patterns. Ecologists working below the population scale (Fig. 1) typically study the behavior of individual plants in terms of evolutionary fitness. A wide range of scientific methods and techniques are used by plant ecologists because ecology typically involves a synthesis of information across many levels of the physical, chemical, biological, and atmospheric sciences (Fig. 1), including the agricultural disciplines. However, ecological analyses have only recently become scientifically and quantitatively rigorous. Indeed, the first plant ecologists were naturalists—students of the natural history of plants.

Prior to the twentieth century, ecological studies were centered on descriptions of what species lived in various habitats, and on describing life-styles and habits, or natural history. Because people instinctively classify their world, this work resulted in structural and behavioral classification systems for individual plants, plant associations (communities), and plant formations (biomes). As ecologists began to formally define their discipline, however, the complexity of ecological systems was realized, as was the need for greater scientific and numerical rigor. The result is a modern ecology with a much greater emphasis on

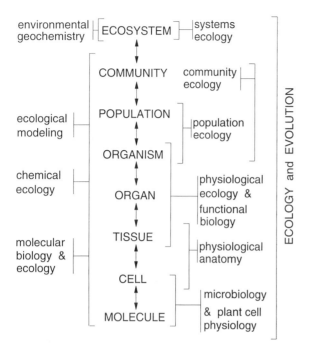

FIGURE 1 Levels of biological organization, showing the many different levels addressed by different ecological subdisciplines. Overall, the study of ecology encompasses all levels of organization and is often intimately intertwined with the study of evolution.

experimentalism and hypothesis testing—that is, manipulative experiments involving control (reference or "check") treatments and statistical analyses—as opposed to the naturalistic observations characteristic of the early ecologists. However, because of the complexity of ecological systems, the many species involved, and the rapidity with which many native ecosystems are disappearing, the classical naturalist's approach can be defended. Nevertheless, our intent is to discuss the work of modern plant ecologists, along with their structural and behavioral classification systems, in order of increasing biological and spatial scale.

II. The Diversity of Modern Plant Ecology

A. Plant Ecological Methods

Ecologists use many research methods, including observing patterns of distribution and abundance, measuring individual and population growth rates and fluctuations, determining community compositions, measuring physiological processes and various attributes of the physical environment, and performing classical and molecular studies of the genetic makeup of individuals and populations. Ecological experiments range from very manipulative studies that can involve drastic treatments like removing a particular species from an area to more natural experiments where hypotheses are tested based on patterns observed in naturally occurring systems.

As an example of the difference between a natural experiment and a manipulative experiment, consider an investigation into the effects of soil fertility on plant growth. In the natural case, one could measure soil fertility level and growth rate in various environments and test for a correlation. Using a manipulative approach, fertilizer could be added in varying levels to plots that are otherwise similar. In both cases a correlation may be obtained, but in principle, the manipulative experiment has the advantage of controlling other factors that contribute to plant growth. In the natural experiment, a correlation might arise because both soil fertility and plant growth rate are correlated with some other unmeasured factor. One advantage of the natural experiment, however, is that large-scale experiments can be conducted with less effort and intrusion.

B. Functional Biology and Physiological Ecology

Many physiological experiments in ecological studies are designed to establish the adaptive significance or functionality of organ shapes, sizes, and physiological processes. Application of the principle of evolutionary optimality—that evolutionary processes will select for optimal growth strategies—is referred to as functional biology, which requires a precise accounting of the metabolic and environmental costs of growth and development options in response to environmental obstacles. Functional plant biologists are essentially physiological ecologists who use an optimization paradigm to construct mathematical models of plant physiology, phenology, and morphology. Physiological plant ecologists are usually interested in measurements per se, particularly in field experiments, in order to gain information about plant behavior, particularly in response to stress. [See PLANT PHYSIOLOGY; PLANT STRESS.]

Efforts to make sense of the diversity encountered within and among plant communities have led to elaborate classification systems for the way plants respond to stress, involving evasion, avoidance (adaptation, acclimation), and tolerance (survival) mechanisms or plant growth strategies. Such strategies involve coping with stresses by making strategic "choices." For an established plant, a choice is made between seed/propagule number and size and the ratio of resources committed to self-preservation and reproduction. Regarding self-preservation, a choice is made between resource (e.g., carbohydrate) storage in roots and stems and resource investment in roots or aboveground canopy structures for further production of reserves or exploitation of resources. The following list describes factors causing and for coping with stress in plants.

List A[1]

I. Water Deficit
   A. Escape (annual and perennial ephemerals)
      i. Fast phenology
      ii. Small plant
      iii. Big seeds, propagules with big storage organs for rapid spring growth
      iv. Shed leaves, develop buds, go dormant (deciduous rain forest)
   B. Avoid
      i. Save and store (succulents)
        a. Few stomata
        b. Water storage
        c. Open stomata during night (CAM)
      ii. Reduce transpiration
        a. Close stomata in strong light
        b. Cuticular barrier
        c. Shed old leaves
        d. Stop leaf expansion
        e. Fold leaves
        f. Orient leaves parallel to sun's rays
      iii. Increase water uptake
        a. Increase root growth or root/shoot ratio
        b. Decrease root/stem resistance
        c. Increase water gradient
          Osmoregulate
          Lower water potential
      iv. Increase transpiration/photosynthesis ratio
        a. C_4 metabolism
        b. CAM metabolism
      v. Absorb dew
   C. Tolerate
      i. Osmotic adjustment
      ii. Smaller cells
      iii. Increase cell wall plastic or elastic extensibility
      iv. Lower permanent wilting point
      v. Wilt leaves, close stomata, go semidormant
      vi. Dehydration tolerance of photosynthetic cells

II. Excess Water
   A. Escape seasonal flooding (see I.A)
   B. Avoid
      i. Initiate root growth in aerated zones
      ii. Increase O_2 transport to roots (aerenchyma formation)
   C. Tolerate
      i. Limit ethylene synthesis
      ii. Modify N metabolism
      iii. Metabolic tolerance of toxins, detoxify toxins
      iv. Anaerobic metabolism, leak ethanol to environment

III. Light
   A. Escape—spring growth in deciduous temperate forests or grasslands (see I.A)
   B. Avoid
      i. Grow upward rapidly (large seeds, reserves)
      ii. Woody stems, perennial behavior (trees, shrubs)
      iii. Grow toward a light gap
      iv. Viny growth habit (lianas)
   C. Tolerate
      i. Increase leaf area, thin leaves, large area/mass ratio
      ii. Larger photosynthetic units
      iii. Minimize leaf reflectance, transmission
      iv. Slow growth

IV. Low Temperature
   A. Escape (see I.A)
      i. Go dormant
          Underground propagules, reserves
          Resistant stems and buds
   B. Avoid
      i. Dehydrate or increase solute concentration, bound water
      ii. Resistant membranes and proteins
      iii. Ice-forming inhibitors
      iv. Enhanced ability to supercool
      v. Leaf folding to avoid back-radiation to sky
      vi. Prostrate growth to take advantage of snow or other mulch
      vii. Store heat in large organs
      viii. Heat from rapid metabolism
   C. Tolerate
      i. Avoid freezing dehydration

V. Heat
   A. Escape (see I.A)
   B. Avoid
      i. Increase transpiration rate
      ii. Increase transmission, reflectance, back-radiation
      iii. Insulation
      iv. Decreased respiration
   C. Tolerate
      i. Metabolic resistance (thermal stability, rapid repair)

[1] Adapted from Levitt, J. (1980). "Responses of Plants to Environmental Stresses. Vol. I. Chilling, Freezing, and High Temperature Stresses. Vol. II. Water, Radiation, Salt, and Other Stresses." Academic Press, New York. Copyright © 1980, Academic Press.

VI. Nitrogen deficiency in soil
    A. Avoid
        i. N$_2$ fixation
        ii. Bigger root system
        iii. Optimize phenology for exploiting N in the soil as it becomes available from decomposition
    B. Tolerate
        i. Redistribution from vegetation to seed
        ii. Decrease level of N in plant parts
        iii. Curtail growth, leaf area
VII. Mineral nutrient deficiency
    A. Avoid
        i. Large root system
VIII. Fire
    A. Avoid
        i. Thick bark
        ii. Underground propagules, reserves
IX. Salt
    A. Avoid
        i. Selective uptake
        ii. Store salts in vacuole
        iii. Exude salts
        iv. Water dilution
    B. Tolerate
        i. Osmoregulation
        ii. Growth reduction

To quantify these strategies, physiological measurements, such as the photosynthetic carbon gas exchange rate, the leaf water content or water potential, or the tissue concentration of a particular mineral nutrient, are usually taken at the individual organ level. The results of such studies are frequently interpreted at the organism level, however, ignoring variation among organs within an individual plant.

Consideration of the iterative, building-block nature of plant development is necessary to properly interpret measurements at the organ level: organs (and organ-level data) are functionally (and statistically) interdependent. A plant shoot is made up of repeated *phytomer* units consisting of a leaf, an internode, a node, and an axillary bud or meristem. Flowers and fruits are made up of modified phytomer units, and a branch phytomer is more likely to produce seed than is a mainstem phytomer. Significant genetic variability can occur among phytomers or series of phytomers (branch modules), or among plants grown from vegetative propagules from the same plant. Such variability results from somatic mutation and has been generally ignored or assumed negligible in most physiological ecology studies. However, the genetic variability among shoot phytomers, by mutations and other means, can lead to natural selection at the phy-

tomer level, that is, the vegetative unit with genetic characteristics that favor it in a particular niche, such as resistance to insects, will survive and replace other clones. This phytomer variation would be particularly advantageous for very long-lived plants that experience many stressful periods and many successively evolved generations of herbivores and pathogens.

C. Population Ecology

As just described, a population of plants can be viewed genetically not only as a group of organisms, but also as a group of phytomers, yet most plant population ecologists (or population biologists) study the factors affecting the observed temporal and spatial variation in the number of individuals comprising a group of plants. The spatial boundaries of a plant population are not always clear, however, as interbreeding between ecotypes or races along with spatially varying edaphic factors or microclimate can result in a continuum of quantitative traits, or the expression thereof. Therefore, populations are typically defined operationally with respect to space. The processes affecting the dynamics of the numbers and locations of individuals, including the alternation of generations as an obvious difference from animal population biology, are the foci of this science.

1. Fecundity, Mortality, and Population Dynamics

The enumerative nature of population biology has nurtured a strong connection between experimental population ecology and mathematical and computer modeling of populations. The simplest population model is the logistic equation

$$\frac{\Delta N}{\Delta t} = r_\mathrm{m} N \left(\frac{K - N}{K} \right),$$

where N is the number of individuals in a population, r_m is the maximum relative rate at which the population (or an individual plant or organ as a population of N organs or cells) can grow over time t, and K, sometimes referred to as the carrying capacity, is the equilibrium population density, with zero population growth rate when $N = K$. Given a fecundity (birth rate) of b and a mortality (death rate) of d, $r_\mathrm{m} = (b-d)$. If N exceeds K, $\Delta N/\Delta t$ becomes negative and the population declines. Real populations often differ from the logistic equation, but, as with all simple models, a discrepancy between the abstract and the real world is informative. In addition, seeds or propagules also can leave one population (emigration, E) to enter another (immigration, I). If the migration rate is independent of N, then both E and I can be added to the preceding equation. However, if migration depends on N, then the effects of E and I should be subsumed by the value of r_m. Clearly, the exact form of this and similar equations is at the discretion of the population biologist, who generally uses such models for explanation and exploration.

Almost any effort to model a real population will require extra complexity to deal with the particulars of that situation. For example, the phenomenon of seed dormancy can allocate portions of future seedlings to different time periods, which, along with environmental variation, helps to create age- and size-structured populations. For population modelers, this complexity leads to the use of matrix models, and for even more realism, environmentally variable matrix parameters are used.

2. Dispersal and Population Structure

Plant migration results from *dispersal,* or the translocation of a seed or propagule away from its place of birth so that it might encounter microsites with adequate resources to grow and reproduce. Spatial dispersal results in *dispersion* patterns that can be random, clumped, or uniform. Certain statistical distribution functions can be used to test the probable existence of certain dispersion patterns, most notably the Poisson (random), negative binomial (clumped), and positive binomial (uniform) distributions. One obstacle to the study of dispersion patterns is that, in practice where quadrat sampling is used, estimates of the dispersion pattern depend on estimates of the mean population density, and vice versa, and both depend on the size of the sampling quadrat chosen, such that an iterative approach is necessary. To avoid this, and for other reasons, many plant ecologists have turned to neighborhood approaches, where distances between plants (rather than plants per quadrat) are measured.

When seeds are dispersed, the term *seed shadow* is used to describe the amount and pattern of seed falling from an individual plant. All else being equal, lighter seeds, taller parent plants, and higher winds result in a wider seed shadow (greater dispersal) and a greater likelihood of seeds reaching suitable growth sites. Plants have also evolved elaborate mechanisms for delivering their offspring to these sites. Field scientists and many active youth are familiar with the "sticktight" seeds of plants like cocklebur (*Xanthium strumarium*), sandbur (*Cenchrus pauciflorus*), beggar's

ticks (*Bidens* spp.), and others that have evolved to cling to animals. Many seeds are dispersed by fruit-eating animals. And farmers and grain haulers know that their machinery must be thoroughly cleaned to limit weed seed dispersal. Farm implements also serve as both propagators and dispersers for the asexually reproducing rhizomes of Johnsongrass (*Sorghum halapense*) and other perennial weeds that otherwise disperse slowly.

Plants hedge their efforts at evolutionary success not only by exploring the landscape, but also by diversifying their genetic investment over time. Plants have evolved elaborate seed and meristem dormancy mechanisms that result in seed banks and bud banks, which allow plant populations to survive severe competition, stress, or disturbance. This is effectively *dispersal in time* and is a method of increasing the probability of finding new microsites for seedling establishment.

3. Interference, Competition, and Density Dependence

Individual plants do not grow in isolation, and any unilaterally or mutually negative interaction between neighboring plants is called *interference*. Mutual interference is referred to as *competition*. When a particular process at the individual plant level is influenced by the presence of neighbors and the degree of crowding, such a process is referred to as being *density dependent*. Interference is thus a density-dependent process. The density dependence/independence concept is often presented as a dichotomy, when in practice it is probably a continuum. In other words, it is hard for one to imagine an ecologically significant process that would not be affected at some density. However, density-dependent analyses have produced substantial ecological insight.

D. Community Ecology

Whereas physiological ecologists and population biologists tend to focus on the behavior of the individual species, community ecology is the study of the interactions between groups of plant and animal species that are associated together as a unit, such as a plant community. Such interactions include *commensalism* and *mutualism,* the terms used respectively for unilaterally and mutually positive interactions, with no negative interactions. The typical positive/negative relationships are *parasitism* (animal–plant and plant–plant) and *herbivory.* These terms are not just ecological jargon: studies of such interactions will lead to discoveries that might facilitate a more ecologically sound and sustainable agriculture.

1. Ecological Niche and Habitat

There is much to learn about the ecological "niches" of our current and potential crops, let alone their pests, commensals, and mutualists. The word *niche,* rooted in the word "nest," is an attempt to conceptualize the temperature, soil, water, and other resource requirements of a single species or race, as well as its growth characteristics, phenology, tolerance to pests and fire, and competitive abilities. The term *ecological niche* was first used during the first quarter of this century, when it was associated with the principle of competitive exclusion, that is, that no two species can occupy the same niche. A niche can be viewed as a "role slot" in a community. In 1957, G. E. Hutchinson, drawing on earlier attempts to mathematicize ecological definitions, explained the niche as an *n*-dimensional response *hyper*volume, where each dimension represents a particular resource-response option. Although it is an effective analogy, the *n*- dimensional space concept is complicated in practice and leads one to a systems analysis approach (see Section II,E). The following list of niche factors is an attempt to outline the environmental factors and plant responses controlling the distribution of plants:

List B[2]

Temperature
  winter cold
    direct damage by degree of cold
    required dormancy and vernalization
    unseasonal temperatures (without cold hardening, etc.)
    production–respiration balance (evergreens)
  summer heat
    lethal temperatures
    length of warm season
    sufficient warmth
    production–respiration balance
  diurnal variations
    sudden cold (without cold hardening)
    required thermoperiodism
  simultaneous extremes of different plant parts
  annual regime and phenology or growth stage
Moisture
  drought
    atmospheric (high saturation deficit, rapid dehydration)

soil (waterholding capacity relative to potential evapotranspiration)
    physiological (reduced uptake at low temperatures, pests, etc.)
    length of seasonal dry period(s)
    required dormancy and estivation
  surfeit
    rotting
    flooding and reduced aeration
  alternation of drought and surfeit
  sufficient degree and length of wet period(s)
  diurnal aspects involving saturation deficit, leaf morphology, transport, and water supply
  soil-moisture regime and phenology or growth stage
  vulnerability to disease due to drought or surfeit
Wind
  increased moisture loss and desiccation
  mechanical damage
  effects on surface temperatures
Insolation (solar ray interception)
  effects on soil and plant temperatures
  sufficient insolation for photosynthesis
  solar declination
  photoperiodism
Soil
  depth
  texture
  nutrient content
  waterholding capacity
  available water capacity or rooting depth
  depth to water table
  parent material
  mobility
  pH
Topography
  soil and surface heterogeneity
  degree of slope and direction
  ocean and lake effects
  elevation
  drainage
  snow accumulation
Fire
  frequency, type
Geological history
  frequency of disruptions (drought, fire, volcanic activity, flooding, etc.)
  recolonization time since last disruption
  equilibrium time since last disruption
Biological interactions
  competition for resources, positive interactions
  allelopathy
  life-history or stress response strategies
Pollution, pests

[2] *Source:* Box, E. O. (1981). "Macroclimate and Plant Forms: An Introduction to Predictive Modeling in Phytogeography." Dr. W. Junk Pub., The Hague. Copyright © 1981, Dr. W. Junk Pub.

Some of this information for a particular plant can be inferred inductively from the geographical distribution of a species and associated topographical and climatic information; some can come from what is known about other similarly behaving or related species. The term habitat is closely related to niche, but whereas niche refers to a species' role in the community, habitat refers to the place where that role can be played. The relationship between niche and habitat can be one-to-one or one-to-many, that is, more than one habitat night contain a particular niche. In the former case, many such species are threatened or endangered.

2. Stress, Competition, and Disturbance: Plant Strategies

Plants become stressed when some component of the environment is not at the optimum level for plant growth. The Plant Life Form classification systems (see Box, 1981) are based on morphological characteristics that are adaptations to various kinds of stress. Rate of acclimation to a stress and the rate of recovery affect the severity of the damage. Frequently a stress in one component of the system translates into a stress in another component, such as a secondary nitrogen stress resulting from a primary water stress. Pest or herbivore stresses result in physiological stresses, with pests pruning or killing different plant parts and clogging vascular tissue.

Fire, set by lightning or human beings, is a major disturbance affecting vegetation. Severe drought and the accumulation of dry fuel contribute to the likelihood and severity of a fire. Plants adapt with thick bark, rootstocks or stems that can survive and regenerate, and specialized fruits that open and release seed only when exposed to the heat of a fire. Grasslands are more susceptible to fire than forests, mainly because of the frequency of droughts. Erosion is another form of disturbance/stress that may temporarily increase after fire until new growth forms a protective cover.

Because plants compete for resources such as light, water, mineral nutrients, CO_2, and O_2, their success depends on resource and other stress levels (such as associated with wind, temperature, pests, herbivores, and fire) and on the frequency of disturbance. At the end of the 1970s, J. P. Grime devised a triangular scheme (Fig. 2) for classifying plants based on the frequency of disturbance and the level of stress. Because the *high stress + frequent disturbance* areas are uninhabitable, the rectangle defined by the two axes of Fig. 2 is reduced to a triangle. The three major classes of plants resulting are the *ruderals,* which capitalize on disturbance, the *competitors,* and the *stress tolerators.*

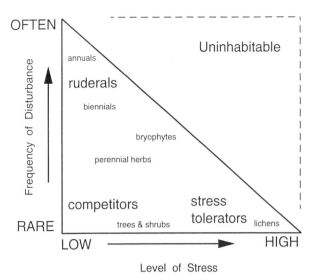

FIGURE 2 Grime's triangular *ruderal–competitor–stress tolerator* scheme (see Section II.D.2). [Redrawn from Grime, J. P. (1977), representing three generalized strategic responses to stress and disturbance.]

Large annual weeds like velvetleaf (*Abutilon theophrasti*) and crops like corn (*Zea mays*) are classified as competitive-ruderals, whereas slower-growing, smaller annual weeds and crops fit the ruderal category. Competitors include trees and shrubs, and stress tolerators include desert succulents.

3. Community Classification, Biodiversity, and Vegetation Processes

The descriptive nature of the early ecological work led to various attempts to classify the forms and functions of organisms and communities, and many of these systems have remained useful. Plant communities can be classified according to species composition, with emphasis on the more numerous (dominant) ones. Communities can be classified according to edaphic and climatic factors (Fig. 3), with azonal (apparently out-of-place) communities existing within zones depending on variations in soils, such as stony, sandy, salty, nutrient-deficient, poorly drained, or wet soils. Then there are altitudinal zones, which may or may not duplicate latitudinal zone temperatures. Topography (slope, direction of slope with respect to the sun and winds bearing rains, surface drainage), soil characteristics, fire, and pests (or herbivores) also play important roles in controlling site characteristics and the diversity of vegetation.

The methods for determining plant diversity within a community, or *species richness,* are similar to those for quantifying dispersion (Section II,C,2). And just

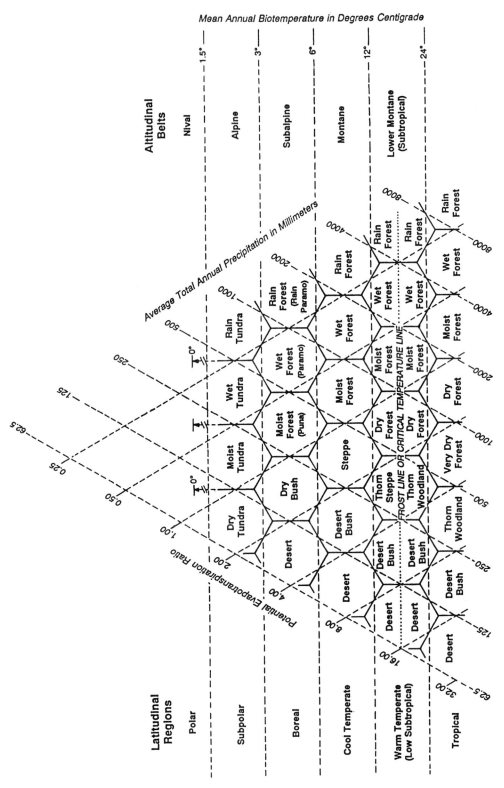

FIGURE 3 A graphical life zone chart for our planet. "Considered horizontally, the diagram shows the relative climatic positions of the sea-level or basal life zones on the surface of the planet. Considered vertically, the chart shows the vertical extensions of the basal life zones" corresponding to latitude or elevation. "The life zones or first order ecosystems are defined by mean annual values of the three major climatic factors of the environment, namely, heat, precipitation and moisture. However, additional factors such as day length, seasonal variations of radiation and atmospheric pressure actually differentiate the life zones within each series of altitudinal and basal life zones with the same biotemperature, precipitation, and potential evapotranspiration ratio ranges." [From Holdridge, L. R. (1964). "Life Zone Ecology." Provisional Edition, Tropical Science Center, San Jose, Costa Rica.]

as dispersion patterns can be assessed at any level of spatial precision, biodiversity can refer to diversity within communities, among communities, or at the global level. Biodiversity within communities has implications for ecosystem stability, and ultimately for global life-support. The many values of biodiversity are indisputable (Table I).

Ecological succession and biodiversity are intertwined in that both result from the effects of the many growth strategies deployed over time and space. After disruption by erosion, development, landslides, agricultural activity, fire, blowing sand, flooding, severe drought, or pests, a bare area is populated first by annual plants that can reproduce quickly, followed by woody perennials. This is ecological succession. Using the successional *climax* model, the plant community reaches a stable state dictated by the local environment. (M. J. Crawley has discussed the Gleason vs. Clements paradigm conflict and the controversy surrounding the term "climax.") However, preservation ecologists were surprised when what appeared to be climax vegetation in a new preserve collapsed from old age and degenerated into a highly degraded ecosystem, fairly common in the geographical area. In the Americas, it now seems that the Amerindians manipulated ecosystems considerably with fire, raising the question of what the pre-Columbian climax vegetation might have been.

4. Plant–Animal Interactions

Many plant ecologists study animals, microbes, and even viruses to develop a greater understanding of the biotic interactions within plant communities. Many of the positive interactions discussed in Sections II,C,3 and II,D result from plant–plant, plant–animal, or plant–microbe coevolution (Section II,G). The importance of microbes as mediators of plant–animal interactions is just now being addressed. Plant–herbivore and plant–pathogen interactions are at the heart of modern biological weed control studies.

E. Ecosystem Processes and Systems Ecology

Cohabiting organisms (biotic factors) constituting a community, together with the physicochemical (abiotic) features of their environment, comprise an ecosystem. Systems ecology is the application of systems analysis techniques, such as those used typically in engineering disciplines, to ecosystem studies. Mathematical models of all relevant biotic and abiotic processes, such as population dynamics or energy or material flow, are at the heart of systems ecology. When emphasis is placed on the experiments needed to determine such equations, the work is usually referred to as process ecology. Ecosystem studies typically involve both systems ecology and process ecology.

Ecosystem studies would appear to be all-encompassing, trying to "do it all." Indeed, ecologists and agriculturists often comment that their system under study is so complex that one could not hope to predict with models what is determining the size and structure of the various populations in a community. They are probably right, although nothing is

TABLE I

The Various Values and Benefits of Preserving Biodiversity

Value Category	Explanation
Economic	As yet unidentified species may provide valuable food, fiber, drugs, or other products for human use
Evolutionary potential	The genetic diversity contained in all species constitutes the basis for future natural evolution and artificial breeding/selection programs
Natural laboratory	Natural ecosystems with their full complements of species represent the museums and laboratories for the study of the Earth's natural history
Aesthetic	Natural landscapes and wild species provide many amenity and recreational values to the public
Ethical	Humans have a moral responsibility to be stewards of the natural environment and protect all species
Ecosystem integrity/function	Diversity must be maintained in order to preserve critical ecosystem services and the integrity of the Earth's life-support system.

Source: Pitelka, C. F. (1993). Biodiversity and Policy Decisions. *In* "Biodiversity and Ecosystem Function." (E-D. Schulze and H. A. Mooney, eds.). Ecological Studies No. 99, pp. 481–493. Springer-Verlag, Berlin. Copyright © 1993, Springer-Verlag.

certain in science. User-friendly computer software is available for generating and solving ecosystem models based on sophisticated mathematical relationships. The underlying mathematical principles can be instructive, but as a computer model becomes complex, understanding tends to vanish. Also, despite the advantages of taking an ecosystem approach in which one pulls information from many different, very focused disciplines and integrates it into an ecosystem synthesis, the modeler can quickly become focused on the model and miss important information gaps or subsystem interactions. Ecosystems are very complex—one can always point to gaps in a model. Nevertheless, critics of ecological modeling should consider the value of the explorations, explanations, and predictions that only an explicit modeling approach can support.

1. Plant Ecophysiology

Plant ecophysiology is centered on studies at the organism level (Fig. 1) and is concerned chiefly with the effects of environmental, phenological, morphological, and physiological processes on mass and energy exchange through a plant—or its role in the ecosystem. Physiological plant ecology (Section II,B), on the other hand, consists of physiologically based studies of a plant's niche—its role in the community. Plant morphology plays an important role in controlling major physiological processes, and phenology quantifies morphogenesis. Heat sum models or detailed temperature response equations are used to describe temperature effects on phenological (developmental) rates. Various techniques are used, such as controlled environment facilities (phytotrons), field sites at different altitudes on a mountain or in different climates or on different soils, rain-shelters, (mini)rhizotrons for root studies, and weighing lysimeters and fumigation chambers, to study the effect of environment on soil–plant behavior. Instruments are now commercially available for measuring most of the relevant soil, plant, and atmospheric processes to make long-range estimates of biomass production, water supply and demand, sunlight energy dissipation, and mineral nutrient cycling. Systems approaches are often applied, and in any systems analysis, a project should begin with a master list of important factors that affect the processes being considered, such as List A or B (Sections II,B and II,D).

Plant ecophysiology is synonymous with plant process ecology, which was developed as a discipline by agricultural scientists, particularly agricultural meteorologists, crop and plant physiologists, and crop ecol-

ogists. The use of Ohm's Law analogs for describing materials (water, CO_2, ions) and energy flow through soil–plant–air systems (Table II), as well as other biophysical theory for energy exchange and physiological processes, contributed greatly to the development of this discipline and of ecosystem process ecology in general. In an ecophysiological modeling study, information from the many classification systems available (soil, climate, plant morphology, phenology and physiology, stress response mechanisms) are used to set (usually time-varying) rules for the boundary conditions for the biophysical theory (Table II).

Assumptions have to be made when such biophysical theory is used, and in real ecosystems such assumptions often break down. Soil cracks, animal burrows, and plant anatomy affect material and energy flow in the soil–plant system; wind and air expansion with heating and associated buoyancy affect materials exchange rates between the canopy surface and the atmosphere. Such factors, of course, can be accounted for operationally, and theoretical work is ongoing.

2. Bioclimatology

Bioclimatology is a comparatively new field that can be viewed as the inverse of plant ecophysiology. Whereas ecophysiology examines the effects of the environment on the plant, bioclimatology examines effects of plants on the environment, or specifically on the climate. We now have the field instruments and the computing power to integrate plant ecophysiology with bioclimatology. A major problem with this approach is scaling from the leaf to the global level, but teams of physiologists, ecologists, atmospheric scientists, and geographers are rapidly researching the scaling problem.

3. Crop Ecology and Agroforestry

Most of the work done by crop ecologists is, in fact, crop systems/process ecology, with the added consideration of human goal orientation and management. Indeed, owing to their regular plant spacings and uniform genotypes, such systems are easier to quantify and represent a better fit to the systems analysis tools borrowed from the engineers than do natural ecosystems. Crop ecology includes studies of the biological-, geophysical-, atmospheric-, economic-, and sociological-subsystem interactions in a managed ecosystem. Agroforestry is a broad term, usually depicting the study of mixed annual/perennial cropping systems, and often limited to indigenous or otherwise

TABLE II

Transport Equations for Different Processes Where Flow = Constant × Density or Force Gradient

Process	Driving force or gradient	Current or flow	State variable	Equation with transport constant
Hydraulic or pneumatic	Pressure (P) N/m^2	Flow (Q) m^3/s	Volume (V) m^3	$Q = dV/dt = KP$ (Poiseveille's law)
Thermal conduction	Temperature (T) °C	Heat flow (q) J/m^2s	Energy (U)	$q = -T$ (Fourier's law)
Diffusion	Concentration (N) q/m^2s (q/m^3)	Mass flow	Concentration (c)	$N = -D_m\, c/x$ (Fick's law)
Water flow in soil	Hydraulic (dh/dx) m^2 (m^3/m^3)	Water flow (Q) m^3/s	Concentration (h)	$Q = K\, dh/dx$ (Darcy's law)
Gas transfer	Concentration g/m^4 or mol/m^4	Mass flow (N) g/m^2s	Concentration (c)	$N = K_V c$

Source: Jorgensen, S. E. (1986). "Fundamentals of Ecological Modeling, Developments in Environmental Modeling 9." Elsevier, The Hague. Copyright © 1986, Elsevier.

ecologically appropriate crop and livestock species. [*See* AGROFORESTRY.]

4. Ecosystem Energetics and Biogeochemical Cycles

Green plants, the foundation of the pyramid of productivity, are about 1% efficient at capturing solar energy and producing chemical energy to drive ecosystem processes. As energy is transferred from producer to primary consumer (herbivore) to secondary consumer (carnivore) in the *productivity pyramid,* or *food chain,* much of it is lost as heat. This has led some to launch an attack on the meat-producing industries, because, for example, one-third of the corn grown in the United States is eaten by cattle and chickens, and society could obtain its dietary calories more efficiently by eating plants rather than plant-eaters. However, much of the grazing lands for these cattle will support only pasture plants that humans cannot eat, so eating grass-fed livestock makes ecological sense in these areas, and some amount of grazing may be necessary to maintain biodiversity. However, widespread overgrazing and erosion of public lands is unjustified.

Whereas energy is lost as heat at each stage of consumption in the production pyramid, matter is conserved as it flows within and among ecosystems. Water flows through each (open) ecosystem, carrying nutrients (and contaminants) to support ecosystem processes. One can speak of water, carbon, and nutrient cycles, most notably the nitrogen and phosphorus cycles. At the global scale, these are closed systems, but local ecosystems are open systems, an often-ignored fact.

F. Global Ecology

Global classification systems have been developed for soils, topography, and climate (Table III) and are discussed elsewhere in these volumes. Global plant formations can be classified into realms, regions, and provinces depending on current climatic conditions and geological history (isolating or mixing factors such as continental drift, land bridges, mountain ranges, deserts, rivers, oceans). There are 4 realms and 30 provinces globally, with 10 provinces in North America. The Holdrige system (Fig. 2) is based on evaporative demand, rainfall supply, and air temperature and was simplified by R. H. Whittaker. Between 1900 and 1920, the Danish ecologist C. Raunkiaer devised perhaps the most widely referenced life-form classification system based on where, during an unfavorable period, a plant maintains its perennating (overwintering) buds: in the air, at the soil surface, or below ground. Raunkiaer's system corresponds closely to climate-based classification.

The rapid growth of the human population and the related natural resource consumption have led to a loss of natural ecosystems and have degraded most of the natural areas that remain. The burning of fossil fuels has led to an increase in atmospheric CO_2 from 343 ppm in 1970 to 357 ppm today. Many efforts are currently under way to define how the future rise in CO_2 will affect biodiversity, as plant species have

TABLE III
Climate, Soil Type and Vegetation Classification Systems

Class	Climate	Soil type	Vegetation
I	Equatorial	Equatorial brown clays, ferrallitic soils, latisols	Evergreen tropical rain forest, little seasonal variation
II	Tropical, with summer rains and winter drought	Red clays or red earths—savanna soils	Tropical deciduous forest or savanna
III	Subtropical arid desert, little rain	Sierozems or syrozems, saline soils	Subtropical desert, rocky landscapes
IV	Mediterranean, winter rain, summer drought	Mediterranean brown earths, fossil terra rosa	Sclerophyllous woody plants, frost-sensitive
V	Warm temperate, summer rainfall, mild maritime	Red of yellow forest soils, lightly podzolic	Temperate evergreen forest, somewhat frost-sensitive
VI	Cool temperate, short period of frost	Forest brown earths and grey forest soils, lightly podzolic	Broadleaf deciduous forests, bare in winter, frost-resistant
VII	Continental, arid-temperate, cold winter	Chernozem, Castanozem, Burozem to Sierozem	Steppe to desert with cold winters, frost resistant
VIII	Boreal, cold-temperate with cool summer and long winters	Podzols (raw humus bleached earths) to frost	Boreal coniferous forest (taiga), very resistant
IX	Polar, arctic and Tundra antarctic, very short summers	humus soils with heavy solifluction	Treeless tundra vegetation, usually on permafrost soils

Source: Walter, H. and Breckle S.-W. (1985). "Ecological Systems of the Geobiosphere, 1. Ecological Principles in Global Perspective." Springer-Verlag, Berlin. Copyright © 1985 Springer-Verlag.

various photosynthetic pathways and respond very differently to increased CO_2. However, the rise in CO_2 is also expected to cause an increase in the mean global air temperature and a redistribution of rainfall, which might be the controlling factors for vegetation CO_2 responses. Very little work has been done on the rate of development (and the demand for carbon) as affected by high temperatures, water stress, and increased CO_2. Recent work by well-known plant ecologists and physiological ecologists suggests that this oversight is being addressed. [*See* AIR POLLUTION: PLANT GROWTH AND PRODUCTIVITY.]

The climate change hypothesis predicts changes in global vegetation distribution patterns. Soils have evolved to reflect the rainfall patterns, and climate change could result in strange and stressful combinations of edaphic and climatic regimens. And although physiological knowledge will help us understand some of this change *as it happens,* environmental physiology alone is inadequate for *prediction* at the global scale. To make such predictions, ironically, some pioneering ecologists are returning to the classical descriptive paradigm. When combined with the knowledge engineering advances of the computer scientists, this approach is likely to prove most fruitful.

G. Evolutionary and Molecular Plant Ecology

The theory of evolution arose from the work of naturalists and is the unifying theme of modern biology, including plant ecology. Whereas global ecology is conducted at the largest spatial scale, evolutionary ecology addresses the largest temporal scale. Simultaneously, however, evolutionary ecology addresses the lowest biological scale—the gene—and all levels in-between. Indeed, much of our ecological information is interpretable only in an evolutionary context (Fig. 1). We began this snapshot of modern plant ecology with a discussion of functional biology, which can be viewed as the intersection of evolutionary and physiological ecologies. In addition, much of population and community ecology, and some would argue the majority of *true* ecology, is concerned with evolutionary questions. The reader is referred to the work of J. L. Harper, S. E. Kingsland, and R. E. Ricklefs. [*See* EVOLUTION OF DOMESTICATED PLANTS.]

Systematics involves the recognizing, comparing, classifying, and naming of organisms. In principle, the taxonomic groupings of plant species into genera, family, orders, etc., reflect evolutionary relationships; species that more recently shared common ancestors are considered to be more closely related. Although systematics is technically a discipline separate from ecology, systematics is a tool inseparable from ecological work because it is essential not only to identify correctly the species with which one is working, but also to advance our understanding of evolution. The emerging herbicide-resistant weed (and herbicide-tolerant crop) phenomenon is an ominous example. [*See* HERBICIDES AND HERBICIDE RESISTANCE.]

Modern molecular techniques, including restriction fragment length polymorphism and DNA fingerprinting, are allowing ecologists to address questions that were unimaginable a short time ago. Genetic variation among individuals and populations, loss of genetic diversity, and even individual parentage can be determined with these techniques. In addition, microbial communities that are associated with plants are being studied with techniques that allow identification and quantification of organisms that cannot even be cultured. The future of molecular ecology likely holds major advances in the understanding of rhizosphere ecosystems, soil microbiology, and many other areas.

III. Applied Plant Ecology

With the advent of remote sensing and molecular ecology, plant ecologists have extended their observational powers to both scalar extremes. The challenge for the future of applied ecology is to match the observational powers with analytical and educational advances. In this respect, the distinction between basic and applied ecology becomes vague.

A. Sustainable Agroecosystems and Ecological Economics

Sustainable agriculture is defined in the National Agricultural Research, Extension, and Teaching Policy Act of 1977 as a system of site-specific production practices that will meet five criteria: (1) satisfy human food and fiber needs, (2) enhance environmental quality, (3) make the most efficient use of nonrenewable resources and integrate, where appropriate, natural biological cycles and controls, (4) sustain the economic viability of farm operations, and (5) enhance the quality of life for farmers and society as a whole. The recent proliferation of articles attempting to define sustainability, and questioning whether sustainable development is an oxymoron, is evidence of the nebulous nature of the concept. At the root of this fuzziness is the need for an ecological–economic compromise. [See Sustainable Agriculture.]

The new field of ecological economics is intimately related to the concept of ecological sustainability. Ecological economics differs from its contributing fields, ecology and economics, in that it takes a broad and long-term view of a system so that both the economic and the ecological subsystems are managed for sustainability. Ecological economics is transdisciplinary and problem-oriented, aimed at meeting the needs of the existing human population while protecting natural resources, biodiversity, and the quality of life in general. Such work inevitably requires knowledge of human psychology and sociology, and methods for dealing with human values.

B. Preservation, Restoration, and Reconstruction

Because of human activities over the last 500 years, most of the pre-Columbian North American landscape has been radically altered but in the past 40 years there has been renewed enthusiasm for what can be achieved by preservation or reconstruction. There also is growing emphasis on using native plants in managed landscapes because of easy maintenance and low water use. To some extent, education has stimulated action.

1. Inventories, Endangered Species, Plant Propagation, and Ecosystem Preserves

Some ecologists argue that all ecosystems are dynamic and always have been, and therefore no effort should be made to preserve a contemporary natural ecosystem for future generations. Nevertheless, species and races must be preserved, and setting aside natural ecosystems is an economical way—the only feasible way in most cases—to achieve this. But first there must be an inventory of what natural ecosystems have been preserved and what remains in private hands that might be destroyed soon. The optimal area per preserve and the number of similar preserves necessary to save most of the species and races involved need to be worked out.

The denser the human population, or the higher the potential agricultural productivity of the land involved, the more difficult it is to maintain a natural ecosystem. Ecosystem restoration or reconstruction is a relatively new activity, going back some 40 or 50 years. Practical expertise in restoration is accumulating rapidly and is being documented in the scientific literature. Various governmental agencies and privately financed wildlife preservation groups are coordinating their activities to achieve a common goal. Many volunteers are involved. The effort can be labor-intensive, and better ways of creating and preserving natural ecosystems are still being developed by professionals and volunteers. Landscapes around private homes and public areas, including transportation right-of-ways, might be used more to preserve native species. Horticulturally improved cultivars of some native plants are available; this effort might be extended to a wider range of species. Hobbyists, who often grow the wild species along with the cultivars, should be recruited for the development and maintenance of ecosystems in publicly owned areas, which can be labor-intensive and therefore expensive.

Natural ecosystems are complex, with a mix of plants and animals below and above ground that is hard to reconstruct. It is best to preserve or restore wherever possible, and to reconstruct only if the site is adjacent to a natural system that will allow species to migrate to the reconstructed area. When restoring or reconstructing a site, the seed used should be collected from another site as close by as possible to maintain local races. The successful preservation of ecosystems in areas with a large human population and a poor economy must take into account the food, fiber, and space needs of the population, but no more or less so than when evaluating the materialistic desires in a rich economy. Compromises must be made. For example, where a forest is needed to maintain rainfall for a rain forest in a large geographical area, one might opt for managed forests with limited biodiversity in selected area to get the needed rainfall for the entire ecosystem.

2. Seed Preservation, Botanical Gardens, and Arboretums

Public seed banks are maintained for many agricultural crop species and their wild relatives. Commercial horticultural seed and plant sources also maintain seed collections, particularly those that supply native species. This phenomenon needs to be fostered and exploited with public funds for the preservation of endangered species. New technology is evolving for preserving meristems or embryos at low temperatures, which might be used for preserving seeds for longer periods of time. For some species like wild cottons (*Gossypium* spp.), perennial plants can be maintained for many years in a frost-free area in a garden (arboretum) for occasional seed collection. Botanical gardens maintain living collections of plant species and races. Nonetheless, seed banks, botanical gardens, or private collections are subject to eradication because of catastrophic events like vandalism, pest outbreaks, fires, hurricanes, or civil wars, or because of the sudden loss of motivated stewards: ecosystem preservation is essential. [*See* PLANT GENETIC RESOURCES; PLANT GENETIC RESOURCE CONSERVATION AND UTILIZATION.]

3. The Louisiana Iris: A Case Study

Some of the wild species of the Louisiana Iris (*Iris* spp.) grew formerly in wetlands within the New Or-

leans city limits but have long since disappeared. The *Iris* seed floats on water, which allowed three species to come together in Abbyville, Louisiana, where three rivers converge, producing colorful hybrids in a very small geographical area, which is now drained and planted in sugarcane. In addition, some of the remaining wild species in the area were recently eradicated when a road was widened. Meanwhile, hobbyists have greatly increased the colors, the numbers of flowers per plant, and vigor by doubling the chromosome number (tetraploids). Potted flowering plants now sell well in New Orleans, returning the germ plasm to the city in a somewhat modified form. The commercial suppliers and hobbyists maintain collected species in their fields or gardens and are still collecting from new locations, but there is no way of knowing what genes were lost with the wetlands. This story can be repeated over and over for other native American species, with variations; cactus species offer another good example.

4. The North American Tallgrass Prairie: A Case Study

The North American grassland biome pushed east during a warm period 8000 to 3000 years ago. The tallgrass prairie, which covered much of the middle of the United States and extended into Canada (Fig. 4), can be subdivided based on soil characteristics. Some of the prairie included numerous potholes created by the last glacier, which have been drained. Hill prairies developed on loess hills, and sand prairies developed on dunes near rivers. There are other texture distinctions like gravel prairies. Prairies developed on soils underlain by calcium carbonate or soils flooded by calcium-rich water (fens); some of these soils were difficult to plow and have remained in grass. The rare types of prairie, such as calcareous fen, sand, and hill prairies, are being saved. The prairie–forest intergrade, or savanna or barrens, is also an important part of the complex, with its unique species.

The Illinois prairie in recent times had enough rainfall to support trees, but fires lit by lightning or by Amerindians for hunting and other purposes kept the tree populations down. Fire was also used to maintain railroad right-of-ways when the railroads prospered; this helped maintain prairie remnants. Partial burning each year is important in the creation and maintenance of a prairie. Much of the original prairie was a wetland that was drained for agricultural production. Because of the productivity of such areas, little of the original ecosystem is left: of the 9 million ha of high-quality

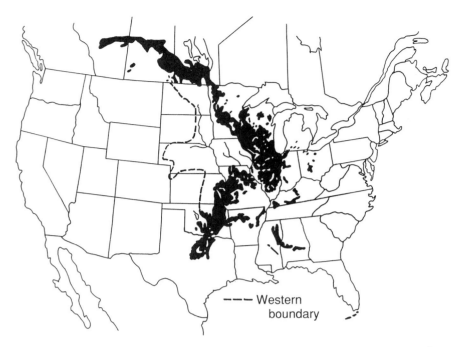

FIGURE 4 North American tallgrass prairie, with western boundary. Area in black is the transition zone between forest and prairie. Above 10 trees per ha in this zone, the prairie is called savanna. The approximate western boundary shown is a transition zone between shortgrass and tallgrass prairie. [From Anderson, R. C. (1982). The Eastern Prairie-forest Transition—an overview. *In* Proc. Eighth North American Prairie Conference, Department of Biology, Western Michigan University, Kalamazoo. (R. Brewer, ed.). pp. 86–92.]

prairie in Illinois in 1820, about 1200 ha remain. Quality prairies still exist in abandoned cemeteries, and some farmers or ranchers have maintained quality prairie remnants. Such areas are disappearing fast, but volunteer organizations are also growing fast and are trying to preserve much of what is left.

Scientists at the University of Wisconsin reconstructed a tallgrass prairie near the university grounds in Madison; this effort quickly galvanized volunteer groups and individual hobbyists to preserve quality ecosystems, to restore others, and to reconstruct new ones. The Nature Conservancy and wildlife groups interested mainly in hunting and fishing have played an important role in this effort. In the past 40 years, efforts to set aside a large preserve met with considerable resistance from agricultural interests, but recently the Nature Conservancy bought and is in the process of restoring a 12,000-ha ranch in Oklahoma that was never plowed. Progress is being made to establish a national park in Kansas. The Nature Conservancy is coordinating the activities of many agencies and volunteers, and the entire effort is undergoing rapid growth. An effort is also being made to identify, propagate, and reintroduce endangered species to preserves where they will not obstruct agricultural and other development. As private remnants are still dis-

appearing rapidly, preservation activists will be kept busy for some time.

C. A Sociopolitical Atmosphere of Urgency

There is a growing awareness that biodiversity needs to be not only protected but protected *now*. As with many environmental and ecological problems, however, public education occurs slowly. Literature syntheses about all the species involved are needed, and advancing computer technology should contribute to solutions concerning the urgent data collection, information storage, communication, and education problems.

Agriculturists now or will soon face rules and regulations on how to manage their land with respect to drainage, erosion, water quality, and biodiversity. It is likely that biodiversity will be achieved in preserves that do not disrupt other human activities; however, agriculturists should be informed about the ecosystems that prevailed before farming started, as well as about reconstruction, restoration, and preservation efforts in progress. Such a background will allow individuals to participate constructively in determining a sound policy for preserving species and ecosystems. Agriculturists could also help considerably in developing an inventory of what native species and ecosystems still exist and in preserving as much of this as possible.

Community ecologists in general are concerned with a large collection of plant and animal species in a complex ecosystem, whereas crop ecologists work with a few species growing in a spatially and temporally managed system. Weed and forage ecologists, however, work with more complex species mixtures than do other crop ecologists, and horticulturists work with yet a wider range of species than field crop production scientists. Forage ecologists and entomologists study the effects of herbivory on plant behavior. Agroecological research is frequently cited in ecological monographs; in fact, many such monographs are strongly based on agroforest ecological research often written by agroforest authors. All of these activities suggest that many of the boundaries between *pure* plant ecology and the production-related disciplines are historical and political. In the future, as these boundaries fade even further, applied ecological and environmental efforts should become more coordinated and more effective. Crop scientists need to be open to the paradigms and methods of ecologists, as they have much to offer. Indeed, our ecological, environmental, and economic problems demand humility, sincerity, and open-mindedness on all sides.

The continual increase in atmospheric CO_2 concentration could lead to increases in regional temperatures and drought severity. Advances in energy-related technology leading to less use of carbon sources and in methods for controlling the growth rate of the human population do not seem to be forthcoming. Fuel conservation and other conservation efforts are limited by public education, or "ecological salesmanship."

Much ecological research, especially that into the effects of pollution on climate, and of climate change on plant productivity and biodiversity, has become policy driven. Because of limited resources for research, the ends should be clearly defined, and a top-down, problem-solving systems approach, including the ecological–economic or value subsystem, is essential to increase agroecological research efficiency and effectiveness. Ecologists are being called on increas-ingly by governing and regulatory bodies to put their theory to use and develop ecological and environmental solutions, which inevitably involve a short-term economic compromise. Public education that is founded in basic research is vital, as people are more likely to compromise if they understand the need to do so.

Bibliography

Box, E. O. (1981). "Macroclimate and Plant Forms: An Introduction to Predictive Modeling in Phytogeography." Dr. W. Junk Pub., The Hague.

Crawford, R. M. M. (1989). "Studies in Plant Survival, Ecological Case Histories of Plant Adaptation to Adversity." Blackwell Scientific, Oxford.

Crawley, M. J. (1986). "Plant Ecology." Blackwell Scientific, Oxford.

Ehleringer, J. R., and Field, C. B. (1993). "Scaling Physiological Processes." Academic Press, San Diego.

Grime, J. P. (1977). "Plant Strategies and Vegetation Processes." John Wiley & Sons, New York.

Groombridge, B. (1992). "Global Biodiversity, Status of the Earth's Living Resources." Chapman & Hall, London.

Gutschick, V. P. (1987). "A Functional Biology of Crop Plants." Timber Press, Portland.

Harper, J. L. (1977). "Population Biology of Plants." Academic Press, London.

Hoelzer, A. R., and Dover, G. A. (1993). "Molecular Genetic Ecology." IRL Press at Oxford University Press, Oxford.

Kingsland, S. E. (1985). "Modeling Nature: Episodes in the History of Population Ecology." University of Chicago Press, Chicago.

Loomis, R. S., and Connor, D. J. (1992). "Crop Ecology, Productivity and Management in Agricultural Systems." Cambridge University Press, Cambridge.

Ludwig, J. A., and Reynolds, J. F. (1988). "Statistical Ecology" (includes software). John Wiley & Sons, New York.

Moore, D. M. (1982). "Green Planet, the Story of Plant Life on Earth." Cambridge University Press, Cambridge.

Pearcy, R. W., Ehleringer, J., Mooney, H. A., and Rundel, P. W. (1991). "Plant Physiological Ecology, Field Methods and Instrumentation." Chapman & Hall, London.

Ricklefs, R. E. (1990). "Ecology." W. H. Freeman, New York.

Walter, H. (1985). "Vegetation of the Earth and Ecological Systems of the Geo-biosphere." Springer-Verlag, Berlin.

Plant Gene Mapping

NEVIN DALE YOUNG, *University of Minnesota, St. Paul*

Glossary

Genetic marker Specific location on a chromosome, defined by a naked eye polymorphism (NEP), protein, or DNA sequence, whose inheritance can be monitored in a mapping population

Isozyme Protein existing in various forms that can be resolved by electrophoresis; isozyme polymorphisms can be used as genetic markers

Linkage map Description of genetic markers and genes showing their linear order, relative location, and the recombinational distance between them

NEP Naked eye polymorphism; a type of genetic marker whose inheritance can be monitored without the need for specialized biochemical or molecular techniques

RAPD Random amplified polymorphic DNA; a type of genetic marker that is defined by differences between individuals in the sites that are primed in a polymerase chain reaction by an arbitrary oligonucleotide sequence

RFLP Restriction fragment length polymorphism; a type of genetic marker that is defined by differences between individuals in the array of fragments generated by restriction enzyme digestion of genomic DNA and is observed by hybridization with a cloned DNA sequence

Gene mapping is the process of locating a gene's position relative to other genes and genetic markers, primarily through linkage analysis in a segregating population. In linkage mapping, distances are based on recombinational frequency. Gene mapping can also involve determination of a gene's physical location by cytogenetic analysis or electrophoresis of very large pieces of DNA. In physical mapping, distances are based on observable chromosomal position and/or the number of nucleotides between loci.

I. Theoretical Basis of Gene Mapping

The most common way to map a plant gene is by linkage analysis. First, a cross is made between two genetically compatible but nonidentical parents. Progeny from such a cross (often a backcross, F2, or recombinant inbred population) segregate for the parental alleles inherited by each individual. Early research in genetics demonstrated that the closer together two genes were on a chromosome, the more likely they were to be inherited together. This established the essential conceptual basis for linkage mapping. Statistical analysis of the inheritance pattern for segregating traits, in turn, provided a basis for gene mapping.

In linkage mapping, the proportion of recombinant individuals out of the total mapping population provides the information for determining the genetic distance between loci. Consider a plant with two linked genes, A and B, of genotype, AB/ab (meaning that the A and B alleles are on one homologous chromosome, while a and b are on the other homologue). This individual produces nonrecombinant gametes of genotype, AB and ab, and recombinant gametes, Ab and aB. If this individual is crossed with another individual where the genotype of progeny can easily be inferred (such as genotype, ab/ab), the frequency of progeny arising from the recombinant gametes (Ab and aB) indicates the genetic distance between A and B.

Encyclopedia of Agricultural Science, Volume 3   Copyright © 1994 by Academic Press, Inc.   All rights of reproduction in any form reserved.

To describe genetic distances, scientists use the unit centimorgan (cM), which is equivalent to 1% cross-ingover (recombination during meiosis between two linked genetic markers). Thus, two loci that are 5 cM apart will be observed to recombine (approximately) 5% of the time. A similar strategy for estimating genetic distance can be extended to a cross involving three or more loci. However, as the number of loci investigated goes up, the number of calculations increases rapidly and computer programs based on maximum likelihood methods are essential.

At very short genetic distances, double (and higher order) crossovers occur infrequently and recombination frequency is a good estimate of genetic distance. As distances increase, the likelihood of a multiple crossover also increases and the true genetic distance becomes greater than the observed rate of recombination. For this reason, algorithms for converting recombinational frequency to genetic distance are required, especially for distances greater than 5 cM. In some algorithms (such as that of Haldane), no effect of interference between crossovers is considered. In other algorithms (Kosambi), the effect of interference between crossovers is included in the conversion calculation. [See PLANT GENETIC ENHANCEMENT.]

II. Genetic Markers

A. Naked Eye Polymorphisms

In early mapping studies, almost all genetic markers were naked eye polymorphisms (NEPs). This included variants in plant stature, disease response, photoperiod, the shape or color of flowers, fruits, or seeds, and many other easily observed visual mutations. There are two major reasons for continued interest in NEP genetic markers, the most important being that they represent actual phenotypes of importance to agriculture. By contrast, the other types of genetic markers in use (see below) are important only as arbitrary loci for use in linkage mapping and often do not correspond directly to specific plant phenotypes. The other desirable aspect of NEPs as genetic markers is they can generally be scored quickly, simply, and without specialized lab equipment. For example, several thousand maize kernels can be screened for variation in color or shape by a single person in a matter of minutes. Other types of genetic markers require considerably more effort to analyze, even with just a few hundred progeny.

Typically only a few NEP factors can be analyzed in a single population, with the number of segregating factors constrained by the difficulty in determining the phenotype for several distinct NEPs in a single plant. In the early days of plant linkage mapping, two and three factor crosses enabled scientists to determine which NEPs were linked to one another and their linear order. Since each cross segregated for only a few markers, data from several separate crosses were required to construct linkage groups. As long as two or more loci were common between different mapping populations (crosses), the linkage order and distances could be integrated among several experiments. The need to integrate mapping data from unrelated crosses continues today with protein and DNA marker linkage maps developed in separate labs.

B. Protein-Based Genetic Markers

With the discovery that variant forms of proteins could be differentiated by gel electrophoresis, genetic markers based on biochemistry were developed. In some cases, there are many protein variants of an enzyme; these are referred to as isozymes. Variants in the molecular weight or isoelectric point of structural proteins, especially seed storage proteins, can also be used as genetic markers. In general, protein markers offer several advantages over NEPs. Because proteins are a biochemical product of genes, they reflect differences in gene sequence more directly than do NEPs. Moreover, only a small amount of plant material is required to perform an isozyme analysis and often the material can be isolated when the plant is very young (or even directly from seeds). By contrast, NEPs generally can only be determined at a specific stage in development—frequently mature plants. Finally, monitoring variation in one protein usually does not interfere with the ability to assay other protein polymorphisms, which means that many different protein loci can be followed in the same cross. However, only a relatively small number of protein polymorphisms may exist between two parents and this limits the total number of protein loci that can actually be scored in a given cross.

C. DNA Markers

In 1980, scientists studying human genetics observed that variation in the pattern of DNA fragments generated by restriction enzyme digestion of genomic DNA could be used as a genetic marker. Soon afterward, this strategy was extended to plants and DNA markers now represent the primary method of genetic mapping. There are several different types of DNA markers, including restriction fragment length polymorphisms (RFLPs) and random amplified polymorphic DNA (RAPDs), but they all share certain properties in common.

First of all, DNA markers are associated with polymorphisms in the actual nucleotide sequence of DNA. Moreover, scoring one DNA marker generally has no effect on the scoring of another and there is essentially no limit to the number of DNA markers that are available for scoring in a single mapping population. This means that linkage maps with many hundreds or even thousands of DNA markers can be constructed with just a single mapping population. Consequently, highly saturated DNA marker maps have been constructed for most important plant species in just a few years. Examples include common bean, lettuce, loblolly pine, maize, potato, rice, soybean, tomato, wheat, and many others. Nonetheless, factors such as the total amount of plant DNA material available and the overall level of DNA sequence polymorphism between two parents influence the number of DNA markers that can reasonably be scored.

D. Restriction Fragment Length Polymorphisms

RFLPs are based on restriction endonucleases, enzymes that recognize and cut DNA at short, specific sequences of nucleotides. Changes in the locations of restriction sites between individuals, either due to basepair substitutions or insertions and deletions, provide alternate alleles that can be used in segregation analysis and genetic mapping. Restriction digestion of a typical plant genome leads to thousands of fragments, which are impossible to analyze all at once. For this reason, genomic or cDNA clones are used to visualize complementary (homologous) regions of the genome of individual plants. To do this, a plant's DNA is first digested with a restriction enzyme, separated by gel electrophoresis, transferred to a membrane, and hybridized with a radiolabeled DNA clone. Only those DNA fragments on the filter that are complementary to the cloned sequence bind to the labeled clone and become visible after autoradiography. If the restriction sites on these fragments differ between the parents, one or more DNA band length polymorphisms appear, which can then be treated as a genetic marker for linkage mapping.

E. Random Amplified Polymorphic DNA

Random amplified polymorphic DNA (RAPDs) provide the same type of information about changes in DNA sequence as do RFLPs. The key difference between RFLPs and RAPDs lies in the method by which changes in DNA sequence are detected. With RAPDs, variation in DNA sequence is observed as differences in the ability to bind to short oligonucleotide primers used in the polymerase chain reaction (PCR). These oligonucleotide primers are synthetically produced, random DNA sequences, approximately 10 nucleotides in length. If two plants differ in their ability to bind an oligonucleotide primer, they also show differences in the DNA molecules produced by PCR. The presence or absence of a given DNA molecule product between two plants can then be used as an allele for genetic mapping. Because RAPDs are based on PCR, relatively small amounts of DNA are required for this type of analysis. Other differences between RFLPs and RAPDs lie in the type of information they provide; RFLP markers tend to be codominant while RAPD markers tend to be dominant/recessive. RAPDs are also more sensitive to experimental conditions, sometimes making RAPD markers more difficult to reproduce consistently (Table I).

F. Other Types of DNA Markers

Because RFLP analysis is technically difficult and RAPD markers tend to have problems of reproducibility, there is considerable effort to convert these markers into a new type of marker, known as a sequence tagged site (STS). An STS is any site on the genome that is unambiguously defined in terms of flanking primers for PCR amplification. Sequencing beyond the ends of a RAPD primer makes it possible to create improved primer sequences that are more reproducible for PCR. Similarly, sequencing the ends of an RFLP clone can also create an STS marker by providing the information to synthesize primers that tag the location of the RFLP locus. While the amount of work to create STS markers can be great, they have one especially desirable feature. The only thing that is required to use an STS marker in a mapping project is knowledge of the primer sequences. For this reason, a new lab that is beginning genetic analysis only needs to know the sequence of the primers for each STS marker to synthesize the appropriate oligonucleotides and begin mapping.

TABLE I
Properties of Different Genetic Markers

	NEP	Isozyme	RFLP	RAPD	STS
Effort to generate	Simple	Moderate	Moderate	Simple	Difficult
Ease of use	Simple	Moderate	Difficult	Moderate	Moderate
Number per cross	<10	<30	Unlimited	Unlimited	Unlimited
Plant material required	Intact plant	Little	Medium	Little	Little
Reproducibility	High	High	High	Moderate	High
Dominant/codominant	Dominant	Codominant	Codominant	Dominant	Codominant

DNA markers can also be based on hypervariable sequences that give higher levels of polymorphisms for use in linkage mapping. Typical RFLP and RAPD markers show limited variation between parents, especially in naturally inbreeding plant species. This limits the number of useful markers that can efficiently be mapped in any given cross. To circumvent this problem, scientists seek highly variable DNA markers that provide high levels of useful sequence polymorphisms in any cross. Micro- and minisatellites (also called simple sequence repeats, SSRs, or variable number of tandem repeats, VNTRs) tend to be hypervariable. In a microsatellites, short di, tri-, or tetranucleotide sequences are repeated many times, with the precise number of repeats frequently differing from one individual to the next. Minisatellites are organized in a similar way, though the repeat unit is longer. The differences in the number of repeat units among individuals provides the variation for using these types of sequences as highly polymorphic genetic markers. By flanking microsatellites or minisatellites with PCR primers that amplify through the hypervariable region, DNA sequences likely to be polymorphic in nearly any cross can be generated.

III. Applications of DNA Markers

A. Linkage Mapping with DNA Markers

Whichever method is used to uncover differences in DNA sequence, DNA markers can be used to construct a genetic map in the same way as with NEP or protein markers. First, a genetic cross is made between parents with contrasting alleles for several DNA markers and the progeny are analyzed with those markers. DNA markers that tend to be coinherited among the progeny are linked, with the frequency of crossingover providing the information for determining map distance. To map a gene of interest, the gene is treated as just another genetic marker and

its recombinational frequency with DNA markers is used to construct a linkage map. A few of the many agriculturally important genes that have been mapped with DNA markers are shown in Table II.

For successful linkage mapping with DNA markers, sufficient DNA sequence polymorphism between parents must be present. Naturally outcrossing species tend to have high levels of DNA polymorphisms and virtually any cross that does not involve related individuals provides sufficient polymorphism for mapping. In naturally inbreeding species, levels of DNA sequence variation are generally lower and finding suitable DNA polymorphisms can be more challenging. Sometimes mapping in inbreeding species requires that parents be as distantly related as possible, which can be estimated based on geographical, morphological, or isozyme diversity. The requirement for sufficient DNA sequence polymorphism may preclude the use of DNA markers in some narrow-based crosses, such as between different cultivars of the same species.

Several different kinds of genetic populations are suitable for linkage mapping with DNA markers. The simplest are F2 populations derived from F1 hybrids and backcross populations. For most plant species, populations such as these are easy to construct, although sterility in the F1 hybrid may limit some combinations of parents, particularly in wide crosses. The major drawback of F2 and backcross populations is that they are ephemeral, that is, seed derived from selfing these individuals will not breed true. This limitation can be overcome to a limited extent by cuttings, tissue culture, or bulking F3 plants to provide a constant supply of plant material for DNA isolation.

A better solution is the use of inbred populations that provide a permanent mapping resource. Using recombinant inbred lines derived from individual F2 plants is an excellent strategy. The recombinant inbred lines are created by single seed descent from sibling F2 plants through at least five or six generations. This process leads to lines that each contain a

TABLE II

A Few of the Agriculturally Important Characters Mapped with DNA Markers

Trait	Gene	Organism	Notes
Simple characters			
Bacterial speck resistance	Pto	Tomato	Gene cloned based on map position
Downy mildew resistance	Dm	Lettuce	Basis for bulked segregant analysis
Hessian fly resistance	H23	Wheat	RFLPs near two unlinked resistance genes
Oat stem rust resistance	Pg3	Oat	RAPD marker identified with near isogenic lines
Potato virus X resistance	Rx	Potato	Associated with segregation distortion
Photoperiod sensitivity	—	Rice	Identified RFLP for use in breeding
Phytophthora rot resistance	Rps	Soybean	RFLPs linked to six resistance loci
Rice blast resistance	Pi	Rice	Identified with near isogenic lines
Root-knot nematode resistance	Mi	Tomato	High resolution linkage mapping with near isogenic lines
Supernodulation	nts	Soybean	Also associated with nitrate-tolerant and autoregulatory nodulation
Tomato mosaic virus resistance	Tm2	Tomato	Also mapped with PFGE
Complex characters			
Fruit soluble solids	—	Tomato	High resolution mapping and comparison among environments
Heat shock protein synthesis	—	Maize	Three to eight QTL control up to 60% of variation
Maturity	—	Maize	One locus controls nearly 50% of variation
Plant height	—	Maize	Three to seven QTL control up to 73% of variation
Powdery mildew resistance	—	Mungbean	Oligogenic resistance controlled by three QTL
Production of 2-tridacanone	—	Tomato	Trait is associated with insect resistance
Seed weight	—	Mungbean	Major locus maps to same position in cowpea
Yield	—	Maize	Mapped QTL control up to 61% of variation

different combination of genomic segments from the original parents, providing a basis for linkage analysis. However, because several generations of breeding are required to generate a set of RIs, this process can be quite time consuming. Moreover, some regions of the genome may tend to stay heterozygous longer than expected from theory. Finally, obligate outcrossing species are much more difficult to map with RIs because of the difficulty in selfing plants.

B. Targeting Specific Genomic Regions with DNA Markers

Genome mapping is often directed toward producing a comprehensive genetic map covering all chromosomes evenly. This is essential for effective marker-assisted breeding, quantitative trait analysis, and chromosome characterization. However, there are situations in which specific regions of the genome hold special interest. One example is where the primary goal of a research project is map-based cloning. In this case, markers that are very close to a target gene and suitable as starting points for chromosome walking are needed. Therefore, the goal is to generate a high-density linkage map around that gene as quickly as possible. While the construction of a complete genome map by conventional means eventually leads to a high-density map, special strategies for rapidly targeting specific regions have also been developed.

One approach for targeting specific regions is based on near isogenic lines (NILs). Over the years, breeders have utilized recurrent backcross selection to introduce traits of interest from wild relatives into cultivated lines. This process leads to the development of pairs of NILs: one, the recurrent parent and the other, a new line resembling the recurrent parent throughout most of its genome except for the region surrounding the selected gene(s). This introgressed genomic region, derived from the donor parent and often highly polymorphic at the DNA sequence level, provides a target for rapidly identifying clones located near the gene of interest. NILs make it easy to determine the location of a DNA marker relative to the target gene, in contrast to typical genetic mapping where it would be necessary to test every clone with a complete mapping population to determine whether it mapped near the target gene.

An alternate strategy to target specific genomic regions is to select the individuals from a segregating population that are homozygous for a trait of interest and pool their DNA. In the pooled DNA sample, the only genomic region that will be homozygous is the region encompassing the gene of interest, which can then be used as a target for rapidly screening DNA markers. This means that any trait that can be scored in an F2 population can be rapidly targeted with DNA markers. Used in conjunction with RAPD markers, it is possible to identify large numbers of DNA markers in a region of interest in a short time. Pooled DNA samples can be generated based on homozygosity for a DNA (or protein marker) as well as NEPs. In this way, any genomic region of interest that has been previously mapped in terms of DNA or protein markers can be rapidly targeted with new markers. This may be especially useful in trying to fill in gaps on a genetic map.

C. Parallel Genome Mapping

One of the most powerful aspects of genetic mapping with RFLP markers is the fact that markers mapped in one genus or species can often be used to construct parallel maps in related, but genetically distinct, taxa. For this reason, a new mapping project can often build on previous mapping work in related organisms. Examples include a potato map constructed with tomato markers, a sorghum map constructed with maize markers, and a turnip map constructed with markers from cabbage.

Not only does a pre-existing map provide a set of previously tested DNA markers, it also gives an indication of linkage groups and marker order. In the case of tomato (*Lycopersicon*) and potato (*Solanum*), only five inversions involving complete chromosome arms differentiate the two maps. Similar conservation of linkage order has been observed between sorghum (*Sorghum*) and maize (*Zea*). Even organisms as distant as rice (*Oryza*) and maize have been shown to have a remarkable degree of linkage order conservation, although many conserved linkage blocks from rice are duplicated in maize. In cases like these, markers can be added to a new map in an optimum manner, either by focusing on markers evenly distributed throughout the genome or by targeting specific regions of interest. However, RFLP clones may hybridize in multiple taxa, yet show little conservation in linkage group or marker order. Even though the tomato and potato maps are nearly homosequential (syntenic) in marker order, both differ significantly from the linkage map of pepper (*Capsicum*), despite the fact that all were constructed with the same RFLP markers.

D. Mapping Quantitative Trait Loci

With the advent DNA markers, it became possible to carry out detailed analysis of genes that underlie complex, multigenic characters. Genes controlling complex traits are known as quantitative trait loci (QTL) and they can be mapped relative to DNA markers using simple statistical tests. To locate QTL, DNA markers throughout the genome are individually tested for the likelihood they are linked to a QTL. First, a population is generated from two parents that differ significantly in the trait of interest. The progeny are then split into groups according to their genotype at each DNA marker locus. The mean and variance of the trait for each subgroup is then calculated and compared among subgroups. A significant difference between the means of the subgroups indicates that there is a relationship between the DNA marker and the trait of interest—in other words, the DNA marker may be linked to a QTL. Table II shows some of the agriculturally interesting characters that have been mapped in terms of DNA markers. In contrast to such single factor approaches for QTL mapping, alternative strategies based on maximum likelihood and interval mapping have also been developed, although these methods generally give results comparable to those of single factor analysis.

IV. Cytogenetic Gene Mapping

An alternate strategy for plant gene mapping is based on cytogenetic analysis. Cytogenetics is the science in which cellular events, especially those relating to chromosome structure, are correlated with transmission genetics (Mendelian inheritance). Cytogenetic analysis provides insights about the number and microscopic appearance of chromosomes, as well as their behavior during mitosis and meiosis. Cytogenetic mapping focuses on actual microscopic characterization of chromosomes and chromosome variants. In cytogenetic mapping, individuals that have an unusual number of chromosome or chromosomes with rearrangements can be quite useful. By studying the inheritance of genes when such individuals are used as parents in a cross, the location of genes can be assigned to a specific physical structure (chromosome or chromosome segment) as seen under a microscope.

One example of cytogenetic mapping is the use of translocations to locate genes to specific segments on chromosomes. Translocations occur when chromosomes break and the ends rejoin in abnormal combinations. If a plant that is heterozygous for a translocation is crossed with another plant of normal karyotype (chromosome morphology), the translocation can generally be treated as if it were a genetic marker because it causes semisterility, a trait like any phenotype in linkage analysis. This orients genetic markers in terms of the translocation point. Repeating this process with a second translocation involving another breakpoint on the same chromosome can then be used to localize markers to specific physical segments.

Another example of cytogenetic mapping involves aneuploids, lines that have a missing or extra copy of a chromosome or chromosome segment. Analyzing the DNA from an aneuploid with an RFLP clone can identify the clone's chromosome location. If the RFLP is located on a missing chromosome (nullisomic), the corresponding DNA band on the autoradiogram will also be missing. If the RFLP is on a chromosome of altered dosage (monosomic or trisomic), there will be a change in the relative signal on the autoradiogram.

In a related strategy, substitution lines (lines with known chromosomes or chromosome arms substituted with homoeologous segments from alien species) can also be used to locate DNA markers to specific chromosomes. In cereal species where the use of this approach is most common, probing DNA from a complete set of substitution lines with an RFLP clone can potentially identify the chromosome location of that clone. This is because restriction digested DNA from the line with the appropriate chromosome substitution tends to show a different restriction fragment pattern compared to the other substitution lines.

V. Physical Mapping by Electrophoresis

A. Pulsed Field Gel Electrophoresis

A very different strategy for physically mapping genes is the use of gel electrophoresis of very large DNA molecules. Normal agarose gel electrophoresis separates DNA molecules up to 50,000 bp only, but even markers that are just 1 cM apart are likely to be physically separated at least one million basepairs apart in most higher plants. With this in mind, a special type of gel electrophoresis, known as pulsed field gel electrophoresis (PFGE), is required to characterize physical associations between genes. PFGE includes different technical variations, such as contour-clamped homogeneous electric field (CHEF), field inversion gel electrophoresis (FIGE), and transverse alternating field electrophoresis (TAFE). In each case, large DNA molecules from 1 million to nearly 10 million basepairs in length can be separated, visualized, and used for nucleic acid hybridization studies.

The different PFGE techniques are all based on the same theory. Above a certain size, the gel exclusion properties of agarose cause all DNA molecules to migrate at roughly the same rate in a uniform, linear electric field. By alternating the direction or angle of an electric field, larger DNA molecules take longer to reorient and move, providing a basis for separating very large molecules. For example, FIGE alternates the field 180° every few seconds, with the time in the forward direction longer than that in the reverse. CHEF and TAFE both alternate between fields that are at angles approximately 120° apart. In all these cases, altering the direction of the electric field makes it possible to separate very large molecules. In the case of CHEF, molecules as large as 10 million basepairs can be separated in a few days.

B. Comparisons between Linkage and PFGE Maps

In the few cases where information is available, the relationship between genetic and physical distance varies dramatically according to location on a chromosome. In general, recombination is inhibited near centromeres, in heterochromatin, and in regions introgressed from wild relatives. Markers elsewhere in the genome appear to undergo relatively higher levels of recombination. This has the practical effect of making linkage maps appear to have many DNA markers clustered near centromeres or in heterochromatin. In other parts of the map, markers are separated by large gaps, even after many hundreds of markers have been placed on the map. In tomato, for example, the ratio between genetic and physical distance is known to vary from 43,000 bp per centimorgan all the way up to 600,000 bp per centimorgan. Despite the nonuniform distribution of markers in terms of recombination frequency, it is possible that the physical distance between DNA markers on a genetic map is more uniform.

Bibliography

Helentjaris, T., and Burr, B. (1989). "Development and Application of Molecular Markers to Problems in Plant Genetics." Cold Spring Harbor Laboratory Press, Cold Spring Harbor, NY.

O'Brien, S. J. (1993). "Genetic Maps: Locus Maps of Complex Genomes," 6th ed. Cold Spring Harbor Laboratory Press, Cold Spring Harbor, NY.

Phillips, R. L., and Vasil, I. K. (1994). "DNA-Based Markers in Plants." Kluwer, Dordrecht, the Netherlands.

Poehlman, J. M. (1987). "Breeding Field Crops," 3rd ed., Chapt. 3. Van Nostrand Reinhold, New York. New York.

Tanksley, S. D., Young, N. D., Paterson, A. H., and Bonierbale, M. W. (1989). RFLP mapping in plant breeding: new tools for an old science. *Bio/Technology* **7,** 257–264.

Plant Genetic Enhancement

EVANS S. LAGUDAH, RUDI APPELS, *CSIRO, Division of Plant Industry*

Glossary

Bulked segregant analysis Technique in which DNA samples from several individuals of an F_2 segregating progeny that share a common feature, such as one or another allele at a single locus, are bulked together, and the bulked DNA samples compared for DNA sequence differences

Conventional crossing Controlled transfer of pollen from one plant to the stigma of another plant to produce a sexual hybrid

DNA marker Defined DNA sequence that frequently varies in length between individuals of a species, after cleavage by a restriction endonuclease or *in vitro* DNA amplification by the polymerase chain reaction

Gene Combination of DNA segments that together comprise a unit that has an effect on the appearance, or phenotype, of an organism by coding for either an RNA or protein product

Gene pool Total genetic variation present in a population, including landraces, induced mutants, breed-

ing stocks, and obsolete and current commercial cultivars

Transformation Transfer and stable integration of a gene into a plant cell, or plant parts, from any organism and subsequent regeneration of whole plants expressing the introduced gene

Wild relative Noncultivated forms of a crop plant that are capable of forming sexual hybrids with the domesticated form so that the resulting offspring express partial or complete fertility

Plant genetic enhancement is the process of altering the genetic constitution of plants, usually crops, in order to increase their value either by improving yield or by improving the quality of the product recovered from the crop. Substantial gains have been made in genetic yield improvement from cultivated forms of crop species through conventional plant breeding. Genetic enhancement from sources other than cultivated forms is discussed in this chapter. Traditionally, it involved exploiting germ plasm from wild, weedy species or landraces to introduce new agronomic characters. A major commitment of plant breeding programs, for example, is to introduce new disease resistance genes into crops to combat the constantly changing populations of pathogens in the field. Resistance to disease is one source of yield improvement. The rapid expansion in scientific knowledge about how plants grow is allowing many other aspects of yield to be targeted for genetic enhancement. It is now also possible to target specific end uses for which the product of the crop is used, and thus add value to the crop.

I. Overview of the Challenges in Plant Genetic Enhancement

The value of wild relatives as sources of new germ plasm is well established in breeding programs for cereal improvement. On a worldwide level, however, the efficiency with which wild germplasm is utilized for introducing disease resistance and other agronomic characters into elite cultivars varies greatly. A common problem encountered with the use of alien species in crop improvement is the introduction of undesirable genetic material accompanying the target gene being introgressed. Irradiation with γ rays can be used to achieve a reduction in the amount of alien genetic material and has been successful in producing modified chromosomes for use in breeding. In addition, in special cases, genetic crossing-over has been used to reduce the amount of unwanted chromatin. An alternative way for genetically modifying plants that is now becoming more generally available is to use the techniques of molecular biology to transform plants. [See Cultivar Development; Plant Genetic Resources; Plant Genetic Resource Conservation and Utilization.]

II. The Three "Gene Pools" Used for Modifying Crop Plants

The crop scientists J. R. Harlan and J. M. J. De Wet proposed the concept of gene pools to describe the wide range of crosses that are carried out with the aim of enhancing genetic variation in a breeding program for crop improvement. Although the gene pool concept does not constitute a formal taxonomic classification system, the grouping of genetic material into primary, secondary, and tertiary gene pools can be useful for breeding programs. Because the concept of gene pools can readily incorporate new molecular biological procedures involving specific genes, combined with DNA transformation, it forms a useful basis for discussing the various ways of genetically enhancing plants (Fig. 1). [See Evolution of Domesticated Plants.]

A. Primary Gene Pools

The primary gene pool is synonymous with traditional concepts of biological species. They are characterized by high crossability, completely homologous chromosome pairing, high fertility of hybrids, normal gene segregation, and a relative ease of gene transfer. The primary gene pool contains wild or weedy races as well as the range of cultivated forms that occur within the biological species.

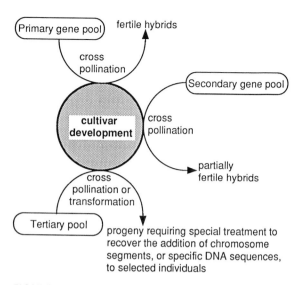

FIGURE 1 The gene pool concept [J. R. Harlan and J. M. J. De Wet. (1971). Toward a rational classification of cultivated plants. *Taxon* **20**, 509–517.] Schematic representation of the process of cultivar development using the total genetic variation available for crop improvement.

B. Secondary Gene Pools

The secondary gene pool consists of all biological species that cross with the crop, but with only partial pairing of homologous chromosomes at meiosis. Hybrids are partially sterile and growth vigor may be low in some types. Difficulties are often encountered in recovering desirable lines in advanced generations. Gene transfer is possible using methods that overcome the sterility barriers that separate the biological species. The secondary gene pool often includes species that may be acceptable to some taxonomists as existing within the generic limits of the taxa.

C. Tertiary Gene Pools

The tertiary gene pool is characterized by species that can be forced to cross with a crop but result in hybrids that are highly sterile and exhibit complete absence of chromosome pairing at meiosis. Some of these species may possess chromosomes that are homeologous with the cultivated crop so that the overall gene order will be similar even though no meiotic chromosome pairing occurs. Although gene transfer is not possible with standard techniques, it can be achieved using methods that induce chromosome pairing and genetic recombination. Hybrids are usually produced by overcoming pre- and postzygotic incompatibility barriers; examples of such techniques include chromosome doubling, embryo rescue, and the use of bridging species. Hybrids from crosses between species spanning wide taxonomic boundaries in the Gramineae have been reported. Species within the tertiary gene pool classification usually constitute the extreme outer limit of the potential genetic resource for crop improvement, using conventional plant breeding methods. Current developments in molecular biology, however, provide for a major expansion of the tertiary gene pool by using DNA transformation to overcome many of the biological barriers encountered in carrying out conventional wide-crosses. Although new problems related to the level of expression of genes can arise, the technology is available for modifying the DNA sequence prior to transformation in order to enhance expression.

III. Genetic Enhancement from the Primary Gene Pool

A primary objective in most plant breeding programs is the genetic enhancement of breeding lines with a high productive performance, desirable resistance to insect pests and diseases, and quality traits required to meet processing and nutritional standards. A significant component in the attainment of these objectives is the availability of germ plasm containing a wide range of useful traits. Large collections of germ plasm are available for most cultivated crop species and are maintained at national and international genetic resource centers. For example, worldwide rice collections are kept at IRRI in the Philippines, maize at CIMMYT in Mexico, and root and tuber crops at IITA in Nigeria.

Selection of desirable varieties or landraces tested for adaptability and performance at a specific locality within a germ plasm collection can be considered as a form of plant genetic enhancement in the primary gene pool at an elementary stage. However, conventional methods of plant genetic enhancement rely on sexual hybridization of selected parental lines followed by selection from the segregating progeny. The new genetic combinations obtained from intercrossing different varieties form the bases of genetic enhancement employed in developing cultivars of crop plants in commercial agriculture.

The absence of desirable variants within varieties, landraces, or induced mutants sometimes necessitates a search for new genetic variants in the weedy or wild forms of the primary gene pool for plant genetic enhancement. A classic example of genetic enhancement from wild relatives in the primary gene pool is in the control of potato cyst nematode (*Globodera rostochiensis*). Continuous cultivation of potato on the same land resulted in rapid increases of potato cyst nematode populations and consequently led to the demise of the potato-growing industry in many European countries in the late 1930s. It was not until 1952, when after screening more than 1200 accessions of cultivated and wild potato for reaction to *G. rostochiensis,* that resistant plants were identified. Of the six resistant accessions identified, four belonged to the wild primary gene pool species, *Solanum tuberosum* ssp. *andigenum.* The close relationship with cultivated potato as well as the single-gene basis (*H1*) for the cyst nematode resistance enabled fertile hybrids to be easily produced and a rapid selection for resistance in potato improvement. Genetic enhancement of the potato crop through the *H1* gene has had a significant impact on the potato industry in many countries where the potato constitutes a major crop.

Sorghum breeding utilizes wild germ plasm for improving commercial cultivars, with 21% of releases containing at least some wild germ plasm in their parentage. A specific "conversion" program for making wild germ plasm available for sorghum breeding has been ongoing in the United States since 1965. The aim of the conversion program is to remove genes controlling tallness and day-length sensitivity without loss of genetic variability for a wide range of other agronomic characteristics. The gene responsible for resistance to biotype C greenbug, originating from *Sorghum bicolor* race *virgatum,* has been reported to be present in commercial sorghums. [*See* SORGHUM.]

Hordeum vulgare ssp. *spontaneum* is the wild progenitor of barley (*Hordeum vulgare*) and has been extensively collected in the Middle East. A program to screen accessions of *H. vulgare* ssp. *spontaneum* from Israel, Iran, and Turkey for resistance to leaf scald, caused by *Rhynchosporium secalis* (Fig. 2), found resistance to be very common. In addition, isozyme electrophoretic variation was also widespread among the *H. vulgare* ssp. *spontaneum* accessions and this provided an opportunity to assess the possibility of using isozymes as biochemical tags for scald resistance genes. Recent studies have focused on the identification of DNA markers that are more closely linked to the disease resistance gene. Markers of this type can be converted to markers that can be assayed by an *in vitro* amplification technique, polymerase chain reaction (PCR) in order to allow their routine application to plant breeding programs. [*See* BARLEY.]

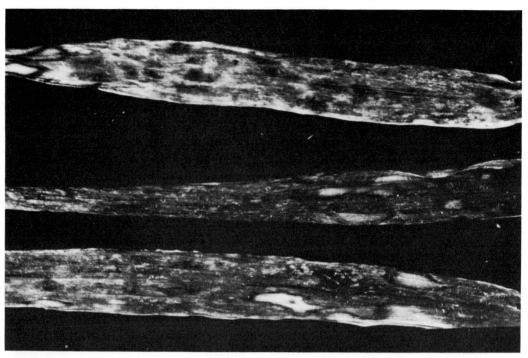

FIGURE 2 Leaf scald damage caused by the fungus *Rhynchosporium secalis* on barley. (Photograph kindly supplied by J. Burdon and D. Abbott.)

Some useful agronomic traits have been transferred from the wild forms of the primary gene pool of *Pennisetum* into cultivated pearl millet. Examples of agronomic traits include rust resistance, cytoplasmic male sterility, and fertility restorer genes from *P. glaucum* ssp. *monodii*.

IV. Genetic Enhancement from the Secondary Gene Pool

A logical progression in conventional methods of plant genetic enhancement involves the use of secondary gene pool species when the levels of useful genetic variation in the primary gene pool are considered inadequate. Additional resources are often required to produce and maintain hybrid offspring between a cultivated crop and its secondary gene pool species. These additional resources often involve methods of overcoming postfertilization barriers of sexual hybrids and subsequent cytological selection. Justification for these extra resources in genetic enhancement are usually determined by the significance of the trait under investigation.

Because of the relatively high genetic divergence between a secondary gene pool species and a culti-

vated crop, a backcross breeding procedure is commonly used to genetically enhance the cultivated crop by removing most of the genome constitution of the secondary gene pool species while retaining the trait under selection. Backcrossing in this situation refers to the crossing of the cultivated, secondary gene pool species hybrid to the cultivated crop. Successive backcrosses are carried out in subsequent generations, while maintaining the cultivated crop as the recipient parent and selecting for the desirable trait transferred from the secondary gene pool species. This technique has been used in several genetic enhancement studies to transfer new traits into commercial crops. The tomato crop is a notable example, in which all varieties in use today with resistance to the root knot nematode (*Meloidogyne* species) are derived from the wild species *Lycopersicum peruvianum*. Other examples of traits transferred from *Lycopersicum* species into cultivated tomato are tobacco mosaic virus resistance, from *L. peruvianum*, and resistance to the fungal pathogens *Cladosporium* and the bacterium *Pseudomonas syringae* pv. tomato, from *L. pimpinellifolium*.

The transfer and expression of apomixis (vegetative propagation through seeds) in hybrid plants with desirable heterozygosity has become a breeding objective because desirable gene combinations would allow the fixation of heterosis and provide a source of true-breeding, seed-propagated hybrids. In the transfer of the apomictic gene from *Pennisetum squamulatum,* double-cross hybrids, each involving autotetraploid pearl millet and *P. purpureum* and *P. squamulatum* (secondary and tertiary *Pennisetum* gene pool species, respectively), were employed in obtaining the obligate apomictic backcross-derived line. Apomictic backcross progeny derived from *P. glaucum* (4x) × *P. squamulatum* (2n = 54) tended to produce plants with high male sterility. Improved pollen fertility was achieved by using trispecific hybrids from sexual 42-chromosome pearl millet × *P. purpureum* hybrids crossed to apomictic 41-chromosome pearl millet × *P. squamulatum.* The use of the secondary gene pool species, *P. purpureum,* as a bridging species represented a significant contribution to the transfer of apomictic gene(s) through the male gametes.

Triticum tauschii is the donor of the D genome of hexaploid wheat (AABBDD), as well as a secondary gene pool species, and was the genome added most recently in the evolution of bread wheat. An extensive collection of the diploid grass *Triticum tauschii* has been characterized for traits controlling disease resistance, plant nutrition, and grain processing in the United States, Canada, Japan, and Australia. The relatively narrow genetic base of bread wheat in the D genome,

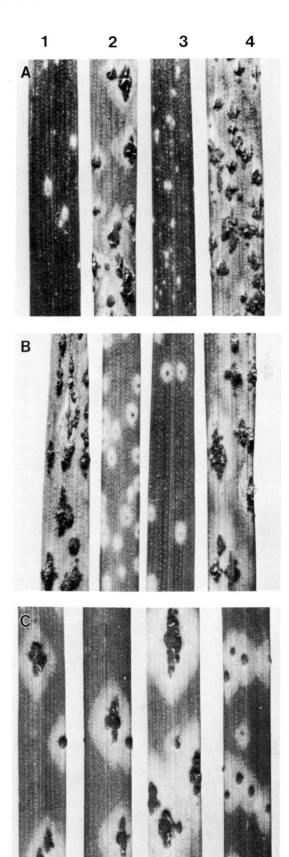

FIGURE 3 Types of interaction between strains of *Puccinia graminis* f. sp. *tritici,* the causal agent of stem rust, on four wheat varieties. Resistance occurs as a result of interaction between a resistance gene in the host wheat variety and infection by an avirulent strain of the rust pathogen. An attack caused by a virulent strain results in a susceptible response from the host plant. (1) Wheat variety W1656 with the resistance gene *Sr36;* (2) wheat variety Vernstein with *Sr9e;* (3) wheat variety Combination III with both *Sr36* and *Sr9e;* (4) Chinese spring variety with neither gene. The genes *Sr36* and *Sr9e* are derived from secondary gene pool species, *Triticum timopheevi* and *T. durum,* respectively. (A) Infection types caused by pathotype 116-2,3,7, which is virulent for *Sr9e* and avirulent for *Sr36.* The infection types on the wheat varieties (from left to right) show a low (resistance), high (susceptible), low, and high response, respectively. (B) Infection types caused by pathotype 34-1,2,3,4,5,6,7 (70-L-5), which is avirulent for *Sr9e* and virulent for *Sr36.* The infection types on the wheat varieties show a susceptible, resistant, resistant, and susceptible response, respectively. (C) Infection types caused by pathotype 116-4,5 which is virulent on both *Sr9e* and *Sr36.* All the wheat varieties are susceptible. [Photographs kindly supplied by R. A. McIntosh; source: R. A. McIntosh and N. H. Luig. (1973). Recombination between genes for reaction to *P. graminis* at or near the *Sr9e* locus. *In* "Proceedings, 4th International Wheat Genetic Symposium, Columbia, Missouri. College of Agriculture, University of Missouri, Columbia."]

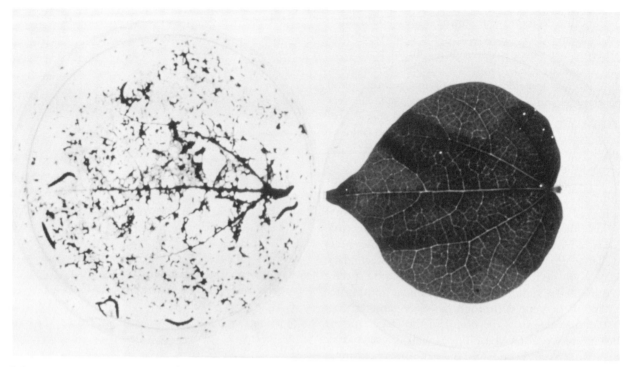

FIGURE 4 Transgenic cotton protected from the insect pest *Helicoverpa armigera*. Leaves taken from nontransgenic (left) and transgenic plant expressing the *Bt* toxin (right) attacked by young larvae. Insects on the transgenic leaf eat only a pinhead-sized piece of leaf before dying as a result of the toxic effects of *Bt* [Photograph kindly supplied by D. Llewellyn; source: D. Llewellyn. (1993). Genetic engineering of insect resistant crop plants. *In* "Proceedings, 15th Australian Society of Sugar Cane Technologists." Cairns, Watson Ferguson and Company, Brisbane.]

in particular, has resulted in the collection of *T. tauschii* accessions being screened for a number of agronomic characteristics, with the view to increasing this genetic base. Exploiting the wide genetic variation in *T. tauschii* and its homologous relationship with the D genome of *T. aestivum* has stimulated the construction of the molecular genetic linkage map of the D genome, based on *T. tauschii*. Several agronomic traits have been mapped in relation to other markers in the genetic linkage map, and these markers are being used to facilitate the introgression of agronomic traits from *T. tauschii* into wheat. Several wheat cultivars have been developed using disease resistance genes from other secondary gene pool species, such as *T. dicoccum* and *T. timopheevi*, for example, rust and powdery mildew resistance (Fig. 3). *Triticum dicoccum* has contributed a particularly durable source of rust resistance (*Sr2*) that is now utilized in several commercial bread wheats. Attempts are being made to develop a DNA marker linked to *Sr2* in order to include and facilitate its selection in breeding programs for rust-endemic wheat-growing areas. [*See* WHEAT BREEDING AND GENETICS.]

V. The Tertiary Gene Pool and DNA Transformation

Species within the tertiary gene pool classification usually constitute the extreme outer limit of the potential genetic resource for crop improvement, using conventional plant breeding methods. The tertiary gene pool of pearl millet is composed of species with varying chromosome numbers of $x = 5, 7, 8,$ and 9; these species are disproportionately distributed throughout most of the tropical and subtropical regions. Varying levels of partial homology between hybrids of pearl millet and other tertiary gene pool species, such as *P. setaceum, P. squamulatum, P. orientale,* and *P. schweinfurthii,* have been reported. Most of the tertiary gene pool species of wheat are found in the *Agropyron* (wheat grass), *Secale* (rye), and *Hordeum* genera. Rust resistance and in some cases yield increases have been derived from chromosomal segments from rye and *Agropyron* present in several commercial wheat cultivars. In the case of *Sorghum bicolor,* Australian native sorghums are currently being investigated as a tertiary gene pool.

The confounding effects of pairing between chromosomes from the same species in certain interspecific/generic hybrid makes it difficult to establish precise relationships at the level of chromosomes within the tertiary gene pool species. The use of species within these gene pools, for pearl millet improvement, is facilitated by a knowledge of the degree of species relatedness and potential recombination. Unlike wheat and its wild relatives, for which chromosome banding patterns have been used successfully to complement genome homology characterization, there is very little use of this cytological application within the gene pool of pearl millet. The isolation of species-specific dispersed repeated DNA sequences has been used for elucidating species relationships in the *Pennisetum* gene pool, as well as for identifying chromosome segments carrying genes conferring apomixis.

Current developments in molecular biology provide for a major expansion of the tertiary gene pool by using DNA transformation to overcome many of the biological barriers encountered in carrying out wide-crosses. Although new problems related to level of expression of genes can arise, the technology is available for modifying the DNA sequence prior to transformation in order to enhance expression. In the case of a gene that originates from a nonplant source (e.g., animal or lower eukaryote), codon usage is an important variable. Any single amino acid in a protein product of a gene can be coded for by several triplets, owing to redundancy in the genetic code, and the respective tRNAs are not equally abundant in plants and animals. Modifying the DNA sequence of a gene from an animal source to code for the same sequence of amino acids but utilizing triplet codons suitable for plants can have major effects on increasing the expression of the protein product in a plant cell. Another level of modification relates to the properties of the gene product formed from the introduced gene; for example, a mutant form of the enzyme 5-enolpyruvylshikimate-3-phosphate synthetase was found to be much less sensitive to glyphosate and thus to confer a more effective resistance to this herbicide. Similarly, a mutant acetolactate synthetase gene from tobacco results in resistance to sulfonyl urea herbicides.

Modifying the region carrying control of transcription signals, and signals determining the cellular location of the expressed product, can also enhance the expression of a gene in a plant cell. It has been shown that the expression of foreign genes in tobacco leaves could be enhanced by altering both the 5′-upstream and 3′-downstream regions. However, the most dramatic effects on expression, as measured by accumulation of protein product, were obtained by targeting the protein product to the endoplasmic reticulum after adding an appropriate signal peptide and the "SEKDEL" sequence to the protein; the "SEKDEL" sequence is an oligopeptide thought to be responsible for retaining a protein in the endoplasmic reticulum.

Although there are now many examples of the genetic transformation of plants and animals, it is useful to examine one in detail. The introduction of *Bt* insect toxin genes, from the bacterium *Bacillus thuringiensis* var. *kurstaki,* has been found to protect a range of crops against insect attack (Fig.4). The proteins encoded by the *Bt* genes are highly toxic to the larvae of lepidopteran insects because they disrupt the midgut cells of feeding larvae. A gene coding for one of the *Bt* proteins has been isolated and introduced into various species of plants such as cotton. The steps in the process of introducing a *Bt* gene into cotton are summarized in Fig. 5. [*See* Cotton Crop Production.]

In addition to introducing a new gene to improve a character such as disease resistance, DNA transformations have also been carried out for the purpose of specifically preventing the production of a gene

BT gene coding for *crylA* protein cloned and sequenced

Gene resynthesized chemically so that protein is made more efficiently in plants as a result of appropriate codon usage

Construct vector with new BT gene plus plant promoter sequences in Ti plasmid for transformation of cotton using *Agrobacterium tumefacians*

Introduce BT gene into leaf cells by co-cultivating with *Agrobacterium tumefacians* carrying the Ti plasmid plus BT gene. A gene conferring resistance to the drug kanamycin is introduced at the same time to allow selection of transformed cells

Tissue culture of transformed cells to select for kanamycin resistance and regeneration of plant from selected cells.

Cross transformed plant with cotton cultivars to improve agronomic characteristics.

FIGURE 5 Schematic outline of the process of transgenic cotton development to express the bacterium *Bacillus thuringiensis* protein with toxic insecticidal action.

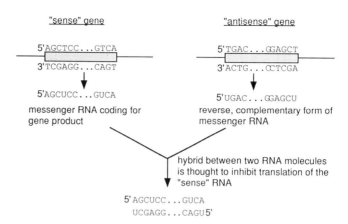

"sense" gene

5'AGCTCC...GTCA
3'TCGAGG...CAGT

↓

5'AGCUCC...GUCA
messenger RNA coding for
gene product

"antisense" gene

5'TGAC...GGAGCT
3'ACTG...CCTCGA

↓

5'UGAC...GGAGCU
reverse, complementary form of
messenger RNA

hybrid between two RNA molecules
is thought to inhibit translation of the
"sense" RNA

5'AGCUCC...GUCA
UCGAGG...CAGU 5'

FIGURE 6 Schematic outline of gene inactivation through the antisense RNA method.

product. The principle of the technique is based on the observation that nucleic acid hybridization between a messenger RNA (mRNA) and its complementary RNA sequence prevents the translation of the mRNA (Fig. 6). A successful example of this approach is the reduction of the amount of ethylene-forming enzyme (EFE) that is produced in tomatoes and carnations. Because ethylene is a phytohormone that is required for senescence to occur in fruit and flowers, the result of reducing ethylene production in tomato plants is to delay ripening of the fruit, whereas in carnations there is a delay in senescence of the flowers. These effects have important commercial implications.

VI. *Arabidopsis* as a Source of Genes for Crop Improvement

The small crucifer *Arabidopsis thaliana* is rapidly becoming a model plant of central importance to the genetic improvement of crop plants (Fig. 7) because it provides a well-defined route for recovering genes of agronomical significance. This is illustrated with a specific example shown in Fig. 8. An important character of plant oils that determines their physical and nutritional characteristics is the number of double bonds in the hydrophobic carbon chain, that is, the degree of unsaturation. The enzymes responsible for introducing the double bonds, in reactions requiring oxygen and an electron donor, are called fatty acid desaturases. Mutational analyses in *Arabidopsis* defined a mutation *fad3* that reduced the conversion of linoleic acid (two double bonds) to linolenic acid (three double bonds). Detailed genetic mapping defined DNA markers close to, and flanking, the *fad3*

FIGURE 7 *Arabidopsis thaliana* plants grown under normal soil (left) and sterile nutrient–agar medium (right). *Arabidopsis thaliana* has major advantages for the study of plant molecular biology: (a) generation time is approximately 3 weeks and can be completed in a test tube as shown here; (b) its five chromosomes are thoroughly mapped at a genetic level; (c) it has a relatively small genome of approximately 100,000 kb; (d) a complete coverage of the genome in the form of DNA clones is available; (e) transformed plants are readily produced; (f) transposable element mutagenesis is possible; and (g) it contains many, if not all, of the genes essential for growth and development of any crop plant. (Photographs kindly supplied by Joanne Burn.)

recovery of a mutation causing reduced accumulation of linolenic acid and increase in linoleic acid (one less double bonded carbon). Mutation called *fad3*.

↓

detailed genetic mapping links *fad3* to DNA markers

↓

DNA markers flanking *fad3* are used to screen a library of large, random DNA segments cloned in yeast artificial chromosomes (YACs)

↓

A single YAC clone, presumed to contain the *fad3* gene, is used as a probe to screen a library of clones produced from the messenger RNA of developing seed of oil-accumulating crop plant (*Brassica* sp.)

↓

Putative clone for the wild-type *fad3* equivalent in *Brassica* sp. is used to transform *Arabidopsis* carrying the *fad3* mutation. Complementation proves that new clone from *Brassica* sp. codes for an enzyme determining the number of double-bonded carbons in a fatty acid chain.

FIGURE 8 Application of the detailed molecular genetic analysis information available in *Arabidopsis thaliana* to identify a useful agronomic gene for edible oil improvement in *Brassica* species [V. Arondel, B. Lemieux, I. Hwang, S. Gibson, H. M. Goodman, and C. R. Somerville. (1992). Map-based cloning of a gene controlling omega-3 fatty acid desaturation in *Arabidopsis*. *Science* **258**, 1353–1354.]

gene and these markers then provided probes to select DNA from a library of cloned DNA segments that contained the *fad3* gene. [*See* PLANT GENE MAPPING.]

The next step in the analysis was to use the cloned *Arabidopsis* DNA segment carrying the *fad3* gene as a probe to select clones representing messenger RNA (cDNA clones) from developing seeds of the oil-accumulating *Brassica napus*. The putative clones carrying the equivalent of the wild-type *fad3* gene from *B. napus* were then proven to be correct by showing that they could complement the *fad3* mutation in the appropriate stock of *Arabidopsis*. The isolation of a crucial gene in controlling the composition of oil in *B. napus* was thus achieved by utilizing the very extensive molecular/genetic data base established for *Arabidopsis*.

VII. DNA Marker-Assisted Plant Genetic Enhancement

The molecular analysis of genes involved in important developmental processes in eukaryotic genomes is hampered by the absence of well-defined gene products controlling these agronomic traits. Although one avenue for cloning genes when the product is unknown is the use of transposable elements, this is not

possible in many crop plants. An alternative approach is to utilize naturally occurring, or induced, mutants and to identify closely linked DNA sequences to the target trait. Molecular genetic markers permit a detailed analysis of plant genomes by allowing the construction of linkage maps, which can provide entry points for studying specific regions of the genome. [*See* TRANSPOSABLE ELEMENTS IN PLANTS.]

Techniques frequently used in identifying molecular markers defining DNA target sequences are DNA hybridization and the polymerase chain reaction (PCR). Restriction fragment length polymorphism (RFLP) obtained from sequence variation is revealed in DNA hybridization as changes in the length of DNA fragments after cleavage by restriction enzymes. RFLP-based markers have been widely reported in several plant, animal, and human genetic studies in the construction of genetic maps and in the identification of mono- and polygenic loci controlling

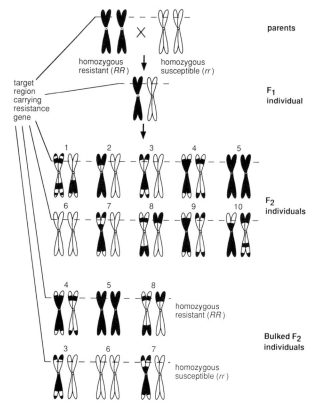

consider one of the chromosomes, carrying either the resistance (*R*) or the susceptible (*r*) alleles, from individuals involved in a cross

FIGURE 9 Schematic illustration of the bulked F₂ segregant analysis. In theory, molecular marker differences detected between the pooled DNA samples from the homozygous resistant (RR) and susceptible (rr) individuals would either be linked or contain part of the gene locus determining the resistant/susceptible phenotype.

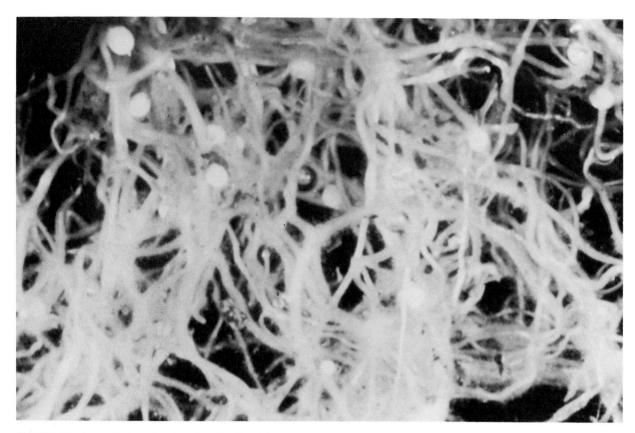

FIGURE 10 Roots of wheat showing white females of the parasitic nematode *Heterodera avenae,* the causal agent of cereal cyst nematode. (Photograph kindly supplied by R. Eastwood.)

disease and other agronomic traits. In crops such as rice, tomato, lettuce, and wheat, RFLP markers targeting introgressed regions carrying viral, bacterial, fungal, nematode, and insect resistance genes have been identified and there are attempts to use some of these markers in breeding programs. However, the main drawback with RFLP analysis is that it involves many steps that are quite difficult to automate.

DNA amplification products obtained from PCR analysis using random primers (RAPD) have been proposed as an alternative tool in targeting DNA sequences for genetic characterization and mapping. The random primers used in these studies were based on 9- to 10-base oligonucleotides. Genetic maps incorporating RAPD-based markers have been reported in lettuce, tomato, and *Arabidopsis*. Near isogenic lines have been used to identify RAPD-based markers linked to resistance genes for *Pseudomonas syringae* in tomato, downy mildew in lettuce, bacterial blight in rice, and stem rust in oats. PCR conditions employed for RAPD analysis permit the detection of several loci using a single primer, and

the relative ease of the assay enables several primers to be tested in a short time period. These factors make RAPDs a preferred option to RFLP analysis when a genome-wide scan aimed at targeting a specific trait is required, but where previous linkage information is unavailable.

In the absence of near isogenic lines, bulk segregant analysis is very useful for the identification of DNA sequences linked to a targeted region (Fig. 9). Bulk segregant analysis is a process by which DNA from selected individuals in a segregating progeny homozygous for specific alternate alleles is pooled and then examined for molecular marker differences. This technique was first tested in lettuce to identify new DNA markers tightly linked to downy mildew resistance. Sometimes the high levels of repetitive DNA sequences in a plant species can prevent the recovery of DNA markers characterizing alternate alleles. Removal of the repetitive DNA sequences prior to screening for molecular marker differences has been successfully applied to finding DNA markers closely linked to the cereal

TABLE I
Applications of Molecular Markers

Trace agronomic characters, in a breeding program, that are
    difficult or expensive to measure
Increase the efficiency of backcrossing
Provide markers for the pyramiding of disease resistance genes
Allow the genetic dissection of complex traits (QTLs)
Verify the existence of a true F_1 hybrid from a difficult cross
Assess germ plasm diversity
Trace resistance genes for diseases that are sporadic
Characterize pathogen populations and carry out early diagnosis
    for the presence of a pathogen
Identify varieties for plant variety registration purposes as well
    as quality control in defining mixtures of lines
Basic plant biology

cyst nematode resistance gene transferred from
the wild wheat *Triticum tauschii* to bread wheat
(Fig. 10).

VIII. Future Prospects

In recent years, rapid progress has been made in
the construction of molecular genetic linkage maps
of many crop plants. The improved DNA technology means that many avenues have been cleared
for applying this to the genetic enhancement of
plant crops. The various applications are listed in
Table I.

The combination of this progress in the application of DNA technology with the ability to transform all the major crops grown in the world today
means that the end products from crops can be
more readily modified to suit the changing demands
from customers.

Bibliography

Baum, M., Lagudah, E. S., and Appels, R. (1992). Wide
    crosses in cereals. *In* "Annual Review of Plant Physiology
    and Plant Molecular Biology," Vol. 43, pp. 117–143.
    Annual Reviews Inc., Palo Alto, CA.
Freeling, M., and Walbot, V. (eds.). (1994). "The Maize
    Handbook" Springer - Verlag, New York/Berlin/Hei-
    delberg.
Gatehouse, A.M.R., Hilder, V. A., and Boulter, D. (eds.).
    (1992). "Plant Genetic Manipulation for Crop Protec-
    tion." C.A.B. International, Wallingford.
Gustafson, J. P., Appels, R., and Raven, P. (eds.). (1993).
    "Gene Conservation and Exploitation." Plenum Press,
    New York.
Koncz, C., Chua, N., and Schell, J. (eds.). (1992). "Meth-
    ods in *Arabidopsis* Research." World Scientific, Sin-
    gapore.
Murray, D. (ed.). (1991). "Advanced Methods in Plant
    Breeding and Biotechnology." C.A.B. International,
    Wallingford.
Paterson, A. H., Tanksley, S. D., and Sorrells, M. E.
    (1991). DNA markers in plant improvement. *In* "Ad-
    vances in Agronomy," Vol. 48, pp. 39–90. Academic
    Press, San Diego.

Plant Genetic Resource Conservation and Utilization

TE-TZU CHANG, *International Rice Research Institute (retired)*

Glossary

Accession Cultivar, landrace, or sample from a wild population registered and maintained in a genetic resources collection
Bulk Growing and maintaining genetically different plants in a population without separation or selection
Clone Group of vegetatively propagated and genetically identical plants
Collection Collected sample; an assemblage of collected samples
Cultivar Cultivated crop variety; the international term for variety
Gene Basic functional unit of inheritance located in a chromosome; a character may be governed by one or more genes
Genebank Storage center for genetic resources; synonym: plant genetic resources center
Genetic diversity Extent of genetic variability in a population or species
Genetic erosion Disappearance of genetic resources from farmer's field, natural habitat, or genebank
Genetic vulnerability Condition that results when a crop is genetically uniform and unable to adapt genetically to a pathogen, pest, or environmental hazard

Genetic resources Sum total of genetic material in a species and its close relatives; genetic information embedded in the cells of a plant; synonym: germplasm
Landrace Population of genetically heterogeneous plants found in traditional agricultural systems resulting from long years of farmer's selection and natural selection under local conditions
Population Group of organisms of the same species occupying a particular geographic area and capable of interbreeding with one another
Sample Individuals taken from a population to represent it
Wild and weedy relatives Naturally occurring plants that have a common ancestor with a crop species but not domesticated

Plant genetic resources (PGR) are the total array of genetic materials that can perpetuate a plant species or the populations within it regardless of the species' immediate value to humans. When a species of economic importance is concerned, its genetic resources (germplasm) include the cultivated forms (cultivars for crops), related wild species in the genus or related genera, and the natural hybrids between cultivated and wild-growing forms. Products of research (genetic stocks) and breeding (improved cultivars, hybrids, inbreds, and breeding stocks) constitute more recent components of plant germplasm.

I. What Are Plant Genetic Resources? Why Conserve?

Plant genetic resources are traditionally propagated in the form of seeds, plants, or plant parts. With the advent of modern cellular biology, genetic resources can be propagated from tissues, cells, and even DNA sequences. Plant genetic resources can be modified by natural and artificial means to alter, and more frequently enhance, the usefulness of a species to humans. While PGR is renewable, it is also subject to destruction by natural forces and human neglect or abuse. [*See* PLANT GENETIC RESOURCES.]

Why should humans conserve the vast number of plant species and their populations while many forms at present appear valueless? The reason for PGR conservation extends beyond their association with human society since recorded history began: since the early existence of human beings, plants have furnished the bulk of naturally occurring materials for food, fiber, feed, shelter, and fuel; plants have also served as a powerful force in maintaining a stable ecosystem in which humans, animals, plants, aquatic organisms, and microbes interact and derive their life-supporting elements. Plant genetic resources have been crucial in supporting the growth of human society and helping humans advance from a hunting–food-gathering status to a farming society and then to the industrialized–urban society. PGR conservation ensures the present, and future generations will continue to derive maximum benefits from plants, the most important segment of the earth's biological resources. The direct connection between PGR and the world's food security becomes more apparent as human population growth continues at an unabated pace. The ever-increasing demand for food is the primary motivating force for human societies to properly manage the conservation activities and to protect the dwindling resources. [*See* WORLD HUNGER AND FOOD SECURITY.]

The life-supporting plants contain 250,000 flowering species of higher plants and 750 conifers. At least 3000 species have been used for human food at one time or another. Through human exploitation and dispersal of the most productive or preferred species, only seven major food crops (wheat, rice, corn, barley, potato, sweet potato, and cassava) are widely grown and consumed in urban areas, each exceeding 100 million tons in annual production. Soybean, sugarcane, sorghums, millets, and oat are the other major crops, though hundreds of minor plant species are consumed by cultivators in traditional agro-ecosystems. The continuing reduction in the diversity of the food crops, coupled with the rapid displacement of numerous traditional genetic resources (landraces) by improved cultivars and the disappearance of many wild relatives have irreversibly narrowed the genetic base on which food supplies are built. As crop breeding programs make rapid advances in productivity, the cultivars become fewer in number and more genetically uniform and related—a factor leading to the

genetic vulnerability of major crops to widespread pest damages and environmental disturbances. [*See* CULTIVAR DEVELOPMENT.]

As many as 50,000 tropical plant species are threatened by extinction. Meanwhile, the recent recognition that the human environment is continually degraded by the widespread destruction of forests and disappearance of many forest species has helped to heighten the human concern for biological conservation. It is now well recognized that once a plant species becomes extinct, either through displacement or human neglect, the genetic resources are lost forever. Only conserved genetic resources will help to fill the gap.

Though plant exploration in and introduction from foreign or remote lands was initiated to enrich the locally available plant genetic resources since the time of the Egyptian dynasties (4500–3500 B.C.), large-scale efforts to search, acquire, and study plant genetic resources, notably the economic species, began as recently as the 1920s when the Russian botanist N. I. Vavilov and co-workers amassed about 200,000 samples and varieties by the time of World War II. Vavilov's work aroused worldwide interest in crop diversity patterns. Concerns for the vanishing crop resources were voiced in the United States during the 1930s. Refrigerated seed-storage facilities came into use in the late 1950s. Systematic programs of field collection, preservation, evaluation, and documentation were begun in the 1950s by U.S., U.K., and Australian workers and on an international scale by the International Agricultural Research Centers (IARCs) during the 1960s and the following decades. Rice was the pace setter. Many national agricultural research programs have added PGR activities in recent years. However, rapid and accelerated loss of irreplaceable crop genetic resources has already occurred in farmers' fields and natural habitats of wild relatives as a result of successful plant breeding and environmental disturbance. The deteriorating situation, coupled with the genetic erosion inside PGR centers (genebanks), lent urgency to the need to rescue much threatened plant genetic resources, to properly manage the conserved materials, and to derive maximum benefit from the existing genetic resources collections. Meanwhile, the world's arable land area also continues to shrink.

The time to capture all the PGR in various areas of the world is definitely past, but there is sufficient room for all concerned to conserve whatever materials are available, to care for the fragile resources, and to exploit their genetic potentials. Different nations and all sectors of society should work together to conserve and use the major biological heritage of human-kind—plant genetic resources.

II. Evolution of Biological Diversity and Its Loss

A. Evolution of Diversity

Of the 250,000 flowering plants found today, a vast number of them probably have stayed genetically similar since prehistoric times if they remain in the same habitat. For many other species that have been dispersed to other sites and undergone human care (domestication), remarkable changes have occurred. These changes are particularly true for those crop plants that have been brought to new habitats following domestication and have experienced many cycles of natural and artificial (human) selection. Since agriculture began little over 10,000 years ago in several regions of the world (Southwest Asia, China, and Meso-America), the rate of genetic differentiation and subsequent diversification was markedly elevated when farmers attempted to exploit the genetic variability in the plants for greater productivity and/or for specialized uses. The process also narrowed the targets of human selection and production to a small number of flowering plant families: Gramineae (cereal grains), Leguminosae (pea, beans, soybean, peanut), Solanaceae (tomato, potato, tobacco, eggplant), Cruciferae (cabbage, turnip, radish, rape), and Rosaceae (berries, apple, pear, plum).

Though sudden changes (mutations) in the genetic makeup of a plant have been the primary source of new variations, natural hybridization (crossing) between related plants in a community produce hybrid progenies that lead to a wide range of variants following segregation and recombination. The differentiated forms became fixed in their original or new habitat following selection. New forms are thus born, extending the range of eco-geographic adaptation. Such processes can be repeated over time and place. An outstanding example of repeated hybridization between primitive cultivated species (cultigens) and their wild relatives in producing an advanced cultigen is bread wheat (*Triticum aestivum*): crosses between two diploid species gave rise to tetraploid forms, one of which hybridized with another diploid wild species to produce the hexaploid bread wheat. Less dramatic in chromosome buildup but more widespread in latitudinal spread is the cosmopolitan Asian rice (*Oryza sativa*): it had its distant origin in the Gondwana super-continent, became differentiated as a semi-aquatic plant in the humid Asian tropics, spread to temperate areas in five continents, diversified into three ecogeographic races, and further differentiated into 100,000 cultivars under different hydro-edaphic-cultural regimes. Rice cultivation now extends from 53° N to 42° S in latitudinal span.

B. Loss of Genetic Diversity and Consequences

It took centuries or even millennia for a crop species to produce the level of diversity that enabled the crop to become established at new habitats. But it only took less than two decades to dissipate the genetic diversity in rice and wheat during the evolution of the "Green Revolution"—the breeding and extensive use of semidwarf high-yielding varieties. Even though the yield levels have been dramatically elevated by the semidwarf wheats and rices, many pest problems and environmental stresses persist in production areas. Control of these yield-destabilizing factors requires a continuous supply of genes, both known and unidentified. Recent deteriorations in the environment have produced new factors for the scientists to counter with: intensified UV radiation, acid rains, and new pollutants. [*See* RICE GENETICS AND BREEDING; WHEAT GENETICS AND BREEDING.]

Well-known examples of genetic vulnerability associated with varietal uniformity are: (1) the Irish famine of the 1840s associated with the potato late blight disease, (2) the southern corn leaf blight epidemic in the United States during 1970–1971, and (3) cold injury to the Bezostaja wheat cultivar in the former U.S.S.R. in 1972. The hybrid rices of China grew from experimental plantings in 1970 to 17 million ha in 1992, but the hybrids have identical sterile cytoplasm and semidwarfing gene. Only a rich collection of genetic resources and judicious deployment of major genes and cytoplasms can maintain and fortify the genetic diversity in crops.

III. Exploration, Introduction, and Collection

A. Processes of Obtaining Genetic Resources

Acquisition of plant genetic resources is accomplished by one or more of the following operations.

1. Exploration, Collection, and Introduction
This is the earliest and most active way of obtaining genetic resources from remote areas or foreign lands.

The objectives of the mission are generally plant-specific and the task is performed by one to a few collectors. Field collections later evolved into broader and more systematic missions that involved scientific planning, team approach, and even international collaboration.

2. Introduction by Exchange

It is a common practice for PGR centers to have materials in one collection transferred into another by mutual consent. This practice accounts for many of the duplications in different genebanks. The sharing of samples collected as above also led to extensive duplication.

3. Introduction by Deposit

Materials in one collection may be deposited in another PGR center for safekeeping or for duplicate storage.

Recent systematic collecting missions have involved advance planning, extensive consultations, thorough preparations, and international teams. Some of the large-scale collecting projects, such as peanut and rice, were preceded by international workshops. Pooling of financial, physical, and human resources is essential. Training of local field collectors helps to enlarge the capability of the collecting teams.

B. Field Collecting Techniques

Foremost in planning collecting missions is the prioritization of target-plants by the extent of known genetic erosion, potential richness in local genetic resources, and the feasibility of local travel and obtaining support. In the case of re-collecting, information obtained from previous trips and collected materials is vital in choosing sites for collection. Alternative plans are needed before the actual execution begins so as to cope with unforeseen developments. Collectors' knowledge about the plants of interest, dedication to the mission, resourcefulness, ability to endure physical hardships, and capability to deal with local customs, religious beliefs, and cultural peculiarities are essential requisites.

Seed or plant-part samples may be drawn from farmer's field, natural habitat of wild forms, storeroom on farm, or the local market. Random (nonselective) lots are generally taken by a variety of sampling methods covering a site. When a collector encounters interesting materials, several samples should be separately collected as selective samples and so marked. For heterogeneous populations, a bulk

sample is preferred to selective sampling. For wild forms that flower over an extended period, repeated sampling may be needed.

After sampling in the field, sufficiently complete information should be recorded on the species and vernacular names of the plant, geographic location of site, date, grower's name and address, eco-agronomic characteristics of the site, breeding history, or special features of the sampled plants, local cultural practices and name of collector—these constitute the important passport data for each sample. Each sample should bear a unique code number.

IV. Global Survey of Plant Genetic Resources Collections

Surveys show that through worldwide collecting efforts the holdings of national and international genebanks now total about 2,500,000 accessions. However, much duplication exists between genebanks conserving the same crop, especially wheat, sorghum, and soybean. The total number of distinct or unduplicated accessions may be around 1,050,000. About 35% of the unduplicated world holdings, which may grow to 1,300,000 by 1996, will be cared for by the nine commodity-based IARCs with assistance from the International Board for Plant Genetic Resources (IBPGR) (later renamed as International Plant Genetic Resources Institute).

The current status of the major crops being conserved with respect to size and genetic representation is summarized in Table I. Among the crops, the com-

TABLE I

Conservation Status of Major Crops

Crop	Total accessions in genebanks	Distinct accessions	Wild accessions	% Cultivars uncollected
Wheat	410,000	125,000	10,000	10
Grain and oil legumes	260,000	132,000	5000	30–50
Rice	250,000	120,000	5000	10
Sorghum	95,000	30,000	950	20
Maize	100,000	50,000	1500	5
Soybean	100,000	30,000	7500	30
Common potato	42,000	30,000	15,000	10–20
Yams	8200	3000	60	High
Sweet potato	8000	5000	550	>50

Source: Chang, T. T. (1992). Availability of plant genetic resources for use in crop improvement. *In* "Plant breeding in the 1990s" (H. T. Stalker and J. P. Murphy, eds.), pp. 17–35. C.A.B. International, Oxon.

prehensiveness of coverage by existing collections may be ranked as: small grains (wheat, rice > maize) > coarse grains (sorghums, millets) > grain and oil legumes (beans, pea, peanut) > soybean > potato > yam > sweet potato. Wild species are generally deficient in collections, except for the relatives of wheat, legumes, potato, and tomato. Maintenance and evaluation are two general constraints in genebanks.

Among the major national genebanks, the United States leads with 557,000 accessions for all crops; followed by China, 400,000. The former U.S.S.R. holds 325,000 accessions, mostly in the Vavilov Institute. The rice collection of the International Rice Research Institute (IRRI) and the cereals and legumes held by the International Crops Research Institute for the Semi-Arid Tropics (ICRISAT) each amounts to 86,000. India has assembled 76,800 accessions, while the International Maize and Wheat Improvement Center (CIMMYT) and the International Center for Agricultural Research in the Dry Areas (ICARDA) each has 76,000 samples.

The above numbers are mere statistics that do not indicate the true genetic diversity in a given collection or the extent of genetic variability in a given crop represented by the holdings. Unknown numbers of accessions have expired in seedstores before the advent of modern refrigerated facilities. The fate of holdings in small and undersupported genebanks is doubtful. Discontinuity in program, departure of experienced staff, disappearance of bulky records, lack of viability monitoring, and shortage of competent workers have all added to the serious genetic erosion in many genebanks. The security of collected genetic resources hinges on adequate and sustained support. Human neglect can easily lead to the loss of large segments of genetic resources.

Currently, 25 national genebanks are operating in the developed countries (DCs), 25 in the less-developed countries (LDCs), and 10 in international and regional centers. The operation of PGR centers requires continued funding, adequate facilities, and competent staff who face a complicated, demanding, and nonglamorous service task.

V. Conservation Methods and Systems

Conservation of plant genetic resources should extend beyond mere preservation so that opportunities for multiplication, evaluation, dissemination, further evolutionary changes, and enhanced use can be pro-

vided. Therefore, a systematic approach should be taken at the start of program planning.

Conservation is implemented under two major approaches: (1) *ex situ* conservation by storing seeds, plants, and plant parts outside the home habitat of the collections, usually in a PGR center, and (2) *in situ* conservation by maintaining plant communities in natural environments. For many crop plants, both approaches can be used as complementary means.

A. *Ex Situ* Conservation

Ex situ conservation is accomplished by refrigerated storage inside genebanks for seed crops, field nurseries for vegetatively propagated plants (clones), deep-freeze storage of seeds/tissues in liquid nitrogen (cryopreservation), or *in vitro* culture of plant tissues and seed. Botanic gardens and arboreta offer another sanctuary for a small number of exotic introductions. Of these methods, cold seed storage is extensively used and most cost effective. Collections need to be multiplied in quantity so that they can be preserved.

Seed storage under cold and dry conditions is widely adopted with recent advances in refrigeration technology, insulation materials, seed drying devices, and airtight seed containers. Three levels of storage conditions and associated seed moisture contents are widely used: (1) Long-term storage ($-20°$ to $-10°C$, low RH) and air-tight containers for the base collection. This is the comprehensive set to be preserved; samples are withdrawn only to replenish exhausted or expired seeds in the active collection or when lowered seed viability requires regeneration. (2) Medium-term storage (around $0°C$, low RH) for the active collection. This is a second set of the base collection, which is widely distributed for use, entailing periodic cycles of regeneration. (3) Short-term storage (above $5°C$, RH under 60%) for the working (breeder's) collection. This consists of frequently used and regenerated breeding stocks, breeding lines, and modern cultivars. Backup (duplicate) collections for safekeeping at another genebank are also placed in long-term storerooms.

Seeds high in starch content such as the cereals are well suited for subfreezing storage after drying to low moisture content (ca. 5%) and sealing in airtight containers. Such seeds are considered to manifest the orthodox storage behavior. Seed viability may last more than 50 years. Other seeds have nonorthodox storage behavior are termed recalcitrant. Those high in oil or protein content require slow and low-heat drying and above freezing temperature storage. Peanut, soybean, onion, and lettuce are in this group, and seed longevity seldom exceeds 20 years. Another group of nonorthodox seeds consists of large seeded species, aquatic plants, and most tropical fruit trees. Their seeds cannot tolerate drying and therefore are not storable around or below 0°C. A new method is to store excised embryos of tropical species in liquid nitrogen and later to regenerate them *in vitro*.

Stored seeds need to be periodically checked for viability level so they will not expire during storage. When viability has dropped below a chosen level (say 80%), the seeds need to be planted for another cycle of multiplication (regeneration or rejuvenation). Care must be taken during regeneration to minimize possible changes in the structure of a population, mechanical mixtures, and human errors so that the genetic identity of the sample (accession) and its genetic integrity are maintained. For many cross-pollinated species, artificial pollination and/or isolation are necessary. Regeneration is a critical yet tedious and expensive operation which has led to the collapse of certain large collections.

Vegetatively propagated plants such as fruit trees, small berries, root crops, some industrial crops, and forest species are perpetuated vegetatively as clones in a variety of ways. Potato and yam tubers can be stored at moderately low temperatures inside dark rooms but only up to a year. They have usually been maintained in field plantings which involved the risk of virus infections and the need to eliminate the virus. Fruit trees are field grown in fenced areas or inside protected structures. Their clones are distributed as budwood or dormant scions. Cloned materials require more frequent regeneration and higher costs than seed crops, but their genetic integrity is seldom altered during propagation.

Many plant species are adapted to *in vitro* culture which can serve the dual function of maintenance and mass propagation. Moreover, virus-free clones can be readily obtained by heat or chemical treatment of stock materials, though subsequent tests (virus-indexing) are needed to ascertain a virus-free condition. Currently clones of potato, cassava, sweet potato, banana, and plantain are maintained and distributed by three IARCs (CIAT, CIP, and IITA) as *in vitro* cultures. *In vitro* maintenance is aided by lowered temperature, minimal cultural medium, and addition of growth retardants.

In recent years, small seeds, embryos, anthers, pollens, protoplasts, and other cell cultures have been successfully treated and stored in liquid nitrogen at −196°C. This method is termed cryopreservation and

has advantages for *in vitro* cultures by avoiding periodic subculture and transfer of stock cultures. It offers an alternate technique for long-term preservation without electric consumption and refrigeration. However, it requires periodic refilling of the liquid nitrogen tanks and the tanks are bulky and space-wasteful. Moreover, the protocol for prefreezing and thawing still needs to be developed for many species.

B. *In Situ* Conservation

In contrast to *ex situ* or static conservation, maintaining plant communities in their home habitat as natural reserves is considered by many biologists as a more dynamic form of conservation by allowing continuous evolution within natural environments. This approach can be used for landraces by growing them in original ecological niches under traditional agro-ecosystems and for wild species preserved in natural plant communities. This method is particularly suited for forest trees. Thus, *in situ* conservation provides a broad genetic base, maintenance of population structures, stability of population numbers, and opportunities for adaptive expansion. However, *in situ* projects require a high degree of technical expertise to monitor and manage the community structure and full cooperation from local residents in protecting the reserves, especially when *in situ* sites are nearly all in remote areas. Therefore, only limited efforts have been devoted to *in situ* projects. On the other hand, *in situ* and *ex situ* approaches could be used to complement each other.

C. Genetic Resources Systems

For large national collections scattered over many institutions, a national plant genetic resources system is essential to effective collaboration and efficient coordination. The United States has the largest number of federal and cooperating state agencies engaged in PGR activities and also the largest holdings (ca. 557,000 accessions). The USDA and collaborating agencies operate four regional plant introduction stations, the National Seed Storage Laboratory, eight national clonal genetic resources repositories, the National Arboretum, Interregional Research Project for three crops, genetic stock collections, National Plant Quarantine Center, and the National Genetic Resources Laboratory which administers the Plant Introduction Office, Genetic Resources Information Network, and National Small Grains Collection and coordinates the various crop advisory committees.

The National Plant Germplasm System (NPGS) was established in the early 1970s to coordinate the large and diffuse network. The NPGS has undergone reorganizations, but the day-to-day coordination and collaboration among crop workers are sustained by the crop advisory committees. These committees are composed of specialists drawn from federal agencies, state universities, and private seed and food processing companies. They exist for most economically important crops and meet frequently. Such committees provide crop-specific advice to the NPGS on operational needs and communications and coordination among workers of common interest.

On the international scene, eight commodity-oriented IARCs cover a wide range of conservation and dissemination activities, exerting an important role in international genetic resources exchange and use. While the IBPGR located in Rome has no physical facilities or genetic resources collections of its own, it has served as a prime coordinator and catalyzer of PGR activities worldwide, especially in collecting, documenting, and training. IBPGR has fostered a network of base collections in which national and international centers participate. Over the years the IARCs have assisted many national centers in upgrading their PGR activities. In recent years forestry research centers (ICRAF, CIFOR) have joined the international consortium.

VI. Characterization, Evaluation, Documentation, and Dissemination

A. Characterization

Characterization is the systematic recording of selected morphological and agronomic characters that not only identify an accession but also indicate potential agronomic usefulness. The time-consuming recording process may be accomplished when the accessions are being multiplied in the field or nursery. To facilitate recording, comparison, and data processing, data can be entered according to internationally standardized descriptors (characters, traits) and descriptor-states (values, codes) on a decimal scale. Only a few crop collections have adequate characterization data, however.

B. Evaluation

A user, whether a breeder or researcher, needs to know the geographic origin, biotic (disease and insect) reactions, abiotic (environmental stresses) responses, quality, and other agronomic features of conserved accessions before he or she will ask for the genetic resources materials. Often the user is discouraged by the lack of evaluation data or the inability to find the past records that are scattered and poorly accessible. The lack of passport data or evaluation data thus becomes a bottleneck in genetic resources use.

A preliminary, often visual, evaluation of new accessions can be made at the time of the first growout. An experienced PGR scientist or plant breeder will readily notice promising or unusual material and place such entries in trials.

After seedstock or plant material is built up, a systematic evaluation program performed by teams of diverse disciplines will reveal the potential value of the conserved accessions. In the mid-1970s, IRRI organized the massive Genetic Evaluation and Utilization Program to explore the hidden potentials in the rice collection by multidisciplinary teams, each charged with a major breeding objective, e.g., drought resistance, insect resistance, or nutritional quality. As many as 38 items per accession may be evaluated. The evaluation data together with the passport and characterization data (50 items) helped numerous rice researchers to search for desired combinations of traits and ask for the appropriate seed samples. The breeders choose parents from a delimited pool of genetic resources in their crosses, thus raising the probability of obtaining the desired combination of traits. In collaboration with national programs, IRRI also organized and coordinated a large number of international nurseries that extend the testing to many sites, often in areas with endemic stresses. Both unimproved (landraces) and elite (improved) genetic resources are included in the international testing program. Such collaborative evaluation has not only stimulated rice breeding activities but also enhanced rice research worldwide. Similar nurseries of elite materials of bread wheat have provided the vehicles for the successful Green Revolution in wheat.

Core collections, representative subsets of a base collection, have been recently advocated by some workers to cope with the difficulty of dealing with larger genetic resources collections. About 10% of the accessions may be drawn by different sampling techniques to form a core set which facilitates initial evaluation or study. Preliminary findings can then be used to determine which eco-geographic sectors of the base collection can be studied more intensively for specific targets.

C. Documentation

An efficient documentation system helps to link collecting and conservation with evaluation, research, breeding, management, and other practical functions of a PGR center. Development of standardized descriptors and descriptor-states has eased the entry and processing of voluminous data related to collecting, characterization, and evaluation and the exchange of data. Microcomputers are now affordable to most genebanks and have sufficient capability to handle the files, records, retrieval, cataloging, statistical analyses, and managerial practices. IRRI has developed and distributed a microcomputer-based software program (IRRIGEN) to help national rice genebanks in setting up related data files toward the establishment of a database. A daunting job for many genebanks is to locate and enter the voluminous backlog of data into their files. For a multi-crop PGR center, the challenge is even greater. National databases should be linked to a large international database, such as IRRI's.

The USDA now has the Genetic Resources Information Network (GRIN) in place. It is capable of linking all affiliated stations in data entry, collation, analysis, and retrieval. Japan also has a national network in its PGR center. On a regional basis, European workers have the best developed crop databases for crops specific to the region.

D. Dissemination

Genetic resources dissemination is the ultimate function of PGR centers. The capability of a genebank to supply genetic resources and related information upon the request of interested users is a mark of the efficacy in carrying out its mission. More often than not, national genebanks find it difficult to provide the desired genetic resources due to either national policy restrictions or problems in producing sufficient and viable seeds or clones. Plant quarantine restrictions also sometimes constrain genetic resources flow.

Among PGR centers in the world, the USDA has the largest annual distribution of crop genetic resources worldwide (ca. 200,000 samples), followed by IRRI's International Rice Genetic Resources Center (c. 50,000). International genetic resources exchange has been greatly elevated during the past three decades by the IARCs, all of which have genetic resources dissemination capability.

In recent years the international flow of genetic materials from the genetic resources-rich countries (LDCs) in the south to the genetic resources-poor countries (DCs) in the north encountered a short-lived slow down. Precipitated by socio-political activists, some of the LDCs claimed national sovereignty over the ownership of their unimproved genetic resources and demanded compensation as "farmer's rights" for their use in the DCs by private seed companies. The contention arose as a response to the rising adoption of plant breeder's rights and patents related to biotechnological innovations. Heated debates in FAO (Food and Agriculture Organization of the U.N.) sessions have continued for several years. Many nations later endorsed the FAO's proclamation entitled "International Undertaking on Plant Genetic Resources" and agreed to develop cooperative efforts through its Plant Genetic Resources Commission. Genetic resources exchange then resumed. It remains a great challenge for different countries to develop workable schemes of genetic resources control and funding while protecting the opportunity of farmers in the LDCs to be benefitted by the combined use of genetic resources, breeding, and biotechnology.

VII. Use of Genetic Resources

The enormous and rich plant genetic resources available today must be effectively used to further improve human well-being, especially that of the future generations, and to justify and it is hoped to recover the costs and efforts invested in conservation, evaluation, and research. Full exploitation of potentials in PGR should be the ultimate objective in the long series of conservation operations. If not properly used, conserved genetic resources will remain as "museum collections." All plant scientists should share the responsibility in channeling conserved genetic resources into productive use.

A. Genetic Resources Transfer and World Economy

The world has seen numerous instances of effective and profitable use of plant genetic resources, especially the foreign introductions in the DCs. Most of the major food crops and industrial crops grown in any country of the world have their origin in foreign lands following introduction and genetic improvement. The productive agriculture in North America, Australia, and Europe derived its strength from introduced plant genetic resources. The agricultural economies of the world are heavily based on transferred genetic resources and are interdependent on PGR.

The significant and continued growth in crop productivity during this century can be readily traced to genetic improvements based on an ever-broadening genetic resources base. Landraces of wheat collected in Turkey have enabled U.S. breeders to develop a large number of disease resistant cultivars that established the wheat industry in the hard-red winter wheat areas during the 1920s, and also provided multiple resistances to wheats in the Pacific Northwest. More recently, potent semidwarfing genes from Japan and Taiwan have triggered the "Green Revolution" in wheat and rice, respectively. Semidwarf cultivars now occupy slightly over 50% of wheat and rice lands in the LDCs. They are also rapidly expanding in the DCs. Cytoplasmic pollen sterility found in a wild rice plant on Hainan Island of China has led to 17 million ha of hybrid rices now grown in China. Another weedy wild form, *O. nivara,* has furnished the only source of resistance to the destructive grassy stunt virus disease. Other outstanding instances of tapping rare and useful genes, mainly for disease resistance from the wild relatives, may be cited from potato, oat, tomato, peanut, tobacco, vegetable crops, and many others. Recent advances in biotechnological techniques have broadened the horizon to use novel genes in distantly related species or genera as the sterility barriers in distant transfers can now be overcome.

Estimates on dividends derived from genetic resources use are yet to be compiled. One report places the contributions to U.S. agriculture at about $1 billion annually. Increased global production from the semidwarf rices and wheats is severalfold in magnitude.

B. Why the Underuse of Genetic Resources?

In spite of the notable successes, there is a common criticism that the vast stores of plant genetic resources have been little used, particularly by the breeders. Causes for the underuse may be traced to: (1) breeders are mainly interested in agronomically desirable materials as they cannot afford to work with obscure, sorry-looking, or difficult-to-handle genetic resources, and (2) potentially useful genes in unimproved genetic resources have not been fully identified and made known to the breeders and related researchers. The chain of events in using exotic genetic resources is: acquisition → conservation–multiplication–dissemination → evaluation/research → prebreeding or enhancement → breeding. Wherever a gap appears in the chain, whether it be genetical material or scientific data, the subsequent steps cease to

function. The major gaps are evaluation/research, dissemination of technical information, prebreeding, and communication among different disciplines. Therefore, more intensive use of exotic genetic resources depends on increased efforts in all these areas. Genetic engineering will lend a powerful hand to bridge the genetic barriers that have prohibited the tapping of genetic potentials in the vast pool of genes in the plant kingdom. New molecular innovations will enable plant scientists to use genes beyond the related plant genera and even those in the microbes and the animal kingdom. Therefore, PGR scientists and their associates should search beyond the usual gene-pools for potentially useful genetic resources. The plant breeder will remain the master technician in producing the final product.

Genetic resources workers need to also exert greater efforts in becoming a more effective partner in crop improvement projects by sharing their knowledge, expertise, and genetic resources with other disciplines. Thus, they can share a part of the credit in crop breeding. The crop advisory committees will remain as the vehicle in providing teamwork and guidance toward well-defined objectives.

All of the intensified and accelerated multidisciplinary activities will call for greater outlays of funds and personnel than the present levels. The 55 million U.S. dollar funding level for PGR programs of the whole world in 1982 is woefully inadequate—a 500 million U.S. dollar annual requirement has been proposed. The human resources also need to be augmented, for conservationists, evaluators, and breeders alike.

VIII. Future Directions

PGR programs have made impressive advances during the past three decades in spite of a belated start. A growing concern for environmental protection and the recent North vs South debates in the FAO sessions have brought greater public recognition of the importance of PGR. Products of its use such as the semidwarf wheats and rices have staved off the spectre of massive food shortages in the mid-1960s and the early 1970s, though nutritional level in the LDCs was only marginally raised. Continued gains from conserving, researching, and using plant genetic resources will undoubtedly be realized and even increased. But time is running short if PGR activities and their scope remain at the present level. It is therefore imperative to focus on the areas for future endeavor.

A. Limited and well chosen field collecting or recollecting of crop plants is needed to fill in some of the gaps in present genetic resources holdings. Wild species should receive increased focus than they have to date. Enhanced feedbacks from evaluation/research activities will aid acquisition efforts of the future.

B. Contents of major genetic resources collections should be re-examined to assess their true representativeness and diversity, and to reduce the redundancy within a collection. Duplicates in different genebanks should be consolidated to develop leaner base collections, while duplicate storage sites should be provided. Such information will aid future acquisition and conservation efforts.

C. Genebank facilities should be continually upgraded and management practices made more efficient. Examples are nondestructive monitoring of seed viability; use of core sets for distribution and initial screening; expansion of cryopreservation of difficult-to-store materials.

D. Systematic evaluation and prebreeding must be expanded and interlinked by improved communications. Genetic resources workers should participate more fully in crop improvement and biotechnology projects.

E. Potentials in exotic genetic resources need to be innovatively exploited to enhance crop productivity at affordable production costs, to reinstate genetic diversity in major crops and to counter changes in the ecosystems due to climatic change and pollution.

F. Increased funding for PGR and related activities is imperative. The private seed industry should contribute to the funding.

G. More genetic resources workers should be trained to better manage the myriad operations which are not taught in universities.

H. Farming communities and garden clubs should be enrolled to help conserve minor crops not included in national PGR centers.

I. Conservation of forest trees needs to be expanded by both *in situ* and *ex situ* schemes. Research and development of conservation technology must be accelerated. International leaderships should be established.

J. LDCs and DCs must fully collaborate and pool resources to expand *in situ* conservation, to facilitate rejuvenation of exotic stocks, to upgrade conservation centers in the LDCs, and to expand the training of PGR workers, breeders, and biotech workers.

K. Intellectual property protection systems should be modified to minimize restrictions on genetic resources exchange and use among nations. Disputes over the ownership, control, and use of PGR must be resolved.

L. Interinstitutional and international collaboration should be further strengthened to broaden the scope of conserving and using PGR.

As human population growth keeps burgeoning and all natural resources continue to dwindle, there is little room for complacency. All sectors of human society must join hands in caring, sharing, and using PGR, our common biological heritage.

Bibliography

Brown, A. H. D., Frankel, O. H., Marshall, D. R., and Williams, J. T. (eds.) (1989). "The Use of Plant Genetic Resources." Cambridge Univ. Press, Cambridge.

Chang, T. T. (1992). Availability of plant genetic resources for use in crop improvement. *In* "Plant Breeding in the 1990s" (H. T. Stalker and J. P. Murphy, eds.), pp. 17–35. C.A.B. International, Oxon.

Chang, T. T., and Vaughan, D. A. (1991). Conservation and potentials of rice genetic resources. *In* "Rice" (Y.P.S. Bajaj, ed.), "Biotechnology in Agriculture and Forestry," Vol. 14, pp. 531–552. Springer-Verlag, Berlin.

Chang, T. T., Goodman, M. M., and Krugman, S. K. (1985). Plant genetic resources—key to future food production. *Iowa State J.Res.* **59,** 323–539.

Consultative Group on International Agricultural Research 1992. "Partners in Conservation: Plant Genetic Resources and the CGIAR System." 2nd ed. IBPGR, Rome.

Frankel, O. H., and Bennett, E. (eds.) (1970). "Genetic Resources in Plants—Their Exploration and Conservation." Blackwell, Oxford.

Kloppenburg, J. R., Jr. (ed.) (1988). "Seed and Sovereignty." Duke Univ. Press, Durham.

Knutson, L., and Stoner, A. K. (eds.) (1989). "Biotic Diversity and Genetic Resources Preservation, Global Imperatives." Kluwer, Dordrecht.

National Academy of Science (1972). "Genetic Vulnerability of Major Crops." NAS, Washington, DC.

National Research Council (1993). "Managing Global Genetic Resources: Agricultural Crop Issues and Policies." National Academy Press, Washington, DC.

Plucknett, D. L., Smith, N.J.H., Williams, J. T., and Anishetty, N. M. (1987). "Gene Banks and The world's Food." Princeton Univ. Press, Princeton.

Plant Genetic Resources

GEOFFREY HAWTIN, *International Plant Genetic Resources Institute*

Glossary

Accession Individual entry in a genebank
Active collection Collection of accessions which are immediately available for multiplication, distribution, and use; often maintained under short- or medium-term storage conditions
Base collection Set of accessions, each of which is distinct and, in terms of genetic integrity, as close as possible to the original sample; since such collections are intended to be permanent, they are usually maintained under long-term storage conditions
Biological diversity (biodiversity) Total biological variability within living organisms and the ecological complexes they inhabit; biodiversity exists at the ecosystem, species, and genome levels
Ex situ **conservation** Conservation of genetic resources outside their original habitat, for example, in seed, *in vitro*, or field genebanks
Gene pool All the genetic information encoded in the total gene composition of a population of organisms that are capable of intercrossing; generally refers to the total genetic diversity of a species and those related species with which it is possible to form hybrids
Genetic erosion Loss of genetic resources, for example, as a result of the replacement of traditional with modern varieties, and the loss of habitats
Genetic resources Genetic material of plants, animals, and other organisms that is of value as a resource for present and future generations of people
In situ **conservation** Conservation of genetic resources in their natural habitat, in the wild, or in farmers' fields; the latter is also referred to as on-farm conservation
Regeneration Growing of a sample of plants from an accession to renew its viability

The innumerable potentially useful genes and gene combinations contained in both cultivated and wild species of plants are a fundamental resource for the development of more productive and stable crop varieties, new products, and improved farming and forestry systems. However these resources are at risk as a result of the loss of habitats, and through the replacement of traditional varieties with new ones. Over the past few decades, worldwide attention has been given to conserving plant genetic resources through the development of genebanks and nature reserves, and by encouraging farmers to conserve their own traditional varieties. Techniques have been developed for conserving seeds, plant tissues, and even pollen, and it is now possible to conserve many species for decades at a time without having to regenerate them. Many different institutions are involved throughout the world in conserving plant genetic resources. International treaties and agreements have been drawn up to promote conservation, and to govern access to genetic resources by individuals that need to use them.

I. Threats to Plant Diversity

It has been estimated that there are more than 250,000 species of multicellular plants in the world. They are fundamental to the earth's life support system and to human survival. Vegetation drives many of the processes of the biosphere and provides food, fuel, shelter, medicine, and many other products. Concern is growing over the increasing loss of plant species, through the loss of habitats and land degradation.

Of equal concern but less obvious in its immediate effects is the loss of diversity within species—intraspecific genetic variation. This is the diversity that enables species to survive and to adapt to different environments, new pests and diseases, and changing climates.

Since the birth of agriculture some 10,000 years ago, farmers, gardeners, foresters, and plant breeders have relied on genetic variability in plants to develop types better suited to their needs. They have bred an immense array of genotypes adapted to widely varied environments and that provide an enormous range of useful products. Breeders have used the natural genetic variation within the species concerned as well as the introduction of genes from closely related, often wild, species. Genetic engineering techniques are making it increasingly possible to introduce useful genes from almost any source. [See PLANT GENETIC ENHANCEMENT.]

As land use patterns change, and new crops and modern varieties replace traditional ones, crop genetic diversity is increasingly put at risk. Local knowledge about the uses and characteristics of traditional varieties is also threatened. Recognizing the magnitude of the problem and its potential consequences for the future of agriculture, there is a growing effort worldwide to conserve plant genetic resources for future generations.

II. Centers of Diversity

The world is not uniformly endowed with plant diversity. The tropics and subtropics are home to far more plant species than are the temperate regions. The Russian plant breeder and geneticist Nicolai Ivanovic´ Vavilov identified, in the 1920s and 1930s, eight regions that were particularly rich in diversity of cultivated plants and their wild relatives. These regions, he postulated, were where agriculture originated and where crops were first domesticated. Although subsequent researchers have revised the concepts and re-

gions proposed by Vavilov, his theory remains a major contribution to our understanding of crop evolution. The eight centers of diversity proposed by Vavilov, and some of the species believed to have been domesticated in them, were:

1. China: soybean, adzuki bean, peach
2. India/southeast Asia: rice, eggplant, mango, banana, coconut, cucumber, black pepper
3. Central Asia: wheat, rye, pea, pear, apple, walnut
4. Near East: durum wheat, barley, chickpea, lentil, melon, fig, grape, pistachio, alfalfa
5. Mediterranean: oats, lettuce, cabbage, olive
6. Ethiopia: teff, finger millet, coffee
7. Mexico/Central America: maize, common bean, pepper, sisal, squash
8. South America: potato, sweet potato, cassava, common bean, tomato, peanut, rubber, pineapple, papaya, cocoa

Significant diversity occurs for many species in more than one region, and secondary centers of diversity can exist away from the probable center of origin, for example, Ethiopia for wheat, barley, chickpeas, and lentils. Regions other than those defined by the eight Vavilovian Centers are also now recognized as centers of diversity and origin of several important crops, for example, oil palm in West Africa and sunflower in North America.

III. Assessment of Genetic Diversity

To conserve the genetic diversity of a species efficiently, it is desirable to obtain information about the extent and distribution of its intraspecific variation. Taxonomic descriptions of species, distribution data in flora, and data on genebank and herbarium specimens all provide useful sources of information. New biochemical and molecular genetic techniques are enabling genetic diversity surveys to become ever more accurate. Assessment of the genetic variability in *ex situ* collections provides a guide to gaps that must be filled. Assessment of genetic diversity in the field provides information for planning collecting expeditions and for locating *in situ* reserves.

IV. Conservation Strategies

Optimum strategies for conserving a gene pool depend on factors such as the nature of the storage organs or propagules (seed, tubers, etc.); reproductive biology; the extent and geographic distribution of genetic diversity; the availability of suitable storage facilities; and financial and human resources. [*See* Plant Genetic Resource Conservation and Utilization.]

Genetic resources can be conserved either *ex situ* or *in situ*. Conservation in *ex situ* genebanks insures that the stored materials are readily accessible; can be well documented, characterized, and evaluated; and are relatively safe from external threats. This method has the disadvantage that these plants can no longer evolve under natural or human selection. *In situ* conservation enables many more species to be conserved, and under conditions that allow them to continue to evolve. The main drawback of this method is that the materials cannot be characterized and evaluated as easily and are more susceptible to hazards such as extreme weather conditions, pests, and diseases.

Species having seed that can be dried can be effectively conserved in cold stores. However, for species that do not produce such orthodox seed or that are propagated vegetatively, *ex situ* conservation must take place through collections maintained in field genebanks, or as tissues in *in vitro* genebanks. Methods are also available for storing the pollen of many species.

In situ conservation methods range from fully protected nature reserves to areas in which only a few simple precautions are taken in their management. The conservation of landraces by farmers on their own land constitutes a special case of *in situ* conservation.

Crop gene pools generally comprise both domesticated and wild forms. There may be considerable intra-gene-pool variation in reproductive biology, type of storage organs, and so on. Thus the optimal conservation strategy for a given gene pool may be a combination of different methods, each covering a different part of the gene pool so the total is conserved in the most cost-effective way possible. Such strategies are often referred to as integrated conservation strategies.

V. Types of *ex Situ* Collections

Ex situ collections serve many purposes. Base collections are assembled for the long-term conservation of maximum genetic diversity. These collections may be national, containing germ plasm of national relevance, or international, covering entire crop gene pools on an international basis. These collections may be located at a single institute or dispersed among several. The collections are normally housed in long-term storage facilities, that is, under conditions that minimize the need for regeneration, and thus limit opportunities for genetic change. For added security, base collections should be duplicated in at least two locations, preferably in different countries.

Because long-term storage is relatively expensive, base collection accessions are normally only of sufficient size to insure the maintenance of their genetic integrity, and to provide samples for periodic viability testing. A minimum of 1000 viable seeds per accession is recommended; however, more heterogeneous samples (frequently the case in allogamous species) should be larger to represent their genetic variability adequately. It may be impossible to meet this minimum quantity for vegetatively propagated species and those with very large seeds.

Small accession size makes base collections generally unsuitable as sources of germ plasm for evaluation, distribution, and use. Separate collections with larger sample sizes are thus commonly assembled. Such collections, referred to as active collections, may include all or a subset of the base collection. Since accessions in active collections tend to be used up more rapidly than they lose viability, they are often maintained under more economical short- or medium-term storage conditions.

Sets of germ plasm may also be assembled for specific purposes, for example, those required for immediate use in plant breeding. Such collections are often referred to as working collections.

VI. Collecting

Collecting plants from the wild, and cultivated forms from farmers, is an activity as old as agriculture. An obelisk at Karnak in Upper Egypt depicts 275 different plants brought to Tuthmosis III from Syria in 1450 BC. Many new crops and other plants were brought from the Americas to Europe from the late 15th century onward; during the colonial period, new species were introduced to many countries, especially in the tropics, and the development of botanical gardens in the 18th and 19th centuries contributed to a massive worldwide movement of plants.

In this century, special efforts have been made to collect species and intraspecific plant diversity of agricultural importance. About 2.5 million accessions of crop plants and their wild relatives are now held in *ex situ* collections globally. Although this represents an enormous resource for agricultural development, much potentially useful diversity remains uncollected.

Collecting requires careful planning, whether to collect plants with particular characteristics, to fill gaps in existing collections or to rescue endangered forms. Sampling strategies should take account of the ecogeographic distribution of known variation and the geographic origin of accessions already collected. Knowledge of habitats and terrain, weather conditions, and the phenology and breeding system of the target species all contribute to successful planning. The optimal sampling frequency depends on the objectives of the expedition: specific traits or maximum diversity within taxa. When the distribution patterns of variation within and between populations is unknown, topographic and climatic data can help guide sampling strategies.

The size of samples to be collected depends on the nature of the material (e.g., whether seed or vegetative), the population size, and the extent of intrapopulation diversity. Random sampling within populations is commonly practiced to insure broad coverage of the gene base, but may be supplemented by selective sampling to guarantee the inclusion of individual phenotypes.

Collecting orthodox seeds is generally straightforward, although drying in the field may be needed. Recalcitrant and short-lived seeds, however, may require modified storage environments in transit and timely transportation to a germ plasm handling facility.

Vegetatively propagated species pose particular problems for collecting. Sampling can be slow, underground storage organs can be difficult to locate, and materials can be bulky and are often short-lived or prone to damage. Vegetative materials also carry a greater risk of disease and pest transfer and require phytosanitary inspection before being grown in a new location. For many species, *in vitro* collecting, for example through excising shoot apices or embryos in the field and transporting them on a culture medium, offers a practical solution to some of these problems.

The Food and Agriculture Organization (FAO) of the United Nations has developed the International Code of Conduct for Plant Germplasm Collecting and Transfer. This code recognizes national sovereign rights over biodiversity, and that even traditional landraces are the products of conscious breeding and selection by farmers and rural communities. The Code, approved by FAO member states in 1993, complements the Convention on Biological Diversity and the International Plant Protection Convention. It outlines procedures for granting collecting permits and describes the responsibilities of collectors, sponsors, curators, and users of the material collected.

VII. Seed Genebanks

Most arable and temperate forage seeds dry naturally during maturation and are orthodox, that is, they can be dried to low moisture contents and stored at subzero temperatures. Some seeds that do not normally dry during maturation, for example, certain *Citrus* species, are also orthodox. However, the seeds of many aquatic and large-seeded woody species such as chestnut, coconut, mango, and cocoa are recalcitrant, that is, they lose viability as they are dried. Since they have to be kept moist, subzero storage results in freezing injury. Most tropical recalcitrant seeds are damaged even at temperatures below 10–15°C. and cannot be maintained in cold stores. Certain other species, for example, coffee and papaya, are intermediate: drying to about 10% moisture content increases their longevity, but even at this moisture content deterioration occurs more rapidly at freezing than at ambient temperatures.

Cold stores provide the most convenient, safe, and economic method for conserving orthodox seeds *ex situ*. Under optimal conditions, the viability of many species can be maintained for well over 100 years. Longevity in storage depends on the initial seed viability, the presence or absence of pathogens, and the care taken to avoid mechanical and physiological damage. The two most important factors are the moisture content and the storage temperature. Although further research is needed on the relationship between them, especially at very low moisture contents, these factors largely influence seed longevity independently.

Seed longevity increases with decreasing seed moisture content. Although the effect varies among species, the benefit becomes greater for each successive reduction in moisture content until a threshold value is reached. Decreasing the seed moisture content of many nonoily orthodox seeds from 9% to 8% increases longevity by a factor of about 2, whereas reducing it from 6% to 5% increases longevity threefold. In sesame, reducing seed moisture content from 5% to 2% increases longevity by a factor of 40.

FAO and the International Plant Genetic Resources Institute (IPGRI) have jointly published standards for seed genebanks that recommend drying to a moisture content of 3–7% (depending on species) through the use of a desiccant or dehumidified drying chamber. Drying at 10–25°C and at 10–15% relative humidity is recommended. Seeds should be dried as soon as possible after reception. The drying period required depends on the size and other characteristics of the seed, their quantity, the initial moisture content, and the relative humidity of the drying chamber.

Relative responses to reductions in storage temperature are similar among most orthodox seeds. Longevity is increased by a factor of about 3 if storage temperature is reduced from 20°C to 10°C; by 2.4 from 10°C to 0°C; by 1.9 from 0°C to −10°C; but only by 1.5 from −10°C to −20°C. FAO/IPGRI recommend subzero storage for base collections, with the preferred standard being at least −18°C. This standard was chosen because domestic deep-freezers, which are cheap, readily available, and perfectly adequate for small collections, operate at −18 to −20°C. Techniques for cooling below −23°C are more costly. However, increased attention is being given to cryopreserving orthodox seeds using liquid nitrogen at temperatures down to −196°C.

Containers for storing seed samples in cold stores should be sealable, moisture proof, and able to withstand storage for long periods without leaking. Glass jars, metal cans, and sealable laminated aluminum foil packets offer effective alternatives. They should be well labeled and the seal should be tested regularly. It is preferable to store individual accessions in multiple containers for extra security.

The viability of samples stored in base collections should be regularly monitored. An initial viability test should be carried out as soon as possible after receipt of an accession. The frequency of subsequent testing will depend on the initial viability, the storage conditions, and the species concerned. Under good conditions, accessions with a high initial viability are normally tested about every 10 years. Viability is usually assessed by standard germination tests but other procedures (such as the topographical tetrazolium test) may be required to determine whether ungerminated seeds are nonviable or dormant.

VIII. Field Genebanks

Clonally propagated crops such as cassava, yam, potato, taro, banana, and plantain and many fruits such as apple, pear, *Citrus,* and *Prunus* species are normally conserved vegetatively, either because they do not produce viable seed (e.g., triploid desert bananas) or because specific genotypes would be lost in the production of conventional seed (e.g., cassava). Such crops, as well as recalcitrant-seeded species, are often conserved in field genebanks in which collections are maintained in the open and managed to conserve maximum variability.

Accessions in field genebanks can be easily observed, characterized, and evaluated. They are also readily available for distribution, which is particularly important for perennial species that are difficult to multiply quickly. However, field genebanks are expensive to establish and maintain, especially for short-lived clonally propagated crops (e.g., potato) that must be replanted frequently. These genebanks are also vulnerable to hazards such as pests, fire, and adverse weather conditions.

Space and cost considerations can severely limit both the number and the size of accessions that can be maintained in field genebanks, placing a particular constraint on the conservation of heterogeneous accessions. Close spacing of materials can increase the number of individual plants per unit area. However, it can also reduce opportunities for taking observations and for producing propagules for dissemination.

Management techniques for field genebanks depend on the specific conservation objectives and the nature of the species conserved. For example, when production of fruit or seed of self-incompatible species is desired, many individuals of each accession and special pollination control procedures might be needed. In some situations, environmental modification might be required, for example, to control photoperiod or to protect sensitive accessions from frost. All field genebanks must be protected from pests and diseases. Systemic infections in particular can be very expensive or even impossible to eradicate.

IX. *In Vitro* Genebanks

An alternative to field genebanks for the conservation of vegetative materials is their maintenance as tissues in *in vitro* genebanks. Such collections may be safer and less expensive to maintain. *In vitro* culture originated as a means of rapid clonal propagation, using nutritional and environmental conditions that are conducive to fast growth. However, such conditions are unsuitable for conservation because of a need for frequent transfer of the cultures to fresh media; this is a labor intensive operation with attendant risks of contamination. Thus techniques are being developed to reduce the need for frequent subculturing. Two main approaches are being followed: the manipulation of conditions to slow culture growth and the use of ultralow temperatures to essentially stop all metabolism. [*See* PLANT TISSUE CULTURE.]

A common approach to slowing culture growth rate is to reduce temperature. For rapid propagation, most temperate species are grown between 20 and 25°C, whereas tropical species generally grow best from 25 to 30°C. Lowering temperatures to 6–10°C for temperate and to 15–25°C for tropical species can extend the period between subculturing to 1–2 years.

An alternative is to modify the culture medium. The addition of chemicals such as osmotic regulators or hormones can slow growth. In some situations, combining low temperature with chemical retardants has proven optimal. Other techniques have also been tried with mixed success, including mineral oil overlays, reduced oxygen tension, and defoliation of shoots. Whatever method is used, the cultures are subject to stresses that can affect their health and ability to re-establish.

Although slow growth methods have been attempted for both unorganized (callus) and organized (shoot) tissues, the latter has so far proven more effective. Callus cultures appear more susceptible to physiological damage and frequently retain a reduced growth rate when returned to normal culture conditions. In addition, unorganized cultures are more prone to genetic instability generated *in vitro*, known as somaclonal variation.

Despite advances in slow growth techniques, these methods can only be recommended for short- and medium-term conservation. The second approach to *in vitro* conservation, the cryopreservation of tissues using liquid nitrogen, offers the prospect of a reliable long-term conservation option.

Cryopreservation has been used for many years in microbiology and animal cell culture; however, less attention has been given to plants. Higher plant culture systems vary enormously in their complexity, culture requirements, and response to freezing and thawing. Except for cell suspension cultures, no generalized protocol for cryopreservation appears possible. Nevertheless, cryopreservation has been successfully applied to calluses, protoplasts, suspension cultures, pollen, meristems, zygotic embryos, somatic embryos, or seeds of more than 50 crop species.

Cryopreservation procedures can be broken down into several stages: pregrowth, cryoprotection, freezing, storage, thawing, and regrowth. Pregrowth treatments can include hardening by exposure to cold or, for cell suspensions, passing through a medium containing mannitol, sorbitol, or proline to help dehydrate them. Cryopreservation of cells at the exponential phase of growth is generally the most successful, when they are small, have relatively small vacuoles, and have a low water content.

Cryoprotection through the use of chemicals such as dimethyl sulfoxide (DMSO), glycerol, mannitol, and sorbitol is often carried out to help maintain cell viability during freezing. Cryoprotectants help reduce ice crystal formation and moderate solute concentration.

Depending on the material to be preserved, freezing should be slow, rapid, or even ultrarapid through direct immersion in liquid nitrogen. For cell suspensions, a widely followed protocol involves slow cooling to −35°C, holding this temperature for about 40 min, followed by rapid cooling in liquid nitrogen. This allows the extracellular medium to freeze first, causing extraction of water from the cells and reducing the possibility of ice damage when cellular freezing eventually occurs.

Following storage, either by immersion in liquid nitrogen at −196°C or suspended in the vapor phase (approx. −150°C), the specimens are normally thawed rapidly by immersion in warm water (34–40°C) for 1–2 min. Thawed specimens are generally very susceptible to injury and must be treated carefully. Factors such as the composition of the regrowth medium and the physical environment during early regrowth are important determinants of successful regrowth.

Recent advances in artificial seed technology offer prospects for improved *in vitro* conservation methods. For example, the encapsulation of embryos and meristems in a protective nutritive casing prior to drying and freezing can increase survival rate significantly.

X. Pollen Storage

Although not widely used in conservation programs, the storage of pollen can usefully supplement other strategies. Pollen is relatively easy to collect and transport, occupies little space, and can be collected when plants are flowering but do not yet have viable seeds.

A drawback to the use of pollen is that, to produce plants for characterization and evaluation, it must first be used in artificial pollinations. This process is time consuming and, if the pollen parent is highly heterozygous, many pollinations may be needed to produce an adequate population. In addition, homozygous recessive alleles in the pollen parent are only expressed in the F_2 generation when hybridization is made to plants having homozygous dominant alleles at those same loci.

An alterantive, and one that allows for the immediate expression of recessive genes, is to culture haploid or dihaploid plants. Unfortunately, reliable techniques for the production of such plants are unavailable for many species.

The longevity of pollen of many species can be prolonged by desiccation and freezing. Storage in liquid nitrogen is possible for many crops. A common procedure is to immerse a vial of pollen directly in liquid nitrogen. Thawing is normally carried out by immersion in a warm water bath.

XI. Maintenance of Diversity in Collections

All collections require periodic regeneration. For seeds, standards have been published by FAO/IPGRI that are designed to insure that viability does not fall below acceptable levels while maximizing the period between regeneration cycles. This period depends on the longevity of the stored seed and the rate at which stocks are depleted as seeds are removed for viability testing, for reconstituting active collections, or for other purposes.

Initial germination capacity depends on the environment during seed production and procesisng, its physiological state at harvest, and inter- and intraspecific genetic differences. For most field crops such as cereals, initial viability in excess of 85% can be expected. This value may be less for certain vegetables and oilseeds and less again for some wild or forest species. The FAO/IPGRI guidelines suggest that regeneration be carried out when viability falls to 85% of the initial value.

In regenerating accessions, sufficient plants, preferably at least 100, should be used to maintain genetic integrity. As much as possible, all sources of selection pressure should be removed (e.g., through careful control of pests and diseases) and the contribution of all plants should be equalized (e.g., through harvesting the same number of seeds from each plant). Special techniques are needed in some species to control pollination.

Active collections should be regenerated no more than two or three times before being reconstituted from base collection accessions. Thus, if regeneration is needed every 15 years, new seed must be taken from the base collection every 30–45 years to insure maintenance of the original genetic identity.

XII. Documentation and Information

Effective management of plant germ plasm collections requires that they be well documented. Most genebanks handle four main types of data: passport, characterization, management, and evaluation data.

Recording data on individual accessions begins in the field and includes the collecting institute, the collector's name, the collecting number and date, the cultivar name, the location from which the sample was collected, the site details (topography, soil type, natural vegetation, etc.), and the conservation status. These data are entered into the germ plasm information system in conjunction with details of the scientific name (genus, species, subspecies, botanical variety), and each sample is given a unique accession number in the collection. Data entered on accessions acquired from other institutions should ideally include (in addition to collecting data) the original accession number, the donor's name and institution, the pedigree, the acquisition date, the accession size, and details of past regeneration. These descriptors are referred to collectively as the passport data.

Characterization data provide a record of highly heritable characters that can be easily observed and are expressed in all environments. Such data constitute a standardized record which, with the passport data, helps provide a unique identity for each accession. To obtain comparable data on accessions in different collections, lists of standardized descriptors have been published by IPGRI for about 70 agricultural and horticultural crops. These are widely used in genebanks around the world.

Management descriptors provide information needed for the effective management of collections. They include accession number, location in storage, date of entry into storage, initial germination percentage, date and result of last germination test, date of next test, initial moisture content, quantity of seeds in storage, duplication at other locations, details on past regenerations, details of samples taken, and mode of reproduction.

Evaluation data may or may not be kept by a genebank. However, ready access to evaluation data greatly enhances the potential use of germ plasm collections. Such data might include yield, harvest index, phenology, growth habit, pest and disease reaction, tolerance of environmental stresses, and nutritional or processing quality. Data generated in more than one environment are particularly useful.

Most genebanks use computers to manage their data. Database management software, either tailor-made or adapted from commercially available sources, is becoming ever more user-friendly. Modern database management systems make it increasingly possible for genebanks to share data electronically.

Information needed by conservationists and users of plant genetic resources is not confined to accession level data. Inventories of collections, for example, can assist in the location of sources of germ plasm. IPGRI publishes directories of germ plasm collections of more than 75 agricultural and horticultural crops held in institutions around the world.

XIII. Access and Use

Materials in germ plasm collections must be accessible to individuals that need them. This requires that they be well documented and that sufficient propagules be available to enable samples to be dispatched expeditiously. When materials are sent internationally, adequate phytosanitary precautions are needed. Many countries enforce strict plant quarantine regulations, but even where these do not exist special care is required to avoid the inadvertent distribution of diseases, pests, or weed seeds along with the germ plasm. FAO and IPGRI, in collaboration with other specialist institutions, publish a series of guidelines on the safe movement of germ plasm of various crops. These guidelines list the main phytosanitary threats, provide details on pest and disease identification, and indicate how they can be eliminated.

Base collections of particular gene pools can comprise many tens of thousands of accessions. For example, the rice germ plasm collection maintained by the International Rice Research Institute (IRRI) in the Philippines contains about 80,000 accessions. The International Center for the Improvement of Maize and Wheat (CIMMYT) in Mexico houses a collection of more than 70,000 accessions of wheat. Such large numbers present breeders with an enormous challenge in their search for new genes.

Recognizing this problem, many genebanks are now developing separate subsets of their collections known as core collections that are constituted to include the maximum genetic diversity within a manageable number of accessions. Core collections aim to provide an efficient means of accessing the diversity of a gene pool, through perhaps no more than a few thousand accessions containing among them most of the known and suspected genetic variation. Accessions not included in the core collection are retained as reserves for future use. Core collections can be constituted from a knowledge of the ecogeographic origin of accessions, and the extent and distribution

of known genetic diversity. As the core collection is evaluated for more traits and in different environments, a body of data is built up that makes the collection ever more useful.

Many different organizations are concerned with the conservation and use of plant genetic resources. These might include government-supported genebanks, plant introduction and variety registration services, plant quarantine agencies, national and international agricultural research institutes, government and private plant breeding programs, university departments, development agencies, forestry departments, environmental agencies, environment and development-oriented nongovernment organizations (NGOs), private sector seed companies and processors, and farmers' organizations. An effective national program requires mechanisms to insure that these various groups can participate in conservation efforts and can have access to the genetic materials and information they need. Coordination mechanisms can range from a rudimentary information system handled by a single agency to a national committee or commission in which all key interest groups are represented.

XIV. *In Situ* Conservation

Ex situ genebanks, although essential for agricultural crops, do not necessarily provide the most effective conservation mechanism for wild relatives, forest species, medicinal plants, or other potentially useful plant species. Although there is growing interest in the *in situ* conservation of genetic resources, most current *in situ* programs target the preservation of ecosystems (often areas of outstanding natural beauty) or of particular species (generally endangered animals) rather than intraspecific genetic diversity, which is the primary concern of most plant genetic resources programs.

Options for *in situ* conservation range from nature reserves from which all human intervention is excluded, through national parks in which economic activities with a potential to disturb the natural ecosystems are carefully regulated, to the implementation of special management regimes in areas used primarily for agriculture or forestry. The choice of conservation option for a particular gene pool depends on factors such as the specific conservation objectives, the biology of the target species, its ecological requirements, the distribution of its genetic diversity, the geographic and ecological characteristics of the habitats in which it is found, and social and economic feasibility.

Extensive biological, ecogeographic, and social research is needed to locate suitable areas for *in situ* conservation, and for the development of effective management systems that may have to take into account economic, social, and conservation objectives.

A global program to develop *in situ* conservation areas for ecosystems was launched by the United Nations Economic, Social, and Cultural Organization (UNESCO) in 1971, and was called the Man and Biosphere (MAB) program. It aims to "develop a basis for the rational use and conservation of the resources of the biosphere." One theme of the MAB program is the "conservation of natural areas and the genetic materials they contain." The program promotes the establishment of Biosphere Reserves, large areas in which a core zone representative of a particular ecosystem is left undisturbed and surrounded by a buffer zone in which limited economic activity is allowed, but with due regard for environmental concerns. Approximately 250 Biosphere Reserves have been established around the world.

The on-farm conservation and mangement of landraces, or farmers' traditional varieties, represents a special case of *in situ* conservation. The enormous genetic diversity still found on farms today, especially in developing countries, attests to the primary role of farmers as creators, users, and custodians of plant genetic resources. Although many farmers' traditional varieties have been lost, much diversity remains. Collecting this diversity for safe storage in genebanks before it is eroded further continues to be the top priority of many genetic resources programs. In parallel with such efforts, increased attention is being given to studying the ways in which farmers and rural communities have traditionally conserved and managed genetic resources and how, in many cases, they continue to do so today. Such studies are providing information of potential value for the design of improved *ex situ* conservation strategies, as well as for improving on-farm systems of conservation, breeding, and management.

Formal plant breeding—whether private or public, national or international—concentrates mostly on crops that are grown on large areas. Minor crops and varieties adapted to specific environments tend to be neglected. Support for on-farm management of genetic resources helps farmers in marginal areas and the producers of minor crops meet their own needs. Assisting farmers to improve their own indigenous varieties rather than replacing them offers an opportunity to achieve local agricultural development objectives while contributing to the conservation of genes for specific traits and adapted gene complexes.

Researchers have argued that since farmers grow those varieties that best meet their needs, the fact that so many traditional varieties are still grown only indicates that nothing better is yet available. Although this may generally be true, it is also true that farmers often retain specific varieties because they have certain desired traits, such as a particular flavor or cooking quality, that are not found in commerical varieties and perhaps never will be. The increasing interest of many European and North American farmers in traditional, or "heirloom," varieties of fruits and vegetables is an example of farmer conservation.

XV. Traditional Knowledge

Like genetic resources, traditional knowledge about local varieties and wild plants is also rapidly eroding. As societies and agricultural systems change, in many cases members of the younger generation do not acquire the knowledge of their elders. However, such knowledge, much of it held by women, provides invaluable information about the characteristics and uses of local varieties and species and their role in traditional agriculture. Indigenous knowledge about wild species can lead to the development of new commercial medicines and other products. While respecting the ownership rights of such knowledge, it is imperative that major efforts be made to capture it before it is lost.

Interviewing farmers at the time of germ plasm collecting can give rise to much valuable information. Local cultivar names, for example, can provide clues to special traits, uses, and even origin. The recording of such information in conjunction with the passport data can significantly enhance the usefulness of collections.

Information about the structure, organization, and function of local societies is essential for the design of effective *in situ* conservation programs. Local customs and values, decision-making processes, land ownership patterns, and access rights to common lands all help determine what modes of conservation are feasible.

XVI. The Global System

The terms "genetic resources" came into common usage at the FAO International Technical Conference in 1967. The scientific principles underlying strategies and methodologies for collecting, conserving, evaluating, and documenting genetic resources were comprehensively addressed for the first time. The conference was followed by a number of seminal publications by Sir Otto Frankel and others.

At that time, it became clear that the conservation and exchange of plant genetic resources required international approaches. Many countries with the greatest diversity lack the resources to conserve and use them adequately. Conversely, many countries with the best conservation facilities and advanced scientific research institutions have less indigenous genetic diversity.

In 1983, FAO developed a voluntary legal framework, the International Undertaking on Plant Genetic Resources, aimed at formalizing arrangements for the conservation, access, and use of plant genetic resources. At the same time, the Commission on Plant Genetic Resources was established by FAO as an intergovernmental forum to address issues of common concern and to monitor the implementation of the Undertaking. Subsequent interpretations of the Undertaking recognized the validity of Plant Breeders' Rights and the parallel concept of Farmers' Rights—the collective rights of farmers to compensation for their conservation and crop improvement efforts. While recognizing plant genetic resources as the common heritage of humanity, the Undertaking also recognizes national sovereign rights and responsibilities. To date more than 120 countries have agreed to adhere to the Undertaking, have joined the Commission, or both.

During the 1970s and early 1980s, the Consultative Group on International Agricultural Research (CGIAR) established a number of International Agricultural Research Centers to address problems limiting food production in developing countries. A major focus was on the genetic improvement of food crops. Genebank facilities were built and germ plasm collections were assembled. By 1990, 11 centers collectively maintained the world's largest international germ plasm collection, comprising more than 500,000 accessions held in trust for humankind. [See CONSULTATIVE GROUP ON INTERNATIONAL AGRICULTURAL RESEARCH; INTERNATIONAL AGRICULTURAL RESEARCH.]

In 1974, the CGIAR established the International Board for Plant Genetics Resources (IBPGR) administered by FAO. IBPGR has since become IPGRI, an independent international center under the auspices of the CGIAR. IPGRI focuses on science, technology, and information, and works with FAO to assist the development of national plant genetic resources programs and to support international linkages and networks. Agreements were signed between IPGRI and many national genebanks for the maintenance of base collections of important crops under specified conditions, including open access internationally. Subsequently, FAO and IPGRI have jointly supported the development of FAO's International Network of *Ex Situ* Base Collections, in which countries and institutions formally agree to adhere to certain standards in the maintenance of their base collections.

In addition to United Nations and CGIAR institutions, several other organizations are working internationally to promote the conservation and use of plant genetic resources. These include international and regional scientific institutions and international NGOs such as the World Conservation Union (IUCN), the World Conservation Monitoring Center (WCMC), the Worldwide Fund for Nature (WWF), the Rural Advancement Foundation International (RAFI), and Genetic Resources Action International (GRAIN). Many of the stronger national programs are also very active internationally.

At the United Nations Conference on the Environment and Development (UNCED), held in Rio de Janeiro in 1992, more than 150 nations signed the Convention on Biological Diversity (CBD). The Convention, negotiated under the auspices of the United Nations Environment Program (UNEP), came into force at the end of 1993. It aims to promote the conservation and use of biodiversity, and regulates the terms and conditions under which biological materials, and the benefits arising from their exploitation, are to be shared. The CBD supersedes the International Undertaking and now provides the legal framework governing international relations with respect to biodiversity. Negotiations aimed at harmonizing the International Undertaking with the CBD are currently underway, and FAO and the Commission on Plant Genetic Resources are taking the lead in the further development of the CBD as an effective instrument for the international regulation of the conservation and use of plant genetic resources.

Bibliography

Board on Agriculture, National Research Council (1994). "Managing Global Genetic Resources: Agricultural Crop Issues and Policies." National Academy Press, Washington, D.C.

Engels, J. M. H., and Tao, K. L. (1994). "Genebank Standards." Food and Agriculture Organization, and the International Plant Genetic Resources Institute, Rome.

Frankel, O. H., and Bennett, E. (1970). "Genetic Resources in Plants—Their Exploration and Conservation." Blackwell Scientific Publications, Oxford.

International Board for Plant Genetic Resources (1991). "Elsevier's Dictionary of Plant Genetic Resouces." Elsevier, Amsterdam.

Plant Pathology

GEORGE N. AGRIOS, *University of Florida*

Glossary

Mycelium Strand (hypha) or mass of hyphae that make up the body of a fungus

Nucleic acid DNA or RNA that make up the genetic material (genes) of living organisms

Parasite Organism living on or in another organism (host) and obtaining its food from the latter

Pathogen Living entity, usually a microorganism, that can cause disease

Phloem System of tubelike plant cells through which sugars and some other organic molecules are carried from leaves to other parts, particularly fruit and roots, of the plant

Probe Radioactive or otherwise tagged segment of a nucleic acid (DNA or RNA) used to detect the presence of, or to identify, a complementary nucleic acid

Prokaryote (also prokaryotic) Primitive microorganism whose genetic material (DNA) is not organized into a nucleus separated from the cytoplasm by a membrane

Serological reaction Binding of a protein, often present at the surface of a pathogen, with its specific antibody and the production of a visible or otherwise detectable reaction product that is used for the detection and identification of the pathogen

Virulence Relative ability of a given pathogen to cause disease

Xylem System of tubelike plant cells through which water and mineral nutrients are carried upward throughout a plant

Plant pathology is the science that studies plant diseases and their control. Basically, plant pathology studies the living entities and the environmental factors that cause disease in plants, the mechanisms by which they cause disease, the methods of their multiplication and spread, and the ways by which they can be detected and diagnosed, and subsequently avoided, reduced, resisted by, or eliminated from the plant. Plant pathology, therefore, is for plants what medicine is for humans and veterinary medicine is for animals. The scientists studying plant diseases are known as plant pathologists.

I. Introduction

Plant disease is the condition of a plant in which pathogenic microorganisms or adverse environmental factors interfere with the plant's normal functions. Normal plant functions include water and mineral nutrient uptake and translocation, photosynthesis, food translocation, utilization and storage, cell division and enlargement (i.e., growth), and reproduction. Interference with such functions is expressed by the plant as symptoms in the form of root rot, wilt, gall, canker, leaf spot, fruit rot, blight, and others (Fig. 1).

As with medicine and veterinary medicine, plant pathology is studied primarily at the postgraduate level; the vast majority of plant pathologists are hold-

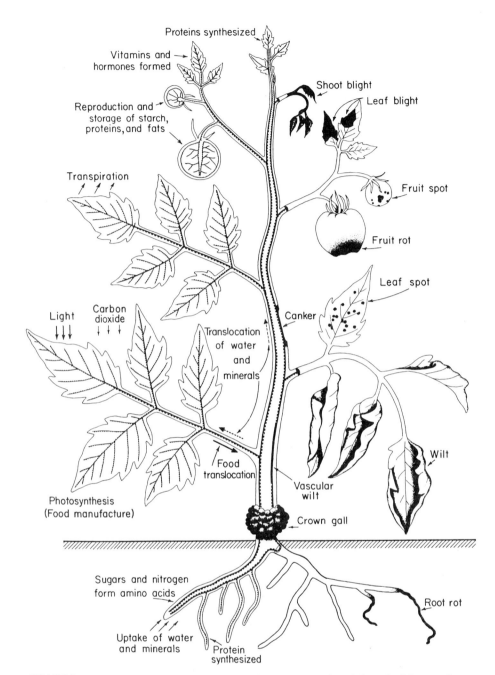

Proteins synthesized

Vitamins and hormones formed

Reproduction and storage of starch, proteins, and fats

Transpiration

Light

Carbon dioxide

Photosynthesis (Food manufacture)

Translocation of water and minerals

Food translocation

Shoot blight

Leaf blight

Fruit spot

Fruit rot

Leaf spot

Canker

Vascular wilt

Wilt

Crown gall

Sugars and nitrogen form amino acids

Uptake of water and minerals

Protein synthesized

Root rot

FIGURE 1 Schematic representation of the basic functions in a plant (left) and of the interference with these functions (right) caused by some common types of plant diseases. [From Agrios, G. N., (1988). "Plant Pathology," 3rd ed., p. 5. Academic Press, San Diego.]

ers of a doctorate (Ph.D.) degree in plant pathology, while a few may hold an M.S. degree and fewer yet may hold only a B.S. degree in plant pathology.

Plant pathologists are employed as researchers at universities, at state and federal agricultural experiment stations, and in large agrichemical and biotechnological industries; as professors who teach plant pathological or biological courses in colleges and universities; and as extension plant pathologists at universities and experiment stations who assist county agents, growers, or homeowners with specific plant disease diagnoses and control recommendations. Plant pathologists are also employed as private consultants (plant doctors) who undertake, for a fee, to monitor plant diseases in the fields of growers and to offer effective control recommendations to the grower(s).

Plant pathology is a scientific discipline that is recognized worldwide. There is an International Society of Plant Pathology, and many multinational, national, and state societies, such as the American Phytopathological Society and the Florida Phytopathological Society. These scientific societies aim at promoting the study and control of plant diseases in general and of the plant diseases in their geographical area in particular. Many such societies also try to improve communication among their members by publishing newsletters, scientific journals, and books in which their members publish the results of their work (research, extension, or teaching) and, concurrently, read and learn from the work of other plant pathologists. For example, the American Phytopathological Society, which has about 4500 members, publishes three journals (*Plant Disease* for somewhat applied research, *Phytopathology* for primarily basic research, and *Molecular Plant/Microbe Interactions* for advanced molecular research), and a monthly newsletter called *Phytopathology News*. Also, through the APS Press, the American Phytopathological Society publishes various kinds of books concerning plant diseases as well as visual aids (slides) and computer software on plant disease diagnosis, development, and control.

II. Economic and Social Importance of Plant Diseases

Plant diseases damage plants and plant products in the field and in storage and result in plant yields that are reduced in quantity and/or quality. Plant diseases, therefore, cause reduced availability and higher costs of plant products needed for food, feed, fiber, and shelter, and for beautification of the landscape. Some plant diseases also endanger the health of humans and animals consuming diseased plant products.

Humans became aware of plant diseases in the early stages of transition from nomadic to the agrarian way of life. Once humans began to collect grain, legume, and fruit tree seeds and to plant them, they noticed that, because of diseases, many seeds often did not germinate, plants did not always grow as well or yield as much, and that the produce often was less and of much lower quality. The quality often continued to deteriorate rapidly until the product rotted and became inedible or was completely destroyed before it could be used.

Early humans, of course, did not know that these effects on seeds, plants, and produce were caused by plant diseases. Most of them attributed these effects to growing the plants in the wrong soil, to bad weather, or, more commonly, to their gods or related spirits who, for their own reasons, interfered with normal plant growth and food production, not infrequently as a punishment of the humans for some imaginary wrongdoing. Plant diseases, however, have continued to take their toll on food production and to cause various degrees of hunger and famine throughout the evolution of mankind from the initial days of agrarian societies to the industrial societies of yesteryear and the computerized societies of today. In the days of the Old Testament, the plant diseases of blasting and mildew were considered among the great scourges of mankind, along with human disease and war. The ancient Greeks, as with so many other things, were the first to begin to study plant diseases, and the Romans created a special god, Robigo, as the god responsible for, and able to protect them from, the devastating rust diseases that destroyed large proportions of their grain crops each year (Fig. 2). In order to please Robigo so that he would not send the dreaded rusts to destroy their grain crops, the Romans held the annual spring festival Robigalia in which all sorts of sacrifices, including that of sheep and red dogs, were offered to appease Robigo. Numerous references to the ravages of plant diseases appeared in the writings of many contemporary historians in the following 2000 years, but little knowledge was added regarding their causes or their control.

In addition to the diseases causing hunger and famine, some plant diseases became notorious because of the poisonous effects they have on humans and animals. Such was the case of ergot, a disease of rye and wheat, which made affected grain, flour, or meal poisonous to humans and animals that ate it. In the Middle Ages, countless humans and animals in north-central Europe developed "ergotism" or "holy fire," a disease characterized by, progressively, hallucinations, mental aberrations, feeling of burning skin, miscarriages, gangrene of hands and feet, and eventual crippling or death. All these symptoms were caused by alkaloid chemicals contained in the fruiting body of the fungus that causes the ergot disease of rye and wheat (Fig. 3). When such fruiting bodies were mixed or ground together with the grains and consumed by humans and animals, they produced the symptoms described earlier. Actually, many grains, legumes, and other foods and feeds, which become infected or contaminated with certain fungi in the field or storage,

FIGURE 2 Symptoms on wheat leaves and kernels caused by two wheat rust fungi. (A) Rust spots on wheat leaves caused by the fungus *Puccinia recondita*. (B) Wheat kernels from a healthy plant (above) and from a plant infected with the stem rust fungus, *P. graminis*. The kernels from infected plants are smaller, are shrivelled, and have only a small fraction of the nutritional value of healthy kernels. [Photos courtesy USDA; from Agrios, G. N. (1988) "Plant Pathology," 3rd ed., pp. 454 and 458. Academic Press, San Diego.]

contain extremely toxic substances (mycotoxins) which cause a wide variety of severe symptoms and diseases, called mycotoxicoses, in poultry, sheep, pigs, cattle, horses, and humans that consume such foods.

One of the best known plant diseases that had important sociological effects is the late blight of potato. Potatoes were introduced into Europe from South America. By the 19th century its cultivation had spread throughout Europe and it had become a staple crop of the poor in many northern European countries. In Ireland, peasants came to depend on potatoes

for most or all of their food. It so happened that some of the imported potatoes were infected with a fungus that, somewhat late in the growing season, attacks and kills leaves and young stems of potato in the field, causing what is known as the late blight disease of potato. The same fungus also attacks potato tubers which rot, partially or entirely, while still in the soil or after harvest (Fig. 4). Both the potato plant and the late blight fungus grow best in cool, wet weather. When in the summers of 1845 and 1846, cool, rainy weather prevailed in northern Europe for a few

FIGURE 3 Ergot of rye caused by the ergot fungus, *Claviceps purpurea*. The kernels have been replaced by long, hornlike structures of the pathogen that contain the chemicals responsible for "ergotism." [Photos courtesy USDA; from Agrios, G. N. (1988) "Plant Pathology," 3rd ed., p. 392. Academic Press, San Diego.]

weeks, the fungus multiplied and spread rapidly, killing potato leaves and stems in its path. Spores of the fungus also reached and infected the potato tubers in the soil or when they were dug up. Some tubers rotted in the soil and were discarded while others appeared to be healthy, were harvested, and were stored in bins or in large dug-outs in the ground. Unfortunately, many of the latter tubers were also infected with the fungus which, while in storage, continued to grow and to spread into the infected and the previously healthy tubers. When the peasants opened up the bins later in the fall and winter they found that their potatoes had been changed into a smelly, rotten mass. Panic, hunger, and starvation followed in many rural areas of northern Europe, and many people died or emigrated to the United States. However, because of the prevailing political and economic conditions at the time, starvation and famine were much worse in Ireland. Hundreds of thousands of Irish peasants died from starvation and more than 1.5 million Irish emigrated to the United States in those 2 years to avoid death from starvation. The social implications of late blight of potato for Ireland, and for states such as Massachusetts, that received many of the Irish immigrants, were monumental at the time of the occurrence and can hardly be overlooked even now. [*See* POTATO.]

The impact of the potato late blight epidemics of the mid-1840s to Europe was so great that several scientists, some of them botanists, some physicians, studied the problem with intensity and perseverance. All of these scientists had to overcome the prevailing belief in "spontaneous generation," according to which microbes such as fungi and bacteria appeared as a result of disease and disintegration of living plant

FIGURE 4 Exterior and cross-section view of potato tubers infected with the late blight fungus, *Phytophthora infestans*. [Photo courtesy Department of Plant Pathology, Cornell University; From Agrios, G. N. (1988). "Plant Pathology," 3rd ed., p. 307. Academic Press, San Diego.]

or animal tissue rather than being the causes of disease and tissue decay. By 1861, however, the German botanist Anton deBary proved experimentally that the fungus *Phytophthora infestans,* the name meaning infectious plant destroyer, was the cause rather than the result of the potato late blight disease. For this and several other important contributions, deBary is considered the "father" of plant pathology. It should be noted that deBary's experimental proof that a fungus was the cause rather than the result of a disease preceded by 2 years Louis Pasteur's well-publicized experiments with bacteria and his proposal in 1863 of the "germ theory" of disease.

Over the centuries, plant diseases like the cereal rusts and cereal smuts caused frequent, widespread, and severe losses in grain crops. In a few instances this resulted only in smaller profits. Considering, however, that most grains throughout the world have been used by the farmers themselves and the other locals in the area as the basic food for survival, loss of grain to disease almost always resulted in partial hunger or total starvation of multitudes of people worldwide (Table I). Numerous other plant diseases have had important social implications. For example, coffee rust in the coffee plantations of the British colonies of Southeast Asia destroyed all coffee trees in the late 1800s and changed the British into tea drinkers. The brown spot disease of rice caused a devastating famine in Bengal in the early 1940s, while the southern corn leaf blight disease destroyed almost one billion bushels of corn in 1970 in the United States. Some diseases, like chestnut blight and Dutch elm disease, threaten to wipe out entire plant species (American chestnut and American elm, respectively) from the face of the earth. Plant diseases like citrus canker are so dreaded by citrus growers and citrus

producing countries that they prefer to eradicate tens of millions of citrus trees around points of new infection in the hope of preventing the establishment of the disease in their area. Many tree diseases, such as tristeza of citrus and fire blight of apples and pears, can reduce yields and usually kill trees within a few years from infection. Most diseases, however, primarily reduce yields, cut profits, or result in hunger. Plant diseases also increase control costs that must be incurred if the diseases are to be kept from causing even greater losses. A rough idea of the worldwide losses of produce caused by diseases, and by the equally important insects and weeds, is presented in Table I.

Control of plant diseases, even when possible, carries with it its own economic and social costs beyond the monetary costs involved in purchasing and applying control equipment and materials. Plant diseases are often controlled by treating plants with certain chemicals (fungicides, bactericides, nematicides) that act as medicines against the specific microorganism that causes the disease. Many of these chemicals, in addition to being toxic to microorganisms causing plant disease are also toxic to humans and animals, some of them being capable of causing birth defects, cancer, and other ailments. Some of these chemicals can also be washed off and carried by the rain into streams and lakes where they may poison fish and other animals and, moreover, may be washed by rain downward into and through the soil and may reach, and contaminate, underground water reservoirs that supply water for human consumption. These findings have resulted in a drastic reduction in the number of chemicals that can be used to control plant diseases and in the number and frequency of application of the chemicals that are still available for use. [*See* FUNGICIDES; NEMATICIDES; PEST MANAGEMENT, CHEMICAL CONTROL.]

III. Causes of Plant Disease

Plant diseases are caused by the same kinds of microorganisms (viruses, mollicutes, bacteria, fungi, nematodes, protozoa; Fig. 5) and by the same kinds of environmental factors (too low or too high temperatures, insufficient or excessive water, insufficient oxygen, soil acidity or alkalinity, air pollution, nutrient deficiencies or excesses) that cause diseases in humans and animals. Microorganisms that cause disease are called pathogens. Some plants may also become infected by certain parasitic higher plants which obtain

TABLE I

Percentage of All Produce Lost, in Roughly Equal Amounts, to Diseases, Insects and Weeds by Continent or Region

Continent or region	Percentage of produce lost to diseases, insects, and weeds
Europe	25
Oceania	28
North and Central America	29
USSR and China	30
South America	33
Africa	42
Asia	43

From Agrios, G. N., "Plant Pathology," 3rd ed. (1988). Academic Press, San Diego; adapted from Cramer, H. H., "Plant Protection and Crop Production" (1967), Leverkusin.

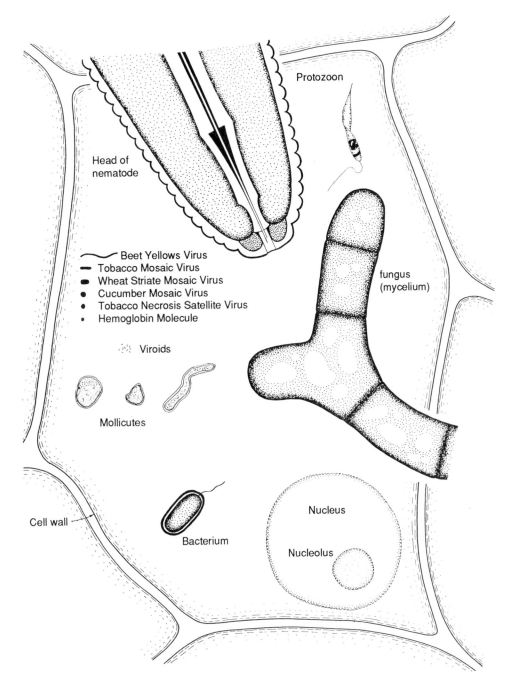

Head of
nematode

Protozoon

~ Beet Yellows Virus
— Tobacco Mosaic Virus
◖ Wheat Striate Mosaic Virus
● Cucumber Mosaic Virus
∙ Tobacco Necrosis Satellite Virus
∙ Hemoglobin Molecule

Viroids

Mollicutes

fungus
(mycelium)

Cell wall

Bacterium

Nucleus

Nucleolus

FIGURE 5 Schematic diagram of the shapes and relative sizes of certain plant pathogens in comparison to a plant cell. [From Agrios, G. N. (1988). "Plant Pathology," 3rd ed., p. 7. Academic Press, San Diego, CA.]

part or all of their nutrients from the first. Diseases caused by pathogenic microorganisms are called biotic, infectious, or contagious diseases and spread from one plant to another because the pathogens that cause them can spread from plant to plant. Diseases caused by environmental factors are called abiotic or noninfectious, affect only plants subjected to the par-

ticular environmental factor involved, and cannot be transmitted from plant to plant. Biotic diseases are much more common and cause greater losses than abiotic ones. Plant damage caused by plant-eating animals (herbivores), or by insects and mites, is not considered disease and therefore is not studied by plant pathologists.

Diagnosing a plant disease (i.e., identifying the cause of a plant disease) is essential if one is to know how to go about avoiding or otherwise controlling the disease. Some plant diseases are easy to diagnose because they induce characteristic symptoms on a plant part, for example, apple scab, corn smut, powdery mildew of rose, just as some human diseases do, for example, warts, chicken pox, mumps, etc. Other plant diseases are much more difficult to diagnose because the symptoms of many of them are similar. Diagnosing such diseases involves finding and identifying the pathogenic microorganism responsible for the disease. This is often very difficult because pathogenic microorganisms are usually very small, and must therefore be examined with a microscope or even an electron microscope, and often they must be separated or distinguished from several nonpathogenic microorganisms with which they coexist. In recent years, several new quick and dependable diagnostic tools and techniques have been developed, for example, selective culture media, serological tests, and tests that use DNA segments (DNA probes) that are complementary to DNA portions of specific pathogenic microorganisms. If a disease is caused by an environmental factor, then, of course, no pathogen can be found and the cause of the disease is usually identified either by some characteristic symptom it causes on the plant or by examining the weather conditions, atmospheric and soil contaminants, and the cultural practices and possible accidents in practices occurring before and during appearance of the plant disease.

IV. Environmental Factors That Cause Plant Disease

Insufficient or excessive amounts of several environmental factors, which go beyond certain normal ranges of these factors for the various types of plants, can cause disease to plants. The most common such environmental factors and some of the diseases they cause are given below.

Temperatures below 5°C reduce growth and may injure many vegetables, annual ornamentals, and tropical plants, while below-freezing temperatures damage or kill most annual plants and many perennial plants. Much less common is sunburn injury from high temperatures at the sunny side of fleshy fruits such as apple and pepper.

Insufficient moisture in the soil (drought) is probably the most common cause of yellowing, poor growth, and low yields of crops. Low relative humidity in the air, especially if combined with wind, often causes excessive loss of water in plants and results in poor growth, sometimes bud or flower drop, and withering of plants and seeds. On the other hand, excessive moisture in the soil (flooding) interferes with availability of oxygen to the roots of terrestrial plants. Plants in flooded areas quickly appear diseased and usually die within a few days or a few weeks.

Air pollutants include ozone, sulfur dioxide, and other toxic organic and inorganic substances produced primarily in the exhausts of automobiles and factories. Air pollutants cause various types of injury to plants ranging from leaf spots to malformations, yellowing, and severe stunting of entire plants. Several of these compounds, especially sulfur and nitrogen dioxides, dissolve in rain drops in which they are converted into sulfuric acid and nitric acid, two of the strongest acids, and fall to the ground as "acid rain" or snow where they cause adverse effects on plants and other organisms. Nutritional deficiencies cause frequent and widespread diseases in plants. Affected plants grow poorly, appear yellowish or reddish, blossom and fruit sparingly and irregularly, may show necrotic areas on leaves, stems, or fruit, and sometimes die. On the other hand, when nutrient elements, toxic minerals, excessive herbicides, and other pesticides are present at concentrations above those that can be tolerated by plants, the plants may turn yellow, then brown and may die within a few days. [See AIR POLLUTION: PLANT GROWTH AND PRODUCTIVITY.]

V. Biotic or Infectious Agents That Cause Disease in Plants

The most important biotic agents that cause diseases in plants are pathogenic members of the microorganisms known as viruses, mollicutes, bacteria, fungi, nematodes, and protozoa (Fig. 5), and the parasitic higher plants. These pathogens have characteristics and properties that are very much the same as those that cause diseases in humans and animals with one main difference: they infect only plants (although a few of them also infect the insect vectors that transmit them from plant to plant).

There are of course many kinds of plant pathogenic viruses, bacteria, etc. Each such virus, bacterium, fungus, and so on, may be able to infect and cause disease in only one species of plant, for example, tomato or apple. Many pathogens, however, often infect several species of plants each, although usually these plants are fairly closely related taxonomically, for example, tomato and pepper, apple and pear, cabbage and cauliflower, etc.

A. Viruses

Viruses are submicroscopic, noncellular pathogens. They infect and survive by using the cellular functions of their hosts, which may be from single-cell to multicelled microorganisms to all types of plants and animals, including humans. [*See* PLANT VIROLOGY.]

Most plant pathogenic viruses are either rod shaped or polyhedral (Fig. 6), or variants of these two basic structures. Plant viruses contain a single or a double strand of either RNA (ribonucleic acid) or DNA (deoxyribonucleic acid) surrounded by several dozen to several hundred molecules of protein. In each virus, the protein serves to protect the nucleic acid (RNA or DNA), while the nucleic acid carries in it the information, in the form of genes, that allows the virus to infect plant cells, to reproduce (replicate) itself, and to produce proteins that help it move through the plant and, if applicable, to be transmitted by its vectors. Viruses can be seen only with the electron microscope. Their detection and identification, therefore, depend on their ability to cause characteristic symptoms on certain host plants, transmissibility by certain vectors (for example, aphids or nematodes), and serological reactions with antisera, or reaction with nucleic acid probes prepared to known viruses.

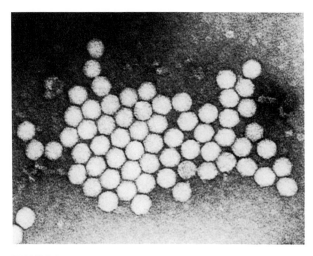

FIGURE 6 Particles of an isometric plant virus (tobacco ringspot virus).

When a plant virus infects a plant, it replicates and spreads systemically throughout the host. Young plants become quite thoroughly invaded by the virus within a few weeks, while large trees may take one or more years to invade.

Plants infected with viruses develop a variety of symptoms including leaf mosaics, yellowing or ring spots, leaf, fruit, and stem distortions, and sometimes leaf and stem necrosis. As a result, infected plants are often stunted, produce less, decline, and may die. Viruses are transmitted from plant to plant through vegetative propagation (for example by cuttings, grafting, tubers, etc.), by natural root grafts, by sap, and through the pollen and seeds of some plants. A few viruses are transmitted by specific nematodes, fungi, and mites. The most common and most important vectors of plant viruses, however, are several types of insects, particularly certain aphids, leafhoppers, and whiteflies.

Similar yet unlike the typical viruses, viroids are very small, naked, single-stranded, circular RNAs that can multiply and cause disease in plants. Viroids cause several devastating diseases of crop plants.

B. Prokaryotic Organisms: Mollicutes and Bacteria

Prokaryotes include the mollicutes and bacteria and are the smallest and simplest cellular microorganisms that cause disease in plants. Their bodies consist of single cells composed of cytoplasm surrounded only by a membrane in the mollicutes, but by a membrane and a rigid cell wall in bacteria. The cytoplasm contains genetic material (DNA) which is not organized into a membrane-bound nucleus. Most bacteria and many mollicutes live off dead organic matter, but some infect, parasitize, and cause diseases in plants, while others infect and cause disease in animals and humans.

1. Mollicutes

The plant disease-causing mollicutes are microscopic or submicroscopic, and their shapes range from spheroidal to filamentous or helical. They have no flagella and do not produce spores. They reproduce by budding and fission.

The plant mollicutes grow internally in infected plants, primarily in young tissues, in the sap of a small number of phloem cells (Fig. 7). They cause many destructive diseases of trees, such as pear decline and coconut lethal yellowing, but also of annual plants, such as aster yellows of vegetables and ornamentals.

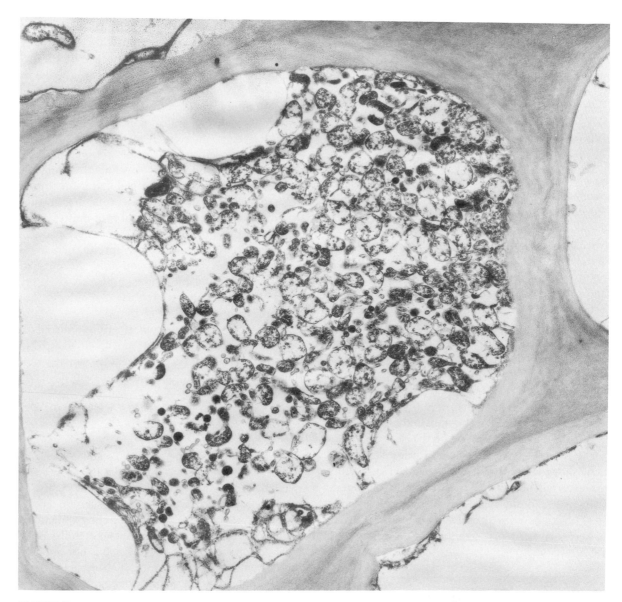

FIGURE 7 Cross-section of an aster phloem cell containing numerous mollicutes of various shapes and sizes. (Photo courtesy J. F. Worley.)

Mollicute-infected plants usually develop gradual, uniform yellowing or reddening of the leaves, smaller leaves, shortened internodes, stunting, excessive numbers of shoots, and greening or sterility of flowers, produce greatly reduced yields, and finally die back, decline, and die.

Plant mollicutes are transmitted by vegetative propagation, and by certain leafhopper, psyllid, and planthopper insects. Plant mollicutes also grow and multiply in their insect vectors.

2. Bacteria

Most plant pathogenic bacteria are microscopic single-celled rods (Fig. 8). They do not form spores but multiply rapidly by fission, doubling their numbers every 20–30 minutes as long as food supplies last. Some bacteria have no flagella, while others have one or several flagella. The bacterium *Streptomyces*, however, consists of branching filaments and produces chains of spores.

In addition to the chromosomal DNA, bacteria frequently have smaller circular chromosomes called plasmids that can move or be moved between bacteria. One such plasmid, known as the Ti (for tumor-inducing) plasmid, can also move from bacteria into plants, where it is incorporated into the plant chromosomes, causes plant cells to divide and enlarge, and results in the formation of galls. The Ti plasmid has

FIGURE 8 Plant pathogenic bacteria (*Pseudomonas syringae*) isolated from a cherry tree canker. (Photo courtesy H. R. Cameron.) [From Agrios, G. N. (1988). "Plant Pathology," 3rd ed., p. 566. Academic Press, San Diego.]

formed the basis for genetic engineering and has been used as a vehicle for transferring new genes into the genome of plants.

Plant pathogenic bacteria infect all types of crop plants and cause symptoms such as leaf spots, blights, soft rot of fruit, root, and other storage organs, wilts, overgrowths (galls), scabs, and cankers. In diseases that appear as localized symptoms, such as leaf spots or galls, the bacteria exist and multiply between the cells. In wilts (Fig. 9), the bacteria multiply and move in the xylem of plants, which they clog or destroy, and the plant wilts and dies.

Plant pathogenic bacteria are spread from plant to plant or to other parts of the same plant by rain splashes, run-off water, insects, and humans, and by infected or contaminated transplants and seeds. Plant pathogenic bacteria can be identified by their ability to grow on certain nutrient media, infection of certain host plants, serological tests with specific antisera, comparison of their fatty acids, and analysis of specific DNA fragments.

C. Fungi

Fungi are usually microscopic, mostly filamentous, branched, spore-bearing organisms that lack chlorophyll (Fig. 10). Most fungi live off dead organic matter. A few fungi infect and cause diseases in humans and animals. However, more than 8000 species of fungi cause disease in plants. Most plant diseases are caused by fungi. Some plant pathogenic fungi can grow and multiply only in their host plants; others can grow and multiply on dead organic matter as well as on living host plants. The elongated, branched,

filamentous body of the fungus is called mycelium, and the individual branches or filaments are called hyphae. Hyphae are usually from 0.5 to 5 μm in diameter and from a few microns to several centimeters long. Fungi reproduce by means of spores, each consisting of one or a few cells (Fig. 10).

Most plant pathogenic fungi specialize in the plants and even in the plant organ they infect, for example, tomato leaves, apple fruit, etc., but some fungi infect many kinds of plants and most of their organs. Fungi, therefore, may cause leaf spots (Fig. 11A), blights (Fig. 3), rusts (Fig. 2), smuts (Fig 12A), mildews, cankers, scabs (Fig. 12B), fruit rots, root and stem rots (Fig. 11B), seed rots, galls, wilts, tree diebacks and declines, etc. These symptoms are usually accompanied by stunting and reduced yields by the plants, and may be followed by death of part of or the whole plant.

Plant pathogenic fungi penetrate plant tissues and draw nutrients from them. Some cause only localized infections, as in leaf spots, whereas others may spread through most of the plant. Some fungi grow only in certain types of cells, such as the xylem vessels.

Plant pathogenic fungi spread from plant to plant generally in the form of spores carried by wind, by running or splashing water, by insects, humans, and other animals, and on or in seeds and transplants. Spores may be carried for a few millimeters on the same leaf, to fields many hundreds of miles away.

D. Nematodes

Nematodes are small, usually microscopic, wormlike animals. Some live freely in water or in the soil, while others parasitize and cause disease in humans, animals, or plants. Plant-parasitic nematodes are usually 300–1000 μm long by 15–35 μm wide and, because they are more or less transparent, they can only be seen with the microscope (Fig. 13A). The females of some species have pear shaped or round bodies. All plant-parasitic nematodes have a hollow stylet or spear which is used to puncture plant cells.

Female nematodes lay eggs which hatch and produce four larval stages before they differentiate into adult males and females. The female can then produce fertile eggs with or without mating with a male. A life cycle from egg-to-egg may be completed within 3 to 4 weeks at optimum temperature but takes longer in cooler temperatures.

Nematode infections cause symptoms mostly on roots of plants which may exhibit galls (Fig. 13B),

FIGURE 9 Tomato plant wilted as a result of invasion of its xylem by the tomato wilt-causing bacteria (*Pseudomonas solanacearum*). (Photo courtesy USDA.)

lesions, excessive branching, injured tips and eventually root rots. Infected plants also exhibit reduced growth and yellowing of foliage, reduced yields, and poor quality of products. Certain nematodes, however, attack and cause symptoms on leaves, stems, and seeds rather than roots. Most plant-pathogenic nematodes occur in the upper 15 cm of the soil. Some feed superficially on roots and underground stems (Fig. 13A), while others attach themselves to roots where they remain and feed for life. Some survive entirely in the tissues of the plants they infect.

Nematodes move through the soil very slowly but are easily spread by anything that moves and can carry particles of soil, for example, farm equipment, irrigation, flood or drainage water, animal feet, dust storms, and nursery plants. Nematodes can be identified from the morphological characteristics of their bodies.

E. Protozoa

Protozoa are mostly one-celled, motile microorganisms. A small number of protozoa species cause disease in humans and animals, while a few species infect and cause disease in plants. The protozoa that cause disease in plants have one or more flagella. Their bodies are usually long, oval, or spherical. Protozoa reproduce by longitudinal fission. There are few plant diseases caused by protozoa. Such diseases include the phloem necrosis of coffee, hartrot disease of coconut palm, and sudden wilt of oil palm.

F. Parasitic Higher Plants

Many higher plants live as parasites on stems or roots of other higher plants and cause disease on them. Some, such as the mistletoes and the broomrapes,

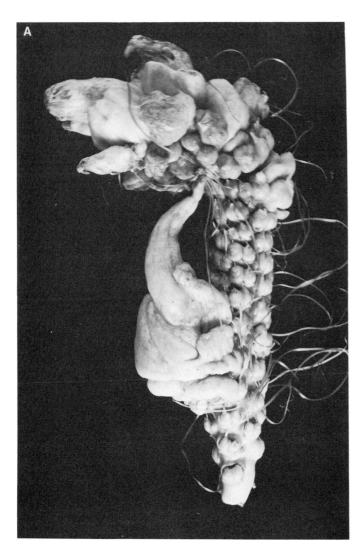

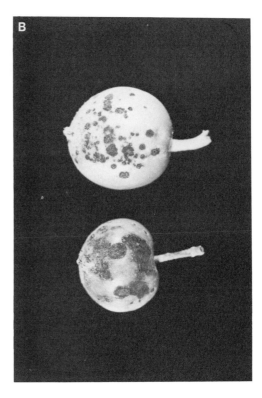

FIGURE 12 Symptoms caused by plant pathogenic fungi. (A) Corn smut disease caused by the fungus *Ustilago maydis*. Infected corn kernels enlarge and form irregular galls which are filled with masses of fungal mycelium and spores. (B) Apple scab symptoms on young fruit caused by the fungus *Venturia inaequalis*.

Even within the same host/pathogen system, however, such as tomato and the fungus that causes tomato leaf mold, certain populations (races) of that fungus infect certain tomato varieties but not others, while other fungus races affect different sets of tomato varieties. Frequently, there is an overlap in the varieties attacked by two or more races of the pathogen and, conversely, there is an overlap in the pathogen races to which two or more plant varieties are susceptible. Whether a plant variety is resistant (or susceptible) to infection by a specific race of the pathogen is often controlled by the presence (or absence) in the plant of at least one specific gene for resistance to that pathogen race. Conversely, the ability of a pathogen to infect one but not another variety of a plant is due

to a specific gene for virulence it has against the variety it can infect while it has a gene for avirulence towards the variety it cannot infect. In many host/pathogen systems, resistance or susceptibility of a plant variety is controlled by one, two, or more genes, as is the virulence or avirulence of the pathogen races involved. In many host/pathogen systems, however, the genetics of disease induction and resistance is not known and may be controlled by several to many genes.

Knowledge that resistance or susceptibility to many diseases is genetically inherited has been very useful for breeding crop varieties that resist disease in the presence of the pathogen. Unfortunately, pathogens multiply so rapidly and so profusely that, as they

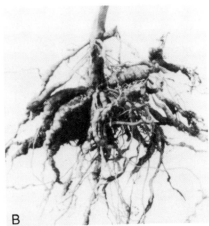

FIGURE 13 Plant diseases caused by nematodes. (A) A plant parasitic nematode (*Trichodorus christiei*) feeding on the surface cells of a young root. (B) Massive swellings and galls caused on a tomato plant root by the root-knot nematode (*Meloidogyne* sp.). (Photo A courtesy B. M. Zuckerman; photo B courtesy USDA.) [From Agrios, G. N. (1988). "Plant Pathology," 3rd ed., p. 744. Academic Press, San Diego.]

multiply, new genetic characteristics appear and new races develop that can infect the plants of varieties that were previously resistant to the earlier races. This "breakdown" of earlier resistance to disease necessitates a constant effort to find and incorporate into susceptible plants new genes for resistance. The advent of genetic engineering promises to make possible the detection, isolation, transfer, and expression in susceptible plants of genes for resistance to disease against specific pathogens; it also promises to greatly enhance our ability to produce plants not only with greater resistance but also at a greater speed so that we can always stay ahead of the new pathogen races that would bypass such resistance.

FIGURE 14 Strands of the parasitic higher plant dodder (*Cuscuta* sp.), parasitizing and smothering alfalfa plants. (Photo courtesy USDA.) [From Agrios, G. N. (1988). "Plant Pathology," 3rd ed., p. 611. Academic Press, San Diego.]

VIII. Development of Plant Disease Epidemics

It is obvious that diseases caused by environmental factors, such as freezing injury, toxic chemicals, etc., develop wherever these adverse environmental factors occur at levels that affect plants, but such disease agents do not spread from diseased plants onto healthy plants not exposed to such adverse factors.

In plant diseases caused by pathogenic microorganisms, however, the pathogens spread from diseased to healthy plants, infect them, and make them diseased. In the newly diseased plant, the pathogen multiplies and from there it spreads to still other healthy plants, which it infects. The stages in the infection, reproduction, and spread of a pathogen to a new host, on which it can start a new infection, comprise a disease cycle. Disease cycles may be repeated as rapidly as every few days. When a plant disease (really, the plant pathogen) spreads, and affects many plants quite severely over a large area, the disease is said to have become an epidemic.

After a plant or plant part has been killed by a pathogen, the pathogen is forced to survive in the dead plant tissues or in the soil, in its growing (vegetative) form, as spores (fungi), or as eggs (nematodes). Several pathogens, such as viruses, mollicutes, protozoa, and some bacteria, cannot survive in either dead plant tissue or the soil. Such pathogens survive either in their crop hosts, if they are perennial, or in wild hosts from which they are transmitted to their cultivated hosts the following spring. All pathogens can survive in vegetative propagating materials such as grafts, tubers, corms, etc., and several can survive in or on seeds.

Usually, very small populations of pathogens survive while crop plants are absent. After the first plants of the season become infected, new pathogen populations are produced in these plants and are capable of infecting new plants. Because a single infection by a pathogen can produce hundreds, thousands, or millions of pathogen individuals, and this can be repeated many times during a growing season, the total numbers of a pathogen near the end of a growth season are vastly larger than at the beginning, and they often cause severe epidemics.

Pathogens spread from plant to plant, within the same or different fields, in various ways. Most severe plant diseases that develop into epidemics, however, are caused by fungi that are spread as spores carried by wind, by fungi and bacteria spread by windblown rain splashes, and by viruses, fungi, bacteria, and mollicutes spread by their specific insect vectors.

Plant disease epidemics develop when three things exist concurrently in the same area: a virulent pathogen, a susceptible host plant, and favorable environmental conditions, particularly favorable moisture and temperature with light winds blowing in the direction of more healthy plants, or the presence of large numbers of appropriate insect vectors. The amount of disease produced is sometimes portrayed by the area of the so-called "disease triangle" (Fig. 15), the sides of which represent the size and virulence of the pathogen population, size and susceptibility of the host

plant population, and the sum total of favorable environmental factors. The prediction and forecasting of a disease epidemic depend upon knowledge of the size of the three factors and of the circumstances under which they interact. Such knowledge is extremely useful because it allows us, when possible, to take early, precautionary or preventive control measures, and thereby prevent the development and the losses that would be caused by the predicted epidemic.

IX. Control of Plant Diseases

Control of plant diseases in most farms and gardens is through application of pesticides (fungicides, bactericides, nematicides) on the plants or in the soil. Control of plant diseases is helped greatly by planting only healthy (pathogen-free) or chemically treated seed, transplants, etc., in soil that is free or has low numbers of the important pathogens that attack that plant. If a variety resistant to one or more of the serious pathogens of the plant is available, seeds, transplants, etc., of that variety should be preferred for planting. Seeds or transplants should be planted in fertile, well-drained soil at a time when they can germinate and grow well without delays. Crops planted in a field should be rotated with other unrelated crops, for example, corn with beans. Diseased plants or plant parts should generally be removed from a field or garden as soon as detected. Diseased and healthy plants should not be handled one after the other, and wounding plants and their produce should be avoided. Harvested produce should be stored in clean, pathogen-free containers, and if possible, refrigerated to slow down development of existing infections and to prevent new infections from getting started.

In recent years considerable efforts have been made to develop "biological controls" of plant diseases, that is, to find and employ microorganisms that are pathogenic or antagonistic on the pathogens that cause disease in plants. To date only a few biological controls are available against plant diseases but numerous ones are at various stages of development. [*See* Pest Management, Biological Control.]

A more promising development in plant disease control is the use of molecular techniques that allow genetic engineering of plants so that susceptible plants can be transformed to disease-resistant plants. Plants genetically transformed so they express the coat protein gene of one of several viruses have been shown to be resistant to that virus. Other viral genes, especially mutant defective genes of those that normally code

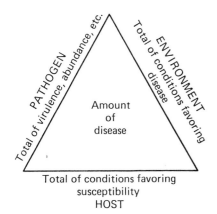

FIGURE 15 The disease triangle, depicting the role of pathogen virulence, host susceptibility, and favorable environment on the amount of disease that is produced. In the presence of all three factors (susceptible hosts, virulent pathogens, and favorable environment), as each of them increases the amount of disease produced also increases. [From Agrios, G. N. (1988). "Plant Pathology," 3rd ed., p. 43. Academic Press, San Diego.]

for proteins indispensable to the virus, when transferred into plants also prevent infection of the plant by the respective virus. Some plants have also been made resistant to fungi by transforming the plant with a gene for an enzyme that disintegrates the cell wall of pathogenic fungi and thereby prevents them from infecting the plant. Molecular techniques are also expected to soon enable us to locate and isolate genes for resistance to certain pathogens from wild or unrelated hosts in which they are found and to transfer and express these genes in susceptible, commercially desirable crop plants which would then become resistant to that pathogen.

X. Concluding Remarks

Plant pathology, like human medicine and veterinary medicine, has made significant contributions and remarkable progress in the last 150 years. Plant pathology proved that plant diseases were the result neither of the wrath of God(s) nor of spontaneous generation but rather of infections of plants by specific pathogenic microorganisms. Plant pathology has contributed to the study of fungi, bacteria, viruses, nematodes, and the other microorganisms that cause disease, and has developed mechanisms, materials, and procedures by which these microorganisms can be managed or controlled. These contributions have helped to significantly protect our crop plants from plant disease, thereby not only reducing or eliminating famines caused by destruction of produce by plant diseases but resulting, instead, in more and better food, feed, fiber, ornamentals, and wood products being available for use by humans throughout the world. Through these accomplishments, plant pathology has improved everyone's quality of life.

Bibliography

Agrios, G. N. (1988). "Plant Pathology," 3rd ed. Academic Press, San Diego, CA.
Alexopoulos, C. J., and Mims, C. W. (1979). "Introductory Mycology," 3rd. ed. Wiley, New York.
Goodman, R. N., Kiraly, Z., and Wood, K. R. (1987). "The Biochemistry and Physiology of Plant Disease." Univ. of Missouri Press, Columbia.
Goto, M. (1992). "Fundamentals of Bacterial Plant Pathology." Academic Press, San Diego, CA.
Horsfall, J. G., and Cowling, E. B. (eds.) (1977–1980). "Plant Disease," Vols. 1–5. Academic Press, New York.
Matthews, R. E. F. (1992). "Fundamentals of Plant Virology." Academic Press, San Diego, CA.
Schumann, G. L. (1991). "Plant Diseases: Their Biology and Social Impact." APS Press, St. Paul, MN.

Plant Physiology

FRANK B. SALISBURY, *Utah State University*

Glossary

Enzyme A protein molecule that speeds up the rate of a chemical reaction, usually in the cells of living organisms (sometimes in digestive tracts, secreted to outside, etc.); enzymes act when the so-called substrate molecule(s) in the reaction briefly (for a fraction of a second) attach to an active site on the enzyme; each enzyme reacts with only one or a few kinds of substrate molecules and is not permanently changed by the reaction

Meristem Tissues in plants made up of cells that are capable of cell division; apical meristems occur at the tips of roots and shoots, lateral meristems (cambium) occur between wood and bark, and intercalary meristems occur at the bases of leaves and in stems between points of leaf attachment (the nodes)

Phloem Tissue in which inorganic and organic molecules (e.g., products of photosynthesis) are transported from regions of high concentration (sources) to regions of lower concentration (sinks: growing regions, storage organs, etc.); solutions in phloem cells (sieve tubes) are under pressure; phloem usually is located in bark (i.e., to the outside of xylem tissue)

Photoperiodism Plant responses to the relative lengths of day and night, for example, in short-day plants, flowering may be promoted by decreasing day lengths (e.g., a day with fewer than 15 hr of daylight for cocklebur); there are various responses besides flowering, and plants may detect decreasing or increasing day lengths or be insensitive to day length

Photorespiration Respiration promoted by light; occurs in C-3 but not C-4 or CAM plants (see *photosynthesis*)

Photosynthesis Process in which carbon dioxide and water are used to produce carbohydrates (especially the sugar sucrose) and amino acids, which are eventually used to synthesize all other compounds, including proteins and fats; oxygen (from water) is released in the process; variations among plants include: C-3 plants (first product has three carbon atoms), C-4 plants (four carbons), and CAM (crassulacean acid metabolism, which occurs in succulents and is similar to C-4)

Phytochrome Proteinaceous plant pigment that exists in two forms (P_r = red absorbing; P_{fr} = far-red absorbing, active form) and that absorbs the light responsible for numerous plant responses

Phytohormone (plant hormone) Organic compound synthesized in a plant that controls or influences developmental processes at very low concentrations; sometimes it is transported from one part of the plant to another, where it exercises control; in other cases, it may function in the cells where it is made

Transpiration Evaporation of water from leaves or other organs of plants

Vernalization Promotion of early flowering by exposure of moist seeds or later stages (depending on species) to a period of low temperatures (close to the freezing point of water)

Water potential Measure of the activity of water in a part of a system compared with the activity of pure water at atmospheric pressure and the same temperature; indicates the tendency for water to diffuse

Xylem Tissue in which sap (a dilute water solution) moves from the soil to leaves, pulled to the leaves under tension as water vapor transpires from the leaves, and held together by forces of cohesion between water molecules; xylem is usually located to the inside of phloem tissue; mature xylem tissue is wood

Plant physiology is the subfield of botanical science that studies how plants function, that is, how plants work. It is of necessity based on knowledge of plant structures—cells and larger anatomical features—the study of which forms independent botanical subfields (cytology and anatomy). Plant physiology studies movements of water and dissolved substances into and throughout the plant; the biochemical processes within plants; how plants grow in size and increase in complexity, often in response to such specific environmental inputs as light or gravity; how plants otherwise adapt to environmental features, including those that might be stressful; and especially during recent years how plants control all of these processes at the molecular (genetic) level (often called biotechnology).

I. The Plant Cell

In basic structure, plant cells are very similar to animal cells (Fig. 1): each plant cell is surrounded by a membrane and contains a viscous material called cytoplasm that in turn contains a number of other important structures along with the controlling nucleus (not part of cytoplasm). Important structures in the cytoplasm common to both plants and animals include bodies (mitochondria) that combine oxygen with substrates such as sugars and fats to release carbon dioxide and water along with energy in a usable form, a network of folded membranes (endoplasmic reticulum) that moves substances within the cell and performs other functions such as fat synthesis, tiny bodies (ribosomes) that synthesize proteins, special structures (Golgi bodies) involved in secretion, and minute fiberlike structures that form a cytoskeleton. [See PLANT CYTOSKELETON.]

In addition to these fundamental cellular structures, plant cells are characterized by three highly characteristic features: green bodies (chloroplasts) suspended in the cytoplasm, a cell wall just outside the membrane, and a typically large vacuole that is filled mostly with water that has many substances dissolved in it. The plant cell, especially the vacuole, absorbs

water by osmosis (see the following), which produces pressure that is exerted against the cell wall. This wall can withstand high pressures on the order of those in an inflated bicycle tire or the water in a culinary water supply. This expansion of the cell against its wall gives form to the soft parts of plants, a phenomenon that becomes quite apparent when a plant loses enough water so that it wilts. In trees and other woody plants, the majority of the plant body may consist of dead cells filled only with water, their walls strengthened not only by cellulose but by lignin, a substance that confers much rigidity.

This brief description of plant cells has emphasized typical living cells, but many plant structures are based on highly specialized cells. Plant physiology studies the functions of these cells and cell structures and more.

II. Plant Water Relations: Osmosis

Because the absorption of water by plant cells is so critical to plant life, and because plants lose much water by evaporation into the atmosphere (transpiration), plant physiologists have studied the so-called water relations of plants for well over a century.

Basic to this study is the process of osmosis, which is the movement of water across a membrane in response to a difference in the activity of the water molecules on opposite sides of the membrane. (Movement of dissolved substances across membranes is *not* called osmosis.) The movement is by diffusion (based on random motions of individual water molecules) from the site of highest water activity to that of lowest activity. The measure of activity is based on concepts from the science of thermodynamics; it is called water potential. Technically, water potential is a measure of the potential of the molecules to do work (their free energy or chemical potential) compared with the potential of pure water at atmospheric pressure and the same temperature as the water being considered.

The presence of dissolved substances in water reduces the water potential to some value below zero, that is, solutions at atmospheric pressure have a negative water potential. The component of water potential caused by dissolved substances (solutes) is called osmotic potential. Pressure applied to the solution (e.g., as in a cell absorbing water against the restriction of its wall) increases the water potential of the solution.

Thus the substances dissolved in a plant vacuole (and in cytoplasm) produce a negative osmotic potential within the plant cell, but as water is absorbed,

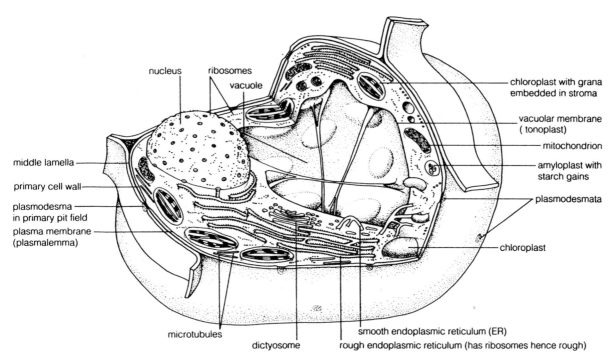

FIGURE 1 A generalized plant cell. The drawing is based on young cells that still have a small vacuole. Mature cells are often much longer and with a vacuole that occupies more than 90% of the cell volume. The cellular organelles (e.g., bodies in the cytoplasm, which is everything within the cell membrane except the nucleus and the vacuole) are drawn much as they appear in electron micrographs. [Drawing by Cecile Duray-Bito. From F. B. Salisbury and C. W. Ross. (1992). "Plant Physiology," 4th ed. Wadsworth, Belmont, CA. Used with permission.]

pressure builds up until the water potential for the cell as a whole equals that outside the cell. Water molecules then continue to move across the membrane, but the movement in equals the movement out (dynamic equilibrium).

Other factors may also influence water potential, notably the negatively charged surfaces on large protein molecules, cell-wall substances, and clay particles in soils; these effects on water potential are collectively referred to as matric potential. Concepts of water potential are extended to water in the soil and to water vapor in the atmosphere. (Please note that water-relations terminology has undergone much revision during the past century. Terms such as osmotic pressure and diffusion pressure deficit, related or equivalent to osmotic potential and to water potential, have been used in the past.)

III. Transport across Membranes

As with the concepts of water movement, transport of dissolved substances (solutes) across membranes and throughout a plant can be understood by apply-

ing principles of thermodynamics. Individual ions (electrically charged atoms or molecules) and neutral molecules in solution will individually diffuse in response to gradients in their activities (i.e., from points of high activity to points of lower activity). This is greatly complicated, however, by the fact that cell membranes typically either do not allow passage of these substances—or the membranes actively transport certain ions or molecules into cells against an activity (usually concentration) gradient.

If dissolved substances moved as freely across membranes as do water molecules, their concentrations would always be the same on both sides of the membrane, and osmosis would not be possible. Osmosis depends on the fact that plant (and animal) membranes use energy to pump certain dissolved substances into the cell and to exclude others. The movement in must greatly exceed the movement out (or prevention of entry) if the concentration of dissolved substances inside the cell is to increase sufficiently to produce a negative osmotic potential and thus make osmosis possible. The process of active movement of dissolved substances into cells is also important because some of those substances are essential for specific metabolic functions. For example, several critical molecules in plants contain an atom of iron. Clearly, unless iron is absorbed by cells there will be no growth of green plants. The essential elements for plant life are discussed in following sections.

IV. The Ascent of Sap

Understanding how sap can move to the top of tall trees proves to be especially challenging to plant physiologists. They realized during the nineteenth century that such movement could not be explained by resorting to principles of operation of simple pumps. One such pump creates a vacuum at the top of a standing pipe, and water (e.g., in a well) will be forced into this pipe from below by the pressure of the atmosphere pushing on the water outside the pipe. But there is only enough atmospheric pressure to push water in such a pipe to a height of about 10 m, whereas many trees reach heights of 30 m or more, and the tallest trees may be over 100 m tall. A pump that pushes the water from the bottom could raise water to heights limited only by the power of the pump and the strength of the pipes holding the water. Do the roots of plants have an analogous pump (called root pressure) capable of forcing water to the tops of tall trees? If they did, drilling into the trunk of such a tree should allow the release of pressure so that water ran out of the drilled hole much as it runs out of a water tap. Although weak root pressures have been observed, water seldom runs out of a hole drilled into a tree trunk; indeed, water may be sucked into such a hole. Furthermore, it is possible to cut a tree off near its base and suspend it in a bucket of water (as was done by a German scientist in the late 1800s) and still observe the ascent of sap from the bucket to the leaves. This will occur even if the bucket contains poisons that kill the cells in the trunk so that they cannot act as miniature pumps along the way. (When the poisons reach the leaves and kill them, water flow stops.)

Near the end of the nineteenth century and at the beginning of the twentieth century, scientists proposed a mechanism in which evaporation (transpiration) of water from leaves created a tension in the water columns extending all the way down to the roots. Thus as water left the leaves, it was pulled up almost as though it were a thin steel wire extending all the way to water in the soil below. The water column was held together by cohesive forces between water molecules. This mechanism was called the cohesion theory for the ascent of sap. It met much opposition, at least until the 1950s, because if one attempts to pull water in a pipe to heights above those supported by atmospheric pressure at the bottom of the pipe, the tension within the water typically causes a bubble of water vapor to form, a process called cavitation. If water is pulled to the tops of tall tress, what prevents cavitation in the trunks of these trees?

The answer gradually became apparent as the idea was evaluated during much of the twentieth century. Our current understanding of the phenomenon depends completely on our knowledge of tree anatomy, especially the anatomy of wood (called xylem tissue; Fig. 2). To begin with, if cavitation does occur in one cell, the vapor cannot spread to other cells. The pores in cell walls are too narrow to permit the passage of a vapor bubble; that is, the surface forces on such a bubble are too strong to allow it to be distorted small enough to pass through the pores in the wall—although liquid water with its solutes passes readily through these pores. Many xylem cells also have check-valves that consist of a pit and a torus. When the pressure in one cell exceeds that in its neighbor, the torus is pushed into the pit, closing it (Fig. 3). Actually, cavitation occurs in many cells in the trunk of a transpiring tree, but enough noncavitated cells remain for the sap to continue to move. Furthermore, in summer when the tree is growing, a layer of living cells between the wood and the bark (the cambial layer) produces new cells by cell division; those to the outside become bark cells and those to the inside become xylem cells filled with sap and ready to take part in sap movement.

A more subtle factor of xylem anatomy appears to be the fact that greater tensions can exist in water without cavitation when the water is confined in very small tubes. This was shown in the 1950s with glass tubes of various diameters.

It is interesting to contemplate that water under tension is in a so-called metastable state, a state that is unfamiliar to our everyday senses. We see that water under tension at the macroscale normally cavitates, which is the same as saying that it boils even at room temperatures when the pressure is low enough. Yet metastable water under tension exists in the vast quantity of living wood in all the trees of the world, as well as in the xylem elements of all other smaller plants. It is something that must be visualized rather than experienced.

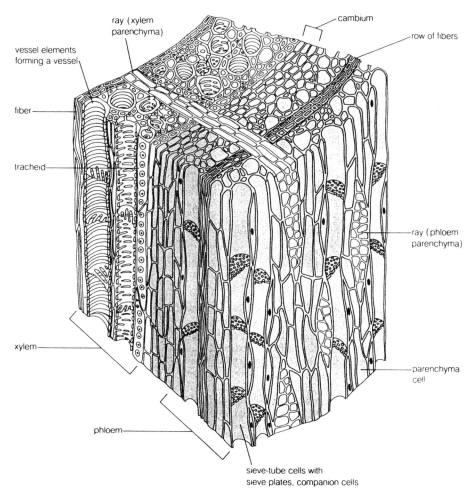

ray (xylem parenchyma)

cambium

row of fibers

vessel elements forming a vessel

fiber

tracheid

ray (phloem parenchyma)

xylem

parenchyma cell

phloem

sieve-tube cells with sieve plates, companion cells

FIGURE 2 A three-dimensional drawing of a woody stem showing xylem (wood) to the inside of a layer of cambium (dividing cells) and phloem (part of bark) to the outside. [From F. B. Salisbury and C. W. Ross. (1992). "Plant Physiology," 4th ed. Wadsworth, Belmont, CA. Used with permission.]

V. Movement of Products from Source to Sink

Plants photosynthesize in cells that contain chloroplasts. Many products are produced, but sucrose (common table sugar) is the most common of these. Many of these products (collectively called assimilates) are moved from where they are produced (the source) to growing tissues such as stem and root tips, flowers, fruits, and seeds, and to storage organs such as tubers (all acting as sinks). The movement of assimilates, called translocation, occurs in specialized tubes called sieve tubes that consist of individual sieve cells (or elements) connected to each other end to end across a perforated plate called a sieve plate (see Fig. 2). These cells are part of the phloem tissue, which

occurs in bark. The sieve elements are alive with a thin layer of cytoplasm around their periphery and a large vacuole in the center—unlike the dead xylem tubes in which sap ascends to the top of trees. Although sieve elements have no nucleus, each sieve cell has a so-called companion cell next to it with all the typical organelles including a nucleus. What companion cells do is not well understood, but they are apparently essential if sieve cells are to function.

How do assimilates move through the sieve tubes from source to sink? As with the cohesion theory for the ascent of sap, a theory for phloem transport was proposed (in 1926 by E. Münch in Germany). It is called the Münch pressure-flow theory, and like the cohesion theory it met considerable resistance for many decades. It is now widely accepted.

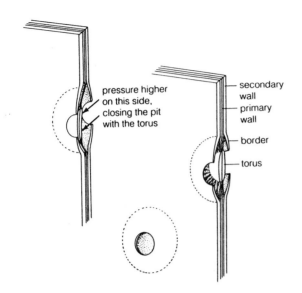

FIGURE 3 Diagram of a bordered pit from a transporting cell in the xylem of a pine tree. If pressure on one side exceeds pressure on the other side, the high pressure pushes the torus so that it plugs the hole, shutting of flow. [From F. B. Salisbury and C. W. Ross. (1992). "Plant Physiology," 4th ed. Wadsworth, Belmont, CA. Used with permission.]

In its modern form, the theory proposes that assimilates are actively moved across membranes from photosynthesizing cells (or perhaps from storage tissues where they are being mobilized instead of stored) into seive tubes (called phloem loading). This buildup of solutes produces a negative osmotic potential. Because water outside the sieve tubes has a more positive water potential, it moves osmotically into the sieve tubes, creating high pressure there. At the sink end of the system, assimilates can be actively removed from the sieve tubes (phloem unloading). These assimilates might be used to support growth, or they may be stored in developing fruits, seeds, tubers, etc. In any case, as the assimilates are removed from the sieve tubes, osmotic potential in those tubes decreases, and the pressure in the tubes that is transmitted from the source cells makes the water potential in the sieve tubes less negative than that outside those tubes, so water moves out osmotically. This maintains the flow from source to sink.

Much of the controversy was based on observations that different substances may move in opposite directions in the same phloem tissue at the same time. It was also suggested that the holes in the sieve plates between sieve elements were too small to allow sufficiently rapid flow of the rather viscous phloem solution. Careful studies have shown that different substances may flow in opposite directions in the same

phloem tissue but most likely not in the same sieve tubes. An interesting tool to study these processes is an aphid that inserts its stylet into an individual sieve tube. The aphid can be anesthetized and its stylet cut off after it has been inserted; fluid will continue to move out of the cut stylet, sometimes for several days, showing that fluid in sieve tubes is indeed under pressure. Such studies have shown among many things that flow through the sieve plates is rapid enough to account for phloem transport.

An alternative hypothesis held that substances were moved actively in each sieve element by a streaming of cytoplasm around the periphery of the element. This would allow different substances to move in a single phloem element at the same time. Much effort was expended to see if cytoplasm does stream around sieve cells as it does in many other living plant cells. The conclusion is that such cytoplasmic streaming, as it is called, does not occur in sieve cells. When it does occur in other plant cells it would surely speed transport of solutes from one part of the cell to another.

VI. The Mineral Nutrition of Plants

As long ago as the middle of the nineteenth century, Julius von Sachs (sometimes called the father of plant physiology), developed a system to grow plants with their roots immersed in a solution containing mineral elements rather than growing them in soil. This procedure is now called hydroponics. It provided a way to discover which elements were essential for normal plant growth and how much of each element had to be supplied. When these experiments were first carried out, it was already known that plants require hydrogen and oxygen from water (H_2O) and carbon from the carbon dioxide (CO_2) in the atmosphere. In the earliest experiments it was possible to grow plants in a solution that apparently contained only six elements: nitrogen, potassium, calcium, magnesium, phosphorous, and sulfur (in decreasing order of quantities required). It was also known (as early as 1840 when Justus von Liebig published the first book on agricultural chemistry) that traces of iron were also essential for plant growth.

During the early part of the twentieth century, as chemical purification procedures improved, it became apparent that other elements were also required for plant growth, these in very tiny quantities. Highly sophisticated purification techniques had to be developed to show that plants would not grow in the absence of boron, manganese, zinc, copper, nickel, and molybdenum. A healthy plant contains 60 million times as many atoms of hydrogen as molybdenum.

Some elements were especially difficult to study because they are so universally present in the environment. One of these was chlorine, which (in the form of sodium chloride) is present in dust particles, not to mention perspiration and other excretions from the humans who were studying its essentiality for plants. Thus, although it is required in even larger quantities than iron, it was the early 1950s before it was shown to be essential for plants. Sodium, which is required in rather large quantities by animals, is apparently essential only for certain kinds of plants (those that photosynthesize by the C-4 pathway, see Section VII). Boron, on the other hand, is clearly essential for plants but may not be essential for animals. Several other elements are required by animals (iodine, selenium, nickel, chromium) but may or may not be essential to plants; their essentiality for plants has not been demonstrated.

Although the traditional approach is to attempt to grow a plant in the complete absence of a suspected element (failure indicates the essentiality of that element), another evidence for essentiality is the presence of the element as part of some enzyme or other compound that is known to play an essential role in plant metabolism. Thus chlorophyll contains an atom of magnesium, and proteins and many other important compounds in plants and animals contain relatively large quantities of nitrogen as well as the carbon, hydrogen, and oxygen that form their basic structures. These would be sufficient reasons for essentiality of magnesium and nitrogen.

Studies of plant mineral nutrition have had a profound effect on agriculture. Knowledge gained has placed fertilizer application onto a sound scientific basis and has greatly increased the yields of agricultural crops.

VII. Photosynthesis

Plants are nearly unique among living organisms because they combine carbon dioxide and water to produce carbohydrates and ultimately all the other compounds of life in the process called photosynthesis. Pure oxygen is released as a by-product of the process, enriching our atmosphere with the gas essential for our own livelihood at the same time that the carbon dioxide that we exhale, too much of which can be harmful, is removed. The study of photosynthesis is

an extremely important part of the science of plant physiology. [See Photosynthesis.]

Photosynthesis proves to be extremely complex. There are at least fifty separate chemical reactions, and each one of these (with a very few exceptions) is controlled by a special protein, an enzyme. The role of enzymes, each one of which is a protein, cannot be overemphasized. Indeed, a fairly good definition of life states that life is a collection of chemical reactions, each controlled by its own enzyme. These wonderful proteins consist of tens to hundreds, even thousands, of amino acids attached to each other in long chains that are folded in such a way that a surface is produced on the protein where some substrate molecule or molecules can exactly fit, leading to some chemical change: perhaps a breakdown of the molecule or a combining of two molecules. There are thousands of different enzymes, probably as many as two thousand acting in any given cell at any given moment.

Most of the fifty or more steps in photosynthesis are controlled by enzymes. The first steps are photochemistry in which light energy is absorbed by chlorophyll and other molecules, exciting those molecules to a state where the energy from the light can be stored in the chemical bonds of certain molecules. Early in photosynthesis, water is broken down, releasing some of the oxygen into the atmosphere and moving the hydrogen across membranes in a way that produces highly energetic molecules abbreviated as ATP and NADPH. The energy of these molecules, with the help of enzymes, is then used to combine a molecule of CO_2 with a phosphorylated 5-carbon sugar (a sugar molecule with two phosphate groups attached). This produces a 6-carbon molecule that immediately breaks down to form two 3-carbon molecules. The enzyme that controls this reaction, often referred to as rubisco, is almost certainly the most abundant protein on earth, making up a considerable portion of the protein in green leaves.

The two 3-carbon molecules are then cycled through a complex series of reactions called the Calvin cycle to regenerate the 5-carbon phosphorylated sugar and eventually to produce 6-carbon sugars (after six turns of the cycle) that typically combine to produce the common table sugar, sucrose (consists of two 6-carbon sugars: glucose and fructose). Many other biochemical steps follow, including the synthesis of amino acids (which requires addition of nitrogen); the amino acids in turn combine to form protein, including more enzymes. Even the DNA (deoxyribonucleic acid) that constitutes the genetic material and ultimately controls the synthesis of all the different enzymes is ultimately a product of photosynthesis.

During the 1960s it was discovered that some plants carry out photosynthesis in a somewhat more specialized way. Certain cells in the leaves of such plants combine CO_2 with a 3-carbon molecule to produce a 4-carbon phosphorylated acid, which in turn is pumped into other cells in the leaf where the CO_2 is released to enter into the Calvin cycle. The 3-carbon molecule that is left over moves back to the first cell, where it combines with another molecule of CO_2. The end result is that CO_2 is concentrated in the photosynthesizing cells where the Calvin cycle takes place. This process works extremely well at warm temperatures and high light intensities; it also has advantages where nitrogen and water may be limiting. Plants that use only the Calvin cycle are called C-3 plants because the first product of CO_2 fixation is a 3-carbon molecule, whereas plants that transfer the 4-carbon molecule to photosynthesizing cells are called C-4 plants. Some C-4 crops are maize, sugarcane, sorghum, and a number of tropical and subtropical grasses that are used as forages. Most crops are C-3 plants.

A real advantage enjoyed by C-4 plants is that they avoid a process known as photorespiration. In this process, which is driven by light, oxygen molecules (O_2) compete with CO_2 molecules for the active site on the rubisco enzyme. The result is that photosynthesis in C-3 plants becomes less efficient at high temperatures and high light where more O_2 is produced. In C-4 plants, the concentrated CO_2 in photosynthesizing cells effectively competes with O_2 to prevent photorespiration.

A third kind of photosynthesis is called CAM (crassulacean acid metabolism) and occurs in succulent desert plants such as cactus and aloe. The only agricultural CAM plant of any significance is the pineapple. CAM photosynthesis is very similar to C-4 photosynthesis except that the fixation of CO_2 is separated from the Calvin cycle not in space but in time. In these plants, CO_2 is fixed at night when humidities are higher and temperatures are lower, resulting in less transpiration from the plant. When the sun comes up, the pores on the surface of the leaves called stomata (singular stoma) close, preventing further transpiration. In the light of day, CO_2 is released from the 4-carbon molecules and fixed in the Calvin cycle.

VIII. Respiration

Respiration in all living things involves a breakdown of organic molecules with the release of energy in a form that can be used to sustain the many activities of life. In a few organisms (yeast, certain bacteria), the process goes only part way and does not use oxygen; in most living organisms, oxygen is eventually combined with hydrogen to produce water, and carbon dioxide is released—the end results being just the opposite to those of photosynthesis. Like photosynthesis, however, respiration is a highly complex process involving at least fifty enzymatically controlled reactions or steps. Various organic molecules can enter into respiration, including proteins and fats, but the process is usually discussed in terms of the breakdown of carbohydrates, specifically the simple sugars glucose and fructose.

These sugars must first be "activated" by combination with phosphate. A series of chemical reactions breaks these activated sugars down into phosphorylated 3-carbon compounds with the release of some energy in the form of ATP (the same high-energy compound produced in the first steps of photosynthesis) and NADH (related to the NADPH of photosynthesis). If no oxygen is present, the 3-carbon molecules may be converted to alcohol (e.g., in yeast) or to lactic acid (in the muscles of an exercising athlete who uses energy faster than oxygen can be absorbed). Up to this point, the process is called glycolysis. If O_2 is present, the 3-carbon molecules release a molecule of CO_2, and the remaining 2-carbon molecules enter into a cycle called the Krebs cycle, the first product of which is citric acid. The citric acid (a 6-carbon acid) is gradually broken down to release two CO_2 molecules, and hydrogen is carried to still another series of chemical steps (the cytochrome system) where they are eventually combined with oxygen to form water and where several molecules of ATP (five times as much as in glycolysis) are produced.

Plant physiologists have studied respiration in plants as long as they have studied photosynthesis. It proves to be remarkably similar to respiration in animals although there are a few unique features. Both plants and animals have still another pathway of breakdown that involved 5-carbon sugars (the pentose phosphate pathway), and perhaps this pathway is somewhat more important in plants than in animals. Furthermore, plants have a cyanide-resistant pathway that can utilize oxygen in the presence of cyanide but without the toxic effects experienced by animals. Cyanide inhibits the cytochrome system in most living organisms, but many plants can bypass this. When such a bypass occurs, heat may be released although little ATP is produced. This occurs in certain voodoo lilies and their relatives that become quite warm to the touch; the heat volatilizes bad-smelling compounds that attract flies to these interesting plants. The flies then pollinate the flowers.

IX. Other Plant Metabolism

Although photosynthesis and respiration are extremely important to plant growth and thus to plant physiology, we should note that plants can synthesize thousands of products besides those that take part in photosynthesis and respiration. Many of these have important consequences for human beings, and some agricultural crops are grown because of these compounds. Tobacco (nicotine), coffee and tea (caffeine), coca (cocaine), poppies (opium), and many other plants may be grown commercially because of the specific drugs that they produce. Of course the food plants upon which most agriculture is based are also characterized by specific compounds that give them flavor as well as nutritional qualities. Plants are able to synthesize the amino acids required by humans and other animals, as well as all the vitamins. The vitamins often activate enzymes in both plants and animals, but though the plants are capable of synthesizing them, animals usually lack this capability.

A significant portion of plant biochemistry is devoted to the study of sulfur compounds as well as nitrogen compounds. Many fats and fatty acids and their derivatives are important, not only in human nutrition but also in industry. Some plant oils, for example, are used as lubricants; others are used to produce paints, varnishes, resins, etc. Natural rubber is an important plant product. All of these things and many more can be made by the complex, enzymatically controlled reactions of plants and provide a valid province for plant physiologists to study. [See PLANT BIOCHEMISTRY.]

X. Plant Growth and Development

The wonderful thing about living organisms is that they create themselves, and plants are no exception. The study of how plants increase in size (growth) and in complexity (development) is a fascinating part of plant physiology. It necessarily involves careful attention to plant anatomy, and because of the uniqueness of plant cell walls, growth and development proceed differently in plants than in animals. The cells are cemented to each other, so the form that they collectively produce as the plant grows depends on how they elongate and increase in diameter.

Higher plants, including the flowering plants but virtually all leafy land plants, have two primary organ systems, the root and the shoot. In seed plants, these are clearly discernable in the embryo. As the seed sprouts (germinates), both the roots and the shoots grow by adding cells at their respective ends by cell division followed by cell enlargement and elongation, often with branching, to produce the complex root and shoot systems that we readily recognize.

The dividing mass of cells at the shoot and root tips are called apical meristems, and they often have the potential to grow indefinitely. The root meristems produce a mass of intricately branched roots, and the stem meristems produce stems and leaves. In many plants, especially the woody plants, there are also lateral meristems that are layers of cells surrounding stems or roots and dividing to produce wood with its xylem to the inside and bark with its phloem to the outside. Grass leaves also have meristems at their base so that they can be cut off near the tips (by a grazing cow or a lawn mower) and continue to grow. Some plants have meristems between the nodes (points of leaf attachment) that function for awhile to cause the stem to elongate.

There comes a time in the life of nearly all plants when the shoot apical meristems (or branch apical meristems) stop producing stems and leaves and begin to produce flowers or other reproductive structures (cones in coniferous trees such as pine and spruce). This conversion from the vegatative (only stems and leaves) to reproductive (flowers, cones, etc.) provides a fascinating area for plant physiologists to study. The basic control systems (located ultimately in the genetic material) must be switched off and on in special ways that change the direction of growth of the meristems. Sometimes this seems to be internally controlled and sometimes it is a response to environmental factors such as day length and temperature.

Some plants germinate, grow, and produce flowers from all their meristems in one year, after which they die; these are called annuals. A few others grow vegetatively the first year and then (often in response to the low temperatures of winter) produce flowers and die in the second year; these are biennials. Others, such as shrubs and trees, never convert all their meristems to flowers, maintaining both vegetative growth and flower production from year to year; these are perennials. [See FLORICULTURE.]

XI. Plant Hormones and Growth Regulators

Beginning in the 1920s, plant physiologists became aware that plants produce compounds that can act at extremely low concentrations to control or influence various aspects of growth and development. Some of these compounds were apparently produced in one part of the plant and transported to another part where the control was exercised, much as hormones in animals are made in certain glands and transported to other parts of the animal's body where they have their effect. We now know that some hormones act in the same cells where they are made.

Thus the science of plant hormones (phytohormones) came into being. By now, vast amounts of information have been accumulated, although we still know virtually nothing about how the hormone or suspected hormone actually works. For example, it was assumed by most plant physiologists until the early 1980s that hormones in plants exercise their control as their concentration in various plant organs (sometimes the same organ where they were made) changed—much as happens with animal hormones. Now some plant physiologists are seriously considering the possibility that hormone concentration may change little or not at all, but that the sensitivity of the plant to the hormone may change in response to genetically controlled developmental programs within the plant or even in response to the environment. Of course it is possible that some hormones act as their concentrations change and others act as sensitivity to them changes. Time will tell about the relative importance of these two possible mechanisms.

Five groups of hormones have been most widely studied, although a number of others certainly exist and some of those have been studied at least in a preliminary way. The five groups are auxins, gibberellins, cytokinins, abscisic acid, and ethylene.

A. Auxins

The first phytohormone to come to the attention of plant physiologists in the Western world causes stem cells to elongate and is called auxin. The primary auxin of plants is thought to be indoleacetic acid (IAA), although there are certainly other auxins as well. During the 1930s and 1940s, it was shown that auxins can influence several aspects of plant growth and development as well as the elongation of young stem cells. (Mature stem cells have lost their sensitivity to auxin.) Examples include stimulation of rooting, inhibition of bud break (although this proves to be more complex than was originally thought), and promotion of fruit development (with the potential of producing seedless fruit when pollination is not allowed to occur and auxins are artificially applied).

Some auxinlike plant growth regulators (compounds that can act like phytohormones but do not occur naturally in plants) include important herbicides like 2,4-D and MCPA.

B. Gibberellins

Gibberellins were discovered in the 1920s by Japanese plant pathologists in cultures of a fungus that causes rapid elongation of rice stems so that the stems fall over (called *foolish seedling disease* in Japanese). These compounds came to the attention of the Western world first in the mid-1950s, and they are now known to be important phytohormones in all higher plants. They are complex molecules that consist of four or five rings with various side chains; well over 80 have now been identified and their structures determined. Only certain ones are active in causing stem elongation, and much is known about the enzymatically controlled transformations that occur among these compounds, ultimately producing the ones that influence plant growth. As with the auxins, these compounds also influence several responses besides stem elongation, but typically not the same phenomena as those influenced by the auxins. For example, they may cause otherwise dormant seeds to germinate, and they may induce flowering in plants that otherwise require long days and/or low temperatures (see following discussion). They have had some application in agriculture, such as the promotion of elongation in grapes so that the grapes are not only larger and longer but farther apart in the bunch, which helps them to resist certain fungi.

C. Cytokinins

Cytokinins were discovered in the 1950s. Their primary activity is thought to be the control of cell division, but they also produce several responses, including stimulating dormant buds to grow and directing the transport of assimilates and other compounds to parts of the plant where cytokinins are most concentrated (as demonstrated by applying a cytokinin to a spot on a leaf and observing the movement of substances toward that spot). They also delay the natural aging of plant tissues.

D. Abscisic Acid

Abscisic acid, discovered in the 1960s, sometimes shuts down certain plant processes. It may inhibit bud growth and germination, although the evidence is somewhat controversial. One important function is the closure of stomata when the plant is under water stress. As a leaf dries, the concentration of abscisic acid increases, and stomatal closure then reduces transpiration. Drying roots can also synthesize or release abscisic acid, which is transported to the leaves, thus closing the stomata.

E. Ethylene

Ethylene is a gas that also is produced by all higher plants and plays a role in several plant responses. Most noticeably, applied ethylene causes plant leaves to curl, as was observed as early as 1900 on trees growing near gas lights where ethylene was a contaminant in the natural gas. Ethylene causes flowering in pineapple and some other bromeliads, and it may play certain positive roles in plants as well as being a hormone that is involved primarily in plant responses to stress. Although the natural hormone is a gas, an ethylene-releasing substance (ethephon) is commercially available; it breaks down in neutral or alkaline solutions to release ethylene.

F. Other Hormones

Since the 1930s there has been evidence that flowering is induced by a special hormone, but attempts to isolate such a compound have so far failed. The evidence remains strong, however (see Section XIII). For many decades there has also been evidence that plants that respond to touch by folding their leaves (e.g., the sensitive plant) do so as a hormone is released by the touch and quickly transported to other leaves. Such compounds have now been isolated and identified; they are called turgorins. Compounds called brassinosteroids occur in plants and have powerful effects at extremely low concentrations; among other things, they may increase plant sensitivity to auxin. Much study is needed on these and other possible phytohormones.

XII. Growth Responses to Light and Gravity

Two growth responses that plant physiologists have studied for well over a century are the bending of stems toward light (called phototropism) and bending of stems away from and roots toward the earth's center of gravity (called gravitropism, formerly geo-tropism). It was the study of phototropism that led to the discovery of auxin. When grasses germinate, including cereals like wheat and oats, the first leaf that comes up through the soil is protected by a tubular, almost transparent sheath called the coleoptile. Charles Darwin, working with his son Francis, discovered in the 1870s that this was a good object to use for studies of phototropism because the coleoptile curved readily in a darkened room toward a light source off to one side. Frits Went in the Netherlands in 1926 removed several tips from oat coleoptiles under dim red light (toward which they do not bend) and placed these tips on and around a small block of gelatin. After some time, he discarded the coleoptile tips and placed the gelatin to one side of a coleoptile from which the tip had been removed. The side under the block grew more than the other side, causing the coleoptile to curve in the direction away from the block.

Went reported that the tips and then the block contained a stem growth substance, which he named auxin. He suggested that in normal phototropism, auxin is moved in the coleoptile tip toward the side away from the light, causing that side to grow more and thus accounting for bending toward the light. This hypothesis has been examined in many experiments since then, Although widely accepted, it has faced serious challenges during the late 1980s and early 1990s. Some researchers report that the amount of auxin, as measured by sophisticated modern methods, does not change in the coleoptile tip but rather that an inhibitor builds up on the side toward the light causing that side to grow less than the other side. (Went's initial experiment remains valid because the gelatin block must have contained more auxin than inhibitor.) It is interesting that after more than 100 years of study, we still do not have a definitive answer as to how plants grow toward the light! (At the same time Went announced his results, N. Cholodny in Kiev published the same hypothesis; hence, it is usually called the Cholodny–Went hypothesis.)

Another phase of the problem includes how the coleoptile absorbs the light in the first place. The first law of photochemistry says that for light to be effective it must be absorbed. What is the pigment in the almost transparent coleoptile that absorbs blue and ultraviolet light and not other wavelengths, eventually leading to the bending? Whatever it is, it must be present in very small amounts because the coleoptile certainly is not bright orange-yellow as it would be if a blue-absorbing pigment were present in large amounts. The best evidence suggests that the pigment is a flavoprotein, although carotenoids might also play a role if only by reducing some of the light that gets to the dark side of the coleoptile.

Went and his colleague Herman Dolk applied the transport hypothesis to gravitropism. It was suggested that plant stems and roots respond to gravity as more auxin accumulates in bottom cells of a stem or root laid on its side, promoting the growth of bottom tissues in stems (and thus causing an upward curvature) and inhibiting the growth of bottom tissues in roots (downward curvature). This hypothesis has also been examined by hundreds of experiments since the early 1930s when it was proposed, and again plant physiologists do not have a definitive understanding of how gravitropism works. Movement of auxin may indeed occur, but it also seems possible (and is supported by strong recent evidence) that gravistimulation causes a change in tissue sensitivity to auxin such that bottom tissues become more sensitive to the auxin already present while top tissues in a horizontal stem or root become less sensitive. Again, time will tell.

There is also controversy concerning the perception of gravity by the plant. Something within the cells must settle (or float) in response to gravity, conveying information about the direction of the gravitational stimulus. Around 1900 it was proposed that the settling objects within the gravity-detecting cells were starch grains in bodies called amyloplasts. Starch is more dense than the cytoplasm, so amyloplasts do indeed settle to the bottom of cells in stems or roots turned on their sides. Alternative proposals appeared during the intervening decades, but the idea seemed to stand the test of time until the 1980s, when mutant *Arabidopsis* (mouse-ear cress) plants were discovered that made no starch but responded to gravity anyway! These plants respond much more slowly, however, so it may be that starchless amyloplasts are settling slowly within the perceiving cells and accounting for the gravitropic response. Time will tell!

There are many other plant responses to light besides photosynthesis and phototropism. Study of the several responses is an important subfield of plant physiology. For example, many seeds germinate only after they have absorbed water and been exposed to light. Furthermore, seedlings grown in the dark are much different in appearance from comparable seedlings grown in the light. The dark-grown seedlings have elongated stems, reduced root systems, and tiny unexpanded leaves (in most cases) and are nearly colorless. Seedlings exposed to light are much shorter, have expanded leaves and larger root systems, and

are usually a healthy green color. In the 1940s and 1950s, it was shown that these responses are all under the control of a single pigment called phytochrome. This pigment absorbs orange-red light most effectively, and by doing so it changes to a form that best absorbs light of longer wavelengths in the so-called far-red part of the spectrum, just visible to the human eye. (Even longer wavelengths are called infrared.) It is the far-red-absorbing form of phytochrome (P_{fr}) that accounts for the responses just described. This ubiquitous pigment accounts for many other responses as well, such as development of color in apples and more rapid growth of maturing plants (as contrasted with seedlings) in the shade of other plants. (Leaf shade is enriched in far-red light.) As in the other cases described here, there are many complications that cannot be discussed in this abbreviated space; for example, though phytochrome responds to low light levels applied over relatively short intervals, there are comparable responses to much higher light levels applied for longer times (the so-called high irradiance reaction: HIR).

XIII. Temperature, Day Length, and Plant Clocks

Plants also respond in developmental ways to changes in temperature and to the relative durations of light and darkness in the daily cycle, which implies that plants have the ability to measure time—that they have a biological clock. As in the cases already discussed, these plant phenomena provide numerous opportunities for study by plant physiologists.

Of course the rate of growth of these "cold-blooded" organisms is strongly influenced by temperature: growth increases with increasing temperature up to an optimum (usually between 25 and 40°C, depending on species), but there are also many interesting developmental responses. Seeds of many species (including agriculturally important fruit trees and many native plants), for example, may require a period of low temperature before they will germinate. They must have absorbed water, and the most effective temperatures are usually a little above the freezing point of water. This prevents many seeds from germinating precociously in autumn; the cold temperatures of winter are required before germination will occur. Dormant buds of deciduous plants often respond in a similar way. Such treatments applied to seeds or buds are collectively called prechilling.

A somewhat similar response to a low-temperature treatment, of moist seeds of some species and of mature plants of other species (e.g., the biennials mentioned earlier), is the promotion of early flowering. This response is called vernalization. Winter cereals not exposed to low temperatures, for example, may form flowers after about 14 weeks in the field; with a vernalization treatment, flowers may appear within 7 weeks. Sugar beet plants (biennials) grow vegetatively during their first year and respond to the cold of the first winter by sending up a flowering shoot (a process called bolting) during the second summer. If the sugar beet plants are never exposed to the low temperature, they never bolt but remain vegetative indefinitely (e.g., for years in a greenhouse experiment).

After the sugar beet plants have been vernalized, they are also promoted in their flowering by exposure to long days; if the days are short, flowering is delayed. Plant response to day length is called photoperiodism; it was discovered by USDA scientists in 1920 (and by a French scientist a few years before, although his work was largely overlooked). The USDA scientists found that many plants were promoted in their flowering as days became longer (e.g., cereals after vernalization, black henbane, radish), others by days that were getting shorter (e.g., chrysanthemums, soybean, cocklebur, rice), and some plants seemed to be little or not at all affected by day length (e.g., tomatoes, cucumber, sunflower). These three groups were called long-day plants, short-day plants, and day-neutral plants. (Note that it is not the absolute day length that is important. Cocklebur plants flower when days are *shorter* than about 15 hr; henbane plants flower when days are *longer* than about 12 hr—both flower when days are between 12 and 15 hr.)

Much study has been aimed at understanding photoperiodism. One question concerns how the plants detect the difference between day and night. The phytochrome system is clearly involved. Short-day plants may be thought of as responding to long nights (although the day length also plays a role), and if this night is interrupted with a brief interval of light (sometimes only a few seconds if the light is bright enough), flowering will be inhibited. Orange-red light is most effective, and far-red light reverses the effect just as it does in the phytochrome responses mentioned in the previous section. Long-day plants are promoted in their flowering when a dark period is interrupted with red light, although it usually takes a little more light to influence long-day plants than short-day plants. Many authors and a few scientists

have assumed that the phytochrome system is actually measuring the length of the light or dark periods. Strong evidence suggests that this is not the case. The phytochrome system is the link between the environment and some more subtle biological clock, which we will discuss in a moment.

It was shown in the 1930s that leaves respond to the length of day although it is the apical meristems that actually become reproductive. Thus some kind of signal must be transmitted from the leaves to the meristems. This is one piece of evidence that a flowering hormone (called florigen) is involved, but as mentioned earlier, such a hormone remains to be isolated and studied although much effort has been expended in attempts to do so.

The biological clock has many manifestations besides its role in photoperiodism (and perhaps also in a temperature response called thermoperiodism). For example, many plants display a daily up-and-down movement of their leaves. Typically, the leaves are more or less horizontal during the day and vertical at night. If these plants are placed in a perfectly uniform, dark environment, the leaf movements will continue for several days (also in a constant light environment although for a shorter time). The intervals between the most vertical position are not exactly 24 hr; the intervals may be more like 25 to 26 hr. Thus, when the clock is allowed to run freely, it usually runs slow. In nature, the clock is reset every day by the change from dark to light (dawn) and from light to dark (dusk). Similar responses occur in virtually all plants and animals, and studies of the biological clock are extensive. (Although initial discovery of the clock, in the 1920s, was through leaf movements of plants, the clock is perhaps less studied now by plant physiologists than animal physiologists.)

XIV. Plants under Stress

An important phase of plant physiology examines how plants respond to environmental conditions that are not optimal for their growth. This subfield, called stress physiology, has especially important applications in agricultural science. In many parts of the world, the rains may fail or never be frequent enough to support a crop (requiring irrigation). Some crops resist the drought better than others. Many millions of hectares of the earth's surface now suffer from too much salt in the soil. Some crops are much more resistant to this salt than others (but the salt must ultimately be reduced if the soils are to become highly productive regardless of crop resistance). In other areas of the world, low temperatures provide a serious stress when they occur during the growth period, for example, the spring frosts that may freeze developing fruit buds in orchards. High temperatures (e.g., in deserts), ultraviolet light, and mechanical stresses caused by wind or human or animal movements are other stresses studied by plant physiologists. Perhaps the most interesting recent discovery in this field is that many plants exposed to low temperatures, drought, or high salt in the soil produce a special set of so-called stress proteins. The exact role of these proteins remains to be determined, although some suggestions have been made. [*See* PLANT STRESS.]

XV. The Molecular Biology of Plants

During the 1980s and especially the early 1990s, plant physiologists have increasingly applied the techniques of molecular genetics (often called biotechnology) to many problems of plant physiology. The basic approach is to isolate a gene that controls some response, study the protein that might be synthesized under the control of that gene, and then determine what role the protein is playing (e.g., what enzymatic reaction it might control). This approach has advanced many frontiers of plant physiology, and it is now likely that over half the papers presented at national meetings devoted to plant physiology report results of studies that use techniques of biotechnology. This trend will continue; indeed, it may well revolutionize the entire field if it has not already. [*See* PLANT BIOTECHNOLOGY: FOOD SAFETY AND ENVIRONMENTAL ISSUES; PLANT GENETIC ENHANCEMENT.]

Bibliography

Briggs, W. R., Jones, L., and Walbot, V. (eds.). (1993). "Annual Review of Plant Physiology and Plant Molecular Biology," Vol. 44. Annual Reviews Inc., Palo Alto, CA. [See other volumes as well.]

Galston, A. (1994). "Life Processes in Plants." Scientific American Library, New York.

Salisbury, F. B., and Ross, C. W. (1992). "Plant Physiology," 4th ed. Wadsworth, Belmont, CA.

Tiaz, L., and Zeiger, E. (1991). "Plant Physiology." Benjamin/Cummings, Redwood City, CA.

Wilkins, M. B. (ed.). (1984). "Advanced Plant Physiology," 2nd ed. Pitman, London.

Plant Propagation

DAVID W. BURGER, *University of California, Davis*

Glossary

Adventitious organ Organ that arises from an unusual, abnormal, or unexpected location, for example, roots growing from leaves, petioles, or stems

Chimera Genetically distinct tissues growing adjacent to one another; from Greek mythology—a fire breathing monster with a lion's head, goat's body, and serpent's tail

Clone Population of plants derived asexually from a single individual and maintained using asexual means, thus making all members genetically identical; also, any single member of the asexually propagated population

Embryogenesis *De novo* formation of an embryo from unorganized cells

Differentiation Development of many cells or tissues from one cell that acquire and maintain specialized functions

Fully imbibed seed Seed that has absorbed the necessary water to initiate the germination process

Interstock Stem piece inserted between the scion and the rootstock

Meristem Undifferentiated tissue made up of cells that divide to form specialized tissues and organs

Micropropagation Aseptic process in which asexual means are used to reproduce many whole plants from cells, tissues, or organs of a single plant on culture media *in vitro*

Organogenesis *De novo* formation of an organ from unorganized cells

Propagule Propagation unit, such as a seed or cutting, used to reproduce a whole intact plant

Scarification Any mechanical or chemical technique used to soften the seed coat for improved water uptake and gas exchange

Scion Detached branch or bud comprising the upper portion of a graft or bud union that ultimately forms the stem and branches

Somatic Nonreproductive, vegetative tissue such as leaves, stems, and roots arising through mitosis

Totipotency Demonstrated capability in plants of a single isolated cell to develop into a whole plant

Propagation is the reproduction of new plants from existing plants or plant parts by sexual or asexual means. The fundamental objectives of plant propagation are (1) to increase the number of a particular plant and/or (2) to perpetuate the essential character of the plant. When the flower or inflorescence is involved as an essential organ in the increase and there is an exchange of genetic material (e.g., sperm, eggs), propagation is sexual. When there is no exchange of genetic material, propagation is asexual or vegetative. The propagation of plants in agriculture includes seed and seedling production, the production of bulbs and other modified stems and roots, the production of plants growing on specially selected rootstocks by grafting and budding, the production of plants growing on their own roots resulting from layering or cuttings, and the production of entire plants from plant cells, tissues, or organs by tissue culture.

I. Sexual Propagation

Seed (sexual) propagation is the primary method by which most plants naturally reproduce, and is one of the most widely used propagation methods for agricultural crops. Seed propagation is a common means of propagating self-pollinated plants and is often used for those that are cross-pollinated. The initial step in seed propagation is the planting of the seed, but the seed itself is the end result of a long process of growth and development usually involving pollination, fertilization, zygote formation, and embryo development.

A. Seed Development

The sexual cycle includes the development of male (pollen) and female (ovule) structures. On flowering, pollen grains are transferred to the stigma where they germinate if suitable conditions exist (related plant, receptive stigma, etc.). Two generative nuclei move through the pollen tube toward the ovule housed in the embryo sac and join with egg nuclei contained within the ovule. Fertilization occurs when one of the generative nuclei (male gametes) fuses with the female egg cell, creating a zygote combining the genetic material of the male and female parents. The zygote ultimately develops into an embryo (embryogenesis). The other generative nucleus fuses with polar nuclei in the embryo sac resulting in a triploid endosperm, a tissue utilized as a nutritive food source during germination. The nucellus, a diploid maternal tissue, is sometimes also associated with the endosperm as a storage tissue. The integuments surrounding the embryo sac become the seed coat (testa). The embryo consists of a shoot apical meristem (plumule), an embryonic axis, at least one cotyledon (seed leaf) attached to the axis, and a root apical meristem (radicle).

Normally, pollination and fertilization must both occur to produce a single embryo in a viable seed. In some plants (citrus, mango), however, the nucellus is stimulated to form somatic embryos during fertilization, resulting in one or many embryos of maternal origin growing beside the single zygotic embryo, resulting in polyembryony. Apomixis, seen in Kentucky bluegrass (*Poa pratensis*), is similar to polyembryony except that embryos form from unfertilized eggs. Embryos in seeds produced from nucellar polyembryony or apomixis have not undergone any exchange of genetic material and are, therefore, genetically identical to the maternal parent or to clones of the maternal plant.

B. Seed Anatomy

A seed consists of three basic parts: an embryo, food storage tissues, and a seed coat. The embryo and food storage tissues are products of fertilization. The seed coat is usually several cell layers thick and can be very hard as a result of cell wall sclerification. The seed coat

of a mature seed protects the embryo from mechanical damage, provides a means for long-term storage, and facilitates dispersal. It can, however, also present a barrier to germination depending on its physical or chemical properties.

C. Seed Dormancy

Seeds may or may not be ready to germinate when they are formed. Many seeds have a protective mechanism (dormancy) to prevent them from germinating during suboptimal times of the year. Seed dormancy may be maintained by physical properties of the seed coat, chemical properties of the seed itself, physiological properties of the embryo, or some combination of these factors. [*See* DORMANCY.]

Seed coat dormancy is usually due to a heavily sclerified, impermeable seed coat. Hard seed coats can be softened using hot water, fire, or concentrated acid, or by mechanical abrasion using a file or sandpaper. The goal of these scarification treatments is to make the seed permeable to water and oxygen and/or to make it possible for growing embryos to overcome any physical resistance of the seed coat so the germination process can commence.

Dormancy caused by chemical inhibitors in the seed can be broken with a moist, cool treatment of the seed called stratification. Stratification treatments mimic the cool/cold winters seeds normally experience in nature when they are produced in the summer or fall and germinate the following spring. To stratify seeds, fully imbibed seeds are mixed with a material (e.g., peat moss, sand, vermiculite, perlite) that will stay moist, but will also provide adequate aeration. The mixture is placed in a plastic bag and stored at temperatures between 0° and 10°C for varying lengths of time (usually 4–10 wk) depending on the species. During this stratification period, seeds may go through an after-ripening stage in which the embryo fully develops or internal chemical changes take place that allow germination to occur when the temperature rises.

The growth regulator gibberellic acid can be used to substitute for a stratification treatment for many seeds. Gibberellic acid has been shown to overcome internal physiological dormancies and stimulate germination in seeds with dormant embryos.

Leaching can also be used to overcome dormancies caused by chemical inhibitors found in the seed coat or surrounding fruit tissues. Leaching can be done by soaking seeds in water that is changed at least every 12 hr. Preferably, seeds can be washed under running water for 1–2 days.

All these dormancy breaking treatments mimic strategies nature has employed to prevent premature germination in the fall and hasten germination in the spring. Alternate freezing and thawing, soil-borne microorganisms, and passage of seeds through the digestive tracts of animals can gradually soften hard seed coats; heavy spring rains can leach out germination inhibitors and changing light intensities and day lengths can speed germination.

D. Seed Physiology

Plants take varying lengths of time to complete their life cycle from one seed generation to the next. Annual plants complete their life cycle in one growing season and then die. Biennials require two seasons to complete their life cycle. The first season is characterized by vegetative growth and development; during the second season, usually after a cold period, biennial plants flower and form seeds. Perennial plants live for more than two seasons and can live for hundreds of years. They complete the reproductive seed-to-seed cycle each year, yet always maintain the ability to grow vegetatively.

Seeds of annuals, biennials, and perennials all go through similar processes during germination. Several conditions must be met before a seed will germinate. First, the seed must be viable. It must have a living embryo and the embryo must be developed to a stage from which it is capable of growing. Next, conducive environmental conditions must exist around the seed: water must be available, the temperature must be within the range appropriate for the seed, and oxygen must be available. The appropriate light level must be present which may mean, for some seeds, no light. Finally, no internal or external barriers to germination can be present; the seeds must not be dormant. Seed germination proceeds continuously, but has well-defined stages (see Table I). Variations in the environmental parameters important to seed germination (water, temperature, oxygen, and light) can positively or negatively affect the rate of germination and the germination percentage.

E. Seed Handling

Recent advances in seed preparation and coating technologies have improved the uniformity of seed germination for many plant species and have enhanced our capability to handle seeds mechanically. Seeds are now routinely graded using various physical properties such as density, length, size, shape, and weight to

TABLE I

Stages of Seed Germination and Important Characteristics of Each Stage

Stage 1 Imbibition of Water	The moisture content of the seed increases as water is taken in by diffusion; seed tissues soften and swell; tremendous pressures can be generated, usually capable of breaking the seed coat; triggers the rapid synthesis of new enzymatic proteins and reactivates stored enzymes
Stage 2 Digestion and Translocation	Stored complex materials (carbohydrates, proteins, fats) are enzymatically converted to simpler substances and are translocated to the embryo
Stage 3 Seedling Growth	Cell division and enlargement commence in the shoot and root apical meristems; the radicle emerges growing downward and the shoot begins growing upward

separate seeds from other tissues and inert materials. Seeds may be altered (e.g., de-tailing of marigold seeds) and then coated with special films or graphite to facilitate easier handling by automated seeding machines. Small seeds may be coated with manufactured clay materials, creating pellets that may include nutrients or fungicides to enhance germination success. Pelleting is performed to increase seed size and to improve the uniformity of a seed lot; however, pelleting may prove a barrier to water if the coating is too hydrophobic and to oxygen if it is too hydrophilic. Film coatings are usually applied to seal in fungicides and provide dust-free handling. Coatings come in a variety of colors so seed lots can be color-coded, making them less prone to animal consumption since they are not recognized as food. Graphite coatings provide a smooth exterior surface and add weight. These coatings are particularly helpful when seeds are used in mechanical seeders (Fig. 1).

Seeds can be germinated before they are sown. Pre-germinated seeds with radicles 1–2 mm in length that are embedded in a viscous liquid to protect from desiccation may be used. This process is called fluid drilling and can improve germination rates and reduce germination times.

Another advance to ease seed handling and planting is the seed tape, in which seeds are attached to or embedded in a water-soluble tape at the recommended spacing. The tape is laid in the furrow at the proper depth and covered with soil. Seed tapes are

FIGURE 1 Automated seeder placing seeds into rows of cells in a plug tray.

readily available for vegetables and, although expensive, reduce the labor involved in seed handling.

Seed quality is an important aspect for the effective use of seeds for plant propagation. Here, quality is defined as the purity of the seed lot and the viability of the seeds. Purity is expressed as a percentage of other materials in the seed lot. Inert materials, seeds from other species and noxious weed seeds should constitute a very small percentage of the seed lot. High-quality seed packages provide information such as botanical name, purity, germination percentage (viability), germination times, collection date and location, and planting instructions.

Seed viability is an important measure of seed quality and is determined by cutting open a sampling of seeds from a seed lot (the cut test), by direct germination tests, or by indirect viability tests. Cut tests show whether seeds contain intact, fully developed embryos. Germination may be tested directly by sowing a small known number of seeds from the seed lot in moistened paper towels or filter paper and counting germinated seeds. Viability may be measured indirectly using tetrazolium chloride (2,3,5-triphenyltetrazolium chloride or TTC). Viable seeds soaked in a solution of TTC will turn red if they are viable. Living tissue converts colorless TTC into formazan, an insoluble red compound. Samples from seed lots can be treated with TTC and the number of red-colored versus noncolored seeds counted. A very simple test for seed viability is to float the seeds on water. Often seeds that float have poorly developed or aborted embryos whereas those seeds with complete embryos and dense endosperm sink. This test is not always effective, since all small seeds tend to float due to surface tension and those seeds with "wings" or "tails" often will not sink even when viable.

F. Seed Sowing and Planting Techniques

Seeds may be planted directly in the field or they may be started in a protective environment (e.g., greenhouse, cold frame) and transplanted to the outdoors. In either case, seed bed preparation must provide a moist environment in which seeds are in intimate contact with the soil. Seed beds also need to supply adequate aeration and nutrition and be free of disease, insects, weeds, and any other soil-borne problems. Seeds must be planted at the correct time of year depending on the species. Planting depths also vary with species. Some plants require complete darkness to germinate; therefore, they should be planted to a depth 2–4 times their diameter. Seeds requiring light to germinate can be placed on the seed bed surface and lightly shaken or raked in. Sowing rates (e.g., seeds per unit area) also depend on the species and can affect the quality of germinated seedlings. Viability percentages should be included in the calculations to determine the desired seedling density.

G. Diseases Associated with Seed Germination

The environment for seed propagation is quite suitable to the development of pathogenic organisms, especially fungi. Warm temperatures, readily available moisture, and young succulent plant tissues can often result in devastating losses due to disease. Damping-off is a term used to describe the rapid death of germinating seeds and young seedlings and is characterized by stem necrosis at the soil surface. Several fungal organisms cause damping-off, including *Pythium ultimum*, *Rhizoctonia solani*, *Botrytis cinerea*, and *Phytophthora* spp. These organisms are usually soilborne; however, infected plant tissues and seeds may also contribute to the spread of damping-off problems. Control measures for damping-off include the complete elimination of viable pathogenic organisms from the germination environment, heat pasteurization or fumigation of germination media, and careful control of the environment around germinating seeds. [*See* PLANT PATHOLOGY.]

II. Asexual Propagation

A. Comparison between Sexual and Asexual Propagation

As the name implies, asexual propagation is a means of increasing the numbers of plants without any exchange of genetic material. The resulting plants are theoretically identical to the plant from which they originated, based on the concept of totipotency. Therefore, a population of asexually or vegetatively propagated plants will be quite uniform. Depending on the plant being propagated, sexual and/or asexual propagation methods may be selected (Table II).

B. Methods of Asexual Propagation

Specialized vegetative structures can be utilized to asexually propagate plants. Naturally detachable

TABLE II

Comparisons of Sexual and Asexual Propagation Methods and Benefits of Each

Sexual propagation	Asexual propagation
Variability in seedling populations	Uniformity of clones
Less expensive	Requires skill
More convenient	Faster and easier
Some plants cannot be separated, divided, grafted, or rooted	Some plants produce no viable seed
Seeds can be stored for long periods of time	Takes advantage of rootstock characteristics
Less likely to transmit diseases	

structures (e.g., bulb, pseudobulb, corm) can be separated from parent plants to begin new plants. Other organs (e.g., rhizome, tuber, tuberous root, tubercles) may require cutting or dividing to produce new propagules. In either case, the resulting plants are genetically identical to the plant from which they came—the parent plant. Separating and dividing techniques have been practiced since the beginning of the agricultural age.

1. Asexual Propagation by Separation

Bulbs are specialized, underground organs consisting of a short, fleshy stem axis enclosed in thick, fleshy scale leaves. Tulips, daffodils, lilies, onions, and garlic are examples of bulbs. Pseudobulbs are specialized organs produced by many orchid species consisting of an enlarged, fleshy section of stem made up of several nodes. Corms (e.g., gladiolus and crocus) are swollen underground stems enclosed in dry, scale-like leaves and are predominantly made up of stem tissue with distinct nodes and internodes. All these underground organs are useful in propagation by separation.

2. Asexual Propagation by Division

Asexual propagation by division occurs with the use of rhizomes, tubers, tuberous roots, and tubercles. Rhizomes are specialized stems that grow horizontally at or just below ground level. Rhizomatous plants (e.g., bearded iris, bamboo) are propagated by dividing the many attached rhizomes into singular pieces containing a stem axis and associated roots. Tubers (e.g., potato, caladium, Jerusalem artichoke) are swollen underground stems that function as storage organs. The "eyes" of a tuber are nodes or small buds subtended by leaf scars. Tuberous roots (dahlia, sweet potato, cassava) and tubercles (begonia) are similar to tubers except that tuberous roots are below-ground swollen root tissue rather than stem tissue and tubercles occur above ground rather than below ground. Plants with tuberous roots or tubercles are propagated by dividing these structures into smaller sections, each containing a shoot bud.

3. Asexual Propagation by Layering

Layering is a long-practiced method of asexual propagation in which adventitious roots are induced to form on plant stems while they are still attached to the parent plant. Layering is the natural method of asexual propagation in some plants (e.g., black raspberries, trailing blackberries). For plants in which layering is a natural occurrence, specialized organs such as runners, stolons, offsets, or suckers may make layering possible. Layering is performed for various reasons. (1) It may be the natural asexual propagation method worth exploiting to maximum advantage. (2) Other asexual methods may not be successful or possible with some plants. (3) Relatively large plants can be produced in a short time. (4) Layering does not require extensive facilities or equipment; however, it does require considerable labor and time. The several different layering techniques include tip layering, simple layering, compound layering, air layering, mound or stool layering, and trench layering. The procedures for each type of layering technique differ, but basic similarities exist among them. First, all layering methods involve the manipulation of a branch that is still attached to the parent plant. The branch may be bent, wounded, or girdled in some way to help induce the formation of roots. Next, the area where roots are to form is exposed to an environment conducive to root formation (i.e., buried in soil or wrapped in a moist substrate). Finally, once roots have formed, the new offshoot is removed from the parent plant and transplanted to new soil to grow on its own.

4. Asexual Propagation by Budding and Grafting

Budding and grafting involve human intervention to combine two or more pieces of plant tissue so they will grow and develop as one plant. Generally, two pieces are used: a scion that develops into the aboveground portion of the plant and a rootstock that becomes the root system. In grafting, scions are usually whole or relatively large sections of a branch whereas in budding, the scion is a single bud. Budding and grafting techniques have been in practice for more than 3000 yr. The ancient Chinese practiced the art of grafting and Aristotle wrote quite knowledgeably on the topic. Modern propagators practice budding and grafting for several reasons. (1) Budding and grafting may be the only effective choice to asexually propagate some plants. (2) Scion characteristics (e.g., flower/fruit quality, branching habit, disease tolerance/resistance) and rootstock characteristics (tolerance to soil conditions such as low aeration, salinity, alkalinity, disease, and insect resistance) can be chosen to best suit the environment in which the plant will grow. (3) Budding and grafting provide the opportunity to change an established field from one scion to another without removing the established root systems. (4) Damaged plants can be repaired. Rootstocks can influence the quality and/or characteristics of the scion, such as yield and fruit quality.

Not all rootstocks and scions of a particular genus can be budded or grafted. When anatomical, physiological, or biochemical barriers exist between a particular scion and rootstock they are said to be incompatible. In some cases, when a scion and rootstock do not join together successfully because of some kind of incompatibility, an interstock may be used between them. Several factors can affect the success of a particular scion/rootstock union including incompatibility, temperature, moisture, season, disease, and oxygen. The many different types of budding methods include T-budding, inverted T-budding, patch budding, and chip budding. Grafting methods include whip grafting, splice grafting, side grafting, cleft grafting, wedge grafting, bark grafting, approach grafting, inarching, and bridge grafting.

5. Asexual Propagation by Cuttings

Asexual propagation by cuttings is the most common type of vegetative propagation. A cutting is a part of the plant that is treated to stimulate the regeneration of missing organs (Fig. 2). The formation of adventitious organs is the basis for cutting propagation and involves the concept of totipotency. With stem cuttings, adventitious roots may develop from preformed root initials or they may be the result of cellular dedifferentiation and redifferentiation phenomena in the stem. The many types of cutting other than stem include leaf (consisting of the leaf blade with or without the petiole), leaf-bud (consisting of the leaf and part of the stem containing an axillary bud), and root. To realize a new plant, stem and leaf bud cuttings must initiate a new root system, root cuttings must initiate a new shoot system, and leaf cuttings must initiate both a new shoot and a new root system. Adventitious organ formation is not an uncommon natural occurrence. Brace roots on corn and aerial roots on English ivy (*Hedera helix*) are examples of naturally occurring adventitious roots.

The four types of stem cuttings are herbaceous, softwood, semi-hardwood, and hardwood. These categories reflect the kind of wood used to make the cutting. Each type of cutting has distinctive characteristics and rooting responses (Table III). Generally, herbaceous and softwood cuttings root easily in a relatively short period of time, whereas semi-hardwood and hardwood cuttings take longer to root and may have reduced abilities to form adventitious roots. This is a broad generalization with many exceptions, since rooting potential is an inherently unique characteristic for each species. Stem cuttings containing the stem tip or apical meristem are called tip

FIGURE 2 Rooted stem cutting of walnut (*Juglans hindsii* × *J. regia* 'Paradox').

or terminal cuttings, whereas those not containing an apical meristem are called basal or subterminal cuttings. Root formation may differ depending on whether the cutting is terminal or subterminal.

Asexual propagation with leaf cuttings is possible with some plants such as African violet, *Sansevieria*, and begonia. Since most leaf cuttings contain no meristem, *de novo* synthesis of a meristem is required through dedifferentiation and redifferentiation. The new plant will have the genetic makeup of the leaf cells forming the new meristem. If the leaf is not genetically uniform (as in the case of a chimera), the unique characteristics of the plant may be lost through leaf cutting propagation. An example of this is the *Sansevieria* clone "Laurentii" that has a yellow leaf margin. The yellow margin is genetically distinct from the rest of the leaf. Leaf cuttings made from this plant will lose the yellow leaf margin attribute since only leaf cells in the middle area of the leaf undergo *de novo* meristem formation and form new shoots.

Plant growth regulators can be used to enhance the formation of adventitious roots in stem cuttings. Auxins are the most commonly used growth regulators for this purpose. Indole-3-acetic acid (IAA), a naturally occurring auxin, can enhance rooting but the synthetic auxins indole-3-butanoic (indole-3-butyric) acid (IBA) and naphthaleneacetic acid (NAA) have been shown to be more effective. These compounds are mixed with talc and are applied as a powder to the base of stem cuttings before they are placed in a rooting medium, or they may be dissolved in a solvent and applied as an aqueous solution (Fig. 3). NAA and IBA are the active ingredients in most commercially

TABLE III
Cutting Types, Characteristics, and Rooting Responses

Characteristic	Cutting types			
	Herbaceous	Softwood	Semi-hardwood	Hardwood
Description	Succulent stems from nonwoody plants	New, tender growth on woody species	Partially matured wood from current season's growth	Dormant, leafless stems or evergreen conifers
Time of year	All year	Spring–early summer	Summer–early fall	Late fall–early spring
Cutting length	<15 cm	5–15 cm	5–15 cm	5–75 cm
Relative humidity	Humidity tent, mist or fog	Humidity tent, mist or fog	Humidity tent, mist or fog	Direct field planting, "stool" beds
Examples	Begonia, coleus, chrysanthemum, sugar cane	Lilac, pyracantha, crepe myrtle	Citrus, camellia, rhododendron	Walnut, grape, fig, willow, poplar
Comments	Easy, quick, most susceptible to disease	Most root in 2–6 wk; bottom heat may improve rooting	Most root in 2–12 wk; wounding, auxins enhance rooting	Roots form after callus; bottom heat may help; inexpensive

FIGURE 3 Application of auxin as a liquid to a group of stem cuttings. Spraying the solution onto the cuttings instead of dipping cuttings into the solution prevents the spread of disease and contamination of the solution.

available powders, talcs, or liquids. IBA can also be obtained as a potassium salt that is freely soluble in water. If solutions are used, cuttings can be treated in high concentrations (1,000 to >10,000 ppm) for short periods of time (5–60 sec, known as a quick dip) or they can be treated in low concentrations (10–500 ppm) for longer periods of time (1–24 hr, known as a slow dip or soak). Once treated, the cuttings are placed in a flat containing a pasteurized rooting medium usually composed of a combination of peat moss, vermiculite, perlite, and/or sand. The flat is then placed on a rooting bench under mist. Bottom heat may be provided, if necessary, to accelerate the rooting process.

Other techniques may be used instead of or in combination with growth regulators to enhance adventitious root formation. Wounding of the cut stem base stimulates root formation in some plants. Stems are wounded by cutting into the stem with a sharp knife, by scraping the outer surface of the stem, by splitting the base of the cutting, or by slicing away a portion of the stem base. Wounding may improve rooting by improving water or growth regulator penetration into the stem or by reducing physical barriers in the stem that may inhibit the development and emergence of newly formed adventitious roots. Shading parts of

the plant where cuttings are to be taken with opaque tapes or Velcro™ strips, an etiolation technique known as banding or blanching, can improve rooting in some plants.

Several factors can influence the formation of adventitious roots from cuttings. The stock plant from which cuttings are obtained must have no nutritional deficiencies, have no disease infections or insect infestations, and be under no environmental stress (e.g., water, temperature, light). Stock plants can be managed to maximize rooting success by careful irrigation, fertilization, and pruning methods. Pruning can be done to increase axillary bud break, thus increasing the number of cuttings available. Plants must be fertilized, but not so cuttings are rapidly growing when they are taken. A thorough watering can be done before cuttings are taken to insure that the cuttings are not water stressed. Avoiding stock plants that are in flower is usually best. In some plants, cuttings from flowering shoots do not root as well or as fast as those from nonflowering shoots.

The environment in which cuttings are rooted plays an important part in the overall success of cutting propagation. Adequate medium moisture and relative humidity, light, and temperature must all be provided to root stem cuttings. Adequate moisture is probably the most important factor, especially at the beginning of the rooting process. Water must be available at the cut stem to allow for water uptake there. At the same time, oxygen must be available for adventitious root formation. Therefore, the rooting medium must be able to hold water well while providing adequate aeration. Rooting media composed of ingredients having large particle sizes (e.g., perlite, vermiculite, coarse sand) can supply the required moisture and air for the rooting process.

Important interrelationships exist between light and temperature during the rooting process. Photosynthesis is important for the cuttings to meet the heavy respiratory demand of the rooting process. However, excess light and its associated heat may increase transpiration above what the cutting can withstand and can be detrimental to rooting. Ideally, the temperature of the rooting medium should be higher than the ambient temperature by 10°F to 15°F. This differential soil/air temperature can help optimize rooting since higher temperatures around the cutting base will enhance adventitious root formation and low temperatures around the leaves will reduce evaporation, transpiration, and respiration of the aboveground parts.

High relative humidity must be maintained around cuttings as they root to reduce the water lost through

FIGURE 4 Propagation benches covered with polyethylene to increase the relative humidity around propagules.

transpiration. Relative humidity can be increased by placing the cuttings in plastic-covered (Fig. 4) enclosures, by using high-pressure fog systems, or by using low-pressure mist systems (Fig. 5). Mist systems are most commonly used to increase relative humidity. The mist system may be controlled by time clocks that activate the system at regular intervals. A combination of two clocks is often used: one clock controls the frequency and length of each mist application and the other clock turns the first clock on at dawn and off at dusk. The time settings must be changed manually by the propagator throughout the year. A common setting for the mist clock is for mist to be applied for 5 sec every 5 min.

Devices that simulate leaf evaporation can also be used to control mist systems. These "electronic leaves" are based on the weight of water (balance type) or the conductivity of water (electrode type). The balance-type mist controller is composed of a screen attached to the end of a lever that activates a mercury switch (Fig. 6). As water evaporates from the screen (the simulated leaf), the lever rises until it reaches a point at which the mist is turned on. Mist accumulates on the screen, causing it to go down and turn the mist off. The electrode-type controller is composed of a platform containing two electrodes separated from one another. As long as a water film is present on the platform, completing the circuit, the water remains off. When the circuit is broken (by water evaporating from the platform surface), the mist system is activated and remains on until the cir-

cuit is completed again. The electrode mist controller cannot be used on mist beds spraying deionized water. Both types of electronic leaves are more environmentally sensitive than time clocks and can help conserve water. Mist cycles will occur based on the need of the cuttings rather than on time. However, electronic leaves require maintenance to prevent excessive salt buildup when tap water is used. New mist controlling devices are incorporating photocells to override the time clock during the night or to fully control the mist applications.

III. Micropropagation—Tissue Culture and *in Vitro* Culture

Micropropagation, tissue culture, or *in vitro* (i.e., in glass) culture are terms used to describe the newest development in asexual propagation. The aseptic culture of plant cells, tissues, and organs has been researched and practiced since the early 1950s. The application of tissue culture to propagation started in the late 1960s and is today an important means of propagating many plants. Tissue culture has three unique characteristics. (1) The plant is in a sterile environment. (2) The part of the plant (explant) placed *in vitro* is isolated from other tissues and organs. (3) The environment (temperature, light, culture media) is under strict control. [*See* PLANT TISSUE CULTURE.]

FIGURE 5 Typical overhead, low-pressure mist system for the vegetative propagation of stem cuttings.

A. Required Equipment, Materials, and General Procedures

Micropropagation requires specialized equipment and highly skilled technicians (Fig. 7). An autoclave (or pressure cooker) is required to sterilize tools, glassware, and solutions used in the culture of plant tissues. A near sterile environment in which to work is also needed. Most micropropagation laboratories use a laminar flow hood for this purpose (Fig. 8). This hood takes in room air, forces it through a series of highly efficient filters to remove almost all particulates, and distributes the filtered air over a work sur-

FIGURE 6 "Electronic leaf" used to control the frequency of mist applications based on the evaporation of water.

face. This process gives the technician confidence that he or she is working in a very clean environment and reduces the risk of airborne organisms contaminating culture media. Other standard laboratory materials and equipment are also necessary and include a pH meter, dissecting microscopes, high-quality (deionized or distilled) water, and a microwave oven.

The culture environment is usually outfitted with fluorescent and/or incandescent lights to provide intensities at 100 ftc (50 μmol·m^{-2}·sec^{-1}). Temperature is also carefully controlled to maintain a nearly constant temperature in the 20–28°C range.

The general procedure for micropropagating plants is as follows. (1) The explant to be placed *in vitro* is chosen (axillary buds are most often the starting material). (2) The explant is treated with disinfestants, such as bleach or ethanol, for 10–30 min to kill any surface microorganisms (fungi, bacteria). (3) The disinfestant is rinsed from the explant with sterile water. (4) The explant is placed on the culture medium (the culture medium had been prepared previously and was distributed into a culture vessel and sterilized). (5) The explant in the culture vessel is placed in the culture room in the light or dark, depending on the needs of the explant. The goal is to induce the explant to proliferate many growing points (shoots) so they can be dissected apart and placed back into culture. The increase in the number of new plants depends on the rate of proliferation. If the rate is high, many thousands of plants can be propagated from a single original explant in a short period of time.

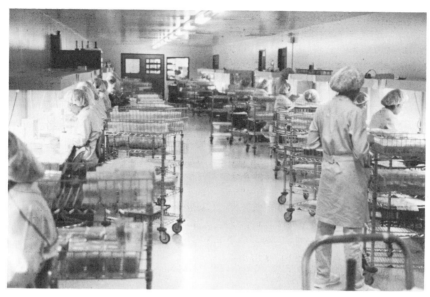

FIGURE 7 Commercial tissue culture laboratory producing micropropagated plants.

B. Tissue Culture Media

The different culture media used in micropropagation are myriad. The presence of macronutrients, micronutrients, vitamins, amino acids, a carbohydrate source, growth regulators, and some kind of support are common to all media. Many micropropagation media used today are modifications of the Murashige–Skoog medium developed for tobacco tissue culture in the early 1960s. Most modified media

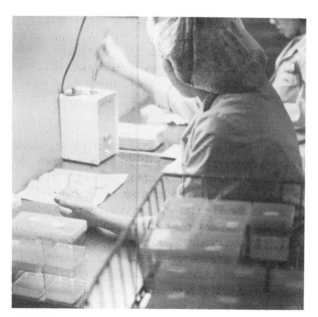

FIGURE 8 Tissue culture procedures require a nearly sterile environment, such as this laminar flow hood, in which to work.

differ in the types and concentrations of growth regulators (cytokinins, auxins) added and the type of carbohydrate source (sucrose, glucose, fructose). Auxins and cytokinins are added to culture media to induce organ formation. Cytokinins such as kinetin (6-furfurylaminopurine), BAP (N^6-benzylaminopurine), 2iP [6-(3-methyl-2-buten-1-ylamino)-purine], zeatin, and thidiazuron (N-phenyl-N-1,2,3-thiadiazol-5-ylurea) are added to stimulate the formation and multiplication of shoots. Auxins such as IAA (indole-3-acetic acid), IBA (indole-3-butanoic acid or indole-3-butyric acid), NAA (α-naphthaleneacetic acid), 2,4-D (2,4-dichlorophenoxyacetic acid), and NOA (naphthoxyacetic acid) are added to stimulate the formation of adventitious roots.

C. Shoot Multiplication, Organogenesis, and Embryogenesis

Micropropagation of plants may proceed via existing meristem multiplication, *de novo* organogenesis, or *de novo* embryogenesis. Depending on the initial explant and its response to culture conditions, existing meristems may be stimulated to grow and proliferate additional meristems; this is an example of multiplication. When new organs are formed in the explant or from tissues derived from the original explant, organogenesis is responsible. Embryogenesis occurs when explants undergo cellular dedifferentiation and subsequent redifferentiation to form somatic embryos. These embryos are genetically identical to the parent explant, assuming no mutation events occurred during the differentiation steps. From a propagator's standpoint, multiplication is the preferred response since it generally involves no callus stage at which cells are totally unorganized and are more likely to undergo mutation. An unorganized callus stage is usually involved in organogenesis and embryogenesis. Therefore, these developmental processes are characterized by higher rates of mutation.

Acknowledgments

The author acknowledges the help of Patricia A. Kiehl in producing this chapter.

Bibliography

George, E. F., and Sherrington, P. D. (1984). "Plant Propagation by Tissue Culture: Handbook and Directory of Commercial Laboratories." Exegetics, Hants, England.

Hartmann, H. T., Kofranek, A. M., Rubatzky, V. E., and Flocker, W. J. (1988). "Plant Science: Growth, Development and Utilization of Cultivated Plants," 2d Ed. Prentice Hall, Englewood Cliffs, NJ.

Hartmann, H. T., Kester, D. E., and Davies, F. T., Jr. (1990). "Plant Propagation: Principles and Practices" 5th Ed. Regents/Prentice Hall, Englewood Cliffs, NJ.

Janick, J. (1986). "Horticultural Science," 4th Ed. Freeman, New York.

Plant Stress

ASHTON KEITH COWAN, *Schonland Botanical Laboratories, Rhodes University*

Glossary

Abiotic factors All physical factors that influence plant growth and development; included are soil type, soil composition, soil moisture content, soil conductivity and pH, nutrient status, salinity, temperature, precipitation, light quality and quantity, humidity, and atmospheric composition

Apoplast Continuum of nonprotoplastic material including the cell wall and intercellular spaces throughout the plant body

Biotic factors Biological factors that influence plant growth and development, including other plants, pathogenic organisms, animals, and humans

Chemical potential of water Component of an aqueous solution which is independent of the concentration of water and dependent only on the mole fraction of water, that is, the number of moles of free water in solution

Turgor Outward pressure exerted by the protoplast on the plant cell wall; outward directed pressure arises from the swelling of the protoplast brought about by the uptake of water by osmosis

Solute potential (Ψ_s) Component of water potential that takes into consideration the concentration of solutes in a cell; when confined by a membrane, a solution exhibits a pressure equal in magnitude to osmotic pressure (π) but opposite in sign ($\pi = -\Psi_s$)

Water potential (Ψ_w) Measure of the energy available in aqueous solutions to facilitate the movement of water across semipermeable membranes during osmosis; the sum of osmotic potential, matrix potential, and pressure potential describing the principle of water conduction in plants, that is, that water tends to move along gradients of decreasing water potential

Stress has been defined as any environmental factor capable of inducing a potentially injurious strain in plants. A more general definition of stress has also been proposed: the excessive pressure from some adverse force that inhibits the functioning of normal systems. The problem with such definitions is that stress is perceived physically as a factor or adverse force. More correctly the term "stress," when used in plant biology, should refer only to the physiological state induced by injurious strain or inhibition of functioning. For example, water deficit is not a stress. Only when plants display a measurable physiological or biochemical response, deviating from the norm, is water deficit perceived as stressful. This metabolic deviation and the subsequent decline in productivity constitute stress. Clearly, any definition of plant stress must therefore consider the stressed state or syndrome induced by extreme physiological and/or environmental pressures. Even so, stress-inducing factors and the stress syndrome are not mutually exclusive and caution is needed when attempting to include the agents responsible in a definition of plant stress. The reasons for this are many and varied. For example, factors that induce a state of stress seldom act individually in the natural environment. Also, a given set of stress factors may cause responses that are temporally dependent on the developmental stage of the organism. Furthermore, factors that might normally be considered detrimental to plant growth and development may in fact be a necessary requirement for full morphogenic expression and survival of the species. In the context of these restrictions, the term "plant stress" is defined as a situation in which normal growth and developmental processes are compromised by extreme abiotic and biotic pressures such that plants, tissues, and cells enter into new and/or different physiological states.

I. The Plant Stress Syndrome: A Centralized System of Responses

Plants will display the stress syndrome when pressure from abiotic or biotic factors (or both) exceeds some theoretical maximum level, that is, when the pressure can no longer be tolerated and normal functioning capitulates. This is best illustrated using a simple dose–response curve. Figure 1 shows that when plants are responding maximally (i.e., close to optimum), the stress syndrome will not be evident. However, an increase in abiotic or biotic pressure beyond saturation not only results in responses that are less than optimal but may be sufficient to induce the stress syndrome. Note that in a no-stress situation, plants seldom function at full genetic potential or optimum. Ideally, maximal response is maintained at about 10–15% below optimum; the differential between maximal response and full genetic potential represents a window of tolerance (see Fig. 1). The term "tolerance" is used in this context to describe all mechanisms by which plants may in some way perform better in response to

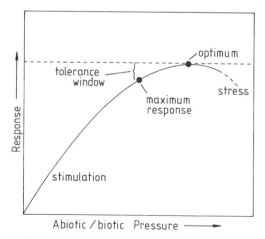

FIGURE 1 Simplified dose–response curve illustrating the relationship between increased abiotic/biotic pressure and the magnitude of the metabolic/physiological response.

increasing physical and/or environmental pressures. The biochemical and physiological processes occurring within this tolerance window are of major interest to agriculturalists, plant physiologists, and molecular biologists. Attempts are therefore constantly being made to understand and integrate these processes to select, breed, and engineer plants that show sustained (or enhanced) productivity under otherwise unfavorable conditions. [See PLANT PHYSIOLOGY.]

Why study plant stress syndrome? Researchers now generally agree that several aspects of different types of plant stress have common features and that plant responses share common components indicating general principles. For example, plants growing in low resource environments display reduced photosynthetic rates, a decreased capacity to acquire nutrients, and low productivity (measured as dry matter accumulation). The same is true of plants exposed to severe physical and/or environmental pressures. In addition, plants respond to many perturbations by changing their hormonal balance, frequently producing more abscisic acid (ABA) and ethylene (ETH) and less cytokinin (CK). Also, the generation of toxic oxygen radicals occurs as a result of increased temperature and light regimes, metal and air pollution, and water deprivation. Such events in plant metabolism are likely to exert a major influence at the cellular level, thus redirecting regulatory mechanisms toward acclimation and adaptation. Acclimation is by nature rapid and is the process sensitive to changes in abiotic and biotic signals. The receptors involved in perceiving adverse pressures (or conditions) must therefore function efficiently to insure optimal response. Some altered cellular characteristics include intracellular compartment pH (and associated proton gradients), redox potentials, electrical potential gradients across plasma and organellar membranes, and calcium ion (Ca^{2+}) concentration, all of which seem to be unifying signals that bring about ordered changes in cell metabolism.

Investigators have well established that genes govern the synthesis of enzymes, which in turn control the chemistry of the cell. Thus, acclimation to change, and adaptation (i.e., the long-term ecophysiological strategy of plants), must be the product of alterations in gene expression and/or enzyme synthesis. Acclimation is catalyzed by flux redistributions that occur in metabolism during adverse conditions and must therefore be regarded as a complex function of the sensitivity coefficients of all metabolic steps. The concept of a plant stress syndrome is preceded, therefore, by the differential expression of all or part of a set of genes (enzymes) induced by extreme abiotic and biotic pressures. In this regard, perhaps the best studied examples are the expression characteristics of ABA-regulated genes. The accumulation of ABA is a characteristic response of plants exposed to severe water deficit, hypersalinity, and wounding. Furthermore, genes induced by application of ABA also accumulate following treatment with NaCl, low temperature, and wounding. The level of expression, however, is not always the same, implying differential regulation of gene expression in response to prevailing abiotic and biotic pressures. Thus, even at the genetic level, a centralized system of physiological response appears evidently responsible for co-ordinating the plant stress syndrome.

II. Adaptive Strategies: A Common Physiological Basis

Adaptive strategies occur as consequences of the resetting of the direction of cellular metabolism, and enable plants to withstand sustained adverse conditions. With respect to plant water deficit, the term "strategy" has been defined as a combination of mechanistically linked responses and characteristics that constitute a particular type of behavior during periods of water stress. It is nevertheless possible to extend this definition to include all abiotic and biotic factors that induce the plant stress syndrome. As shown in Fig. 2, a common feature of hyperosmotic conditions, water

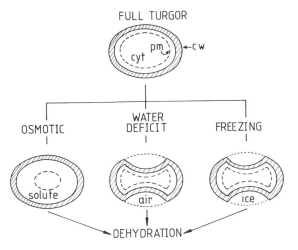

FIGURE 2 A diagrammatic representation of osmotic-, water deficit-, and freezing-induced transition of cells from full turgor to dehydration. Abbreviations: cw, cell wall; cyt, cytosol; pm, plasma membrane.

deprivation, and freezing is a change in volume of the cytoplasm and associated distortion of the plasma membrane. These are not mutually exclusive processes and are the result of dehydration, that is, the elimination or removal of chemically combined water. A similar phenomenon might also be expected for plants exposed to elevated temperature and/or solar radiation, particularly when the rate of water loss by evapotranspiration exceeds that of water acquisition or supply by or from the roots. Similarly wounding, which results in localized alterations in turgor, facilitates dehydration since the water potential outside nonwounded cells or tissues will be more negative than that inside. Thus any change in the chemical potential of apoplastic water will impact on the turgor-sensing mechanism of individual cells, tissues, and organs.

III. Turgor Maintenance and Dehydration Tolerance

Continuity of liquid water exists from the soil through the plant to the liquid–gas interface at the evaporating surfaces within the leaf. Within tissues, all living cells contain a soluble phase containing dissolved organic and inorganic solutes; this soluble phase is separated by membrane-bound compartments. When conditions are optimal for growth and development, individual cells and intracellular compartments exist in a state of equilibrium at or near full turgor. Cell turgor is the result of the intake of water by osmosis, that is, the movement of water along gradients of decreasing water potential. Any change in solute concentration and/or the availability of water will therefore alter the chemical potential of cellular water, resulting in loss of cell turgor. To overcome the effect of fluctuations in the apoplastic microenvironment and to prevent loss of turgor, cells possess the ability to osmoregulate. Essentially, osmotic adjustment contributes to the maintenance of turgor and hence the hydration of cells, tissues, and organs.

The relationship between osmoregulation and hydration is dependent on the elasticity of cell walls and is defined as

$$\frac{\mathrm{d}P}{\mathrm{d}\Psi} = \frac{\varepsilon}{\varepsilon - \pi},$$

where P is the rate of change in turgor potential, Ψ is the rate of change in water potential, π is the osmotic potential, and ε is the modulus of elasticity. Thus the rate of loss of turgor with a fall in water potential is less when cell walls are more elastic (i.e., at low values of ε). Cell enlargement is a characteristic of growth processes; the rate of cell expansion depends on several interrelated factors including cell wall extensibility. Young expanding cells, tissues, and organs are therefore more sensitive and will have higher osmoregulatory capacity. Thus, even after turgor is lost, cells with greater ability for osmotic adjustment will have higher water contents at low water potential values. Maintenance of hydration through osmoregulation is clearly an adaptive strategy enabling plants to both tolerate and avoid periods of water deficit.

IV. Effect of Solutes on Hydration Equilibria

Over 20 years ago it was proposed that failure to accumulate ions into the protoplast would result in the development of low apoplastic solute potentials leading to cellular dehydration. This idea has now become known as the Oertli hypothesis. Solute potential (Ψ_s) is directly proportional to the number of solute molecules in a given volume:

$$\Psi_s = \frac{-nRT}{V},$$

where n is the number of solute molecules in a solution of volume V. A decrease in apoplastic volume of free water will therefore impact directly on apoplastic solute potential. For example, slow cooling results in ice initiation in apoplastic water, because of the lower solute concentration and because of ice nucleators such as dust and bacteria. Thus, a reduction in the availability of free water results. Similarly, during periods of water deficit, water is first lost from the cell wall; this reduction in apoplastic water increases solute concentration. Since accumulation of nonreacting solutes influences chemical equilibria where water is a reactant by shifting the equilibrium toward dissociation, changes in solute potential will impact on processes such as protein hydration and will facilitate the direct interaction of solutes with protein and macromolecular structures. Since all enzymes are proteins, metabolism is inextricably linked to solute potential and hence to perturbations in turgor.

V. The Turgor (Volume)- Sensing Mechanism

A sensory mechanism demands stimulus–response coupling. All available evidence suggests that this pro-cess is intimately linked to volume change and the associated distortion of the plasma membrane. Phosphorylation of plant nuclear proteins is known to be involved in active metabolism. Protein phosphorylation and/or dephosphorylation may represent the most important component in the integration of external and internal stimuli. In general, the coupling of the stimulus to the response involves a cascade of discrete steps:

1. perception of stimulus
2. change in membrane activity
3. transient increase in concentration of second messengers
4. activation of second messenger-dependent enzymes
5. phosphorylation of proteins and return to homeostasis.

Unfortunately, the presence of a turgor-sensing mechanism in plant cells remains equivocal. Nevertheless, two likely candidates involve alterations in plasma membrane chemistry and volume-induced changes in the concentration of regulatory ions and molecules.

Changes in the plasma membrane following loss of turgor include ultrastructure, lipid composition, permeability, lipid metabolism, and ion transport systems. These membrane phenomena occur rapidly (within minutes) in response to volume change and are largely transient in nature, consistent with a turgor-sensing mechanism. The molecular basis for change appears to result from lipid metabolism and the subsequent release of second messengers. For example, in stomatal guard cells inositol 1,4,5-trisphosphate (IP_3), a product of hydrolysis of phosphoinositide 4,5-bisphosphate (PIP_2), causes a transient increase in cytosolic Ca^{2+} concentration and stomatal closure. The causative agent appears to be IP_3, which is released from the membrane by phospholipase C (PLC)-catalyzed hydrolysis of PIP_2. A schematic illustrating these events is given in Fig. 3. Activity of PLC has an absolute requirement for Ca^{2+}. Alterations in phospholipid metabolism and hence membrane composition must therefore depend on the availability of Ca^{2+}.

Volume changes necessarily lead to alterations in cytosolic concentrations of molecules and ions within the cell, which may affect key enzymes in metabolism. In this regard, a decrease in volume will increase the intracellular concentration of Ca^{2+}, inorganic phosphate (P_i), and hormones such as ABA. Additionally, dehydration induces the formation of free radicals that catalyze de-esterification of membrane

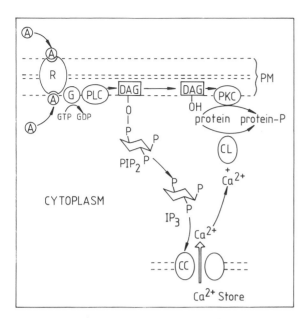

FIGURE 3 Simplified overview of the main components involved in stimulus–response coupling. Abbreviations: A, agonist; CC, calcium channel or pump; CL, calmodulin; DAG, diacylglycerol; G, GTP-binding protein; IP_3, inositol 1,4,5-triphosphate; PIP_2, phosphoinositide 4,5-bisphosphate; PKC, protein kinase C; PLC, phospholipase C; PM, plasma membrane; R, receptor protein.

phospholipids, causing a transition from liquid-crystalline to gel phase. Furthermore, a decrease in cytoplasmic volume is usually associated with rapid but transient intracellular acidification, a process that might be expected to alter the partition coefficient of metabolites between cytosol and membrane. Clearly, the turgor-sensing mechanism must involve interplay between the plasma membrane and cytoplasmic factors, which together facilitate signal–response coupling. For example, ABA is known to interact with receptors in the plasma membrane of stomatal guard cells; various membrane receptors are autophosphorylatable and movements of Ca^{2+} via membrane-localized Ca^{2+} channels are modulated by phosphorylation.

Protein phosphorylation involves a minimum of three proteins and two reactions:

$$\text{Protein} + n\text{ATP} \rightarrow \text{protein-P}_n + \text{ADP} \quad (1)$$

$$\text{protein-P}_n + H_2O \rightarrow nP_i + \text{protein}. \quad (2)$$

Reaction 1 is catalyzed by protein kinases and reaction 2 by phosphoprotein phosphatases. Such a reversible process is referred to as a cascade, and functions to regulate and coordinate metabolic networks. These systems not only amplify signals but control flux re-

distributions in metabolism by acting in additive, opposite, or synergistic ways.

VI. Integration of Metabolic Responses

When cells are exposed to conditions that induce a state of dehydration (e.g., water deficit, freezing, hypersalinity), a cascade of events is triggered by volume change, brought about by alterations in the chemical potential of water. Similarly, nutrient stress causes a substantial reduction in water uptake and/or loss, which alters the ionic status of cells. In comparison, the influence on photosynthesis of high light intensity appears to be directed to the thylakoid membranes, resulting in accumulation of inactive photosystem units and chlorophyll depletion. The subsequent reduction in levels of ATP and NADPH retards the rate of photosynthetic carbon reduction and productivity. Volume-induced changes, on the other hand, include altered cytoplasmic pH (transient acidification) and increased cytosolic P_i concentration, altered $NADH:NAD^+$ ratio, and transport of P_i into the chloroplast in exchange for triose phosphates. High levels of P_i in the chloroplast also arise from reduced photophosphorylation. Together these events represent components of the signal transducing mechanism that ultimately controls the direction of carbon partitioning within the cell.

There is a large difference in redox potential between the $NADPH:NADP^+$ ratio in the stroma and the $NADH:NAD^+$ ratio in the cytoplasm. Maintenance of this redox potential is achieved via the triose phosphate–3-phosphoglycerate translocator and the malate–oxaloacetate shuttle. Changes in cytoplasmic pH and $NADH:NAD^+$ ratio can therefore be transduced into changes in chloroplast pH and $NADPH:NADP^+$ ratio and vice versa. These changes will impact on metabolism and compartmentalization of regulatory compounds, and on cell sensitivity to such regulators. Plant hormones are perhaps the best studied examples of regulatory molecules.

It is now well established that plant hormones have the capacity to either activate or repress gene expression. Plant hormones also influence membrane properties and can therefore contribute to the regulation of protein phosphorylation. When abiotic and biotic pressures exceed tolerance capacity, levels of ABA and ETH rise whereas CK levels fall. Little direct information is available on changes in auxin (IAA) and gibberellin (GA) levels in response to adverse pressures. Even so, there is increasing appreciation of the concept that plant responses are not correlated only with the concentration of endogenous growth regulators but also with the state of sensitivity of the cells and/or tissues to those substances. Since a major function of plant hormones is control of growth and development, any change in hormonal balance and/or sensitivity is therefore likely to influence plant morphogenesis.

Characteristically, plant hormones play a role in processes such as cellular differentiation and morphogenesis; seed development, dormancy, and germination; root and shoot development; and leaf, flower, and fruit development. Although our understanding of the molecular mechanisms involved remains incomplete, we are beginning to appreciate the finer details of the mode of action of many plant growth-regulating substances. In this regard, physicochemical properties of both plant growth regulators and plant tissues have been shown to determine the distribution and redistribution of plant hormones within and between cells and tissues. The major contributing factors include the partition coefficient of the hormone concerned and the membrane conductances of neutral and charged phytohormone species. Only CK and the ETH precursor 1-amino-cyclopropane-carboxylic acid (ACC) are distributed evenly among cellular compartments. ABA is the only hormone that distributes along pH gradients, whereas all other phytohormones show little or no redistribution after changes in compartmental pH. Thus, only ABA is principally capable of being a stress messenger in higher plants. [See PLANT CYTOLOGY.]

A. Abscisic Acid

A major physiological role of ABA in higher plants is control of stomatal aperture. Stomata are responsible for regulating rates of photosynthesis and transpiration to optimize carbon gain against water loss. An increase in levels of ABA facilitates stomatal closure and hence retards water loss under unfavorable conditions. In general, stomatal aperture is controlled by osmotic adjustments in the two opposing guard cells surrounding each stoma, and is directly related to the apoplastic concentration of ABA. An increase in apoplastic ABA levels results from dehydration-induced release of ABA from leaf mesophyll cells and production of ABA by roots in drying soil, which moves via the transpiration stream and accumulates at or near the guard cells.

The sites of ABA action in guard cells appear to be on the outer surface of the plasma membrane.

Here, ABA prevents stomatal opening by arresting H^+ extrusion and K^+ influx, and initiates closure by facilitating the release of K^+, Cl^-, and malate. ABA may therefore influence the H^+–ATPase translocating machinery associated with plasma membranes by binding to one or more receptor proteins. Alternatively, ABA could interact directly with membrane phospholipids to alter membrane permeability.

Volume changes are associated with altered metabolism of inositol phospholipids, which may influence the cytoplasmic Ca^{2+} concentration (see Section V). Although ABA may activate Ca^{2+}-permeable ion channels in the plasma membrane, ABA also initiates release of Ca^{2+} from intracellular stores (e.g., endoplasmic reticulum membranes and vacuoles). This event is supported by the demonstration that, in IP_3-treated guard cells, a decrease in stomatal aperture was accompanied by an increase in free cytosolic Ca^{2+}. Although this finding suggests a role for IP_3 in ABA-induced changes in Ca^{2+} concentrations, there is at present no information to suggest that ABA causes hydrolysis of phosphoinositides to yield IP_3. Nevertheless, the following sequence describes the possible events occurring in cells following exposure to volume-induced pH-coupled redistribution of intracellular ABA or elevated apoplastic ABA levels:

1. activation of inward-directed currents providing a driving force for K^+ efflux
2. stimulation of inositol phospholipid turnover and a rise in cytosolic Ca^{2+} concentration
3. inactivation of Ca^{2+}-sensitive and inward-rectifying K^+ channels
4. activation of Ca^{2+}-dependent and voltage-gated anion currents for anion efflux
5. rise in cytoplasmic pH
6. activation of outward-rectifying K^+ channels for K^+ efflux

B. Ethylene and Polyamines

A change in endogenous levels of ETH and alterations in the rate of ETH synthesis form part of the acclimation process developed by plants to cope with sustained adverse abiotic and biotic pressures. In all probability, the increase in ETH acts synergistically with changes in levels of other phytohormones to bring about inhibition of growth, epinasty, stomatal closure, and senescence and abscission of leaves, flowers, and fruits. These responses clearly have survival value. For example, abscission of leaves and fruits helps minimize water loss during periods of excessive water shortage by reducing the evaporating surface of the plant. Also, ETH-induced formation of adventitious roots increases the survival potential of plants exposed to waterlogging. These examples illustrate that the influence of ETH is exerted only after plants display the stress syndrome.

ETH synthesis is linked to the methionine cycle; the conversion of S-adenosyl methionine (SAM) to ACC is the first committed step in ETH biosynthesis. The synthesis of polyamines also depends on SAM. Polyamines inhibit conversion of ACC to ETH by reducing synthesis of ACC synthase and scavenging free radicals involved in the catalytic conversion of ACC to ETH. Thus it might be expected that any change in levels of polyamines would directly influence levels of ETH.

Several studies have shown that polyamines and polyamine synthetic enzyme activities are higher in plants exposed to unfavorable conditions. For example, nutrient deficiency, heavy metals, low pH, low temperature, and hyperosmotic conditions all contribute to elevated endogenous polyamine levels. An increase in polyamine levels does, however, appear to be a short-term response. Since polyamines are products of intermediary nitrogen metabolism, a stress-induced decline in nitrogen assimilation will consequently lower the available endogenous nitrogen for polyamine synthesis. Likewise, SAM also serves as the precursor to several osmolytes and membrane phospholipids. Thus a relatively small flux redistribution via the methionine–SAM pathway will be transduced into a relatively large change in rate and capacity of ETH biosynthesis.

C. Carbon Flux and Reallocation

The redirecting of carbon flux is coordinated to meet the demands of increased abiotic and biotic pressures. Strong evidence has been obtained to show that membrane disruption under adverse conditions is the consequence of an uncontrolled increase in free radicals that degrade membranes through lipid peroxidation. Furthermore, oxygen uptake can cause cells to produce superoxide radicals. These are normally converted to hydrogen peroxide by superoxide dismutase (SOD), and then broken down to water and molecular oxygen by catalase. When the system is saturated, however, mutual interaction of free radicals may give rise to the extremely reactive singlet oxygen (O_2^-) and hydroxyl radicals ($OH^.$):

$$O_2 \rightarrow O_2^{.-} \rightarrow HO_2^. \rightarrow H_2O_2 \rightarrow OH^. \rightarrow 2H_2O.$$

If not scavenged immediately, these toxic oxygen radicals will cause irreparable damage to the cell. Thus, a large part of the carbon budget is allocated to producing compounds that will prevent the occurrence of free radicals (e.g., osmolytes such as proline, glycine betaine, glycerol, and certain carbohydrates) or, alternatively, to producing compounds that scavenge free radicals directly (e.g., carotenoids, ascorbic acid, and glutathione).

All osmolytes (compatible solutes) function to maintain protein and membrane integrity; this is usually achieved by raising the temperature requirements for protein denaturation, promoting subunit interaction, and preventing inhibition of enzyme activity induced by inorganic solute accumulation. Interestingly, almost all compatible osmotic solutes accumulate within the cytosol of plant cells, which suggests that an additional function of these compounds is maintenance of cytoplasm–nucleus interaction and hence contribution to regulation of the cell cycle (i.e., the time of DNA replication relative to nuclear division). Chloroplasts have also evolved defense mechanisms to cope with inordinate increases in toxic oxygen radicals, generated through lipid peroxidation and the Mehler reaction (where molecular oxygen substitutes for NADP as the terminal electron acceptor in photosynthetic electron transport). In general, ascorbate (via ascorbate peroxidase) and glutathione (via glutathione reductase) protect by preventing lipid peroxidation from entering into a propagation phase, by preventing enzyme thiol group oxidation, and by scavenging free radicals in the aqueous phase of the cell.

One of the major injurious responses to a sustained water deficit is photoinhibition because of the coincident effect of high photon flux density. Under these conditions, the quantity of light absorbed by the photosystems exceeds the capacity of the chloroplast to utilize products of the photochemical reactions. Destruction and/or uncoupling of the photosystems results in accumulation of chlorophyll triplets. These excited states of chlorophyll can facilitate production of singlet oxygen and hydroxyl radicals. In addition to the enzymes SOD and catalase, chloroplasts have evolved the xanthophyll cycle to cope with increased oxygen toxicity (Fig. 4). This cycle involves two enzymes located at the thylakoid membranes which convert zeaxanthin by NADPH or NADH and O_2 to violaxanthin via antheraxanthin, thus dissipating the excess energy harmlessly as heat. The reaction is reversible and has been linked to the synthesis of ABA in plants exposed to water deficit and/or hyperosmotic

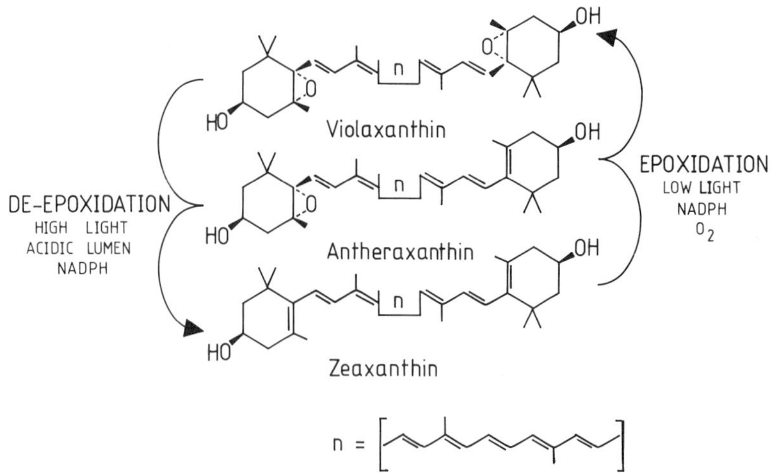

FIGURE 4 The light-regulated interconversion of thylakoid-localized carotenoids with NADHP and O_2 in the xanthophyll cycle.

conditions. Clearly coordinated regulation of carbon allocation within the cell is necessary for acclimation and will contribute to the long-term adaptive strategy of the plant.

D. Gene Expression, Protein Synthesis, and Enzyme Activation

An increase or decrease in concentration of hormones, or other regulatory chemicals, is achieved by gene activation, enzyme synthesis, and hence alterations in both anabolic and catabolic enzyme activities. Also, short-term changes in metabolite flux and compartmentation and long-term ecophysiological adaptation are consequences of altered carbon allocation patterns, both controlled by an increase or decrease in levels and/or activities of the necessary enzymes. For these changes to occur, alterations in the expression of the genetic program in plants by abiotic and biotic stimuli is needed. Remember that normal ontogenic changes are also governed by the genetic blueprint of a species and, for the most part, these changes in ontogeny occur in response to environmental signals such as day length, and seasonal variations in temperature and precipitation. Thus, it should not be surprising to discover that many proteins induced as part of the stress syndrome are also produced at different stages during the life history of an organism.

Although the exact function of many stress-related proteins identified to date remains to be elucidated, several interesting observations deserve mention. The largest family of stress-related proteins characterized to date is that of the dehydrins. This family includes the LEA (late embryogenesis) group 2 proteins and RAB (ABA responsive) proteins, and is characterized by the amino acid consensus sequence KIKEKLPG which may be repeated many times within the complete dehydrin protein sequence. Messenger RNA transcripts for many of these proteins accumulate in a number of species and in a variety of tissue types including leaves, roots, callus, seeds, and intact seedlings. Furthermore, these proteins occur in both angiosperms and gymnosperms, and have recently been detected in cyanobacteria, the proposed progenitors of higher plant chloroplasts. Even more interesting is that these proteins accumulate in response to a wide variety of abiotic and biotic stimuli including drought, wounding, low temperature, heat shock, ABA treatment, and exposure to hyperosmotic conditions. It therefore appears that this particular family of proteins has been highly conserved during evolution.

One feature that has emerged, particularly from studies on ABA-regulated gene products, is that the proteins are readily soluble, basic, very hydrophilic, and stable even after boiling. It has thus been concluded that these proteins are neither enzymes nor structural proteins. In fact, their solubility and likely occurrence in the cytosol and stroma suggest that they function to solvate membranes and proteins during periods of dehydration.

Obviously when prevailing conditions retard photosynthetic carbon fixation, metabolism and reallocation of cellular carbon must involve mobilization of the stored starch reserve. A combination of reduced triose phosphates and increased P_i levels within the chloroplast is sufficient to activate both starch phosphorylase and phosphofructokinase, two of the enzymes required for starch metabolism. The end result is a net increase in triose phosphates, which are exchanged for cytosolic P_i. Under conditions of elevated endogenous ABA and/or cellular dehydration, activity of cytoplasmic fructose-1,6-bisphosphatase is inhibited, preventing sucrose synthesis. Thus, triose phosphates are freely available for respiratory and biosynthetic processes. Whether any of the enzymes required for biosynthesis form part of the dehydrin family of stress-related proteins is still uncertain. However, several proteins induced by anaerobiosis are known to be enzymes of glycolysis, and at least one protein induced by dehydration has been identified as phosphoenol pyruvate carboxylase, a major enzyme of CAM photosynthesis.

Many enzymes in metabolism are regulated entirely or in part by the free Ca^{2+} concentration, and thus are subject to control through signal–response coupling. Enzymes that are modulated in this way include NAD kinase, β-glucan synthase, Ca^{2+}– and H^+–ATPase, quinate:NAD^+ oxidoreductase, and protein kinase. The most important Ca^{2+}-modulated enzyme is the ubiquitous eukaryotic protein calmodulin. This protein binds Ca^{2+} and undergoes a conformational change that allows it to interact with and alter the activity of calmodulin-dependent enzymes. The pigment phytochrome, a protein-containing chromophore, also mediates many metabolic processes and does so by regulating membrane permeability, particularly to Ca^{2+}. Phytochrome exists in two interconvertible forms—Pr (red-light absorbing) and Pfr (far-red-light absorbing)—and interconversion allows phytochrome to mediate responses that are promoted by red light and inhibited or reversed by far-red light. Researchers know that Ca^{2+}, calmodulin, and phytochrome all play roles in the phosphorylation of nuclear proteins and hence enzyme activation.

VII. Physiological Responses: From Cell to Whole Plant

The physiological response displayed by plants to excessive abiotic and biotic pressure is determined by both genotype and phenotype of the species. For growth and development, carbon dioxide diffuses through the stomata and enters the chloroplasts where it is fixed by photosynthesis into simple and complex carbohydrates. Similarly, some species of plants fix atmospheric nitrogen which is eventually incorporated into amino acids, proteins, and polypeptides. Continued synthesis of large complex molecules from simple gases, minerals, and various metals requires the presence of sufficient water and will ultimately give rise to larger and more complex cells, tissues, and organs. Photosynthesis and water (nutrient) acquisition are linked by the vascular system, which is responsible for controlling the movement and disbursement of nutrients, organic solutes, and water among cells, tissues, and organs. [See PHOTOSYNTHESIS.]

The vascular system, a continous network of xylem and phloem, forms the link between sources and sinks. Demand for resources is usually exercised by sinks, which include all developing regions of the plant. Young growing areas are more sensitive to fluctuations in the environment and will perceive adverse conditions more readily. In response to perturbation, demand for nutrients by sinks will be reduced. Consequently, an imbalance between source and sink will be established and communicated from sink to source. Since sources are usually mature, fully functional organs, perception of the sink signal will alter solute allocation patterns. Depending on the nature, intensity, and duration of the signal, development may switch from vegetative to reproductive, or senescence and abscission of the source may be induced to insure survival.

Crop production (yield) has been inextricably linked to crop transpiration and has been discussed in terms of water use efficiency, that is, the amount of carbon fixed per unit of water lost. However, it is becoming increasingly clear that stomata regulate the amount of carbon dioxide for photosynthesis against available water. A limitation in supply of CO_2 is transduced into mechanistic down-regulation of energy transfer within the photosystems and hence reduced rates of carbon fixation. The presence of a vascular network means that fluctuations in shoot physiology can be transduced to roots and vice versa, providing plants with the ability to adjust and optimize resource acquisition and allocation in concert with prevailing conditions. Central to this process is source–sink chemical communication. In this regard, ABA regulation of stomatal guard cell movement and hence conductance, coupled with ABA-induced dehydrin protein synthesis and increased vigor, suggests that genetic differences in ABA accumulation may be the determinant for selecting high-yield crops with increased tolerance capacity for cultivation in low or poor resource environments.

Bibliography

Alscher, R. G., and Cumming, J. R. (1990). "Stress Responses in Plants: Adaption and Acclimation Mechanisms." Wiley–Liss, New York.

Blatt, M. R., and Thiel, G. (1993). Hormonal control of ion channel gating. *Annu. Rev. Plant Physiol. Plant Mol. Biol.* **44,** 543–567.

Cote, G. C., and Crain, R. C. (1993). Biochemistry of phosphoinositides. *Annu. Rev. Plant Physiol. Plant Mol. Biol.* **44,** 333–356.

Cowan, A. K., Rose, P. D., and Horne, L. G. (1992). *Dunaliella salina:* A model system for studying the response of plant cells to stress. *J. Exp. Bot.* **43,** 1535–1547.

Davies, W. J., and Jones, H. G. (1991). "Abscisic Acid Physiology and Biochemistry." Bios Scientific Publishers, Oxford.

Hartung, W., and Slovik, S. (1991). Physicochemical properties of plant growth regulators and plant tissues determine their distribution and redistribution: Stomatal regulation by abscisic acid in leaves. *New Phytol.* **119,** 361–382.

Hetherington, A. M. and Quatrano, R. S. (1991). Mechanisms of action of abscisic acid at the cellular level. *New Phytol.* **119,** 9–32.

Jones, H. G., Flowers, T. J., and Jones, M. B. (1989). "Plants Under Stress," Society of Experimental Biology Seminar Series, Vol. 39. Cambridge University Press, Cambridge.

Kreeb, K. H., Richter, H., and Hinckley, T. M. (1989). "Structural and Functional Responses to Environmental Stresses." SPB Academic Publishers, The Hague.

Levitt, J. (1980). "Responses of Plants to Environmental Stresses," 2d Ed., Vols. 1 and 2. Academic Press, New York.

Liljenberg, C. S. (1992). The effects of water deficit stress on plant membrane lipids. *Progr. Lipid Res.* **31,** 335–343.

Mooney, H. A., Winner, W. E., and Pell, E. J. (1991). "Response of Plants to Multiple Stresses." Academic Press, San Diego.

Smith, J. A. C., and Griffiths, H. (1993). "Water Deficits: Plant Responses from Cell to Community." Bios Scientific Publishers, Oxford.

Plant Tissue Culture

MARY ANN LILA SMITH, *University of Illinois*

I. The Chemical Microenvironment
II. The Physical Microenvironment
III. Facilities and Aseptic Methods
IV. Techniques
V. Methods of Experimental Evaluation
VI. Implications for Agricultural Science

Glossary

Chemical microenvironment Includes the water, supporting inorganic (macronutrients and micronutrients) and organic chemicals in the medium, medium pH, and the physiochemical conditions (gaseous composition and humidity) in the headspace of each containment vessel

In vitro From the Latin translation "in glass," the term refers to cultivation of whole organisms, cells, or tissues under *aseptic conditions*, on an *artificial medium*, in an *enclosed vessel* (typically glass or plastic). As a result of the limited physical microenvironment and influence of the growth regulators, the cultivated material usually exists on an extremely *small scale*. Synonymous with the term "microculture"

Physical microenvironment Unique set of conditions that exist *inside* a tissue culture growing vessel, including headspace temperature, incident light at the culture surface (irradiance, spectra, and duration), air movement, physical boundaries (dimensions) of the culture vessel, agitation rate in liquid suspension cultures, electrical stimulation, or the gel strength and physical characteristics of solidified culture medium (which determine physical resistance to growth and gel water availability); these conditions are in part dictated by the rigorously controlled parameters set in a tissue culture growth room or chamber, but are also influenced by the metabolism of the plant material and the ventilation/construction of the vessel/ closure, which governs exchange between the inter-

nal microenvironment and the growth room environment

Regeneration Restoration or redifferentiation of a whole plant or plant organ from dissimilar or undifferentiated tissues *in vitro;* for example, plants can regenerate via *organogenesis,* a morphogenic response that results in production of roots, shoots, or entire plants from a disorganized mass of callus cells; alternatively, plants can regenerate via *somatic embryogenesis,* the formation of somatic (vegetative, nonsexual) embryos on the surface of callus or another cultured plant organ

Plant tissue culture, also known as the *in vitro* culture or microculture of plants, concerns the growth and development of plants, plant cells, plant organs, or plant tissues in an aseptic, strictly controlled chemical and physical microenvironment. As is the case for *in vitro* culture in animal (vertebrate and invertebrate) systems, plant tissue culture dictates that material is cultivated within enclosed vessels which maintain the sterile, usually defined nutrient medium and gaseous environment. Usually, the small-scale and limited internal space within tissue culture vessels results in a miniaturization of the material, hence the term "microculture." Plant tissue culture comprises a range of technologies integral to both the commercial plant production industries and research investigations. In commercial production, plant tissue culture emerges as an advantageous means of mass propagating economically valuable clones and new crop introductions, with concurrent gains in production timing, product uniformity, efficiency, availability of clean, virus-free material, and flexibility in response to market demands. In research, plant tissue culture facilitates the engineering and selection of elite, superior genotypes, and serves as a vehicle for in-depth investigation of physiological or biochemical processes. Fi-

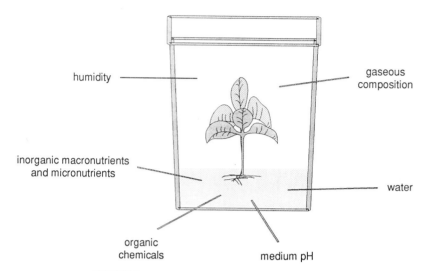

FIGURE 1 The chemical microenvironment *in vitro*.

nally, the ability of plants to synthesize valuable secondary metabolites *in vitro* has recently shifted from a research curiosity into the arena of intensive commercial interest, as the consumer demand for natural products and valuable natural medicinals has escalated.

I. The Chemical Microenvironment

A. Medium Composition

Medium components routinely include purified water, inorganic salts (macronutrients and micronutrients), a carbohydrate source, vitamins, hexitols, and growth regulators; they *may* also include amino acids, antibiotics, antioxidants, adsorbents, various natural (undefined) complexes which aid in growth and development, and a gelling agent, such as agar or gellan gum (Fig. 1).

1. Water

Water, which will comprise about 95% of the nutrient medium, must be available in high-quality purified grade: preferably pretreated (distillation or reverse osmosis), and then distilled or deionized just prior to use in preparing a medium formulation. Untreated tap water contains contaminating solutes that would otherwise interfere with the chemistry of the other nutrients added to the medium.

2. Macronutrients and Micronutrients

The six major elements required for plant cell growth are each provided in the inorganic salts of culture media. Nitrogen (N) may be supplied as nitrate or ammonium ions, or more typically a combination of the two. (Some nitrogen may also be supplied in amino acids—an organic source.) Phosphorus (P) is supplied as potassium phosphate and sodium phosphate. Potassium (K) is included in potassium nitrate, potassium chloride, or potassium phosphate forms. Calcium chloride or calcium nitrate satisfies plant requirements for calcium (Ca) in the medium, while magnesium sulfate provides both the essential magnesium (Mg) and sulfur (S) elements.

The required micronutrient elements are iron (Fe), copper (Cu), zinc (Zn), manganese (Mn), molybdenum (Mo), and boron (B). Iron, which may be the most influential of the micronutrients in terms of cultured plant response, is supplied as a chelate, typically NaFeEDTA or sequestrene Fe. Other micronutrient elements are included in some medium formulations, but are not proven as essential to support culture growth.

Several commonly used media formulations featuring different concentrations and molecular sources of these macro- and microelements have been devised for plant tissue culture, largely based on empirical trials with a broad range of plant species. These are available commercially as premixed powdered formulations, or more commonly, the published formulae can be prepared from inorganic salt stock solutions.

3. Carbohydrates

A sugar source is normally supplied for metabolism *in vitro*, because cells in culture are not generally *auto-*

trophic (capable of supplying their own carbohydrates via photosynthesis). Sucrose and glucose are the most common sources, usually at about 3% of the medium volume; other sources have proven optimal for some plant genotypes. In other specialized tissue culture systems (e.g., protoplast culture), sugars in much higher concentration may be added not as metabolic agents, but as an osmotic agents.

4. Vitamins

Thiamine (B_1), nicotinic acid, and pyridoxine (B_6) are routinely incorporated into the nutrient formulation, and function as catalytic agents (coenzymes) in enzyme reactions. Other vitamins that are added in more complex formulations include biotin, folic acid, ascorbic acid, pantothenic acid, vitamin E, riboflavin, and *p*-aminobenzoic acid. The actual plant *requirements* are not well established, but the supplemental vitamins appear to support growth *in vitro* especially when the population densities (of cells) are low.

5. Hexitols

These components enhance *in vitro* growth beyond their simple carbohydrate identity, and demonstrate a vitaminlike action. For this reason, *myo*-inositol (a commonly included hexitol) is often discussed as a vitamin in the plant tissue culture literature. Hexitols also have a role in cell wall production and membrane stabilization.

6. Growth Regulators

These organic chemicals, which may mimic or modify the action of inherent plant hormones, are applied in very small concentrations to elicit an *in vitro* response. The growth regulators instrumental in plant tissue culture include the *auxins, cytokinins, gibberellins, ethylene,* and *abscisic acid.* The two classes of growth regulators with the most overwhelming influence on performance in plant tissue culture are the auxins (which cause cell elongation, swelling of tissues, cell division, formation of adventitious roots, the inhibition of adventitious and axillary shoots, xylem differentiation, and embryogenesis in suspension cultures) and the cytokinins (which promote adventitious shoot proliferation and retard aging). Very often, when a plant tissue culture researcher is faced with the task of culturing plant material without a recorded history of *in vitro* performance, the first step in experimentation will be an empirical screening of auxin/cytokinin formulations and concentrations, to determine an optimal balance to maximize growth.

7. Miscellaneous Additives/Natural Complexes

Other tissue culture medium components that are required in specialized applications include amino acids, antibiotics, antioxidants, adsorbents, and various undefined complexes of biological origin. Amino acids provide a more readily available N source than the inorganic salts, and are useful in morphogenesis and embryogenesis research. Antibiotics are provisional additives to help alleviate persistent problems with contamination, in particular if the plant *in vitro* harbors internal pathogens, but they are not recommended for long-term use. Antioxidants or adsorbents alleviate inhibition due to natural exudates from the plant material in culture. The final category—undefined complexes of biological origin—embraces quite a potpourri of additives which, although largely undefined, may effect dramatic stimulation of performance *in vitro*. Some of these additives used in plant tissue culture include coconut endosperm, orange juice, tomato juice, V-8 (mixed vegetable) juice, yeast extract (natural source of vitamins), malt extract, banana puree, potato extract, casein hydrolysate (weak acid/protein mixture), and fish emulsion.

8. Gelling agents

The nutrient medium may be solidified with a gelling agent in order to support plant material in an upright orientation (as for a shoot culture), to facilitate handling and storage of media, and to avoid hyperhydricity (a glassiness or "water-logged" hyperhydration of some plant material which occurs readily on liquid medium, although it may also occur on gels in high humidity, in the presence of high salt or cytokinin concentrations, or when material is maintained on wet gel surfaces). The connected pore spaces/solid matrix of the gels permits diffusion of liquid medium nutrients to the plant and simulates the structure of a soil system. Agar (a polysaccharide extract from seaweed or red algae) is the most common gelling agent used *in vitro;* as a natural product it is undefined, not completely inert, has impurities, and has adsorptive properties (may remove harmful metabolic products from the plant during culture). A substitute gellan gum (self-gelling hydrocolloid) may be used alternatively to form a more transparent rigid gel. Several other gelling agents *or* physical support systems have been introduced to support *in vitro* cultures. Since the gelling agent/gel strength and/or use of alternative supports (filter paper bridges, membrane rafts, or screens over liquid medium) have a profound effect on the physiochemical phenomenon of water availability, and on physical resistance to plant growth, they are also considered components of the physical microenvironment of plant tissue culture.

B. Medium pH

Medium pH (a measure of solution acidity) influences the solubility and availability of nutrient ions, the growth response of plant cells or organs, the stability of some growth regulators and vitamins, and (in gel-solidified medium) the ability of the gelling agent to form an adequate solid matrix. The pH is routinely adjusted after preparing a medium formulation to a specified range (5.2–5.8). However, pH will decrease slightly after autoclaving, and media will further acidify during the culture cycle as nutrients are exhausted and metabolic by-products accumulate. A buffer may be added in media to moderate pH fluctuation during culture.

C. Composition of Vessel Headspace

Chemical factors in the headspace of a tissue culture vessel are not as well defined or easily regulated as in the medium, but according to recent research, these chemical factors have tremendous influence on the performance of the plants *in vitro* (Fig. 1).

1. Humidity

Humidity in the interior of the vessel headspace is quite high—typically approaching 90–100%, which suppresses transpiration/transport through the plant's vascular system. As the gel concentration in solidified medium increases, the relative humidity (RH) decreases. Treatments which effectively lower the RH in the headspace will increase the transpiration of *in vitro* cultures (at least during early stages of the culture growth cycle), and enhance the ability of the microcuttings to survive the subsequent acclimatization to *ex vitro* production.

2. Gaseous Composition of the Headspace

Gas composition within a vessel is dictated by both the absorption/generation due to plant metabolism and gas exchange with the outside culture growth room. Typically, unless the closure of a culture is tightly sealed with a wax wrap or other material, there is some gaseous exchange between the headspace of a vessel and the air in a culture growth room. When exchange is limited, the ambient internal levels of ethylene and other gasses tend to accumulate within the vessel headspace, especially in callus cultures. Eth-

ylene may also be released from the synthetic growing vessels and closures which contain the cultured plant material. Elevated atmospheric ethylene in sealed vessels invariably inhibits *in vitro* growth and morphogenesis, and has been implicated in reacquisition of the juvenile status in plants. In light of the profound effect of vessel closure on both the humidity and gaseous composition of the headspace, new research efforts are aimed at redesign of growing receptacles, and incorporation of gaseous exchange mechanisms into the tissue culture system. Photoautotrophic tissue culture production systems (featuring photosynthetically functional plants) require enrichment with supplemental CO_2 gas and/or forced ventilation systems to support maximum productivity. These points are discussed further under the subject of the physical microenvironment, below.

II. The Physical Microenvironment

A. Headspace Temperature

In vitro cultures are more temperature sensitive (tissues lack the protective adaptations of intact plants) than counterpart plants *in vivo*. Therefore, the need to maintain consistent temperature and constantly monitor temperature in the growth chamber, and to examine the parallel temperatures within the vessels, is far more critical than for most other plant culture systems. In the *in vitro* growth facility, temperature is usually maintained near 25°C throughout the day/night period, with temperature fluctuation ideally less than ±2.0°C. Significantly different temperature regimes, however, have been determined optimal for production of some specific crops *in vitro*; for example ranges between 32 and 35°C are required for certain tropical plants, microtuber formation in culture is maximized at 20°C, and alternating temperatures are required to stimulate *in vitro* bulb production, break dormancy, or regulate internode length. As a consequence of the light source's radiative heating of the vessel, the interior vessel atmospheric temperature is about 1°C higher inside the vessel when lights are on. The temperature of the plants themselves are between 0.1 and 1°C higher still than the inside air temperature, because of the usually low air velocity within the vessel and negligible transpiration (Fig. 2).

B. Incident Light

Lights are typically supplied in a growth room on a 16/8 hr photoperiod, although for some systems (cal-

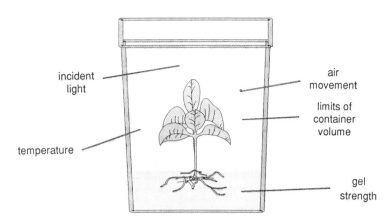

FIGURE 2 The physical microenvironment *in vitro*.

lus culture or prerooting treatments), the cultures may be maintained 24 hr in darkness. In most tissue culture schemes, light irradiance is relatively low ($\sim$30–100 μM m^{-2}s^{-1}) to avoid a "greenhouse effect" of extreme temperatures as lights reach the culture vessels. Actual light penetration into the headspace is modified by culture vessel arrangement and vessel/closure characteristics. The relatively low photosynthetic photon flux (PPF) reaching the internal headspace of the vessel is not normally problematic, as cultures with carbohydrate in the medium do not rely on vigorous photosynthetic activity, and photomorphogenic events (such as organ production) require only low levels of irradiance. However, for photoautotrophic micropropagation systems (in which supplemental CO_2 is supplied to the headspace, and the carbohydrate is eliminated from the medium), a PPF up to five times higher can be successfully introduced to stimulate vegetative productivity, *and* will result in more successful transition of the plant from *in vitro* to *ex vitro* environments.

Several morphogenic events *in vitro* appear to be related to the spectral quality of the light reaching the cultures. Fluorescent lamps (which supply a broad wavelength range) are sometimes supplemented with other sources to enrich portions of the spectra, or, alternatively, narrow-band filters have been used to exclude certain wavelengths from reaching the plants. Photochemical degradation of medium components upon interaction with certain wavelengths can also be circumvented using the latter strategy. Recent innovations in light placement have demonstrated that sideward/lateral lighting of cultures (rather than conventional overhead lighting) can amplify growth and quality by allowing uniform PPF to reach all levels of the plant canopy, not just the top. This strategy is facilitated by the introduction of cool light emitting

diode (LED) lamps or vertically distributed optical fibers in place of conventional lighting systems. Space utilization is more efficient, and plant production is enhanced.

C. Vessel/Closure Design

Vessels and closures provide the outside limits of the physical microenvironment, effectively insulating their contents (plant material, media, and headspace atmosphere) from the contamination and low RH of the *in vivo* environment. They impose the boundaries of the internal growing area. At the same time, vessels and closures are exposed and responsive to the environmental parameters set for the larger growing chamber, allowing the chamber set points to temper the physical regime within each culture's individual microenvironment. The vessel and closure design have enormous influence on how significantly the physical chamber specifications for air movement, temperature, light are perceived in the headspace surrounding the culture. Vessel and closure also regulate chemical microenvironmental parameters of gaseous headspace composition, and humidity.

Culture rooms feature mechanisms for forced-air ventilation, to ensure adequate circulation, velocity, and turbulence of air in the chamber. The density of culture vessel spacing or presence of other obstructions in the chamber will in part modify these air movement parameters. Air movement *within* a culture vessel headspace, however, may range from nearly stagnant air (tightly capped, wrapped vessels) to systems which feature excellent circulation of *sterile* forced air within the headspace. Vessels which accommodate the plants include a broad range of containers including culture tubes, petri dishes, glass mason jars, fabricated vessels of various sizes, and customized growing chambers; closures may involve simple foil or wax wraps, cotton plugs, or custom-designed polypropylene fitted lids. On the lower end of the air movement range, tightly wrapped culture vessels afford maximum protection against possible contamination or desiccation. However, there is little exchange between the culture room atmosphere and the headspace. Unfavorable gaseous accumulation or excessive humidity may result. Alternatively, other vessel closures have been constructed with grooved lids or flexible rib designs to fit more loosely and permit better exchange between the headspace microenvironment and the culture room. A further innovation in this direction is the introduction of gas-permeable filters in the lid design, to facilitate enhanced gaseous exchange. Gas-permeable films (which permit CO_2 supplements to reach the headspace) are ideal for photoautotrophic micropropagation of plantlets with functional stomates. Finally, on the opposite end of the range, some customized, automated microculture systems now feature sterile, forced-air ventilation systems for free flow of gasses within the culture headspace. Bioreactors designed for large-scale liquid fermentation similarly feature automated, sterile forced aeration systems.

Variables of vessel size, volume of growing space within a vessel, and surface areas of container and lid surfaces are significant factors which not only have repercussions for humidity and air exchange, but also influence temperature and light levels in the headspace. Temperature in the headspace is related to the ratio of lid and side surfaces to the bottom vessel surface, and the heat absorption of vessel walls. Vessel construction and composition regulate how much irradiance reaches the growing surface, and the spectral transmittance of vessel materials differ from each other. The distribution of light in the headspace is linked to the vessel/closure combination.

D. Gel Strength

Gels used to solidify the culture medium have some chemical influence (impurities and adsorptive properties) as described previously. They further modify the physical microenvironment by offering some physical impedance to unrestricted growth. For callus, dry weight increase tends to be inversely related to the concentration of the gel. While growth of undifferentiated callus on hard gels is inhibited, morphogenesis may be stimulated by this simple physical stimulus. Medium hardness/physical resistance increases linearly with gel (agar) concentration. Gels also appear

to exert a strong effect on the water availability *in vitro*. An agar-solidified medium will have a lower (more negative) water potential than its liquid counterpart. Gels contribute to the matric potential of the medium, and effectively restrict water movement as compared to liquid culture systems. Although these gel matric potential negative effects seem to be very small (not measurable by conventional instruments) they are quite influential when measured in terms of plant performance.

E. Other Factors

The degree of culture *agitation/rotational speed* for liquid suspension cultures is a physical factor that not only regulates aeration of the medium, but also affects the turbulent shear force encountered by the plant cells. Because plant cells have rigid cell walls and a large vacuole, they are more sensitive to shear than animal or microbial cells cultured in a similar environment. The unique vulnerability of plant cells to physical shear stress led to development of modified bioreactor designs in order to maintain cell viability. However, recent research has determined that some plant cells are relatively shear resistant (less sensitive than previously surmised). *Pulsed electromagnetic field stimulations* are additional physical factors that have promoted superior *in vitro* plant performance. Environmental disturbance induced by magnetic fields has promoted physiological and morphological changes *in vivo,* related to the polarity of plant organs and polar transport of plant organs. Similarly *in vitro,* electrical stimulation has been proposed as a physical regulation mechanism with strong commercial promise for control of plant growth. Electrically stimulated callus tissue and *in vitro* plantlets respond with modified growth rate and morphogenesis (increased rhizogenesis and more prolific shoot production). In particular, electromagnetic treatment of shoot cultures favors development of plants with enhanced ability to survive acclimatization to the *ex vitro* environment.

III. Facilities and Aseptic Methods

A plant tissue culture facility includes three distinct areas, which are physically separated to promote efficiency and maintain cleanliness. A *preparation lab,* where the nutrient medium is processed, houses the equipment required to purify water, weigh and combine media components, mix, dispense, and sterilize media, store heat-sensitive chemicals, and wash glassware. A *sterile transfer chamber* contains a hood where all manipulation of the plant material (introduction to *in vitro* culture, subculture transfer, etc.) takes place (Fig. 3). In a laminar flow hood, for example, air is forced through a high-efficiency particulate air (HEPA) filter, which screens out microorganisms on route to the sterile working surface. The closures on a sterile culture vessel can safely be removed under the hood in order to insert and maneuver the sterile tools used in the subculture process, without contaminating the aseptic microenvironment. A *growth chamber* or growing area contains the well-lighted, temperature-controlled shelves in a clean, air-filtered room for maintenance and growth of *in vitro* cultures (Fig. 4). Rotator or shaker culture instruments are similarly housed in the growth chamber.

Aseptic technique requires deliberate sterilization of all working surfaces and tools prior to the transfer (subculture) session, cleansing and/or sheathing of the operator's hands and arms, and assurance that all plant material to be introduced to culture is free of microorganisms. Surfaces are usually disinfected with chemical antiseptics or bleach solution. Tools such as forceps or scalpels used to dissect plant material for culture are sterilized chemically and/or heat sterilized. All receptacles, media, water, or other instruments

FIGURE 4 A growth room for *in vitro* cultures includes controlled irradiance, temperature, and air velocity.

used in the transfer process must similarly be sterilized by autoclaving, filter sterilization, electromagnetic radiation, or use of germicidal antimicrobials or other cold-sterilization methods.

The plant material to be introduced from nature into the *in vitro* environment is called the *explant*. Explants can range from entire organisms or organs (seeds, seedlings, bulb scales, roots, etc.) or can be excised as small segments from a plant (isolated meristems, shoot tips, anthers, etc.). After detachment from the parent plant, the explant is subjected to surface treatment to eliminate intrinsic microorganisms. Although electromagnetic radiation treatments have proven successful for preparation of some material, the most common procedures involve chemical treatment of plant tissues with disinfectant solutions (dilute sodium hypochlorite, alcohols, hydrogen peroxides, calcium hypochlorites, or other solution). Explant preparation for recalcitrant plant material (especially material with difficult to cleanse bud scales, waxy surfaces, or surface trichomes) may require sonication, use of a surfactant, and a series of sequential disinfecting solutions prior to achieving a successful, aseptic culture free of contaminants. In other cases, some of the tissues of an explant must be stripped and discarded during the sterilization process: for example, the seed coat of some species, or the rough bark layers of a woody explant. An explant is finally trimmed to remove tissues damaged during the procedure, then inoculated on (or partially submerged in) the surface of a sterile nutrient medium, maintaining the same polarity (orientation) that existed in nature.

The new growth *in vitro* from the original explant may occur as organized tissues (division, shoot formation, etc.), or the explant may be induced to produce

FIGURE 3 A laminar flow hood for sterile transfer of cultured tissues.

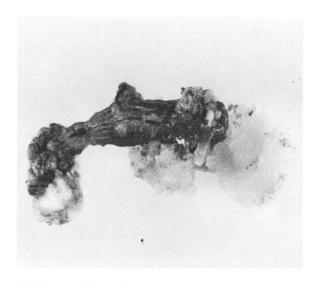

FIGURE 5 Callus production from the cut ends of a grape stem explant.

FIGURE 6 Micropropagated shoot cultures.

unorganized callus (Fig. 5). New growth from the explant provides starting material for initiating and maintaining each of the plant tissue culture systems described below. After a culture growth cycle, the material may be subcultured by dividing or excising plant parts for transfer to fresh medium. Normally, a plant can remain in the same vessel until growth has exceeded the physical limits of the headspace and/ or the nutrient medium has been depleted, but timing may be regulated to maximize the productivity of the culture and maintain optimal quality. The length of time a culture can be sustained (prior to subculture or re-entry into the *ex vitro* environment) varies enormously depending on the type of culture system.

IV. Techniques

A. Commercial and Production Methods

1. Micropropagation

Micropropagation, a technique for rapid, clonal mass multiplication of plants *in vitro*, is the most striking application of plant tissue culture, and the technique with the strongest practical impact on the commercial plant production industries. Induction of multiple shoots from axillary buds cultured *in vitro* (usually on cytokinin-supplemented medium) results in asexual production of identical, highly uniform clones (Fig. 6). Shoot cultures can be subcultured on a 3- to 6-week rotation to successively increase the number of plantlets in a short amount of time. Alter-

natively, microcuttings produced by the end of the subculture cycle can be delivered as rooted or un-rooted microplants to conventional growers, who will process them through final stages of production. Cultures can be mass-produced, and production timing scheduled to match market demand. Micropropagation has provided a mechanism for growers to rapidly produce a bulk supply of a new crop introduction, accelerate the introduction of unique genotypes, and reproduce certain plants that are recalcitrant to conventional propagation techniques. Micropropagation in the industrial arena has emerged as a highly efficient, predominantly automated set of production steps which provides uniform plant product to progressive growers of agronomic crops, foliage plants, horticultural fruits and vegetables, nursery crops, or forestry crops. Although traditional growers and producers were initially hesitant about dealing with the micropropagated plantlet, this reluctance has been largely outweighed by realized savings in timing and space requirements and the guarantee of healthy, clean, disease-free plantlets.

Beyond commercial implications, micropropagation techniques are used to preserve and maintain vegetatively propagated plant germplasm collections on a global scale. The micropropagated germplasm repositories have reduced requirements for controlled environment space (as compared to conventional field or greenhouse-maintained collections) and therefore lower costs are involved. The healthy, disease-free status of the reserved plant material can be maintained long term. Stock maintained in micropropagation blocks is readily available in response to requests for rapid bulk-up multiplication and international transport and exchange. Several techniques to further enhance long-term storage potential and maintenance of microcultures (cold storage or cryopreservation) are used in tandem with the micropropagation techniques to maintain collections. Micropropagation has added advantages toward preservation of endangered species, which may not propagate efficiently by conventional methods, and/or may be difficult to maintain in germplasm repositories. [See PLANT GENETIC RESOURCE CONSERVATION AND UTILIZATION; PLANT GENETIC RESOURCES.]

2. Meristem Culture

Meristemming or meristem culture relies on excision and axenic culture of a plant's meristematic dome and leaf primordia to establish a virus-free, clonal replicate of the original plant. The technique is similar to micropropagation, but uses a smaller (<0.5 mm) explant in order to achieve a clean clonal replicate of the original plant. Because the meristem may grow more rapidly than viruses in the vascular system, and because the normal vascular connections and cell-to-cell connections (plasmodesmata) are not yet formed in a meristematic dome, it is possible to rescue a virus-free plant using meristem culture techniques. The donor plant may also be heat treated prior to explant excision, in order to inactivate some viruses and guarantee very rapid tip growth. Successful meristem cultures can next be mass-propagated using a micropropagation routine. Meristem culture is a practical procedure for eliminating virus infestations from some plants that are routinely propagated by clonal methods, and can be a component in virus-indexing programs for plant production.

3. Micrografting

Micrografting, which combines the productivity of grafting with the economics and efficiency of micropropagation, is a practical method for virus elimination, circumventing lengthy, fruitless juvenile phases for orchard crops, or rejuvenation of mature tissues. For micrografting, both members of the graft union can be selected from micropropagated plant cultures, or a microcutting could be grafted on a seedling rootstock which is clean and disease-free. The technique involves splicing of a desired scion and rootstock plant *in vitro*, using delicate microscalpels to accomplish excision and transfer of a shoot apex to a decapitated rootstock, and requiring significant manual dexterity on the part of the grafter. Grafting unions heal while the stock/scion combination is maintained in the aseptic *in vitro* environment, and potential graft incompatibilities may be detected more rapidly than *in vivo*.

4. Callus and Suspension Culture

Callus induction is the formation of non-organized, proliferating, aggregated cell masses from organized plant tissues. In nature, callus is a wound tissue produced as a reaction to tissue injury. In plant tissue culture, callus is induced (usually on an auxin- or auxin/cytokinin-supplemented medium) from a wide range of explants. Callus is maintained either on gel-solidified medium (Fig. 7) or on some support over liquid nutrient medium. Some types of callus readily revert back to organized structures (regeneration), but other cell types do not exhibit regenerative capacity. The status (in particular, the maturity) of the original explant may have a profound influence on the later ability of the callus cells to regenerate into entire plants or organs.

Callus cells range in color from green or white to yellow–brown, and can be generated either as a highly friable, loosely aggregated mass or as a compact, nodulated aggregate of cells. When isolated from the parent tissue and grown separately, callus divides more rapidly than differentiated plant cells. Callus is typically subcultured to fresh medium while still growing rapidly, and eventually will maintain a predictable growth rate and uniformity.

Suspension cultures are initiated by transferring loose, friable callus masses to agitated liquid culture systems (Fig. 8). The callus cells ideally should have been maintained for several subcultures in the unorganized cell state, and exhibit a fairly consistent growth rate and uniformity. Plant cells have a minimum inoculation density requirement at the initiation stage. If an insufficient number of cells are inoculated to initiate a suspension culture, growth will slow to a lag phase and death may occur. Single cells or small aggregated cell masses disperse evenly in the medium, multiply while suspended, and in general will grow more

FIGURE 7 Callus culture of *Ajuga reptans*.

FIGURE 8 Suspension cultures of *Ajuga reptans.*

rapidly than the same cells grown as callus. The suspension-cultured cells are more homogeneous than either differentiated plants or callus masses, and can be more uniformly controlled. Cells in suspension cultures typically multiply at an exponential rate until culture medium and volume become limiting, then cells will slow down to stationary phase. Transfers to maximize the productivity of suspension cultures must occur before the stationary phase is reached.

Both callus and suspension culture systems are potentially unstable as compared to the more uniform clonal techniques of micropropagation and other methods using differentiated plant tissues and organs. Cytological deviations occur spontaneously during even routine, carefully regulated cell culture cycles. Variant cell types may be selectively favored during the culture cycle, and eventually become a dominant cell type. As a result, the tendency for suspension and callus cells to exhibit cytological heterogeneity is a recognized phenomenon, which may be advantageous (when callus is used to select for superior cell types) or detrimental (when regenerated plants from callus culture exhibit unexpected deviations in expected traits). Somaclonal variation is the term reserved for genetic variation in plants regenerated *in vitro,* often the consequence of genetic instability.

5. Regeneration (Organogenesis and Somatic Embryogenesis)

Regeneration can spontaneously occur from callus cultures or directly on an excised organ of a plant (Fig. 9). Organogenesis is the production of roots or shoots. Root induction in callus culture occurs in a medium with a high auxin-to-cytokinin ratio, while adventitious shoot formation is favored by the opposite growth regulator balance. In many cases after induction of the organ, further regeneration of an entire, differentiated plant will require transfer of the culture to a growth regulator-free medium.

Somatic (asexual) embryos regenerate from somatic cells or tissues, and form an organized bipolar structure. The regeneration can occur directly from the explant somatic cells, or indirectly (with an intervening callus stage). The strategy for somatic embryo induction from callus may involve an auxin treatment (which induces formation of embryogenic cells) followed by transfer to growth-regulator-free medium (for embryo and plantlet regeneration). Somatic embryogenesis is a technique with considerable commercial application, since the somatic embryos can be encapsulated as synthetic seeds. This innovation allows for clonally produced germplasm to be handled via automated techniques, using some strategies borrowed from conventional seed propagation technology.

6. Rhizogenesis and the *in Vitro* to *ex Vitro* Transition

Microcuttings produced by any of the commercially applicable methodologies previously described need to be rooted and acclimatized to the *ex vitro* growing environment, before the practical benefits of plant tissue culture production can be a realized advantage to the plant production industries. The quality of the root system is vital to plant quality, and will dictate survival during later stages of the plant growth cycle. The procedures used to initiate roots and form a new adventitious root system have pivotal influence on the anatomical structure and overall morphology of the plant as it makes the transition into the greenhouse or field setting (Fig. 10). Microcuttings can be excised from shoot cultures, transferred to auxin-rich or growth-regulator-free rooting medium, and rooted while still in the culture environment. The rooted cultures are subsequently removed from culture and transferred to a transitional acclimatization environment (with temperature, humidity, and light irradiance set points intermediate between the microculture physical microenvironment and the *in vivo* greenhouse conditions). Alternatively, the microcuttings can be removed from culture and rooted *ex vitro* under mist or high-humidity tents, with acclimatization occurring at the same time. In most production schemes, the *ex vitro* rhizogenesis option has proven to be the most efficient for commercial production schemes, and for woody plant microcutting production, the *ex vitro* option results in a far superior root system and overall im-

 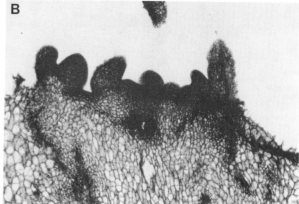

FIGURE 9 (A) Shoot regeneration from pecan embryos *in vitro*. (B) Adventitious buds arising from pecan callus *in vitro*.

proved plant quality during later stages of production and use.

7. Mass Cultivation, Bioreactors, Automation, and Robotization

Although each of the *in vitro* techniques described so far are categorized as commercial and production methods, the economic feasibility of plant tissue culture hinges on innovations to increase efficiency, reduce costs, and introduce practical improvements to a production scheme. Micropropagation has erased many previous barriers to production, and has proven to be a cost-competitive means for mass cultivation of plants. Micropropagation technology has introduced a powerful tool for phenotypic improvement and delivery of uniform, clean, disease-free plant material in exceptionally short turn-around times. These advantages have led to improvements in the way plants are shipped and marketed, and have altered the seasonal nature of the plant production industries.

While most commercial micropropagation is performed on gel-solidified media, new successes with liquid micropropagation systems and/or bioreactor-based large-scale micropropagation systems have been achieved. In these liquid medium systems, the plant material is rapidly mass-produced in the form of meristemoid clusters, bud clusters, or somatic embryos, which are mechanically separated, and then transplanted *ex vitro* for further growth in the greenhouse. Bioreactors are also an integral part of the synthetic seed strategy described earlier. The bioreactor systems in particular are characterized by rigorous grower control over both the physical and chemical microenvironments. When this intensive control over the plant product is coupled with enhanced opportunity for mechanization and automation, the industry benefits from reduced labor costs and increased productivity.

Automation and robotization have gradually permeated nearly every phase of microculture production on a large commercial scale. The technologies and engineering tactics include systems for routine media preparation/vessel handling, vessel washing, *in vitro* manipulation of plant cultures during the transfer steps, and automatic evaluation of product quality. Finally, automation and robotization have aided the inventory, packaging, and shipping processes in commercial operations. Robotic systems for plant manipulation can use machine vision algorithms to guide robotic tool performance, and/or lasers to cut plant tissues during the transfer process.

B. Breeding, Selection, and Investigative Research Methods

1. Haploid Culture

Isolated anthers, pollen grains, or unfertilized ovules can be induced *in vitro* to produce haploid individual plants. The technique is of critical interest to breeders, who can identify desirable traits in the expression of single-copy genetic information. Breeders are then able to immediately obtain homozygous pure-breeding parental lines (doubled haploids) for controlled crosses *in vivo*. Most of the success in this technique has been achieved with anthers. [*See* PLANT GENETIC ENHANCEMENT.]

2. Embryo Culture/Embryo Rescue

Embryos can be isolated from immature seeds and nurtured *in vitro* to produce viable plants, even when the same embryos would spontaneously abort in nature. Embryo culture may be applied to crops with

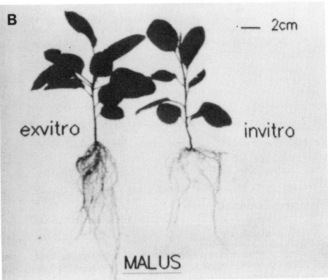

FIGURE 10 (A) *In vitro* rhizogenesis from a micropropagated pecan shoot. (B) Comparison between *in vitro* and *ex vitro* root morphology for apple microcuttings.

unusually poor germination success, or to seeds with complex dormancy requirements, as a means of propagation. When the technique is applied to recover the progeny of a unique, yet incompatible, sexual cross accomplished in nature, the term *embryo rescue* is used. Embryo culture requires exacting dissection of the tiny immature embryo and provision of complex nutrient medium to simulate conditions in the ovule. Alternatively, larger mature embryos can be rescued from ripe seed to eliminate seed inhibition to germination. Unlike other methods, embryo culture does not usually require direct disinfestation of the explant, but rather pre-sterilization of the seed prior to dissection.

3. Protoplasts (Somatic Hybridization and Electroporation)

Protoplasts are plant cells with enzymatically removed cell walls, bounded by a plasmalemma (Fig. 11). Culture of protoplasts provides a vehicle for several physiological and genetic studies of plants, but protoplasts also are particularly interesting due to the potential for plant improvement by inducing genetic change in the cultured cells. Plant protoplasts are quite susceptible to *in vitro* stress-induced genetic change

(somaclonal variation) and, consequently, the plantlets eventually regenerated from protoplasts can exhibit considerable variation in phenotypic traits. Protoplasts can be isolated from plant leaves or other organs, or alternatively from callus and suspension culture cells. Release of the individual protoplasts from the cell walls is accomplished in an enzyme solution supplemented with high sugar concentrations to osmotically stabilize the cells. After isolation and purification, protoplasts remain in a high osmotic medium until cell walls are regenerated. The cell divisions eventually lead to callus and/or somatic embryo formation as described previously.

Somatic hybridization is the technique of fusing the vegetative cells to produce a polynucleate cell. This is a valuable option for joining germplasm from sexually incompatible plants. Fusion is deliberately induced by treatments (electrofusion or fusion induced by chemical agents) that join cell membranes, and then selective techniques for isolating the desired somatic hybrids are conducted. Electroporation is a related technique which takes advantage of the wall-free status of protoplasts. Electrical field strength and pulse duration are adjusted to produce minute pores

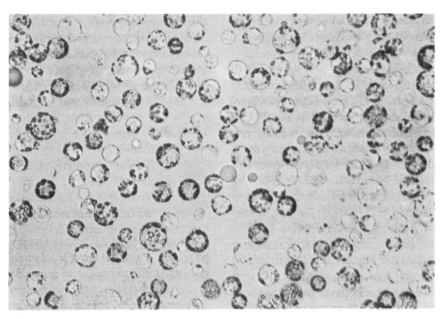

FIGURE 11 Protoplasts isolated from a birch shoot culture.

in the protoplast membranes. The pores are used as the entry port for molecules of genetic information that would normally be excluded from the cell. This is an additional means of engineering unique genetic combinations using *in vitro* methods.

4. Biotechnology/Gene Transfer Systems

Plant tissue culture, since it features sterile plant tissues preacclimated to manipulation in culture, provides excellent donor material for several other transformation procedures. Biolistics (microprojectile bombardment) is a technique which uses accelerated particles to deliver genetic material directly into plant cells in culture. *Agrobacterium*-mediated transformation takes advantage of the fact that this bacterium is able to insert genetic information directly into plant tissues. Both systems accomplish transfer of foreign DNA into plant cells and circumvent the complex culture and regeneration problems inherent to protoplast systems.

5. Screening and *in Vitro* Selection

Plant tissue culture can be used on both the cell-level and the whole-plant level to analyze responses to environmental stress, and to selectively isolate superior genotypes with potential to perform well under natural field conditions. Cell cultures (callus and suspensions) can aid in isolation and study of cell-level responses to environmental stress (high salt, water deficit, herbicides or pesticides, a pathogenic toxin, or ion toxicity). Cell cultures can be evaluated for

tolerance to these factors by gradually introducing gradients, stepwise increasing concentrations, into the nutrient medium over the course of a series of subcultures. The cells which are able to survive and proliferate despite exposure to the selective agent *may* possess a cellular trait which will be advantageous in a regenerated plant; experimental results on this front have been mixed. In other approaches, whole plants or shoot cultures have provided the test plants for screening and selection *in vitro*. These systems permit evaluation of whole-plant responses to stress, yet eliminate many of the ungoverned variables that hinder similar research in the field. Microculture screens using differentiated plants are an efficient intermediate screen to evaluate regenerated plants (from cell culture experiments) in terms of whole-plant potential, prior to expensive replicated field trials (Fig. 12).

Superior genotypes can be isolated under selection pressure as described above, or selected by chance given the tendency for somaclonal variation in undifferentiated cultures. Alternatively, the *in vitro* cultures can be subjected to deliberate mutation treatments in the hope of generating novel germplasm. Deliberate mutation can be induced chemically or by irradiation.

6. *In Vitro* Methods in Physiological/Biochemical Research

Plant tissue culture yields a unique research model for in-depth analysis of direct plant responses to environmental conditions that would be obscured in com-

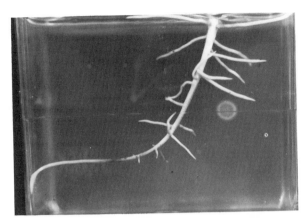

FIGURE 12 Colonization of an *in vitro* soybean root with *Phytopthora* fungus, as observed directly through the walls of the tissue culture vessel. Clear examination of root–pathogen interaction is not possible in a soil environment.

plex *in vivo* growing environments. Callus cultures have been intriguing subjects to examine the cell-level receptors or ultrastructural changes under reproducible, defined, and adjustable physical and chemical microenvironmental conditions. Microscopic studies of callus have elucidated methods of cytodifferentiation and clarified genetic and structural research questions. Cell, callus, and protoplast culture in phytopathology have defined host–parasite interactions and virus entry into cells under controlled conditions with a clarity that would never be possible in nature. Root cultures have provided a unique experimental system for investigating the precise requirements of plants for vitamins, growth requirements, plant nutrition, and hormone physiology.

C. Secondary Metabolites from Plant Tissue Cultures

Plant cells from familiar horticultural and agronomic crops are actually an underestimated resource for many valuable secondary products, including rich natural pigment extracts, flavors, aromas, and pharmaceutical compounds. As today's consumers have become more aware of (and resistant to) artificial food additives and synthetic chemicals, research has shifted to explore these presumed safe, practical alternatives derived directly from plants. Harvest and extraction of natural plant extracts directly from field crops are restricted by problems with variation in quality, chemical instability, seasonal restrictions, and higher costs than synthetic alternatives in most cases. In some cases, endangered species (e.g., plants indigenous to tropical rain forests) are disappearing

before their full value as sources of medicinals can be exploited. In other cases, the plants which produce valuable secondary products are slow-growing, woody specimens, which are available in insufficient supply, and are difficult to harvest. The compound taxol, extractable from the endangered Pacific Yew tree, is a case in point. Although there is good evidence that taxol is an effective chemotherapeutic agent, the natural plant stands are not able to supply sufficient chemical for testing, much less commercial production.

However, plants *in vitro* can be stimulated to produce and accumulate secondary metabolites, which can provide a producer with continuous supply of high-quality product. In many cases, the products are harvested from cell cultures, whereas in other cases, differentiated organs (roots, shoots, embryos) must be cultivated to achieve synthesis of the secondary metabolite. *In vitro* production of secondary metabolites also offers the potential for rigorous control, to tailor a cell line to produce a specialized product. Elite cell lines designed for this application can be created using many of the techniques described previously (biotechnology or selection), and then multiplied using asexual plant tissue culture methods. Once research has identified each of the exacting conditions required in the chemical (nutrient) and physical microenvironment to stimulate metabolite production, these cultured cells can be scaled up for production in large-volume bioreactors, which increases production efficiency and helps lower costs.

V. Methods of Experimental Evaluation

Satisfactory evaluation and quantification of experimental results or production efficiency in plant tissue culture is a particular challenge, due to the nature of the *in vitro* systems. Growth, developmental, or production changes that occur *in vitro* will be reflected by changes in culture growth rate, cell survival, three-dimensional volume of cell masses, color, morphology, number of entities, or density. The difficulties of *in vitro* measurement arise because the plant material exists within the sterile, humid, protected physical and chemical microenvironment inside a culture vessel. Any venture to directly measure the culture usually requires intrusion on the delicate microenvironment. At best, the culture is disturbed and altered in the measurement process, and at worst the culture must be terminated at the time of measurement.

Most standard techniques for monitoring/evaluating *in vitro* performance require that cultures are sacrificed at the time of measurement. Analysis of performance in callus or differentiated plant cultures is based on fresh (FW) or dry weight (DW) measurements. FW measurements can be carefully made using aseptic techniques on sterile surfaces in the laminar flow hood; however, this measurement (which is dependent on the water content of tissues) is not a thorough measure of growth, and the culture microenvironment is still invaded during the measurement. Cell numbers in aggregated suspensions or callus are assessed microscopically (using a hemocytometer) after treating the tissues with chromic acid. The chromic acid treatment separates cells from one another. Cell number in fine cell suspensions and protoplast cultures is determined by sampling without chromic acid pretreatment. Alternative measurements for suspension-cultured cells or cells in bioreactors include packed cell volumes (after centrifugation), mitotic index, and total protein. Analysis of secondary products (using for example, high-performance liquid chromatography) may be done concurrently. Protoplast or single-cell culture development is commonly assessed using a plating efficiency (PE) calculation, where PE = (number of colonies/initial number of units) × 100. The estimates can be used to prepare growth curves.

Indirect measurements of cell culture performance (light absorbance, radioactive tracer counts, redox potential, electrical impedance, ATP concentration) have been substituted for the above, or used to augment the measurements. Very often visual assessments of relative growth or culture appearance have been used to evaluate treatment differences. The visual observations can be made through the transparent or translucent walls of a culture vessel without invading the culture microenvironment; however, they are subjective. Computer-generated mathematical models have been constructed to predict culture performance with more accuracy for both cell and whole-plant systems.

Machine vision (microcomputerized image analysis) is a technique which has particular merit for evaluation of plant tissue cultures, since it permits direct, objective, innocuous measurement of both cell and differentiated plant cultures. The machine vision measurement, which is conceptually straightforward and simple, evaluates the photometric (spectral) and morphometric (spatial) visible characteristics in the *image*

of a sample. The image is captured using a video camera, the culture can be viewed through the walls of the vessel without intruding on the microenvironment, and the measurement can be adapted for both micro- and macroscopic sample images. Machine vision measurements applied to cell cultures have proven to be exceptionally sensitive, and accurate correlations between image analysis calculations and conventional destructive measurements (FW, DW, cell number, etc.) have repeatedly been established in diverse plant tissue culture systems. The combination of machine vision assessment and automation in bioreactor production has particular merit, to manage control of each of the complex physical and chemical cultural parameters toward enhanced productivity and lower production costs.

VI. Implications for Agricultural Science

Plant tissue culture, with its broad range of applications, has had a significant impact on the agricultural sciences both as a vehicle for basic scientific investigation and as a means for development of superior plant genotypes. Many of the gains have been realized most efficiently via the partnership of *in vitro* technologies and traditional *in vivo* plant production and improvement strategies. For example, *in vitro* investigative methods have facilitated in-depth understanding of plant membrane function and cell-wall receptors. This insight provides clues for development of more effective herbicides for field application, or for designing genotypes with superior yield or disease resistance. Tissue culture techniques for mass-propagation of superior clones have revolutionized the foliage plant production industries, and have had growing impact on forestry, agronomic, and horticultural plant industries especially as methods for acclimatization of microplants have been improved. In particular, the *in vitro* technologies have been a boon to strategies for preservation and reintroduction of rare and endangered plant species. Screening and selection coupled with other strategies for *in vitro* genetic improvement have afforded complementary techniques for traditional plant breeding trials. Finally, the emerging gene transfer and *in vitro* plant manipulation techniques have permitted testing and introduction of novel genotypes which have been engineered specifically to withstand several of the climatic or environmental obstacles that have traditionally limited agricultural plant productivity throughout the world.

Bibliography

Debergh, P. C. and Zimmerman, R. H. (1991). "Micropropagation. Technology and Application." Kluwer, Dordrecht.

Payne, G. F., Bringi, V., Prince, C., and Shuler, M. L. (1992). "Plant Cell and Tissue Culture in Liquid Systems." Oxford Univ. Press, New York.

Smith, R. H. (1992). "Plant Tissue Culture. Techniques and Experiments." Academic Press, San Diego.

Stafford, A., and Graham Warren. (1991). "Plant Cell and Tissue Culture." Open Univ. Press, Buckingham.

Torres, K. (1989). "Tissue Culture Techniques for Horticultural Crops." Van Nostrand Reinhold, New York.

Plant Virology

ROBERT M. GOODMAN, THOMAS L. GERMAN, *University of Wisconsin*

JOHN L. SHERWOOD, *Oklahoma State University*

GLOSSARY

Gene Segment of DNA (or in some viruses RNA) that encodes the information to specify the amino acid sequence of a protein and the information about how the coding sequence should be expressed; genes are the unit of heredity in all living things, transmitting genetic information from one generation to the next; fidelity of transmission of information by genes is the basis of predictability of inheritance, and errors in transmission are an important source of variation on which natural selection operates during evolution

Infection Process by which a pathogen becomes established in a host; infection is necessary for a virus to carry out its life cycle, making virus-encoded proteins and copies of its nucleic acid and assembling complete progeny virus that can be spread to other cells and other hosts

Inoculation Process by which a pathogen is introduced into a host; inoculation may be successful or unsuccessful depending on whether infection takes place

Nucleic acid Chemical polymers that genes are made of; nucleic acids are composed of a backbone polymer of sugars, either ribose or deoxyribose, linked together by phosphate bonds; attached to each sugar is one of four bases, two of which are in a class called purines and two in a class called pyrimidines; in deoxyribonucleic acids (DNA), the purines are adenine and guanine and the pyrimidines are thymidine and cytosine; in ribonucleic acids (RNA), the purines are adenine and guanine and the pyrimidines are uracil and cytosine; a special form of RNA, called

messenger RNA, is used to make proteins by a process called translation; in a process called transcription that uses DNA as a template, viruses containing DNA make a messenger RNA before protein synthesis can occur

Protein Polymers of amino acids that do the biochemical work of living cells; 20 different amino acids are found in proteins; each protein gets its unique properties and character from the primary sequence of the amino acids, which in turn determines in part how the protein folds into the three-dimensional shape that gives it its function; many proteins are enzymes, or catalysts for biochemical reactions, but others can play structural roles, such as in virus coat proteins

Serology Use of antibodies for detection and study of antigens such as plant viruses. Foreign proteins, and to a lesser degree nucleic acids, when injected into an animal can stimulate the immune system of the animal to produce antibodies that specifically bind the material injected. Serum from the blood of an immunized animal, containing the antibodies, is collected and used to detect and differentiate viruses from each other and from host proteins

Plant virology is the study of the viruses that infect plants. Viruses are *infectious nucleoproteins*. Infectious means that they infect and multiply in a living host, and nucleoprotein means that viruses are composed of nucleic acid and protein. Viruses must infect a host cell to multiply and spread. Viruses are as a rule specialized for a specific host, but all living organisms that have been well-studied in biology, including bacteria, insects, humans, and plants, can be hosts for specific viruses. When a host is infected by a virus, the effects can range from debilitating chronic disease or death to mild or even no symptoms. All viruses use either DNA or RNA to encode genetic information.

Virus genes are similar in structure and function to the genes of the host cell. Viral proteins, as well as new viral nucleic acid, are produced by the host during infection. One or more of the viral proteins form the coat, enclosing and protecting the viral nucleic acid so it can be transmitted to a new host. Other proteins are involved in the process of infection and multiplication of the virus, but they do not possess the protein machinery necessary for life independent of a living cell.

Many, even very well educated people are surprised to learn that plants can be infected by viruses. In fact, there are many viruses that infect many different plants, both cultivated and noncultivated. In cultivated crops, viruses cause diseases that result in important problems for agriculture. Plant viruses have much in common with the viruses that infect animals. These similarities were discovered when the molecular structure and organization of the viruses were investigated. Infected plants do not sniffle or sneeze, but that is because plants do not respond to virus infections in the same way that animals do. The symptoms of virus infections have much more to do with the biology of the host than with that of the virus. [See PLANT PATHOLOGY.]

I. Discovery and Nature of the Plant Viruses

The first known written record of a plant virus infection is in a poem written by the Japanese Empress Koken in A.D. 752. The poem refers to the striking yellow pattern of symptoms on the leaves of a Japanese honeysuckle. The virus involved is now known to be a member of the Geminivirus group, one of 34 families and groups of plant viruses.

Other early records of plant virus infections are paintings and drawings by seventeenth-century European artists of tulips displaying showy striped patterns on their flowers. Varieties with these striking features (Fig. 1), which are symptoms caused by virus infections, were highly prized and speculatively priced in the horticultural industry of the time. The result was an early example of speculation and subsequent collapse in a marketplace.

Though the striking symptoms of some plant diseases caused by viruses have been recognized for centuries, knowledge about viruses themselves is less than 100 years old. Viruses are too small to be seen with a light microscope. Viruses also cannot live on their own; they depend on infection of a host. Unlike some organisms that also require an intimate association with a host for growth but possess their own biochemical machinery needed to make proteins and for generation of energy through respiration, the viruses are completely dependent on their host for these functions.

A. The Physical and Chemical Nature of Plant Viruses

By the late nineteenth-century, many of the plagues and disorders of common farm animals, humans, and plants were known to be caused by bacteria or fungi. But there were some diseases, among them some notably conspicuous ones of plants, for which the tools and methods of late nineteenth-century science failed to explain the cause. One of these diseases was the tobacco mosaic disease (*Mosaikkrankheit* in German). The Russian bacteriologist Dimitri J. V. V. Iwanowski and the Dutch scientist Marinus Willem Beijerinck showed that the sap from a mosaic-diseased plant, when passed through a filter that would retain all known bacteria, caused infection when rubbed on healthy plants. Other investigators soon showed that numerous other plant diseases were caused by similarly "filterable viruses."

The choice of the tobacco mosaic virus (or TMV) for these early experiments was fortuitous. TMV is one of the most highly infectious plant viruses and is transmitted to a new host by rubbing with the sap of an infected plant. Although many other plant viruses can be transmitted in this way, few are as highly infectious, and most (unlike TMV) are transmitted in nature by insects (the first report was by the American virologist L. O. Kunkel in 1922). Some plant viruses cannot be transmitted experimentally at all with sap from infected plants.

TMV is also atypical in that infected plants contain enormous amounts of the virus. Virus makes up 10% of the total dry weight of an infected plant (or 1% of the wet weight). TMV-infected plants were thus ideally suited for attempts to isolate pure preparations of the virus. Success came in the 1930s with methods developed earlier for isolation and purification of proteins, which was also pioneered with plants, by James B. Sumner in 1922, who studied the enzyme urease from jack bean. Wendell M. Stanley reported the crystallization of a "protein" from TMV-infected plants in 1935. Frederick W. Bawden and Norman Pirie working in England reported the next year that purified TMV contained not just protein, but also nucleic

FIGURE 1 Color breaking of tulips. (Courtesy of Bulb Research Centre, Lisse, The Netherlands.)

acid, that is, it was a nucleoprotein. The distinction between "protein" and "nucleoprotein" was critical, because the nucleic acid of the virus carries the genetic information. But until the 1950s, this point was not at all clear. Indeed, Stanley shared the 1946 Nobel Prize in Chemistry with Sumner and with John Howard Northrop, also a protein biochemist, while the contributions of Bawden and Pirie went unrecognized even though the English group got the right answer about the chemical nature of TMV.

B. The Structure and Function of Viruses

The English biochemist Roy Markham in the early 1950s studied a spherical plant virus (not unlike a microscopic soccer ball) that produced some protein coats that were noninfectious. Physical measurements suggested that these were "empty" shells, lacking nucleic acid, but were otherwise just like the nucleic-acid-containing ones. Markham's conclusion was that the nucleic acid must be contained inside the protein shell.

This idea was extended by James D. Watson and Francis Crick for TMV. Work done in the 1930s, including some of the earliest images seen in the electron microscope, had hinted that TMV was rod shaped. Watson and Crick suggested that virus particles consisted of numerous identical protein subunits arranged in a symmetrical array. A general theory for the structure of simple viruses, based largely on the work with plant viruses, was put forward by D. L. D. Caspar and Aaron Klug in 1960. The model presented for TMV was based largely on data from work by Rosalind E. Franklin. Rod-shaped simple viruses (Fig. 2) have their protein subunits arranged in a helical array, with every internal subunit just like every other subunit, and in identical relation to its neighbors. The nucleic acid is held in intimate structural contact with the helical protein array. In the spherical viruses, such as that studied by Markham, the arrangement of coat protein subunits in the simplest cases has cubic symmetry.

These structural ideas have held up, and recent studies of plant virus structure at the atomic level by sophisticated X-ray crystallography have elegantly confirmed the early models. The structural details of the simple plant viruses have been found to be a conserved feature throughout the world of viruses and are a further example of the fundamental way in which work with plant viruses has been seminal in the development of biology. Nevertheless, there are numerous variations on these simple structural motifs. The different shapes or morphologies found among the plant viruses are illustrated in Fig. 3.

C. The Role of Virus Nucleic Acids

Twenty years after Bawden and Pirie showed that nucleic acid was present in TMV, the specific biologi-

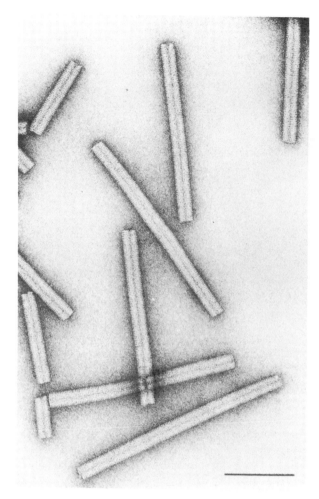

FIGURE 2 Soil-borne wheat mosaic virus, an example of the simple rod-shaped viruses of which TMV is also an example. [Reprinted with permission from R. I. B. Francki, R. G. Milne, and T. Hatta. (1985). "An Atlas of Plant Viruses," Vol. II. CRC Press, Boca Raton, FL. Copyright CRC Press.]

cal role of the nucleic acid was demonstrated. In 1952, Alfred Hershey and Martha Chase reported on work with a bacterial virus that the nucleic acid (DNA in the case of the bacterial virus used) entered the cell but the protein did not. Three years later, in 1955, Heinz Fraenkel-Conrat and Robley Williams reconstituted active TMV from isolated protein and nucleic acid components. The next year, several workers in Germany and the United States (including Fraenkel-Conrat) independently reported that deproteinized TMV nucleic acid (in this case RNA) was infectious, and inoculation with the RNA alone resulted in the production in infected plants of virus containing protein and nucleic acid. Taken together, the results reported for TMV in 1955 and 1956 clearly established the genetic role of the nucleic acids and the protective role of the protein coat in the virus life cycle. These

and many subsequent studies showed that chemical treatments that abolish infectivity of the naked RNA do not abolish that of the intact virus particle, and digestion of the proteins of plant viruses with enzymes that do not destroy nucleic acids do not abolish infectivity.

Among the simplest plant viruses, of which TMV is the classic example, the nucleic acid consists of one strand of nucleic acid, on which genes are encoded, including that for the protein coat. More complex viruses exist. The complexities can be in the number or structure of the nucleic acids, the number or structure of the coat proteins, and in the presence of other components in the mature virus particle. One common variation in the plant viruses is that the full complement of genes in the virus is distributed among more than one nucleic acid molecule. These are called multicomponent viruses. Another variation is that the virus coat, called the capsid, in some viruses can consist of more than one type of protein. Some of these other proteins in some viruses have other biological activities in addition to serving the structural role of enclosing and protecting the nucleic acid. Yet another variation is that the virus outside the infected cell is enveloped within a lipid membrane, which is primarily derived from a membrane from the cell in which the virus is produced.

Different plant viruses contain either DNA or RNA, which is either single stranded or double stranded. In the case of the vast majority of plant viruses, the nucleic acid found in the viruses is single-stranded RNA. Most RNA plant viruses, whether single stranded or double stranded, have linear RNA genome segments, but the double-stranded and single-stranded DNA viruses of plants all have circular genomes.

D. The Genetic Organization and Life Cycles of Viruses

During infection, a virus must enter a susceptible host cell and partially reprogram the cellular machinery of gene expression and metabolism for virus production. This it must do, however, in a manner that does not kill the cell, or at least not until the infection cycle is completed. This requires a collection of genes whose function is finely tuned to this elegant level of parasitism.

Every virus encodes the genes needed to infect the host cell and to produce the virus coat protein. Every virus also needs to have genes to make the necessary enzymes to make more copies of the virus nucleic acid. Some viruses encode their own enzymes for

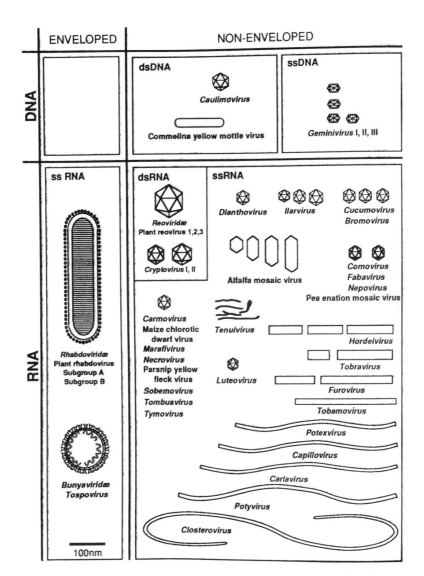

FIGURE 3 Families and groups of viruses infecting plants. A summary of the major groups of plant viruses grouped according to virus morphology and nature of the nucleic acid they contain. [Courtesy of Springer-Verlag; from R. I. B. Francki, C. M. Fauquet, D. L. Knudson, and F. Brown (eds.). (1991). Classification and nomenclature of viruses: Fifth Report of the International Committee on Taxonomy of Viruses. *Arch. Virol.*, Suppl. 2, p. 62.]

nucleic acid synthesis; others encode a portion of the necessary enzymes for this virus replication function and obtain the rest from the host cell.

Other functions encoded by virus genes, beyond those for the coat protein and the replication machinery, have to do with functions related to the host cell that the virus infects. In the very simplest cases, such as in TMV, one nonstructural protein (i.e., it is not a part of the virus particle) acts to make possible the movement of the virus from cell to cell in the plant. Such "movement" proteins are found in many plant viruses. Numerous viruses are transmitted in nature by insects or other invertebrate animals that feed on plants. These "vectors" have a specialized biological

relationship with the virus as well as with the plant. In at least some important virus families, a virus gene encodes a protein that is required for the specificity of the vector relationship. Several groups of plant RNA-containing viruses produce the viral encoded proteins by first producing a long, multifunctional protein that is "processed" into individual proteins that serve their various purposes in the virus life cycle. These large proteins (called polyproteins) are processed by specialized enzymes, called proteases, that are encoded by the virus.

In the more complex plant viruses, other biological functions are encoded by virus genes. These can include enzymes that modify the nucleic acid, proteins that control the expression of other virus genes, and proteins that specifically interact with the host cell, for example, to produce cellular structures in which virus assembly takes place or to produce symptoms.

II. Relationship of Plant and Animal Viruses

A. Origins and Evolution of Viruses

We do not know whether viruses evolved from different progenitors or from a common progenitor. But we do know that viruses with nearly identical molecular organization have hosts as divergent as bacteria, plants, and animals. All viruses have the same basic biological problems to solve. Their proteins and nucleic acids must be synthesized and assembled, the virus must be protected from biological or physical degradation, and they must be spread within a host and between hosts. Because there is a remarkable similarity in the biological processes that occur in all living cells, it makes sense that the survival mechanisms developed by viruses be similar regardless of the host organism to which a virus is specialized.

The factors that drive evolution are variation and selection. Variation occurs as the result of point mutations or when deletions or insertions of genetic sequences are caused by errors in the synthesis of genomic material. Variation in the sequence of bases in nucleic acids arises from chemical, physical, or biological perturbations that cause mistakes in the replication of DNA or RNA. These mistakes can alter the function of the protein produced from the altered genetic code, and such changes are then acted on by selection. Other variation arises as the result of exchange of regions of large segments, for example, between two viruses that infect the same cell. Presumptive evidence for evolution based on exchange of large segments of genes comes from the analysis of viruses that infect both plants and insects.

B. Organizational Themes among the Viruses

The architectural principles of icosahedral, rigid, or filamentous rod-shaped structures described earlier for the plant viruses apply to all viruses. Thus, viruses infecting organisms from every kingdom have the same physical features. Polio virus of humans, cowpea mosaic virus of plants, and black beetle virus of insects are essentially identical in structure because they are based on the same underlying principles of molecular architecture.

There are also common features among unrelated viruses at the genetic level. The majority of plant viruses encode their genes on one or more single-stranded RNA molecules that are directly used by the infected cell's protein synthesis machinery to make proteins. Such viruses are called "plus" (or +) sense RNA viruses. There are also many (+) sense RNA viruses known that infect animals. Among the (+) sense RNA viruses of plants, insects, and higher animals, there is a high degree of similarity between the regions that encode the enzyme required to produce more viral RNA (polymerase). The fact that genetically related viruses are found in hosts as diverse as mammals, birds, mosquitos, and plants suggests that they diverged from a common progenitor or that extensive gene exchange has occurred among viruses of plant and animals and possibly bacteria.

That a common progenitor may account for the commonalities of viruses is further supported by the linear arrangement of the genes on the genomes of (+) sense RNA viruses. The linear arrangement is conserved among the viruses that infect both plants and animals and is irrespective of the morphology of the virus particle. Besides the polymerase, viral replication requires other gene products, which include capping enzymes, helicases, small genomic-associated proteins (VPg), and proteases. These proteins make functional modifications of the viral nucleic acid or are involved in the replication of the virus. When these genes are present, the arrangement of the genes along the RNA is conserved among virus groups. For example, both the picornaviruslike group and the alphaviruslike group include viruses that would seem to be extremely diverse. Polio virus, a spherical-shaped virus and serious human pathogen, and potato virus Y, which has filamentous particles and is a type member of one of the most agriculturally important plant virus groups, are both picornaviruslike in gene organization. The arrangement of the genes involved in replication in all picornaviruses is the same (Fig. 4A). TMV, a rod-shaped plant pathogen, and sindbus virus, a spherical animal pathogen, are both members of the alphavirus group. All alphaviruslike viruses share the same linear gene arrangement (Fig. 4B), which differs from that in the picornaviruslike viruses.

C. The Organization of Virus Gene Expression

The genomic expression strategies are different between, but consistent within, each group as well. For

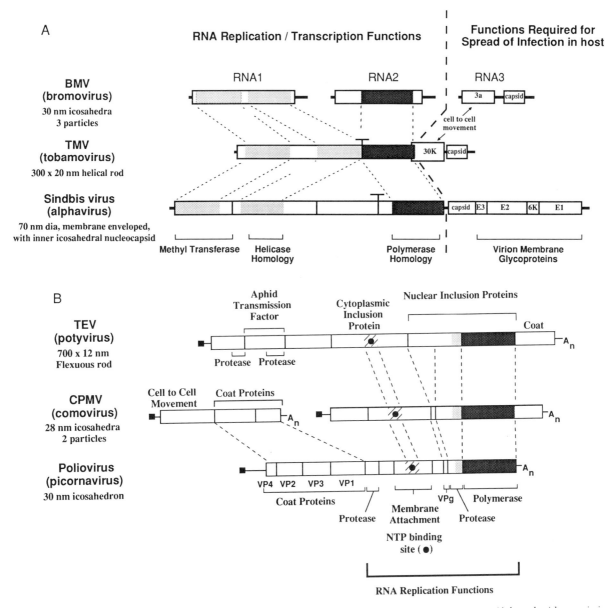

FIGURE 4 An alignment of the organization of (A) picornaviruslike and (B) alphaviruslike RNA genomes. (Adapted with permission from Paul G. Ahlquist.)

example, the picornaviruslike viruses all express their genomes through polyprotein processing by virus-encoded proteases. The viruses in the alphaviruslike group make extensive use of subgenomic messenger RNAs to regulate the expression of various portions of the viral genomic nucleic acid. These details suggest that the viruses within these groups have evolved from a common progenitor or exchanged genetic information to arrive at a common solution to the problems involved in virus survival. Similar genome organizational themes are shared among structurally different viruses from different hosts and are found

among the negative-stranded RNA viruses, double-stranded RNA viruses, and DNA viruses. An organizational scheme based on the composition, organization, and expression strategies of viral genomes is considered by many virologists to be the best way to create virus categories to reflect true evolutionary relationships.

III. Plant Viruses in Modern Agriculture

The effects of virus disease can range from subtle reductions in plant growth to abnormal effects on plant foliage, flowers, or fruits to plant death. The ultimate goal in dealing with plant viruses in agriculture is to manage the spread of, and control the losses resulting from, virus infections. To these ends, methods to detect and identify viruses are critical. Different viruses, transmitted in nature by different means (sap transmission, insect or nematode vectors, *via* pollen or seeds), can cause similar symptoms. Yet because of the means of transmission, the approach to managing spread will differ greatly from virus to virus. Also, symptoms of virus disease may mimic symptoms produced by other biotic or abiotic stresses. Thus it is very important to be able to diagnose accurately the cause of a virus infection.

A. Diagnosis

Methods for identification of a virus rely on detecting its biological activity, the unique structures or molecules produced in a host during virus infection, or the virus or its components. The classic way to identify a virus is based on host range (the plant species that can and cannot be infected) and symptoms. An unknown virus is experimentally transmitted to a range of different plants (called indicator hosts), and 2 or more weeks after inoculation, the presence or absence of infection and the pattern of symptoms are systematically recorded. A well-chosen collection of indicator hosts can be used to identify many viruses because viruses infect different hosts and on those infected produce different and characteristic symptoms. This method is primitive but effective. It requires little more than a greenhouse where unintended spread of viruses can be prevented, the capacity to grow a supply of different plant species, and some prior knowledge about what plants to grow, how to perform the inoculations, and what to look for in the way of symptoms. However, the method is not very precise and it is also impractical for some viruses that are difficult to transmit experimentally. So for the past several decades more precise and discriminating methods of diagnosis have been developed.

Direct observation of viruses is limited by the size of viruses. The light microscope can, however, be used to view inclusion bodies produced as a result of virus infection. These can be characteristic of virus groups and even of specific viruses. Inclusion bodies consist of aggregations of virus particles, virus-encoded protein, or a combination of virus and host material. They range in appearance from crystalline to amorphous.

The electron microscope permits the examination of viruses themselves and also more refined examination of inclusions. Electron microscopy is used in diagnosis for examination of freshly expressed sap in "leaf-dips" or by investigation of virus host interaction in fixed and sectioned tissue. Although many viruses have the same physical appearance (Fig. 3), information about host range, symptoms, and physical structure can often allow the diagnostician to narrow the identification of a virus to a small number of possibilities.

Serology has long been used for detection of virus proteins. Of the techniques used to detect plant viruses, serology is most commonly used when many samples need to be assayed because the production of antibodies is inexpensive and the testing methods are amenable to automation. Both polyclonal antibodies (PAB) and more recently monoclonal antibodies (MAB) have been used in a variety of serological assays. The enzyme-linked immunosorbent assay (ELISA) is commonly used for detection of viruses or for determining taxonomic relationships. Because of their greater specificity, MABs are used to discriminate among closely related viruses that differ in fine details of antigenic structure. Because different viruses in the same taxonomic group can share epitopes, it is also possible to use MAB or PAB produced using one virus for detection of related strains of the same virus. Serology using antibodies has been combined with electron microscopy in a method called immunoelectron microscopy. Virus or cellular structures associated with infection can be seen at the same time that an antibody specific for a virus is also detected. These structures are seen with the aid of heavy metal stains that are used to enhance the contrast between cellular structures and viruses in the cell. The location of the virus to which the antibody is specific is revealed by the prior labeling of the antibodies with gold particles, which can be seen in the electron microscope.

Virus-specific nucleic acids can also be detected. Nucleic-acid-based methods are an increasingly popular approach to identification of viruses involved in plant diseases. These methods rely either on detection of specific types or forms of nucleic acids not typically present in healthy cells or on the detection of virus-specific sequences of either RNA or DNA that correspond to the nucleic acid sequences of the virus. Infection of plants by most RNA viruses results in the accumulation of a double-stranded RNA (dsRNA) molecule representing both the (+) and (−) sense of the virus genome. Detection of dsRNA molecules of expected sizes (which differ from virus to virus) can itself give strong evidence of the involvement of a virus in a disease. But the most specific and informative detection methods are those that specifically detect the virus nucleic acid sequences.

Nucleic acid hybridization detects the viral nucleic acid directly by binding a labeled probe designed to be complementary to the virus nucleic acid in the sample. In a common application, a DNA or RNA copy of a portion of the viral genome is labeled using a chemical- or radioisotope-labeled nucleotide and commercially available labeling kits. A sample of tissue, or an extract containing the nucleic acids, from a plant is fixed to a cellulose or nylon sheet (filter) along with suitable controls (such as a sample from a plant known to be healthy). The labeled copy of the nucleic acid specific to the virus being tested for is used to "probe" the filter. After washing to remove any probe that did not bind to the sample, the filter is "developed" to reveal whether the probe bound or not. Specific binding of the probe to the filter indicates that the sample contained nucleic acid sequences specific to the virus, and therefore perhaps was infected by the virus.

Polymerase chain reaction (PCR) is a new and powerful method that can be used to amplify a specific segment of DNA in a large enough quantity to be detected. PCR requires that two virus-specific primers be designed based on the nucleotide sequence of the viral nucleic acid. Since most plant viruses have RNA as the genome, a cDNA is first made using the viral RNA as template and one of the primers to be used in the PCR to prime the synthesis of the cDNA. The cDNA produced is then used in the PCR reaction with the two primers. The amplification of the region of virus sequence between the primers results in the production of a specific fragment of virus sequence that is then detected by the routine method of gel electrophoresis. Thus, no radioactive materials or serological materials are needed. In addition, sample preparation is relatively rapid, primers can be ordered and quickly obtained commercially, and the reagents needed for cDNA synthesis and the PCR are generally commercially available in comprehensive kits. As nucleic acid sequences of viruses are obtained, the PCR will have greater utility for detection of viruses. However, the extreme sensitivity of the PCR may not be entirely beneficial. Theoretically, the PCR can detect a single copy of a piece of nucleic acid; even pieces of nucleic acid from broken noninfectious virus particles might be detected.

B. Transmission

Viruses of the greatest importance in agriculture are transmitted by a vector such as nematodes, mites, fungi, and insects. In addition, virus-infected materials are moved in commerce by shipment of virus-infected seeds or living plant tissues. Viruses are not disseminated directly by wind and rain as other plant pathogens are, but wind and water directly influence the movement of viruses by affecting the movement of vectors of plant viruses. A few plant viruses can be transmitted by pollen and seeds from the infected plant. Although transmission of viruses by sap is a common method used in research, it is of importance in nature only in certain cases, such as when the host plant is vegetatively propagated (*e.g.*, potatoes and some ornamentals) and transmission *via* sap can occur as cuttings are made. [*See* PLANT PROPAGATION.]

The association of a virus with a vector is generally quite specific, that is, certain species of vector transmit only certain viruses. Hence, the number of viruses transmitted by the few species of nematodes, mites, and fungi that transmit viruses is probably less than fifty. Of the insects, members of the Homoptera that feed by sucking and the Coleoptera that feed by chewing are the most important plant virus vectors. Most of the insects transmit virus in a nonpersistent fashion. The viruses are taken up as the insect feeds on epidermal or mesophyll cells, retained in the foregut, and transmitted when the insect feeds again. The virus is lost after a few feedings and the insect must acquire the virus again to continue to be able to transmit. Although this appears to be a simple process, the nonpersistent transmission of viruses by insects is the result of a specific biological relationship among the virus, the insect, and the plant, and it is not completely understood. Not all viruses that are present in plants are transmitted in this fashion, so it is not just a result of the virus being carried in the plant sap acquired during feeding.

Some viruses are circulative in the insect vector. Once acquired, the virus moves from the food canal into the insect tissue and then to the salivary system before transmission to a healthy plant on which the insect feeds. The mechanism by which a virus is selectively translocated or excluded from translocation from the food canal appears to involve the proteins located in the virus coat. Some circulative viruses are also propagative, that is, the insect is a host for the virus as is the plant. These viruses have a similar route of movement through the insect as the circulative viruses, but they also infect and multiply in the insect.

The insect is generally able to transmit the virus through its life and in some cases through the egg to progeny. Thus, propagative viruses are pathogens of both insects and plants.

C. Control

Once a method for detection of the virus and some understanding of how it is moved is developed, then one of several control strategies can be evaluated to develop the appropriate tactics to control the virus diseases. For diseases caused by plant viruses, the tactics fall generally into five areas: therapy, resistance, protection, exclusion, and eradication.

Therapy is the elimination of the pathogen from the host. This has been done with a number of viruses in a number of crops, particularly those that are vegetatively propagated, by tissue culture, temperature treatment and/or chemical treatment. One example is the elimination of the potyvirus, potato virus Y (PVY), from potatoes. The success of using therapy to eliminate PVY from potato is based on the observation that plants are not uniformly infected with virus. The apical meristem is in many cases free of virus. In other cases, the application of certain chemicals, primarily nucleotide analogues, or extremes in temperature, may retard virus replication so the virus does not become established in the newly developed tissues. If the meristem can be aseptically removed and cultured *in vitro* to produce new plants, virus-free plants may be obtained that are used as a stock for subsequent propagation of additional plants. The plants may be virus free, but are not virus resistant. Thus, once the virus-free plant is obtained it must be prevented from becoming infected again. The process of freeing planting stock from known virus infections, whether seeds or vegetatively propagated plant parts like potato tubers, is often used commercially. Virus-free planting stock is then subject to "certification," often by a government agency or a cooperative. Certification is a cornerstone of plant disease control in potato production worldwide and is also used in many quarantine systems to restrict the movement of virus-infected seeds.

Resistance is an important approach for control of plant viruses in agriculture. Over the past 80 years of plant breeding, many successful efforts have been made to breed crop varieties that are tolerant of, or resistant to, plant virus infection or to the feeding activity of the vectors of plant viruses. In a few cases, such as TMV resistance in tomato, bean common mosaic virus resistance in beans, beet curly top virus resistance in sugar beet, and turnip mosaic virus resistance in lettuce, genetic resistance provided by single or a small number of resistance genes has been very durable. Nevertheless, one of the major problems with the use of simply inherited (usually single-gene) resistance is that resistance-breaking strains of the virus can arise in the natural population of viruses and be favored by natural selection.

Genetic resistance to viruses in plants is generally considered in three categories. When a plant is a non-host to a virus, the plant is said to be immune. The term immunity was first used in plant pathology before the true nature of the immune system in higher animals was discovered. When used in this context, immunity does not imply the involvement of an immune system such as in animals, but rather the complete lack of ability of the plant to be infected by the virus. The terms resistance and tolerance are used to denote different degrees of reaction to infection. Resistance results when infection of the plant is localized or limited to a few infected cells and virus does not spread beyond these initially infected cells, or when no visible reaction is observed in the plant. When infection remains localized, resistance of the whole plant is effected because the virus is limited to those cells that as a result of being infected collapse and die. Such reactions are called local lesions or the hypersensitive response and are characteristic of the response of certain *Nicotiana* spp. and cultivars of tobacco to tobacco mosaic virus. Tolerance denotes an infection during which the plant is nevertheless able to continue to function in a manner similar to that of an uninfected plant, even though the virus infection is systemic.

Protection is the prevention of establishment of the disease by the introduction of a prophylactic barrier between the pathogen and the host. Both chemical and biological barriers have been used. Spraying crops with mineral oils lessens the transmission of many potyviruses by their aphid vectors. However, the foliage must be uniformly and frequently sprayed to be certain new growth is covered because the protective effect is not systemic in the plant. Another tactic directed against the vectors of potyviruses (and other viruses nonpersistently transmitted by aphids) is the use of reflective mulches between the rows of a crop. The idea is to confuse the visual signals used by insects to locate their feeding sites by reflecting ultraviolet light from plastic strips coated with a reflective material.

A biological protection tactic that has been used is cross protection. Cross protection occurs when prior infection of a plant by one strain of a virus (typically a strain causing only mild or no symptoms) prevents damage when a plant is subsequently exposed to infection by a related but more severe strain of the same virus. Cross protection was first demonstrated experimentally in the 1920s. Although cross protection is effective and was described decades ago, it has been used to control only a few virus diseases. The mechanism of cross protection has been the subject of much research, but the events that occur in the host that control the outcome of the interaction between virus strains have yet to be fully explained. Citrus tristeza closterovirus, tomato mosaic tobamovirus, and papaya ringspot potyvirus are the only disease examples in which cross protection has been earnestly investigated for disease control.

Exclusion is the prevention of entry of a pathogen into an area. Quarantines to prevent entry of pests into new geographical areas are generally temporary at best. However, the prevention of establishment of a disease where the pathogen is seed transmitted by using pathogen-free seed has been a successful disease control tactic. Serological assays have been used to develop virus-free seed stocks through testing programs, which have facilitated control of the seed-transmitted lettuce mosaic potyvirus and barley stripe mosaic hordeivirus viruses (BSMV).

Prevention of establishment of a virus source by planting virus-free seed helps to delay or prevent the disease from developing. If the only source of virus is from the crop itself, the disease may be completely avoided although the vector may be present. If there are alternative hosts for the virus among the plants in an area where the crop is grown, either other crops or noncultivated plants, the development of the disease in the crop may be delayed depending on vector activity. In addition to virus-free seed, virus-free material-derived from tissue culture, physical, or chemical treatments are used.

Eradication is the removal of a source of the virus that may come from outside the field (*e.g.*, weeds) or from inside the field (*e.g.*, infected trees in an orchard). Once a virus is established it can be difficult or impossible to eradicate, particularly with a virus that has a wide host range and/or an active vector. A classic example is the attempt to control a virus that is transmitted by mealybugs, the cocoa swollen shoot badnavirus (CCSV) in Ghana. In an attempt to control CCSV in Ghana, from 1946 to 1980 nearly 179 million affected cocoa trees were destroyed. This did not control the disease and in 1983 there remained 39 million trees still affected by the disease. A factor that contrib-

uted to the failure of this procedure may have been the late and incomplete effort, because the disease was first noticed in the 1930s and the eradication program was not begun in earnest until 1946. Additionally, there were interruptions in the implementation of the program owing to social and political factors. The implementation of any eradication program requires excellent detection methods, comprehensive surveys, and efficient destruction of the infected material.

D. Mixed Infections

The foregoing discussion about virus detection and control has ignored the fact that, in nature, plants may be infected by more than one virus. This is an important issue because in some cases infection of a plant by more than one virus at the same time can result in disease of a greater magnitude than either virus would be expected to cause alone. In some cases, the mixed infection results in a much more severe reaction by the plant than would be expected from a merely additive effect of the two viruses. This phenomenon is called synergism. One of the classic studies on plant viruses was that of Kenneth M. Smith in England in the 1930s on potato mosaic disease. Potato mosaic is a severe disease that results when two distinct viruses, now called potato virus X (PVX) and potato virus Y (PVY), are present in a mixed infection. PVY is aphid transmitted, causes mild symptoms, and is typically present in low concentration in singly or mixed infected plants. PVX is not aphid transmitted and causes mild to inapparent symptoms, though it does reduce yields. PVX is present in much higher concentrations in mixed infections than in single infections.

Corn lethal necrosis disease is another important example of a synergistic interaction resulting from a mixed infection. It occurs when plants are infected by maize chlorotic mottle virus (MCMV) and one or more PVY-related viruses that infect corn. Either virus alone produces a mild mosaic, but MCMV infection with either potyvirus results in losses up to 70%. In the United States, this disease seems to be limited to small areas of Nebraska and Kansas.

IV. Future Prospects

Genetic engineering of plants has opened up a new era for research on plant viruses and as a result offers new approaches to the possible control of plant virus infections. Research on plant viruses and their infec-

FIGURE 5 Genetically engineered tomato with tolerance to infection by tomato spotted wilt virus. All plants were inoculated with the virus. The plants on the right were engineered with the gene encoding the virus coat protein; those on the left are controls showing typical virus symptoms.

tions using the tools of molecular biology has profoundly influenced our ability to ask critical questions about virus infection, replication, and variation, as well as about virus movement within the plant and transmission from plant to plant. All of this work not only offers new insights into the mechanisms of infection, but also promises new approaches to disease control. That this statement is true has already been proven in fewer than 10 years after the first genetic engineering of plants. [See PLANT GENETIC ENHANCEMENT.]

Plants genetically engineered to express genes derived from plant viruses as if they were normal plant genes often prove to be resistant to the virus from which the genes were taken. The coat protein genes of more than 20 different plant viruses have been used to make engineered plants in which the result was tolerance or resistance to infection by the virus donating the coat protein (Fig. 5). The phenomenon is

not limited to use of the coat protein genes. Partial sequences taken from the genes that encode virus replicases when expressed in genetically engineered plants also protect the plants from virus infection. Another genetic engineering strategy that has shown some success is to express a gene designed to produce an RNA sequence that is of the opposite polarity of (or complementary to) a virus gene. Yet other strategies based on the interaction of plant viruses with defective interfering (DI) RNAs or unrelated, naturally occurring small RNAs (satellite RNAs) are also being investigated for their potential to create virus-resistant crops.

We must remember that approaches based on a single, simple mechanism of interference with virus infection, whether achieved by conventional breeding or molecular biology, can be readily overcome by corresponding changes that arise from normal rates of mutation in the virus population. If resistant crops are planted widely, selective pressure is placed on the virus population and the result can be selection for the resistance-breaking mutants. Genetic engineering of resistance may increase our ability to create truly durable resistance in crops, because it is based on detailed molecular information about the mechanisms of resistance. Thus, the possibility exists of engineering multiple traits that differ in their mechanisms of action. Such engineering strategies, if deployed with intelligent attention to epidemiological knowledge about virus spread and rates of mutation, offer promising new tools for controlling plant virus diseases in an environmentally and economically attractive package, the genetically engineered seed.

Bibliography

Agrios, G. N. (1988). "Plant Pathology." Academic Press, New York.

Francki, R. I. B., Fauquet, C. M., Knudson, D. L., and Brown, F. (eds.). (1991). Classification and nomenclature of viruses: Fifth report of the International Committee on Taxonomy of Viruses. *Arch. Virol.,* Suppl. 2.

Hiruki, C. (ed.). (1991). Structure and function of plant virus genomes. *Canad. J. Plant Pathol.* **13,** 139–195.

Hull, R., and Davies, J. W. (1992). Approaches to nonconventional control of plant virus diseases. *Crit. Rev. Plant Sci.* **11,** 17–33

Kay, L. E. (1986). W. M. Stanley's crystallization of the tobacco mosaic virus, 1930–1940. *ISIS* **77,** 450–472.

Matthews, R. E. F. (1991). "Plant Virology," 3rd ed. Academic Press, New York.

McLean, G. D., Garrett, R. G., and Ruesink, W. G. (eds.). (1986). "Plant Virus Epidemics: Monitoring, Modelling and Predicting Outbreaks." Academic Press, New York.

Plumb, R. T., and Thresh, J. M. (eds.). (1983). "Plant Virus Epidemiology: The Spread and Control of Insect Borne Viruses." Blackwell Scientific, London.

Popcorn

KENNETH ZIEGLER, *Iowa State University*

Glossary

Flakes Popped product of popcorn enjoyed by consumers

Hull Outside covering of an unpopped popcorn kernel; consumers know it as the part of popcorn that gets stuck in their teeth; also called pericarp or seed coat

Lodged popcorn Popcorn plants not remaining erect until the popcorn crop is harvested

Popcorn conditioning Procedures to ensure that popcorn kernels are at the right moisture content for popping: drying or rewetting kernels, and then sealing them in moisture-proof containers or storing them in environmentally controlled storage areas near 70% relative humidity and 70°F (21°C) for 4 to 6 weeks, until all kernels reach an equilibrium moisture content of 13.5 to 14.5%

Popcorn kernels The raw product of the popcorn commodity used for any purpose except planting

Popcorn processing Procedures to ensure consumers a quality product, such as cleaning and sorting the unpopped kernels for size and color; for distributors of popped popcorn, these procedures depend upon the type of popcorn product marketed, for example, caramel coating for caramel covered popcorn

Popcorn seed Popcorn seeds used for planting

Popping expansion Term interchangeable with "popping volume," is a measure of the volume of flakes produced by a known weight of unpopped kernels; standard measuring units are cubic centimeters per gram (cm^3/g)

Popcorn, a specialty type of corn (*Zea mays* L.), is believed to be one of the oldest types of corn and has been referred to as the original cereal snack food. When kernels of popcorn are heated, they explode and produce large puffed flakes. This characteristic separates popcorn from all other types of corn. Consumers use the term "popcorn" to refer to either the unpopped kernels or to the popped flakes. Growers and processors also use the term "popcorn" to refer to the crop growing in the field. Thousand-year-old corn kernels similar to modern-day popcorn have been found by archeologists in Peru and in the state of Utah in the United States. The first European explorers of the New World wrote that popcorn was eaten and used for decorations by the Native Americans. The early Native American practice of parching corn to make it more palatable may have led to the discovery of popcorn. The true genetic origin of popcorn, however, is one subject in the continuing debate about the genetic origin of all types of corn. Today, popcorn is a well-known snack food; it accounted for 3.9% of the 1991 $31.032 billion U.S. snack food market.

I. The Popcorn Plant

Popcorn is corn, so the best way to describe a popcorn plant is to compare it with a dent-corn (commercial field-corn) plant. At first glance, a field of mature popcorn looks similar to a field of any other type of corn. But as one examines individual plants, some differences become apparent (Fig. 1). A popcorn plant grows and develops like a dent-corn plant. However,

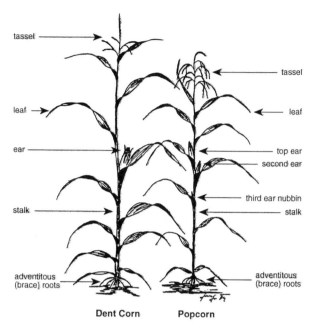

FIGURE 1 Plant type differences between generalized dent corn and popcorn hybrid plants. (Modified from Ziegler and Ashman, 1994).

FIGURE 2 Ears of dent corn and popcorn. (A) Dent-corn ear and cross-section showing kernels and cob. (B) Yellow-popcorn ear and cross-section showing pearl-type kernels and cob. (C) White-popcorn ear and cross-section showing pearl-type kernels and cob.

popcorn seedlings are slower emerging from the soil than dent-corn seedlings. In early growth stages, popcorn plants are difficult to differentiate from dent-corn plants. At this stage of growth, popcorn leaves may be narrower and popcorn plants may have more tillers (secondary plant-stalks emerging from one root system). The most noticeable visual differences between popcorn and dent-corn plants occur when tassels emerge. The popcorn tassel has many more branches and the ends of the tassel branches droop, giving the tassel a "weeping willow" appearance. Because popcorn tassels have more tassel branches, popcorn produces more pollen than dent corn. The popcorn plant generally produces more than one ear, whereas the dent-corn plant generally produces only one ear. Popcorn ears are also more slender (Figs. 1 and 2). Most present-day popcorn hybrids produce at least two good ears and, under optimum growing conditions, may produce a third ear. Usually, however, the third ear does not fully develop. Even though Fig. 1 indicates that dent corn is taller than popcorn, some popcorn hybrids are taller than some dent-corn hybrids. The popcorn plant in Fig. 1 has adventitious (brace) roots; however, they develop only when popcorn plants are grown under favorable conditions. The figure also shows no tillers on the popcorn plant but many popcorn hybrids have tillers, unlike most dent-corn hybrids. [*See* CORN CROP PRODUCTION; CORN GENETICS AND BREEDING.]

Popcorn plants generally have weaker stalks and roots and are more susceptible to pests than are dent-corn plants. Work is in progress at Purdue University, Iowa State University, the University of Nebraska, the University of Missouri, and at private companies to improve the strength of popcorn stalks and roots and to increase the popcorn plant's resistance to pests.

All pests that attack dent corn also can attack popcorn, but two of the most economically severe pests of popcorn plants are the second-generation of the European corn borer (ECB) and the stalk-rotting organisms. There is little natural resistance in popcorn germplasm to the second-generation ECB. Larvae of the second-generation ECB damage the popcorn plant by tunneling into the stalk and ear shank and by feeding on the leaf sheath, leaf collar, and ear. This damage not only causes yield loss but, under some growing conditions, causes poorer popping quality. The stalk-rotting organisms are a disease complex composed of several fungal or bacterial pathogens, or both, depending upon the year and the location. The most damage is usually caused by two species of fungi: *Fusarium moniliforme* (Sheld) and *Gibberella zeae* (Schw). Infection by these pests causes yield loss. Infected plants have poorly filled ears, lodged stalks, or both. Quality is poorer because infected plants lodge and die prematurely. Lodged plants result in reduced quality by exposing ears to deterioration on the ground and by slowing natural drying. Premature death results in poorly formed and filled kernels, which in turn reduce popping quality. The combination of ECB and stalk rot pests causes even more damage. Heavy ECB infestation decreases the tolerance of plants to stalk rots, and a heavy infection of stalk rots, specifically anthracnose [*Colletotrichum graminicola* (Ces.) G. W. Wils.], can decrease the popcorn plant's tolerance to ECB.

II. The Popcorn Kernel

Popcorn ears and kernels also visually distinguish popcorn from other types of corn (Figs. 2 and 3). All corn kernels have three main parts (Fig. 4): the pericarp, or seed coat or protective covering of the kernel; the endosperm, or food-storage tissue composed primarily of starch; and the embryo, or respiring tissue, which develops into a new popcorn plant. Within all corn there is a gradation in the relative

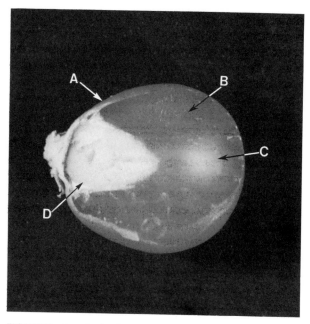

FIGURE 4 Longitudinal section of a yellow-pearl popcorn kernel. (A) Pericarp. (B) Hard, translucent endosperm. (C) Soft, opaque endosperm showing through translucent endosperm. (D) Embryo.

amounts of hard (translucent) to soft (opaque) endosperm in the kernels. Flint corn and popcorn have the highest proportion of hard endosperm, dent-corn kernels have less, and floury corn kernels have none. The hard endosperm in dent-corn kernels (Fig. 3) appears somewhat clear, whereas the soft endosperm appears either opaque or light yellow. Sweet-corn kernels have a brittle, hard endosperm. When comparing all types of corn germplasm, the ratio of hard to soft endosperm in a kernel can determine popability. Within 100% popcorn germplasm, this ratio varies, but, when comparing different popcorns, other factors are more important to popability than the ratio of hard to soft endosperm. Popcorn has been described as a small-kerneled flint corn selected for its ability to explode and to produce a flake when heated. Popcorn kernels have the thickest pericarp of all types of corn. This thick pericarp provides enough strength to confine the internal pressure until the pressure reaches a high enough level for flake formation when the kernel bursts. There are two popcorn-kernel shapes: rice and pearl. Pearl-type kernels are spherical with a smooth top, whereas rice-type kernels are long and slender with a sharply pointed top (Fig. 3). Generally, rice-type kernels are associated with white popcorn, and pearl-type kernels with yellow popcorn. Because of intercrossing of the two types in breeding

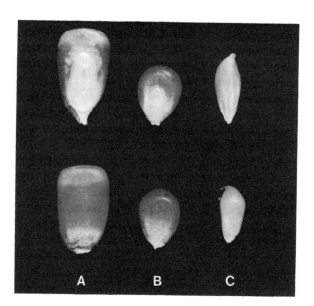

FIGURE 3 Kernels of dent corn and popcorn. (A) Dent-corn kernels. (B) Yellow-pearl popcorn kernels. (C) White-rice popcorn kernels. Upper kernels show embryo side, and lower kernels show the opposite side.

programs, however, both shapes are found in either yellow or white popcorns. The commercial trend has been toward increased use of the pearl types. Although popcorn kernels can be found in many different colors, only the white and the yellow are commercially important. Some red and black popcorns are sold commercially for popping, and an even smaller amount of variably colored popcorn is sold as novelty-popping popcorn. [*See* CORN PROCESSING AND PRODUCTS.]

III. Types of Popcorn

Popcorn itself is a specialty corn, but even within popcorn there are two different categories. One category contains types defined by kernel characteristics, and the other category contains types defined by processor needs. Types as defined by kernel color (yellow, white, other) and kernel shape (pearl, rice) were discussed in the previous section. Another kernel characteristic that is commercially important and defines types of popcorn is kernel size. Yellow popcorns move through commercial channels in three different kernel sizes. There is no industry standard for kernel size, but a commonly used measure is the number of kernels in 10 g, which defines kernel size as follows: 52 to 67, large; 68 to 75, medium; and 76 to 105, small. Small, yellow kernels are generally preferred by home consumers because small kernels tend to produce tender flakes with few hulls. Vendors prefer the medium to large kernel sizes, because they tend to produce large, tougher flakes which reduce breakage during handling, even though the flakes can have more hulls. Medium-kernel popcorns are used by both end-users. Different popcorn hyrids produce different kernel sizes.

A second category of types of popcorn is determined by popcorn processors and consumers. Two different flake types, butterfly and mushroom, are commercially important (Fig. 5). Different hybrids produce different flake-types. Hybrids that produce butterfly flakes are referred to as butterfly hybrids, whereas hybrids that produce mushroom flakes are called mushroom, ball, or caramel-corn hybrids. The butterfly hyrids produce flakes irregular in shape, with many "wings." This type of flake is associated with tenderness and freedom from hulls; but it will not withstand much handling, because the wings break off easily. Popcorn is purchased by weight and sold by volume. Popcorn with butterfly flakes are generally

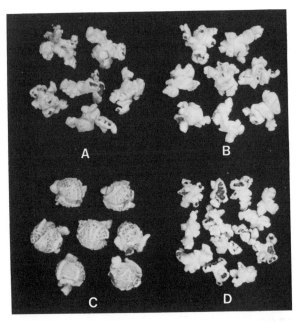

FIGURE 5 Popcorn flake-types. (A) Yellow butterfly-flakes. (B) White butterfly-flakes. (C) Yellow mushroom-flakes. (D) Small, white, butterfly-flakes from black-kernel popcorn.

undesirable to vendors who package popcorn. Product shrinkage is inevitable when packaging butterfly flakes, but some vendors will sacrifice a degree of product shrinkage for tenderness and freedom from hulls. Butterfly flakes work best when popcorn is to be eaten directly by the consumer, such as in the home or in movie theaters. Vendors in the popcorn flavor-coating industry prefer mushroom-type flakes, which are round with few "wings" that can break off during coating procedures. Popcorn hybrids are visually rated on the percentage of mushroom flakes in a popped sample. Some hybrids produce no mushroom flakes (100% butterfly), but no hybrid produces 100% mushroom flakes. Experimentally, some popcorn germplasm has produced 95% mushroom flakes, but no commercial hybrids have yet achieved this large of a percentage.

The acceptance of microwave popping has resulted in yet another type of popcorn: one that performs well under microwave popping conditions. The industry has identified a few popcorn hybrids that perform well when microwave popped, and these are called microwave popcorns. All popcorn hybrids will pop in a microwave, but some hybrids generate a more acceptable product than others.

IV. The Popping Phenomenon

Any type of corn kernel with an intact pericarp and the right amount of moisture will explode when heated. Other types of corn, namely flints and certain dents, depending upon the amount of hard endosperm in the kernels, can produce a small flake. These small flakes, however, would never be acceptable to popcorn consumers. Popcorn is the only type of corn that produces large flakes. Two events occur when popcorn pops: explosion and flake formation. Each popcorn kernel is a sealed vessel with moisture on the inside. The explosion results from heating the moisture until the pressure generated blows the kernel apart (bursts the pericarp). The buildup of internal pressure provides the driving force necessary to expand the hard, translucent endosperm into a flake. Popping usually occurs near 177°C (350°F), equivalent to 2.5 t/cm² (135 psi) on the inside of the kernel. At the moment of explosion, the superheated water in the kernel turns to steam and blows apart the thermoplastic endosperm. The individual starch granules of the hard endosperm do not explode, but, upon release of internal steam pressure, are gelatinized and expanded by heat. Subsequently they are dried and left in a three-dimensional network. The soft endosperm or soft starch granules, found on the inside of the popcorn kernel, undergo little change during the explosion aside from spreading further apart. Even though this soft endosperm seems to have little if any role in the popping phenomenon, it may have some effect on flake formation, because, even after major improvements in increased flake size, all popcorn kernels still contain some soft endosperm. If the soft endosperm was unnecessary for flake formation, selection for increased flake size should have eliminated it from present-day improved popcorn.

The commercial measurement of flake size is called popping expansion or popping volume. Expansion is the most important quality trait of popcorn, because most other quality traits are positively associated with expansion, and because vendors purchase popcorn by weight and sell by volume. Over the years, several different methods have been used to measure expansion, but the measure used today is cubic centimeters of flakes produced per gram of unpopped kernels (cm³/g). The industry's standard measuring instrument is Cretor's Metric Weight Volume Tester (MWVT), an instrument which pops a 250-g sample of popcorn kernels. Popped flakes are funneled into a 13⅓-cm (5¼-in.) diameter, graduated cylinder marked to reflect cm³/g. Commercially acceptable popping expansion for yellow popcorn is 40 cm³/g or greater. White popcorns and mushroom popcorns generally have smaller popping expansions than yellow popcorn. The smaller expansion of white popcorn hybrids results from poorly expanding germplasm, whereas that of mushroom hybrids results from the round flakes packing more closely together in the graduated measurement cylinders.

Popping expansion is controlled by both genetic and environmental factors. Genetically, it is controlled by four or five major genes with heritability estimates ranging from 0.62 to 0.96. Thus, popping expansion is a heritable trait that popcorn breeders can continue to improve. Of interest to popcorn growers and processors are the environmental factors affecting popping expansion. Some environmental factors are under the control of growers and processors, whereas other environmental factors are not. Different growing conditions in different years produce popcorn with different abilities to expand. Although weather conditions cannot be controlled, other factors can. A popcorn hybrid's optimal expansion can be achieved only with an intact pericarp, a nondefective endosperm, and a moisture content of 13.5 to 14.0%. Anything adversely affecting these three factors will reduce popping expansions. Growers and processors must be careful not to use procedures that could damage the pericarp. They also must be careful not to use procedures, such as fast drying, that could alter or damage the endosperm. Changing the moisture content too fast during conditioning can cause stress cracks to develop in the hard endosperm, which decreases popping expansion. Processors and sellers of unpopped popcorn kernels need to condition popcorn to the optimal moisture content and then package it in moisture-proof containers. Attention to detail in these procedures is necessary to provide consumers with a consistently acceptable product.

It is difficult to define just what makes an excellent popcorn product, because, to a great extent, excellence is a matter of taste. Consumers generally demand a popcorn product that has an attractive appearance, has a pleasant flavor (taste) and texture, and has few hulls and few unpopped kernels. Snack-food consumers are increasingly demanding healthful snack foods that are quick and easy to prepare, a niche that popcorn fills. Plant-breeding and research and development efforts continue to improve all the quality characteristics of popcorn.

V. Popping Methods

There are three main ways to pop popcorn—in heated oil, in heated air, or in a microwave oven. Regardless of the method used, a uniform heat source is critical. All popcorn kernels must pop in 1 to 3 min to prevent scorching of the flakes that popped early. There are large, commercial poppers that are continuous-flow, in which the popped flakes are removed from the heat source. But even in these poppers the kernels should explode in 1 to 3 min to produce the best flakes.

Oil popping requires a heat source that is applied to oil in a pan. As the oil heats up, it uniformly disperses the heat to all kernels in the pan. The best oil poppers have either a stirring or a shaking mechanism that evenly coats all kernels with the oil. The oil becomes a part of the popped product and imparts flavor, as it either adheres to or is soaked up by the flakes. Oil also provides a surface on the flakes for salt or for salt-based flavorings to adhere. This method of popping is associated with movie theaters and with many types of home poppers.

Hot-air popping holds unpopped kernels in a hot-air flow. This flow can be directly over an open flame or in air that has passed around and through electrical coils. Poppers of this type generally have some method of removing flakes from the air flow before they become burned or scorched. Uniform heating under these conditions is supplied by a rotating or shaking container, which holds the unpopped kernels, or by the hot air itself, which mixes the unpopped kernels as it flows through them. The popped product from this type of popping is 100% popcorn. The flakes are dry because of the drying effect of the heated air. It may be difficult for salt or salt-based coatings to adhere to the flakes. This type of flake is excellent for consumers on a restricted fat-intake diet and also works well for candy coating, because there is no additional oil to consider before the candy coating is applied.

The newest method of popping popcorn, the microwave oven, has, in a relatively short time period, become widely used. Most of the prepackaged bags of microwave popcorn, however, represent a combination of microwave and oil popping. The only truly microwave popcorn bags are those containing no oil. The oil in prepackaged microwave bags provides the same benefit to popping that it provides to regular oil-popped popcorn—uniform heat dispersion throughout all kernels. Microwave ovens do not provide a uniform enough heat source to pop all kernels at nearly the same time, so microwave packages have been developed that transfer heat uniformly to all portions of the popping sample. Most prepackaged microwave bags contain a heat pad to heat the oil, spreading the heat uniformly throughout the sample. Prepackaged bags without the heat pad rely on the explosive action of the kernels to move unpopped kernels around in the bag. In addition to microwave bags without oil, different shapes and sizes of bowls have been developed for use in microwave ovens to dry-pop popcorn. The popped product from prepackaged microwave bags with oil and various flavorings tastes similar to popcorn popped in an oil popper. The popped product from the microwave, without the addition of oil or other flavorings, is similar to that of hot-air popcorn, except that the microwave popped popcorn is moister, even to the point where salt or salt based flavorings will adhere to the flakes, if applied soon after popping. Microwave popcorn is usually used in homes or in vending-machine areas.

VI. Growing Popcorn

U.S. commercial popcorn production is centered in the U.S. Corn Belt. Nebraska, Indiana, and Illinois are the major producing states, with Iowa, Kansas, Missouri, and Ohio also being significant producers. Other states produce some popcorn, but these seven states generally lead in amount of commercial popcorn produced. In 1992, the Popcorn Institute reported that its members, who represent about 85% of U.S. popcorn sales, harvested an annual average of 215,000 acres (87,000 ha) for the years 1989–1991, producing 693 million pounds (3.15 million q) of popcorn per year.

Because popcorn is a type of corn, growing popcorn is similar to growing dent corn. There are, however, a few differences. A large proportion of popcorn is produced on contracted acres. Popcorn processors enter into a contract with growers to produce popcorn. This guarantees a price that the grower will be paid for the crop and has a stabilizing effect on supply and demand. Unlike growing dent corn, where the grower selects the hybrids to plant, popcorn processors generally select popcorn hybrids to plant. A few acres each year are grown on the open market (not contracted). Growers wanting to try open-market growing can obtain information about hybrids from popcorn seed companies and from the hybrid popcorn performance-trial bulletin published yearly by the Department of Botany and Plant Pathology (Agricultural Experiment Station, Purdue University, West Lafayette, IN 47907). Growers producing open-market popcorn assume the risk of finding a market for their crop. This can be difficult in a good popcorn-production year, when carryover supplies are large. Growers wanting to try contracted acres should contact popcorn processing companies to see if any contracts are available.

Growing popcorn is quite similar to growing dent corn until harvest, at which time the crops are handled differently. Commercial popcorn can be either ear-picked or combine harvested. Some contracts will state which method to use. If the machinery is operated well, both methods can produce a quality product. The main concern of either type of harvesting method is to eliminate as much kernel damage as possible. Ear-picked popcorn can be harvested at 25% moisture or less if a drying system is available to dry the ears to a safe storage moisture. Because of the greater potential for kernel damage by a combine, popcorn should not be combine harvested until it is below 16 to 18% moisture.

Once the popcorn is harvested, it is either delivered to a processor or put in some farm storage facility. Popping expansion is affected by harvesting and handling procedures and by the moisture history of the popcorn before popping. The most critical aspect of conditioning and processing popcorn is controlling moisture content. Once harvested, the crop should be dried to 14 to 15%, cleaned, processed, and sealed in moisture-proof containers at 13.5 to 14.0% moisture until popped. Because of these constraints, popcorn storage is critical to the delivery of a high-quality product. Shelled popcorn at 14.5% moisture can be safely stored from harvest until the next spring, but popcorn should be dried to 12.5 to 13.5% moisture for longer term storage. Control of stored-grain insects and rodents is necessary to maintain the quality of stored popcorn.

VII. The Popcorn Industry

The U.S. popcorn industry is more than 100 years old. It became a commercial industry in the late 1800s. In 1991, for the first time ever, sales of unpopped popcorn were greater than one billion pounds (4.54 million q). Although small compared with commodity industries such as dent corn and soybean, the popcorn industry is composed of many diverse entities. The overall industry consists of groups involved with the production and preparation of raw popcorn (unpopped kernels) and groups involved in the use of raw popcorn. The first group includes university popcorn-breeding programs, seed companies, growers, pesticide producers, herbicide producers, fertilizer producers, popcorn processors, and packaging companies. The second group includes brokers, truckers, popper manufacturers, microwave companies, prepop companies, coating companies (both salt- and sugar-based coatings), home consumers, companies packaging popcorn for home consumption, entertainment poppers (theaters, sports stadiums, fairs, etc.), gourmet popcorn companies, and novelty popcorn markets (colored popcorn, very small kernel and flake popcorns, and microwave popcorn on-the-cob, etc). Individual involvement levels in the popcorn industry can be in one or in any combination of these groups. Some large popcorn companies are involved at most levels, beginning at the plant-breeding research level and continuing to the marketing and selling of the finished product to consumers. The Popcorn Institute (401 North Michigan Ave., Chicago, IL 60611-4267) is the only trade association representing popcorn processors. Its membership represents approximately 85% of U.S. popcorn sales. Members are popcorn processors and/or popcorn seed companies. Activities of the Institute include projects to improve popcorn growing and processing technology, liaison with various government regulatory agencies, and generic marketing and promotion programs to spur sales and consumption of popcorn. In conducting these activities, the Institute is involved in three major areas: (1) It has developed an ongoing program entitled "The Popcorn Institute Seal of Quality Performance." This Seal is not a warranty, but represents a well-designed, statistically structured evaluation program for popping machines. Poppers displaying the Seal have undergone stringent laboratory testing for the production of a high-quality popped product. Poppers displaying the Seal assure the consumer that they are high-quality poppers. (2) The Popcorn Institute has also instituted procedures by which research money is generated. This money is used to fund popcorn breeding programs at Purdue University, the University of Nebraska, Iowa State University, and the University of Missouri. (3) The Institute's marketing and promotion program is an ongoing effort to promote popcorn consumption, which has been increasing steadily at about 9 to 10% per year for the past 10 years.

VIII. Summary

Popcorn has traditionally been the leading specialty corn item exported from the United States, with the annual amount steadily increasing. For U.S. agriculture, popcorn represents a viable alternative crop, especially for dent-corn and/or soybean farmers in the Corn Belt. As an alternative crop, popcorn has in its favor the fact that it is an established industry. Other positive aspects of using popcorn as an alternative crop include a consistently growing demand for popcorn, the similarity of popcorn production practices to dent-corn production, its potential function in a corn–soybean rotation sequence (at this writing, acres of popcorn are not included in the corn base for a farm), the established crop-delivery and handling systems, the availability of information on the best procedures for handling and storing the harvested crop, and the established marketing system of contract acres, which decreases the economic risks usually associated with growing alternative crops. Minor drawbacks of growing popcorn as an alternative crop are the limited number of available contract-acres, popcorn's susceptibility to pests and poorer stalk quality compared with dent corn, and the potential bad effects of certain growing conditions, such as early frost, which may reduce the quality of the crop below acceptable contract provisions, making the crop unmarketable as popping corn. A few markets for a damaged popcorn crop are available (popcorn kernels can be fed to hogs or used in pigeon feed), but these markets are small and tend to fill up fast.

Popcorn is a unique type of corn. For more than a century, this snack food has delighted U.S. consumers. It is now being introduced into other countries, where acceptance has been slow, but is gradually increasing. Popcorn has withstood the test of time and cannot be considered a "fad" commodity. Consumers will continue to enjoy popcorn for many generations to come.

Acknowledgments

Contribution from the Iowa Agric. and Home Econ. Exp. Stn., Ames, IA. Journal Paper No. J15588. Project No. 3059, supported in part by a grant from the Popcorn Institute.

Bibliography

Alexander, D. E. (1988). Breeding special nutritional and industrial types. In "Corn and Corn Improvement" (G. F. Sprague and J. W. Dudley, Eds.). American Society of Agronomy, Madison, WI.

Matz, S. A. (1984). Snacks based on popcorn. In "Snack Food Technology," 2nd ed. AVI, Westport, CT.

Rooney, L. W., and Serna-Saldivar, S. O. (1988). Food use of whole corn and dry-milled fractions. In "Corn Chemistry and Technology" (S. A. Watson and P. E. Ramstad, Eds.). American Association of Cereal Chemists, Inc., St. Paul, MN.

Ziegler, K. E., and Ashman, R. B. (1994). Popcorn. In "Specialty Corns" (A. R. Hallauer, Ed.). CRC Press, Boca Raton, FL.

Ziegler, K. E., Ashman, R. B., White, G. M., and Wysong, D. B. (1984). Popcorn production and marketing. Cooperative Extension Service, Purdue University, West Lafayette, IN. A fact sheet for the National Corn Handbook Project NCH-5.

Postharvest Physiology

GRAEME HOBSON, *Horticulture Research International*

Glossary

Climacteric Rise in ethylene evolution and respiration coincident with the onset of ripening in some fruits (classified as climacteric rather than nonclimacteric); peaks in ethylene and carbon dioxide (CO_2) production are commonly found during flower senescence

Development Period between flower set (fertilization) and full-opening (flowers), and ideal ripeness (fruits) and commercial picking stage (vegetables)

Ethylene Hydrocarbon gas, C_2H_4, ethene, fairly insoluble in water, that acts as a hormone in plant development; it is produced in small quantities during normal growth, and its concentration rises with tissue wounding, senescence, flower fertilization, climacteric fruit ripening and invasion by pathogens; some flowers and fruit become much more sensitive to its effects toward the end of development

Growth Increase in cell number or size of a plant organ

Harvest Removal of an immature, mature, or the ripe organ from a plant for use, completion of development, consumption, or processing

Maturation Occurs toward the end of development and includes preparations by the organ for flower opening, ripening, and seed viability; when mature, an organ can be removed from the parent plant and development can be continued to give a product of generally the same quality as if harvest had been postponed

Ripening Programmed change at the end of maturation during which color, texture, composition, and physiology alter to some degree, the rate of which is dependent on the species in question and on environmental conditions; it is the final phase of development for fruits, and is complete when acceptability is at a maximum

Postharvest physiology is that branch of biology that deals with flowers, food, and utility crops once they have been harvested and until they are used. The commodities are still living, and methods have been devised to control respiration, postpone deterioration, and limit loss of quality. Individual crops at a number of stages of development have been studied in order to minimize damage during harvest, grading, storage, and distribution. The products should be as free from crop-protective chemicals, pest-infestation, and pathogen invasion as possible. In essence, postharvest techniques are designed to exert control over the rate of development and deterioration of crops for economic, nutritional, and consumer-led benefits.

I. The Plant Cell

A typical cell (see Fig. 1) consists of a fairly rigid wall composed of pectin, cellulose, and hemicellulose, sometimes with suberin or lignin as well. The middle lamella, rich in pectic substances, binds adjacent cells together. Lining the wall is the plasmalemma, inside of which is the cytoplasmic layer. This is separated from the vacuole by the tonoplast. Cell walls are fairly permeable to water and solutes, and communicating lacunae, where the walls of adjacent cells do not

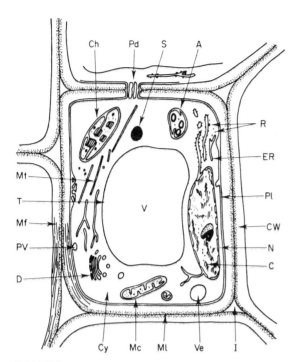

FIGURE 1 Diagram of the components of a typical plant cell. A, amyloplast; C, chromatin; Ch, chloroplast; CW, cell wall; Cy, cytoplasm; D, dictyosome; ER, endoplasmic reticulum; I, intercellular space; Mc, mitochondria; Mf, microfibrils; Ml, middle lamella; Mt, microtubule; N, nucleus; Pd, plasmodesmata; Pl, plasma membrane; PV, plasma vesicle; R, ribosomes; S, spherosome; T, tonoplast; V, vacuole; Ve, vesicle. (After A. W. Robarts (1974). "Dynamic Aspects of Plant Ultrastructure." McGraw-Hill, New York).

touch, form a system of channels through the tissue. The semipermeable properties of the membranes bounding the cytoplasm restrict the movement of solutes but allow water through. The resulting wall pressure (turgidity) together with the middle lamella binding substances is largely responsible for the strength of flower, fruit, and vegetable tissues. The cytoplasm contains an assortment of organelles with specialized functions:

- the nucleus contains genetic information in the form of DNA;
- mitochondria break down organic acids and trap the energy as ATP;
- chloroplasts contain the pigment chlorophyll, an essential component of the photosynthetic apparatus for fixing carbon dioxide (CO_2) using visible light energy;
- chromoplasts are the remains of chloroplasts in which carotenoids are concentrated, and which form the yellow and red pigments of many flowers and fruits;
- the Golgi complex is concerned with cell wall synthesis;

- the endoplasmic reticulum together with ribosomes is involved in protein synthesis, directed by messages from the nucleus. [*See* PLANT CYTOSKELETON; PLANT PHYSIOLOGY.]

II. The Essentials of Flowers

A flower is formed at the end of a shoot, and houses one or more sets of sexual reproductive organs. The bloom may be solitary or contain an arrangement of blooms in an inflorescence. The part of the head where the floral parts are attached is the receptacle. Most flowers contain two sets of sterile appendages, identified as sepals (forming the calyx) and petals (combined, these give the corolla), and these join the receptacle below the reproductive parts, the stamens (male) and carpels (female). The carpels are often folded lengthways and differentiated into a lower part, the ovary (containing one or more ovules), a middle connecting "style" and an upper "stigma" that receives the pollen (see Fig. 2). When pollen grains of a compatible species land on the stigma, a germ tube from each grows down the style to an ovule where fertilization takes place. This often precipitates senescence of the petals (or their abscission), swelling of the receptacle, and maturation of the seeds. If steps are taken to prevent fertilization, or to provide additional metabolites for the petals, vase life of the blooms can be very much extended. [*See* FLORICULTURE.]

III. The Essentials of Fruits

While one aspect of fruits stresses its edibility, botanically it comes from the expansion of the ovary of an angiospermous flower. Thus, while nuts, grains, and legumes fall within this description, fleshy fruits often arise from the growth and expansion of other structures. Instances of this are shown in Fig. 3. The selection of seedless cultivars of certain fruits (such as citrus and grapes) represents a move toward the elimination of unnecessary features for plant economy and consumer interest. Some of the characteristic attributes of fleshy fruits during ripening can be listed as follows.

A. Differential Response to Ethylene

1. Climacteric fruit—ripening changes are initiated and co-ordinated by a rising synthesis of and sensitivity toward ethylene. Respiration also peaks at about the same time as the hormone. Typical fruit showing

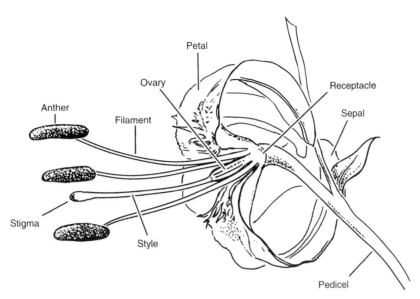

FIGURE 2 Section of a lily (*Lilium henryi*) flower showing the sepals attached to the receptacle below the petals. The gynoecium consists of the ovary, style, and stigma, while the stamen is made up of both filament and anther. (After P. H. Raven and R. F. Evert (1981). "Biology of Plants." Worth, New York).

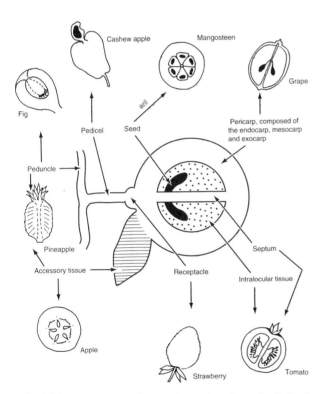

FIGURE 3 Flower parts that can expand to form the flesh of various fruits.

such behavior are apples, avocados, bananas, mangoes, melons, and tomatoes. The continuous exposure to ethylene usually accelerates the onset of ripening.

2. Nonclimacteric fruit—ethylene has little permanent effect on metabolism, and other growth hormones are thought to control the rate of ripening and coloration. Ethylene output and respiration fall continuously following fertilization, and exposure to the hormone does not precipitate ripening. Examples in this class are cherries, cucumbers, citrus, lychees, and strawberries.

3. Intermediate behavior—fig, passionfruit, pineapple, and watermelon.

B. Color Change

Ripeness is generally indicated by an alteration in the pigments in the skin or underlying tissue. Anthocyanins (red to purple) and carotenoid pigments (yellow to red) replace chlorophylls toward the end of development.

C. Flavor Change

The organic acids in a typical fruit, often mainly citric and malic, are offset by an increase in soluble sugars, often sucrose with some glucose and fructose, sometimes the latter two alone. These arise from the degradation of starch within the fruit or from the parent plant. Phenolic components tend to fall in concentration with ripening. The taste components are modulated by the synthesis of small quantities of volatile components during ripening that together give a characteristic flavor profile.

D. Texture Modification

Tissue softening accompanies ripening to a lesser or greater extent through loss of cell turgor, erosion of the middle lamella intercellular cement, and thinning of the cell walls.

E. Abscission Layers

Fruits often fall from the plant, bush, or tree as ripening progresses through the formation of a corky layer of cells in the pedicel. Such abscission can be precipitated by ethylene.

A summary of some of the terms used in describing fruit development is given in Fig. 4.

IV. The Essentials of Vegetables

Salad and cooking vegetables are derived from a diverse range of plant tissue as is illustrated in Fig. 5, and they are drawn from an equally wide range of plant families. However, they do naturally fall into

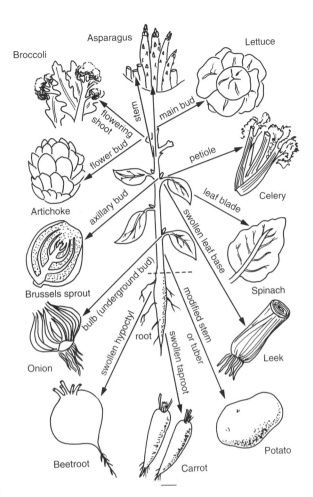

FIGURE 5 Source of tissues that expand to form a number of common vegetables (After Wills, R. B. H., McGlasson, W. B., Graham, D., Lee, T. H., and Hall, E. G. (1989). "Postharvest." BSP Professional Books, Oxford).

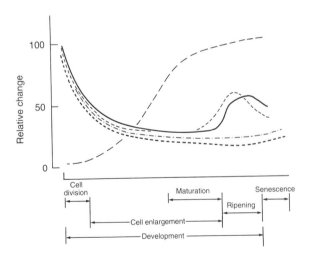

FIGURE 4 Typical growth, respiration, and ethylene evolution patterns in climacteric and nonclimacteric fruits. (Growth (— —), respiration of climacteric (——) and nonclimacteric (– – –) fruits, and ethylene production by climacteric (- - -) and nonclimacteric (-·-·) fruits.

three main groups: seeds (with or without their pods); bulbs, roots, tubers, rhizomes, and corms; and leaves, stems, buds, and flowers. The potato is a modified underground stem, while a sweet potato is simply a swollen root. Ginger is a rhizome or underground stem, while taro is a condensed form of rhizome or corm. Fruits such as cucumber, tomato, and aubergines are normally regarded as vegetables (as they are most often consumed with meat or fish), but botanically this is incorrect. Seeds and pods if harvested fully mature have low moisture contents, respire slowly and verge toward dormancy. However, some seeds such as peas, snap and slicing beans, and sweet corn are picked when still immature and metabolically very active (Table I). If they are to be preserved by freezing, then rapid processing is necessary to prevent serious loss of quality.

TABLE I
Stage of Development When Vegetables Are Normally Harvested

Immature at picking stage	
Legumes:	snap, lima and French beans garden and mange-tout peas
Cucurbits:	cucumbers, soft rind squashes
Solanaceous:	aubergines, bell peppers
Others:	okra, sweet corn
Mature at picking stage	
Cucurbits:	muskmelons, watermelons, pumpkins, hard-rind squashes
Solanaceous:	tomatoes, red peppers

Bulbs, roots, and tubers are storage organs that contain meristems and energy reserves that would sustain growth of the plant if returned to suitable field conditions. When harvested, their metabolic rates are low (about 5 ml CO_2 kg^{-1} hr^{-1} at 5°C or less) and after harvest steps can be taken to prolong this condition by using modified-atmosphere storage (potatoes) or drying followed by low-temperature storage (onions).

Those vegetable crops derived from above-ground parts of the main plant often contain photosynthetically active tissue. Hence, the bright green color associated with fresh vegetables will show signs of yellowing after a few days in darkness at ambient temperatures unless steps are taken to slow deterioration and loss of water. In addition, the texture may

suffer, nutritive value fall, and off-flavors become prominent. A comparative summary of the typical behavior of flowers, fruits, and vegetables at points toward the end of development when they might be harvested is given in Table II.

V. Composition of Fruits and Vegetables

This short section is included to emphasize that in common with flowers, the major constituent of fruits and vegetables is water (see Table III). Whereas the steady transpiration rate of cut flowers can be maintained by water uptake by the stem, the best way to prevent edible produce from excessive water loss is to store at high relative humidity. This is increasingly difficult to ensure as the storage temperature rises. The attractiveness of fruits and vegetables will begin to be eroded when 1–2% of its harvested weight is lost, and this represents a loss in returns as well. Most edible produce will become unsaleable if allowed to lose 3–10% of its original weight, and in almost all cases, unlike flowers, this loss is irreplaceable. Some fruits such as the grape have a waxy layer over the cuticle as do several species of Brassica, and its re-

TABLE II
Comparison of Flowers, Fruits, and Vegetables after Harvest

	Flowers	Fruits	Vegetables
Respiration rate	Reaches a peak at opening, often with a second peak toward senescence	A slow fall in nonclimacteric fruit; a peak in climacteric species	Generally a single peak prior to senescence
Chlorophyll	Slow fall in the sepals and leaves associated with the flowering shoot due to low light and senescence	A slow decline in such fruit as the avocado and citrus to a rapid fall where fruit change color (tomato and banana)	Slow fall in green species due to low light levels and senescence
Ethylene	Rise with senescence slightly after the respiration peak associated with inrolling of the petals and wilting, triggered by fertilization; the perianth may be shed	Slow reduction in nonclimacteric fruit; a peak with climacteric species	Small increase with the onset of senescence
Texture	Loss of turgidity, color, and cell integrity leading to browning; receptable growth	Nonclimacteric fruit usually show a steady decline, but there are exceptions such as the strawberry and the cherry. In climacteric fruit, there is a wide range from apples (slow) to bananas (rapid)	Steady decline due to desiccation, senescence, substrate use for respiration and secondary thickening
Dry matter	Rapid loss of water increases dry matter but substrates are used up rapidly with increased respiration	Increases due to transpiration matched by respiratory losses (see Hardenburg et al.)	Same as for fruit
Protein loss	Usually rapid	Usually slow	Usually rapid for leafy vegetables, but very slow for storage organs

TABLE III
Composition of Some Fruits and Vegetables on a Fresh Weight Basis

Commodity	Dry weight (percentage)	Starch and sugars (percentage)	Organic acids (percentage)	Fiber (percentage)	Vitamin C (mg 100 g^{-1})
Apple	15	13	0.90	0.9	5
Grape	20	16	0.65	0.5	4
Tomato	6	3.5	0.54	0.5	21
Potato	22	19	0.50	0.5	20
Lettuce	5	2.0	0.19	0.5	15
Cucumber	4	2.0	0.25	0.6	8
Beans (string)	10	6.1	0.16	1.0	19
Broccoli	11	4.4	0.33	1.5	113

moval or damage greatly increases transpiration. With care, it is possible to harvest produce when the water-content is close to a maximum and the core temperature is at a minimum. Table III shows that, of the carbohydrate present in fruits, almost all of it is in the form of simple sugars, whereas starch or other polymeric carbohydrate forms the bulk of storage vegetables. Some starch is also present in unripe fruits and leafy vegetables. Most ripe fruits contain some sucrose (cane sugar) with the reducing sugars glucose and fructose there as well. However, cherries, grapes, and tomatoes for instance contain only trace amounts of sugar as sucrose, and the rest is reducing sugar. Dietary fiber, both soluble and insoluble, confers beneficial effects on human health, and this is one of the reasons why the proportion of fruits and vegetables in Western diets is rising. These commodities contribute substantially to dietary vitamin C, and importantly to vitamin A and folic acid requirements. Vitamin C contents of fresh produce are reduced by unsympathetic or prolonged storage, and while new potatoes have 30 mg 100 g^{-1}, storage of 7–9 months reduces this to 8 mg, while extended cooking will decrease this figure even further.

The appearance, color, texture, and (to a lesser extent) the aroma of produce are used by the consumer to assess the likely flavor and the potential shelf-life of a commodity when contemplating purchase. After some time of storage, preparation, and perhaps cooking, the eating texture, ripeness, and flavor components can be assessed more directly. All flowers, fruits, and vegetables produce a range of low-molecular-weight products—mainly esters, alcohols, acids, and carbonyl compounds—that are volatile at ambient temperatures. In flowers they act to attract insect pollinators, while in fruits and vegetables they contribute to the characteristic flavor profile, complementing the taste components which are the

sugars, organic acids, and, to a lesser extent, the salty and bitter principles. The taste and flavor attributes, assessed by the tongue and nose, respectively interact with the other senses to give the consumer a total signal about the desirability and quality of the produce. Postharvest technology has as one of its more difficult tasks the design of methods whereby changes in the appearance, composition, texture, and aroma can be slowed down. The last one is by far the most elusive.

VI. Temperature Effects

Product storage is used to provide a more efficient marketing strategy and has as one of its tenets that deterioration in quality during this period should be minimized. Commodities destined for storage should be harvested at optimal maturity and as far as possible be free from mechanical, physiological, and pathological damage. As indicated in Fig. 6, one of the most

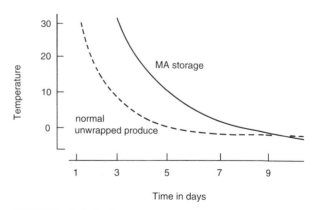

FIGURE 6 Relation between temperature and the length of possible storage for both unwrapped and film-encased (thus modifying the atmosphere) produce that deteriorates rapidly, e.g., mushrooms.

effective methods for preserving harvested crops is to store at temperatures just above the freezing point of the tissue. This slows the onset of senescence, cuts respiratory heat production, moisture loss, spoilage, off-flavor development, and re-growth, but for each commodity a minimum temperature for storage has been found below which irreversible damage will be sustained. A description of the different options for low-temperature storage will be given later. [*See* GRAIN, FEED, AND CROP STORAGE.]

A. Flowers

Figures suggest that there is a 20% loss of cut flowers during marketing, and to keep flowers in good condition, many growers, marketing organizations, and florists have one or more rooms at 1° to 4°C, with the lower temperature being used for longer-term storage. However, higher temperatures (7° to 10°C) are required for holding tropical or cold-sensitive flowers such as orchids and *Anthurium* species. Because of the large surface area of flowers, they quickly assume the store temperature, but water loss is a major difficulty. Maintaining stores with relative humidity of 90–95% is recommended for ensuring good bud-opening, floret development, and freedom from wilting. "Dry pack" storage where flowers are kept without water at close to 0°C is an effective way of maximizing the length of time they can be kept for some (*Gladiolus*, carnation, and *Strelitzia*) but not all species (*Freezia*, *Dahlia*). Infiltration with a warm preservative solution in good-quality water and then the use of moisture-proof boxes help flower longevity. Holding the flowers in keeping-solutions throughout marketing and home use postpones petal senescence and improves flower quality for a wide range of species. Carnations continuously kept in preservative lasted 16.9 days against 6.8 days for control flowers with stems in tap water. Commercial formulations usually contain sugar, a biocide to inhibit microorganism blockage of the water-conducting vessels (8-hydroxyquinoline citrate or a quaternary ammonium compound) and an acidifying agent. Alternatively, pulsing media containing more carbohydrate than is normal for vase solutions can be used before distribution, and the benefits persist even when no subsequent additions are made to the vase water. Bud opening solutions also contain relatively high concentrations of carbohydrate and a biocide, and stems are steeped in such mixtures until the flowers begin to open. Light is of additional benefit.

If flowers senesce prematurely, then one or more of the following causes may apply.

1. Depletion of respiratory substrates—while refrigeration slows down respiration and tissue metabolism, on return to normal temperatures carbohydrates will be used up by the flower head at an increased rate (see Table IV). These can be supplemented by translocation from the vase water and inclusion of a mild bactericide helps to keep the conducting vessels free from blockage. Recutting the stem often helps.
2. Stemborne or surface diseases—these abbreviate the vase life, but prompt refrigeration, control of the relative humidity, stem-shortening, and preservative solutions in a fresh container should maximize resistance.
3. Normal senescent changes—some flowers such as roses and snapdragons should be harvested at the bud stage to have adequate market-life, and then refrigerated promptly. Flowers in bud are often easier to handle, less fragile, and resist adverse conditions well such as the effects of pollutant gases containing ethylene. Nevertheless, maturity at cutting should be such that, on opening, the flower will still be of good quality.
4. Wilting—resaturation after cutting and high humidity in storage and preservative solutions should reverse an excessive loss of moisture.
5. Chilling—while too high a temperature promotes senescence, fading, and the "blueing" of roses, for example, conditions that are too cold may discolor the petals or inhibit subsequent flower opening.
6. Bruising—mechanical damage increases respiration and wound ethylene; both reduce vase-life, while the latter can lead to floret shattering.
7. Ethylene—this natural hormone is symptomatic of senescence, and vase-life can be ended prematurely if flowers are exposed to ethylene from aerial pollution, rotting vegetation, or the proximity of ripening fruit. Sensitivity rises steeply with temperature, and this can

TABLE IV

Respiration Rate and Heat Evolution of Carnation Flowers at Various Temperatures

Temperature (°C)	Respiration rate (mg CO_2 kg^{-1} hr^{-1})	Heat evolution (kcal tonne^{-1} hr^{-1})
0	9.7	91
10	30.0	281
20	239.0	2235
30	516.0	4825

Source: Hardenburg, R. E., Watada, A. E., and Wang, C. Y. (1986). "The Commercial Storage of Fruits, Vegetables, and Florist and Nursery Stocks." USDA, ARS, Agriculture Handbook No. 66, Washington, DC.

be counteracted by either high concentrations of CO_2 (up to 7%, the basis of modified-atmosphere systems) or the incorporation of minute quantities of silver in a translocatable form (the thiosulphate, usually known as STS). However, silver is a heavy metal pollutant and its use in flower preservation is discouraged. Not all flowers respond.

B. Fruit

There is a rather narrow temperature range (5°–30°C are the common extremes) over which fruit can be ripened successfully to meet a particular market demand. If the temperature is too high, shrivelling and abnormal coloration are risks. The lower limit for some fruits is the freezing point of the tissues, but quite a number of species that come from the warmer parts of the world show chilling injury below about 10°C. Low, but not chilling, temperatures are commonly used to delay the onset of ripening, and these conditions obviously also postpone the autocatalytic production of ethylene and decrease the sensitivity of the tissues to the hormone. The aim of many marketing schemes for distant destinations involves the harvest of mature fruits, their transport at temperatures just above those inducing chilling with ventilation to avoid ethylene build-up, and then treatment with ethylene at an appropriate time in specialized ripening rooms. This scheme is successful for a number of fruits, especially bananas and tomatoes. In general, there is an inverse relation between the respiration rate of unripe produce and the length of time they can be successfully stored. Product-life is thus an integral of the storage temperature and the time for which it is held. Most fruits can withstand a short while below their chilling temperature, especially if the ripening process is mostly complete.

C. Vegetables

These generally show no marked respiratory increase toward the end of development, and can be divided into three main categories:

1. Seeds and pods—if these are harvested fully mature as is the practice with cereals, many types of beans and peas, and pulses, then they have a low metabolic rate because of their restricted water content. There is, therefore, no necessity to refrigerate. In contrast, all seeds consumed as fresh vegetables such as legumes and sweet corn have a high metabolic activity because they are harvested at an immature stage. Generally, seeds are sweeter and more tender when not fully

developed, but deteriorate quickly even when stored at 0°C. As the seeds age, sugars are converted to starch, the moisture content decreases, and the amount of fibrous material increases.

2. Bulbs, roots, and tubers such as onions, carrots, and potatoes are storage organs that contain carbohydrate reserves which are mobilized when the growth of the plant is resumed. Onions are usually field-dried, but forced-air can be used to reduce the moisture to about 10%. Long-term storage is achieved at 2°–4°C at 90% relative humidity. The sprouting suppressant maleic hydrazide is being phased out, and irradiation is not completely effective in this direction so reliance is on low temperature, perhaps with atmospheric modification in the future. Storage vegetables may be left in the ground in mild climates and harvested as required. Potatoes may be gathered into cold stores (controlled temperature) or field-clamps (ambient), but below 10°C the texture is increasingly poor with increased sugar content. Atmospheric modification is another possibility.

3. Flower buds, stems, and leaves such as broccoli, asparagus, and spinach have high metabolic rates and need to be cooled rapidly. Refrigeration combined with modified-atmosphere storage (described subsequently) in small packs can be used. Broccoli, for instance, survives well in an atmosphere of 2% oxygen, 10% CO_2, and 88% nitrogen, and when sealed in a semipermeable plastic film, can be kept for a week or more. An increasing use is being made of refrigerated storage for keeping Dutch White (Winter) cabbage. While low temperature alone is adequate for keeping this commodity for up to 6 months, additional storage-life can be obtained by using controlled-atmospheres (2.5–5% oxygen and similar amounts of CO_2) to assure year-round supplies.

VII. Storage Techniques

The orderly and efficient marketing of produce depends on the maintenance of a balance between supply and demand. As consumer purchases often vary with the time of year, the day of the week, and the local weather, short-term storage of various types of produce is becoming more common. Although local supplies can satisfy demand over an ever-longer period through the growing of cultivars having a range of maturity dates, storage during transport to distant markets may necessitate an interval of a few days to a month or more between harvest and arrival at the final destination. Appropriate techniques for each commodity must be devised so that quality and product-life are not eroded overmuch. Stored lines must also compete with fresh produce on the international market.

Storage has the aims of slowing the deterioration of freshly harvested produce economically and reproducibly, commonly using some combination of high or low temperatures and atmospheric modification. The well-being of the commodities must be supported so that they can resist attack by pathogens, excessive moisture loss, and the onset of some physiological ripening disorders. An indication of appropriate cooling methods for flowers, fruits, and vegetables is available, as are the recommended storage temperatures for a wide range of produce. The anticipated storage life of fruits and vegetables within specific temperature ranges has been listed as well (see Hardenburg *et al.*, also Kader).

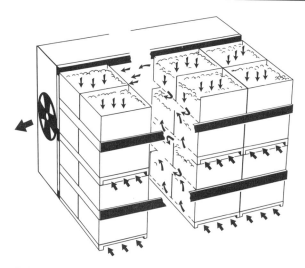

FIGURE 7 Pattern of air-flow in a forced-air cooling system, where cold air is moved across or through produce stacked in containers against a perforated wall behind which air is kept under slight vacuum. The air is then recooled and may be rehydrated before being returned to the store.

A. Temperature and Humidity Control

Many methods for cooling produce have been developed, and the appropriateness of each depends on the commodity in question and the length of storage envisaged. The containers involved, the rate of cooling needed, and the humidity requirements also affect the choice of system.

1. Room cooling—cold air from a refrigeration unit is circulated round a store, sometimes with partial or complete venting to the outside atmosphere. Heat transfer is rather slow, and unless precautions are taken, there may be excessive water loss by the produce. Humidity control or a limited temperature drop on the cooling coils (preventing icing up) avoids excessive drying.
2. Forced-air cooling—this method of cooling uses a system whereby containerized produce is stacked against a perforated vertical panel. A slight pressure gradient pulls cold air from the room through vents in the storage boxes and over the commodities in question and through the panel where the air is recooled and returned to the room (Fig. 7). This intimate mixing of chilled air and product achieves rapid cooling. By humidifying the cold-air, water-loss can be minimized. A range of commodities can be accommodated, but ethylene build-up should be monitored.
3. Hydrocooling—water at a constant low temperature is used to remove the field-heat from produce. Commonly, a bath of cold water kept at a constant temperature is used in which the commodity is immersed for a set period, and the produce is removed by a conveyer belt, dried, graded, and packed. Chlorine is often added to the water for sterilization, and salt may be dissolved to help the

produce to float. Alternatively, a cold-water drench directly onto the produce can be used. While the method of cooling is particularly efficient with compatible crops, water may be drawn into fruit such as tomatoes that is not removed on drying. This may make subsequent storage of this crop more difficult.

4. Vacuum cooling—if a crop that has a large surface area and a high water content is placed in a container which is then evacuated, the very low air pressure inside causes the water contained by the commodity to boil at ambient temperatures (Fig. 8). This quickly removes the field-heat, and is particularly useful for crops such as lettuce, celery, and mushrooms. Weight losses are about 1.8% per 10°C fall in temperature, but not all need come from the produce if a fine spray of water is added during operation.
5. Ice-bank cooling—air is passed through ice-chilled water before passing over the produce. The cool air is thus saturated with water vapor and this minimizes desiccation of the product. There is also no risk of the commodity freezing, and the weight loss is only 1.3% per 10°C temperature reduction. The capital cost is not so great as with vacuum cooling, and it is more versatile.

B. Controlled Atmosphere (CA) Storage

This term implies that gases are removed or added to air to form an atmosphere which is kept constant in composition and in which certain commodities can be stored at low temperature, if necessary, for long periods. A modern system, based on the pioneering work of Kidd and West in the 1930s, combines re-

FIGURE 8 Vacuum cooling is an efficient way to remove field heat from leafy vegetables or salad crops. The illustration shows produce packed in boxes being pushed into a chamber which is then sealed and evacuated. Water evaporating from the surface of the commodities reduces the temperature without any danger of freezing, sometimes in as little as 20 to 30 min.

frigeration with levels of oxygen and carbon dioxide of 1 to 5%. These conditions slow many of the ripening processes in apples and pears so that storage disorders such as apple scald are minimized and the growth of decay organisms is inhibited. Storage rooms need to be relatively gas-tight and some means must be available whereby ethylene levels can be kept consistently low, that is less than 1 μl per liter of atmosphere, for the preservation of pome fruit and nonclimacteric produce such as flowers and vegetables to avoid premature senescence.

Originally, the system relied on the respiration of the produce to increase the CO_2 concentration to 5–10% at the expense of oxygen. Then oxygen levels lower than 10% were found to be beneficial, while some apple and pear cultivars were damaged by CO_2 accumulating to more than 3%. The burning of hydrocarbon fuels in associated generators can be used in conjunction with natural respiration, and the excess carbon dioxide is removed by incorporating an alkali,

activated charcoal, or by the use of a molecular sieve. Atmospheres of 2–3% oxygen and 2–5% CO_2 are commonly used with pome fruit, and the tendency is toward using ultralow oxygen concentrations (<1% oxygen, <1% CO_2) with dessert apples. This is said to reduce spoilage and maintain tissue firmness. In addition, some means should be incorporated whereby ethylene levels can be kept within limits. Ethylene-absorbers (potassium permanganate on kieselguhr, known as Purafil), high-temperature catalytic burners, or ultraviolet radiation can be used to oxidize ethylene and other hydrocarbons in the volatile fraction.

C. Modified-Atmosphere Storage

Rather than artificial adjustment of the storage atmosphere, the respiration of the commodity itself is used to reduce oxygen and increase CO_2 concentration. Depending on the temperature of storage and the basic respiration rate of the produce, alteration of the atmosphere inside a sealed container takes place rather slowly, eventually forming an equilibrium mixture of gases. Because the respiration of harvested crops is inhibited by oxygen concentrations of less than ambient, also by the build-up of CO_2, the deterioration rate of many types of produce is slowed down.

This method of storage is highly adaptable. Whole rooms can be filled with produce and sealed, and then the atmosphere allowed to adjust. Alternatively, shipping containers can be fitted with polyethylene liners to restrict gas exchange, or pallet loads can be shrouded in film so that the contents are exposed to an atmosphere low in oxygen. Diffusion windows may be fitted to plastic covers on storage bins, or, at consumer level, to individual packs of produce. This ensures that the produce is never in danger of becoming anaerobic. Punnets filled with produce and over-wrapped with one of many types of film (sometimes microperforated) are gradually being accepted as a useful way of preserving and protecting cut flowers, fruits, and vegetables. The application of waxes or other surface coatings can modify gas exchange, but they are applied mainly to reduce water loss and for cosmetic purposes rather than for extending product-life.

MA storage is a versatile and inexpensive technique, but it has associated disadvantages. Many commodities have a high water content, and transpiration under sealed conditions leads to an atmosphere high in humidity—ideal conditions for the growth of spoilage organisms. It is difficult to design a plastic film that is sufficiently permeable to water vapor, resistant to the ingress of oxygen, yet able to retain sufficient CO_2 to restrict respiration of the produce. The commodity in any one pack may be of variable maturity, and the temperature at which small packs are kept add to the complications of the system. However, pallets of prepacked strawberries are overwrapped with plastic and flushed with pure CO_2 in the "Tectrol" system, and this treatment preserves both quality and freedom from spoilage. In addition, film-lined boxes or individual consumer-packs of flowers benefit from modifications that allow respiration to take place at a reduced rate, especially when flushed with an appropriate gas mixture prior to sealing.

D. Hypobaric Storage

In this process, perishable commodities are maintained at low temperature and pressure while being continuously ventilated with humidified air. Conditions used for flowers involve a temperature close to 0°C and 40–60 mm Hg. The longevity of flowers is often doubled compared with normal low-temperature storage, and it is thought that restricted oxygen and ethylene slow respiration very much. The capital cost of the apparatus is high.

E. Irradiation

The use of γ-radiation has been explored as a means of controlling decay, insect infestation, and deterioration in flowers, fruits, and vegetables. Dosages are usually less than 1.5 kilograys (kGy), as damage to most tissues occurs above this level. Low doses of about 100 Gy can inhibit the sprouting of potatoes or onions with relatively little effect on subsequent quality. The use of chemical sprout inhibitors such as 3-chloroisopropyl-N-phenylcarbamate and maleic hydrazide is under legal restriction in some countries. While the use of radiation to limit the deterioration of fresh produce is probably not a practical proposition on the grounds of cost, adverse side effects, and consumer resistance, as an effective disinfestation treatment involving 50–200 Gy, it has some potential in the future. [See FOOD IRRADIATION.]

F. Chemical and Physical Treatment

The waxing of fruits is extensively practised, mainly on citrus, but occasionally on deciduous crops and on vegetables as well. Fungicides are often incorporated along with solvent-soluble or water-soluble waxes. The benzimidazole group of fungicides, thiabendazole, benomyl, thiophanate methyl, and carbendazim are compounds usually utilised for their antifungal activities. This group of related compounds is usually used on citrus, but is effective with bananas and peaches as well. Gray mold on table grapes is controlled by the slow release of sulfur dioxide, while the development of biological control methods for postharvest diseases through antagonism between a harmless microorganism and the pathogen is receiving much attention. [See CITRUS FRUITS; FUNGICIDES.]

Climacteric fruit are often harvested when they have just reached maturity, transported while still firm and underripe, and then stored until fit for market. Provided that fruits are at a correct stage of maturity when picked, cantaloupe melons, papaya, and tomatoes will ripen on their own to give a good-quality product. Bananas and melons can, however, be ripened with ethylene even when immature, and 10 μl 1^{-1} for 1–3 days at about 20°C and in an atmosphere low in CO_2 would start off the ripening process. Citrus that has not received sufficient cold or has been left unharvested for too long will tend to re-green. Ethylene at 50 μl 1^{-1} at 25–30°C should convert chlorophyll to carotenoids in a few days.

VIII. Conclusions

Centers of production and the final consumer are often far apart. Assuming that the commodity has been well grown and harvested at an appropriate time, much can be achieved by the distributor during marketing to preserve, or even enhance, the quality that the grower has put into the product. He will have been encouraged to use the minimum of crop-protective chemicals, and the crop will have to have been graded, cooled, packed, stored, and delivered at an appropriate time, place, and core temperature. The commodity may well have had to be prepacked and labeled with appropriate information, and even with suggestions for continued storage until use. There is now an increasing awareness by all points in the chain of distribution that flowers, fruits, and vegetables are alive and that they respire, produce heat and carbon dioxide, and are sensitive to adverse conditions of storage. Growers and marketing provide the consumer with an ever-increasing range of commodities from which to choose. It is up to all concerned to ensure that postharvest technology is correctly applied so that the appearance and quality of fresh produce are as close to its potential as possible.

Bibliography

Hardenburg, R. E., Watada, A. E., and Wang, C. Y. (1986). "The Commercial Storage of Fruits, Vegetables, and Florist and Nursery Stocks." USDA, ARS, Agriculture Handbook No. 66, Washington, DC.

Kader, A. A. (1992). "Postharvest Technology of Horticultural Crops." Publication 3311, University of California, Division of Agriculture and Natural Resources, Oakland, CA.

Monselise, S. P. (1986). "CRC Handbook of Fruit Set and Development." CRC Press, Boca Raton, FL.

Salunkhe, D. K., Bhat, N. R., and Desai, B. B. (1990). "Postharvest Biotechnology of Flowers and Ornamental Plants." Springer-Verlag, New York.

Seymour, G. B., Taylor, J. E., and Tucker, G. A. (1993). "Biochemistry of Fruit Ripening." Chapman and Hall, London.

Weichmann, J. (1987). "Postharvest Physiology of Vegetables." Marcel Dekker, New York.

Wills, R. B. H., McGlasson, W. B., Graham, D., Lee, T. H., and Hall, E. G. (1989). "Postharvest." BSP Professional Books, Oxford.

Potato

C. R. BROWN, *USDA-Agricultural Research Service, Washington*

Glossary

Autotetraploid State of having four identical sets of chromosomes, contrasted to diploid where two sets are present

Axenic culture Growth of potato in a sterile medium in a closed container that excludes recontamination by other organisms

Callus Disorganized cell clumps that develop in tissue culture and give rise to differentiated shoots under the right conditions

Clone Line of organisms derived from single individual and produced by vegetative (asexual) propagation

Heterozygosity State of having multiple forms of genes; genetic diversity is characterized by heterozygosity in individuals and among individuals in populations; potato breeding emphasizes maintenance of heterozygosity to achieve high vigor

Meristem Growing tip of stem in which resides all the tissues of the plant before further growth and differentiation

mRNA "Messenger" ribonucleic acid polymers synthesized in the cell nucleus which transport the genetic code to the cytoplasm and serve as the template for protein synthesis; mRNA molecules are the intermediaries between the genetic code and the protein product of the gene

Parenchyma storage cells Undifferentiated cells in which starch is stored in potato tuber; distinct from specialized cells such as periderm and vascular tissues

Seedpiece Portion of tuber used as seed to plant the crop

Seed tuber Tuber used as seed to plant the crop

Stolon Underground stem which grows horizontally in darkness a portion of which is modified to become a tuber

True seeds Sexually derived seeds produced in fruits

Tuber Underground stem (stolon) which has expanded and become a starchy storage organ

The potato plant consists of leafy, herbaceous aboveground foliage, and underground roots and stems (stolons). Stolons are initiated at below-ground basal locations on the main stems and grow diageotropically (horizontally) in the darkness of soil, until tuberization or radial expansion of the stolon tip commences (see Fig. 1). The stolons swell (tuberize) and become filled with starch during the growing season, at which time they are referred to as tubers. The tubers are harvested as food for human and animal consumption, for extracted starch, and for ethanol production. Besides being consumed directly or converted to a processed product, tubers are used as propagules or seed to establish a new crop. After a dormancy period, upon planting below the soil surface, axillary buds on the tubers grow, giving rise to aboveground stems. Some potato varieties flower and form fruits containing sexual true seed, which when germinated can give rise to potato seedlings. Clonal propagation perpetuates the genetic characteristics of the potato unchanged, while sexual propagation does not. Sexual propagation results in large genetic variation of the progeny, none of which will be genetically identical to the parents.

The potato belongs to the family Solanaceae, which includes food crops such as tomato, pepper, eggplant, pepino, naranjilla, and the nonnutritive physiological stimulants of tobacco and belladonna. Potato is in the

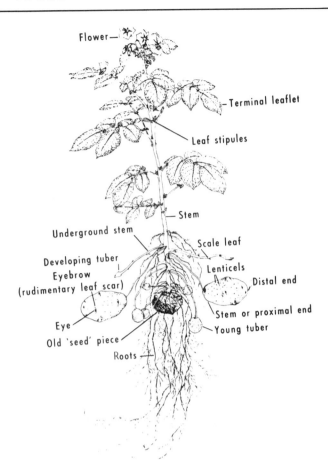

Flower

Terminal leaflet

Leaf stipules

Stem

Underground stem

Scale leaf

Developing tuber

Lenticels

Eyebrow
(rudimentary leaf scar)

Distal end

Stem or proximal end

Eye

Young tuber

Old 'seed' piece

Roots

FIGURE 1 The potato plant showing seed tuber, foliage, inflorescence, roots, stolons, and immature and mature tubers.

genus *Solanum*, subgenus *Potatoe*, section *Petota*. This section is further divided into non-tuber-bearing, *Estolonifera,* and tuber-bearing, *Potatoe,* subsections.

Potato originated in the Andean Cordillera of South America, probably in the vicinity of Lake Titicaca where today the greatest concentration of diversity of native cultivars is to be found. Having been domesticated by native South Americans and under cultivation for millennia, the potato was discovered by Spanish conquerors and transported to Spain around 1570. By the turn of the century it had diffused to Italy, Germany, The Low Countries, and England. Although Spanish Chroniclers had observed and documented that it was an important food in the Andes, this knowledge seems not to have convincingly spread throughout Europe for some time. The potato was regarded as a botanical curiosity and with some suspicion by herbalists of the time due to its membership in the "Nightshade" family, a group of plants reputed for poisonous or neurologically potent effects. Nevertheless, within a hundred years of introduction, ac-

counts were appearing of its use as a food, usually in the context of nutrition for the poorest segments of society. This process of adoption continued in accelerated fashion, undoubtedly accompanied by genetic evolution of a higher yield through selection of sexually generated seedlings from true seed with earlier maturity, until potato became a major item of the diet in many areas of Europe.

I. The Tuber

The tubers originate from tips of stolons. Being a modified underground stem, it contains dormant true buds (eyes) formed at the base of a rudimentary leaf with detectable leaf scars (eyebrows). Like other stems the tuber has lenticels for aeration, which are very enlarged in hypoxic environments. The buds (eyes) are found in a spiral pattern on the tuber, and tend to be more plentiful at the apical end. The most apical bud exerts partial dominance over the other buds, which are activated as the tuber ages, or as successive apical buds are removed or killed by adverse factors. [*See* ROOT AND TUBER CROPS.]

The outer layer of the tuber, called the periderm is 6 to 10 cells thick. The cells of the periderm harden as the tuber matures forming a protective barrier. Pigments are found in the periderm or in the cortex, the region inside the periderm that extends to the vascular tissue (see Fig. 2). The remainder of the tuber comprises a vascular ring and an outer and inner medulla. The inner medulla extends toward each eye, forming a continuous tissue connecting all the eyes. Starch content varies between 10 and 25% of fresh weight. Starch grains are formed within storage parenchyma cells by organelles called amyloplasts. Size of grains and density of starch may differ between tissues within tubers and this pattern may differ between potato genotypes. The greatest concentration of starch granules is found near the vascular tissue. Starch is composed of amylose and amylopectin in a 1:3 ratio. Amylopectin is a large, highly branched polymer of approximately 100,000 glucose residues while amylose is composed of about 5000 mainly unbranched residues.

II. Agronomy

The potato crop is established by planting seed tubers. The production of seed tubers is a specialized endeavor, with particular attention paid to preventing the infection of seedstock by pathogens. By virtue of

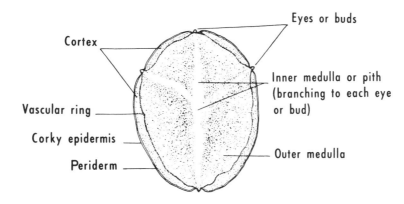

FIGURE 2 Cross-section of tuber showing outer and inner tissues and vascular system.

the vegetative propagation, viruses and some bacteria, fungi, and nematodes are carried in a systemic fashion or by tuber surface contamination from one vegetative generation to the next. The common method of establishing pathogen-free potato seedstock is to place shoots in axenic culture. It is possible to eradicate viruses by application of heat to tissue cultured shoots or whole plants in nonsterile pot culture, excision of meristematic tissue while under treatment, and use of antiviral reagents in the growth medium. Verification of freedom from pathogens is performed by testing tissue serologically for viral pathogens, inoculation of tissue extracts to indicator host species, and use of nucleic acid hybridization tests (potato spindle tuber viroid).

An innovative variation in potato production is to use the sexual true seed from the fruit as a source of the crop. True seed may be used to plant a crop for the marketplace or as a starting point of a seed program where several cycles of vegetative propagation would be carried out to produce a tuber seed crop. True seed has the advantage of excluding transmission of nearly all systemic pathogens.

Seed potatoes are produced typically in colder, upland environments which are isolated from commercial production, have short growing seasons, and have relative freedom from flights of aphids. Avoidance of aphids is necessary to prevent introduction and spread of the two most important viruses of potato: potato virus Y (PVY) and potato leafroll virus (PLRV).

The vast majority the world potato crop is planted from uncut seed potatoes. The seed tuber is kept in storage after harvest until the next planting season. Following a dormancy period, at the end of which the eyes begin to sprout, the seed is exposed to light (chitting) to green the sprouts and break apical dominance. In North America and certain other countries, most notably Argentina and Uruguay, seed tubers are cut into seedpieces prior to planting. Ideally, the cut surfaces are allowed to form a corky wound tissue before the seed is planted, but in practice the time between cutting and planting is usually too short for this to occur. The cutting must be carried out to optimize seedpiece size and to minimize blind seedpieces, i.e., those lacking eyes. Cutting of seed permits economical use of oversize potatoes and eliminates apical dominance permitting greater control of the number of buds which give rise to growing stems. But, it introduces an additional avenue for contamination of seed by disease organisms, particularly those that cause seedpiece decay and ringrot bacteria (*Clavibacter michiganense* ssp. *sepedonicus*).

Potatoes are planted at rates of 1600 to 2800 kg/ha of seed depending on variety, seed size, and spacing. During the crop growth one practice is required that addresses the need for the plant to have ample underground space for proper development of tubers. Thirty to forty days after planting, when the stems have emerged from the soil, it is necessary to raise soil up around the base of the plant, a procedure called hilling-up or bed-shaping.

III. Growth of the Potato Plant

Genetic control of time of tuber initiation is quite marked and cultivars are classified as early, intermediate, and late maturing according to their responses. Classification of maturity is an interaction of time of tuberization initiation and rate of bulking (increase in weight of tubers) (see Fig. 3). Early maturity is characterized by early onset of tuber formation and a faster accumulation of foliage and tuber weight. The later maturing cultivar is slower in both categories,

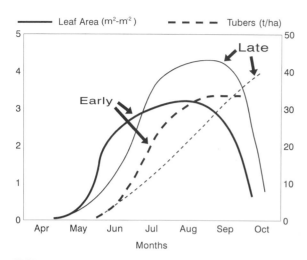

FIGURE 3 Diagram of the development of leaf area (meters per square meter of ground surface area) and tuber mass (kilograms per hectare) in early and late maturing potato varieties.

but the continued development of new foliage and additional tuber weight gain later during the growing season can lead to a much greater final yield for late maturing cultivars.

By virtue of its evolution and domestication at low latitudes, the potato is a short daylength adapted crop. Tuber formation takes place only if the daylength is equal to or shorter than the critical daylength. The critical daylength of late maturing cultivars is shorter than that of early ones. In practical terms, daylength is a consequence of the latitude and season at which a crop is grown. The tuber initiation of a late maturing variety (short critical daylength) occurs earlier in the growth cycle when exposed to a reduced daylength whereas an early maturing cultivar (long critical daylength) is less affected by this change. For example, native cultivars from the low latitudes of South America tuberize poorly, or not at all, at high latitudes because they are growing at daylength regimes that exceed their short critical daylength. In contrast, cultivars selected for adaptation to high latitudes (long critical daylength) typically develop a small amount of foliage, and tuberize very early under short day conditions, resulting in poor yield. Due to these considerations, varieties usually do not perform well at daylengths that are very different from that in which they were selected. The tuberization initiation response is a continuously varying character which interacts with other factors. While tuberization is favored by shortage of nitrogen and phosphorus, high solar radiation, and low temperatures, it is, conversely, delayed by long days, abundant nitrogen, and high temperatures.

Most of the tuber dry weight arises from current photosynthate and bulking continues as long as foliage is photosynthetically active. The mean tuber size is an inverse function of the number of tubers initiated. Tuber dry matter yield is directly related to the amount of foliage developed, up to a limit, which in turn relates to the quantity of light intercepted by the foliage. Bulking can be slowed or stopped by frost, disease, very high day and night temperatures, or high water deficit. The potato crop may cease growth due to the physiological senescence of the foliage, disease, or agronomic practices at the point where optimal tuber size and specific gravity required for the intended utilization have been achieved. Artificial termination of foliage growth is accomplished by desiccation with chemicals, mechanical crushing with a roller, or stripping of foliage away from the crown of the plant by mechanical pulling or beating. Maturation of tubers promotes tuber skin set (suberization) reducing mechanical damage at harvest, and release of tubers from the stolons.

Comparative productivity and dollar values for potato and other major crops are presented in Table I. Potatoes produce a considerable amount of dry matter per unit area, and are surpassed in this and energy production per hectare per day only by sweet potatoes. Even though protein is only 2 and 10% of the weight of raw and dried potato, respectively, the magnitude of total yield results in protein production that is relatively high per unit land area per day, ranking fourth among the crops listed. Potato is usually a high value and high input crop. This is reflected by the fact that potato is third in monetary value produced per day per unit land area, expressed in U.S. dollars, ranking well above the grains and cassava.

IV. Utilization

Specific gravity of potato tubers refers to the density of the tubers with respect to water. Potatoes are slightly denser than water and usually range between 1.060 and 1.090. Specific gravities in the upper range are more desirable from a quality standpoint. Variety selection and appropriate agronomic management are combined to target specific gravities of 1.080 and above for processing. The influences of yearly climate, locality, and agronomic treatment cause variability in specific gravities. However, processors usually pay premiums over the contract price for higher specific gravities. The consumer's satisfaction with the culinary quality of home cooked potatoes, or a

TABLE I

Productivity and Value of Potato Compared to Other Major Crops

Crop	Dry matter kg/ha/day	Energy Mcal/ha/day	Protein kg/ha/day	Value $/ha/day
Cabbages	12	29	1.6	27.5
Tomatoes	8	25	1.3	25.3
Potatoes	18	54	1.5	12.6
Yams	14	47	1.0	8.8
Sweet potatoes	22	70	1.0	6.7
Rice (paddy)	18	49	0.9	3.4
Peanuts	8	36	1.7	2.6
Wheat	14	40	1.6	2.3
Cassava	13	27	0.1	2.2

Source: Horton, D. E. (1988). Potatoes: truly a world crop. *SPAN* **30,** 116–118.

processed product, is closely correlated with specific gravity of the raw product.

Potatoes may be stored for long periods of time (3 to 9 months) to provide a steady supply for processing and fresh market. Dormancy is a postharvest rest period during which sprouting of the tuber eyes does not occur. The dehydration and conversion of starch to sugars which accompany sprouting are undesirable for stored potatoes intended for fresh market or processing. Genetic differences in length of dormancy have been noted. The most utilized varieties may not have long dormancy. Furthermore, the long periods of storage of potatoes necessitate a lengthening of dormancy by artificial means. Length of dormancy can be increased by storage at low temperature (4°C) and sprouting can be prevented by application of sprout inhibiting chemical treatments applied to foliage during crop growth in the field or to tubers in storage. [*See* DORMANCY.]

Ideally tubers are kept at nearly constant temperatures between 4.4° and 10°C. The heat the tubers have accumulated in the field and heat of respiration must be removed. The tuber is a watery modified stem which must be protected from dehydration; thus, storages are kept at high relative humidities. This may require moisture supplementation through mechanical humidification in arid climates. Potatoes stored at the lowest temperatures accumulate reducing sugars (sucrose, glucose, fructose) which impart a brown color to fried products. Therefore, potatoes for chip and french fry processing are stored at minimum temperatures of 10° and 7.3°C, respectively. Potatoes stored at lower temperatures can be "reconditioned" to lower reducing sugars by storage at 10°C for 3 weeks or longer. Considerable effort is being expended to select new "cold chipping" varieties that can be stored at 4.4°C, but do not accumulate reducing

sugars, and thus can be fried directly out of cold storage.

The harvested tuber may be marketed as a raw vegetable or it may be processed. Most processed potato is fried before consumption. Potato chips (crisps) are thin slices of tuber which are fried and packaged as a dry nonrefrigerated snack food. French fries [*pommes frites,* chips (U.K.)] are strips that are precooked, stored frozen, and fried in fat before consumption. In addition, potatoes are dehydrated to provide flakes or granules for dry products which are rehydrated during cooking. Extraction of potato starch, used as a thickener in cooking, is an important industry in The Netherlands and Japan. In Poland, Ireland, and the former Soviet Union a portion of the potato harvest is used as animal fodder. Also a small tonnage of potato is fermented to obtain ethanol in Northern Europe.

The largest amounts of potato are produced by the former Soviet Union, Poland, and China (see Table II). The per capita consumption of China is about a quarter of that of Poland and the former USSR. Of the production 53 and 27% is used as animal fodder in Poland and the former USSR. The Netherlands allocates 54% of its production for industrial starch production. Otherwise potato is predominantly a food for human consumption. Production statistics for Peru are included to illustrate the relatively minor contribution to world production made by an area where potato originated as a crop.

V. Nutritional Aspects of Potato

Potato is approximately 80% water, 17% carbohydrate, and 2% protein in its raw and cooked states. As a consequence, the energy content of raw potato

TABLE II

Potato Production and Consumption in the Highest Producing Countries and Peru Where the Potato Originated

Country	Area (10^3 ha)	Production (10^6 tons)	Yield (t/ha)	Per capita consumption (kg)
Soviet Union	6866	77.8	11	107
China	4005	46.7	12	26
Poland	2219	36.3	16	118
Germany	741	16.3	24	50
United States	506	15.4	30	52
India	748	9.9	13	10
Peru	192	1.6	8	65

is considerably less than that of raw (uncooked) cereals and legumes. When cooked, the latter staples change their composition by absorbing large quantities of water whereupon the energy content of potato compares favorably to cereals and *Phaseolus* beans. Wheat bread and maize tortilla are considerably richer in energy because of the much lower water content per edible portion. Potato dried to a moisture content of 12% compares favorably to cereals such as rice, corn, wheat, and sorghum, in content of crude protein and carbohydrate. Potato has a high-quality protein due to its balanced amino acid composition, and may partially compensate for lysine deficiency in diets where cereals predominate. It is an excellent source of water-soluble vitamins, including the B group vitamins and ascorbic acid. Contrary to common perception it is a poor source of energy unless fried. Boiled potato provides less energy (kcal per edible portion) than sweet potato, cassava, and plantain (Table III). If consumed alone, a considerable quantity of potato must be eaten to satisfy energy requirements. This becomes limiting in small children who may not be able to ingest sufficient volume. However, in diets that are not deficient in calories, potato can provide a significant amount of protein.

A 100-g portion of potato (one small tuber) can supply 5% of the energy requirement, and 10% of the protein requirement of children ages 3–5 years. Potato's low energy density may be a dietary advantage in societies where obesity is a common form of malnutrition, as long as it is consumed without energy enriching amendments.

VI. Disease and Pest Problems

A. Insects

Colorado potato beetle *Leptinotarsa decemlineata* is the most damaging leaf-chewing insect of potato. Adults overwinter in the soil, emerging in the spring as soil temperatures rise. Mating occurs, egg clusters are laid, and larvae hatch and pass through four instars followed by pupation. Population build-up can be very rapid in warmer growing seasons. The larvae are voracious consumers of foliage, stripping the stems of leaves when in sufficient numbers. Insecticides and rotation of fields are the only effective control measures. Rotation and spatial separation of succeeding potato crops are effective in reducing initial populations due to the sedentary nature of infestations and the difficulty that emerging populations have in colonizing distant fields.

Potato has been genetically engineered to synthesize a protein toxin from *Bacillus thuringiensis* that effectively kills the Colorado potato beetle. Advanced breeding clones possessing glandular trichomes derived from the wild species *Solanum berthaultii* also have effective resistance to the beetle.

Potato tuber moth, *Phthorimaea operculella,* is a serious pest of potato in warm arid regions. Hatching larvae burrow into leaves and mine them. Larvae can enter tubers, often destroying the eyes, while predisposing the tuber to rot while in the field and during storage. High infestations can make the crop uneconomical to harvest. Control measures include pesticide applications to foliage and seed tubers (seed potatoes) and application of agronomic measures that seal the soil to protect newly formed tubers from infestation. The highly pungent foliage of the Peruvian plant muña (*Minthostachys* spp.) was mixed with stored potatoes in pre-Columbian times to control insect damage and has been shown to be very effective in controlling the spread of moth damage in stored tubers.

Potato is a host to many species of aphids. Usually the only economic damage is caused by the viruses that aphids vector. In this regard the most important aphid worldwide is the green peach aphid (*Myzus*

TABLE III

Comparison of Composition of Potato Prepared in Various Ways and Foods Derived from Other Food Crops

Food item	Energy (kcal)	Moisture (%)	Crude protein (g)	Fat (g)	Total carbohydrate (g)	Ascorbic acid (mg)
Potato						
Boiled in skin	76	79.8	2.1	0.1	18.5	16
Boiled, peeled	72	81.4	1.7	0.1	16.8	4–14
French fries	264	45.9	4.1	12.1	36.7	5–16
Chips	551	2.3	5.8	37.9	49.7	17
Other crop sources						
Rice[a]	135	67.9	2.3	0.3	28.0	0
Tortilla[b]	210	47.5	4.6	1.8	45.3	0
Bread[c]	278	32.7	8.7	1.6	55.7	0
Sweet potato	116	70.2	1.4	0.4	27.4	26
Cassava	145	62.6	1.1	0.3	35.2	36
Plantain	127	64.5	1.2	0.2	33.3	22
Beans[d]	118	69.0	7.8	0.5	21.4	0

Note: All values are for a 100-g edible portion.

[a] Boiled, white.

[b] Lime-treated maize.

[c] White wheat flour, unenriched.

[d] *Phaseolus vulgaris,* boiled.

persicae) due to its ubiquity and important role as vector of PLRV and PVY. Most control of aphids is therefore directed at early intervention to prevent virus introduction and spread. This is most crucial in tuber seed production where selection of locality may rest on relative freedom from natural aphid infestations. Application of insecticides and artificial vine killing are used to prevent introduction of viruses by aphids into seed crops.

B. Nematodes

Two genera of nematodes cause the greatest damage in potatoes. These are the potato cyst nematodes (*Globodera* spp.) and the root-knot nematodes (*Meloidogyne* spp.). *Globodera* spp. are important in the higher latitude temperate production areas or in alpine production areas in low latitudes. Two species, *G. rostochiensis* and *G. pallida,* originate from Andean regions of South America. Juveniles enter the roots inciting the formation of a syncytium or specialized feeding cell. Those juveniles developing into females after fertilization deposit eggs internally. Upon dying the female's body becomes a cyst containing several hundred eggs. The penetration and feeding are damaging to the roots, resulting in a much reduced root system and impaired plant vigor. Substantial losses are possible. Potato cyst nematodes are highly specialized on *Solanum* species. Cysts are resistant to drying, surviving 20 to 30 years in the soil. In years when nonhosts are cultivated, only a small proportion of eggs hatch. Normal rotations are therefore usually not effective in eradicating the nematode. Traditional Andean rotations of planting potato one year followed by 6 years in pasture or fallow are somewhat effective. [*See* PLANT PATHOLOGY.]

Root-knot nematodes occur in warmer, lighter soils. Although they are less of problem than potato cyst nematodes worldwide, much of the expansion of potato culture into warmer production areas of the developing world will encounter root-knot nematodes in the soil. Second stage juveniles invade roots, settling in the vascular tissue, and inciting formation of giant cells which facilitate feeding. The females are parthenogenetic, laying thousands of eggs in a gelatinous matrix that is secreted outside the root tissue. Some species of root knot cause severe root galling (*M. incognita*) while others restrict their physiologically and economically important damage to the tuber (*M. chitwoodi*). *M. hapla* and *M. chitwoodi* are exceptional in their adaptation to colder production areas.

Control of both nematodes may be accomplished by soil fumigation. Crop rotation is potentially more effective against *Meloidogyne* spp. owing to the perishable status of eggs in soil, even in the presence of nonhosts. However, *Meloidogyne* species have a broad host range, making it difficult to select economically viable nonhost rotation crops. Efforts are under way to utilize resistance to *Meloidogyne* species derived from wild *Solanum* species. A single gene resistance, H_1, to *G. rostochiensis* has been extracted from *S. tuberosum* ssp. *andigena* and incorporated into many modern varieties while multigenic partial resistance derived from wild species is being developed for *G. pallida*. [See NEMATICIDES.]

C. Fungi

The most important fungal pest of potato is late blight caused by the fungus, *Phytophthora infestans*. Historically this has been the most damaging disease of potato on a worldwide basis. Symptoms start with irregular lesions on leaves developing into leaf and stem collapse. Unprotected susceptible cultivars can be completely destroyed in short periods of time. Infection of tubers while still in the soil is a serious potential problem often not apparent until tubers are in storage and beginning to rot. Protectant and systemic fungicides have been used effectively against late blight. These are often capital intensive in situations of high blight pressure as they must be applied numerous times during the growing season on a regular schedule. The large quantity of fungicide used to control blight is an environmental issue in many countries.

Single gene resistance derived from the wild Mexican species *S. demissum* was incorporated into many potato varieties in the years between 1930 and 1950, but new virulence genotypes of the pathogen rapidly overcame this type of resistance. Recent host resistance work has focused on development of horizontal resistance which consists of many distinct components of partial pathogen inhibition involving the action of many genes. There are several wild species sources of horizontal resistance. Late maturing cultivars generally suffer less damage than early maturing ones.

Another blight disease of worldwide importance is called early blight or target spot caused by *Alternaria solani*. It is controlled by contact fungicides while resistance has not been effectively exploited.

Potato is susceptible to a number of fungi which colonize and cause blemishes on the skin. Fungicides have limited effect in controlling these diseases. *Rhizoctonia solani* and *Helminthosporium solani* each cause superficial skin disorders called black scurf and silver scurf, respectively. Powdery scab, *Spongospora subterranea*, has a worldwide distribution. Damage is characterized by thickened corky skin lesions or pits which result in unmarketable tubers. Some resistance in breeding clones has been identified.

Verticillium dahliae is a soil-borne fungus which invades roots and spreads in the vascular tissues of potato causing wilting and plant death. It is one of the single most important causes of crop yield loss in certain warm arid irrigated production areas. It is particularly important in western North America. Soil fumigation, rotations of 2 years or longer in alfalfa, corn, or small grains, and host resistance are effective countermeasures. Late maturing cultivars suffer less damage than early maturing ones.

Storage rots are caused by *Fusarium* and *Phoma* spp. (dry rots) *Phytophthora* spp. (water rots) and *Pythium* spp. (leaks). Together they account for serious losses in temperate potato growing regions where tubers must be stored for part of the year. [See FUNGICIDES.]

D. Bacteria

Four bacteria cause serious problems in potato. The skin blemish, common scab, is caused by *Streptomyces scabies*. Genetic resistance is the most effective means of reducing damage. Ringrot disease is caused by *Clavibacter michiganense* ssp. *sepedonicus*. There are often no typical symptoms making it difficult to reliably identify infection by visual inspection. Plants may be dwarfed or severely wilted. Tuber symptoms are characterized by rotting of the vascular ring. It is found almost exclusively in potato growing areas that cut whole tubers to obtain seedpieces, the main mode of transmission being through contact of cut surfaces with contaminated cutting implements. The bacterium will not persist in the soil past 1 year of fallow or rotation with a nonhost crop, provided volunteer potatoes are controlled. Control is achieved by assuring that seedstock is free of the pathogen. Seed programs have a zero tolerance for ringrot.

Erwinia carotovora causes distinct diseases on both the vine and the tuber of potato. Blackleg derives its name from the black base of the stems that develops, followed by wilting and plant death. Tuber soft rot, as the name implies, is a condition where the tuber breaks down into an almost liquid mass, accompanied by a strong rotting odor. The stem rot generally develops from a rotting seedpiece. Control of seed piece decay through the use of whole uncut seed, suberized seed, fungicide/bactericide application, and optimum planting conditions all contribute to blackleg control.

Tuber soft rot is managed by minimizing bacterial contamination and managing storage conditions so as to avoid free moisture on the tubers and poor aeration.

Pseudomonas solanacearum is a bacterium limited to the subtropics and tropics. It causes a disease called bacterial wilt. Infection results in wilting and sometimes in plant death. Tubers from infected plants frequently rot in storage. It is ubiquitous and persistent in soil, often being harbored in decaying root systems of previous plantings or in the roots of living weeds. The pathogen is difficult to eradicate, because mild symptoms, transmission in tubers, and contamination of clean soil by infected seed contribute to its durability in the production system. As no genetic resistance is known that is effective in hot growing areas where bacterial wilt is most prominent, cultivation methods are sometimes used to reduce damage. One of these is minimum tillage to diminish release and activation of the bacterium from living and dead roots of other hosts. A second method utilizes superheating of bare fallow soil by exposure to direct sunlight at the hottest time of the year, a process called solarization. It has been found, also, that bacteria-free seed is effective in reducing the incidence of wilted plants in problem fields.

E. Viruses

There are approximately 30 viruses that infect potato. Three are of major importance. Potato virus X (PVX) is a contact transmitted virus causing mild mosaic symptoms. Potato virus Y (PVY) is transmitted by many species of aphids and can be transmitted mechanically. It is nonpersistently transmitted, i.e., it is acquired easily by the aphid feeding on an infected plant, but the aphid is only able to transmit to another plant for a short period of time. Symptoms of PVY range from very mild mosaic, to severe mosaic with leaf distortion, to necrosis of leaves, growing points, and plant death. Potato leafroll virus (PLRV) is transmitted in a persistent fashion by the green peach aphid (*M. persicae*). Aphids require some time (1 hr) to acquire the virus, but retain the ability to transmit the virus for long periods of time (several weeks to lifetime of aphid). [See PLANT VIROLOGY.]

Genetic resistance to all three viruses is available. Single gene resistances to PVX and PVY, derived from *S. tuberosum* ssp. *andigena,* and several wild *Solanum* species, are available in breeding programs and have been incorporated into a number of varieties. Partial resistance to PLRV is available in certain cultivars. Ironically, the most widely grown varieties are

not virus resistant. Therefore, techniques of hygienic seed production are essential to produce virus free seed. Axenic culture is widely used to eradicate virus, and maintain the seedstock in an environment that prevents recontamination. Axenic cultures are maintained in laboratories, but the plants must be transferred to the field to start the repeated multiplication cycles that are necessary to build up quantities of tuber seed sufficient to supply seed for the commercial potato crop. Seed production is carried out in geographic locations that permit isolation from commercial potato, and in climatic circumstances that disfavor aphid populations (cold, short growing seasons). For this reason they are often in upland or mountainous locations and/or at high latitudes with short summers.

VII. History of Potato Germplasm

After its introduction into Europe the potato became crucially important in Ireland. Potato was grown on the poorest soils, yet yielded well when planted on "lazybeds," raised strips of soil, fertilized with animal manures. Potato outyielded grain and meat production per unit area, and may have been a causal factor in a population explosion. The population of Ireland increased from 1.5 million to 9 million between 1790 and 1845. Ominously, the Irish diet became monopolized in large degree by potato.

Despite the fact that scientists of the day knew where it had originated, only a few, if any, new potato introductions had occurred up until two and a half centuries after introduction into Europe. The few new types that appeared were difficult to maintain because they produced tubers late in the season, and they became infected with virus and were lost. The value of breeding earlier maturity in varieties was recognized and this was attempted by selecting sexual progeny. Furthermore, all varieties declined in yield, due to accumulated tuber-borne disease and replacement of old clonal cultivars with new sexually produced seedlings was a method of restoring vigor. Yet potato breeding at that time was basically recycling a static gene pool.

The narrow genetic base meant the potato crop was vulnerable to disease. In 1845, late blight disease, incited by the fungus *Phytophthora infestans,* appeared in Ireland. The varieties in Ireland, foremost among these was "Lumper," had no resistance and either succumbed in the field or rotted in storage. The utter dependence of the Irish diet on potato resulted in massive starvation and emigration. It is estimated that a million people starved and two million emigrated.

The potato late blight epidemic was the most spectacular example of widespread crop failure due to a narrow genetic base. Late blight heightened awareness of the need to develop resistant cultivars. The Reverend Chauncey Goodrich of New York State obtained potato clones of South American origin through the Consulate in Panama in 1851. The single clone that he kept out of these introductions was named Rough Purple Chili. Successive sexual cycles stemming from seedlings grown from open-pollinated seedballs led to Garnet Chili and Early Rose. Goodrich saw the potato as healthful food, which could uplift humanity if varieties of sufficient hardiness could be bred. To him, potato breeding had more the character of a crusade than a hobby. But the difficulty of working with germplasm introduced from the area of origin was great. He discarded all but a few clones after considerable labor over many years, doubting whether even these were worthy. Although Reverend Goodrich believed his efforts were a failure, the impact of his breeding has been tremendous due the use of Early Rose as a universal parent. Russet Burbank, the most important processing potato in North America, is a somatic mutant of Burbank, which was a progeny of Early Rose. Thus, Rough Purple Chili was a great-grandparent of Russet Burbank, and an ancestor to more than 100 North American varieties and more than 300 European varieties. On the other side of the Atlantic Ocean, a counterpart of sorts to Chauncey Goodrich, William Paterson of Dundee, Scotland, commenced breeding effort in 1853 with the express purpose of obtaining greater blight resistance. Although it is debatable if blight resistance was obtained, one of his varieties, Paterson's Victoria, was a great success, and, over time became an ancestor to the most important varieties of the early 20th century.

The last quarter of the 19th century was characterized by considerable private interest in potato breeding. Generally, breeders only made crosses between established varieties. As a consequence, the nature of the gene pool changed very little. Breeders did not keep breeding lines that might have had important characters yet lacked sufficient performance for commercial success. The origin of varieties was often a secret, and there were cases of varieties being reintroduced under new names for profit motives.

An important breakthrough in resistance to late blight was discovered in the Edinburgh Botanical Garden in 1910. The entire tuber-bearing collection of *Solanum* spp. was killed by late blight with the exception of one accession from Mexico. This accession was called *S. x edinense*. Later, it was determined that it was a hybrid of cultivated potato with the wild Mexican species *S. demissum*. Late blight resistant breeding lines were developed in Germany and the United States using *S. demissum*, becoming the first intensive use of a wild species to achieve a specific improvement of a trait in potato breeding history.

Starting in 1925, scientifically planned collecting expeditions were conducted, beginning with Soviet scientists under the guidance of N. I. Vavilov. Usually collections were restricted to certain geographic areas. Wild species and native cultivars were collected. Over time, scientists from many countries carried out exploration and collection often in multinational teams. The expeditions led to the establishment of germplasm banks in Europe and the United States. Evaluation of entries in these banks for resistance to pests and pathogens led to the intensive utilization of certain wild species as sources of resistance. By the mid-20th century, this work was well under way in Europe. Only more recently has this been the case in the United States and Canada. About this time, it was realized that the late blight resistance from *S. demissum* was not durable. There was a gene-for-gene correspondence between the potato host resistance and *Phytophthora* pathogen virulence. Blight immune cultivars developed from *S. demissum* became completely susceptible when a new virulence genotype of the fungus appeared. Since 1950, there has been a general search for "durable" sources of resistance to *Phytophthora*. "Durability" implies that resistance is based on multigenic impedance of infection and disease progression at several points in the pathogenesis process.

Although efforts to develop late blight resistance have been the most extensive, several other serious disease and pest problems have been reduced by breeding for resistance using exotic germplasm. These include cyst nematode, PVX, PVY, and wart (a disease caused by the fungus *Synchytrium endobioticum*, characterized by growth of large galls on tubers and the base of the plant).

Extensive collection in the last 50 years has permitted the elaboration of a considerable body of biosystematic knowledge. The approximately 180 species known, span a polyploid series ranging from diploid to hexaploid (Fig. 4). The majority of the wild species are diploid and many of these are directly crossable to cultivated diploid potato. Furthermore, awareness of the diversity of native cultivars has expanded. It is known, for example, that eight species of potato are

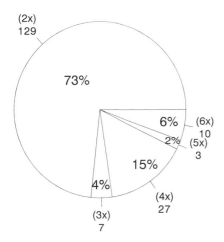

FIGURE 4 Ploidy composition of wild potato species in number and percentage of total.

cultivated. There are about 5000 cultivars maintained in the collection at the International Potato Center. As seen in Table IV, the vast majority of cultivars are classified as *S. tuberosum* ssp. *andigena,* a tetraploid.

The second largest group are the diploids. Triploids and pentaploids are few in number, and two of these, *S.* x *juczepzukii* and *S.* x *curtilobum,* are specialized as frost-resistant cultivars by virtue of being hybrids with the wild species *S. acaule.* These cultivars are grown at the very highest elevations of potato culture in the Andes of South America. They are high in glycoalkaloid content, requiring processing to remove these bitter compounds to make the edible dried product *chuño. Chuño* captured the attention of the writers of the first commentaries, the early Spanish conquerors, for storability, transportability, and nutrition.

TABLE IV

Composition of World Collection of Cultivated Species Maintained by the International Potato Center (CIP, 1988)

Ploidy	Species	Number	Percentage
2x	*S.* x *ajanhuiri*	26	0.5
2x	*S. goniocalyx*	99	2.0
2x	*S. phureja*	43	0.8
2x	*S. stenotomum*	449	8.7
2x	2x hybrids	147	2.8
3x	*S.* x *chaucha*	235	4.5
3x	*S.* x *juzepczukii*	75	1.4
4x	*S. tbr* ssp. *andigena*	3789	73.3
4x	*S. tbr* ssp. *tuberosum*	264	5.1
5x	*S.* x *curtilobum*	38	0.7

In modern times, native cultivars are found primarily in rural locations. To a certain extent, they are in competition with modern day bred varieties, yet their appeal is still retained because they are prized for their culinary traits. Thus, a rural household may grow a modern variety, *papa aguanosa* (watery potato) or *papa blanca* (white potato), for sale to city markets because the yields are higher, whereas the old traditional varieties, *papa de regalo* (gift potato) or *papa de color* (colored potato), will be retained for home consumption or sale of surplus in the local markets.

A detailed examination of two mixed fields in southern Peru identified 26 distinct varieties, 5 of which were diploids, 3 were triploids (not bitter types), and the remainder were tetraploids (*S. tuberosum* ssp. *andigena*). Another study near Cusco in southern Peru found that a field with native traditional cultivars may have from 10 to 30 varieties, averaging 20. Out of 28 mixed potato fields sampled, 79 cultivars, which could be further separated into 164 subcultivars, were distinguished. Out of the 79 cultivars, one-third were cosmopolitan in being found across three distinct regions, whereas the rest were restricted in distribution, in certain cases to only one or two fields over the whole sampling area. Potato folk nomenclature reflects considerable sophistication in differentiation by morphology and use in the household economy. Comparisons of biochemical genetic markers and local names indicate a high degree of correspondence between farmer identification and genotype, although occasionally similar appearing but genetically different varieties are classed as one variety by farmers. Potato varieties are classified according to their method of preparation as being more suitable for boiling, in soups, as mashed potatoes, for specialized yellow flesh dishes, or bitter potatoes for processing into *chuño* (frost-resistant cultivars). Cultivar names, often in the *Quechua* or *Aymara* languages, are descriptive of: (1) appearance, e.g., "cat's face," "black girl," "llama's tongue," "puma's paw"; or (2) function, e.g., "potato for fever" or "potato for weaning of children from mother's milk."

VIII. Genetic Improvement

The potato of commerce is an autotetraploid, having four sets of homologous chromosomes which have homology and pair with equal affinity in meiosis. Potato clones are highly heterozygous. This means that crosses between two clones produce an array of progeny that differ from each other in many traits.

There is no way to recover exactly parental genotypes, through backcrossing. Potato suffers inbreeding depression, which necessitates the avoidance of selfing, backcrossing to the same recurrent parent, or mating of related parents. In population terms autotetraploids have a very small recovery of recessive genotypes. For example the cross of AAaa × AAaa, produces 1/36 aaaa progeny. This is considerably less than the 1/4 aa progeny that would be produced by the cross Aa × Aa in a diploid cross.

The most important characters in breeding programs are generally controlled by many genes, with low heritability. These include yield, tuber size, specific gravity, maturity category, tuber shape, skin texture (russeting), fry color, sugar accumulation at low storage temperatures, scab resistance, horizontal late blight resistance, early blight resistance, *Verticillium* resistance, field resistance to viruses, length of tuber dormancy. Certain characters are controlled by single genes: Immunity to PVX and PVY, cyst nematode resistance, flesh color, and skin color.

The manipulation of ploidy is an important part of potato breeding. A portion of potato germplasm exists in diploid form. Some of the diploid material is derived by extracting haploids from tetraploids, while the rest is derived from cultivated diploids of South American origin. Diploids may be crossed directly with tetraploids if the gametes of the diploid are not reduced in ploidy. These clones are said to produce "2n gametes," because pollen and/or eggs have twice the number of chromosomes as a normal "1n gamete." Diploids which produce 2n gametes and are crossed to tetraploids (4x), transmit high levels of heterozygosity to the tetraploid progeny which have greater uniformity and higher yielding capacity than tetraploid progeny from 4x–4x crosses. This trait, 2n gametes, also permits the direct crossing of diploid wild species, which comprise the bulk of wild species, directly with tetraploid materials.

There are approximately 180 species of cultivated and wild potatoes spanning five ploidy levels (see Fig. 4). Besides the diploid (2x = 24) species which comprise 73% of the group, there triploids (3x = 36), tetraploids (4x = 48), pentaploids (5x = 60), and hexaploids (6x = 72). Many wild species are directly crossable with cultivated forms, a circumstance which set the stage for extensive use of wild species in variety development as has already been mentioned. There exist, however, species which are difficult or impossible to hybridize sexually. To solve this, a technique called "somatic" hybridization has been successfully employed. Somatic hybridization is accomplished by breaking up leaf tissue of two different species into single cells (protoplasts) and axenically culturing them in mixture. Various treatments are applied to induce the single cells from the different species to fuse into one hybrid cell. The hybrid cells are then cultured with special conditions that result in the regeneration of plants with all the genetic information of the two species. The somatic hybrids are often crossable to cultivated potato and serve as the bridge necessary to transfer wild species genes into a breeding program. Tomato, the non-tuber-bearing wild potato *S. brevidens,* and the tuber-bearing *S. bulbocastanum* have been somatically hybridized with cultivated potato.

Breeding programs generate a large number of new genotypes through crosses. The progeny are seedlings which are typically planted in the field as single plants. Selection is performed at harvest based on tuber type and freedom from defects. Upon selection each seedling is established as a clone and will be vegetatively propagated in future cycles. Each year clones are tested in larger plots and at more localities in order to obtain quantitative data on traits that are strongly affected by environment such as yield, tuber size, and specific gravity. It can take from 10 to 15 years to release a new variety from the time the cross is made.

Biotechnology has entered the arena of potato improvement through introduction of foreign genes (transformation) into potato plants by use of *Agrobacterium tumefaciens.* This bacterium can insert portions of its own genetic information into plant chromosomes. The transferred DNA is stably incorporated and includes genes cloned from other sources that have been specifically designed by molecular geneticists to be expressed in plants. Potato has proven to be amenable to genetic improvement through transformation technology because it is easily manipulated in tissue culture. Normal plants can be recovered in high frequency from callus tissue which develops after transformation occurs. Also, the effects of gene insertion on fertility and sexual reproduction are of no consequence in a vegetatively propagated crop. [*See* PLANT GENETIC ENHANCEMENT.]

Several types of genetic engineering strategies have been successful. Expression of a gene from a potato pathogen has been used to interfere with the disease process. This so-called "pathogen-derived resistance" has been effective against PVX, PVY, and to a lesser extent PLRV. In all cases plant-expressible versions of the gene encoding the coat protein of the viruses have been inserted into the plant. It has been shown that at least the mRNA of the coat protein is expressed and that coat protein is recoverable when the PVX and PVY genes are expressed. Responses range from complete freedom from infection to a lowering of the virus titer.

A second strategy is embodied in the expression of genes from the soil bacterium *Bacillus thuringiensis* that code for proteins that are toxic to the Colorado potato beetle and potato tubermoth. Expression of these proteins in plants has been shown to be extremely effective in the control of these important insect pests. Thus, the expression of traits encoded by genetic information derived from other organisms is completely feasible in potato and suggests nearly limitless possibilities.

A third strategy is designed to reduce the expression of a gene by impairing the translation of the gene product from mRNA. This is accomplished by inserting a gene where the product of transcription is an RNA complementary to the mRNA transcript that one wishes to suppress. Translation will be impaired by the formation of duplex RNA structures. This has been shown to be effective in the suppression of an enzyme responsible for the synthesis of amylose (granule bound starch synthase) and in the suppression of PLRV concentration in infected plants.

Bibliography

Harris, P. (ed.) (1992). "The Potato Crop: The Scientific Basis for Improvement." Chapman and Hall, London, U.K.

Hawkes, J. G. (1990). "The Potato: Evolution, Biodiversity, and Genetic Resources." Smithsonian Institution Press, Washington, DC.

Hooker, W. J. (ed.) (1981). "Compendium of Potato Diseases." American Phytopathological Society, St. Paul, MN.

Horton, D. (1987). "Potatoes: Production, Marketing, and Programs for Developing Countries." Westview, Boulder, CO.

Li, P. H. (ed.) (1985). "Potato Physiology." Academic Press, Orlando, FL.

Ross, H. (1986). "Potato Breeding-Problems and Perspectives." Verlag Paul Parey, Berlin, Germany.

Rowe, R. C. (ed.) (1993). "Potato Health Management." American Phytopathologic Society, St. Paul, MN.

Salaman, R. N. (1985). "The History and the Social Influence of the Potato." Revised impression (J. G. Hawkes, ed.). Cambridge Univ. Press, Cambridge, U.K.

Woolfe, J. A. (1987). "The Potato in the Human Diet." Cambridge Univ. Press. Cambridge, U.K.

Poultry Processing and Products

ALAN R. SAMS, *Texas A&M University*

I. Poultry Processing
II. Poultry Meat Characteristics
III. Processed Poultry Meat Products

Glossary

Aging Process in which a freshly processed carcass is held undisturbed and refrigerated for a period of time (at least 4 hr) to allow rigor mortis to develop and thereby prevent toughening during deboning

Chickens Are called Cornish hens, broilers (or fryers), roasters, and hens at 4, 6–8, 9–16, and 52+ weeks of age, respectively; the predominate commercial form is a broiler

Meat In poultry, "meat" generally refers to striated, skeletal muscle

Poultry Any domesticated avian species used for food; while the primary types are chickens and turkeys, additional poultry can include ducks, geese, quail, pigeons, and pheasant; chickens are the only type of poultry with a significant, commerical, egg industry

Scalding Technique of soaking a freshly killed bird in hot water to loosen the feathers for easier picking

Slaughter Can refer either to the specific act of killing or to the entire process of transforming a live bird into a ready-to-cook carcass

Stunning Process of rendering an animal unconscious prior to killing

Turkeys Commonly called fryers, roasters (or "young" hen/tom), and hens/toms at 9–16, 16–24, and 52+ weeks of age, respectively; the predominate commercial form is a roaster

Value-added products Products which involve some form of additional processing beyond the whole ready-to-cook carcass; the consumer convenience imparted by the additional processing is an intangible asset for which consumers are willing to pay

Viscera Gastrointestinal tract and associated organs, reproductive tract, heart, and lungs; some viscera is edible by humans (heart, liver, gizzard)

The field of poultry processing and products is generally concerned with converting live poultry into food products for human consumption. Poultry can refer to any domesticated avian species used for food and the products can range from a whole, ready-to-cook carcass to extensively processed items such as frankfurters or fried nuggets. Studies in this field are a complex combination of biology, chemistry, engineering, and economics. While human food is the main objective of poultry processing and products, related fields include processing waste water management, nonfood uses, and pet and livestock feeds. The presentation of specific, numeric processing conditions are for guideline purposes only, and it should be remembered that different poultry processors will vary in their procedures.

I. Poultry Processing

A. Introduction

Today's commercial poultry is extremely uniform in appearance and composition. Highly managed breeding, incubation, rearing, and nutritional regimens have created a market-aged bird that is virtually indistinguishable from its siblings. This uniformity has allowed poultry processing plants to develop into highly automated processing facilities with an efficiency that is unmatched by other livestock processors. With line speeds of 140 chickens/minute, uniformity, automation, and efficiency are recurring themes and have been keys to the success of poultry processing. Most poultry in the United States is processed at plants with at least some degree of vertical integration. This is a system in which the same corporate entity (e.g., company, cooperative, etc.) owns/manages several (or all) steps of the production process from the breeder through the processing plant (Fig. 1). This ensures maximum efficiency and uniformity. Although there are some minor differences, broiler and turkey processing are quite similar. Because of this similarity and the fact that far more head and kilograms of broilers are produced annually in the United States, broiler processing is the focus of this article and is diagrammed in Fig. 2.

B. Feed Withdrawal, Catching, and Transporting

Poultry processing begins at the growout farms, where the birds are reared to a market weight of 1.5–4.0 kg live weight at 6–9 weeks of age, depending on the sex and intended use of the carcass. Males tend to be larger than females and birds to be sold as whole or as parts tend to be smaller than those produced for boneless meat production. Although the growth rate and condition of the bird at the time of processing depend largely on genetics, nutrition, and management practices, consideration of the conversion of a live bird into salable meat normally begins with preslaughter feed withdrawal. Feed is withdrawn from the birds at a time from 8 to 10 hr prior to slaughter to allow adequate time for clearance of intestinal contents from the gut. It is important that the

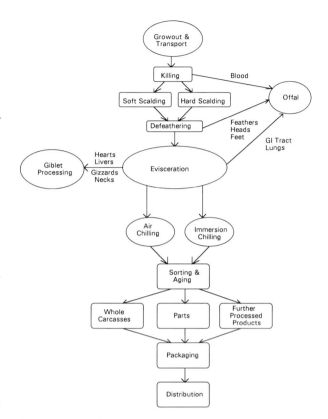

FIGURE 2 Poultry processing flow chart.

birds continue to have full access to water to prevent dehydration and the associated loss in body weight and carcass yield. Gut emptying reduces fecal and microbial contamination of the carcass during processing if the bird defecates during slaughter or if the intestinal wall ruptures during evisceration. Although there is a wide variation around the ideal 8- to 10-hr feed withdrawal duration, it is important to achieve a compromise between the intestinal clearance and the body weight loss that occurs if the bird experiences excessive fasting. Another adverse effect of excessive fasting is the accumulation of fluid and watery feces in the gut once it is empty. This would restore the potential for carcass contamination. [*See* POULTRY PRODUCTION.]

Although machinery to catch poultry exists, the majority of poultry in the United States is still caught by hand. The machines may be more efficient in terms of manpower, but they still require a major change in operating procedures and even more investment of capital. At some time during the feed withdrawal period, allowing sufficient time to complete gut emptying prior to slaughter, the birds are corralled into one area of the house. This brief confinement makes

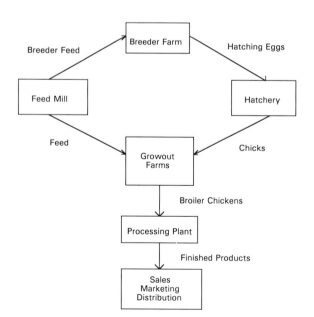

FIGURE 1 Organization of a vertically integrated broiler company.

it easy for catchers to grab the birds by both feet and place them into transport coops. Broiler coops vary in size and design but generally are rectangular in shape and hold from 10 to 50+ birds. The coops are manufactured as fixed stacks or as individual coops which are stacked during loading on a truck for the 1- to 3-hr ride to the processing plant. Once at the plant, most processors use mechanized unloaders which dump an entire stack of coops simultaneously onto a conveyor belt. The conveyor transports the birds into a room where they are hung by their feet on a shackle conveyor line to be slaughtered. A common problem with both manual and mechanized catching and unloading is inducing carcass defects such as bruising and broken bones. This problem can be minimized with proper training and supervision of personnel.

Scheduling is an important consideration when planning any poultry processing operation. Feed withdrawal, catching, transportation, and unloading all need to be closely synchronized with the processing schedule to avoid excessive feed and water withdrawal that may occur if the birds are not processed as soon as possible after arriving at the plant. Scheduling considerations become even more important during hot weather when the weekly evaporative weight loss by birds awaiting unloading can exceed 5000 kg for just one plant.

C. Slaughter

Slaughter is the series of events involved in humanely killing the bird. The first step in humane slaughter is "stunning" to render the bird unconscious prior to killing. With exceptions for religious practices, most countries with commercial poultry industries require stunning before killing. While there are many ways to achieve unconsciousness, the most widely practiced and accepted method is electric shock in a saline solution. While hanging by their feet, the birds are conveyed over a saline-filled trough so that just their heads are submersed. The saline solution is charged so that an electrical current of 20–40 mA and 30–60 V AC flows through the bird to the shackle system which serves as the ground. A proper stun will produce 1–2 min of unconsciousness during which the legs are extended, the wings are tight against the body, and the neck is arched. While stunned, the bird is moved to the killing machine to initiate bleeding. In addition to humane slaughter, there are other benefits to be gained from proper stunning such as calming for improved killing machine efficiency, more complete blood loss, and better feather removal during picking. Inadequate stunning can result in carcass defects such as incomplete bleeding, while excessive stunning can cause shattered clavicles (wishbones) and ruptured arteries, particularly in the thigh.

Within seconds after stunning, the shackle conveyor moves the bird to the killing machine. The killing machine usually consists of a rotating circular blade against which the birds neck is directed. This blade then severs the jugular veins and, depending on the machine design, the carotid arteries. The machine and shackle line must be so adjusted as to prevent the cut from being too deep or superficial. If the spinal nerve cord is cut, the resulting nervous stimulation "sets" the feathers and makes picking more difficult. Conversely, if the cut is too shallow there will be insufficient bleeding and the residual blood causes engorged vessels and can discolor the skin. In some plants a second stun, called a "poststun," is applied to the bleeding bird to extend the unconscious period, help squeeze the blood out of the bird, aid in feather removal, reduce struggle during bleed-out, and improve meat tenderness.

Once the neck has been cut, the bird is allowed to bleed for 1.5–2 min, depending on the plant. During this period the bird loses about 50% of its blood which eventually causes brain failure and death. If the blood loss is insufficient to cause death or if the neck cut is missed altogether, the bird may be still alive at the end of the bleeding period when it enters the scalder. In this case, the blood rushes to the skin surface in response to the scald water's heat, imparting a bright-red off-color to the carcass, a condition known as a "cadaver" by the industry.

D. Scalding and Picking

Feathers are difficult to remove in their native condition due to their attachment in the follicles. To loosen them, the carcasses are submersed in a bath of hot water which serves to denature the proteinaceous structures holding the feathers in place. Two particular combinations of time and temperature have become industry norms and have quite different effects on the carcass. Scalding at 128°F for 120 sec is called "soft scalding" or "semi-scalding" and loosens the feathers without causing appreciable damage to the outer skin layers. Because it leaves the waxy, yellow-pigmented layer of the skin intact, soft scalding is the preferred scalding method for producing fresh poultry with a yellow-pigmented skin exposed. If the skin will not be exposed or is not pigmented with xanthophylls from the feed, the carcasses are usually scalded at 140–142°F for 45 sec, a process called "hard scald" or "subscald." Because it loosens the outer skin layers, this is a harsher procedure than soft scalding; however, it allows easier feather removal than milder scalding conditions. Once loosened the outer skin layer and its associated pigmentation are removed with the feathers by the abrasion of the mechanical pickers. This is a major concern in poultry markets where the yellowness of chicken skin is used as a marketing tool. The loss of the waxy, outer skin layer may be beneficial, such as for the processor whose product is destined to be coated and fried. Coatings generally adhere to the skin better in the absence of the waxy layer.

The actual removal of feathers can be done manually but is usually accomplished by passing the bird through a series of picking machines. These consist of rows of rotating clusters of flexible, ribbed, rubber "fingers." While rapidly rotating, the fingers rub against the carcass and the abrasion pulls out the loosened feathers. By combining a series of these rotating clusters of fingers, each directed at a different region of the carcass, the whole carcass is picked. Picking machines adjusted too close to the bird may cause skin and bone damage while those machines that are too distant may not adequately remove the feathers. An added benefit from the pickers is that the mechanical abrasion may help squeeze some residual blood from the carcass. However, if the blood is squeezed into the wings, it may accumulate in the tips and cause skin discoloration.

Three final steps prior to evisceration are singeing, pinning, and head removal. Birds have hairlike structures called "filoplumes" that are singed off by rapidly passing the defeathered carcass through a flame. Immature feathers, called "pin feathers" which are too small to be removed by the picking machines, are manually plucked. Finally the head, which was partially severed by the neck cut created to allow bleeding, is cut or pulled off of the neck. The neck remains with the carcass to be inspected as salable product.

E. Evisceration

Evisceration, the removal of inedible viscera, has become a coordinated series of highly automated operations. While the specifics of evisceration vary from plant to plant and from one equipment manufacturer to another, three basic objectives are accomplished: (1) the body cavity is opened by making a cut from the posterior tip of the breastbone to the cloaca (anus);

(2) the viscera (primarily the gastrointestinal tract and associated organs, reproductive tract, heart, and lungs) is scooped out; and (3) the edible viscera or "giblets" (heart, liver, gizzard) are harvested from the extracted viscera, trimmed of adhering tissues, and washed with water. Although opening the body cavity and extraction of viscera are usually accomplished by three or more machines, giblet harvesting is frequently still done manually.

The overall dependence of the evisceration process on machinery underscores the importance of machinery maintenance and adjustment for bird size. Poorly adjusted machines are a frequent cause of ruptured intestines, torn skin, and broken bones. While technically edible, the preen gland is also usually removed during evisceration because of its objectionable sensory characteristics.

F. Inspection and Grading

No discussion of poultry processng would be complete without a mention of inspection. Although this inspection process varies between countries, it generally involves inspection of (1) the processing environment for sanitation, (2) both the exterior and interior of the carcass for freedom of disease, and (3) the packaging for accuracy of labels. While these objectives are maintained, the governmental and corporate inspection agencies and systems are dynamic and in a constant state of evolution to ensure the maximum degree of safety in the food supply. Any mention in this text of sampling size or frequency used today would likely be obsolete by publication. It will suffice here to say that in addition to constant monitoring of the carcasses being killed and eviscerated, corporate quality control and government inspectors sample the finished product (whole carcass, individual part, or fabricated products) on a regular basis (e.g., hourly) for compliance with existing regulations.

In addition to the required wholesomeness inspection, most processors also choose to have their fresh poultry products graded for eating quality (broken bones, fleshing, discoloration, etc.) and uniformity by federal government representatives. Some states and companies also have their own grading systems. Despite its voluntary nature, the virtual universality of grading by the poultry industry has made this reassuring marketing tool practically required for widespread product sale. Consumers will readily pay for the added assurance of quality and uniformity imparted by grading. This allows processors to recover the relatively small cost associated with having their product graded, making grading an effective means of adding value to their product. However, companies producing only further processed products, beyond cut-up parts, do not typically grade their carcasses.

G. Chilling

The primary objective of chilling poultry is to reduce microbial growth to a level that will maximize both food safety and the shelf life. Generally, body temperature is reduced to 4°C or less as soon as possible after evisceration (1–2 hr postmortem). The two most common methods used for chilling poultry are immersion chilling, in which the product is placed in chilled (0–4°C) water, and air chilling, in which the product is sprayed with water and then exposed to circulating chilled air which absorbs heat as it evaporates the water. While immersion chilling is used almost exclusively in the United States, air chilling is used almost exclusively in Europe. If additional chilling is needed (e.g., freezing), air chilling, with or without the water spray, is most commonly used in both markets.

Because they are essentially soaked in water, immersion-chilled carcasses absorb considerable water during chilling. To maximize both chilling rate and water uptake, immersion chillers are usually countercurrent flow in which the carcasses and water flow in opposite directions. This system bathes the carcasses being chilled in water of an ever-decreasng temperature. The amount of water uptake usually approaches the governmentally regulated limit of from 6 to 12% of the prechill carcass weight, depending on the country and product. Because this water uptake is allowed to offset any drip loss that will occur during storage and shipping, the higher uptake limits are used for high drip loss products packaged in drainable containers while the lower limits are used for products in sealed packaging. A frequent problem is that the actual amount of drip loss incurred varies with marketing practices (i.e., temperature, time, location) and results in inaccurate label weights with drainable containers. The added water from immersion chilling also adds to shipping costs and can interfere with the use of the product in further processing. Some advantages to immersion chilling compared to air chilling are that the added water increases the salable product yield and will prevent excessive dehydration during frozen storage that could reduce sensory quality.

The two chilling systems have very different effects on the microbial quality of poultry carcasses. The greater washing action achieved with immersion chilling flushes bacteria from the skin, resulting in a greater reduction in total bacterial load. However, the extensive bird-to-bird contact via the water results in more spreading of pathogenic bacteria to other carcasses in the community immersion chiller than occurs in an air chiller where the carcasses are relatively isolated from each other. This cross-contamination results in immersion-chilled carcasses generally having a greater incidence rate of pathogen contamination than air-chilled carcasses. However, pathogen levels on individual immersion-chilled carcasses are typically lower than for air-chilled carcasses.

H. Aging

The emphasis placed on further processing by the poultry industry has made cutting carcasses into parts and deboning meat common practices. Because meat is toughened when it is cut prior to the development of rigor mortis (i.e., 4 hr postmortem for broilers), poultry carcasses should be stored under refrigeration for an aging period prior to deboning to ensure optimum tenderness. Processors that do not observe this aging period can suffer from tough meat and poor tenderness uniformity. Because a carcass can be cut into parts without disturbing the muscle tissue, this aging period is not critical for parts production and the cutting is usually done immediately after chilling (i.e., 1 hr postmortem). Although not yet adopted by the poultry industry, a similar aging period obstacle has been alleviated in other meat industries with such postmortem treatments as electrical stimulation, high temperature conditioning, and muscle tensioning.

I. Packaging

By definition, all packaging systems for fresh poultry involve a container which protects the product from external contamination. The various poultry packagings differ in their cooling method, storage temperature, and additional factors that extend shelf life. The simplest form of packaging is to place the product in a cardboard or plastic box, fill the box with ice, cover the box, and store it below 4°C. This "ice-packed" poultry has the problems that shelf-life is short (7–10 days) and drip loss can be excessive. A more prevalent packaging system involves placing the product on a styrofoam tray, overwrapping the tray with a clear film, crust freezing the surface of the product in a blast freezer, and storing it at −2°C. This "chill-packed" poultry has the advantages of a longer shelf-life (21 days) and less drip loss. A modified atmosphere packaging system involves vacuum sealing the product in a plastic bag, injecting antibacterial gases into the sealed bag, placing the bag in a box, filling the box with dry ice snow, covering the box, and storing the product at −2°C. Although this system is expensive, it has a shelf-life of 21–28 days. [See FOOD PACKAGING.]

II. Poultry Meat Characteristics

A. Chemical Composition

In addition to the convenience and variety of poultry products, one of the main reasons poultry meat has enjoyed an increasing trend of consumption is because of its nutritional value, lower fat content, and more unsaturated fat type. Poultry meat is rich in high-quality protein and lower in total fat and saturated fat than beef (Table I). While this comparison varies with method of preparation, it is important to note that frying is no longer the predominant method, with broiling and roasting becoming more popular due to the lower fat content they yield. Although currently being de-emphasized in importance for the general consuming public, the cholesterol content of poultry meat is similar to that of other meat types.

Important differences exist in chemical composition between parts of a poultry carcass. Because broiler chickens are normally processed prior to the deposition of significant intramuscular fat (marbling),

TABLE I

Fat, Saturated Fat, and Cholesterol Contents of Raw Poultry and Beef

Meat	Total fat (%)	Saturated fat (%)	Cholesterol (mg/100 g)
Poultry[a] lean and skin	15.1	4.3	75
Poultry[a] lean only	3.1	0.8	70
Beef[b] lean and fat	19.2	7.8	67
Beef[b] lean only	6.2	2.3	59

Sources: United States Department of Agriculture (1979). "Composition of Foods: Poultry Products," Handbook 8-5, Washington, D. C., 20402; United States Department of Agriculture (1990). "Composition of Foods: Beef Products," Handbook 8-13, Washington, D. C., 20402.

[a] Composite of all retail broiler parts. Skin includes adhering fat.
[b] Composite of all grades and all trimmed retail cuts.

the majority of their body fat is outside of the muscles with much of it associated with the skin. This makes the fat easy to separate from the meat prior to preparation and has further improved the appeal of poultry meat. Another difference between tissues is the greater fat content of the dark (leg) meat compared to the light (breast) meat. Although still low relative to other meats, this greater fat content makes dark meat juicier and more flavorful. Fat acts as a lubricant during mastication to improve juiciness and most flavor compounds in animal products are found primarily in the fat.

B. Tenderness

While all chickens (and turkeys) are similar in most respects, differences exist in tenderness that greatly impact the ultimate use of the meat. Broiler chickens are hatched, grown, and processed by 7 weeks of age because their sole function is meat production. The chickens (laying hens) that lay table eggs for human consumption and the other chickens (broiler breeders) that lay eggs to be hatched for broiler production (these eggs are not the same) are generally processed after their egg production abilities are exhausted (>1 yr of age). The meat from these older birds is much tougher than broiler meat because as an animal matures, the connective tissues holding the muscle together become very heat resistant and no longer break down easily during cooking. As a result, meat from these older birds is used in products receiving extreme heat treatments such as the retorting of canned soups or the prolonged cooking of stewing hens. These extreme heat treatments are sufficient to overcome the heat resistance of the connective tissue in older animals.

C. Microbiological Quality

Increasing public concern about food safety has caused the poultry and other food industries to closely examine the microbiological quality of their products. While the loads of spoilage bacteria are generally an indication of general plant sanitation and packaging/distribution efficiency, pathogen incidence is more inherent to the production and slaughter of living animals for food. Knowledge of sources of pathogen contamination, modes of spreading, and methods of elimination are all essential to producing a safe food. Depending on the pathogenic organism, the initial contamination usually comes from either the breeder (an infected hen lays an infected egg which hatches into an infected chick) or from the chick's environment. Sources of environmental contamination include previous infected birds in an inadequately sanitized house, rodents and exogenous birds carrying the organism, and feed containing inadequately cooked animal by-products. One last source of contamination is by contact of a cooked product with an unclean surface (tabletop, utensil, or raw product) or an unclean human. This type of contamination is generally prevented with controlled product flow in a plant and proper sanitation practices by workers. [*See* FOOD MICROBIOLOGY.]

Once present in a flock on the farm, the organism can spread by bird-to-bird contact. At the processing plant, the primary sites of spreading are the community bath environments of the scalder and the immersion chiller. Far more spreading occurs in the plant than on the farm with incidence rates rising approximately 10-fold in the plant alone. Principal sources of plant contamination are processing equipment surfaces, scald water, and immersion chill water. It should be stressed that flocks and plants vary dramatically in their pathogen incidence and that the poultry industry is diligently researching methods of reducing the spread of pathogens between birds during processing.

Despite all the efforts to reduce contamination and spreading, it is not likely that pathogenic organisms will ever be eliminated from poultry production and slaughter. Therefore, methods of reducing bacteria on finished products have received much attention. While various antibacterial chemical dips and sprays have been tested on raw carcasses with only limited success, irradiation of raw or cooked products has proven quite effective in eliminating bacteria from poultry meats. Although this process was recently approved for use on poultry in the United States and such use causes no appreciable sensory deterioration, negative consumer perception of its safety will prevent the poultry industry from rapidly adopting irradiation.

III. Processed Poultry Meat Products

A. Marketing Concepts

The willingness by the poultry industry to continue developing new products to meet ever-changing consumer demands for convenience, lower fat, and versatility has been a main reason for the drastic increases in poultry meat consumption. The importance that consumers have placed on food convenience has caused the poultry industry to shift from producing fresh, whole carcasses to marketing the carcass in the form of parts or highly processed, precooked products. Because the intangible property of convenience is added to the product during this "further processing," these products are called "value-added products." These products are profitable because not only are consumers willing to pay for the added production costs and the added convenience, but also these products are frequently made from relatively low value parts of the carcass (e.g., leg meat). While the simplest form of further processing is cutting a carcass into parts, this topic usually describes products undergoing more drastic changes in form, such as deboning, chopping, forming, coating, and cooking.

B. Formed Products

Formed products are products in which deboned meat is formed into some desirable shape and held there while being cooked to retain the shape. These products are of two general types: (1) those in which the meat is left as whole muscles or only cut into large pieces to retain the texture of intact muscle tissue, or (2) those in which the meat is finely chopped to the extent that it no longer resembles intact muscle tissue. Seasonings, skin and adhering fat, and water-holding binders such as phosphates are usually added to the meat prior to forming into the desired shape (e.g., loaf, patty, nugget). Subsequent production steps for these products vary considerably. Loaves may have a browning glaze applied and are generally roasted or smoked. Whole muscle patties may be grilled while formed patties and nuggets are generally coated with batter and/or breading and partially or completely cooked by frying. Because these products are made with unsaturated poultry fat and are at least partially cooked before distribution, they are prone to lipid oxidation and rancidity. This source of off-flavor development remains a major limitation to the shelf-life of these products.

C. Emulsified Products

Emulsified products are meat products in which the meat has been thoroughly chopped into a viscous fluid with the pasty consistency of batter. During chopping, the muscle proteins coat (i.e., emulsify) the droplets of fat, preventing their coalescence and making them stable while dispersed in the surrounding aqueous environment. Extra fat and/or water may be added during the chopping to enhance the sensory properties of the product. Curing agents and seasonings are other possible ingredients, depending on the product. The resulting batter is pumped or stuffed into a tubular casing, usually made of cellulose, which is then divided into segments or links by applying constricting ties or clips at intervals along the cylindrical product. Frankfurters and bologna are the most common examples of these products which are generally cured, smoked, and cooked.

D. Mechanically Deboned Poultry

Regardless of how efficiently a carcass is deboned, substantial meat remains on the skeleton between small bones and in other inaccessible places such as the vertebrae of the neck and back. The relatively low meat:bone ratio, combined with the difficulty in harvesting this meat, makes these skeletal remains cheap by-products of boneless meat production. Mechanical deboners are machines designed to harvest this meat. Demeated skeletons (called "frames"), backs, and necks are coarsely ground and then forcefully squeezed against a sieve with pores approximately 0.05–0.1 cm in diameter. Soft tissue (mainly meat, fat, and bone marrow) is squeezed through the pores while hard tissue (mainly bone) is retained by the sieve. The pressure needed to separate the soft and hard tissues generates substantial heat; which, when added to the unsaturated fat of poultry and the iron content of the marrow, makes mechanically deboned poultry (MDP) very susceptible to lipid oxidation and rancidity. In an attempt to slow rancidity development and prolong shelf life of this unstable product, MDP is usually used or frozen immediately after production. Its pastelike texture and high protein functionality make MDP useful as a binder in formed products and as the main ingredient in emulsified products.

E. Future Developments

One type of product currently in development is a meat analog product made from washed poultry meat. Meat is finely ground and mixed with buffers to remove the soluble protein fraction of muscle tissue. The insoluble protein fraction remaining can then be formed and cooked into any desired shape. These products are high in protein, low in fat, and can be flavored to make them inexpensive substitutes for other meats like lobster and crab.

The demand for healthier snack foods has prompted the development of an extruded snack product made from poultry meat. Meat, binders, and flavoring are mixed together under heat and pressure to cook the ingredients. The pressure decrease when the mixture is then squeezed through a small opening causes the water in the mixture to vaporize, resulting in a puffed-type texture. Depending on the formulation, conditions, and orifice shape and size, these products can have a variety of appearances. Their bland native flavor makes them suitable for dusting with traditional snack food flavors like cheese and seasonings. These products are versatile, high in protein, and low in fat.

Bibliography

Austic, R. E., and Nesheim, M. C. (1990). "Poultry Production," 13th ed. Lea Febiger, Philadelphia, PA.

Bailey, A. J., and Light, N. D. (1989). "Connective Tissue in Meat and Meat Products." Elsevier Applied Science, New York, NY.

Henrickson, R. L. (1978). "Meat, Poultry, and Seafood Technology." Prentice-Hall, Englewood Cliffs, NJ.

Lawrie, R. (1979). "Meat Science," 3rd ed. Pergamon, New York, NY.

Mead, G. C. (1989). "Processing of Poultry." Elsevier Applied Science, New York, NY.

Mountney, G. J. (1976). "Poultry Products Technology," 2nd ed. Haworth, Binghamton, NY.

North, M. O., and Bell, D. D. (1990). "Commercial Chicken Production Manual," 4th ed. Van Nostrand Reinhold, New York, NY.

Parkhurst, C. R., and Mountney, G. J. (1988). "Poultry Meat and Egg Production." Van Nostrand Reinhold, New York, NY.

Stadelman, W. J., Olson, V. M., Shemwell, G. A., and Pasch, S. (1988). "Egg and Poultry Meat Processing." Ellis Horwood Ltd., Chichester, England.

Poultry Production

COLIN G. SCANES, *Rutgers—The State University of New Jersey*

MICHAEL LILBURN, *Ohio State University*

Glossary

Photoperiod Amount of time per day that the bird is exposed to light
Photostimulate To artificially extend the photoperiod
Poult Young turkey, under 8 weeks old

Poultry is produced for the purpose of supplying meat and eggs for consumption. Poultry production is a major component of agriculture throughout the world. This article will discuss the macroeconomics of poultry production; historical, prehistorical, and evolutionary aspects of poultry; the poultry industry; poultry nutrition; poultry health and disease; and poultry reproduction and growth.

I. Macroeconomics

A. World Production and Trade

Poultry production represents a significant feature of the agricultural production of the world. Table I shows estimated regional production figures, and production in selected countries is shown in Table II. It is obvious that poultry meat and eggs make a significant contribution to the nutrition of the population of the world, particularly supplying high-quality protein. [*See* EGG PRODUCTION, PROCESSING, AND PRODUCTS; POULTRY PROCESSING AND PRODUCTS.]

Although much poultry meat and egg is for domestic consumption, there is a considerable world trade in poultry and eggs. The three countries exporting the most broiler chickens account for over 55% of all broiler exports in the world in 1992 (United States, 26%; France, 16%; Brazil, 15%). Importers of U.S. broilers in 1992 included Japan ($130 million), Hong Kong ($127 million), Canada ($90 million), and Mexico ($71 million), with substantial exports to Eastern Europe, the Russian Federation, the Middle East, and the Caribbean. The United States is by far the largest exporter of turkey meat also.

B. United States Production

In the United States, poultry and eggs represented a major component of animal agricultural production in 1992: broiler chicken meat, 9.5 million metric tons; turkey meat, 2.2 million metric tons; and eggs, 2.7 million metric tons. By comparison, beef production was 10.5 million metric tons and pork was 7.8 million metric tons (data from the Agricultural Statistical Board, USDA).

Poultry meat production continues to rise. Broiler production was reported to be 4.6 million metric tons in 1977, 7.4 million metric tons in 1978, and 9.5 million metric tons in 1992. Egg production has declined in the United States since 1945, but is now relatively steady (see Section I,C).

The reasons for the increase in poultry production are consumer demand and the efficiency of the industry. The low price of poultry is the result of the efficiency of the industry and is a major reason for the high consumer demand. Poultry prices have increased 18% between 1977 and 1992 and 168% since 1914. In comparison, livestock prices increased by 35% (1977–1992) and 240% (1914–1992), while vegetables increased 52% (1977–1992) and 936% (1914–1992) (data from the Agricultural Statistics Board, USDA).

TABLE I

Worldwide Poultry Production in 1991[a]

Region	Poultry meat[b] (billion metric tons)	Eggs[a] (billion)
Africa	0.8	10.8
Asia	7.6	239.0
Europe (European Union)	6.8	83.7
Europe (Eastern)	1.7	30.2
Former Soviet Union	3.1	79.2
Europe (other)	0.2	3.4
Middle East	0.9	12.1
Oceania	0.5	3.5
North America	12.7	94.6
Central America	0.1	—
South America	3.4	20
World	38.0	576.7

[a] Based on selected countries. Data from the Agricultural Marketing Service, USDA.
[b] Predominantly broiler chicks.

Poultry production tends to be focused in specific regions of the United States, with production of broiler chickens predominantly in the southern states. In 1992, the top five states for broiler production were Arkansas (15.3%), Alabama (13.5%), Georgia (13.2%), North Carolina (9.9%), and Mississippi (7.5%).

Turkey production is another major feature of agriculture in the southern states but also in the Midwest (where feed grain is produced) and in California. The top five states for turkey production in 1992 were North Carolina (17.4%), Minnesota (15.2%), Virginia (9.8%), Arkansas (7.9%), and California (7.7%).

TABLE II

Examples of Poultry Production in Selected Countries in 1991[a]

	Poultry meat (million metric tons)	Eggs (billion)
Brazil	2.7	13.6
Canada	0.7	5.6
China	5.0[b]	18.5
France	1.8	185
Italy	1.1	15.3
Japan	1.4	42
Mexico	0.8	19.8
Spain	0.9	10.2
United States	11.2	69
United Kingdom	1.2	11.0

[a] Data from the Agricultural Marketing Service, USDA.
[b] 60% broilers and 40% ducks.

A similar situation exists for egg production, with the top five states being California, Indiana, Pennsylvania, Ohio, and Arkansas.

C. Poultry and Egg Consumption

In the United States, the major increase in poultry meat production has met the high consumer demand for low-priced, high-quality meat with a perceived greater healthfulness. Whole dressed chickens are available under different market classes established by the U.S. Department of Agriculture. These include broiler or fryer (~7 weeks old, male or female), roaster (older male or female, ~10–12 weeks old), rock Cornish game hen (male or female, 5 weeks old, up to 2 lb or 1 kg), and capon (young male, less than 8 months old, or technically a castrated male). There have also been increases in the variety of further processed poultry meat products (e.g., chicken nuggets, turkey franks, turkey hams) and poultry meat in fast-food restaurants (e.g., chicken sandwiches, McNuggets, Kentucky fried chicken). Another feature of poultry meat marketing is the premium brands (e.g., Perdue), which are widely advertised and are perceived as having greater quality.

In contrast to the tremendous increase in poultry meal consumption, there has been a decline in egg consumption in the United States. For instance, per capita consumption fell 39% from 411 eggs per year in 1945 to 252 eggs in 1992. This reflects marked decreases in shell or table egg use for home breakfasts (46%) and baking (94%); the former decline reflects changes in life-styles (more women employed outside the home), preference, and perhaps also health concerns. However, there has been a marked fourfold increase in the use of egg products (fried, frozen, etc.), include egg "substitutes" with reduced fat.

II. Historical, Prehistorical, and Evolutionary Aspects

A. Poultry Domestication and the Origin of Poultry Species

Poultry species include representatives of the orders Galliformes (chicken, turkeys, guinea fowl, quail, and pheasants), Anseriformes (duck and geese), and Columbiformes (pigeons and dove). The chicken or domestic fowl is derived from jungle fowl from Southeast Asia. Four species of jungle fowl exist today: the red jungle fowl (*Gallus gallus,* also known previously as *G. bantiva* or *G. ferrugineus*), the Ceylonese jungle fowl (*G. lafayetti*), the gray jungle fowl (*G. sonneratii*), and the black or green jungle fowl (*G. various*, previously known as *G. furcatus*). There is controversy regarding the ancestral origin of the chicken or even whether all breeds have the same origin or phylogeny. One school of thought is that chickens are derived from the red jungle fowl and hence chickens should be referred to as *G. gallus* or as the subspecies *G. gallus domesticus*. Alternatively, chickens may be derived from a combination of ancestral stock, including, and arguably predominantly, the red jungle fowl but also other species of jungle fowl (existing or perhaps extinct). Under these circumstances, it is appropriate to refer to chickens as *G. domesticus*. The domestication and geographical distribution of chickens can be attributed to their use as fighting cocks as well as being sources for eggs and meat, not to mention their hardiness and relative ease in rearing.

The turkey (*Meleagridis gallopavo*) was domesticated in antiquity by Native Americans in what is now Central America and Mexico. The present domestic turkeys may be the result of their introduction to North America via Spain from Mexico and possibly also crossing with wild turkeys.

B. Breeds of Poultry

Breeds or varieties of poultry were developed in specific geographical locations, indeed many breeds are identified by the area in which they originated. Some of the approximately 200 breeds of chickens recognized by the "American Standard of Perfection" are listed in Table III, as are varieties of turkeys. There

are also small or bantam breeds for chickens (~63 different breeds) and ducks. Breeding companies have employed crosses of heavy-meat breeds such as the Plymouth Rock and the Cornish in the development of the grandparent lines of broiler chickens, whereas lightweight lines of egg-producing chickens have been developed from light breeds such as the White Leghorn.

III. Poultry Production

A. Introduction

Historically, each segment of the poultry industry (i.e., layers, broilers, turkeys) acted as a sector industry under which there were independent hatcheries, feed suppliers, producers, and processors. Often the processors would sell their product to retailers. As the poultry industry grew, reliable scheduling and product commitments led to contractual relationships among producers, feed suppliers, and processors, particularly in the broiler industry. The individual producers would own their birds but they would commit to buy from and sell to a specific feed supplier and processor, respectively. These contractual arrangements evolved into the present day arrangement of "vertical integration" (Fig. 1).

Under this type of corporate management, a company owns a hatchery, feed mill, and processing plant. The chicks are produced and owned by the company and are grown on independent "contract" farms that are owned by the individual growers. The contract farm may consist of one or many poultry houses. The

TABLE III

Examples of Breeds of Chickens and Varieties of Turkeys

Breeds of Chickens

American   Delawares, Dominiques, Hollands, Javas, Jersey Black Giants, New Hampshires, Plymouth rocks, Rhode Island red, Rhode Island white, Wyandotte

Asiatic   Brahmas, Cochins, Langshans, Shanghais, Sumatras

British   Australoops, Cornish, Dorking, Orpington (buffs), Redcaps, Scots dumpies, Sussex (light)

Continental   Campines, Crevecoeurs, Faverolles, Hamburgs, Houdans, LaFleche, Polish

Mediterranean   Ancona, Andalusian, Cantalanas, Leghorn (black, white, red, brown), Minorca, Silician Buttercups, Spanish

Varieties of Turkeys   Black, Bronze, Narraganselt, Slate, White Holland, Bourban red, Beltsville small white, and royal palm

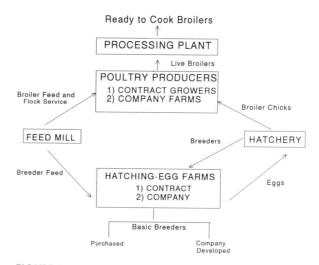

FIGURE 1 Organizational structure of the poultry industry.

contractual arrangement between a poultry company and its producers has alleviated the corporate need for investment in actual rearing facilities and facilitated the availability of bank loans.

B. Broiler Industry

More than 95% of commercial broilers in the United States are produced by vertically integrated companies. The top ten broiler companies are Tyson Foods, ConAgra, Goldkist, Perdue Farms, Pilgrims Pride, Hudson Foods, Wayne Poultry, Foster Farms, Seaboard, and Townsend. The top five and top ten companies account for approximately 46 and 61%, respectively, of all broilers reproduced. A broiler company may have one or many "complexes," each containing a hatchery, feed mill, processing plant, and associated hatching egg and broiler producers. Within a company, a moderate-sized complex (over one million broilers processed per week) might emphasize the production of a particular size of broiler (e.g., small broilers, 1.70 kg; heavy broilers, >2.5 kg). Within a given geographical area, there are often broiler complexes from several companies and there is competition for the best contract producers. Each producer is responsible for day-to-day flock management but a company technical representative visits each flock weekly. This allows for standard vaccination, feeding and management practices, and flock health program.

1. Broiler Breeding Stocks

Most broiler companies purchase day-old breeding stock from commercial broiler breeding genetics companies. Some companies (e.g., Tyson Foods, Perdue Farms) use their own in-house strains for part of their production, particularly if they have a specific market niche (e.g., heavy roasters) for which commercial breeding stock is either unavailable or unacceptable. In the United States, the pullets (females) and cockerels (males) are usually purchased from different breeding companies. Body weight gain during rearing and production in broiler pullets and cockerels is controlled via restricted feeding. Pullets are normally reared under a light-restricted photoperiod (8 to 10 hr) and are photostimulated (>15 hr) at 18–19 weeks of age. They will initiate egg production (5%) at 24–25 weeks and peak in production (80–83%) at 28 to 30 weeks. The normal productive period for a commercial broiler hen is 40 weeks (25 to 65 weeks of age) and she will produce approximately 160–165 hatching eggs resulting in 140–145 chicks. The minimal egg size for setting is normally 52 g.

C. Turkey Industry

1. Introduction

In the turkey industry, independent hatcheries (e.g., Cooper Hatchery Oakwood, OH; Cuddy Farms, Marshville, NC) still play a significant role as suppliers of day-old poults to independent or contract producers, but vertical integration is accounting for an increasing percentage of total turkeys produced. The top ten turkey companies are Butterball Turkey Co. (a division of ConAgra), Jennie-O Foods, Rocco Turkeys, Wampler-Longacre, Carolina Turkeys, Cargill, Bil-Mar Foods, Louis Rich Co., Jerome Foods, and Norbest. The five and ten companies account for 41 and 65%, respectively, of all turkeys produced. Turkeys generally spend the first 5 to 6 weeks in a starter building with a rearing density of 1.5 m^2 per poult. In the past, a significant number of turkeys would be moved from the starter building to an outdoor, fenced range for the remainder of the growing period. Nowadays, however, most turkeys are moved from the starter house to an enclosed growing house and the floor space is increased by 0.28–0.32 m^2 per bird. A normal market age for turkey hens is approximately 14 to 15 weeks of age or 7.3 to 8.6 kg liveweight. Toms are most typically grown for 16 to 18 weeks to a final body weight of 12.7 to 13.6 kg.

2. Turkey Breeding Stock

Turkey breeding stock are also purchased from commercial turkey breeding-genetics companies. Unlike that described for broiler breeders, day-old hens and toms are usually purchased from the same company. Turkey hens and toms used as breeders are allowed free access to feed (*ad libitum* fed) during the rearing and reproductive periods. Depending on strain and sex, body weight gain during the growing period is often restricted or reduced with the feeding of low-protein diets. There is growing interest in restricted feeding of hens, similar to what is done for broiler hens. From 1 to 2 weeks through 30 weeks of age, turkey hens are on a short daily photoperiod (6 to 8 hr light per day) with an intensity of approximately 5 foot-candles (Hybrid Turkeys, Inc., Ontario, Canada). At 30 weeks of age they are photostimulated (the day length increased to 13–14 hr of light per day at 5 foot-candles) and this may be then increased to as much as 17 hr/day. Breeder toms are exposed to a 12- to 14-hr photoperiod from approximately 2 weeks old throughout growth and reproduction, but the light intensity is less than that for hens (2 to 4 foot-candles). Commercial turkey hens are artifically inseminated with semen collected from toms kept in a separate building. After collection, semen is mixed with a commercial diluent but cannot be frozen for later use, unlike practices in the dairy industry. The diluted semen from one tom is normally used to inseminate 1–12 hens.

D. Commercial Layers

1. Introduction

In 1993, there were 235 to 240 million commercial layers in the United States that produced slightly fewer than 170 million cases of eggs (30 dozen/case). Approximately 25% of the eggs went into further processed egg products. Sixty companies with more than one million layers accounted for 74% of the total production. Commercial layers can be divided into two groups, those producing brown or white eggs. Brown eggs have been historically sold in the northeast United States. Strains that lay brown eggs have always been somewhat larger than the White Leghorns that lay white eggs (i.e., 2.0 kg vs. 1.8 kg at end of lay). The egg industry has essentially two types of production units: individual contract farms whose eggs are transported to a central processing plant and "in-line" complexes where the eggs from multiple houses flow to a central, on-farm processing facility.

During rearing, the pullets are reared on contract pullet farms (sometimes in a different state) or in separate pullet rearing houses on the same complex. Pullets are reared under a restricted photoperiod and then photostimulated at approximately 16 to 18 weeks of age, depending on the strain. Some modern layer strains have been selected for early maturity and can be photostimulated as early as 14 weeks of age. Initial egg size will be considerably larger in hens that are photostimulated at 18 versus 14 weeks. Most commercial layers will initiate egg production at about 20 weeks of age and approximately 20% will be force molted (forced cessation of egg production) at the end of the initial production cycle (75 to 80 weeks). This is achieved through a combination of short-term feed restriction and decreased photoperiod. Hens lay in excess of 300 eggs through the initial production cycle and over 400 eggs after two cycles of production (105 weeks of age).

2. Hatching and Incubation

Prior to setting, eggs are held in coolers on breeding farms at temperatures designed to slow down embryonic development and minimize bacterial growth (chicken eggs at 18–21°C; turkey eggs at 13–16°C). Eggs are generally held 3–5 days before setting; eggs held longer than 9–10 days will not hatch as well. The normal incubation time for chickens is 21 days and for turkeys and ducks 28 days. There is considerable variability within a species (i.e., ±24 hr), and this is related to age of breeder, strain, egg size, and incubation environment. In a commercial hatchery, these factors are taken into consideration to maximize total chicks or poults hatched and to minimize overheating and subsequent dehydration of newly hatched poultry. Temperature, air quality, and humidity are critical factors for successful incubation. Most commercial incubators are forced-air units and the optimal incubation temperature is approximately 37–38.5°C. At sea level, forced-air incubators would supply approximately 21% oxygen, which is adequate for normal embryonic development. A by-product of embryonic metabolism is carbon dioxide. If poor ventilation results in >0.5% carbon dioxide in the air of the incubator, hatchability (i.e., the percentage of chicks hatching compared to the number of eggs set) may decline. Turkey and chicken hatching eggs will normally lose approximately 10–11% of their initial (or set) weights over the course of incubation. This represents evaporative water loss and can be influenced by the relative humidity within the incubation environment. A relative humidity setting of 50 to 60% is adequate for normal incubation but may be increased during the latter stages of incubation. The exchange of metabolic gases and moisture occurs across the shell and shell membrane. Poor shell quality can reduce overall hatchability. Eggs are set with the large (or blunt) end up. Although this is not an absolute requirement for hatching, it will greatly improve overall hatching percentage. Hatchery sanitation (cleanliness) is critical to the overall viability of newly hatched chicks and turkey poults. The incubation environment provides an optimal place for bacterial growth and dirty "exploder" eggs can contaminate an entire incubator. Incubator-associated bacteria can increase the mortality of hatched chicks during the first 2 weeks of age.

Over the past 5 years, an innovative piece of equipment has been developed that allows for mechanized ("in ovo") injection vaccinations of broiler embryos at Day 19 of incubation as eggs are being transferred from the incubators to the hatchers (Embrex, Inc., Raleigh, NC). The equipment has been successfully field-tested within commercial broiler hatcheries.

E. Housing and Caging

Most broilers, chickens, and turkeys are reared under total "confinement," although some seasonal turkey growers still keep their birds in fenced ranges for the latter portion of the growing period. Commercial broiler houses are partitioned and for the first 14–18 days, young broilers are "partial house brooded" (i.e., chicks are brooded only in a portion, usually one-third, of the broiler house). This minimizes energy cost and facilitates management during the initial stages of growth. The same principle is true for turkeys with their starter and grower house concept. Most commercial broiler and turkey companies will make seasonal (environmental) adjustments in the number of birds they place in a growing house.

Turkeys and broilers are grown on absorbent litter material. Properly managed poultry houses should have litter moisture levels of approximately 25%. Excess litter moisture can contribute to poor air quality (ventilation problem) within a poultry house. Litter materials include wood shavings, rice hulls, chopped straw, and soybean hulls and will vary depending on geographical region. In many parts of the country, the litter in well-managed broiler houses is completely removed only once a year (sometimes even less frequently) and used in fertilizing fields in compliance with local environmental regulations. In general, turkey houses are cleaned out and disinfected after every flock.

Pan-type feeding systems predominate in most modern broiler and turkey houses. There are normally two or four feed lines within a house, each consisting of numerous pans connected to a tube through which the feed is augered from a main bin. In commercial broiler houses, nipple drinkers (pressurized nipples connected to a main water line) have largely replaced large, bell-type waterers as this greatly facilitates litter management. Turkey houses are still equipped with automatic bell-type waterers, which are hung from the ceiling and can be adjusted to an optimal height as the flock ages.

Almost all commercial layers are kept in wire cages during the production period. Many modern layer houses have a stair-step cage arrangement. The cages often overlay an opening in the floor into which the manure is collected and stored. In a properly managed layer house, the manure is removed every 2–3 years. Some caging systems have a slow-moving belt under the cages, which moves slowly toward one end of the building, where the manure is removed daily. The slow-moving belt also allows for significant moisture evaporation prior to the manure leaving the house. Feeding systems for commercial layers are most often a small troughtype feeder in which the feed is moved down the line by a flattened chain. Nutritionalists

must tailor diets to the feeding system to avoid particle separation of the feed as it moves down the feed line. As with broilers, nipple drinkers are used extensively in commercial laying houses.

F. Processing

The initial processing step occurs in the growing house when feed is withdrawn several hours prior to catching. This minimizes fecal contamination while birds are waiting to be processed. Catching of birds for processing is often done at night under a low light intensity to keep birds as quiet as possible. The processing procedure incorporates humane killing (electrical stunning, exsanguination), followed by feather removal and evisceration. The whole processed carcasses are subsequently chilled in ice water. Most of these steps are highly automated.

The greatest proportion of whole processed turkeys and broiler carcasses are then "further processed," the term used whenever a carcass is subsequently cut into pieces for direct sale or deboned. Further processing may occur within the same processing plant or at a separate facility, especially if cooking is involved. Most deboning rooms, particularly where breast meat is removed from whole carcasses, are kept very cold to minimize microbial contamination of the products.

IV. Poultry Nutrition

A. Requirements and Sources

The energy requirements for a given class of poultry are expressed in terms of kilocalories per kilogram of finished feed (from 2750 kcal in a diet for commercial layers to greater than 3350 kcal for turkeys). The requirements for other nutrients (e.g., protein, amino acids, minerals) are normally expressed as a percentage of the diet. Within a species, "starter" diets have the highest protein levels and this varies among species (e.g., commercial layers, 20%; broilers, 23%; turkeys, 28%). After the starter period, reproductive stock will be switched to a "growing" or "developing" diet that contains 15–16% protein. Near the onset of egg production, hens will be fed a "layer" diet with higher calcium levels (3.2 to 4.0%). The layer diet for commercial layers may also be somewhat higher in protein (18 vs 16%) because of their lower feed intake relative to egg production. [See ANIMAL NUTRITION, NONRUMINANT; ANIMAL NUTRITION, PRINCIPLES.]

B. Broiler Nutrition

The National Research Council (NRC), under the auspices of the National Academy of Science, publishes and periodically updates nutritional recommendations for most classes of commercial poultry. For broilers, the NRC recommendations are chronologically based (e.g., for broilers, 0–3 weeks, 3–6 weeks, etc). Most commercial feeding programs, however, incorporate defined quantities (i.e., weights) of the particular diets: broiler starter feed, 0.8–1 kg; grower, 1.8–2.0 kg; finisher feed, 1.0–2.0 kg. Most of these programs are not in accordance with the chronological changes suggested by the NRC. For example, 0.8 kg of starter feed might last 15 to 18 days, whereas 1 kg may last 18 to 20 days. Commercial broiler feeding programs strive for feed efficiencies of 0.5–0.52 kg gain per kilogram feed consumed and a caloric efficiency of 6000 kcal consumed per kilogram gain. Young broilers (<42 days) will have better feed efficiencies than older, heavier broilers. A starting diet for broilers would contain approximately 22% protein and 3135 kcal/kg. Protein would be gradually decreased and energy increased in subsequent feeds (e.g., finishing diet, 18–19% protein, 3245 kcal/kg).

C. Turkey Nutrition

As for broilers, the NRC recommendations for turkeys are also chronologically based (0–4 weeks, 4–8 weeks, 8–12 weeks, etc.). However, the industry uses feed scheduling similar to what was described for broilers. Normally there are six to seven different feeds that might be used depending on the sex and length of the growing period. Toms that are marketed at 18 weeks have a feed efficiency of 0.36–0.40 kg gain/kg feed. Turkey diets have the highest protein levels found in commercial poultry feeds. Starter diets have approximately 28% protein and this is gradually decreased to 16–17% protein during the latter stages of growth. The energy level in starting turkey diets is somewhat lower than that in broiler diets (2800 kcal/kg), whereas in finishing diets the energy levels may exceed 3300 kcal/kg.

D. Commercial Layer Nutrition

Starting and growing diets for commercial layers contain 20 and 15–16% protein, respectively, and these are normally fed from approximately 0–6 and 6–16 weeks of age. Diets for layers in production are formulated for the consumption of fixed nutrient quantities (e.g., 290–310 kcal, 16–17 g protein, 770 mg lysine, 4.2 g calcium) and the percentages of these nutrients in the diet would vary according to feed intake. Some strains of White Leghorns would consume only 85–105 g per day at peak production, whereas brown egg strains might consume 105–115 g day. Feed consumption is significantly influenced by temperature and energy concentration of the diet.

E. Feed Ingredients

Corn, wheat, and sorghum are the most common cereal grains used in poultry diets in the United States, though barley is used to some extent in the Pacific Northwest. Cereal grains are used as energy sources and can comprise approximately 50–70% of the diet depending on age and species. Cereal grains are not particularly high in protein but could quantitatively still contribute 30–40% of the total dietary protein; they are also low in some essential amino acids (e.g., lysine and methionine). Soybean meal is the single largest protein source for U.S. poultry diets and normally contains either 44 and 48% protein (with or without ground hulls). Rendered poultry and meat by-product meals (50–60% protein) are used extensively as protein supplements but their level of inclusion is rarely higher than 10% of the finished feed. Other common, high-protein ingredients that could make up smaller proportions of most poultry diets are hydrolyzed feather meal (80% protein), corn gluten meal (60% protein) from wet milling of corn, canola and cottonseed meals (35–45% protein), and bloodmeal (85% protein). Rendered fat or vegetable oil is a common energy source but most poultry diets would not contain more than 3–5% added fat. Most dietary minerals (calcium, phosphorus), vitamins, and trace minerals come from dietary supplements. [See FEEDS AND FEEDING.]

F. Feed Formulation

Most commercial diets are generated from "least-cost" feed formulation programs. The nutritional composition and cost of prospective ingredients are known as the "matrix" from which diets are formulated. The specifications for a particular diet, as delineated by a nutritionist, would include nutrient levels or ranges and limits (minimum and maximum) on those ingredients that might be available for use in the diet. The computer then chooses from a myriad of ingredient combinations and arrives at a "solution," the lowest-cost diet that meets the desired ingredient and nutrient specifications. Modern feed formulation programs allow the nutritionist to see where levels of ingredients or nutrients might be manipulated for further cost savings. These programs also have ingredient inventory control options and plant options when formulating for multiple feed mills.

V. Poultry Health and Disease

Poultry, like other animals, are susceptible to diseases that can result in death or reduced production. The development of the poultry industry has depended on healthy chickens and turkeys, and this has been possible because of research into methods to control, and hopefully eliminate, poultry diseases. These techniques include thorough sanitation, vaccination as a prophylactic, agents that kill the disease organisms (antibiotic, insecticides, anthelmintics), monitoring poultry health by farmers and veterinarians (with autopsies and anatomical and biological diagnosis), biosecurity (isolation of poultry facilities) to avoid cross-contamination between people/poultry in adjacent commercial flocks, and the depopulation and disposal on site of infected flocks (as in the case of avian influenza). Poultry diseases can be categorized as environmental diseases or based on the disease organism: virus, bacteria, protozoa, endoparasites, and ectoparasites. [See ANIMAL DISEASES.]

A. Viral Diseases

Viruses that cause diseases in poultry have had devastating effects on the poultry industry in the past. There are now successful vaccination schemes for many of the most serious diseases (Table IV). These have supplemented proper sanitation, monitoring, and biosecurity.

B. Bacterial Diseases

Bacteria not only induce diseases per se, but also may cause opportunistic infections in viral diseases. Bacterial diseases include mycoplasmosis (chronic respiratory disease) due to *Mycoplasma gallisepticum,* Pullorum disease (a salmonellosis) due to *Salmonella pullorum,* and fowl typhoid (a salmonellosis) due to *Salmonella gallinarum,* as well as colibacillosis and staph infections. The primary methods of eliminating bacterial diseases are sanitation, antibiotics, monitoring, and biosecurity.

C. Protozoan Diseases

Poultry are susceptible to several protozoan diseases, the most common being coccidiosis, caused by various species of *Eimeria,* and *histomoniasis* (blackhead) due to *Histomonas meleagridis.* These can be eliminated by treatment with appropriate drugs (e.g., coccidiostats).

D. Endoparasites

Poultry can harbor in their intestines either roundworms (nematodes) or tapeworms (cestodes), both of which will reduce the general health and hence the productivity of the poultry. The problem can be eliminated by sanitation, careful monitoring, and use of anthelmentics.

E. Ectoparasites

Both lice (insects) and mites (arachnids) are examples of ectoparasites that can live on the skin or feathers in poultry. These can become a significant problem that can be alleviated by removing infected birds, sanitation, and use of insecticides.

TABLE IV

Commercial Poultry Vaccinations

Newcastle disease, infectious bronchitis	These are viral diseases that affect the respiratory tract of birds. The objective of the vaccination would be for individual bird protection.
Marek's disease	A viral disease that affects the nervous system of young poultry. The vaccine is administered at hatch and its objective is individual bird protection.
Infectious bursal disease	A viral disease that affects the Bursa of Fabricius in young poultry and renders them immunologically incompetent. Young poultry can be vaccinated for protection but the major route of protection is maternal vaccination for transfer of antibodies to newly hatched chickens.
Reovirus	A viral disease that results in nutritional "malabsorption" and stunting of young broilers. The major mode of protection is via maternal vaccination and passive transfer of antibodies to the chicks (via the yolk).

F. Environmental Diseases

Poultry can be adversely affected by environment-related factors. For instance, poultry feed may be contaminated with toxins, an example being the fungally produced aflatoxins. In addition, temperature (particularly high in the summer) and other environmental conditions can stress either chickens or turkeys, resulting in the production of the stress hormone corticosterone (a glucocorticoid) and reduced growth and depressed immune function. The poultry industry is striving to improve the conditions of poultry to reduce losses of productivity.

VI. Poultry Reproduction and Growth

A. Biology of Female Reproduction and Egg Production

The egg of the chicken or turkey is made up of three major constituents, yolk, egg white, and shell. The solids of the yolk are lipid (two-thirds) and protein (one-third). Yolk precursors are synthesized by the liver under the influence of the estrogenic hormones, including estradiol 17β. These are transported to the developing follicles in the single ovary by way of the bloodstream. Yolk is deposited in the developing ovum in the follicle under the influence of follicle-stimulating hormone (FSH). When the follicle/ovum are mature, it is about 2 cm in diameter and has a yellow appearance due to the accumulated yolk. At this time, the ovum is induced to ovulate owing to hormonal stimulation by luteinizing hormone. Ovulation occurs at a frequency of up to once per day. If eggs are being ovulated daily, this will occur at exactly the same time each day. If not, there is a pattern of later and later ovulation followed by a missed day of ovulation.

The ovulated ovum is picked up by the infundibulum of the single oviduct. If fertilization is to occur, it will occur during the brief time (15 min) in the infundibulum. The proteins of the egg white are added to the ovum during its time ($\sim$3 hr) in the next region of the oviduct, the magnum. The production of egg white protein requires estrogen and, to some extent, also progesterone. The egg then passes to the next region of the oviduct, the isthmus (1 1/4 hr), where the membranes are added. Subsequently the egg spends over 20 hr in the uterus, where the shell is added. More than 2 g of calcium are deposited into the shell, coming from calcium stored in the bones. Female birds have an additional component to their bones, the medullary bone, which is a storage site for calcium laid down under the influence of estrogen and androgens. The second component of the shell, carbonate, is derived from circulating bicarbonate ions in a process involving uterine carbonic anhydrase. After a brief period (less than 30 min) in the vagina, the egg is laid or oviposited. Oviposition is induced again by hormones, including arginine vasotocin and prostaglandins of the E series. The next ovulation of the next ovum from the ovary will occur about 30 min after oviposition of the previous egg.

Spermatozoa are stored in vaginal ducts in the female tract following coitus or artificial insemination. Released spermatozoa pass up the oviduct once the physical obstacle of the egg is out of the way (i.e., following oviposition). Viable spermatozoa can be found for prolonged periods in the vaginal ducts (>1 month).

Development of the poultry egg begins immediately after fertilization in the upper oviduct. In some lines of turkeys, development can be initiated without fertilization. These parthenogenic ("virgin birth") offspring show very high embryonic mortality and few survive past hatching.

B. Male Reproduction

Spermatozoa are produced in the seminiferous tubules of testes in the abdomen of the chicken or turkey. The testes of poultry are large (>8 g in the sexually mature chicken, and each testis is connected to cloaca by a duct, the ductus deferens. Unlike in mammals, there are no accessory glands (e.g., prostate, seminal vesicles) or a penis. There is, however, a small papilla in the cloaca that acts to deliver semen during coitus.

Normal functioning of the reproductive system in male birds requires anterior pituitary hormones; follicle-stimulating hormone (FSH) for spermatogenesis and luteinizing hormone (LH) for testosterone production by the Leydig cells. Male secondary characteristics (in the rooster, the comb, wattle, spurs; in the tom or stag turkey, the snood) are maintained by the male hormone testosterone. Testosterone also acts on the brain to allow male behavior (aggressive and sexual).

C. Artificial Insemination

In the United States, most turkey poults result from artificial insemination (A.I.) for which there is an increasing use in the production of broiler chicks. The rationale for using A.I. includes the physical problems in natural coitus because of the large size of the male and also the advantage of the ability to genetically select superior males. Semen is collected and females are inseminated manually. More than ten females can be inseminated per semen sample obtained from a male. Insemination of the females is repeated every 5 to 10 days. Spermatozoa are stored in vaginal ducts following A.I.

D. Light and Reproduction

In many species of birds, reproduction is restricted to one period of the year. In this seasonal breeding, birds respond to increasing day length, which initially involves changes in the brain. The hypothalamus releases chicken luteinizing hormone releasing hormone I (cLHRH I) and thereby stimulates the secretion of LH and FSH from the pituitary gland. This is known as the photoperiodic control of reproduction. After a time, a stimulatory day length is no longer perceived as stimulatory and the bird is referred to as "photorefractory." A period of time on a nonstimulatory, short day length is needed to "break" photorefractoriness.

This photoperiodic system controlling reproduction is clearly seen in the turkey. Increasing day length is required and is, in fact, used to bring either male or female turkeys into sexual maturity. After a perid, there is the onset of photorefractoriness and egg production declines. In chickens (hens) on a natural day length, egg production shows a marked seasonal pattern. However, if day length is increased above 14 hr of light per day, egg production is maintained at a high level.

E. Behavior and Reproduction

After laying eggs, female turkeys and hens of some breeds (particularly unselected bantams) of chickens show brooding behavior and incubate eggs. This is controlled, at least partially, by the hormone prolactin. These broody chickens and turkeys show the commercially undesirable feature of a precipitous decrease in egg production. Broodyness has been selected out of many chicken breeds.

F. Recycling and Molting

Both egg production and egg quality in chickens decline the longer a hen has been in production. However, if a hen can be induced to have a period when egg production ceases (and also the hen molts), subsequent egg production and quality are increased. This is known as recycling. Techniques to induce a rest from egg production or molts include nutrition deprivation (fasting or limiting critical nutrients such as protein, calcium, and sodium), the inclusion of zinc in the diet, and reducing day length.

G. Biology of Growth

As in virtually all animal species, growth has specific characteristics, and the growth of either chickens or turkeys can be described by a sigmoid curve that reaches a plateau after sexual maturation. Selection for growth rate has in fact increased the overall size/weight of adult poultry but not at the rate at which it achieves the sexual mature size/weight. This gives the net effect of increasing the apparent growth rate to market weight. All tissues do not grow at the same rate, with muscle growth preceding that of adipose tissue. Growth is under hormonal control, with growth hormone, thyroid hormones, and insulinlike growth factors required for normal growth.

Bibliography

Austic, R. E., and Nesheim, M. C. (1990). "Poultry Production." Lea & Lebyer, Philadelphia.

Bell, D. J., and Freeman, B. M. (1971). "Physiology and Biochemistry of the Domestic Fowl," Vols. 1–3. Academic Press, London.

North, M.O., and Bell, D. R. (1990). "Commercial Chicken Production Manual." Van Nostrand Reinhold, New York.

Prices

DAVID A. BESSLER, CARL E. SHAFER, *Texas A&M University*

Glossary

Demand Quantity of a particular commodity which a consumer is willing to purchase as its price varies, assuming all other factors are held constant

Supply Quantity of a particular commodity which is offered for sale per time period as price varies, assuming all other factors are held constant

Market Place (perhaps a physical place but not necessarily) where buyers (demanders) and sellers (suppliers) of a commodity come together to negotiate an exchange at a price

Price is a basic information signal in economic systems. In generic form price represents the rate at which one commodity (good or service) exchanges for another commodity. Usually, price is defined such that one of these commodities is a numeraire, money (dollars, pounds, etc). Price need not, however, be defined in terms of a numeraire, as in barter economies; units of one commodity in exchange for units of another commodity quite adequately defines price. The notion of price exists in free markets, where no constraints exist on entry or exit of buyers and sellers and each is free to negotiate (bid or offer) as he or she wishes. Price is also defined in regulated markets, where some constraints are placed on the behavior of buyers and/or sellers.

I. Functions

Price serves two major functions: first, to ration scarce commodities (who gets what?) and second, to motivate (direct) future production of commodities (what products will be produced?). The manner in which price carries out these functions differs according to the institutional rules placed on the market. Rules range from very little regulation (livestock markets in the United States) to extreme forms of regulation where price is fixed by governmental authority (prices in the former Soviet Union). We describe two forms—free markets with little or no regulation, except legal requirements of enforcement of contracts and type of communication among buyers and sellers (double oral auction versus posted prices), and fixed price markets. The efficiency with which price is able to carry out the above-mentioned functions is determined by the institutional rules placed on the market.

II. Equilibrium Price

In free markets price adjusts according to changes in *supply* and *demand;* the market price is that which just clears the market—the amount of the commodity or service suppliers want to supply (give up title to) equals the amount of the commodity or service demanders want to purchase.

Generally, quantity supplied is a positive function of price, so that suppliers (those who hold title to the commodity) offer more for exchange when price increases; whereas quantity demanded is a negative function of price, as demanders (people who do not have title to the commodity but may want to acquire title to it) offer to buy less when price increases. These are illustrated in Fig. 1. The line labeled dd' represents the demand for the commodity and the line labeled ss' represents the supply of the commodity. If the

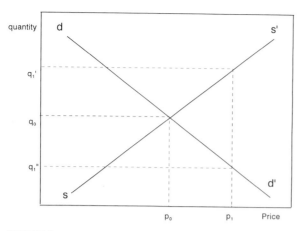

FIGURE 1 Demand and supply lines.

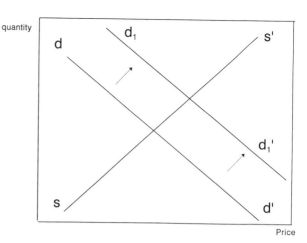

FIGURE 2 Shift in demand due to increased wealth.

quantity suppliers offer at a specific price is greater than the quantity demanders wish to purchase at that price, price will fall, inducing demanders to take more. In Fig. 1 this is illustrated by noting that at price p_1, the quantity supplied (q_1') is greater than the quantity demanded (q_1''). This price (p_1) will not be maintained (is not stable). The reverse will occur if the quantity suppliers wish to supply at a given price is less than the quantity demanders wish to acquire at a specified price. It is only when quantity supplied equals quantity demanded that a particular price is stable. Price arrived at in free markets by this interaction of supply and demand is called an *equilibrium price*—as there will be a tendency for price to return to this level if there is some perturbation to it. In Fig. 1 this is illustrated by price p_0. The equilibrium quantity is given as q_0.

The notion of equilibrium price discussed above has been defined under *ceteris paribus* (others things equal) conditions. Among the other things held equal (constant) on the demand side are income (wealth) of demanders, number of demanders, prices of substitute and complement commodities, preferences of demanders; on the supply side are production technology, prices of inputs to the production process, and number of suppliers in the market. If any of these change, the simple supply and demand model given in Fig. 1 changes. In particular, if wealth of demanders increases, so that each has more money to spend on commodities, including the one under consideration (and if the commodity under consideration is a normal good), the demand line shifts out to the right, so that at each price more of the commodity is demanded. This is illustrated in Fig. 2 by a shift in the demand line from dd' to d_1d_1'. Similar shifts in demand or supply schedules are defined for changes in any of the

variables held fixed in our *ceteris paribus* condition given above.

Price determined by the interaction of demand and supply reflects the marginal valuation of the particular commodity. This explains the seemingly odd fact that a commodity as useful as water (say an acre foot of water) has a price much lower than an apparently less useful commodity—say a diamond. The latter has a more limited supply relative to demand than the former. However, if we placed constraints on the availability of water we would see, rather quickly, that its price would be equal to or much greater than that of a diamond. The marginal utility of a unit of water (at current availability) is much less than that of a diamond, but the total utility of water is far greater.

In free markets those individuals willing to pay the market price will be able to acquire the commodity, while those neither willing nor able to pay the price do without. The entire amount brought to the market (amount suppliers are willing to put on the market) is rationed (allocated) among users by price. If demanders wish to consume more than suppliers offer at a specific price, price will increase. This increase in price brings forth two changes. First, at a higher price suppliers are willing to put more of the commodity on the market and, second, demanders take less of the commodity at the higher price.

If the price arrived at in free market equilibrium is high (relative to the cost of producing the product), suppliers in future periods will have an incentive to bring forth additional supplies of the commodity. If the price is below the cost of production, suppliers will bring forth smaller quantities in future periods. Thus, price serves its second function: it directs (motivates) what and how much is to be produced.

III. Planning Prices

In regulated markets, the extent to which price can accomplish its two functions is limited. In extreme cases where price is held fixed by government authority, both the rationing and direction functions are stifled. For example, in centrally planned economies, where price is not allowed to adjust according to market demand and supply, rationing may be accomplished on a first come first served basis. Once the entire inventory of product is sold at the fixed price, waiting lines form, until additional product is brought to the market. Here time spent in line does the rationing. In such cases, it is common for parallel or black (illegal) markets to form which mimic the behavior of a free market. (Coupons may be used in lieu of price as a rationing device when a shortage exists: e.g., gasoline rationing in the United States during World War II.)

The direction of production is also impaired in fixed price markets. If the price is set above the free market equilibrium price, suppliers will have an incentive to produce more than demanders wish to purchase. Such products will sit in inventory (wait to be sold). If price is set below the equilibrium price, producers will have no incentive to produce additional units of that commodity, even though demanders would be willing to pay more for additional quantities of that commodity. Thus, surpluses and shortages occur only at fixed prices: too high and too low, respectively.

In centrally planned economies, provisions for changing price and production have been devised which rival market prices, at least in theory (Lerner). Prices in these economies were modeled to provide much of the same information as prices from free markets offer. Essentially this model works off of accounting (planning) prices and information interchanges between managers and the central planning authority. The central planner sends information to the managers in the form of a set of price signals; the manager returns information to the central planners in terms of his or her output. If this planned output was greater than needed, a new set of accounting prices would be communicated to the managers. This process continues until the particular set of production levels were realized. Lerner argued in 1944 that such a system could replicate the market pricing system. While the Lerner model was never actually tested, the late 1980s collapse of the centrally planned economies of eastern Europe suggests that the system may have suffered in application.

Accounting prices from central planning authorities have a rigorous interpretation in terms of the output of mathematical programming models. In such models an objective function is defined in terms of alternative activities, each level of that activity contributing positive or negative returns. The objective function is optimized (maximized or minimized) subject to resource constraints. For example, farm planning models may be defined to choose among alternative crop activities, each unit having associated returns and costs. Further, units use certain levels of each of several (potentially) limiting resources (land, labor, capital, water, time, etc.). Associated with the optimal solution to this problem (which will be levels of production of each crop) are shadow prices (shadow meaning not real) which reflect the marginal valuation of the constraining resources. These prices arise naturally in the solution to dual pairs of mathematical programming problems; and while they are not equivalent to the market prices discussed in the remainder of this article, they do reflect similar information, in terms of marginal valuation.

IV. Price in Agricultural Markets

Agricultural prices are prices received for commodities at the farm/ranch gate or at the packing shed door. The demand for these products is derived from the demand by consumers for the final product. Accordingly, farmers and ranchers receive only a share of the price paid by consumers. This difference, the margin, covers transport costs, processing costs, packaging costs, etc. The relative size of the margin differs across commodity groups. For example, the farm value share of the retail beef price is almost 60%; it is only about 7% for bread.

The demand functions (described generically above) facing farmers are believed to be particularly price inelastic. That is, relatively small percentage change in quantity produced (say due to a bad harvest) will result in relatively large percentage change in equilibrium price. The reader may grasp the relevance of this point by understanding that certain commodities have few substitutes. Food (as a composite commodity) is one such commodity. Consumers are likely to give up large portions of other commodities (shoes, entertainment, etc.) rather than do without sufficient levels of food which is quite habit forming and perishable. Weather uncertainty has led to variable production which interacts with the inelastic demand resulting in highly variable prices. For example,

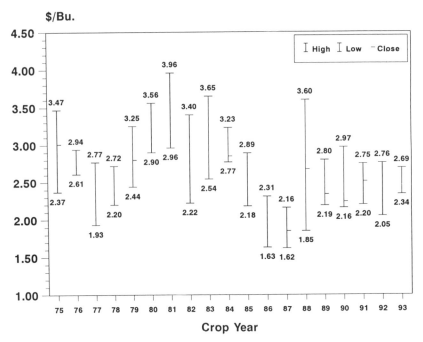

FIGURE 3 December CBOT corn prices high, low, close, and loan rate.

wheat farmers have experienced large year-to-year and within-season price moves. Within-season cash price changes averaged 29% (low to high) during the l981 to 1991 period. Accordingly, market institutions which deal with high price variability, directly or indirectly, have found extensive use in agriculture. Two major market risk-bearing institutions are agricultural commodity futures markets and government programs.

V. Futures Prices

Organized futures markets which allow producers, marketing firms, and consumers to reduce market price uncertainty through hedging with futures contracts have existed in the United States for well over 100 years. Such markets allow pricing of commodities for future delivery by facilitating the trading of futures contracts. The futures contract calls for the delivery of a specific grade of the commodity to a given location by a specific time for a specific price.

Holbrook Working aptly defined hedging as using the futures market as a temporary substitute for the cash market. The maximum decline in December corn futures prices ranged from 46 cents to $1.75 per bushel during 14 of the 19 seasons 1975 through 1993 (Fig. 3). A farmer in the Corn Belt wishing to avoid

such price uncertainty could have priced his or her crop at anytime up to 14 months before harvest by selling December corn futures. Even though the futures transaction may take place months before harvest, delivery is not required until December. Our farmer would probably not deliver on the futures contract, as it calls for delivery in a particular location (Chicago, Toledo, or St. Louis warehouse). Rather, he or she will most likely buy back the contract and sell the harvested crop in the local cash market (at the local elevator in central Illinois). This transaction uses the futures as a hedge against possible adverse price moves in the cash market. The degree of protection depends on the relationship between cash and futures prices: i.e., the basis. Basis, the difference between the local market cash price and the appropriate futures price is less variable than price because the same supply and demand forces which influence the futures price also influence the local cash price at the end of the futures contract's life; the future has become the present. The farmer remains subject to yield uncertainty which may not be particularly easy to forecast but can be partially covered by crop insurance.

The farmer who protected against a drop in his local cash market price by initially selling futures, also "protected" against the benefit of a price increase during the course of the hedge. In 1984, a type of "price insurance" in the form of commodity options on futures contracts became available. By paying a premium, a farmer can buy an option to sell (put option) December corn futures at a specific price (strike price). If price has fallen by the time the crop is ready for sale, the farmer would exercise his or her right to sell futures at the predetermined strike price. If price rises, the producer would not use (exercise) the option but simply sell the crop at the higher cash price level at harvest. The put option provides a price floor (subject to basis risk) but no ceiling against the possibility of higher prices. As with any insurance policy, the benefits come at a cost of a market determined premium which is a significant cost. Similar futures and options market operations can be defined for a user of agricultural products as inputs into other productive activities to protect against unforeseen increases in input prices, say a cattle feeder who feeds corn. In fact, futures are probably used more intensively by marketing/processing firms than by producers.

VI. Government Regulations on Prices

While prices generally perform well in free market environments in determining production and allocating goods and services to consumers, there are environments that economists have identified in which market prices result in less than optimal performance. Pure public goods are not well allocated by free market systems. A pure public good has the characteristic that one person's consumption of the good does not diminish another person's consumption of the good. An example is national defense: construction of a missile defense system for my home against attack from missiles also protects my neighbors. Because of the nonexclusionary property of the good, free markets are generally believed to underinvest in public goods. More directly, such goods have a free rider problem: "I'll not provide the good (and, in fact, may disguise my preferences for the good!) in hopes that you will provide it." A similar problem of nonexclusionary costs results in public "bads" (termed negative externalities). Where such nonexclusionary "goods" or "bads" exist, group action oftentimes replaces the free market. Here clubs or local or national governments carry out the function of the market. Goods having a mixture of public and private good characteristics are provided in a mixture of public and free market allocation. Government intervention in the pricing of agricultural products is an example of a mixed public good. There is some "public good" characteristic to

food production and allocation. [See CROP SUBSIDIES; TARIFFS AND TRADE.]

Government programs designed to influence prices of U.S. agricultural products have been in effect since the 1930s. Many products, including corn, cotton, wheat, barley, soybeans, milk, tobacco, wool, rice, and sugar, are subject to government programs which exist to influence both the level and variability of prices within and between production cycles. Motivations for such actions stem from general societal welfare considerations including provision of an adequate food supply, increased income to farmers, and/or improvement in productivity and resource allocation.

The commodity price received by farmers can be increased by purposely restricting its supply in the market. Such restrictions may include import quotas, taxing imported commodities, or restricting the domestic production of the commodity. The use of each of these tactics depends on the particular characteristics of the market. For example, if a country is self-sufficient in the production of a commodity, import programs, such as quotas or tariffs, will be ineffective in changing the level or variability of that products' price over time. Here production controls may be utilized. Participation may be voluntary or required. In the United States these programs are generally voluntary and operate through acreage set aside programs using a system of target prices and nonrecourse loan programs. In both voluntary and nonvoluntary programs, there will be an tendency for participants to "cheat," so that the monitoring or policing activity is a nontrivial cost of the program. Measures such as export subsidies and food stamps are attempts to support price from the demand side.

Commodity reserve or storage programs can be used to keep prices within narrow (more or less) bands by storing excess commodity in years of surplus and releasing the stored crop in years of shortage. Assuming particular shapes in the demand function and particular patterns of temporal over- and underproduction, such a program can be demonstrated to increase societal welfare, even with nontrivial storage costs.

An interesting historic case of government's establishing "fair" prices for farmers involves parity prices. The initial idea was that fair prices to farmers would be on a par or "at parity" with the prices of things farmers buy. Original or "old" parity prices were computed by simply multiplying the average price of wheat during the 1910–1914 base period by the current index of prices paid by farmers. The 1910–1914 period was selected as a "reasonable" period for relating farm and nonfarm prices. This rather crude method was modified January 1, 1950, to account for relative changes among farm prices since the 1910–1914 period. Nevertheless, the intent remained the same; parity price indicates the farm price necessary to provide farmers with purchasing power per unit of produce equivalent to the costs they pay for inputs and living expenses. U.S. parity prices are still reported; e.g., the December 1992 parity prices for corn and wheat were $5.75 and $8.09 per bushel, respectively. In contrast, cash corn (central Illinois) and wheat (Kansas City) prices were $2.07 and $3.80, respectively. If farm prices were actually supported at the parity level, considerable resource misallocation would ensue.

VII. Efficient Price

The notion of an efficient price relates to its ability to quickly transfer new information to both buyers and sellers resulting in an improved allocation of resources. Thus, prices in an efficient market respond to new information in a timely manner. More directly, in an efficient market prices reflect information such that the marginal benefits from acting on additional information are not greater than the marginal cost of acting on the information. This is the essence of competitive arbitrage.

Notions of pricing efficiency have temporal, spatial, and form dimensions. All can be summed under the heading of "the law of one price": identical commodities should have the same price. Of course, the cost of making commodities identical must be taken into account. Wheat next month is not the same as wheat today—the two commodities are connected by storage. Wheat in Kansas is not the same as wheat in Houston—the two commodities are connected by transportation. And wheat is not the same as bread—the two commodities are connected by processing. Three forms of market efficiency have been defined: *weak, semi-strong,* and *strong.* These forms focus attention on predictability of prices under particular information sets: weak form conditions on past prices (past own prices that is), semi-strong conditions on all publicly available information, and strong form conditions on all information. Under the weak-form market efficiency assumption, prices, adjusted for the costs discussed above, will follow a random walk. Thus, the best predictor of price in period $t + 1$, based on information at period t, is price in period t. This simple model of price behavior is sensible when one considers that in free markets, if one

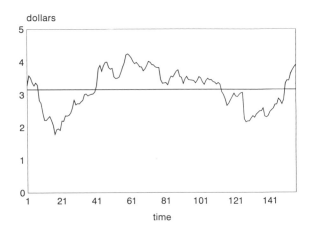

dollars

monthly data (1976 - 1988) in dollars per bushel

FIGURE 4 Prices of central Kansas wheat.

finds recurring patterns in historical prices (in either weak or semi-strong information sets), he can act on that information today. His actions, taken to profit on that information, will drive the price today to just equal the predicted price for tomorrow. So the only difference between today's price and the expected price for tomorrow will be the cost of acting today.

The efficiency arguments suggest particular time series properties of price data. Price data tend to be nonstationary through time. That is, they do not tend to return to their historical mean, but rather, wander (seemingly aimlessly) through time. As an example, we have plotted monthly wheat prices in central Kansas and Houston, Texas, over the years 1976 through 1988 in Figs. 4 and 5. The mean price in both markets is plotted in each figure as well. Note, in both figures price crosses its historical mean infrequently. In Fig.

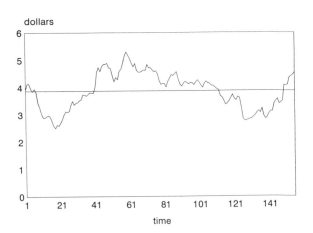

dollars

monthly data (1976-1988) in dollars per bushel

FIGURE 5 Prices of Houston-Port wheat.

4, the number of crossings is four, while in Fig. 5 the number of crossings is five in 156 observations. Both series appear to wander from their historical means for long periods of time. So that the mean price is not particularly helpful as a measure of central tendency.

In Fig. 6 the contemporaneous differences between the Houston-port wheat price and the central Kansas wheat price (price in Houston in period t minus price in Kansas in period t) are plotted. Notice, the number of mean crossings for these differences is considerably greater than for the individual price plots (25). Here the mean price difference is a more meaningful statistic (meaningful as a measure of central tendency). Bessler and Fuller show that these price series exhibit a type of equilibrium relationship. The series move away from their historical "basis" relationship (their mean price difference of $0.70) in the short run, but return to it in the long run. In technical jargon these prices are cointegrated, which means the series are individually nonstationary (appear not to return to their historical mean), but their contemporaneous differences are stationary. Among other things cointegration places a discipline on interpretation of price movements.

The results presented in Figs. 4, 5, and 6 are illustrations of the "law of one price." For short periods of time such prices may deviate from one another, but in the long run they are brought back into their "normal" equilibrium relationship; with the difference reflecting the cost of making the commodities identical. Prices which reflect the "law of one price" will be cointegrated. Engle and Granger present the general model of cointegration, including appropriate tests, test statistics, and critical values.

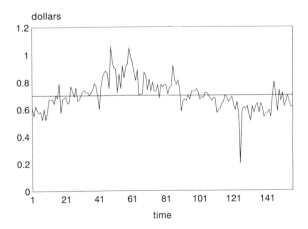

dollars

monthly data (1976 - 1988) in dollars per bushel

FIGURE 6 Differences between Houston-port wheat price and central Kansas wheat price.

Cointegrated prices help reflect the extent to which the market is defined. In agricultural markets we might expect to see cointegrations between futures and cash markets for commodities, between prices of spatially separated commodities (as illustrated above), or between raw and processed commodities (price of raw cotton and the price of gray cloth). Government interventions may serve to effectively block such cointegrations by restricting the "free flow" of information. Thus, if suppliers are prevented from selling certain commodities in a particular country, prices of the commodity in the restricted country will respond to local supply and demand conditions and not global conditions.

In addition to price efficiency, economists have studied other aspects of performance of markets and their prices. Some criteria useful for measuring the efficiency and performance of a pricing system include:

1. Allocation efficiency including short-run market clearance, inventory adjustments, production adjustments, and marketing efficiency.
2. Welfare considerations regarding price level and stability, price dispersion, and sharing of gains or losses within the marketing channel.
3. Competitive considerations on market access, bargaining power, structure, and information flow.
4. Cost considerations including waiting time and transportation and system costs.
5. The level of buyers' and sellers' satisfaction with the operation and results of the system.

VIII. Price Level

The foregoing discussion centers on relative prices. They are defined in terms of exchange for money (the numeraire good). The quantity of money in the economy will define the nominal level of these prices. In periods following substantial growth or contraction in the quantity of money, the price level may change quite rapidly. This is defined as a period of inflation or deflation—a period when the general price level increases or decreases. Historically, such periods have been at the heart of social and political turmoil—witness the late 1800s and the silver coinage (bimetallism) political campaign of William Jennings Bryan in the United States, the hyperinflation in Germany in the 1920s, the great depression of the 1930s, and the inflation during the 1970s in the United States. Each of these periods is characterized in terms of increases or decreases in the general price level. Further, each period experienced contractions (the United States in the late 1800s and 1930s) or expansions (Germany in the 1920s, and the United States in the 1970s) in the quantity of the numeraire good (money).

The equation of exchange is helpful in understanding such periods: $MV = PQ$. Here M is a measure of the money stock, V is a measure of the velocity of money (the number of times per unit of time that a unit of money turns over), P is a measure of the general price level (changes in the general price level are measured by broad-based price indexes, such as the consumer price index or the wholesale price index), and Q is a measure of the number of transactions. For V constant and the economy at full employment (the economy is operating such that all resources are fully employed), an increase in M will result in a increase in P; decreases in M result in decreases in P. During the late 19th century, the United States was on the gold standard. This tied the quantity of money (defined as the amount of currency and bank deposits) to the quantity of gold held by the government. From 1870 to 1896 there were few new gold discoveries to support the rapidly growing U.S. economy. Bordo and Schwartz and Bessler argue that the relatively fixed money stock (M) and an expanding economy (Q) had to result in either increased velocity (V) or a decrease in the price level (P) or both. In fact, Bessler provides empirical evidence that the price level actually fell in response to the contracting money supply.

A similar argument is behind explanations for the Great Depression, the inflation in the United States of the 1970s, and the hyperinflations in Latin America in the mid-20th century.

It is generally thought that changes in money supply do not affect changes in relative prices. That is to say, the traditional view is that money is a veil, which merely covers, but is not a part of the real workings of the economy. However, recent work has at least questioned the short-run relevance of this view. Bordo discusses the short run nonneutrality of money. Differing market institutions are offered as explanations for money showing nonneutral effects. Agricultural products and other basic commodities (metals) are sold in auction-type markets; while other goods and services are often sold in posted price markets or on contracts which cover long periods of time. Pulses in the money supply will find their immediate effect in agricultural and other basic commodity market prices and show lagged (sluggish) effects throughout the rest of the economy. Bessler, however, shows that for price data in Brazil, a country which has a long experience of rapidly changing prices, this differential effect does not appear. He speculates that there market institutions have adjusted to a long history of living with a rapidly changing money supply.

Bibliography

Ardeni, P. G. (1989). Does the law of one price really hold for commodity prices? *Am. J. Agricult. Econom.* **71**, 661–669.

Barnett, R. D. Bessler, and Thompson, R. (1983). The money supply and nominal agricultural prices. *Am. J. Agricult. Econom.* **65**, 303–307.

Bessler, D. A. (1984). Additional evidence on money and prices: U.S. data 1870–1913. *Explorations Econom. History* **21**, 125–132.

Bessler, D. A. (1984). Relative prices and money: A vector autoregression on Brazilian data. *Am. J. Agricult. Econom.* **66**, 25–30.

Bessler, D. A., and Covey, T. (1991). Cointegration: Some results on US cattle prices. *J. Futures Markets* **11**, 461–74.

Bessler, D. A., and Fuller, S. W. (1993). Cointegration between U.S. wheat prices. *J. Regional Sci.* **33**, 481–501.

Bordo, M. (1980). The effects of monetary change on relative commodity prices and the role of long term contracts. *J. Pol. Econ.* **88**, 1088–1109.

Bordo, M., and Schwartz, A. (1980). Money and prices in the nineteenth century: An old debate rejoined. *J. Econom. History* **40**, 61–67.

Buchanan, J. M. (1968). "The Demand and Supply of Public Goods." Rand McNally, Chicago.

Dunham, D. (1983). "Food Costs . . . From Farm to Retail in 1992." United States Department of Agriculture, Economic Research Service, Information Bulletin Number 669, April.

Engle, R., and Granger, C. W. J. (1987). Cointegration and error correction: Representation, estimation and testing. *Econometrica* **55**, 251–276.

Fama, E. (1991). Efficient capital markets:II. *J. Finance* **46**, 1575–1617.

Friedman, M. (1976). "Price Theory." Aldine, New York.

Friedman, M., and Schwartz, A. (1982). "Monetary Trends in the United States and the United Kingdom." Chicago: Univ. Chicago Press, Chicago.

Forker, O. D. (1975). "Price Determination Processes: Issues and Evaluation," p. 5. FCS Information 102. Farmer Cooperative Service." US Department of Agriculture.

Harberger, A. C. (1981). A primer on inflation. *J. Money Credit Banking.* **13**, 1–10.

Intriligator, M. (1971). "Mathematical Optimization and Economic Theory." Prentice Hall, Englewood Cliffs, NJ.

Johansen S., and Juselius, K. (1990). Maximum likelihood and inference on cointegration—With applications to the demand for money. *Oxford Bull. Econom. Statistics* **52**, 169–210.

Lerner, A. (1944). "The Economics of Control." MacMillan, New York.

Samuelson, P. (1971). Stochastic speculative price. *Proc. Nat. Acad. Sci.* 335–337.

Tesar, R. (1986). Agricultural options: Practical usage by a commercial firm. *Rev. Res. Futures Markets.* **5**, 24–34.

Tomek, W., and Robinson, K. (1981). "Agricultural Product Prices." Cornell Univ. Press, Ithaca.

Turnovsky, S. The distribution of welfare gains from price stabilization: A survey of some theoretical issues. *In* "Stabilizing World Commodity Markets." (F. G. Adams and S. A. Klein, eds.), pp. 199–248. Lexington Books, Lexington, MA.

Working, H. (1953). Hedging reconsidered. *J. Farm Econom.* **35**, 544–561.

Production Economics

PETER BERCK, *University of California, Berkeley*

GLORIA HELFAND, *University of California, Davis*

Glossary

Conditional factor demand Input use, x_i, chosen for least-cost production of output, y, when input prices are $w_1, \ldots, w_n$: $X_i(y, w_1, \ldots, w_n)$
Cost function Least amount of money needed to purchase inputs to produce a specified output level, y. When the prices of inputs are $w_1, \ldots, w_n$, the cost function is written $C(y, w_1, \ldots, w_n)$
Production function Agricultural yield expressed as a function of the use of inputs, such as water, nitrogen, labor, land, and equipment. If the physical quantities of inputs are $x_1, \ldots, x_n$ and the physical quantity of output is y, then the production function is mathematically represented as $y = F(x_1, \ldots, x_n)$
Profit function Maximum farm earnings net of production costs as a function of output price, p, and input prices, $w_1, \ldots, w_n$. Profits are total revenues from farming less total costs of farming; the profit function, π, is written as $\pi(p, w_1, \ldots, w_n)$
Supply curve Profit-maximizing output level as a function of output price, p, and the prices of inputs, $w_1, \ldots, w_n$: $Y(p, w_1, \ldots, w_n)$
Unconditional factor demand Input use, x_i, chosen for profit-maximizing production when input prices are $w_1, \ldots, w_n$ and output price is p: $X_i(p, w_1, \ldots, w_n)$

Farmer behavior and firm decisions in the marketplace are examined in production economics. Technical relationships between inputs and farm output are summarized and incorporated into the behavioral analysis. Cost minimization is usually assumed when farmers choose their inputs. When they choose their level of production, profit maximization is assumed. Production economics is used for improving farm management and for estimating the effects of agricultural policies and practices. Statistical and programming techniques have been developed to quantify these effects.

I. The Basic Theory of Production

A. Production Functions

Production economics begins with the production function—a mathematical relationship between the inputs used for farming and the resulting output. The production function F is purely a technical relationship. It summarizes knowledge about the production process but implies nothing about the behavior of farmers or of markets.

Economists assume that the production function has several properties. First, output is nondecreasing in each input. That is, if use of one input increases, all other inputs held constant, then output should increase or, at least, not decrease. Mathematically, the partial derivative of F with respect to each input is nonnegative. In agriculture, it is possible that output will decrease if the input (such as fertilizer or manure) is applied excessively (first derivative becomes negative), but it is usually assumed that such applications are outside of the range of what farmers would use.

Second, each additional unit of input usually adds less to output than the unit before, a phenomenon referred to as diminishing marginal product. Mathematically, the second partial derivative of output, with respect to that input, is nonpositive.

Third, production functions have constant or decreasing returns to scale. A production function has constant returns to scale if doubling all inputs doubles output. (Returns to scale are decreasing if doubling all inputs leads to less than a doubling of outputs. They are increasing if doubling all inputs leads to more than doubling of outputs.) When output is analyzed as a function of physical inputs, such as land and fertilizer, agricultural production functions can reasonably be assumed to have constant returns to scale. In other words, if a farm is doubled in size, output should double. Because, as farm size increases, farmers are often able to specialize and make better use of certain inputs (such as labor and machinery), the production function may have increasing returns over much of the relevant size range. In wheat farming, for instance, the ability to use large implements on large tractors makes it possible for a large farm to use less labor per unit of output than a small farm.

A production function with constant returns to scale can be written on a per-acre basis. That is, if x_1 is the number of acres, then $y/x_1 = g(x_2/x_1, \ldots, x_n/x_1)$. A frequent way to analyze agricultural output is to write separate equations for per-acre production and for acreage in the crop. Acreage is assumed to adjust to changes in profits from one crop relative to profits from an alternate crop.

Pesticides and other damage-control agents do not behave like ordinary inputs in that they do not increase yields directly. Instead, they limit the reduction in yield from pests or other damaging agents. To reflect the difference between damage-control agents and other inputs, a production function can be specified as

$$y = f_1(x_2, \ldots, x_n) + f_2(x_2, \ldots, x_n) \, G(x_1),$$

where x_1 is the damage-control agent and $x_2, \ldots, x_n$ are other inputs. The function $G(x_1)$ is the proportion of the pest damage that is controlled by use of the input x_1; it has a range between zero and one. Any cumulative probability distribution function is a potential choice for a functional form for G. When damage is uncontrolled, the output is f_1. When it is perfectly controlled, it is $f_1 + f_2$.

B. Cost Minimization

In most production processes, different techniques can be used to produce a given level of output. Irrigated crops, for example, can produce the same output with flood irrigation (more water and less equipment) or sprinkler irrigation (less water and more equipment). Economists assume that, for any specified level of output, firms choose the combination of techniques to produce that output at least cost. In the

example above, if water were expensive relative to irrigation equipment, sprinklers would be adopted. If water were inexpensive relative to equipment, flood irrigation would be used.

More technically, firms choose inputs, $x_1, \ldots, x_n$, to minimize the cost of those inputs, $w_1 x_1 + \cdots + w_n x_n$, subject to the constraint that a given level of output, y, is achieved: $y = F(x_1, \ldots, x_n)$. Levels of input use chosen as a function of input prices and output are conditional factor demands, $X_i(y, w_1, \ldots, w_n)$. The cost function is $C(y, w_1, \ldots, w_n) = w_1 X_1(y, w_1, \ldots, w_n) + \cdots + w_n X_n(y, w_1, \ldots, w_n)$.

A conditional factor demand decreases as its own price increases. Effects of other input prices on this curve depend on whether inputs are substitutes or complements. For most inputs, increased output shifts factor demands out; these are termed normal inputs. Inputs whose use decreases when output increases are called inferior inputs.

Other facts about conditional factor demands are derived from economic theory as well. First, $\partial x_i / \partial w_j = \partial x_j / \partial w_i$. That is, the effect of the price of input j on factor i is the same as the effect of the price of input i on factor j. Second, they are characterized by homogeneity of degree 0, which means that doubling all prices with output held fixed does not change the input mix. This assumption implies that only relative prices (the ratio of each input price to one designated "numeraire" input price) matter.

C. Profit Maximization

Each firm engaged in agricultural production supplies only a tiny fraction of the market for that product. No single farm can significantly affect market volume or market price. Such firms are called price takers. A reasonable approximation of their behavior is derived by assuming that they maximize profits. There are two steps to maximizing profits: deciding how much to produce and choosing inputs to produce that amount at minimum cost. Because the input decision problem is summarized in the cost-function analysis, the remaining problem for the firm is to choose the output level to maximize profits, which are revenues less costs, $py - C(y, w_1, \ldots, w_n)$. The y chosen to maximize profits yields the decision rule $Y(p, w_1, \ldots, w_n)$, called the supply function. The supply function gives the profit-maximizing level of output for any set of input and output prices. The profit function summarizes the output and input decisions in the same way that the cost function summarizes input decisions and is written as $\pi(p, w_1, \ldots, w_n)$.

Economic theory predicts that the supply function is nondecreasing in output price and nonincreasing in input prices. In practice, one expects that an increase in the output price will increase the quantity supplied while an increase in an input price will decrease the quantity supplied. Finally, the supply function is homogeneous of degree 0 in output and input prices: If all output and input prices double, production should remain unchanged. Thus, only relative prices matter.

The supply function from profit maximization can be substituted for y in the conditional factor demands. The resulting functions (which give input use as functions of input prices and output price) are termed unconditional factor demands, since they are no longer conditioned on output. Properties of the unconditional factor demands are the same as those of conditional factor demands.

D. Risk

Agricultural firms bear considerable yield and price risks. Weather, world demand conditions, government programs, and pest problems are a few sources of risk in agriculture. A better approximation of farm behavior than simple profit maximization includes a consideration of farmer attitude toward those risks and of the actions farmers can take to avoid those risks.

Risk-averse farmers are willing to take lower expected profits in order to lower the variability of their profits. Consider two possible outcomes for farm income: high income, H, which happens with probability λ, and low income, L, which happens with probability $(1 - \lambda)$. A risk-averse individual prefers a certain income equal to the average, $\lambda H + (1 - \lambda)L$, to the uncertain income stream. In fact, such an individual may be willing to take a payoff lower than $\lambda H + (1 - \lambda)L$ in order to avoid the uncertainty.

Risk aversion can lead farmers to alter their practices in a number of ways. For instance, they may choose to spend more on inputs if they can reduce the uncertainty in crop yield. Instead of growing just the most profitable crop, farmers may choose to grow many crops in the hope that some crops will do well even if others do not.

E. Duality

Although it is most natural to think of the physical relationships of production being summarized in the production function, the same information is contained in the cost and profit functions due to their derivations from the production function. This property of cost, profit, and production functions is called duality. Its practical consequence is that all technical information about production can be recovered from the cost function (which depends upon prices and output) or from the profit function (which depends only on prices).

1. Shepard's Lemma

The problem of minimizing cost for given output makes direct use of the production function and leads to conditional factor demands. Those same factor demands can be found in a second way. The calculus can be used to prove that the derivative of a cost function with respect to input price is the conditional factor demand. Since the same input decisions are found either by working out the minimization problem, using a production function, or by taking the derivative of a cost function, these two functions contain the same information.

2. Hotelling's Lemma

Just as conditional factor demands can be found directly from the cost function, unconditional factor demands can be found from the profit function. The derivative of the profit function, with respect to output price, is the supply function. The negative value of the derivative with respect to an input price is an unconditional factor demand.

F. Multiproduct Model

Most farm operations have more than one output. Hog farms often produce corn and soybeans, cotton farms tend to grow alfalfa, and farms in the developing world frequently include several types of livestock as well as vegetables and grains. The formal analysis for such operations is an extension of the single product analysis presented above. Instead on only one output, y, let $y_1, \ldots, y_m$ represent the m possible outputs. The multiproduct production function, H, is commonly written in an implicit form, such as $H(y_1, \ldots, y_m, x_1, \ldots, x_n) = 0$. The cost function is $C(y_1, \ldots, y_m, w_1, \ldots, w_n)$; the profit function is $\pi(p_1, \ldots, p_m, w_1, \ldots, w_n)$; and so forth. With a suitable choice of functional form, these relationships can be used directly. While there are examples of work at this level of generality, such as a study of expansion effects and substitution possibilities for Canada, there is a tendency, particularly among those interested in farm management, to place more structure on the problem.

The most common method of analyzing a multiproduct system is first to examine the production relationships of each activity (crop or livestock production) individually and then to examine the profit maximizing choice of activities.

Farmers have many reasons for producing more than one crop. First, different soils may be suited to different crops. Second, if labor, capital, or other input requirements for two different crops are complementary, then both can be grown with the same inputs and greater production can be achieved with little increase in cost. Third, producing a variety of crops reduces risk, since factors that may cause one crop to fail may not be factors that will destroy another crop. Even if one crop is destroyed, the farmer can still earn profits from those remaining. Fourth, rotations may be desirable to control pests or maintain soil characteristics. Various programming methods, described below, can be used to solve this allocation of scarce resources among crops.

II. Estimating Production Relationships

The previous discussion has presented the economic theory of production. The next step in analyzing production relationships is data collection and analysis. Data collection and analysis are used to determine empirical production, cost, and profit functions and associated factor demands. These functions are used for policy analyses and farm-management recommendations.

A. Data Sources

Data derived from a range of sources can be used to develop estimates of production, cost, and profit functions. Some of these data sources are described below.

1. Experimental

Agronomic field trials provide data that include different input combinations and their resulting output. Examples include experiments to find responses of various crops to nitrogen, potassium, and water.

2. Farm Survey

The U.S. Department of Commerce, Bureau of the Census, conducts a census of agriculture every 5 years. Every farmer in the United States is asked information on their farm operations, including input applications, irrigation methods and amounts, acreage in different uses, size of farming operations, market value of assets and debts, and other areas. Every year as well, the National Agricultural Statistics Service of the U.S. Department of Agriculture surveys a subset of farmers and collects similar information for specific crops. Other surveys are often conducted by researchers interested in specific issues or regions.

3. County Data

The local government in most states in the United States collects acreage, output, and value of output by crop for the major crops in each county. This information can be used to construct countywide agricultural land-use models.

4. Cost and Return Estimates

Land-grant universities in most states in the United States estimate how much it will cost to produce a crop and what revenues to expect on a representative farm. These estimates are based on a standard set of inputs, management techniques, and level of output, determined by consultation with farmers and local crop experts, and with specified prices for inputs and output. They are used for a wide range of purposes, including determining whether an investment in a particular crop is worthwhile, examining the effects of alternative management practices on profits, estimating the value of land in production of a crop, and analyzing whether the costs on a particular farm are relatively high or low.

In addition, the U.S. Department of Agriculture conducts a Farm Costs and Returns Survey for a range of crops (including all crops involved in federal crop support programs) every 5 years. This survey reports how much, on average, farmers spend for production of each crop and how much they earn.

B. Statistical Techniques and Functional Form

Statistical estimation of crop production functions requires not only data on inputs applied and output attained, but also a class of assumed mathematical forms to relate the inputs and output. A class of functions is determined up to r numerical values, $\alpha_1, \ldots, \alpha_r$, called the parameters of the function. A choice of the form of the function $F(x_1, \ldots, x_n; \alpha_1, \ldots, \alpha_r)$ is necessary to conduct most analyses. Statistical techniques are used to determine the parameters that make the equation fit the data as accurately as possible.

Data never conform perfectly to a functional form chosen to approximate the true production function. Measurement error, omitted variables, and error in the choice of functional form all contribute to the imperfect fit. These discrepancies between actual yield and the yield predicted by a production function are called error terms, ε. The relationship between inputs and output is assumed usually to be of the form

$$y = F(x_1, \ldots, x_n; \alpha_1, \ldots, \alpha_r) + \varepsilon.$$

This form means that, for each observation (year and farm, field, county, or state), the output for that observation is related to the inputs for that observation plus an error term. Assumptions about ε, such as normality, lead to particular types of estimation.

A common choice for an agricultural production function is a quadratic function. With two inputs, the function is written as

$$y = \alpha_1 x_1 + \alpha_2 x_2 + \alpha_3 x_1 x_2 + \alpha_4 x_1^2 + \alpha_5 x_2^2 + a_6.$$

In this function, six parameters need to be determined by statistical techniques. Generalizations of the quadratic include higher order polynomials, polynomials in square roots, and polynomials in logarithms (called translog).

As noted above, diminishing marginal product is commonly assumed in economics and commonly observed in reality. This assumption is incorporated by limiting the choice of forms (or parameters) to those that permit positive first derivatives and negative second derivatives. In addition to polynomials, the following functional forms (given for the two-input case) are frequently used and have these characteristics.

Mitscherlich	$y = (1 - e^{\alpha_1 x_1})(1 - e^{\alpha_2 x_2})$
Von Liebig	$y = \min(\alpha_1 x_1, \alpha_2 x_2)$
Cobb-Douglas	$y = \alpha_0 x_1^{\alpha_1} x_2^{\alpha_2}$
Translog	$\ln y = \alpha_0 + \alpha_1 \ln x_1 + \alpha_2 \ln x_2 + \alpha_3 (\ln x_1)(\ln x_2).$

C. Pope/Just Production Function

The Pope/Just production function generalizes the other functions described by including the possibility that different inputs have different effects on the variability of output. Let the function $g(x_1, \ldots, x_k)$ include only those k inputs that affect yield variability, either positively or negatively. The Pope/Just function is then written as

$$y = f(x_1, \ldots, x_n) + g(x_1, \ldots, x_k)e,$$

where e is the error term. This generalization of the error term can be used to analyze situations where farmers use inputs that are relatively expensive but that reduce risk and reduce use of inputs that increase risk.

D. Estimation of Multicrop Production Functions

The estimation of individual production functions for each crop in a multicrop system is hampered by lack of data. Farmers often keep records of the inputs purchased for the whole farm, but they rarely keep records of which inputs were used for which crops. For instance, a farm might have records on the total purchase of nitrogen fertilizer and on output of soybeans and corn but no records about the amount of fertilizer allocated between soybeans and corn. Without the allocation of the input between the two crops, it is not possible to determine directly the effect of changing fertilizer use on the production of either crop.

One solution is to use an aggregate function, with the underlying assumption that fertilizer is allocated optimally between the two crops. The added information from this behavioral assumption gives the problem enough structure to be solvable.

E. Statistical Techniques and Problems

The most commom technique to determine the parameters, $\alpha_1, \ldots, \alpha_r$, is to find the set of parameters that minimizes the sum over all observations of the error terms squared, ε^2. (The error term is squared to prevent positive and negative errors from canceling each other out without finding a good fit.) This method, called least-squares estimation, is used with an additive error term. Using the least-squares estimation, inputs are assumed not to be related to the error term and parameters that best explain the relationship between the inputs and output are found.

Another class of models are those in which the error term is assumed to be inside of the function, such as $y = F(x + \varepsilon)$. For instance, the von Liebig function with two inputs, when the error term is incorporated, can be written as

$$y = \min(a_1 x_1 + \varepsilon_1, a_2 x_2 + \varepsilon_2).$$

A switching regression can be used to estimate the parameters of this function.

A requirement for consistent estimation of a production function by least squares is that inputs are independent of the error term. For instance, inputs may be related to planting, while the error term might consist largely of weather which occurs after planting has taken place. In this case, the assumption of independence is reasonable. In more managed agricultural systems, however, the input data available often violate this assumption. For instance, the number of pests, which is difficult to count, is an omitted variable and thus part of the error term. If pesticide inputs are applied after the pests are observed, then their use is correlated with pest presence and violates the independence assumption. Similarly, the amount of water applied often correlates with weather (a frequently omitted variable). This violation of assumptions leads to estimators with very poor statistical properties: the estimated parameters do not approach the true parameters even with a large amount of data. Statistical methods for estimating production functions in these circumstances call generally for some measure of the omitted variables—pests or weather in the above examples. An alternative is to estimate the cost function or factor demands, since these forms do not have the same problem.

F. Prediction and Extrapolation

In agricultural work, the major reasons to estimate a production function are to predict yields in new circumstances or to assess the efficacy of changing the input mix. Standard statistical methods would seem to suffice for assessing the effect of a small change in inputs. Typically, investigators try many functional forms and different sets of variables before deciding which combination is "right." As a result, the statistics calculated for a single regression tend to give confidence intervals smaller than the true intervals, because they do not reflect all prior analyses that were rejected. Partly for this reason, predictions from estimated equations tend to be worse than the regression statistics suggest.

G. Estimation of Cost Functions

In managed agricultural systems, where least squares yield inconsistent estimates of a production function, the estimation of conditional factor demands or cost functions provides an alternative. This alternative is also used with data aggregated over many crops or many regions. Because input prices can be assumed not to be correlated with an error term reflecting weather or pest conditions, estimating factor demands avoids the problem of the interrelatedness of inputs and the error term.

A system of factor-demand equations is often estimated together. A typical equation for conditional demands is of the form

$$x_i = X_i(w_1, \ldots, w_n, y).$$

In unconditional factor demands, output price p substitutes for output y. Properties of factor demands, described earlier, are imposed as restrictions on the equations.

H. The Supply Function

To examine the effects of government policies on farm output, economists need to know the relationship among those policies, prices, and farmer behavior—a relationship contained in the supply function. If prices can be considered to be unaffected by the behavior of the farmers whose output is being studied, the supply function can be estimated by least squares. This situation holds if price is not determined with output simultaneously.

Under several circumstances, this assumption is plausible. First, if the supply function is estimated for a small segment of the overall market, such as a small region, farmers can be considered to be price takers. Second, if output price is determined by government policy rather than by market forces, then price truly is determined before output. Third, if it can be assumed that farmers base their planting decisions on the price expected for the crop at the end of the season and if actual price at harvest is related to actual harvest (which is determined by exogenous factors, such as weather and pests), then output is determined before actual price. In this case, price, rather than output, is the dependent variable.

However, if a supply function is estimated for the entire market of an output and if output price and quantity are not set separately, then price is determined by interaction of the supply function with the demand function which reflects consumer desires for the commodity at a range of prices. In this case, information about consumer demand behavior is also necessary to find the supply relationship. Estimation techniques, such as two-stage least squares and instrumental variables, are used to solve this problem.

III. Use of the Basic Theory of Production

The theory of production and the techniques used to develop quantitative estimates of the theoretical relationships are outlined in the previous description. The ultimate reasons for conducting studies of farm production are to improve farm production (the field of management economics) or to examine the effects of government policies (the field of agricultural policy). The use of the theory and quantitative relationships for examining input levels, farm-level resource allocation, regional modeling of agriculture, and agricultural policy analysis is described in the following section. [See GOVERNMENT AGRICULTURAL POLICY, US; QUANTITATIVE METHODS IN AGRICULTURAL ECONOMICS.]

A. Input Intensity

While it is often safe to assume that farmers use levels of inputs that maximize their profits, at times it is useful to examine effects of alternative levels of input use and production technologies, such as irrigation techniques. The choice of functional form for the production function makes a great difference to recommendations for input use. When a polynomial or Cobb-Douglas form is used, recommended input levels will often be less than when a von Liebig form is used. According to empirical evidence, the von Liebig form is closer to being correct for experimental plots, though nonuniformity in fields leads to arguments for the use of other functional forms.

The choice of irrigation technology involves a trade-off between cost and uniformity. Less expensive techniques can lead to parts of the field receiving more water than necessary and other parts of the field receiving far less water. Since excess water often does not adversely affect crop growth, farmers may "over-irrigate" parts of the field in order to get sufficient water to the drier parts of the field. Alternatively, farmers can invest in more expensive technologies that distribute water with great uniformity and reduce water use. A more surprising result is that irrigation equipment can be used to substitute for land! The more expensive technology could be more profitably used by increasing water per acre, which increases yield per acre, and by decreasing the number of acres. [See IRRIGATION ENGINEERING: FARM PRACTICES, METHODS, AND SYSTEMS.]

B. Whole-Farm Models and Planning

Input decisions involve choosing methods to maximize profits for a given crop. In contrast, whole-farm planning involves choosing how many acres to devote to each crop. Depending on the nature of the

research question, researchers use several forms of programming techniques to allocate acreage among crops and practices.

1. Linear Programming

The simplest of these techniques is linear programming which is used to analyze static systems with no uncertainty. The linear-programming analysis begins by determining the fixed inputs available to the farm. Fixed inputs are those that cannot be adjusted quickly or easily, such as the total acreage available, soil types, and major capital equipment. The next step is calculating revenues, purchased inputs, and needs for fixed inputs for each crop on a per-acre basis. Finally, total profits (revenues minus costs of variable inputs per acre) are maximized, subject to the constraints on fixed inputs. If fixed inputs are used similarly for each crop, then only the crop with maximum profits per acre will be grown according to the linear-programming algorithm. More generally, the number of crops that will be grown depends on the number of fixed inputs. For instance, if different land types each grow a different crop more profitably or if both water and land are in fixed supply, then the programming solution will put acreage into more than one crop.

A simple example is the choice of growing corn and growing sorghum, with profits per acre of π_1 and π_2, respectively. With a total of A acres, A_1 is allocated to corn, and A_2 is allocated to sorghum. The linear-programming problem for crop choice and profit maximization is

$$\max_{A_1, A_2} A_1\pi_1 + A_2\pi_2$$

subject to

$$A_1 + A_2 = A.$$

With more complicated production relationships (for instance, with a restricted profit function rather than the simple profit per-acre formulation), nonlinear programming is necessary to solve the problem. In these cases, multiple outputs are more likely than with strictly linear problems.

2. Quadratic Programming

The generalization of the whole-farm planning problem either to a regional model or to a model that accounts for risk involves quadratic programming. Risk is usually incorporated in these models by assuming that a farm operator maximizes profits less a term reflecting the risk aversion of the farmer. In terms of the simple example,

$$\max_{A_1,A_2} A_1\pi_1 + A_2\pi_2 - \lambda \mathrm{Var}(A_1\pi_1 + A_2\pi_2)$$

subject to

$$A_1 + A_2 = A,$$

where λ is a coefficienct reflecting risk aversion and Var refers to variance with

$$\mathrm{Var}(A_1\pi_1 + A_2\pi_2) = A_1^2\mathrm{Var}(\pi_1) + A_2^2\mathrm{Var}(\pi_2) + A_1 A_2 \mathrm{Cov}(\pi_1,\pi_2).$$

This is a quadratic function in the choice variables A_1 and A_2.

3. Dynamic Programming

Many agricultural problems include dynamic aspects. For instance, because soil loss happens over many years, a model to examine its effects must look a number of years into the future. Similarly, pesticide resistance and fertilizer carryover are multiyear problems. Using dynamic-programming methods, time aspects of these issues are incorporated.

In the dynamic-programming framework, farmers are assumed to care about future as well as current profits. At the same time, it is assumed that current profits are more valuable than future profits due to the existence of alternative investment possibilities and uncertainty about the future. In other words, future profits are discounted relative to present profits. The discounted value of future profits, found by dividing future profits by a discount factor related to the interest rate, is termed the present value of future profits. The dynamic-programming problem involves maximizing the sum of present-value profits over time.

Constraints on the maximization problem, such as the effects of practices on soil erosion and, therefore, on soil quality, involve multiple years. These equations are termed "equations of motion" because they describe how variables, such as soil quality or pest resistance, change over time.

Controlling pests over 2 years provides an example of dynamic modeling. Let $s(t)$ be pest population at time t, and let $j(s)$ be the year to year increase in pests. With pest control, v, applied at a cost of w per unit, profits with a given pest level are $\pi_t(p, s) - vw$ in year t. With the discount rate, r, the dynamic problem for a 2-year model is

$$\max_v \pi_1(p_1,s_1) - v_1 w + \pi_2(p_2,s_2)/(1 + r) - v_2 w/(1 + r)$$

subject to

$$s(2) = s(1) + j[s(1) - v_1].$$

Since the function, j, is generally not linear, the prob-lem is not linear in the choice variable, v, and requires nonlinear programming.

When the equation of motion is influenced by a random element, ε, the resulting problem is a stochastic dynamic-programming problem. Since profits are no longer deterministic, expected profits are maximized. In terms of the example,

$$\max_v \mathbf{E}\{\pi_1(p,s) - vw + \pi_2(p,s)/(1 + r)\}$$

subject to

$$s(2) = s(1) + j[s(1) - v_1] + \varepsilon,$$

where $\mathbf{E}$ is the expectations operator. There are computer codes to solve these problems, but computational difficulties increase greatly as the number of equations increases. Thus, the decision to add stochastic and nonlinear elements to a model comes usually only with less thorough modeling of other elements, such as seasonality in labor or number of crops modeled.

C. Regional Models

Supply functions and demand functions (amount of product purchased by consumers as a function of price) can be combined to give equilibrium (quantity supplied equals quantity demanded) models of agriculture on a regional (or even national) basis. These models are used to analyze the effects of such issues as how government policies affect cropping patterns, the optimality of water use rules, and the costs of soil preservation. In California, water is allocated to districts based on political and historical factors. Regional production models are used to determine how much more money could be made in the farm sector if water were allocated to maximize profits. In Iowa, a soil preservation law requires great care in farming very erodible land. A regional model was used to determine the cost of existing regulation and compare it to more and less restrictive regulations. Linear- and quadratic-programming methods are used generally to analyze these types of models.

D. Implications for Agricultural Policy

Policy issues are often centered on farmer response to government-imposed price policies. The aggregate supply curve for a crop can be used to analyze the effects of price changes on production. For instance, if the government lowers the support price for a crop, the supply curve would indicate by how much production would decrease. To determine who gains and who loses from agricultural policy, both aggregate supply and aggregate demand curves are necessary to determine the effects on consumers. In the above case, a drop in price would hurt producers, but consumers would benefit from lower prices. [*See* PRICES.]

A more complicated example is the probable effects of a tax credit for ethanol. Such a credit would increase corn output (requiring an estimate of the supply curve for corn). It would also produce by-products that compete with soybeans, lowering soybean demand. The supply curve for soybeans is also needed to determine the effects on the soybean market. Complicated analyses of this nature are underlain always by the estimated supply relationships of the crops.

IV. Issues in Economics

It is assumed that agriculture is a relatively static industry in the proceeding analysis. In other words, farmers operate in basically the same way year after year and change their behavior only in response to input and output prices. Of course, this assumption is wrong. Farming has changed substantially over time, in response to changes in technology, regulations, and altered market circumstances. These changes are reflected in contemporary work in production economics.

A. Technological Change

Supply curves change over time because of technical change. Technical change involves increases in output above that which can be accounted for solely by changes in input use. Examples include improvements in varieties, invention of the tomato harvester, and better pest management. The challenge is to measure the increases in output (or savings in inputs) caused by new technologies. Three approaches are used: experimental plots, time trends, and proxy variables.

For a discrete event, such as hybrid corn, experimental plots are planted with the old and new varieties. If the new variety has desirable characteristics, such as greater output, pest or drought resistance, etc., then it is a possible improvement. To be a technical change (and a real improvement), it must also increase profits (or decrease risk).

The time-trend approach involves estimating a production function with a trend term—for instance,

$$y = e^{rt} f(x_1, x_2),$$

where t is time and r is the rate of technical change. In this case, the cause of the change is not specified.

The last approach is to use proxies, such as agricultural experiment station expenditures, for new techniques. An example of such a functional form would be

$$y = I^a f(x_1, x_2),$$

where I is the expenditure on research and development. Estimates of productivity that account for research and development show that the returns to research activities are quite high.

Technical progress causes major problems for agricultural price-support activities because it increases output. Price-support authorities find it necessary to purchase ever larger quantities of product, or to restrict output, to continue to support prices. [*See* CROP SUBSIDIES.]

B. Farm Size

Returns to scale have real consequences for farm demographics. Middle-sized farms in the United States are decreasing in number, while larger and smaller farms both seem to be increasing. One of the driving economic forces in this revolution (at least for wheat) is returns to scale. If a farm is operating with increasing returns to scale, it can increase its output more quickly than its costs increase; the farm will expand. If a farm is operating with decreasing returns, it will benefit by reducing its size. The range of output where the farm faces constant returns to scale yields the minimum cost per unit of production and is termed the minimum efficient scale. Many crops, such as wheat, are grown in vast expanses to take advantage of large equipment. The large equipment is less costly per acre if the farm has enough acres to average out the capital costs. [*See* U.S. FARMS: CHANGING SIZE AND STRUCTURE.]

C. Overproduction Trap

Supply curves may not be the same for a price increase and a price decrease because factors of production can easily move into agriculture but not out. Capital (such as tractors), for which farmers paid large sums, cannot be resold at the same price.

For example, the Russian Wheat Deal led to greatly increased wheat prices and the expectation that high prices would continue. In response, farmers purchased considerable quantities of equipment to expand output. When wheat prices subsided, the additional investments could not be converted to other uses and remained in the hands of farmers. The farmers continued to produce high output because they already owned the equipment and had already prepared the land.

This phenomenon is called the overproduction trap. Production economics is used to estimate how fast production will respond to increased prices and to make a separate estimate of how fast production will respond to lowered prices.

Bibliography

Beattie, B. R., and Taylor, C. R. (1985). "The Economics of Production." Wiley, New York.

Chambers, R. G. (1988). "Applied Production Analysis." Cambridge University Press, Cambridge.

Jensen, H. R. (1977). Farm Management and Production Economics 1946–1970. In "A Survey of Agricultural Economics Literature" (L. R. Martin, ed.). Univ. of Minnesota Press, Minneapolis.

Quantitative Methods in Agricultural Economics

H. ALAN LOVE, BRUCE A. MCCARL, *Texas A&M University*

I. Econometric Estimation
II. Operations Research Techniques
III. Usage of Quantitative Techniques

Glossary

Econometrics Field wherein a set of statistical techniques is used to analyze empirical data; generally, econometric analysis aims at quantifying economic theories, testing hypotheses, forecasting future values of economic variables, and quantifying the effects of alternative government, firm, or individual decisions on the economy

Hypothesis test Used to determine whether a specific conjecture is supported by a particular data set; hypothesis tests are formally stated as a null hypothesis that the specific conjecture is true and an alternative hypothesis that the conjecture is not true; a test statistic is constructed from observed data assuming that the null statement is true and the hypothesis is rejected if the calculated test statistic exceeds a critical level for its known distribution

Linear programming Special case of mathematical programming where the objective function is linear as are the restrictions on the decision variables

Mathematical programming Class of problem solving where one is trying to find the appropriate values of a set of decision variables so that an objective is optimized while obeying a set of restrictions on the allowable decision variable values

Operations research Application of mathematics and the scientific method to decision making with the intent of improving the design and operation of the system studied

Regression analysis Statistical tool for quantifying the relationship of one or more independent explanatory variables to a continuous dependent variable; properly implemented, regression analysis can be used to determine a probabilistic statement of the extent to which changes in independent variables are related to changes in the dependent variable; the most commonly used regression technique is ordinary least squares

Simulation Problem analysis method where one develops a computerized model of a system then uses that model to depict the reaction of the system to different operating decisions and/or levels of external stimuli

Simultaneous equations Econometrically estimated multiequation systems involving interdependent variables; in simultaneous equation models, several dependent, or endogenous, variables are jointly determined by a set of explanatory, or exogenous, variables and error terms; each equation may have one or more endogenous variables and one or more exogenous variables

Generally, agricultural economists use quantitative methods for three purposes. First quantitative methods can be used in conjunction with observable data to discover how individuals behave when confronted with various economic stimuli and in cases to test the adequacy of economic theory. Second, quantitative techniques are used to forecast economic trends and the consequences of alternative decisions. Third, quantitative techniques are used to resolve choices among decision alternatives. The methods employed for satisfying the above purposes can be broadly classified into estimation-based, econometric, and operations research (OR)-based techniques. The OR techniques applied are predominantly those based on mathematical programming and simulation.

In this article we provide a brief overview of the most commonly used quantitative techniques. This article is organized into three sections, two sections concentrating on econometric and OR-based techniques and then a third dealing briefly with the way these techniques have been used to address the classes of problems mentioned above. Readers interested in more detail and a guide to the further literature should consult the quantitative methods volume of the Survey of Agricultural Economics Literature.

I. Econometric Estimation

Econometrics is concerned with describing and evaluating the relationship between a target variable, called the dependent variable, and one or more explanatory variables, often called independent variables or regressors. Generally a causal relationship is presumed from the explanatory variables to the dependent variable and an equation is estimated in which the dependent variable is expressed as a function of the independent variables. However, many analyses are concerned with systems of simultaneous equations where independent variables influence a set of dependent variables which also influence each other. In either case, economic theory and empirical observation guide the selection of variables to include in any particular equation.

The basic ingredients in an econometric model are an assumed functional relationship between the dependent variable and the independent variables and an error term which allows deviations between predicted values of the dependent variable and its observed values. A widely used econometric model incorporates a linear functional relationship between the dependent variable and independent variables, along with an additive error term. This model is given as

$$y_i = \beta_0 + \beta_1 x_{1i} + \beta_2 x_{2i} + \beta_3 x_{3i} \\ + \cdots + \beta_K x_{Ki} + e_i, \qquad (1)$$

where y_i is the ith observation of the dependent variable, $x_{1i}-x_{Ki}$ are the ith observations on the independent variables, $\beta_0-\beta_K$ are unknown parameters, and e_i is the ith observation on the stochastic error term or disturbance. The regression problem generally involves finding the β's so that the error terms across the observations are minimized. Alternative functional forms can be used to express the relationship between the dependent variable and the independent variables and, generally, the error term need not be additive.

A model specification is not complete without specifying the characteristics of the error term. There are two main sources of randomness incorporated in the error term: unpredictable randomness in human behavior and the effects of a large number of unimportant omitted variables. It is commonly assumed that the error term has a mean value of zero and an equal finite variance for all observations.

Given data for the dependent and independent variables, the objective of econometrics is to obtain estimates for unknown parameters $\beta_0-\beta_K$. To achieve this goal, certain prerequisite conditions (assumptions) must be met. These assumptions vary depending on the estimator selected, but the following are commonly used.

1. The expected value (mean) of e_i is zero (i.e., $E(e_i) = 0$, for all i, where i indexes the number of observations in the data set).
2. The variance of e_i is equal for each observation (i.e., $V(e_i) = \sigma^2$, for all i).
3. The explanatory variables x_1-x_K are nonstochastic.
4. Each explanatory variable contributes a unique influence on the dependent variable y, considering the whole set of x variables.
5. The model is correctly specified.

The first assumption is not restrictive. The second assumption is some times referred to as homoskedastic errors. Homoskedasticity can be violated in two distinct ways. First, the variances can be correlated across observations such that the probability of a positive deviation in one observation influences the deviation in adjacent observations. When the data are for sequential time periods (called time series data), this is referred to as serial correlation or autocorrelation. Second, the variances can differ by observation depending on the values of some subset of the explanatory variables. This condition is termed heteroskedasticity and is most common in cross-section data. The third assumption is sometimes referred to as fixed regressors. It implies that the explanatory variables come from an experimental process where the investigator sets the level of the independent variables and then observes the outcomes of the dependant variables; a rare occurrence in economic modeling. In fact, the explanatory variables can be stochastic so long as the distribution of the error term does not depend on any explanatory variable. The fourth assumption implies that the data are not perfectly multicollinear; i.e., there is not an exact linear relationship among any subset of the regressors. The fifth assumption implies that the correct functional form is truly linear, that the model contains all relevant variables, but does not include any irrelevant variables, and that the model has an additive error.

Although it is not necessary, another commonly used assumption is that the error term is normally distributed. If sample size is large, normality of the error term is assured through the central limit theorem. In small samples, statistical inference (hypothesis testing) is based on assumed normality of the error term.

Each of these assumptions can be relaxed to obtain estimates of the unknown parameters β_0–β_K, and variance of the error term, σ^2. In addition, the validity of each assumption can be determined using appropriate hypothesis tests.

A. Empirical Estimation

Econometric estimation uses available data plus any additional information about the probability distribution of the error term, to provide estimates of the unknown model parameters which form the predictive equation. The true, but unknown, parameter values are called population parameters. Parameter values obtained from data are called estimates. The criteria for obtaining parameter estimates from data are called the estimation method or simply an estimator.

Since estimation methods employ functions involving data that include the dependent variable, which by assumption contains error, the resultant parameter estimates are random variables with both a mean and a variance. Only by coincidence will an estimated parameter estimate equal its "true" population value, but as random variables, parameter estimates have sampling properties. Evaluating estimator's distributional properties amounts to characterizing their sampling distributions.

1. Desirable Estimator Properties

Unbiasedness means that if a parameter is estimated repeatedly from different data samples, the average value of the estimated parameter will equal the underlying population parameter. An estimator is efficient if it is unbiased and has no greater variance than any other alternative unbiased estimator. Unbiasedness and efficiency are small sample properties that hold for all sample sizes. But, if any assumption discussed in the data generation section is violated, unbiased and, hence, efficient estimators are not available. In such cases it is customary to apply estimators with desirable large sample properties. Consistency ensures that, as sample size expands, the probability that an estimator equals its true population parameter value approaches one. Sufficient conditions for consistency are that bias and variance both tend toward zero as sample size expands. Asymptotic efficiency is

the large sample analogue of finite sampling efficiency. An estimator is asymptotically efficient relative to any other if it is a consistent estimator and the variance of its limiting distribution is smaller than that of any other consistent estimator.

2. Some Commonly Used Estimators

Estimators can be grouped into one of three categories: minimum distance estimators (including the most common technique, ordinary least-squares regression), maximum likelihood estimators, and method of moments estimators. Method of moments estimators simply replace population parameters with their sample counterparts. For example, the sample mean can be substituted for the population mean. Under assumptions 1–5 above, method of moments estimators are unbiased and efficient. However, method of moments estimators need not be unique and they depend on the random variable in question having finite moments. In practice, method of moments estimators are rarely used.

The commonly used ordinary least-squares (OLS) criterion is a minimum distance criterion that produces parameter estimates which minimize squared prediction errors. Given a data sample of T observations, the least squares estimator obtains values for β_0–β_K that minimize

$$S = \sum_i \hat{e}_i^2 = \sum_i (Y_i - \hat{Y}_i)^2,$$

where

$$\hat{Y}_i = \hat{\beta}_0 - \hat{\beta}_1 x_{1i} - \hat{\beta} x_{2i} - \cdots - \hat{\beta}_K X_{Ki}$$

and is the forecasted value. OLS offers two major advantages over alternative estimators. First, it does not require specifying the error term's underlying distribution. Second, if assumptions 1-5 are met, OLS parameter estimates are unbiased and efficient.

Maximum likelihood estimation yields estimates of the unknown parameters that have the largest probability of giving rise to the sample that was actually selected. Maximum likelihood estimation requires assumption of a particular probability distribution function for the error term and computes the probability of the sample occurrence. The resultant equation is the probability function of the dependent variable outcome as a function of the values of the unknown parameters. Maximum likelihood estimates are those resulting from maximization of this function with respect to parameters to be estimated. When the error term is normally distributed, the maximum likelihood estimator is the same as the method of moments and ordinary least-squares estimators just discussed.

Under fairly general conditions, maximum likelihood estimators are consistent and asymptotically efficient.

3. Statistical Inference

Many times the investigator is only interested in estimating parameters for forecasting or simulation. In these cases estimates can be useful in their own right. Other times, the purpose of analysis is to test whether a particular independent variable is a significant factor in the equation for the dependent variable. In addition, more advanced hypothesis tests are allowed that answer the question of whether the observed difference between a function of the estimated parameters and some hypothesized value is real or is due to chance variation. A test statistic is constructed from estimated parameters and their sampling distribution. Two simple test statistics appropriate in small samples when the error term is normally distributed are the t test and F test. The t test is constructed as the ratio of the estimated parameter to its standard error and is used to test hypotheses concerning a single parameter; (e.g., whether β statistically differs from zero). The F test is constructed as the adjusted percent loss in sum-of-squared predicted errors from imposing a hypothesized linear restriction on a group of parameters.

Three more general alternative hypothesis testing procedures are available: Wald, likelihood ratio, and Lagrange multiplier (score) tests. The Wald test can be implemented with any estimation procedure, but the likelihood ratio and score tests require maximum likelihood parameter estimation. The Wald test requires estimating only the unrestricted model and specifying the hypothesized restriction. This can be advantageous when the restricted model is difficult or impossible to compute. The likelihood ratio test requires estimating both the restricted and unrestricted model parameters. The likelihood ratio test statistic is computed as the ratio of the likelihood function evaluated at the restricted parameter estimates and the likelihood function evaluated at the unrestricted parameter estimates and hence is easy to implement when both the restricted and unrestricted models can be estimated. The score test requires estimating only the restricted model, but can be somewhat difficult to compute without specialized computer software. Under the null hypothesis all three test statistics are distributed chi-squared with degrees of freedom equal to the number of parameter restrictions imposed. The null hypothesis is rejected when calculated test statistic values exceed tabled critical values.

B. Violation of Model Assumptions

The assumptions stated above may be violated in a set of data. In this section each model assumption will be considered from three perspectives: (1) How can the validity of each assumption be checked? (2) How are the parameter estimate sampling properties altered if a particular assumption is violated? (3) What alternative estimators can be used to obtain desirable parameter estimates when standard techniques are inappropriate?

1. Homoskedastic Errors

So far we have assumed that the error terms are homoskedastic; the error terms have common variance at each observation. One possible violation of homoskedasticity is heteroskedasticity. Heteroskedastic errors have different variances and generally occur in cross-sectional data. An often-cited example of heteroskedasticity comes from models of household saving. In any particular time period, households can use earnings for consumption or savings. Suppose savings are modeled as a function of income. Low-income households must spend nearly all income on consumption items, but high-income households can consume or save. As a result, high-income households exhibit greater variance in savings than do low-income households. The error term in predicting savings is therefore generally found to be correlated with income. Under heteroskedastic errors, the OLS estimator is still unbiased, but it is not efficient. More importantly, estimated parameter variances from OLS are biased, invalidating significance tests and hypothesis testing.

A number of tests are available for detecting heteroskedasticity. Tests involve examining whether the error terms from the regression are significantly affected by the value of the decision variables either through a second regression of the error term on a function of the independent variables or by examining whether groupings of the error terms for different ranges of the dependent variables are statistically different.

Estimation methods may be altered to correct for heteroskedasticity. Generalized least-squares estimators (GLS) use a three-step estimation process to obtain efficient parameter estimates. First-round OLS estimation is used to compute error terms at each observation and then these errors are squared and regressed on the independent variables. The dependent variable and independent variables are then divided by the estimated variance of each observation and OLS estimation is reapplied to the transformed data. GLS estimators are biased in small samples but they are consistent and asymptotically efficient in large samples.

Autocorrelation of residuals is another phenomenon that causes failure of the homoskedasicity assumption. Autocorrelation occurs when there is correlation between error terms, i.e., when error term e_i is correlated with error terms e_{i+1}, e_{i+2}, . . . and e_{i-1}, e_{i-2}, . . . and so on. One explanation for autocorrelation is that factors omitted from the model are correlated across time periods. Autocorrelation results in unbiased but inefficient parameter estimates and makes hypothesis testing inappropriate. The most widely used test for autocorrelation is the Durbin–Watson test. It is based on OLS residuals and is appropriate for testing first-order autocorrelation—correlation between e_i and e_{i-1}—when the model does not contain lagged dependent variables. Similar tests are available for testing higher-order autocorrelation and for testing autocorrelation in the presence of lagged dependent variables.

Maximum likelihood and generalized least-squares estimators are available that correct for autocorrelation. The GLS estimator involves transforming the data to correct for autocorrelation and then obtaining parameter estimates using OLS. The data transformation requires estimating the correlation between error terms from OLS residuals, transforming the data to eliminate correlation between errors, and reestimating the transformed model using OLS.

2. Stochastic Regressors

Until now, we have assumed that each model contains a single endogenous or dependent variable; all other variables appearing in the model are independent or exogenous. The crucial assumption is that each explanatory variable is independent of the error term. If this condition is not satisfied, the OLS estimator is both biased and inconsistent.

The fixed regressor assumption can be violated in two ways. First, an explanatory variable can be measured with error that is correlated with the disturbance term. Second, the economic system being examined may involve simultaneously determined "endogenous" variables appearing as independent variables. Such variables are necessarily correlated with disturbance terms appearing in each equation. The classic example in agricultural economics models is a supply and demand model where price, quantity supplied, and quantity demanded are simultaneously determined. Consider a simple demand and supply model,

$$q_d = a_0 + a_1 p + a_2 I + e_d, \quad \text{demand function}$$

$$q_s = b_0 + b_1 p + b_2 C + e_s, \quad \text{supply function}$$

and

$$q_d = q_s, \quad \text{market identity,}$$

where p is market price, q_s is quantity supplied, q_d is quantity demanded, I is income, C is production cost, a_0–a_2 and b_0–b_2 are unknown parameters, and e_s and e_d are stochastic error terms. Here p, q_s, and q_d are endogenous variables while I and C are the exogenous variables. Equilibrium price, quantity supplied, and quantity demanded can be obtained as the solution to the system of equations. A change in production cost, C, shifts the supply function causing a change in equilibrium price and quantities. Similarly, realization of an external shock through either error term reveals itself through a change in equilibrium quantities and price. As a result, price and quantities are stochastic variables that are correlated with the error terms in each equation. Both supply and demand are functions of price, a stochastic variable correlated with the error terms; hence, OLS is an inappropriate estimator for the unknown parameters. Hausman's specification test is a very general specification test that can be used for detecting stochastic variables.

Before estimation can be considered, the fundamental question of whether the unknown parameters of interest in a simultaneous equations model are estimatable must be addressed. The question that arises is: Given information about the underlying model structure, is more than one theory, or set of equations, consistent with the same data? If so, the two models are observationally equivalent and model parameters cannot be estimated. This issue is referred to as the econometric identification problem. For linear models, a necessary, but not sufficient, condition for parameter identification in any equation is that the total number of exogenous variables excluded from the equation must be at least as great as the number of included endogenous variables, less one. This result is known as the order condition of identification. A necessary and sufficient condition for parameter identification is given by the rank condition. In simultaneous equations model, each equation must be checked for econometric identification before any estimation technique is applied.

A general method for obtaining consistent parameter estimates when stochastic regressors are a problem is the instrumental variable method. An instrumental variable (IV) is one that is uncorrelated with the error term but is correlated with the explanatory variables in the equation. In the simultaneous equations model above, income would be a good instrument for price in the supply equation and production cost would be a good instrument for price in the demand equation. One special instrumental variable estimator is two-stage least squares (2SLS). 2SLS estimation is achieved by applying OLS in two stages. First, OLS is used to obtain predicted values for each endogenous variable using all exogenous variables in the system as explanatory variables. Second, endogenous explanatory values are replaced with their predicted values and OLS is used to obtain parameter estimates. Two-stage least squares is a biased but consistent estimator; it is not efficient. Asymptotically efficient estimation must account for potential correlation among error terms across equations at the same observation. Systems estimators account for cross-equation correlation of error terms and are asymptotically efficient. Three-stage least squares is an IV type systems estimator. Maximum likelihood estimators are also available for simultaneous equations systems.

3. Multicollinearity

Multicollinearity is the term used to describe the problem where the independent variables are linearly dependent. Such a case causes one to be unable to isolate the effects of those variables and leads to inflated variance in the estimated parameters. The problem is most extreme when each explanatory variable does not exert an independent effect on the dependent variable leading to a situation of perfect multicollinearity where model parameters cannot be estimated. Multicollinearity appears most often in time series data; data that consist of a series of observations on the variables over a number of time periods. Tests and corrective procedures for multicollinearity are available, but generally result in biased parameter estimates.

4. Functional Form

Obviously, the adequacy of an estimated equation is greatly influenced by choice of functional form. There are three important considerations in making such a choice. These are: (a) avoiding excluding explanatory variables that should have been included; (b) inclusion of superfluous explanatory variables; and (c) the exact nature of functional form assumed. Exclusion of relevant explanatory variables results in biased and inconsistent parameter estimates and renders hypothesis tests inappropriate. Including extraneous explanatory variables results in loss of efficiency but does not induce bias in parameter estimates. Inap-

propriate functional form results in biased and inconsistent parameter estimates. Various hypothesis tests are available for detecting whether an explanatory variable should be included in a model and the appropriateness of the assumed functional form.

C. Special Situations

Certain other special situations underlie the use of other estimation methods. Here the use of limited dependent variable and time series approaches is overviewed.

1. Limited Dependent Variable Models

There are many situations where the dependent variable of interest involves a yes or no, 1 or 0, outcome. Such models are often called limited dependent variable or qualitative choice models. For instance, in analyzing farmer participation in government programs we might have data on whether any particular farmer participated, along with explanatory variables for expected returns from participating and expected returns from not participating. Estimation of equations for such types of dependent variables is done using limited dependent variable estimation methods. In such a case, the outcome of the discrete choice is viewed as a reflection of an underlying continuous valued index function

$$y_i^\star = \beta_0 + \beta_1 X_{li} + \cdots \beta_{1k} X_{ki} + e_i,$$

where $y_i^\star$ is an unobservable index representing farmer i's desire to participate in the government program. The estimation technique is based on an assumed distribution and an approach that maximizes the probability of observing the decisions found in the sample. The unknown parameters are typically estimated using either the probit model assuming the errors are normally distributed or the logit model when the errors are assumed to be logisticly distributed. Both models are widely used and in practice and there is little difference in the estimation results.

The regression coefficients from either model cannot be directly interpreted as the marginal effect of the explanatory variable on the dependent variables as is generally the case in linear models. In general the marginal effect of any explanatory variable on the probability of a 1 decision outcome of the dependent variable is

$$\frac{\partial E[y]}{\partial x_i} = f(\beta_0 + \beta_1 X_{li} + \cdots + \beta_k X_{ki}) \cdot \beta_i,$$

Where $f(\cdot)$ if the probability density function of the error term.

2. Time Series Models

The econometric models we have considered so far are known as structural models. Structural models are based on economic theory and are developed to specifically incorporate the underlying theoretical relationship between variables in the model. In contrast, time series models do not presuppose an underlying economic structure and as a result are often called atheoretical to emphasize their loose theoretical basis. The major focus of time series analysis is forecasting. In some situations time series models have been found to produce more accurate forecasts than structural models, particularly in short-term forecasts of a period or two ahead. The basic approach in time series analysis is to predict future values of variables of concern based on recently observed past values. The most commonly used method for estimating time series models was developed by Box and Jenkins. Their approach involves: (1) transforming the data by subtracting period-to-period values until the data are independent of time, (2) identifying appropriate lags to include in the model, and (3) checking the ability of the model to predict within the data sample period. Special techniques are available for predicting future values from Box–Jenkins models.

II. Operations Research Techniques

The field of operations research is a broad one. Virtually all of the techniques in the field have been employed in agricultural economics applications. However, the most common applications involve mathematical programming and simulation.

A. Mathematical Programming

Mathematical programming problems are used to set up the problem of finding the best objective-motivated decision strategy from a set of allowable decisions constrained by scarce resources, minimum requirements, and other considerations. The mathematical representation of this problem is

$$\text{Max} \quad f(x)$$
$$g_1(x) \leq b_1$$
$$g_2(x) \geq b_2$$
$$g_3(x) = b_3$$
$$x \geq 0,$$

Where $f(x)$ is an expression of the objective to be maximized when choosing the decision strategy x.

The set of possible decision x includes all nonnegative values for a vector of possibilities which must be chosen so as to satisfy the constraints expressed in the equations containing the $g_i(x)$ functions. Note that each of the $g_i(x)$ and b_i relations are vectors so that multiple restrictions are defined by each constraint equation. The solution to such a model contains information about the best or optimum values of the decision variables (x) as well as information on whether constraints are restrictive. Duality information is also provided which tells how the objective function would change with changes in the constraint constants (b_i).

Commonly such models depict the maximization of such things as firm profits subject to resource availability limits. However many applications involve objectives which are designed to simulate economic behavior (i.e., maximizing agricultural sector consumers'/producers' surplus yields a solution simulating a perfectly competitive agricultural sector).

1. The Linear Programming Problem

Linear programming (LP) is a special case of the above problem where all the functions are linear. In the LP context, the model becomes

$$\text{Max} \quad CX$$
$$A_1X \leq b_1$$
$$A_2X \geq b_2$$
$$A_3X = b_3$$
$$X \geq 0$$

Here X is a vector of decision variables while C gives the contributions of each unit of the decision variables to the objective function, and A_1 gives the usage of each constraining resource item by each decision variable while A_2 and A_3 are defined analogously. There are two types of assumptions underlying the use of such a formulation. The first set of assumptions are mathematical and indicate that: (a) returns and constraint effects are proportional to the level of the variables; (b) there are no interaction terms between the variables; (c) the X values are continuous with fractional parts allowed; and (d) all parameters are known with certainty.

The second type of assumptions involve the model's ability to portray the actual decision problem. These are that the objective function, variables, and constraints are appropriate.

The linear programming model can be applied to a wide range of problems. Consider two examples. First, suppose one is trying to decide on a least cost livestock diet while satisfying the animals nutrient needs. An LP formulation of that decision problem involves minimization of feed cost (the objective function) while the variables depict amount of ingredients, such as corn and soybeans, purchased. Simultaneously, the constraints impose dietary nutrient (i.e., calorie and protein) requirements that must be met. Second, suppose one is trying to predict how changes in farm program price supports might alter production on farms. In that case, a family of LP formulations might be assembled depicting a number of different farms each containing an objective such as farm profit maximization with variables representing acreage by crops and constraints on farm resources like land, labor, and water. The models then could be used to examine how the acreage mix across the farms varies with policy changes.

Much of the mathematical programming literature reflects attempts to appropriately formulate linear programs for different situations and ways of relaxing the assumptions to investigate the effects of the various model assumptions. The treatment below elaborates on assumption relaxation.

a. Proportionality The assumption in linear programming that the objective function value and the resource usage be strictly proportional to x is commonly relaxed both within and outside of linear programming. The techniques within linear programming deal with cases where there are diminishing returns to scale and basically involve specifying multiple activities which represent the effects of the nonproportional phenomena over different ranges of activity. Thus, one might specify that the return of selling the first two units of any output of the firm is $2.00 while the return of selling an additional two units is $1.50, etc. This involves incorporating additional variables and constraints with resulting discrete jumps in returns and costs. It is also possible to model a continuous change in returns by using more general nonlinear programming techniques. The most common such technique is called quadratic programming (QP). This technique only relaxes objective function proportionality. The QP objective function is modified to look like

$$\text{Max } cx + x'Qx,$$

where $x'Qx$ reflects interactions between variables and negative entries on the diagonal of Q reflect declining returns to increases in x.

Proportionality can also be relaxed in the constraints by using multiple variables or through the use of general nonlinear programming techniques.

b. Additivity The additivity assumption prohibits interactions between variables. Thus, the quantity of one commodity cannot affect the marginal returns or resource use of another commodity. This assumption can be relaxed within linear programming by specifying multiple activities, but its formal relaxation requires nonlinear programming techniques. Quadratic programming formulations are commonly used to allow objective function interactions. Namely, when the off-diagonal elements in the Q matrix are nonzero then there are interactions between variables. Relaxationing of the additivity assumption within the constraints requires the use of general nonlinear programming where, for example, production could be modeled as a nonlinear function of multiple inputs given that the inputs individually and collectively reflect diminishing returns to scale.

c. Certainty The certainty assumption requires that the modeler assert that each and every parameter in the model is known without question. This is quite often unrealistic in that parameters in agriculture are often subject to weather variations and other random events. The most common way of relaxing the assumption involves the parameters in the objective function. Concepts from financial portfolio theory have been adapted where decision makers are assumed willing to reduce average return to achieve a reduction in the variance of return. This is usually modeled within the quadratic programming model where the constant terms reflect average returns to x and the Q matrix reflects the degree risk times the variance covariance matrix of returns.

There are additional models handling other forms of uncertainty. Typically, the models depict reactions to uncertainty such that a conservative strategy is followed (i.e., one might maximize variability discounted returns or be limited by variability discounted while assuming variability inflated resource availabilities along with inflated resource usages). Uncertainty models also reflect varying assumptions about how uncertain events arise. The key differentiating factor involves the interrelation of decision timing and the uncertainty resolution. Nonsequential models assume decision makers commit themselves at the beginning of the year and then discover the uncertain outcomes at the end of the year. Sequential formulations (commonly called stochastic programming with recourse) model a interrelated stream of decisions and uncertainty resolution. For example, when modeling an agriculture operation in a sequential setting, planting decisions are made under planting and harvesting season uncertainty but harvesting decisions are made with knowledge of what happened during planting season.

d. Continuity   The continuity assumption allows any variable to take on any value including fractional values. However, certain variables may only make sense if they take on discrete values. This is commonly the case when one is modeling capital investment where equipment must be purchased in whole units. If this is the case, then integer programming must be used. Integer programming restricts the values of certain of the decision variables to be integer valued prohibiting fractional parts.

e. Appropriate Objective Function   The final commonly relaxed assumption involves the choice of a single appropriate objective function. This assumption is relaxed by using multiobjective programming techniques which define a set of several simultaneous objectives. When using this technique one simultaneously models all the objectives and establishes goal preferences under a assumed utility function structure. The portfolio model discussed above is the most common example of this approach where returns and variance of returns are the two objectives.

2. Computational Mathematical Programming

Mathematical programming is fundamentally an empirical tool and its use requires reliance on computers. Some problems require conventional linear programming solution algorithms while others require adoption of specialized nonlinear, quadratic, or integer programming techniques. Today, solution of such models is approached in one of three fashions. First, one can use traditional approaches where the model is converted into solver format by hand or with a custom computer code and the problem is submitted typically in MPS format to the solver. Second, one may use algebraic modeling systems which do data calculations, formulate the model, interface with a variety of solvers, and can do report writing. Third, one can develop very specialized custom software to allow direct use by decision makers such as agricultural producers.

3. Other Programming Methods

The above review, by its nature, is an overview and does not cover many possible programming techniques. Other techniques include:

a. Analytical programming methods wherein solutions are characterized in an abstract fashion to gain theoretical insight into problem solution properties. In such a setting the theories involving Quality, Lagrangian multipliers, Kuhn Tucker conditions, and Optimal Control are relevant.

b. Dynamic programming and optimal control theory methods which are concerned with optimizing multiperiod problems.

C. Simulation

Another commonly used operations research technique is computer simulation. However, this technique, unlike the others discussed above, does not have a fixed structure. Simulation models may use ad hoc functions, theoretically derived functions, and/or econometrically derived functions. The models may incorporate numerically based or mathematical programming-based techniques. Further the model may simulate certain or uncertain systems. A rigorous discussion of simulation techniques is not possible in this short space; rather the reader is referred to the Survey of Agricultural Economics Literature.

III. Usage of Quantitative Techniques

Quantitative methods are applied in agricultural economics as a means of achieving an end. As such, we close with a discussion of how the methods reviewed are used in the three problem settings mentioned at the beginning of the paper. [*See* MACROECONOMICS OF WORLD AGRICULTURE; PRODUCTION ECONOMICS.]

First, let us consider the use of quantitative methods to discover information about the influence of certain factors. In such cases, econometric techniques are commonly used to fit functions based on observable data which allow one to determine if particular independent variables significantly influence our ability to predict the values of individual behavior and whether the form of these responses conform with economic theory. However it is quite common today to express the econometric estimation problem as a mathematical program.

The second problem posed above involved forecasting economic trends and the consequences of decisions. Econometric techniques are most amenable to forecasting future values of various economic items of interest. For example, constructing a forecast of what the inflation rate will be in the next few months. However, as the time frame gets longer and more variables become involved in the process one needs to go toward econometrically based simulation. Econometrically based forecasts are most appropriate where the forecasted behavior is felt to be reasonably extrapolated from the past observations used in forming the econometric model. Forecasts from programming and noneconometric-based simulations are relevant when the extrapolation of past behavior is not felt to be in order. However, the use of such techniques requires adoption of a considerably larger set of assumptions. Namely, under the econometric technique one must choose independent variables and functional form whereas under the mathematical programming technique all the data must be specified as well as the objectives pursued, the functional forms, the manipulatable variables, and the constraints. Nevertheless, if one assumes that an OR model adequately represents the decision situation, then one can use it to predict the consequences of things that alter model data. This includes forecasts of the consequences in technology, alterations in farm policy, and new equipment investments.

The traditional role of mathematical programming models is widely felt to be in the arena where one is trying to resolve choices among decision alternatives. However this is not the most common use of programming techniques in agricultural economics. Usually models are not adequate enough to directly resolve choices and are used rather as an aid to making decisions and a forecasting tool. Nevertheless when programming models are to be used in such a setting they are commonly supported by econometrically or simulation-derived estimates of parameters.

Bibliography

Boisvert, R. N., and McCarl, B. A. (1990). "Agricultural Risk Modeling Using Mathematical Programming," pp. 1–103. Regional Research Bulletin No. 356, Southern Cooperative Series.

Brooke, A., Kendrick D., and Meeraus A. (1988). "GAMS: A User's Guide." The Scientific Press, San Francisco, CA.

Greene, W. H. (1990). "Econometric Analysis." Macmillan, New York.

Hazell, P. B. R., and Norton, R. D. (1986). "Mathematical Programming for Economic Analysis in Agriculture." Macmillan, New York.

Hillier, F. S., and Lieberman, G. J. (1974). "Operations Research,"2nd Ed. Holden-Day, Inc. San Francisco, CA.

Judge, G. G., Griffiths, W. E., Hill, R. C., Lütkepohl, H., and Lee, T. C. (1985). "The Theory and Practice of Econometrics," 2nd Ed. Wiley, New York.

Maddala, G. S. (1992). "Introduction to Econometrics," 2nd Ed. Macmillan, New York.

Martin, L. R., ed. (1977). "A Survey of Agricultural Economics Literature." Vol. 2. "Quantitative Methods in Agricultural Economics, 1940s to 1970s." (G. G. Judy, R. H. Day, S. R. Johnson, G. C. Rausser, and L. R. Martin, eds.) Univ. Minnesota Press, Minneapolis.

McCarl, B. A., Candler, W. V., Doster, D. H., and Robbins, P. (1977). "Experiences with Farmer Oriented Linear Programming for Crops Planning." *Can. J. Agric. Econ.*. **25**(1),17–30.

Rabbit Production

JAMES I. MCNITT, *Southern University and A&M College*

Glossary

Buck Mature male rabbit
Buckling Immature male rabbit
Doe Mature female rabbit
Doeling Immature female rabbit
Heritability Proportion of the expression of a trait that is a result of genetics
Kindle To give birth
Kit (kitling) Unweaned rabbit of either sex
Senior weight Weight of a rabbit 6 months of age or over in the small breeds and 8 months or over in the large breeds
Stereotypical behavior Behavior which is repetitively performed with little or no variation

Rabbits are kept as pets, for show purposes, and for various commercial products including meat, furs, wool, and laboratory stock. The most common commercial use is for meat. Rabbits are sold as fryers at 8 to 10 weeks of age and as roasters or stewers at later ages. Rabbits of the Rex breed have guard hairs and underfur of the same length which makes the pelts suitable for making garments. Angora rabbits produce high-quality wool. This is harvested by shearing or by plucking and spun into yarn. Rabbits reared for sale to research laboratories may be standard rabbits or may be specially produced and guaranteed to be free of specific pathogens. This is a highly specialized production system which requires skill and business acumen from the producer. Some breeders also sell breeding stock to other commercial producers or, if they are heavily involved in showing rabbits, to other show people. By-products from rabbits include the manure for use as fertilizer and fishworms which grow well in rabbit manure.

I. Breeds and Types of Rabbits

The domestic rabbit *Oryctolagus cuniculus* is classified in the order Lagomorpha in the family Leporidae which includes the rabbits and hares. Lagomorpha differ from the Rodentia because they have three pairs of chisel-like incisors. The genus *Oryctolagus* includes the European wild rabbit and its domesticated descendants. The genus *Sylvilagus* includes the Swamp, Eastern, Desert, Bush, and Marsh cottontail rabbits of North America. Hares are in the genus *Lepus* and are distinct from rabbits because they are born in poorly defined nests above ground, are fully haired, and can run shortly after birth.

The American Rabbit Breeders Association, Inc. (ARBA), recognizes 45 breeds of domestic rabbits. These are divided into groups based on color pattern and, within groups, into varieties depending on color. Some breeds may have only one color whereas the Netherland Dwarf has 23 recognized varieties. The breeds can be separated into two classes on the basis of size with the small breeds having an ideal senior weight under 4 kg and the large breeds having an ideal senior weight over 4 kg. The breeds within each of these classes are listed below.

Under 4 kg	Over 4 kg
(Small breeds)	(Large breeds)
American Fuzzy Lop	American
Angora (English)	American Sable

Angora (French)
Angora (Satin)
Belgian Hare
Brittania Petite
Chinchilla (Standard)
Dutch
Dwarf Hotot
English Spot
Florida White
Harlequin
Havana
Himalayan
Jersey Woolly
Lilac
Lop (Holland)
Lop (Mini)
Mini Rex
Netherland Dwarf
Polish
Rex
Rhinelander
Silver
Silver Marten
Tan

Angora (Giant)
Beveren
Californian
Champagne D'Argent
Checkered Giant
Chinchilla (American)
Chinchilla (Giant)
Cinnamon
Creme D'Argent
Flemish Giant
Hotot
Lop (English)
Lop (French)
New Zealand
Palomino
Satin
Silver Fox

Rabbits are also informally spoken of as small (under 4 kg), medium (4–6 kg), or large (over 6 kg) breeds.

II. Rabbit Physiology and Reproduction

A. Physiological Parameters

A number of the general physiological parameters of the domestic rabbit are shown below. More detailed values may be obtained from the literature. These "normal" values may be affected by the diet, breed, environment, and age of the rabbit.

Life span	5–13 years
Birth weight	20–70 g
Mature weight	1–8 kg
Rectal temperature	38.5–40.0°C
Heart rate	120–330 beats/min
Respiration rate	32–60 breaths/min
Blood volume	55.6–57.3 ml/kg
Food intake	30–50 g/kg/day
Water intake	60–100 g/kg/day
Urine production per day	20–350 ml/kg body wt
Feces production per day	30–45 g/day (80% hard feces, 20% soft)

Dental formula[1] I $\frac{2}{2}$, C $\frac{0}{0}$, PM $\frac{3}{2}$, M $\frac{3}{3}$

B. Reproductive Anatomy and Physiology

The female rabbit (doe) has a duplex uterus with two horns, no uterine body, and two cervical canals. The

[1] I, incisers; C, canines; PM, Premolars, M, molars. Numerator shows number in upper jaw. Denominator shows number in lower jaw.

doe reaches puberty at 3 to 8 months depending on the breed with the larger breeds tending to reach puberty later than the smaller breeds. Rabbits breed year around and ovulation is induced within $9\frac{1}{2}$ to $13\frac{1}{2}$ hr after copulation. Does have irregular cycles of about 18 days with 14 days receptivity and 4 days nonreceptivity although this varies greatly with season of the year and with individual females. Reproductive activity is reduced in the autumn and winter as a result of the shortening daylengths. This can be overcome by the addition of extra hours of artificial light to simulate the long daylength of the summer. [See ANIMAL REPRODUCTION, NONPREGNANT FEMALE; ANIMAL REPRODUCTION, AN OVERVIEW OF THE REPRODUCTIVE SYSTEM.]

Males (bucks) reach puberty at 4 to 8 months and will produce a 0.5- to 2.5-ml ejaculate containing 10 to 100 million sperm/ml. Bucks are capable of multiple daily services. One buck was reported to have served 50 does within an 8-hr period and to have impregnated the 50th doe. Bucks, like does, exhibit reduced fertility in the autumn and winter. [See ANIMAL REPRODUCTION, MALE.]

Fertilization occurs in the ampulla 1 to 2 hr after insemination. The fertilizable life of the spermatozoa is about 30 hr whereas the ova remain fertile for 6 to 8 hr. The embryo moves into the uterus about 3 days after fertilization and implantation occurs 7 to $7\frac{1}{2}$ days after fertilization. Gestation is 30–35 days depending on the breed and litter size. Smaller litters have longer gestations. The young are born in a nest formed of fur pulled from the ventrum of the female mixed with various nest materials including hay, shavings, or other materials supplied in the nest box. Does do not retrieve kits so a box must be provided which will prevent the kits from wandering from the nest and getting chilled on the wire of the cage. The average litter is 4 to 8 kits although litters as large as 17 or 18 are sometimes seen. Kits normally weigh from 20 to 70 g at birth. [See ANIMAL REPRODUCTION, PREGNANCY.]

Doe milk has the following approximate composition:

Protein	13–17%
Fat	9–18%
Lactose	0.2–1.6%
Minerals (ash)	1.6–2.8%
Water	63–74%

This will vary with the stage of lactation and the number of kits suckling. Does normally nurse once daily visiting the nest for only 3 to 5 min. The period gets shorter as the litter ages. Does normally have 8 to 10 nipples. Milk production reaches a maximum (225–250 g/day for New Zealand White does) at about 20 days postkindling and declines thereafter. Kit's eyes open at about 10 days, they begin eating solid food at about 3 weeks and can be weaned anytime from 4 to 8 weeks of age.

Rabbits can be sexed at any time after birth by exposing the sex organs by placing them between the thumb and forefinger which are then spread. The female orifice will be closer to the anus than that of the male and, as they age, the tubular appearance of the penis will become obvious while the vagina of the female takes a more characteristic slit appearance. It is much easier to sex the kits after 4 weeks.

C. Artificial Insemination

Semen can be collected from male rabbits and inseminated into the reproductive tract of the female. Equipment for this technique is available commercially or can be constructed from locally available supplies. All inseminations are carried out using fresh semen, however, because technology for freezing rabbit semen is not yet available. It is necessary to inject the doe to induce ovulation, usually with a synthetic gonadotrophin releasing hormone (GnRH) compound. Artificial insemination can reduce the work required for hand mating in large commercial herds. It also provides a large number of does which have been served at the same time which will, in turn, provide a large number of marketable fryers of the same age. Artificial insemination will not provide an increase in fertility over natural service. Because of the labor involved in semen collection and handling, artificial insemination may involve more work for small herds than natural mating. With clearly defined selection and improvement programs, artificial insemination will increase the number of does which can be inseminated with semen from superior sires.

D. Embryo Transfer

Embryo transfer in rabbits is accomplished by performing a midline ventral laparotomy and using a flushing medium to flush the fertilized ova back through the oviduct. The embryos can also be flushed from the uterus, but larger amounts of flushing medium are needed, there is more trauma and the recovery is less efficient. Before the embryos are transplanted into the recipient, they should be examined under a microscope and any that are abnormal should be discarded. [See EMBRYO TRANSFER IN DOMESTIC ANIMALS.]

To ensure success, the recipient should be at the same stage of gestation as the donor. This can be accomplished by sham-mating the recipient with a vasectomized buck or injecting with LH or hCG to induce ovulation at the time the donor is mated.

The primary reason for embryo transfer is to increase the use of superior females by allowing them to produce ova without having to nurture the embryos. These females are stimulated to produce a large number of eggs (superovulated) and then the ova are harvested and placed into other females to be gestated. The rabbit was used as a vehicle to transport cattle embryos in the early days of development of the technique. The embryos were removed from the donor cow, implanted into a rabbit which was then flown to the location of the recipient cows where the embryos were removed from the rabbit and implanted into the recipients.

E. Breeding Management

Does should be bred when they reach about 80% of the mature weight for the particular breed. That is about 5 months for New Zealand Whites. Bucks tend to mature more slowly but can be used at about the same age or slightly older than does. Fanciers breed their stock to ensure animals of the proper ages for the shows they wish to attend and thus may only breed their does two or three times each year. Commercial fryer producers want to maximize the output from their does so they will try to keep the does breeding throughout the year to produce anywhere from 5 to 11 litters. In the former case, the kits generally are not weaned and are sold directly from the doe's cage 8 weeks after birth. The doe is rebred 6 weeks after kindling, and will produce another litter slightly over 2 weeks after the previous litter has been weaned. In the latter case, which approximates the biological maximum for the doe, the kits will be weaned at 4 weeks of age. The doe, having been bred 1 day postpartum, will kindle a new litter 3 days after weaning. Rebreeding at intervals such as 7, 14, or 21 days after kindling are more common for commercial production.

Ten days after breeding, the developing embryos can be felt in the uterus of the doe. Pregnancy diagnosis is carried out by holding the scruff of the doe's neck with one hand and palpating the abdominal organs with the thumb and forefinger of the other hand. In the medium breeds, the embryos will be the size of marbles at 10 days and will increase to the size of ripe olives by the 14th day. Palpation poses little danger to the embryos as they are protected by the placental tissues: especially the fluid filled amnion.

The pregnant doe should be provided with a nestbox at about day 28 of gestation. The box should be large enough so the doe will readily enter it. If the box is too large, the doe may spend time loafing or defecating in it. Care should be taken that the nestbox is not placed in the area of the cage where the doe urinates and defecates. A box 60 cm long × 20 cm wide × 20 cm deep is suitable for medium-sized breeds such as the New Zealand White. The box may be made of wood, metal, or plastic depending on the preference of the breeder and the materials available. The design will vary to some extent depending on the prevailing climate. In some areas, nestboxes are covered to provide warmth whereas as in warmer areas covers are not needed and may even be detrimental. In the south, nestboxes can be made with bottoms of fine screen. This allows the waste nest material to fall out and decrease in quantity to coincide with the kit's age and development. The nestbox should be provided with an absorbent nest material such as shavings or chopped fine hay. A few hours before kindling the doe will begin pulling fur from her ventrum and mixing it with the nest material provided. Kindling occurs most frequently early in the morning although does will kindle at other times as well. Once the doe has finished kindling, the nest should be checked and any dead kits and other nonliving materials removed from the box.

Because the hormones involved in the birth process are also involved in lactation, does will sometimes nurse the litters during or just after kindling. Once kindling is completed, most does will leave the box and only return once daily to nurse. About 22 hr after the previous nursing, the kits will move to the top of the nest material to await the doe. Once the doe jumps into the box, the kits will rapidly begin searching for nipples and begin nursing. At the beginning of the nursing period, the kits will move frequently from nipple to nipple. After 30 to 60 sec, such movement will slow or cease and the kits will nurse for 60 sec or more after which more movement will begin. The kits move frequently from nipple to nipple so it is important that there not be more kits than available nipples. Litters should be culled to ensure this is the case. Rabbits readily accept fostered kits so if several does kindle on the same day, it will be possible to even out the numbers between the does. After the doe leaves the nestbox, the kits all urinate and carry out a stereotypical digging behavior which serves to fluff and dry the nest material. After about

15–30 min of this behavior, the kits form into one or several clusters and remain quiet until the next nursing episode. Does do not retrieve kits that are pulled out of the nestbox. The nestbox should be checked daily and any dead kits removed. After about 14 days, the kits are able to maintain their body temperatures well and the nestbox can be removed from the cage. [See LACTATION.]

III. Rabbit Genetics

A. Selection for Multigenic Traits

Most characteristics that are of economic importance in commercial rabbit production such as growth rate, feed efficiency, and reproductive efficiency are dependent on a number of genes (i.e., they are multigenic or quantitative traits). In these cases, each gene tends to contribute independently to the expression of the trait so the more genes that are available or the larger the effects of the genes, the more that trait will rely on genetics. This proportion is referred to as the heritability and is an indication of how successful a breeder might be in selecting for that particular trait. The higher the heritability, the more rapid the selection progress will be. In general, heritabilities of traits related to reproduction and survival, such as number of kits born alive and survival to weaning, tend to be low whereas production and growth traits like milk production, litter growth rate, and wool production have medium to high heritabilities. This means that selection programs which emphasize the latter traits will usually be more successful than those which are focused on traits with low heritabilities. [See ANIMAL BREEDING AND GENETICS.]

Breeding systems that are used with rabbits include line-breeding, inbreeding, random mating, line-crossing and cross-breeding. In line-breeding, rabbits of the same line that are deemed superior for a particular trait or traits are mated (e.g., grandsire–granddaughter or first cousin matings). Mating of more closely related individuals is termed inbreeding. Inbreeding has the advantage of exposing any genetic strengths or weaknesses in the line. It is used for this purpose because it exposes and codifies the strengths of the line. Random mating refers to mating of individuals without regard to parentage. Outbreeding includes line-crossing (crossing two unrelated lines within a breed) or cross-breeding (mating of individuals of two different breeds.) Planned matings of two or more breeds each of which contribute specific genetic capabilities are called hybrid matings. The aim of this type of cross is breed complementation in which each breed contributes a specific strength or strengths to the hybrid offspring.

B. Coat Color Genetics

Coat color in rabbits is largely under the influence of a series of five multiple alleles on four chromosomes. The coat color of the wild-type rabbit is referred to as agouti or full color. Individuals in this category have banded hairs including gray, yellow, black, and brown. Rabbits lacking pigment are albinos. Mating of a homozygous agouti to an albino always produces agouti offspring. Agouti has dominant gene expression at each of the five loci. Phenotypes lacking yellow pigment in the coat but possessing a silver gray appearance are referred to as chinchilla. Another common phenotype is the Himalayan in which the coat is white except for the extremities, nose, ears, feet, and tail, which are black. This is due to activation of melanin pigment in the cooler areas of the body. Himalayan rabbits (or Californians which also have the Himalayan color pattern) born in the cooler times of the year will often have a gray cast to the baby coat as a result of having been chilled in the nestbox.

There are five coat color series (A, B, C, D, and E) with the A series (agouti) on one chromosome. The B and C series, which code for brown and color, are on another chromosome and D, dilution of color, is found on a third chromosome. E, the gene series for extension, is found on the fourth chromosome. Each locus has several alleles and their expression, as well as epistatic effects between the loci, will determine the color of the fur. Show persons spend a lot of time and work diligently to obtain the particular coat color they wish for their show animals.

IV. Rabbit Diseases

A. Causes of Disease

Disease is a condition in an animal which is different from the normal. It can be caused by management factors or by microorganisms. There is a wide variety of organisms which cause disease in rabbits including viruses, rickettsia, bacteria, protozoa, fungi, and external and internal multicellular organisms. [See Animal Diseases.]

Perhaps the most damaging and pervasive disease among rabbits is pasteurellosis. This is caused by the bacterium *Pasteurella multocida* which manifests itself in a variety of conditions including skin abscesses,

snuffles, orchitis, vaginitis, and wry neck. Enterotoxemia is caused by proliferation of the bacterium *Clostridium spiroforme* type c in the hindgut. These bacteria produce toxins which kill the rabbits—usually within just a few hours. This occurs more frequently if the rabbits are fed a diet which is high in carbohydrates. This condition causes numerous deaths in rabbits: especially among growing stock at the times of year when the weather is most changeable (spring and autumn).

B. Disease Management

There are very few drugs which have been approved for use in rabbits. Rabbit diseases tend to progress very rapidly, and the first sign of a disease situation often is a dead rabbit. As a result, most rabbit diseases are controlled through management and preventive measures rather than treatment. An individual comes down with a disease when the number of the microorganisms in the environment increases to the point where the microorganisms overcome the natural defenses of the host to resist. To prevent disease, the producer must therefore reduce the number of microorganisms and ensure that the resistance of the host remains high.

Reduction of the number of microorganisms relies on maintaining a high level of cleanliness in the rabbitry. This includes the building and the cages as well as prompt removal of any sick or diseased animals and good ventilation to reduce the numbers of microorganisms which are in the air. Maintaining a high level of resistance in the host means that the rabbits must be well fed and cared for and not stressed by environmental factors which might lower their resistance.

V. Rabbit Housing

A. Requirements for Rabbit Housing

There is no one type of housing which is suitable for all situations, but there are certain requirements that should be met regardless of the location of the rabbitry. The following is a list of requirements for rabbit housing:

The housing should be comfortable for the rabbits.
The housing must confine the rabbits and keep them
from escaping.
The housing must protect the rabbits from predators.
The housing must protect the rabbits from the weather.

The housing should allow easy, comfortable access for the manager.

The housing should be "self-cleaning" or easy to clean.

The housing should be of a reasonable cost, be easy to maintain, and be durable.

B. Environmental Effects on Production

Rabbit productivity is affected by the environment. In hot weather, they eat less and do not grow as rapidly, litter sizes are smaller, and conception rates are lower. Rabbits are also affected by the length of day and, as autumn approaches and the days become shorter, producers who do not provide extra hours of artificial lighting may find that their does will stop breeding. This effect will wane when the days begin getting longer and the does once more begin breeding. Simulation of a constant daylength as long as the longest day in the area will avoid the reduction in reproductive efficiency in the autumn.

C. Types of Housing

Housing for rabbits can be either outdoors or within a building. Outdoor housing is most usual when only a few rabbits are housed and simple hutches can be used. This type of housing provides for excellent ventilation and may be cheaper than indoor housing but suffers the disadvantages of not providing as good protection from weather and predators as indoor housing. In severely cold or wet weather, the producer may also suffer from the effects of the weather with outdoor housing. This may affect the quality of care given the rabbits. Indoor housing generally consists of a number of stacked cages or cages hung from the rafters inside a building. These cages are usually constructed of welded mesh wire which provides the strength needed to support the rabbits and provides the required protection. Cages may be built with the opening on the top or on the front or they may be built in a quonset configuration with the door on the curve. There is no set rule for cage sizes, but a rule of thumb suggests 0.15 m²/kg mature body weight. In some cases cages are stacked two or three high but this configuration makes good management and ventilation extremely difficult. Most commercial producers use a single tier of cages.

Locating a rabbitry must be done with care. Drainage away from the site is required to keep the area dry, but care should be taken that the effluent does not contaminate local water sources. Zoning laws must be met, and care should be taken that extra-territorial jurisdiction will not, at some time in the future, place the unit within municipal limits where it will not be tolerated.

A rabbitry must have good ventilation to reduce the amount of moisture, ammonia, and disease-causing organisms. This can be accomplished by using natural ventilation or by installing fans within the rabbitry to either blow the stale air out (positive ventilation) or suck in fresh air (negative ventilation). In a production situation, rabbits do well at about 15–20°C. They also will grow and produce at higher and lower temperatures but that range is good to aim for. Providing shade and air movement will help keep temperatures down in the summer, whereas blocking drafts will help to warm the unit in the winter. In areas with prolonged freezing in the winter, some form of additional heating may be necessary to keep the water lines from freezing. Rabbits do quite well at temperatures around freezing although their feed efficiency is lowered due to the higher energy cost to keep warm.

Rabbits produce urine and feces. The urine contains urea which reacts with water in the presence of the enzyme urease to produce ammonia which creates an odor and health problem. The feces provide a solid waste problem. These problems have to be controlled and the wastes removed regularly to reduce odors as well as to reduce the number of pathogenic organisms in the rabbitry. There are numerous waste handling systems. Some involve daily washing to remove all the urine and feces from the unit. Others allow the manure to build up under the cages and rely on drainage to remove the urine and keep the ammonia problem in check. There are several automatic manure removal systems on the market, but most producers end up using a shovel to remove the manure from their units because the automatic systems do not do the job satisfactorily.

D. Cage Equipment

Rabbits require large amounts of clean water. For instance, a doe and her litter may require as much as 4 liters of water each day. As with any other livestock species, this should be provided in containers free from contamination and algal growth. It rapidly becomes apparent that with any number of rabbits, the producer should have an automatic system which provides a continuous flow of fresh water to his animals. There are several types of drinkers on the market. Dewdrop drinkers are the cheapest, but they also tend to drip and, if drainage within the rabbitry is poor, can make a mess. Other types of drinkers are more expensive but are less prone to dripping.

In the wild, the doe makes its nest at the bottom of the lowest burrow in the warren. This is composed of fur pulled from the ventrum as well as grasses and other materials carried into the burrow. After the doe kindles, it visits the nest only once per day to nurse and, as it leaves up the burrow, any kits still attached to nipples come free and roll back to the nest. The doe, therefore, exhibits no retrieving behavior. The domestic doe is provided with a box in which to build its nest. Because there is no retrieval, the manager has to observe the litters carefully to be sure that any kits which are out of the box are returned before they become chilled and die.

The dimensions of the box and the materials used for construction depend on the preferences of the manager, the size of the rabbits and the materials available. A plywood box 20 x 40 x 20 cm with a double wire bottom made of a layer of 6.35 mm (1/4 in.) hardware cloth and a layer of 25.4 x 12.7 mm (1 x 1/2 in.) weld mesh is suitable in areas with a mild climate for New Zealand Whites. Other areas may require a lid or a solid bottom—especially in the winter months. Some boxes are constructed of wire and a new cardboard liner is used for each litter. Other boxes may be made of metal or plastic with or without removable bottoms. Smaller breeds will require a smaller box and larger breeds correspondingly larger ones.

Rabbit rations are normally pelleted to stop the rabbits from selecting the portions of the diet they will eat. Pellets can be offered in ceramic or plastic crocks placed in the cage but, with more than just a few rabbits, metal feeders are normally used. One common design is a "J" shape with the lower end of the J in the rabbit's cage and the upper part on the outside. These are mounted on the outside of the cage so the pellets can be added without opening the cage. This allows the rabbits easy access for eating the feed, but they are not able to foul it by walking or defecating in the feeders. Most designs have screen or slotted metal at the bottom of the J. This allows the fines or unpelleted feed material to fall out of the feeder.

VI. Nutrition and Feeding

A. Digestive Anatomy and Physiology

The rabbit has a monogastric digestive system with a functional cecum. The cecal contents (cecotrophes or soft feces) are consumed directly from the anus (cecotrophy or copraphagy). This provides B vitamins and bacterial protein from the bacterial activity in the cecum. The small intestine is about 4 m long, the ceca about 30 cm each, and the colon about 150 cm. Daily food intake on a dry matter basis is 3.0 to 5.5% of body weight. Water intake is about twice dry matter intake. Rabbits eat 25–30 "meals" each day. The young eat throughout the day and night whereas adults tend to eat mostly at dawn and dusk. [See ANIMAL NUTRITION, NONRUMINANT; ANIMAL NUTRITION, PRINCIPLES.]

B. Dietary Requirements

Most rabbit breeders use only one feed for all the stock. Many feed manufacturers offer more than one feed, but the producers choose not to use them. As a result, the feeds are formulated to suit the fryers which require a high fiber feed to prevent enteritis. As a result, does, especially lactating does, are sometimes unable to obtain sufficient energy from the ration so a high-energy supplement is added to increase productivity. [See FEEDS AND FEEDING.]

Rabbits are able to utilize forages, so most diets contain 50% or more alfalfa or other roughage sources. Soybean meal is the most common protein source although small amounts of other protein sources such as cottonseed meal may be used. Carbohydrate sources may include corn, barley, oats, wheat mill feed, or rice by-products. Tallow may be added as a fat source and molasses as an energy source and pellet binder. If good quality alfalfa is used, there is little need to add vitamins to the diet and the only need for minerals is for salt and dicalcium phosphate.

Rabbit feeds are normally ground and formed into pellets about 4 mm in diameter and 5—7 mm long. This prevents the rabbits from selectively eating the portions of the ration they prefer. It is also useful for reducing wastage because of feed becoming caked in the feeding bowls or the feeders.

A common formulation for general use in the rabbitry was developed at the Oregon State University Rabbit Research Center and is as follows:

Alfalfa meal	54%
Soybean meal	21%
Wheat mill run	20%
Molasses	3%
Tallow	1.25%
Salt	0.5%
Dicalcium phosphate	0.25%

This would be a suitable ration for growing fryers.

It can also be used for breeding does, but is somewhat low in energy for lactation so supplementation of nursing does with an energy source such as whole corn might be necessary. Bucks, replacement stock, and nonlactating, gestating does can be fed a diet which contains a high proportion of high fiber feeds such as oats or barley and a much lower level of soybean meal and other protein sources.

C. Feeding Practices

Whatever the method of providing feed for the rabbits, the feeders must be kept clean. Stale feed or feed that has been contaminated by water, feces, or urine will not be consumed by the rabbits and should be promptly removed. Uneaten feed will become caked in the feeders and may mold. In some areas, the feeders for lactating does and growing fryers can be filled with feed every few days whereas in others where it is very humid, feeding must be done on a daily basis. Feed left in the feeder for more than one day becomes soggy and the rabbits will not eat it. Whichever method is used, behavior of the rabbits should be observed carefully during the feeding period as this is a good time to identify those animals which may be sick or getting sick.

The amount fed depends on the production level of the animals. Lactating does should be full fed (e.g., about 125 g for a New Zealand White doe plus an additional 50 g for each kit it is nursing). Smaller does require less feed; larger does more. Where high-fiber feeds are being used, a carbohydrate supplement such as a spoonful of whole corn may be added for the doe. This practice should cease when the kits leave the nestbox and begin consuming solid feed because the high energy feed will cause them to scour. Nonlactating does and breeding bucks of the medium breeds should be restricted to less than 100 g per day. Breeding stock should be handled regularly to ensure they maintain adequate but not excessive body condition. Fryers should be full fed or nearly so.

VII. Rabbit Management

A. Handling

Regardless of the reasons for which rabbits are kept, they are sentient beings and must be handled in a manner which will not cause them pain or undue discomfort. Whenever possible, rabbits should be lifted by placing the hands under the body. They also can be lifted by grasping the scruff of the neck (NOT the ears) with one hand with the other hand under the rump supporting the weight. When being carried, rabbits should be supported by resting them on the forearm.

B. Identification

Rabbits can be identified by tattoos in the ear or by metal tags of the sort made as wingbands for poultry. If the rabbit is to be registered with the ARBA, the right ear is reserved for the registration tattoo. The left ear is available for the individual breeder's marks. These should be devised to convey as much information as possible (e.g., the year of birth, sex, and ownership as well as an individual identification number). Identification should be done at weaning or earlier.

C. Records

Records are critical for the rabbit raiser. These include the records needed for daily operations such as the dates of breeding, days on which the does should be palpated for pregnancy, and when nestboxes should be placed in the cages. Keeping these records in a desk diary/calendar which shows a week at a time will allow making notes of future chores to be carried out. Performance records of individual breeding rabbits will include the number of litters produced, the number of kits in the litters, and their weights as well as the number of the kits born that were weaned and marketed. A useful criterion is the weight of the litter at 3 weeks. Until that time, the kits subsist almost entirely on milk so that figure gives an estimate of the doe's ability to produce kits and its milk producing ability. In a commercial situation, financial records listing expenses and income are critical. These should include all expenses attributable to the rabbit enterprise as well as any income whether it be from sale of live rabbits, rabbit meat, manure, or other products from the unit.

D. Purchasing Stock

Before seeking breeding stock, the breeder should decide why he or she is going into the rabbit business. This will determine the type of rabbits, e.g., meat quality, show quality, or pets, in which the purchaser is interested. When visiting a rabbitry to buy rabbits, the buyer should take a look around the unit. Do the rabbits look healthy? Are there lots of litters of a variety of ages in the cages? Do the rabbits appear to be of the type desired? Are detailed performance records kept? Once the buyer feels comfortable with the rabbitry and the quality of stock available, negotiation of price can begin. This will depend on the use to which the rabbit is to be put. Fancy show stock will sell for whatever the market will bear. Show stock for children in 4-H or other youth programs is often much cheaper although many breeders are very cautious about this as they have had bad experiences with selling rabbits which supposedly were for children and then having to compete against those same rabbits being shown by the parents. Meat rabbit breeding stock will sell for slightly more than $1.00 per week of age so a doe at 10 weeks might sell for $10–$15 whereas the same doe at 20 weeks might sell for $20–$25. Buying the younger stock allows them to become accustomed to the rabbitry before breeding begins but they must be held for 10 weeks until they are old enough to breed. The older stock cost more but there is no rearing cost. They may have problems kindling however if they are not given time to adapt to the rabbitry.

VIII. Marketing Rabbits and Rabbit Products

There are no large, well established markets for rabbits and rabbit products and one of the problems faced by the new producer is locating places where the products can be sold.

Meat processors tend to move in and out of business rather rapidly so it is a good idea to have more than one outlet available. It is important, however, that the producer be loyal to his processors and provide product on a continuous basis. Many processors are forced out of business because their fryer producers sell a large portion of the fryers to temporary markets such as the Easter bunny trade. The processor may then be unable to meet his contractual obligations. Production tends to be highest in the summer when the market for meat is the weakest. At that time, many producers look for additional markets and are unable to find them because they have not supported their processors throughout the year. Prices for fryers are generally set by the processors and tend to fluctuate only slightly during the year.

Markets for laboratory animals and animal products are arranged between the producer and the buyer and it is up to the producer to find buyers for the product. Prices are set by agreement between the two parties. The price for laboratory rabbits is much higher than for meat rabbits because the laboratory rabbit producer must carry an inventory of older animals to cater to last minute orders. Most scientists are not willing to wait 6 months to have a 6-month-old rabbit on which to carry out research. Sales of breeding stock depend on the producer building a reputation for producing high quality animals. Pet stores generally purchase rabbits at about 4 to 6 weeks of age and will usually specify the price they are willing to pay. Producers intending to sell laboratory stock or to sell live animals to other commercial outlets are required to have rabbitries licensed in accordance with the Animal Welfare Act. They should check with the Animal and Plant Health Inspection Service of the United States Department of Agriculture for licensing regulations and procedures.

Rex fur producers normally harvest the pelts, pay to have them dressed (tanned), and then sort them by quality and color and ship them to a furrier who will offer a price for the batch. The market is extremely limited however and furriers are very particular regarding the quality and colors they will accept. It thus sometimes happens that the price offered is less than the producer has invested in the pelts. As a result of this marketing structure, many Rex producers develop their own markets by making a variety of novelty items from the pelts which are then sold wholesale or directly to the consumers.

There is little, if any, large scale commercial spinning of Angora wool in the United States. Producers may market raw wool to spinners who use it to make yarn which is used to make garments for sale at wholesale or directly to consumers at craft fairs and other outlets. There are also markets for spun wool which can be used for making garments for sale. Many Angora producers carry out all the steps from production of the wool to sale of the finished products. [See WOOL AND MOHAIR PRODUCTION AND PROCESSING.]

Bibliography

American Rabbit Breeders Association, Inc. (1991). "Standard of Perfection." American Rabbit Breeders Association, Inc., Bloomington, IL.
Cheeke, P. R. (1987). "Rabbit Feeding and Nutrition." Academic Press, New York.
Cheeke, P. R., Patton, N. M., Lukefahr, S. D., and McNitt, J. I. (1987). "Rabbit Production," 6th ed. Interstate, Danville, IL.
Harkness, J. E., and Wagner, J. E. (1989). "The Biology

and Medicine of Rabbits and Rodents." Lea and Febiger, Philadelphia.

Kilfoyle, S., and Samson, L. B. (1992). "Completely Angora," 2nd ed. Samson Angoras, RR 1, Brantford, Ontario, Canada.

Lebas, F., Coudert, P., Rouvier, R., and deRochambeau, H. (1986). "The Rabbit: Husbandry, Health and Production." FAO Animal Production and Health Series No. 21, Food and Agricultural Organization of the United Nations, Rome.

Lukefahr, S. D. (1992). "A Trainer's Manual for Meat Rabbit Project Development." Heifer Project International, Little Rock, AR.

Manning, P. J., Ringher, D. H., and Newcomer, C. E. (eds.). (1994). "The Biology of the Laboratory Rabbit." 2nd ed. Academic Press, San Diego.

Rangeland

RICHARD H. HART, *USDA-Agricultural Research Service, Wyoming*

Glossary

Ecosystem Unit of the environment populated by a definable group of organisms interacting with a definable physical environment to produce a clearly defined trophic structure, nutrient and water cycles, and biotic diversity

Multiple use Use of an area of land to produce a wide array of material and nonmaterial products while sustaining the basic resources of soil, water, plants, and animals

Native vegetation Consisting predominantly of plant species present before the arrival of European peoples; may include some species introduced accidentally from other areas

Riparian area Areas in which, because of the presence of flowing or standing water, soil and vegetation are different from soil and vegetation in the adjacent area

Stocking rate The number of animals, including humans, occupying a unit area of land for a given period of time

Rangeland is land occupied by grasses, shrubs, or other native vegetation and capable of sustaining one or more uses such as grazing for wild and domestic herbivores, wildlife habitat, water production, and recreation. Forested land is included when it supports grazing or other rangeland uses, but timber production is not usually considered to be a rangeland use.

Areas of land seeded to native plant species are included, but not cropland or seeded pasture.

Rangeland occupies about 36% of the land surface of the United States (Table I), when millions of hectares of grazable coniferous forest in the West and smaller areas of grazable coniferous and deciduous forest in the East are excluded. When these are included, about 483 million ha or 51% of the total area of the United States can be considered as rangeland. Nearly all of U.S. rangeland is concentrated in 19 western states where 18 to 80% of a state's total area may be classified as rangeland. In contrast, only 7% of Maine and 6% of Florida is rangeland, and no other eastern state contains as much as 4% rangeland, when forest land is excluded.

I. Rangeland Ecosystems

Rangeland in the contiguous 48 states includes 11 major ecosystems (Fig. 1). A 12th ecosystem, Arctic tundra, occupies one-third of Alaska, and small areas of alpine tundra occur in mountain ranges throughout the West. A wide variety of rangeland ecosystems, tropical to temperate and humid to arid, are found in Hawaii, but each occupies a very small area. Riparian areas, in which vegetation is affected by the presence of running or standing water, occur throughout all ecosystems. Forage production varies widely among ecosystems and among years within each ecosystem, because of variation in vegetation, climate and soils. [*See* RANGELAND ECOLOGY.]

Tallgrass prairie originally occupied 74 million ha in the United States, extending from the Canadian border to the Gulf of Mexico. Big bluestem (see Appendix Table I for scientific names of plant species) dominates this ecosystem. Little bluestem, yellow Indiangrass, and switchgrass are subdominants, and a long list of other grasses and forbs are present. In Gulf

TABLE I

Total Area, Land in Rangeland, and Rangeland in Federal Ownership in 19 Rangeland States and the United States

State	Total area	All range		Federal range	
	(% million ha)	(% million ha)		(% million ha)	
Alaska	151.9	62	94.2	97	91.3
Arizona	29.5	62	18.3	46	8.4
California	41.1	42	17.3	59	10.2
Colorado	27.0	42	11.3	31	3.5
Hawaii	1.7	24	0.4	31	0.1
Idaho	21.6	44	9.5	66	6.3
Kansas	21.3	31	6.6	1	0.1
Montana	38.1	57	21.7	21	4.6
Nebraska	20.0	49	9.8	2	0.2
Nevada	28.6	80	22.9	92	21.1
New Mexico	31.5	63	19.9	31	6.2
North Dakota	18.3	27	4.9	12	0.6
Oklahoma	18.1	21	3.8	2	0.1
Oregon	25.1	36	9.0	59	5.3
South Dakota	20.0	47	9.4	7	0.7
Texas	69.2	54	37.4	1	0.4
Utah	22.0	55	12.1	78	9.4
Washington	17.7	18	3.2	21	0.7
Wyoming	25.4	75	19.0	44	8.4
Total	628.1	53	330.7	54	177.4
Other 31 states	310.7	1	3.1	28	0.9
Total United States	938.8	36	333.8	53	178.3

Coast prairie, which originally occupied 8 million ha along the north coast of the Gulf of Mexico, silver beardgrass and Texas wintergrass are added to the above species. Annual precipitation decreases from nearly 100 cm in the east to 65 cm in the west. Annual evaporation increases from 100–125 cm in North Dakota to 175–200 cm in Texas. Because of adequate precipitation, nearly level topography, and fertile soils, nearly all of these rangelands have been converted to cropland.

Northern mixed grass prairie (Fig. 2) originally occupied 71 million ha. Tall grasses are replaced by the cool-season mid-grasses needle-and-thread, green needlegrass, and western wheatgrass, and the shortgrass blue grama. At its northern extremity, thickspike wheatgrass and porcupine needlegrass are locally abundant. Dryland sedges are important constituents, as are numerous forbs and a few other grasses.

At its southern margin, this ecosystem changes into the Southern mixed grass prairie with an original area of 42 million ha. The cool-season mid-grasses of the northern prairie are replaced by their warm-season equivalents, including little bluestem and sideoats grama. Precipitation and evaporation are intermediate between those of tallgrass prairie to the east and

shortgrass prairie to the west. Significant areas of mixed grass prairie have been converted to cropland, mostly dryland wheat and irrigated crops, but most remains as grazing land.

Shortgrass prairie is dominated by blue grama, buffalograss on heavier soils, and several species of three-awns. Prickly pear cactus may be locally abundant. Annual precipitation is 25–60 cm, with 70–80% falling during the growing season. This ecosystem is highly resistant to grazing, remaining productive after millenia of wildlife grazing and over a century of livestock grazing. Most of the original 28 million ha remains in rangeland. Areas that have been cultivated are very slow to return to perennial grassland.

Desert grassland (Fig. 3) covers 20 million ha of the Southwest. This is the driest grassland, with 25–50 cm of precipitation and up to 200 cm of evaporation. The dominant grass is black grama; several other species of grama, mesquitegrass, and three-awn are common. Several species of shrubs, including mesquite and creosotebush, often form an open overstory. Shrubs tend to increase at the expense of grasses when these grasslands are overgrazed or when fire is excluded.

Bunchgrass steppe covers 52 million ha over a wide variety of climates in the intermountain West. Precipi-

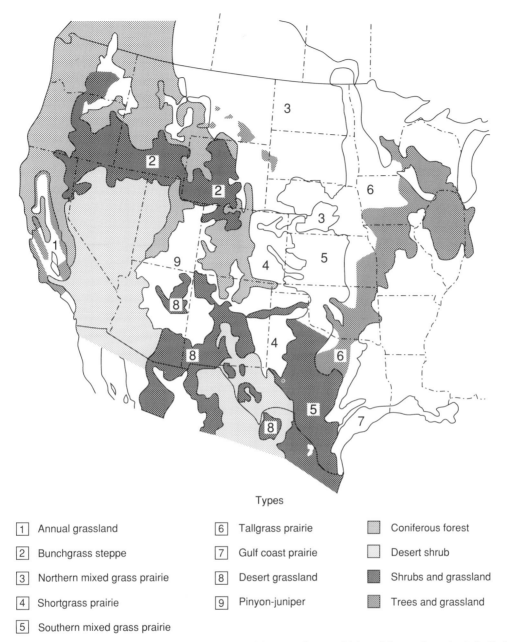

Types

1 Annual grassland   6 Tallgrass prairie   Coniferous forest

2 Bunchgrass steppe   7 Gulf coast prairie   Desert shrub

3 Northern mixed grass prairie   8 Desert grassland   Shrubs and grassland

4 Shortgrass prairie   9 Pinyon-juniper   Trees and grassland

5 Southern mixed grass prairie

FIGURE 1 Major range ecosystems of the contiguous 48 states of the United States. [Adapted from a figure by J. L. Dodd, published by Tetlyanova, A. A., Zlotin, R. I., and French, N. R. (1990). Changes in structure and function of temperate-zone grasslands under the influence of man. *In* "Managed Grasslands" (A. Breymeyer, ed.) pp. 301–334. Elsevier, Amsterdam.]

tation ranges from 20 to 50 cm annually, with much of it falling as snow from September to April. Bluebunch wheatgrass is the dominant grass; Idaho fescue, needle-and-thread, bluegrasses, and Basin wildrye are subdominants. Big sagebrush dominates the landscape in much of the bunchgrass steppe (Fig. 4); shadscale saltbush and several other species of shrubs are present. In southern Idaho and other areas, cheatgrass brome is replacing many of the perennial grasses.

The bunchgrass steppe merges with the Desert shrub ecosystem to the south, occupying 62 million ha. Big sagebrush, black sagebrush, shadscale saltbush, and rabbitbrushes dominate in the north. These are replaced by greasewood, blackbrush, and creosotebush farther south. Bluebunch wheatgrass and

FIGURE 2 Northern mixed grass prairie.

FIGURE 4 Bunchgrass steppe, dominated by big sagebrush. (From an original oil painting "The Southern Medicine Bows" by Linda Lillegraven, Laramie, Wyoming, 1992).

bottlebrush squirreltail are the most important grasses. Cold winters, hot dry summers, and limited precipitation limit plant growth to a brief spring period.

The Pinyon–juniper ecosystem (Fig. 5) is dominated by pinyon pines, Rocky Mountain juniper, Utah juniper, western juniper, and one-seeded juniper. These tree species form open stands on 31 million ha of land characterized by shallow soils, rough topography, and less than 40 cm of annual precipitation. Production of the dominant understory grasses bluebunch wheatgrass, western wheatgrass, blue grama, needlegrasses, and galletagrass is very low. Pinyon–juniper seems to be expanding into adjacent

sagebrush–bunchgrass land, possibly because of overgrazing, control of fire, or climate change.

The California annual grasslands, covering 10 million ha, have replaced the original vegetation of perennial bunchgrasses. Dominant grasses include wild oats, annual bromegrasses, and annual barleys. Forbs are of secondary importance but bur clover and alfilaria furnish abundant forage. Grasses and forbs make most of their growth in the mild wet winters and become dormant in the hot dry summer. In the foothills a great variety of shrubs occur, in open stands at lower altitudes and impenetrable tangles at higher altitudes. Shrubs include snowberry, bearberry, buckthorn, and many others.

FIGURE 3 Desert grassland.

FIGURE 5 Pinyon–juniper rangeland.

Arctic tundra covers 52 million ha in the northern third of Alaska. Vegetation is highly variable but usually includes some or all of the grasses dupontia, meadowsweet, reedgrass, and arcticgrass, as well as sedges, cottonsedges, bulrushes, cattails, and reeds. Various fructicose lichens provide forage for caribou. Alpine tundra, differing from Arctic tundra in the absence of permafrost and exposure to higher levels of solar radiation, is found above timberline in the Rocky and Sierra Nevada Mountains. As in Arctic tundra, vegetation is highly variable, but tufted hairgrass is nearly ubiquitous. Alpinesedges, sedges, and many species of mat-forming forbs are widespread.

Coniferous forest covers 185 million ha in the 19 rangeland states, including nearly all the southern two-thirds of Alaska. The Forest Service recognizes over 40 habitat/community types dominated by ponderosa pine. In the Pacific Northwest, understory grasses include bluebunch wheatgrass, Idaho fescue, spikefescue, pine reedgrass, onespike danthonia, blue wildrye, bluegrasses, and bromegrasses. In the Rocky Mountains, common grasses under ponderosa pine are mountain muhly and Idaho fescue from central Colorado northward, and Arizona fescue from there southward. Under Engelmann spruce and alpine fir at higher elevations, Thurber needlegrass dominates from northern Colorado southward and Idaho fescue and bluebunch wheatgrass from there northward.

Riparian areas are those in which, because of the presence of flowing or standing water, soil and vegetation are different from soil and vegetation in the adjacent area. They occur in all ecosystems along streams, around lakes and ponds, and in marshes and wet meadows. Riparian areas may cover only 1–2% of a watershed, but provide 20–80% of the forage for grazing animals. Riparian vegetation shades streams and keeps water temperatures cooler and more suitable for fish. Vegetation slows runoff, reducing peak flows and prolonging flow after precipitation. Well-managed riparian areas filter out sediment and pollutants, regulate stream flow, provide habitat for wildlife, and improve fish habitat in streams and lakes.

High-altitude wet meadows occupy only about 1.6 million ha in the rangeland states, but provide grazing and hay essential as winter feed for livestock and wildlife. Vegetation on undisturbed meadows includes a wide range of grass and sedge species. Many meadows have been reseeded to cultivated grasses after water control structures were built and fertilizer was applied. Production on such improved meadows may be two to four times that on unimproved meadows.

[See RANGELAND GRASS IMPROVEMENT; RANGELAND SHRUBS; RANGELAND SOILS; RANGELAND PLANTS.]

II. History of Western Rangelands

Meriwether Lewis and William Clark in 1804–1806 recognized the great variability and potential of western rangeland, but Steven Long, after his 1819–1820 expedition, described the Great Plains as "The Great American Desert." For the next several decades, the range was a place to cross to get to somewhere else, or to live in temporarily while resources were extracted. The mountain men trapped the beaver nearly to extinction by the 1840s, and hide hunters almost wiped out the bison after the Civil War. Gold and silver discoveries in the 1840s and 1850s brought temporary population booms to many areas, but these seldom lasted longer than the mineral deposits.

John Wesley Powell recognized that much rangeland was well suited to livestock production but not to crop production. He recommended that the minimum size of a rangeland claim should be 1036 ha (2560 acres), and that boundaries of claims should follow the divides between watersheds. When Congress passed the Homestead Act in 1862 it ignored his recommendations and retained the rectilinear boundaries and 65 ha homesteads as in the East. The area settlers were allowed to claim was increased to 130 ha in 1873 and to 260 ha in 1877 but this was not nearly enough rangeland to provide a living for a family.

Livestock production in the West boomed immediately after the Civil War. The discovery of gold attracted settlers into areas that had been bypassed previously, creating a market for beef. Cattle in Texas had run wild and reproduced in great numbers during the War. The War also produced a great number of adventurous young men, white and black, looking for a way to turn cattle into cash. Initially cattle were driven to market or to unoccupied rangeland farther north; the coming of railroads accelerated cattle movement. By the late 1860s cattle occupied most of the Plains. The open-range livestock industry peaked in the 1880s. Promoters trumpeted the immense profits to be made in cattle, Eastern and European capital flooded the West, and herds increased until much of the range was grossly overstocked.

Federal government land policies were totally inadequate to deal with the situation. No mechanism existed for the government to manage its own land or to turn over adequately sized units of land to private ownership. Stockmen controlled large areas of range-land by acquiring control of water sources, filing claims through "dummy" representatives, or more often illegally excluding anyone who attempted to claim land or water. This practice led them into conflict with the Federal government because they were using Federal land without compensation. The question of proper compensation for the use of Federal land remains with us today.

The Federal government finally began to accept its responsibility for management of forest and rangelands. The Forest Reserve Act of 1891 set aside 19 million ha of National Forests, and President Theodore Roosevelt added over 40 million ha from 1901 to 1907. The Forest Service was created within the Department of Agriculture in 1905 to manage these lands, which by 1988 amounted to 68.3 million ha. Much of this land is not forested by the usual definition; about 14% is grassland and 20% is shrublands and shrub–grasslands.

The 1934 Taylor Grazing Act sought "to stop injury to the public grazing lands (and) provide for their orderly use, improvement, and development" through leases of the public domain to stockraisers. A Division of Grazing was established within the Department of the Interior to manage 32.4 million ha of grazing lands. The Division became the Grazing Service in 1941 and merged with the General Land Office to become the Bureau of Land Management in 1946. The Bureau now manages 71.7 million ha, of which about one-fourth is forest and three-fourths is grassland, shrub, or shrub–grassland.

The Wilderness Act of 1964 defined wilderness as "area[s] where Earth and its community of life are untrammeled by man, where man himself is a visitor who does not remain." Originally 3.7 million ha were set aside. This has since grown to 12.9 million ha. All but 0.9 million ha of wilderness is in the 19 rangeland states.

The National Environmental Policy Act of 1969 required Federal agencies to prepare Environmental Impact Statements and present them for public discussion before implementing any significant management decisions on range and other public lands. The Federal Land Policy and Management Act of 1976 defined procedures for acquiring, reclassifying, or selling public lands; authorized identification of areas of critical environmental concern; contained formulas for leasing public lands and setting lease fees; and ended homesteading in the contiguous 48 states and, after 10 years, in Alaska.

However, very little Federal land was transferred to private ownership after the 1930s. A high percentage of western rangeland remains in Federal ownership in 13 of the 19 rangeland states. Federal ownership ranges from 21% of Washington and Montana rangeland to 92% of rangeland in Nevada and 97% in Alaska (Table I). The proportion is much lower in the eastern tier of rangeland states, from 1% in Kansas and Texas to 12% in North Dakota. Texas never surrendered its public land to the Federal government; the other states did so but most was subsequently homesteaded and removed from Federal ownership. The high percentage of Federal ownership complicates the management of rangeland and exacerbates conflicts among rangeland users.

III. Uses of Rangeland

A. Grazing and Habitat

Historically, the primary use of rangeland has been to provide forage and habitat for domestic and wild herbivores and many other species of animals from mammals to microorganisms. Of the total animal species found in the United States, 84% of the mammals, 74% of the birds, 58% of the amphibians, and 38% of the fishes are represented in nonforested rangeland ecosystems. At least 90% of the total bird, amphibian, and fish species and 80% of mammal and reptile species utilize forest ecosystems. Rangeland also is home for countless species of insects, other arthropods, assorted worms, and other sorts of small to microscopic animals. About 330 threatened or endangered species of animals and 180 species of plants are found on rangeland.

Nationwide, range and pasture provide 83% of the nutrients consumed by beef cattle, 91% of nutrients for sheep and goats, and 72% of nutrients for horses and mules. These percentages are probably higher in the rangeland states, which have a lower proportion of livestock in feedlots than do the eastern states. [See RANGELAND GRAZING.]

The 19 rangeland states are home to 58% of all beef cattle on range or pasture (total beef cattle minus cattle on feed) in the United States (Table II). Not all are on range; considerable numbers are on cultivated pasture, particularly in the tier of states from North Dakota to Texas and in the Pacific Rim states. The 19 rangeland states also harbor 79% of the stock sheep (sheep not in feedlots) and 88% of the goats in the United States. Again, some are on pasture rather than

TABLE II

Numbers of Livestock and Big Game Animals in 19 Rangeland States[a]

State	Beef cows	All range cattle	Sheep	Goats	Horses	Burros, etc.	Pronghorn	Deer	Elk	Moose	Bighorn sheep	Mtn. goats
Alaska	3	6	3	0.2	2	<0.1	0	166	1	150	73	16
Arizona	259	518	220	98.0	39	0.5	8	139	8	0	3	0
California	935	2721	775	26.0	123	2.0	7	1000	3	0	4	0
Colorado	774	1953	455	5.7	72	0.9	60	550	160	0	5	1
Hawaii	75	172	0	4.0	4	0.6	0	0	0	0	0	0
Idaho	530	1194	270	3.0	55	0.7	21	334	115	4	4	3
Kansas	1390	4001	185	4.8	60	0.4	3	175	0	0	0	0
Montana[b]	1328	2189	640	1.5	69	0.6	20	170	18	1	<1	<1
Nebraska	1755	3742	135	3.0	52	0.3	8	120	0	0	0	0
Nevada	290	514	97	1.0	15	0.1	11	155	1	0	6	0
New Mexico[b]	589	1152	473	78.0	53	0.4	2	23	2	0	0	0
North Dakota	872	1547	152	1.0	35	0.1	5	125	0	0	0	0
Oklahoma	1880	4805	105	9.5	91	0.9	0	175	0	0	0	0
Oregon	592	1173	345	9.8	57	1.0	15	712	10	0	1	0
South Dakota	1505	2964	535	1.4	48	0.1	47	350	0	0	0	0
Texas	5210	10593	1890	1257.0	233	3.0	18	4200	1	0	0	0
Utah	325	611	485	1.6	33	0.1	8	500	25	0	4	0
Washington	375	839	83	8.0	52	0.8	0	450	56	1	1	1
Wyoming	650	1133	705	1.0	47	0.3	469	530	63	8	6	0
Total	19337	41827	7553	1514	1140	12.8	702	9874	463	164	107	21
U.S. total[b]	33200	72157	9601	1724	2258	33.8	702	20687	466	203	107	21

[a] Values for animals are 1000s.

[b] Mississippi, Montana, and New Mexico reported only harvested game animals, not totals.

on range. Fifty percent of the horses and 38% of burros and mules, including feral horses and burros, are located in the 19 rangeland states.

Population estimates from state Game and Fish Departments show that most of our herbivorous big game animals are found exclusively or in very large proportion on rangeland. The rangeland states provide habitat for all the pronghorn antelope, bighorn sheep, mountain goats, musk oxen, and caribou in the United States (Table II; the 2000 musk oxen and 466,000 caribou in Alaska and 245,000 large exotic herbivores, nearly all in Hawaii and Texas, are not shown). The rangeland states also harbor 99% of the elk and 81% of the moose, but only 48% of the deer. Game and Fish Departments regard bison as quasi-domestic animals and do not count them, but the American Bison Association estimates that about 80,000 bison still roam the range. Population figures are not available for large predators, but most are necessarily rangeland animals because the range is where their prey is and man, their primary enemy, is not. Thirty percent of the hunting licenses sold annually in the United States are sold in the rangeland states.

B. Water

Thirty percent of the U.S. population lives in the 19 western states. The aridity of the West makes the limited water supply of this region doubly precious. Most of the surface water in the 19 western states originates on forest or rangeland because these ecosystems occupy most of the area of these states. Some of this water is required for instream uses including fish and wildlife habitat, drinking water for wildlife and livestock, hydropower generation, recreation activities, and navigation. Instream uses usually require a minimum flow rate and thus compete with off-stream uses which reduce stream flow. Fish and wild-life habitats are extremely sensitive to reduced flows, because damage may persist long after streamflow is restored. States have been found to have the legal right to reserve unappropriated water and purchase water rights to insure adequate stream flow. [See RANGELAND WATERSHED MANAGEMENT.]

Offstream uses include domestic drinking water, irrigation, and industrial and commercial uses. Most of the water withdrawn for offstream uses, except for irrigation, eventually returns to the stream, although some is lost through evaporation. A larger portion of water used for irrigation is lost through transpiration from crops, gardens, or lawns. Such lost water is considered consumption. In the West, water may be piped from one watershed to another; this represents complete consumption of water from the original watershed and a gain for the second.

Much of the subsurface or ground water also originates on rangeland. Ground water may remain in the ground for a few years or few hundred years before it is removed for use. In many areas of the West, ground water is being removed faster than it is being replenished. [See GROUND WATER.]

The Forest Service estimated water use in the region encompassing all the 19 rangeland states except Texas and Oklahoma. From 1960 to 1985, withdrawals of surface and ground water increased 42 and 92%, respectively, and consumption increased 42%. From 1985 to 2040, groundwater withdrawals are expected to increase another 26% while surface withdrawals increase 36%, and consumption increases 41%. Irrigation was the major consumptive user, accounting for 95% of consumptive use in 1960 with a decrease to 88% expected by 2040. In the same period, domestic and other municipal uses will increase from 3 to 7%. Livestock water will remain steady at 1% of consumptive use, while industrial uses, including power plant cooling, will increase from 1 to 4%. The increased demand for water for nonagricultural uses will necessitate transfer of water rights.

Quality is as important as quantity. Domestic, agricultural, and industrial users are concerned with sediment, salts, and other pollutants in water, as are wildlife and fish habitat biologists. They will become advocates for preventing and mitigating ecological disturbances and rehabilitating areas where previous disturbances have degraded surface and ground water quality.

Vegetation management on upland and riparian areas, control of time and severity of grazing, and trapping snow with structures or vegetation all contribute to increasing the amount and decreasing the variability of streamflow and to increasing quality of surface water. Proper management also improves percolation of precipitation into ground water and improves ground water quality. More efficient irrigation management means less consumptive use of water and better quality of water returned to streams. [See WETLANDS AND RIPARIAN AREAS: ECONOMICS AND POLICY.]

C. Mining and Energy Production

Over half of U.S. coal reserves are located in the 19 rangeland states. The Bureau of Mines reported that, in 1991, the rangeland states produced 400 million tons of coal or 40% of total U.S. production. Nearly 90% of this coal was produced by surface mining. The rangeland states also produced 5.5 million barrels of oil in 1991, 74% of the U.S. total.

Nonfuel minerals valued at over 16 billion dollars were mined in the 19 rangeland states in 1989. These include not only metals but bentonite and other clays, cement, trona (the raw material for baking and washing sodas and many industrial chemicals), salt, limestone, gypsum, and sand, gravel, and stone for construction. Precious metals are only a small fraction of the total mineral value. The sand, gravel, and crushed stone mined in California was valued at 900 million dollars, or 2.5 times the value of the gold mined in that state in 1989.

The minerals removed from rangelands are enormously valuable, but costs and risks of reclaiming surface mined land are also enormous. Mined land in the rangeland states must be reclaimed with one-half to less than one-third as much precipitation as mined lands in Appalachia and the Midwest. Rangeland soils have limited organic matter, natural erosion rates are high, and vegetative succession is very slow. Surface mining impacts rangeland hydrology as well as vegetation. Mining and reclamation must be conducted to minimize damage to surface and ground water quality and quantity.

D. Recreation and Esthetics

In 1986, recreationists logged 15 million visitor days on BLM lands in the West. These lands represent just over 11% of the total land area and 23% of the rangeland of the 19 rangeland states. In 1991, the National Park Service reported 106 million visitor days on National Parks, Monuments, and Historic Sites in the 19 rangeland states, but not all these sites are on rangeland. Add to these figures the 234 million visitor days per year on the forests and rangelands of the National Forests, plus recreational use on private and other public rangeland, and one can see that the total recreational use of rangelands is prodigious.

The Forest Service visitor figures include hunting and fishing use, but the BLM figures do not. The popularity of hunting and fishing on rangelands is indicated by the 4.7 million hunters and 13.1 million fishers licensed per year in the 19 rangeland states.

While it is impossible to put an economic value on the benefits of outdoor recreation to physical, mental, and spiritual health, the value of recreation to the local

economy has been estimated. Estimates include about $7 per day for motorized travel; $14 for camping or winter sports; $20 for hiking, horseback riding, and water-related activities; $23 for fishing; $35 for elk hunting; and $54 for antelope hunting. BLM estimates that those 15 million visitor days have an economic value of $193 million. If each hunter and fisher spends only 3 days per year in the field, and a hunting and fishing day are valued at $45 and $23, respectively, the annual contribution of hunting and fishing to the economies of the 19 rangeland states totals $1.5 billion.

Vicarious or existence value of range cannot be so easily quantified, but will have increasing impacts on range use. Persons who will never set foot on rangeland derive pleasure from the experiences of others on the range, experienced vicariously through books, films, television, and other media. To some it is enough simply to know that rangeland with all its values exists, as part of our national heritage. A National Academy of Sciences report stated:

. . . Rangeland may be far better at producing the stuff of myth and national identity than . . . beef and mutton products. Yet, in the long run, the production and perpetuation of national myth may be one of the most valuable resources harvested from public rangeland.

Future decisions on range management must take these values into account, and may have to give them as much weight as productive or consumptive uses.

IV. Multiple Use Management

Multiple use is the overriding paradigm of range management. All resources must be managed simultaneously and monitored constantly, to provide the combination of uses that best meets the needs of rangeland owners and users. Multiple use does not necessarily mean that every tract of rangeland should be managed for all possible uses. Instead each tract should be evaluated for its suitability for each use and its ability to support that use without deterioration of soil, water, plant, and animal resources. [See RANGELAND MANAGEMENT AND PLANNING.]

Rangeland is managed by manipulating such basic ecological processes as plant succession, water and nutrient cycling, and energy flow. Plant succession refers to changes in occurrence and abundance of different plant species and plant communities as changes occur in the environment or in competitive relationships among species. Water, nutrients, and energy are removed from the abiotic or nonliving part of the environment (soil, water, air, and solar energy) by producers (plants). They then flow through primary consumers (herbivores) to secondary and tertiary consumers (carnivores). Both herbivores and carnivores include many species, ranging from microorganisms to large mammals. Eventually, decomposers (mostly microorganisms) return water, nutrients, and energy to the abiotic part of the environment.

Ecological processes are manipulated directly, or indirectly by controlling when, where, and how many people, livestock, and wild animals use the range. For example, livestock grazing is managed to match the appropriate number and kind of animals with amount and kind of forage available at each season of the year. Managers ensure that animals are well distributed over the land so forage is not over- or undergrazed in any part of the range. Plant species which are particularly desirable to grazing animals or which are less competitive with other species must be protected from overgrazing or they will be replaced by less desirable or more competitive species through plant succession. This is accomplished by reducing stocking rate (the number of animals per area of rangeland), by providing periodic rest from grazing, or by grazing with a different animal species or mix of species. Additional water or nutrients may be provided through irrigation or fertilization, desirable plants introduced through artificial reseeding, or undesirable plants removed mechanically or with herbicides. However, these practices are not often used because of cost and unintended impacts to the environment. [*See* FORAGES.]

Wildlife is managed by improving habitat to provide more forage, shelter, water, or other needs so that larger, healthier populations of game and nongame wildlife can be supported. Hunting and trapping seasons and bag limits are adjusted to maintain the number of animals that the habitat can support. Exotic animals may be removed and native animals that have died out may be reintroduced. [*See* WILDLIFE MANAGEMENT.]

Control of stocking rate is as important in managing humans as in managing livestock and wildlife. Overuse by humans can destroy vegetation or direct succession to less desirable plant communities, drive off wildlife, accelerate soil erosion, and impair water quality.

Maintaining rangeland vegetation in satisfactory condition slows runoff of rain and snowmelt, holds the soil in place, and allows water to infiltrate the soil. This insures a continued supply of ground and surface water of good quality. Dams and other structures may be built to control flow and insure more uniform quantity and quality of water in streams and lakes.

It must be recognized that implementing any management decision about rangeland use, including the decision to allow no human use whatever, will have an impact on rangeland resources, and that these impacts may have results contrary to those desired. Removal of livestock grazing to improve wildlife habitat is detrimental to bird species such as McCown's longspur and the mountain plover, which prefer to nest in heavily grazed grasslands. Fire control and prohibition of timber harvest cause fuel to accumulate and produce uncontrollable wildfires such as the Yellowstone fires of 1988. Hydroelectric power, initially applauded as cheap and nonpolluting, now is castigated for destroying populations of anadromous fish. [*See* RANGELAND CONDITION AND TRENDS.]

Conflicts may be social as well as biological. For example, recreational users protest that livestock producers and timber harvesters pay less than fair market value for the use of range and forest lands. Yet these same recreational users pay essentially nothing for the use of rangelands and enjoy the roads and other improvements installed and paid for by the livestock and timber industries. These are examples of the conflicts that must be addressed in multiple use management.

Bibliography

Bureau of Land Management (1990). State of the public rangelands, 1990. U.S. Department of the Interior, Washington, DC.

Coupland, R. T. (ed.) (1992). "Ecosystems of the World. Natural Grasslands." Elsevier, Amsterdam.

Evenari, M., Noy-Meir, I., and Goodall, D. W. (eds.) (1985). "Ecosystems of the World. Hot Deserts and Arid Shrublands," parts A & B. Elsevier, Amsterdam.

Flather, C. H., and Hoekstra, T. W. (1989). "An Analysis of the Wildlife and Fish Situation in the United States: 1989–2040." USDA Forest Service General Technical Report RM-178. U.S. Government Printing Office, Washington, DC.

Guildin, R. W. (1989). "An Analysis of the Water Situation in the United States: 1989–2040." USDA Forest Service General Technical Report RM-177. U.S. Government Printing Office, Washington, DC.

Heady, H. F., and Child, D. R. (1993). "Rangeland Ecology and Management." Westview, Boulder, CO.

Holechek, J. L., Pieper, R. D., and Herbel, C. H. (1989). "Range Management Principles and Practices." Prentice-Hall, Englewood Cliffs, NJ.

NRC/NAS (1984). "Developing Strategies for Rangeland Management." Westview, Boulder, CO.

Vallentine, J. F. (1990). "Grazing Management." Academic Press, San Diego, CA.

West, N. E. (ed.) (1983). "Ecosystems of the World. Temperate Deserts and Semi-deserts." Elsevier, Amsterdam.

Appendix: List of Rangeland Plant Species, in Order in Which They Appear in the Text

Big bluestem	*Andropogon gerardii* Vitman
Little bluestem	*Schizachyrium scoparium* (Michx.) Nash
Yellow Indiangrass	*Sorghastrum nutans* (L.) Nash
Switchgrass	*Panicum virgatum* L.
Silver beardgrass	*Andropogon saccharoides* Swartz.
Texas wintergrass	*Stipa leucotricha* Trin. & Rupr.
Needle-and-thread	*Stipa comata* Trin. & Rupr.
Green needlegrass	*Stipa viridula* Trin.
Western wheatgrass	*Pascopyrum smithii* (Rydb.) A. Love
Blue grama	*Bouteloua gracilis* (H.B.K.) Lag. ex Steud.
Thickspike wheatgrass	*Elymus lanceolatus* (Scrib. & J.G. Smith) Gould
Porcupine needlegrass	*Stipa spartea* Trin.
Sedge	*Carex* and *Cyperus* spp.
Sideoats grama	*Bouteloua curtipendula* (Michx.) Torr.
Buffalograss	*Buchloe dactyloides* (Nutt.) Engelm.
Three-awn	*Aristida* spp.
Prickly pear cactus	*Opuntia* spp.
Black grama	*Bouteloua eriopoda* Torr.
Curly mesquitegrass	*Hilaria belangeri* (Steud.) Nash
Honey mesquite	*Prosopis juliflora* (Woot.) Sarg.
Creosotebush	*Larrea tridentata* (DC.) Cov.
Bluebunch wheatgrass	*Pseudoregneria spicata* (Pursh) A. Love
Idaho fescue	*Festuca idahoensis* Elmer
Bluegrass	*Poa* spp.
Basin wildrye	*Leymus cinereus* (Scribn. & Merr.) A. Love
Big sagebrush	*Artemisia tridentata* Nutt.
Shadscale saltbush	*Atriplex confertifolia* (Torr. & Frem.) Wats.
Cheatgrass brome	*Bromus tectorum* L.
Black sagebrush	*Artemesia nova* Nels.
Rabbitbrush	*Chrysothamnus* spp.
Greasewood	*Sarcobatus vermiculatus* (Hook.) Torr.
Blackbrush	*Coleogyne ramosissima* Torr.
Bottlebrush squirreltail	*Elymus elymoides* (Raf.) Swezey
Pinyon pine	*Pinus edulis* Engelm. and *P. monophylla* Torr. & Frem.
Rocky Mountain juniper	*Juniperus scopulorum* Sarg.
Utah juniper	*J. utahensis* (Engelm.) Lemmon
Western juniper	*J. occidentalis* Hook.
One-seeded juniper	*J. monosperma* (Engelm.) Sarg.
Galletagrass	*Hilaria jamesii* (Torr.) Benth.
Wild oats	*Avena fatua* L.
Annual bromegrass	*Bromus* spp.
Annual barley	*Hordeum* spp.
Bur clover	*Medicago hispida* Gaertn.
Alfilaria	*Erodium* spp.
Snowberry	*Ceanothus* spp.
Bearberry	*Arctostaphylos* spp.
Buckthorn	*Rhamnus* spp.
Dupontia	*Dupontia* spp.
Meadowsweet	*Filipendula* spp.
Reedgrass	*Calamagrostis* spp.
Arcticgrass	*Arctagrostis* spp.
Cottonsedge	*Eriophorum* spp.
Bulrush	*Scirpus* spp.
Catttail	*Typha* spp.
Reed	*Phragmites* spp.
Tufted hairgrass	*Deschampsia caespitosa* (L.) Beauv.
Alpinesedge	*Kobresia* spp.
Ponderosa pine	*Pinus ponderosa* Laws.
Spikefescue	*Leucopoa kingii* (S. Wats.) Weber
Pine reedgrass	*Calamagrostis rubescens* Buckl.
Onespike danthonia	*Danthonia unispicata* (Thurb.) Munro ex Macoun
Blue wildrye	*Elymus glaucus* Buckl.
Mountain muhly	*Muhlenbergia montana* (Nutt.) Hitchc.
Arizona fescue	*Festuca arizonica* Vasey
Engelmann spruce	*Picea engelmanni* Parry
Alpine fir	*Abies lasiocarpa* (Hook.) Nutt.
Thurber needlegrass	*Stipa thurberiana* Piper

Rangeland Condition and Trend

WILLIAM A. LAYCOCK, *University of Wyoming*

Glossary[1]

Climax (1) Final or stable biotic in a successional series which is self-perpetuating and in dynamic equilibrium with the physical habitat; (2) assumed end point in succession

Decreaser Plant species of the original or climax vegetation that will decrease in relative amount with continued disturbance to the norm (e.g., heavy defoliation, fire, drought); some agencies use this only in relation to response to overgrazing

Ecological status Present state of vegetation and soil of an ecological site in relation to the potential natural community for that site; ratings usually are early seral, 0–25% (similarity to PNC); mid seral, 26–50%; late seral, 51–75%; and potential natural community, 76–100%

Increaser Plant species of the original vegetation that increase in relative amount, at least for a time, under continued disturbance to the norm (e.g., heavy defoliation, fire, drought)

Invader Plant species that were absent in undisturbed portions of the original vegetation and will invade or increase following disturbance or continued heavy grazing

[1]Definitions of terms are the same or slightly modified from those presented by the Society for Range Management (1989). "A Glossary of Terms Used in Range Management," 3rd Ed. Society for Range Management, Denver, CO.

Potential natural community Biotic community that would become established on an ecological site if all successional sequences were completed without interferences by man under the present environmental conditions; natural disturbances are inherent in its development; the PNC may include acclimatized or naturalized non-native species

Range condition (1) Generic term relating to present status of a unit of range in terms of specific values or potentials; specific values or potentials *must* be stated. (2) Present state of vegetation of a range site in relation to the climax (or natural potential) plant community for that site; expression of the relative degree to which the kinds of proportions and amounts of plants in a plant community resemble that of the climax plant community for the site

Range site Area of rangeland which has the potential to produce and sustain distinctive kinds and amounts of vegetation to result in a characteristic plant community under its particular combination of environmental factors, particularly climate, soils, and associated native biota (Synonymous with ecological site)

Range trend Direction of change in range condition or ecological status observed over time; trend should be described as *toward* or *away from* the climax (or PNC) or as *not apparent*

Range Condition has a number of nontechnical and technical meanings, leading to confusion by professionals and nonprofessionals alike. Farm and ranch reporters often use the term to mean the amount of range production at a particular time, especially as related to precipitation or growing conditions. Thus a farm reporter on the radio or in a newspaper might say, "Because of favorable precipitation this spring, range conditions in the state are good," meaning that there is an abundant growth of green grass. This general usage of the term condition is presented to illustrate the problem of multiple definitions and will not be used again.

I. Range Condition

A. Why Do Managers Need Range Condition Information?

The concept of range condition evolved because range managers needed a measure of the current situation on a given area of rangeland in relation to some standard in order to assess management success. In this context, it has been described as the state of "health" of the range. Both "condition" and "health" have become value laden terms and, this will be discussed later. Many people feel that the term, "health," is inappropriate because it equates land or a plant community to a living organism or even a human. [See RANGELAND; RANGELAND MANAGEMENT AND PLANNING.]

B. Technical Definitions

Even in a technical sense, the term "range condition" has more than one meaning. The most recent Glossary published by the Society for Range Management (SRM) indicated that the term has had two historical definitions. The *first* is a generic term relating to the present status of a unit of range in terms of specific values or potentials. An earlier definition by the Society for Range Management was "The current productivity of a range relative to what that range is naturally capable of producing." This definition now is considered too vague because the specific values or potentials were not stated except in the broad generality of "productivity." However, this definition was the traditional one used by range managers for many years.

1. Resource Value Ratings

Within the first definition of range condition is the concept of a "resource value rating" which is the value of vegetation present on a range site for a particular use or benefit. The most common use of a resource value rating is for livestock forage. In this case, the palatability and nutrient quality of each species for livestock production is considered and the most desirable mix of species for maximum livestock production would have the highest resource value rating. However, resource value ratings can be determined and stated for wildlife habitat for a given species of wildlife or for any other use of a specific rangeland area.

2. Desired Plant Community (DPC)

The DPC is a range condition concept being implemented by the Bureau of Land Management. The DPC for a given range site is defined as the vegetation composition which embodies the necessary attributes (species composition, production, structure, and cover) for all of the present and/or intended uses of an area. The DPC must be consistent with the site's documented capability to produce the required community. While this concept is related to the resource value rating described above, it has the advantage of considering all of the uses of a range site, not just a single use such as grazing, in determining the best composition of vegetation to meet those uses while protecting the site and soil from degradation.

The Forest Service uses a similar concept called "desired future condition" (DFC) which is defined as "the specific future condition that meets management objectives as identified in the Forest Plan or Allotment Management Plan. (It is expressed) in terms of future ecological status. . . . , desired soil protection. . . . , and loss of soil productivity."

For both the desired plant community and the desired future condition concepts, success of management is measured by progress in moving the composition of vegetation toward or away from the DPC or DFC.

C. Range Condition Related to Climax

The *second* definition for range condition given by the SRM is "the present state of vegetation of a range site in relation to the climax (natural potential) plant community for that site. It is an expression of the relative degree to which the kinds, proportions and amounts of plants in a plant community resemble that of the climax plant community for the site." This is related to "ecological status," which will be described in detail later and is the definition of range condition most commonly used in the United States.

D. Basis for Traditional Range Condition Concepts

The basic ideas behind the traditional range condition concepts given in the second SRM definition are based on the writings and concepts of succession and climax described in the early part of the 20th century by the classical ecologist, Frederick Clements. In 1919

Arthur Sampson proposed that the succession concept could be applied to rangelands to determine if they had been degraded by livestock grazing by measuring changes in plant species composition. In 1949 a formal procedure based on this concept was published [E. J. Dyksterhuis (1949). *J Range Manage.* **2**, 104–115] and quickly adopted by the Soil Conservation Service, other federal land management agencies, and the range management profession as a whole. This system enabled managers to quantify range condition and led to development of the range site classification system.

These 1949 concepts still dominate the thinking of range managers and are the basis for range condition ratings in the United States. This system assumes that, in a given area, a sequence of communities occur, through plant succession, until a relatively stable community in equilibrium with the environment is reached. This final community is the "climax" which is made up of two main types of plants: "decreasers" and "increasers." With heavy grazing, decreasers diminish in amount or importance in a community because they either are quite palatable to grazing animals and thus are consumed in preference to other plants or they are quite easily damaged by grazing (or both) (see Fig. 1).

Increasers are those plant species which initially increase under grazing pressure, presumably because they either are less palatable or are damaged less by grazing than the decreaser species. With prolonged heavy grazing the increasers will also begin to decrease. Invaders are species not originally present in the undisturbed or climax community but which enter the community after disturbance and then increase with continued heavy grazing or other disturbance. The relative amounts of each class of species, expressed as a percentage of each in the climax, deter-

mine the range condition class of a given sampled area.

The terms decreaser, increaser, and invader are not much used any more because a given plant species may have different palatability to different herbivores. A forb that is quite palatable to sheep might be a decreaser under heavy sheep grazing. If this species is unpalatable to cattle, it may be an increaser under heavy cattle grazing. These terms have been replaced by other terms to describe species such as "desirable" and "undesirable." However, the basic concept of the response of decreasers, increasers, and invaders to grazing still is essential to the understanding of traditional range condition classification based on succession and climax. [*See* RANGELAND GRASS IMPROVEMENT; RANGELAND GRAZING; RANGELAND PLANTS; RANGELAND SHRUBS.]

E. Range Condition Classes

The traditional range condition concepts describe retrogression, or community deterioration, along a gradient or continuum and is measured by a numerical score (Fig. 1). The four condition classes have been traditionally called: "excellent," "good," "fair," and "poor." These terms are applied to describe how close existing plant composition of a sampled rangeland area is to the "climax." These adjectives are still used by the Soil Conservation Service (USDA) to describe range condition. However the terms have caused some of the controversy concerning management of rangelands because they imply value judgments which may create unrealistic expectations about management success and the potential for improvement to higher stages of condition. As will be pointed out later, a fair condition rangeland may be fully meeting the needs for grazing and other multiple uses on a specific area of rangeland.

F. Ecological Status

The public land management agencies, the U.S. Forest Service and the Bureau of Land Management, now use the term "ecological status" to describe range condition. This is conceptually slightly different from that described above because the presumed end point of secondary succession is the "potential natural community" (PNC) instead of "climax." The Society for Range Management defined PNC as "the biotic community that would become established . . . if all successional sequences were completed without interferences by man under the present environmental

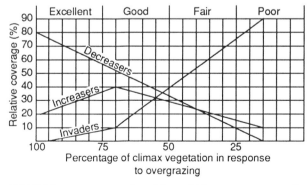

FIGURE 1 The conceptual model on which present range condition standards (presented schematically for any range site) in the United States are based. (From Dyksterhuis, E. J. (1949). *J. Range Manage.* **2**, 104–115.)

conditions." The main difference is that the PNC may include acclimatized nonnative species, whereas "climax," by definition, excludes these alien species.

The condition or state of the vegetation of a sampled area of rangeland is described in relation to the PNC, using the terms "early seral," "mid seral," "late seral," and PNC. Table I compares the terminology used in the two different methods of describing range condition.

G. Determining Range Condition

The percentage similarity to the climax or PNC is calculated by first sampling the vegetation on a range site and then expressing the amount of each plant species as a percent of the total. Originally, amount of plant cover for each species was determined and expressed as a percent of total plant cover (Fig. 1). In recent years, the amount of herbage or biomass of each species expressed as a percent of the total biomass on a range site has been used more often to determine condition. The percent of total composition of each species in the sampled area is then compared to the percentage assumed to have been in the "climax" (or PNC). The more the percentage of each species deviates from the amount in the climax, the lower the total score indicating the range condition or seral stage as shown in Table I.

Table II contains a simplified example of range condition calculations based on two areas on a "common" range site on shortgrass plains in Eastern Colorado. To determine range condition, the percentage of each species present in each area was compared with the percentage in the climax or reference list. If the amount in the sampled area was *larger* than the maximum allowable in the reference list, the amount in the reference list was recorded in the "% allowed" column. If the percentage composition of a species in the sampled area was *smaller* than that in the reference list, the actual percentage in the sampled area was

TABLE I

Comparison of Different Range Condition Terminology

% of "climax" or "PNC"[a]	Range condition class	Ecological status
76–100%	Excellent	PNC[a]
51–75%	Good	Late seral
26–50%	Fair	Mid seral
0–25%	Poor	Early seral

[a] PNC, potential natural community

recorded in the "% allowed" column. Thus, the smallest of the two figures was always recorded. The figures in the "allowed" column were added and the total score indicated the range condition rating as indicated in Table I. This example illustrates how two areas with very different species composition can be classified within the same range condition class.

The methodology used in making up the maximum allowable percentages in range site guides can lead to mistakes in determining range conditions. Often, the maximum allowable quantity allowed in the climax is an average from a composite list from a number of ungrazed reference areas. Because each area has a somewhat different mix of plant species, the amalgamation and averaging of such information ensures that no single site can achieve the maximum values. The author of Table II stated, "Similarities of only 70–80% between replicate sites is a normal characteristic of similarity indexes." Thus "excellent" condition (76–100% similarity) is essentially unobtainable. In the example in Table II, Area 1 was in the best condition of any area known to the range conservationists because it had not been grazed for several decades. Yet, it did not rate excellent because it had less blue grama and more western wheatgrass than "allowed" in the climax list. Area 3 had received heavy prolonged grazing pressure and was in the poorest condition of any unplowed area examined. However, the site guide classified it in the lower end of "good," not in "fair" (or even "poor") condition where it probably should have been rated.

II. Range Trend

Range condition or ecological status ratings taken at a single point in time, no matter how accurately obtained, may not be useful to a manager as the sole source of information. Range trend is the direction that range condition is moving over time. A range site in "fair" (mid seral) condition that has an improving trend often is quite a satisfactory situation and usually will not require any changes in stocking level or management. Even a range site in fair condition that has a stable trend may be maintaining site potential, protecting the soil from erosion and meeting all management objectives. Examples will be presented later. In contrast, a range in "good" (high seral) condition that has a downward trend may be of concern and require management changes. Thus, the combination of both condition and trend is a better indication of range "health" than range condition alone.

TABLE II

Example of How Range Condition Ratings Are Calculated from Two Rangeland Areas in Eastern Colorado[a]

Species (abridged)	Maximum % in "climax" community	Area 1		Area 3	
		% Present	% Allowed	% Present	% Allowed
Blue grama	40	10	10	47	40
Buffalo grass	5	0	0	30	5
Green needlegrass	20	20	20	0	0
Western wheatgrass	35	60	35	0	0
Red three-awn	1	0	0	10	1
Misc. forbs (1% spp. up to 10)	10	5	5	10	10
Other	1	5	1	3	1
Totals	—	100	71	100	57
Range condition class		Good		Good	

[a] *Source:* Wilson, A. D. (1989). The development of systems of assessing the condition of rangelands in Australia. *In* "Secondary Succession and the Evaluation of Rangeland Condition" (W. K. Lauenroth and W. K. Laycock, eds.), pp. 77–102. Westview Press, Boulder, CO.

The amount of actual change (upward or downward) in range condition over time obviously is quite important. Rate of change is also important but often is difficult to determine. Many factors, especially amount and timing of precipitation, have a very pronounced effect on different species on a range site. In a dry year or a series of such years, some species do not grow or may be present in very low amounts. The same species may be quite prominent in a wet year or sequence of years. Other species, such as many shrubs, may fluctuate less in production between years with different amounts of rainfall. Therefore, any accurate determination of range trend should be a comparison of initial range condition measured over several consecutive years with the same multiyear sampling of range condition at the end of the period. Weather conditions during the years sampled at the beginning and at the end should also be similar. Lack of sufficient funds and field personnel in land management agencies often preclude this ideal method of trend determination and trend often is determined from single beginning and end points. This can lead to erroneous conclusions about trend. Even large changes between years with greatly differing precipitation may not represent trend at all, but simply the reaction of the community to the weather.

Trying to determine trend and rate of change may also be complicated by other factors. Rangeland plant communities that may be in a stable state condition may be little influenced by external forces to move

the condition either upward or downward. This will be discussed later. As indicated earlier, trend toward or away from a "desired plant community" or a "future desired condition" indicates the relative success or lack of success of management.

III. Problems with the Current Range Condition Model

A. Criticisms of the Present Model

A great deal of dissatisfaction with the current range condition model, which emphasizes succession and climax, has been expressed by a number of U.S., Australian, and other range scientists. Some of the main "problems" that have been presented are:

1. Climax is the implied goal of management: Acceptance of the view that succession culminates in maximum stability, productivity, diversity, and other presumed desirable qualities leaves little choice but to manage for near-climax, or admit that the management goal is a second-rate, degenerated ecosystem.

2. Similarity to or departure from climax is not necessarily correlated with livestock forage production or any other resource values.

3. Similarity to climax does not provide an adequate basis for predicting response to management.

4. An index of departure of vegetation composition from climax does not provide information needed to identify where site degradation is occurring.

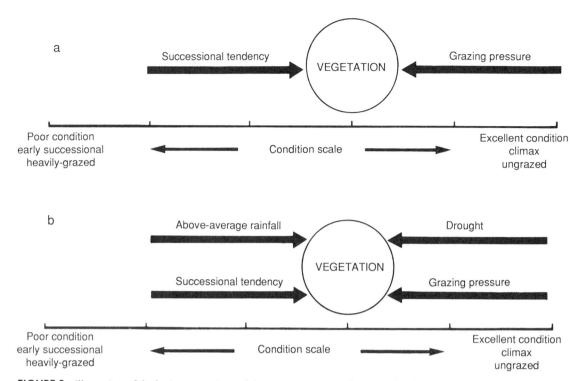

FIGURE 2 Illustration of the basic assumptions of the current range condition model showing: (a) successional tendency and grazing pressure as opposing forces and (b) incorporation of rainfall variability into the model. (From Westoby, M., et al. (1989). *J. Range Manage.* **42,** 266–274.)

5. Climax-based range condition assessment does not work on seeded ranges or forest ranges.

6. Range condition and trend assessment with climax-based concepts have not provided managers with an accurate indication of the state of our rangelands or their management.

7. Soil stability factors and other indicators to site productivity and stability need to be incorporated into the determination of range condition.

The basic, yet generally unwritten, assumptions of the current range condition model are: (1) The model supposes a given rangeland has a single persistent or stable state (the climax) in the absence of grazing; (2) grazing pressure produces changes which are retrogressive; (3) removal of grazing pressure allows succession to proceed toward the climax in a steady, continual process in the opposite direction of retrogression; (4) the stages or communities leading toward the climax are the same as those occurring during retrogression (i.e., there is no hysteresis); and (5) at a given stocking rate, the opposing grazing pressure and successional tendencies are opposite, producing an equilibrium.

In this model, all possible states of the vegetation can be arrayed on a single continuum (see Fig. 2a)

from heavily-grazed, early-successional, poor condition to ungrazed, climax, excellent condition. Trend is the direction of the movement of the composition of vegetation along the continuum. The model also recognizes that vegetation is affected when rainfall varies from year to year. It assumes that drought affects vegetation in a manner similar to grazing and, conversely, that above-average precipitation accelerates the successional tendency (see Fig. 2b). The model assumes that range condition can be modified continuously and reversibly by adjusting stocking rate and giving time for the range to adjust to it through succession. An alternative "state-and-transition" model, which will be discussed later, was proposed as a replacement to overcome the lack of hysteresis and other obvious limitations of the traditional model.

B. Observed Changes Are Not Accurately Predicted by the Model

The current model and procedures for determining range condition generally seem to predict changes caused by grazing well in the more humid grasslands of the Great Plains. These are the vegetation types in which the concepts were developed and it is not surprising that the concepts work reasonably well there. However, the concepts do not work consistently or well in shrub-dominated and some other vegetation types. The range management literature contains a great many references to this lack of applicability.

As early as 1949, the same year the present range condition model was published, other researchers reported that presence or absence of grazing had little effect on species composition on high elevation parks in Colorado dominated by Kentucky bluegrass. The same situation was found in the mixed grass prairie of western Manitoba where a pure Kentucky bluegrass stand also had become established. On shortgrass steppe areas in Colorado, several authors have stated that "there is no ecological basis for visual recognition and separation of vegetation changes induced by grazing from those induced by weather" and "conventional range condition classification on shortgrass plains serves no useful purpose."

Many researchers in rangeland vegetation types dominated by shrubs have long recognized that the prevailing range condition concepts do not seem to apply. On southwestern desert grasslands invaded by mesquite and other woody species, many research papers have reported little or no improvement in range condition following the removal of grazing.

Similar results have been reported in areas of the Great Basin dominated by sagebrush or salt desert shrub communities. Heavy grazing following settlement, along with control of fires, has resulted in large areas of the Great Basin being occupied by dense stands of big sagebrush. Sagebrush dominates the site with little change in species composition, even with long periods of complete rest from grazing such as in exclosures which prevent livestock grazing. Complete protection from grazing for periods ranging from 14 to 45 years has caused no significant change in vegetation composition and thus no improvement in range condition.

Other shrub-dominated vegetation types in the Great Basin react similarly. Studies of winterfat, Nuttall saltbush, shadscale, and black sagebrush communities have shown little difference between grazed and ungrazed areas and researchers have concluded that the concept of grazing succession is not meaningful in these shrublands.

Communities dominated by well-adapted introduced annuals often do not return to the original perennial community even with extended protection from grazing. Dynamics of these systems are not ex-

plained by the current range condition model. The California annual grassland originally consisted of perennial bunchgrasses. After settlement by the Spanish in the late 18th century, a number of well-adapted annuals from the Mediterranean region were introduced and almost completely replaced the perennial grassland communities. On 2.5 million ha of former sagebrush rangelands in southern Idaho and surrounding areas, the vegetation is now almost completely dominated by cheatgrass brome and other annuals. In both of these vegetation types changes in grazing will not result in the return of the perennial species and the only way to replace the annuals is to artificially plant perennials.

IV. Some New Ideas about Range Condition

A. Stable States

The original range condition model introduced in 1949 incorporated ecological thinking at that time. Some of these ecological concepts are now somewhat outdated and the range management profession needs to incorporate newer ecological concepts and information in the determination of range condition. One of these concepts discussed in the ecological literature is that of relatively stable domains or states with thresholds of environmental change between them.

The stable state idea has been incorporated into a "state-and-transition" model in which a "state" is a recognizable and relatively stable assemblage of species occupying a site. A "transition" between states is triggered by either natural events or by management actions or a combination of the two. In order to move from a state into a transition, a community has to cross a threshold which often is difficult or impossible to recross.

A state-and-transition model describing the dynamics of a sagebrush–grass system is shown in Fig. 3. The boxes represent recognizable stable states and arrows represent transitions. States I, II, and III occur in areas without annuals. State I represents the "climax" which occurred without livestock grazing but with occasional fires to kill or thin the sagebrush plants. Transition T1 represents heavy livestock grazing which damages the perennial understory plants and eventually converts the community to a stable, dense stand of sagebrush. As indicated previously, this state can maintain itself with little change even after decades of protection from grazing. Fire (T3),

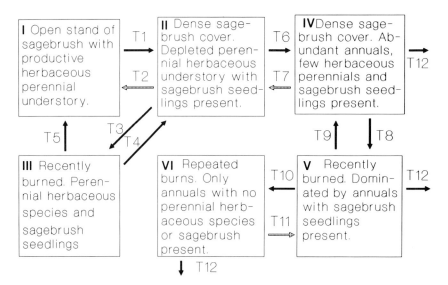

FIGURE 3 State-and-transition diagram for sagebrush–grass vegetation. **Catalogue of transitions:** Transition 1, heavy continued grazing. Rainfall conducive for sagebrush seedlings. Transition 2, difficult threshold to cross. Transitions usually will go through T3 and T5. Transition 3, fire kills sagebrush. Biological agents such as insects, disease or continued heavy browsing of the sagebrush by ungulates could have the same effect over a longer period of time. Perennial herbaceous species regain vigor. Transition 4, uncontrolled heavy grazing favors sagebrush and reduces perennial herbaceous vigor. Transition 5, light grazing allows herbaceous perennials to compete with sagebrush and to increase.

If climate is favorable for annuals such as cheatgrass, the following transitions may occur: Transition 6, continued heavy grazing favors annual grasses which replace perennials. Transition 7, difficult threshold to cross. Highly unlikely if annuals are adapted to area. Transition 8, burning removes adult sagebrush plants. Sagebrush in seed bank. Transition 9, in absence of repeated fires, sagebrush seedlings mature and again dominate community. Transition 10, repeated burns kill sagebrush seedlings and remove seed source. Transition 11, difficult threshold to cross if large areas affected. Requires sagebrush seed source. Transition 12, intervention by man in form of seeding of adapted perennials. (From Laycock, W. A. (1991). *J. Range Manage.* **44,** 427–433.

or some other force that kills adult sagebrush plants, will release the perennial understory from the competition from the sagebrush and allow it to increase with proper grazing management. The sagebrush, which is a natural part of the community, returns over time and the site can return to something resembling State I.

In areas with well-adapted annuals, such as cheatgrass brome, continued heavy grazing (T6) can convert the stand to a thick stand of sagebrush with annuals instead of perennials in the understory. Fire (T8) and repeated fire (T10) can convert the site into a stable community dominated by annuals (State VI). This transition generally is irreversible. Fire return frequency in cheatgrass-dominated communities may be less than 5 years which is frequent enough to prevent the reestablishment of the original perennials including sagebrush. Thus State VI becomes the stable annual community described previously and does not return to any previous state in a time measured by decades or even centuries regardless of grazing pressure.

B. Domain of Attraction

Another model that may conceptually describe range condition somewhat more realistically than the present model is the "domain of attraction" which depicts a vegetation community as a ball or marble in a cup or trough. The boundaries of each cup represent the range of environmental conditions under which that community is stable. Different communities may have different numbers of stable states. Figure 4a represents a community that is *globally* stable because, after all disturbances, it will return to its original configuration. This somewhat accurately represents the "climax" or "successional" range condition model currently used because a disturbed rangeland community will always return to the climax condition (bottom of the cup) via internal feedback mechanisms when the disturbance stops.

Figure 4b represents a community which has multiple stable states, only three of which are shown. It is *locally* stable at Configuration I. If this represents climax or "excellent" range condition, minor grazing perturbations may change the composition of the

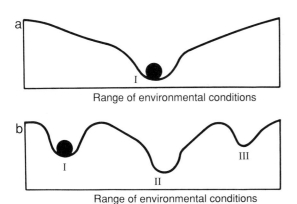

a

Range of environmental conditions

b

I

II

III

Range of environmental conditions

FIGURE 4 Diagrams illustrating the ball and cup analogy to illustrate stability concepts. A community is represented as a black ball on a topographic surface (cup or trough) which represents the range of environmental conditions under which the community is stable. In (a) the community is globally stable because after *all* perturbations, it will return to Configuration I. In (b) the community is locally stable, but if perturbed beyond a certain critical range, it will cross the threshold between Configurations I and II and move to a new locally stable configuration at "II." [From Krebs, C. J. (1985). Ecology: The experimental analysis of distribution and abundance. 3rd Ed. Harper and Row, New York.]

community somewhat (i.e., move the ball up the side of the cup) but, once the disturbance is stopped, the community returns to its original climax state. While the community is with the bounds of Configuration I, current concepts about range condition and the reversibility of change may fit quite well. However, if a grazing disturbance is strong or prolonged enough, the community will cross a threshold between Configurations I and II and move to a new locally stable Configuration II. For a rangeland community in this new stable state, the conventional range condition standards *no longer apply* either to describe the state or identify the forces required to move the community out of this state.

Both the state-and-transition and the domain of attraction models depict the dynamics of rangeland communities in discrete "steady states" with definite thresholds to replace the linear, reversible climax or "successional" range condition model now in use in the United States.

C. Implications for Rangeland Management

Our current range condition model often describes the early stages of deterioration caused by grazing reasonably well for most rangeland communities. The main problem with the current model is that it predicts that reduction or removal of grazing pressure will allow a deteriorated rangeland community to

improve toward "climax" through secondary succession by moving through the same stages observed during retrogression (see Fig. 2). Some more mesic grassland and mountain communities react this way but most of the arid and semi-arid rangeland vegetation types remain stable at one or more lower successional states for long periods of time, even when grazing is removed.

It is important that managers be able to recognize stable states of range condition and when and where they occur. If they do not recognize a stable state situation, they usually assume that reducing the stocking rate will result in an increase in range condition as predicted by the prevailing model. If the vegetation is in a stable state condition, little or no change will take place. Some managers are beginning to realize this. One [A. H. Winward (1991). *Grazier* **273**, 1–4] stated, "There are more areas of sagebrush–grass lands in the western United States being held in a low ecological status in the last decade due to abnormally high sagebrush cover and density than currently is occurring due to livestock grazing."

The transitions between relatively stable states often require a combination of climatic circumstances and management actions. Climatic circumstances which allow favorable transitions, i.e., toward a desired alternate state, could be coupled with the necessary management actions to achieve that objective but only if the manager recognizes the opportunity.

Rangeland researchers are beginning to experiment with alternative models to describe and quantify range condition. Only when additional research is completed and managers recognize the existence of stable states will true changes in range condition concepts take place.

V. Conditions on U.S. Rangelands

All major state and federal management agencies in the United States use the same, climax-based range condition model presented in section I. However, this is where the similarity ends. The three major federal land management and advisory agencies (Soil Conservation Service, Forest Service, and Bureau of Land Management) currently use somewhat different terminology and different field methodology to determine range condition. This makes it difficult to accurately compare condition data from the different agencies. A further complication is that both range condition criteria and sampling methodologies have changed over time in all three agencies which makes comparing trends in range condition over time difficult. The Society for Range Management has been working to standardize terms and concepts among agencies in North America.

Despite the deficiencies and weaknesses outlined above, comparisons of range condition over time are useful. Such comparisons are even more important now because of a current active campaign against livestock grazing on public land in the Western states being waged by certain environmental groups who claim that public rangelands are severely overgrazed and deteriorated. To determine if these claims of overgrazing have any validity, the following sections present trends in range condition over time on rangelands of various ownership in the United States.

A. Bureau of Land Management Rangelands

Table III shows the condition of BLM rangelands from 1936 until 1989. There has been a definite improvement in range condition. Since 1936, the amount of BLM rangeland in the best two conditions has doubled and the amount in "poor" condition has decreased by more than one-half. Despite this relatively clear picture of range improvement since 1936, criticisms continue. The Natural Resources Defense Council (NRDC) concluded that the BLM figures for 1984 and 1989 "do not reveal any significant improvement in range health." This statement indicates little recognition that most BLM and other rangelands have improved to a "stable state" condition in which little improvement takes place in short or even relatively longer periods of time. Because of this stable state situation, the removal of livestock from public lands being called for would result in little or no change in range condition.

The NRDC also manipulated the 1989 BLM data and stated that 68% of the BLM range is in "unsatisfactory" condition. They apparently lumped the

"fair," "poor," and "unclassified" ranges together to get this figure. The unclassified BLM lands included seeded and ungrazed areas and thus should not be lumped with any condition. In addition, the BLM has emphasized that fair condition range often is "a very satisfactory stage for producing high-quality forage, wildlife cover, watershed protection and an aesthetic landscape."

Examples of this include critical mule deer winter ranges which often are classified in fair condition because of the abundance of sagebrush which is critical to deer survival in deep snow. When livestock grazing is removed, prolonged heavy winter use of the shrubs by deer can result in an eventual conversion of these rangelands to a grass-dominated rangeland which is in excellent condition but which makes very poor deer winter range because the available forage is buried deep below the snow cover.

Another example is introduced pasture species (smooth brome, timothy, etc.) which often invade meadows on public land. Because these highly adapted species do not "count" in condition determinations, such sites often rate in fair condition even though the introduced species are highly productive and may protect the site as well as native species. A "low" range condition rating in the examples given above does not mean that the sites are in a deteriorated or nonproductive state. Instead, they usually are productive and are meeting or exceeding management objectives.

B. Rangelands on Private Land

Data from the National Resources Inventory by the Soil Conservation Service show that range conditions on private rangelands has also improved considerably (Table IV). The improvement in condition is quite similar to that on BLM lands, over a shorter period of time.

TABLE III

Condition of Bureau of Land Management Lands, 1936–1989

	Percent of each condition on BLM lands				
Year	Potential natural (excellent)	Late seral (good)	Mid seral (fair)	Early seral (poor)	Unclassified
1936	2	14	48	36	—
1975	2	15	50	33	
1984	5	31	42	18	4
1989	3	30	36	16	14

TABLE IV

Condition of Private Rangelands, 1963–1989

	Percent of each condition on private lands				
Year	Excellent	Good	Fair	Poor	Unclassed
1963	5	15	40	40	—
1977	12	28	42	18	—
1982	3	31	45	17	5
1987	3	30	47	14	6

C. Forest Service Rangelands

No comparisons over time were available for Forest Service rangelands. In 1987, range conditions on Forest Service rangelands were as follows:

Range condition	%
Potential natural community (excellent)	15
Late seral (good)	31
Mid seral (fair)	39
Early seral (poor)	15

These figures are similar to those on BLM land in 1989 and on private land in 1987. An improving trend over time would be expected because the Forest Service has actively managed grazing with the goal of improving range conditions on National Forests for more than 90 years.

VI. Range Condition Trends on U.S. Rangelands

The long-term condition data from private and BLM lands clearly indicate that conditions of rangelands have improved greatly over time. This would imply that trend at the present time is either still improving or, as might be expected from the preceding discussion on stable states, now unchanging or stable. The Society for Range Management, with the assistance of the federal agencies, compiled an "Assessment of Rangeland Condition and Trend of the United States: 1989." The trend figures from this publication are shown in Table V and they confirm the conclusions from the condition data over time, i.e., 84% of the rangelands in the United States are in either an upward or stable condition.

TABLE V

Apparent Trend in Range Condition in the United States

Land or agency	Year	Percent in each category			
		Upward	Stable	Downward	Undetermined
BLM	1986	15	64	14	6
FS	1987	43	43	14	—
Private (SCS)	1982	16	70	14	—
Weighted average[a]		18	66	14	2

Source: Society for Range Management (1989). "Assessment of Rangeland Condition and Trend of the United States, 1989." Public Affairs Committee, Society for Range Management, Denver, CO.
[a] Determined by adding acreage for each land ownership in each class and computing overall percentage.

VII. Conclusions

Based on both the reported conditions and current trend in condition, the rangelands of the western United States are not in unsatisfactory condition but, instead, are in the best condition in this century. Some exceptions to this broad conclusion may be some brush-dominated areas where the Forest Service or BLM have stopped using range improvement practices such as brush control because of budget constraints or environmental concerns. In the absence of naturally occurring fires, species such as sagebrush will gradually and naturally reinvade both grazed and ungrazed areas. As the sagebrush becomes thicker, range condition ratings decline.

As a result of the overall increase in range condition, federal land management agencies, livestock operators, and range management professionals deserve recognition and credit for steadily improving range conditions resulting from successful range management. What is needed now is for both range researchers and range managers to recognize that more realistic concepts are needed to measure and determine range condition and trend. Dialogue is beginning to take place toward this end and future editions of this encyclopedia should have a greatly different discussion of range condition and trend.

Bibliography

Bureau of Land Management (1990). "State of the Public Rangelands 1990." Washington, DC.

Clements, F. E. (1916). "Plant Succession." Carnegie Inst. Wash. Pub. 242.

Dyksterhuis, E. J. (1949). Condition and management of rangeland based on quantitative ecology. *J. Range Manage.* **2,** 104–115.

Friedel, M. H. (1991). Range condition assessment and the concept of thresholds: A viewpoint. *J. Range Manage.* **44,** 422–426.

Lauenroth, W. K., and Laycock, W. A. (eds.) (1989). "Secondary Succession and the Evaluation of Rangeland Condition." Westview Press, Boulder, CO.

Laycock, W. A. (1991). Stable states and thresholds of range condition on North American rangelands: A viewpoint. *J. Range Manage.* **44,** 427–433.

Smith, E. L. (1989). Range condition and secondary succession: A critique. *In* "Secondary Succession and the Evaluation of Rangeland Condition" (W. K. Lauenroth and W. A. Laycock, eds.), pp. 103–141. Westview Press, Boulder, CO.

Society for Range Management. (1989). "A Glossary of Terms Used in Range Management," 3rd Ed. Society for Range Management, Denver, CO.

Society for Range Management. (1989). "Assessment of Rangeland Condition and Trend of the United States, 1989." Public Affairs Committee, Society for Range Management, Denver, CO.

U.S. General Accounting Office. (1988). "More Emphasis Needed on Declining and Overstocked Grazing Allotments." GAO/RCED-88-80. Washington, DC.

Westoby, M., Walker, B., and Noy-Meir, I. (1989). Opportunistic management for rangelands not at equilibrium. *J. Range Manage.* **42,** 266–274.

Wilson, A. D. (1989). The development of systems of assessing the condition of rangelands in Australia. *In* "Secondary Succession and the Evaluation of Rangeland Condition" (W. K. Lauenroth and W. A. Laycock, eds.), pp. 77–102. Westview Press, Boulder, CO.

Winward, A. H. (1991). A renewed commitment to management of sagebrush grasslands. *The Grazier* **273,** 1–4.

Rangeland Ecology

R. K. HEITSCHMIDT, *USDA-Agricultural Research Service, Montana*

Glossary

Agriculture The business of capturing solar energy and transferring it to humankind for their use

Assimilation efficiency Proportion of energy consumed by organism that is actually converted into new biomass (i.e., growth)

Consumer organisms Heterotrophic organisms, primarily animals, that acquire energy by ingesting other organisms

Decomposers The last or final consumer organism, primarily bacteria and fungi, in a food chain

Detrital food chain Food chain consisting of producer being consumed by decomposer organisms

Ecological efficiency Measure of how efficient solar energy is captured and transferred via food chains

Ecological stability In reference to an ecosystem's ability to maintain itself

Ecosystem Any definable assemblage of living organisms and their associated physical and chemical environment

Ecology The study of the interrelationships of organisms and their environment

Food chain Process whereby energy is transferred from producers to consumers

Grazing food chain Food chain consisting of producer being consumed by consumer organisms other than decomposer organisms

Harvest efficiency Proportion of energy available to an organism that is actually consumed

Primary productivity Measure of how much solar energy is captured by producer organisms

Producer organisms Autotrophic organisms, primarily green plants, that acquire energy by fixing solar energy

Rangeland Land on which the indigenous vegetation is predominantly grasses, forbs, or shrubs, and is managed as a natural ecosystem

Secondary productivity Measure of how much of the solar energy initially captured by producer organisms is converted into consumer biomass

Trophic level Refers to steps in food chain whereby producers occupy first trophic level, herbivores second trophic level, and decomposers last trophic level

Rangeland ecology is the study of how rangeland ecosystems function over time and space. It is the study of the interrelationships between rangeland organisms and their rangeland environment. Rangeland ecology is the core science of a multitude of scientific disciplines that collectively are identified as rangeland science.

I. The Structure and Function of Ecological Systems

A. Structural Organization

The organizational structure of ecological systems (i.e., ecosystems) centers around the functional roles of the component parts of the system rather than taxonomic entities. As such, the organizational structure of ecosystems is commonly built around four component parts; one abiotic and three biotic (Fig. 1).

The abiotic component is the inorganic, nonliving component of ecosystems. It defines the physical and chemical environment of the biotic component. The abiotic component includes such things as climatic (e.g., temperature and precipitation), atmospheric (e.g., gaseous composition and radiation), edaphic

ABIOTIC COMPONENT

Radiation   Soils
Climate   Geography
Atmosphere   Fire

BIOTIC COMPONENT

Organism	Function	
Plants	Producers	} Autotrophs
Herbivores	Primary Consumers	
Carnivores	Secondary Consumers	} Heterotrophs
Decomposers	Primary, Secondary, and Tertiary Consumers	

FIGURE 1 Generalized description of the structure of ecological systems. The abiotic (nonliving) component comprises the physical and chemical environment of the biotic (living) component. [Reprinted with permission from Heitschmidt, R. K. and J. W. Stuth (eds.) (1992). "Grazing Management: An Ecological Perspective," p. 12. Copyright 1992 by Timber Press, Inc., Portland, OR.]

(e.g., nutrient concentrations and texture), and topographic features (e.g., slope and aspect), to name but a few of a near endless list of components.

The biotic or living component of ecosystems is conveniently subdivided into three components: producers, consumers, and decomposers. Producers are autotrophic (i.e., self-nourishing) organisms that acquire energy by fixing solar energy. Thus, the green plants that cover the earth's surface, including the phytoplankton that flourish in the earth's waters, are the principle producer organisms of our biosphere. In contrast to the energy acquisition tactics of the producer organisms are the energy acquisition tactics of the consumer components. Consumers are heterotrophic organisms, that is, organisms that attain their energy by ingesting other organisms. In other words, consumers are animals except in rare instances (e.g., the Venus fly trap). Consumer organisms are often classified by the type of foodstuff they consume (Fig. 1). For example, herbivores consume plants (e.g., cattle), carnivores consume other animals (e.g., lions), and omnivores consume both plants and animals (e.g., humans). Decomposers are those consumers, mostly bacteria and fungi, that functionally fulfill the role of last or final consumer of organic matter. Although all consumers act as partial decomposers in that they reduce a portion of their food to simple abiotic substances such as water and carbon dioxide, decomposers are those consumers that specifically ensure that the task of decomposition is complete. [*See* SOIL MICROBIOLOGY.]

B. Functional Attributes

The functional attributes of ecological systems are related to two fundamental phenomena: energy flow and nutrient cycling. Life is energy and the integrity of all ecosystems is dependent on efficient flow of energy through the system and the efficient cycling of the raw materials (i.e., nutrients) required to capture and process this energy.

1. Energy Flow

Food chains are energy processing pathways that determine pattern of energy flow through an ecosystem. These flows are governed by the first two laws of thermodynamics. In their simplest forms, these laws state that although energy can be transformed, it can never be created nor destroyed (1st Law) and that no transformation can be 100% efficient (2nd Law). The impacts of these laws on ecological systems are many, but the major effects are depicted in Fig. 2 wherein amount of fixed energy decreases whenever energy (i.e., food) is transferred from one organism to another (i.e., the predator). This decrease in fixed energy at each step results from the energy expenditures required to simply maintain an organism. This expenditure is commonly referred to as respiration (Fig. 2) which is the sum total of the energy in a system that is required to simply maintain the existing biotic components.

There are two fundamental types of food chains: detrital and grazing. The first step or trophic level in both chains is the primary producers (i.e., green plants) which convert solar energy to chemical energy via photosynthesis. Thereafter, the primary consumer determines the pattern of energy flow. If the primary

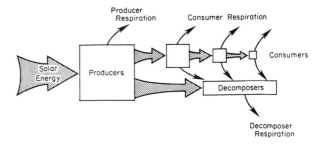

FIGURE 2 Simplified illustration of energy flow through ecological systems. Solar energy is initially captured by primary producers and transferred through at least one consumer feeding level to form the grazing food chain, or directly into the decomposer compartment to form the detrital food chain. [Reprinted with permission from Heitschmidt, R. K., and Stuth, J. W. (eds.) (1992). "Grazing Management: An Ecological Perspective," p. 13. Copyright 1992 by Timber Press, Inc., Portland, OR.]

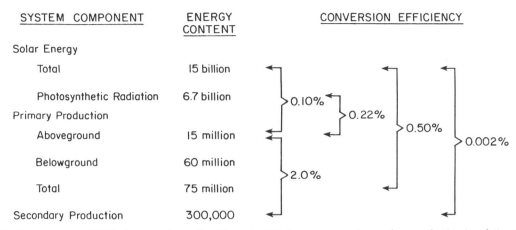

FIGURE 3 Energy content (MJ/ha/year) and transfer efficiencies (%) for primary and secondary production in relation to total and photosynthetically active solar radiation. Energy values are calculated by multiplying primary and secondary production values by 19.7 and 23.5 MJ/kg, respectively. Conversion efficiencies represent the quotient of two energy values at specified locations within the system multiplied by 100. [Reprinted with permission from Heitschmidt, R. K., and Stuth, J. W. (eds.) (1992). "Grazing Management: An Ecological Perspective," p. 17. Copyright 1992 by Timber Press, Inc., Portland, OR.]

consumers are decomposers, the defined food chain is detrital; otherwise, the defined chain is grazing (Fig. 2).

Ecological or conversion efficiencies refers to how efficiently the biotic components of ecosystems capture and process energy. There are many ways that these efficiencies can be calculated, all of which are appropriate depending upon point of interest. For example, primary productivity is a measure of how much solar energy is captured by primary producers and the ratio between total amount of solar energy available and total primary production is a measure of ecological efficiency. Similarly, secondary productivity is a measure of how much primary production was captured by primary consumers and converted into new biomass (i.e., growth), and the ratio between amount of primary production and amount of secondary production is another measure of ecological efficiency. The ratio of amount of solar energy available to the primary producers and amount of secondary production is also a measure of ecological efficiency. Thus, the manner in which ecological efficiencies are calculated will vary depending upon interest (Fig. 3), but in all instances, they are a reflection of the efficiency whereby food chains process energy.

Generally, ecological efficiencies, irregardless of ecosystem or tropic level, are much less than expected. For example, on a worldwide basis, the estimated efficiency whereby green plants capture solar energy is <2%. This apparent inefficiency is the result of three factors. The first limitation stems from the fact that only about 45% of the solar energy reaching the earth's surface is photosynthetically active radiation (PAR), meaning that about 55% of the solar energy reaching the earth's surface is unusable relative to its capture and conversion into chemical energy by green plants. The second constraint is that the inherent physiological capacity of green plants to fix PAR is limited. Maximum efficiencies of solar energy capture for green plants growing under optimal conditions (i.e., irrigated, fertilized, etc.) range from about 10 to 20%. The third major limitation is that nature seldom provides the ideal plant growth or solar energy fixation environment with regards to water, temperature, and soil nutrient regimens. Evidence of the existence of these limitations is reflected in our extensive use of supplemental water (i.e., irrigation) and soil nutrient amendments (i.e., fertilizer) in agronomic systems.

The ecological efficiencies associated with secondary productivity are also often much less than expected. The reasons for these inefficiencies are somewhat analogous to those affecting efficiency of solar energy capture. For example, just as only about 45% of all solar energy is "available" for fixation by green plants, so is only a portion of the energy fixed by green plants available for secondary production. This is because a large portion of the energy fixed by green plants is processed via the detrital rather than grazing food chains. This aspect of secondary production is commonly referred to as harvest efficiency which is

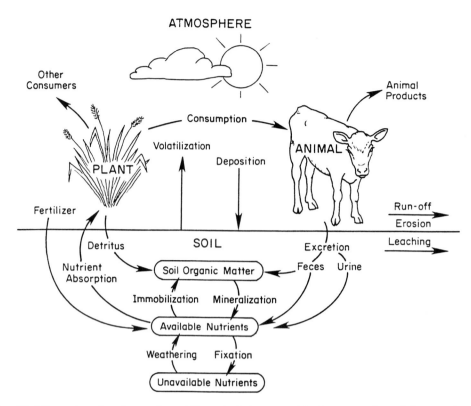

FIGURE 4 Simplified illustration of nutrient cycling within ecological systems. Nutrients move from their respective reservoirs within the abiotic component of the system, into the biotic component, and back into the environment in a cyclic pattern. [Reprinted with permission from Heitschmidt, R. K., and Stuth, J. W. (eds.) (1992). "Grazing Management: An Ecological Perspective," p. 15. Copyright 1992 by Timber Press, Inc., Portland, OR.]

a measure of the proportion of energy available to an organism that is actually consumed by that organism.

Secondary productivity is affected also by assimilation efficiency which is the inherent physiological capacity of a consumer to convert consumed energy into stored energy. Although conversion efficiencies vary in part as a function of animal species and quality of food, inherent maximums are near 10%.

A third factor limiting the overall efficiency of secondary production is efficiency of primary production (i.e., efficiency of solar energy captured). Regardless of how inherently efficient a consumer may be at capturing (i.e., harvesting) and processing (i.e., converting) energy, a consumer's well-being is dependent first on magnitude of primary production.

2. Nutrient Cycling

Nutrient cycling is the second indispensable function of ecosystems. Essential nutrients (e.g., nitrogen, carbon, oxygen, etc.) can be viewed as the abiotic raw materials required to maintain the structures and associated mechanisms required to capture and process energy.

In contrast to energy flowing through an ecosystem via food chains, nutrients are cycled via food chains in concert with natural geochemical processes (Fig. 4). The fundamental cycle revolves around the assimilation of nutrients by the primary producers followed by the sequential reduction of complex organic compounds to simpler, less complex forms by consumers with the decomposers completing the task as final consumers.

II. The Structure and Function of Rangeland Ecosystems

A. Abiotic Component

The abiotic components of greatest interest to rangeland ecologists are climatic, edaphic, and topographic features. This is because of their overwhelming impact on ecosystem function with particular regard to primary productivity.

Rangelands are characteristically located in arid and semi-arid regions. About 80% of U.S. rangelands are located in regions where average annual precipitation is less than 500 mm. Moreover, seasonal and/or annual droughts are common (Fig. 5). For example, in the Great Plains of the United States, about 25% of all years are drought years (i.e., precipitation <75% of average). [See RANGELAND.]

Rangeland soils are generally less fertile than cropland soils. This is primarily because climatic conditions (i.e., aridity) and/or topographic features (e.g., steep slopes) have restricted *in situ* soil development over time. [See RANGELAND SOILS.]

Although the topography of rangelands varies broadly, it is probably the major factor preventing cropping of rangelands. The two major topographic features of greatest interest to ecologists are slope and aspect particularly with regard to their effects on site aridity. Generally, site aridity increases as: (1) slope increases; (2) aspect shifts toward the equator (i.e., south in the Northern hemisphere and north in the Southern hemisphere); and (3) position on slope moves upward.

B. Producers

The primary producer component of rangeland ecosystems normally consists of a multitude of plant species. It is not unusual to find >50 plant species growing on a hectare of rangeland. Moreover, it is not unusual to find this wide array of plant species is both structurally and functionally diverse varying in

growth forms (e.g., grasses, forbs, and shrubs), growth and reproductive strategies (e.g., annuals and perennials), photosynthetic capacities [e.g., cool-season (C_3) and warm-season (C_4) species], and stature (e.g., short-, mid-, and tallgrasses). [See RANGELAND GRASS IMPROVEMENT; RANGELAND PLANTS; RANGELAND SHRUBS.]

This high level of diversity of producer species is a fundamental characteristic of most natural ecosystems including rangelands. It is an important characteristic because it provides functional stability to rangeland ecosystems. This increased stability results because the greater number of producer species present provides a greater opportunity for an ecosystem to "survive" the impact of various disturbance agents, such as natural drought and/or intense levels of defoliation. This is a logical thesis since the "survival" of an ecosystem is dependent first on its ability to capture solar energy. Moreover, for an ecosystem to maintain itself over time, it must have the capacity to repeatedly capture solar energy. The diverse assemblage of primary producers characteristic of most natural ecosystems provides this opportunity.

Because the integrity of an ecosystem is dependent first on capturing solar energy, estimates of primary productivity are critical for the development of appropriate ecosystems management schemes. Unfortunately, accurate estimates of primary production in even simple, single species agroecosystems (e.g., corn field) are difficult to attain because they require a detailed accounting of not only aboveground seed and fodder yields but also belowground root yields, a most difficult parameter to estimate. Moreover, because of the impact of fluctuating abiotic variables on plant growth, biomass accumulation rates (i.e., production) vary widely over time (i.e., among seasons and years) and space (i.e., between fields and across regions). Thus, to accurately assess primary productivity in even the simplest of agroecosystems requires tremendous resources. However, the magnitude of resources required to accurately assess primary productivity in single-species systems is only a fraction of the resources required in a multi-species rangeland systems because of the need to quantify primary production for every plant species present. Yet, despite these difficulties, many primary productivity estimates for rangeland ecosystems are available with most aboveground productivity estimates averaging <3000 kg ha^{-1} year^{-1} (Table I) whereas belowground estimates generally average <5000 kg ha^{-1} year^{-1}. Although these levels of productivity may seem substantial when evaluated on a biomass basis, evaluation

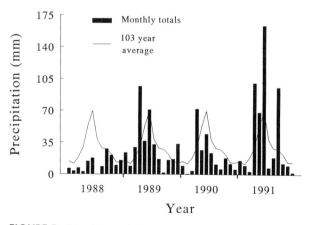

FIGURE 5 Monthly precipitation at Miles City, Montana, from 1988 through 1992 and long-term average. Variation among years in monthly precipitation are typical of most rangelands in that annual and seasonal droughts are common.

TABLE I

Aboveground Herbage Production Estimates for Several North American Rangelands

	kg ha^{-1} year^{-1}
Annual grasslands	3000–4000
Shortgrass prairie	1000–2000
Mixed grass prairie	2000–3000
Tallgrass prairie	4000–5000
Shrub steppe	500–1000
Desert	200–500
Piñon–juniper woodland	200–1000
Southern pine forest	2000–4000
Alpine tundra	500–1000

Source. Lauenroth, W. K. (1979). Grassland primary production: North American grasslands in perspective. *In* "Perspectives in Grassland Ecology" (N. French, ed.), p. 3–24. Springer-Verlag, New York. Holechek, J. L., R. D. Pieper, and C. H. Herbel. (1989). "Range Management Principles and Practices." Prentice-Hall, Englewood Cliffs, NJ.

on an ecological (i.e., energy flow) basis reveals efficiency of solar energy capture in most rangelands is <1% (Fig. 3).

C. Consumers

The consumer component of most natural ecosystems including rangelands is even more diverse than the producer component. The major reason for this is that total number of animal species, on a worldwide basis, grossly exceeds number of plant species. Thus, the probability that consumer diversity will exceed producer diversity in any given ecosystem is considerable.

The functional diversity of the consumer component of ecosystems is also much greater than that in the producer component. This is because many consumer organisms may simultaneously function as a primary (i.e., herbivore), secondary (i.e., carnivore), and tertiarary (i.e., carnivore) consumer. For example, humankind can functionally act as a primary, secondary, and tertiary consumer at a single meal by consuming a vegetable or cereal grain, beef steak, and an egg derived from a grasshopper fed chicken. This is in contrast to the producer organisms of which most only function as primary producers.

Just as diversity in the producer component limits the ease of attaining precise and accurate primary production estimates, so does diversity in the consumer component limit accuracy of secondary productivity estimates. This is largely because total secondary pro-

ductivity in an ecosystem is the sum total of all increases in herbivore mass (i.e., energy) over a specified period of time. Thus, to attain accurate total secondary productivity estimates requires the growth and reproductive performance of every consumer be quantified concurrently, a near impossible task. Still, secondary productivity estimates for targeted herbivores, particularly various species of livestock, are readily available and most have shown levels of ecological efficiency <1% (Fig. 3). [*See* RANGELAND GRAZING.]

III. Ecological Systems, Efficiencies, and Agriculture

To understand range ecology from an agricultural perspective, requires first an understanding of agriculture from an ecological perspective. The objective herein is to briefly outline this perspective.

Agriculture is traditionally defined as the business of producing food and fiber. But basic understanding of the structure and function of ecological systems reveals that agriculture is really the business of capturing solar energy and transferring it to humankind for their use. It can be reasoned then that success in agriculture is closely linked to the employment of management tactics that enhance efficiency of solar energy capture and/or efficiency of harvest and/or efficiency of assimilation.

Examples of such management practices are numerous. For example, common tactics utilized to enhance efficiency of solar energy capture (i.e., primary production) include irrigation, fertilization, and the planting of hybrid seeds. Two examples of tactics used to improve the efficiency whereby solar energy is transferred to humankind (i.e., harvest efficiency) are the use of insecticides and livestock grazing of postharvest residue. Insecticides are often employed to shift the flow of captured solar energy from food chains that do not include humankind (e.g., rangeland forage → grasshoppers → decomposer) to those that do include humankind (e.g., rangeland forage → livestock → humankind → decomposer). This shift is achieved by simply eliminating the competing consumer. Likewise, livestock grazing of postharvest residue works in a similar fashion in that it shifts the flow of energy from a detrital food chain (e.g., corn stalks → decomposers) to a grazing food chain that includes humankind (e.g., corn stalks → livestock → humankind → decomposers).

Essentially all examples of management tactics employed to enhance efficiency of assimilation are related to food quality. In fact, food quality can be defined strictly by its relative effect on efficiencies of conversion in that high- and low-quality foods are those that result in high and low net energy gains to the consuming organisms. For example, rangeland forages are deemed low-quality human foodstuff and high-quality ruminant livestock feedstuff. The reason for this disparity is that ruminants digestive systems are such that they can process range forages in a manner whereby they can derive most of their life-giving nutrients from the forage. This is in contrast to humankinds' digestive systems which are incapable of effectively digesting these same forages. Thus, the assimilation efficiency of range forages is low for humankind and high for ruminants. [*See* ANIMAL NUTRITION, RUMINANT; FORAGES.]

Efficient production of fiber (e.g., cotton, timber, and wool) is also dependent on the efficient capture of solar energy and its subsequent harvest. That is why cotton, for example, is often irrigated and fertilized (i.e., increase efficiency of solar energy capture). But in contrast to food production practices, postharvest processing of fibers is designed primarily to interrupt food chains and prevent consumption of the fiber (e.g., termites consuming wood).

IV. Rangeland Ecology and Management

Ecology, in it broadest sense, is the fundamental science driving managerial decisions in any natural ecological system including rangeland ecosystems. Successful management is related directly to humankind's level of understanding of the fundamental relationships that exist between organisms and their abiotic (i.e., nonliving) and biotic (i.e., other living organisms) environment. This is particularly so when people are functionally viewed as energy processing, nutrient cycling organisms that occupy "influential" positions relative to the shaping of the structure and function of ecosystems.

It is from our basic understanding of the structure and function of rangeland ecosystems and agriculture in general that we recognized that grazing by large herbivores is the principle form of agriculture practiced on rangelands. This is necessarily so because most of the primary producers growing on rangelands do not meet the energy and nutrient requirements of

humankind (i.e., low-quality foodstuff). They do, however, meet the requirements of many large herbivores (i.e., high-quality foodstuff), of which the same meet human requirements. Thus, in contrast to many forms of agriculture, humankind must act primarily as a secondary (i.e., top carnivore) rather than primary (i.e., herbivore) consumer in rangeland agricultural systems.

It is from our basic understanding of rangeland ecology that we recognize that the fundamental challenge in rangeland agriculture (i.e., grazing) is to balance the delicate, often antagonistic relationships between efficiency of solar energy capture and its subsequent harvest and conversion. The fundamental ecological dilemma is that it is not possible to maximize efficiency of solar energy capture, efficiency of harvest, and efficiency of assimilation simultaneously (Fig. 6). This is true regardless of whether the consumers of interest are domesticated (i.e., livestock) or undomesticated (i.e., wildlife) herbivores.

It is also from a basic understanding of rangeland ecology that we recognize that the fundamental principle of rangeland management is to control the frequency and defoliation of individual plants. Focus on

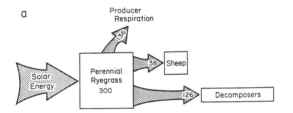

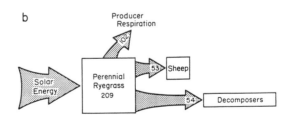

FIGURE 6 Energy capture and flow (kg carbon/ha/day) within (a) leniently and (b) severely grazed perennial ryegrass pasture. A greater amount of solar energy is converted into ryegrass production in the leniently grazed picture, but this grazing regime reduces the relative amount of energy consumed by livestock and increases the relative amount of energy transferred into the decomposer compartment in comparison with the severely grazed pasture. [Reprinted with permission from Heitschmidt, R. K., and Stuth, J. W. (eds.) (1992). "Grazing Management: An Ecological Perspective," p. 16. Copyright 1992 by Timber Press, Inc., Portland, OR.]

individual plants or species of plants is imperative because of the effects that varying levels of defoliation have on the competitive abilities of plants.

Control of the frequency and severity of defoliation of individual plants is a formidable challenge in rangeland ecosystems because: (1) kinds and numbers of plants (i.e., primary producers) are large; (2) most animals (i.e., primary consumers) are selective grazers; and (3) climatic conditions are often highly variable. These three factors in combination ensure that the quantity (i.e., efficiency of solar energy capture) and quality (i.e., assimilation efficiency) of available foodstuff (i.e., forage) will vary dramatically over time (e.g., seasonally) and space (e.g., regionally).

It is also from our understanding of ecological systems that we recognize that the basic concepts presented in this section are as applicable to undomesticated (i.e., wildlife) as domesticated (i.e., livestock) herbivores, to large (e.g., deer) as small (e.g., ants) herbivores, and to natural (e.g., rangelands) as other agroecosystems (e.g., croplands). The only differences are related to scale and management goals. [*See* RANGELAND CONDITION AND TREND; RANGELAND MANAGEMENT AND PLANNING.]

Bibliography

Harrington, G. N., Wilson, A. D., and Young, M. D. (eds.) (1984). "Management of Australia's Rangeland." Commonwealth Scientific and Industrial Research Organization, Australia.

Heitschmidt, R. K., and Stuth, J. W. (eds.) (1992). "Grazing Management: An Ecological Perspective." Timber Press, Portland, OR.

Holechek, J. L., Peiper, R. D., and Herbel, C. H. (1989). "Range Management: Principles and Practices." Prentice-Hall, Englewoods Cliffs, NJ.

Lowrance, R., Stinner, B. R., and House, G. J. (eds.) (1984). "Agricultural Ecosystems Unifying Concepts." Wiley, New York.

Odum, E. P. (ed.) (1959). "Fundamentals of Ecology." Saunders, Philadelphia.

Pomeroy, L. R., and Alberts, J. J. (eds.) (1988). "Concepts of Ecosystems Ecology." Spring-Verlag, New York.

Smil, V. (ed.) (1991). "General Energetics: Energy in the Biosphere and Civilization." Wiley, New York.

Tueller, P. T. (ed.) (1988). "Vegetation Science Applications for Rangeland Analysis and Management." Kluwer, Boston.

Vallentine, J. F. (ed.) (1990). "Grazing Management." Academic Press, San Diego.

Rangeland Grass Improvement

K. H. ASAY, *USDA-Agricultural Research Service, Utah State University*

Glossary

Certified seed Seed that is produced under the supervision of an official certification agency, and is verified to be a particular variety (cultivar), free of weeds and other types of seed, and of high quality; classes of certified seed are foundation, registered, and certified

General combining ability Breeding value of a clonal line based on the performance of its progenies from a series of crosses with several other clonal lines or a genetically broad-based cultivar

Heritability That portion of the total (phenotypic) variance among plants or clones that is due to genetic differences (genetic variance); the phenotypic variance consists of the environmental variance and the genetic variance

Synthetic cultivar Cultivar (or variety) generated by intermating a group of selected clones; the parental clones are usually selected on the basis of their general combining ability and the population is advanced from one generation to the next by open pollination in isolated seed-increase blocks

Variety Also referred to as a cultivar, a population or group of plants that are genetically distinct from other groups in the same species; varieties are often developed through a series of breeding cycles

North American rangelands are noted for environmental extremes associated with variations in soil properties, amount and seasonal distribution of precipitation, and wide variations in temperature. Extended periods of severe water deficit are not uncommon and these effects are often compounded by the grazing management system as well as infestations of insects, disease, and nematodes. Reclamation of depleted rangelands with cultivars of perennial grasses that have been developed for these harsh conditions is a relatively economical means of effecting long-term improvements. Research with range grasses has received considerably less emphasis than other crop species; however, improved cultivars have been developed that have significantly impacted rangeland improvement programs. Genetic progress has been expedited by modern breeding methods and more effective selection criteria. Advances in technology have led to genetic transfer across previously unsurmountable barriers. An expanded genetic base or gene pool generated through plant exploration and the activities of the National Plant Germplasm System have provided more impetus. [*See* RANGELAND.]

I. Introduction

Breeding programs to develop improved grasses for temperate rangelands vary from evaluation and release of superior ecotypes to more elaborate procedures involving wide hybridization, molecular technology, progeny testing, and use of selection indices. Although range-grass breeding programs have received comparatively less support from public and private resources, improved cultivars of perennial cool-season grasses have been a significant component of seeding mixtures used on temperate rangelands of North America. Plant materials developed from interspecific hybrids are beginning to make inroads as exemplified by the crested wheatgrass cultivar Hycrest, developed from a hybrid between an induced tetraploid form of *Agropyron cristatum* and natural tetraploid *A. desertorum* and NewHy, which is a hybrid between quackgrass (*Elytrigia repens*) and bluebunch wheatgrass (*Pseudoroegneria spicata*). Other noteworthy examples of genetic progress include Ephraim and Kirk crested wheatgrass, Bozoisky-Select and Mankota Russian wildrye (*Psathyrostachys juncea*), and Reliant and Manaska intermediate wheatgrass (*Thinopyrum intermedium*). Interdisciplinary research continues to generate more effective selection criteria and screening procedures for water-use efficiency, drought resistance, seedling vigor, forage yield and quality, soil stabilization, and resistance to disease and insect pests. New plant materials from modern breeding programs are expected to conserve and improve the vast rangeland domain as habitat for livestock and wildlife as well as for recreation use and aesthetic beauty.

II. Impact of Basic Research

Effective plant breeding programs, particularly those aimed at developing improved range grasses, must entail interdisciplinary research involving the interaction of physiology, biochemistry, molecular genetics, plant pathology, entomology, animal science, statistics, and other fields of study. Basic information generated from such research leads to more efficient breeding methods and selection criteria. For example, attempts to produce an interspecific hybrid are often stymied by abortion of the embryo relatively early in its development. This problem has been circumvented in many instances with a technique known as embryo rescue. The immature embryo is aseptically removed from the floral structures of the female parent where it would normally abort and transferred to an artificial medium. The proper nutrients, hormones, and other environmental conditions are provided in the laboratory for the embryo to develop into a mature hybrid plant, which then can be used in subsequent crosses.

The rapidly advancing science of molecular biology continues to develop technology that may revolutionize plant breeding programs. This technology is providing the means to precisely map the location of genes on chromosomes, determine genetic relationships among species, and even mark and characterize genes associated with specific traits. Present breeding programs are based on selection for characters that

are the products of genes. These characters are often masked by complex environmental factors and may in some environmental surroundings not even be expressed. Using molecular technology, screening and selection may soon be based on the gene itself (or marker associated with a particular DNA sequence). This would allow the breeder to screen large populations of seedlings in the greenhouse or laboratory for characters that are expressed in mature plants often under specific environmental conditions. The selection process not only would be done more precisely, but the savings in time and resources would be enormous. [*See* PLANT GENE MAPPING; PLANT GENETIC ENHANCEMENT.]

Genetic engineering or direct transfer of genes from one species to another has yet to have a major impact on range-grass breeding programs. Although a considerable amount of technology remains to be developed and some political concerns must be addressed, the procedure offers considerable promise for designing plants to meet specific needs. It is evident, however, that even when genetic transfer is achieved, conventional breeding approaches will be required to convert the initial transformed population into a fertile and genetically stable cultivar.

III. Breeding Methods

The choice of methods used in breeding range grasses is dictated by several factors including long-range objectives, mode of reproduction for the species involved, nature of the genetic and environmental effects on character(s) of interest, type of cultivar grown commercially, and resources available in the breeding program. Because most range grasses are cross-pollinated, various breeding schemes are used to develop improved populations or synthetic cultivars. A typical sequence (Fig. 1) would include the following steps: (1) development of a germplasm base, (2) initial evaluation in source nurseries, (3) replicated progeny tests and/or clonal evaluation trials, (4) crossing blocks to develop experimental strains, (5) strain evaluation, and (6) seed increase and cultivar release. [*See* CULTIVAR DEVELOPMENT.]

A. Development of Germplasm Base

The development of a gene pool or a germplasm base from which breeding populations and improved cultivars can be derived is an essential part of an effective plant breeding program. Use of the most elabo-

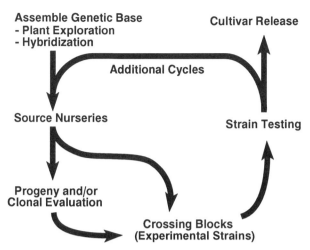

FIGURE 1 Typical breeding scheme used to improve cross-pollinated forage grasses.

rate technology and facilities will do very little to compensate for inadequate genetic resources. The nature of the gene pool required is dependent on the objectives of the breeding program, but in most grass breeding programs, it is developed through plant exploration and hybridization procedures. Promising germplasm also can be obtained from naturalized strains and old existing cultivars. [See PLANT GENETIC RESOURCES; PLANT GENETIC RESOURCE CONSERVATION AND UTILIZATION.]

Because many grasses used in seeding operations on North American rangelands were originally introduced from other continents, foreign plant exploration has been particularly instrumental in providing the essential genetic diversity. Introduction of crop species, including forage grasses, was initiated by the first settlers in the new world. Many species were brought to North America in the ballasts of ships or in livestock feed; others were introduced intentionally by settlers and by plant explorers. These early introductions are still an important component of the genetic resources included in breeding programs, and they have provided a germplasm base for several modern cultivars. Today organized plant exploration has become an integral part of breeding programs.

The National Plant Germplasm System (NPGS), which was organized by the Agricultural Service of the U.S. Department of Agriculture (USDA-ARS) in cooperation with state agricultural experiments stations, and commercial entities in the private sector, coordinates much of the acquisition and preservation of plant germplasm as well as distribution of these plant materials to plant breeders and other research

scientists. Collections in the NPGS consist of more than 380,000 accessions representing some 8700 species. These include essentially all of the crops of interest to agriculture in the United States. Important components of the system include: the National Seed Storage Laboratory at Fort Collins, Colorado, which provides for long-term and back-up storage of collections; four regional plant introduction stations; clonal germplasm repositories at 10 locations in the United States; and other regional repositories for wheat, potatoes, and other crops.

The impact of plant exploration in plant breeding programs is exemplified by the USDA-ARS at Logan, Utah, in support of research involving the development of improved grasses and other forages for rangelands. Since 1972, scientists from the unit have participated in 16 foreign expeditions to Iran, Turkey, the former USSR, China (PRC), Australia, New Zealand, Romania, Pakistan, and Nepal. In addition, several domestic expeditions have been made by the research unit to collect germplasm from range sites in the Intermountain Region as well as the Great Plains of the United States and Canada.

B. Source Nurseries

After a broad germplasm base has been assembled, representative plants of each entry (accession, collection, strain, cultivar, hybrid, etc.) are established in nurseries on a range site that typifies the environmental conditions that the end product (new cultivar) will be exposed to. A source nursery is usually established as spaced plants to permit evaluation on an individual plant basis. Spacing between plants depends on the objectives of the program, the species, and environmental conditions. For example, a highly rhizomatous species would be spaced farther apart than those with a caespitose or bunch-type growth habit. Wider spacings also may be used in particularly xeric sites. In a typical range-grass breeding program, plants would be spaced from 0.5 to 1.5 m apart.

The source nursery should adequately sample the genetic variation of the assembled germplasm base, particularly in terms of genetic traits desired in the end product. The size of the nursery is often limited by available resources, but it usually consists of from 5000 to more than 20,000 spaced plants. Evaluation criteria vary according to the objectives of the breeding program. Because of the large number of plants involved in most source nurseries, evaluations at this stage of the program are based largely on visual observations for traits such as vegetative vigor, reaction to disease and insect pests, response to environmental stress (drought, salinity, temperature, etc.), relative maturity dates, seed yield potential, leafiness, and vegetative spread.

C. Replicated Progeny Tests

Superior lines selected from the source nurseries are then included in progeny tests and/or replicated clonal evaluation trials. The purpose of a progeny test in a typical range-grass breeding program is to determine the general combining ability of the clonal lines. The topcross, polycross, and open-pollinated progeny tests are most commonly used to obtain this information. Topcross progenies are derived by crossing each of the clonal lines (as female parents) with a common genetically broad-based population or cultivar. A polycross progeny is obtained by crossing a maternal clonal line with all other clones in the group. This is accomplished in an isolated crossing block or with a controlled crossing scheme. Different designs are used in polycross crossing blocks, but replicates of the individual clonal lines should be arranged to provide each clone an equal chance of being pollinated by any other clone in the block. The open-pollinated progeny test requires the least amount of time because OP seed lots are harvested directly from selected clones in the source nursery. A degree of error is inherent in the OP procedure due to variation in the pollen source in the field nursery.

After the seed is obtained, progeny tests are usually established on multiple sites. Because fewer lines are involved, more elaborate evaluation criteria are employed. Traits of interest are forage and seed yield, forage quality, stand-establishment vigor, drought resistance, water-use efficiency, and animal performance. Progeny lines also can be evaluated under laboratory conditions for seed quality and seedling vigor.

D. Development of Experimental Strains

Clonal lines with the best general combining ability as determined in the progeny tests are established in isolated polycross crossing blocks to produce experimental synthetic strains. Sufficient clones should be included in the parentage of the strain to minimize inbreeding in subsequent generations. The parentage of most synthetic cultivars comprises from 4 to 20 clonal lines. Equal amounts of seed from each PC line are bulked to form the Syn-1 generation or breeder seed. The Syn-1 is grown under isolation to produce

the Syn-2, which is generally used in evaluation trials. Seed from the polycross blocks also can be used to initiate a new breeding cycle, a process known as recurrent selection.

E. Strain Testing and Cultivar Release

Experimental strains are then evaluated in seeded plots on several range sites for at least 3 years. Evaluation criteria will again depend on the objectives of the breeding program; however, under rangeland conditions, stand establishment, soil stabilization, and persistence under environmental stress are always of major concern. Characters associated with forage quality and animal performance are measured in more detail. Not only are nutritive value, palatability, animal carrying capacity, and daily gains important, but the effect of the grazing animal on plant performance and persistence must be considered as well. During this phase of the program, grass entries should be evaluated in mixtures with other species. This is consistent with the goal of modern range management, which is to establish a balanced and productive ecosystem consisting of several grass, forb, and shrub species. Cultivars generated by the breeding program will represent a component of this genetically diverse system. [See RANGELAND GRAZING; RANGELAND MANAGEMENT AND PLANNING.]

The synthetic strain with the best performance record in the evaluation trials is given a name and released as a cultivar. Release procedures vary, but in general, classes of seed include breeder, foundation, registered, and certified. Breeder seed is produced by the breeder from the polycross crossing block. Foundation seed is produced from breeder seed by the institution or company releasing the cultivar. This seed is then distributed to commercial seed growers who produce registered and certified seed classes. Seed-increase procedures are monitored by a certification agency, usually associated with the state agricultural experiment station, and the state department of agriculture to validate the purity and viability of the seed. In accordance with state and federal seed laws, a label is attached to seed offered for sale with pertinent information relating to cultivar name, viability, and amount of inert material. It is not uncommon for range grass seed to be sold as a particular cultivar, but not certified as such. In those instances, the buyer has no guarantee regarding genetic purity of the cultivar or that the seedlot is even the cultivar that it is purported to be.

F. Breeding for Improved Drought Resistance

Production of the world's food crops is limited more by inadequate water than any other factor. Rangelands of western North America are particularly susceptible to drought-related problems in that periods of water deficit usually occur annually. Poor grazing management often has compounded the effects of drought, and many sites have been depleted of productive vegetation. Seeding with cultivars of perennial grasses that have been bred for resistance to drought and are more competitive with aggressive invader species is a feasible means of reclaiming these areas.

Resistance to drought may be perceived as the ability of a plant to grow satisfactorily in areas subjected to periodic water deficits, or from an ecological viewpoint, the capacity of a plant to survive or persist when its water supply is limited. Three mechanisms have been described that are associated with drought resistance in plants.

1. Drought Escape

Many species escape drought by completing their life cycle during the spring and early summer before soil water becomes a serious limiting factor. Plant breeders often employ this mechanism in developing drought-resistant cultivars of wheat. Perennial grasses often escape drought through rapid seedling development during the early spring or fall when water is available near the soil surface.

2. Drought Avoidance or Endurance with High Internal Water Content

A perennial grass plant that has escaped drought through rapid seedling development then can more effectively resist or avoid ensuing drought conditions by excluding the effect of drought from the plant tissues. The internal water content of the plant is maintained by restricting water loss from the leaves by reducing the size of the stomatal openings in the leaf surface or through the development of a cuticle that is more impervious to water movement. Plants may reduce leaf area when water becomes limiting or have morphological adaptations that reduce the amount of radiation absorbed by tissues. Temperate range grasses also protect their internal tissues from drought through the development of an efficient and extensive root system.

3. Drought Tolerance at Low Water Potential

Many plants are able to maintain turgor during periods of water deficit through osmotic adjustment, increasing the elasticity of the cell walls, or decreasing the size of the individual cells. Mechanisms also function in drought-tolerant plants that protect the protoplasm from the mechanical and physiological effects of dehydration. This enables the plants to endure the effects of drought and recover after the tissues are again hydrated. Although these mechanisms are not fully understood, protoplasmic proteins in drought-tolerant plants are considered to be more resistant to denaturation, coagulation, or hydrolysis.

Survival of plants under drought stress is usually dependent on a combination of escape, avoidance, and tolerance mechanisms at some stage of their development. Most range grasses are perennials and must be able to survive periods of drought stress after annual plants have completed their life cycle. In temperate range grasses, this usually entails various degrees of dormancy, which often is accomplished at the expense of a slower growth rate or reduced forage production. Range grasses must be particularly flexible in order to break dormancy and renew active growth when moisture and temperature conditions become more favorable.

4. Screening for Drought Resistance

Various methods have been devised to artificially impose drought stress on germinating seeds and developing seedlings. Osmotic solutions such as mannitol, polyethylene glycol (PEG), and sodium chloride have been used to subject germinating seeds to drought. In the USDA-ARS research program at Utah State University, crested wheatgrass and Russian wildrye breeding lines were screened in germination vessels that were designed to regulate the soil water content. Soil in the vessels was separated from a solution of water and PEG with a semipermeable (cellulose acetate) membrane. The amount of water that moved across the membrane into the soil was regulated by the concentration of PEG in the osmotic solution. Although the procedure effectively discriminated among the breeding lines, relative differences were not closely related to stand establishment of the same lines in the field. Screening procedures based on seedling recovery after exposure to severe drought in a growth chamber yielded similar results. Significant differences were detected among progeny lines of crested wheatgrass, but results were not closely related to stand establishment under field conditions.

The most effective screening procedures used to date in range-grass breeding programs have emphasized avoidance mechanisms. Drought stress on arid and semiarid rangelands is of particular concern during seedling establishment. Seedlings are more susceptible to drought than established plants so it is crucial that they develop rapidly during the period when moisture conditions are most favorable. Selection for seedling vigor in the field, laboratory, and greenhouse has been effective (see methods described under seedling vigor), and differences have been positively correlated with differences in stand establishment observed under field conditions. This has been particularly true for screening procedures based on seedling emergence from deep seedings and weight of individual seeds.

Measurement of root growth rates has shown promise as a screening procedure for improving the drought resistance of plants. Rate of root penetration in tubes that have been filled with soil or sand has been positively correlated with survival of grasses in rangeland seedings. Hydroponic culture has been effectively used at the University of Nebraska to evaluate root development of sorghum breeding populations. The hydroponic system has distinct advantages over traditional methods involving soil, because the level of stress is easier to control and conditions are more uniform near the individual plant roots. Other methods that may have merit for evaluating grasses for resistance to drought include chlorophyll stability, desiccation and heat tolerance of leaf discs, photosynthetic rate of stress plants, and proline content.

Regardless of the method used to screen individual plants for their reaction to drought, genetic progress must be reflected under actual field conditions. Evaluation of progeny lines and experimental strains on representative range sites is essential to an effective range-grass breeding program.

IV. Breeding for Improved Efficiency of Water Use

Breeding for improved water-use efficiency (WUE), expressed as kilograms of dry matter produced per kilogram of water transpired, should improve the productivity of range grasses on water-limited sites. Because of the time and resources required to measure this trait, however, it has not been used as a selection criterion in grass breeding programs. Recent research with carbon isotopes may provide a feasible method to evaluate plants for this trait. A negative correlation has been found between carbon isotope discrimination (designated as Δ) and WUE in several cool-season species. During photosynthesis in these plants, ribu-

lose bisphosphate carboxylase-oxygenase (Rubisco) discriminates against $^{13}CO_2$. Hence, when the stomates are fully open less $^{13}CO_2$ is fixed in the plant tissues than would be expected on the basis of its concentration in the air. As water becomes less available, the size of the stomatal aperture is reduced. If water loss from the plant is reduced proportionately more by the stomatal closure than the rate of photosynthesis, the intercellular concentration of CO_2 decreases and the plant has less opportunity to discriminate against $^{13}CO_2$. The result is a higher $^{13/12}C$ ratio in the plant tissues and more carbon is fixed in the plant tissues per unit of water lost. A lower Δ value would, therefore, be associated with higher WUE. This relationship has been confirmed by USDA-ARS workers at Logan, Utah. Moreover, the magnitude of the genetic variability and heritability values observed to date indicates that Δ would be a promising breeding tool in the development of crested wheatgrass (and perhaps other cool-season grasses) with enhanced WUE.

V. Breeding Grasses for Grazing during Fall and Winter

Livestock operations in the temperate regions of United States and Canada rely heavily on harvested hay during the late fall and winter months. This obviously is a very expensive practice, and costs of producing hay have increased to the point where it often is no longer cost-effective on a typical western ranch. The development and use of perennial grasses that have the characteristics necessary to extend the grazing season during the late fall and winter are receiving increased attention. Characters of interest include a tall upright growth habit, which would enable the plant to protrude above the snow level. It also would be essential for leaves to remain on the plant and retain their nutritive values. Grasses that are capable of significant growth under cold temperatures, often under the snow, can be used in such a management scheme. Development of such grasses likely will involve genetic modifications in the metabolism of specific carbohydrates (fructans).

The development of improved grass cultivars and hybrids for grazing during the fall and winter has been an objective of the USDA-ARS breeding program at Logan, Utah. Germplasm resources include tall upright grasses such as Altai wildrye (*Leymus angustus*), Great Basin wildrye (*Leymus cinereus*), tall wheatgrass (*Thinopyrum ponticum*), and intermediate wheatgrass (*T. intermedium*). A breeding population was generated from crosses among Altai wildrye, Great Basin wildrye, and mammoth wildrye (*L. racemosus*). This germplasm pool is exceptionally robust and upright, and has demonstrated excellent potential in this breeding program. Russian wildrye also offers considerable promise. Although this species is not exceptionally tall, its leaves remain intact and have excellent curing qualities. Research involving fructan metabolism in crested wheatgrass likely will lead to the development of cultivars that are more productive during the winter and early spring.

VI. Forage Quality Considerations

Because of emphasis placed on establishment and persistence under environmental stress, breeding for improved forage quality or nutritive value has received comparatively less attention in range grasses than other forage crops adapted to the more humid regions. Forage quality can best be evaluated on the basis of actual animal performance; however, animal feeding trials are time consuming and expensive, and are not feasible when a large number forage samples are involved. Several laboratory procedures have been developed to predict animal performance from a relatively small forage sample. These include analyses for specific mineral elements, fiber, and lignin fractions, *in vitro* dry matter digestibility (IVDMD), cellulase solubility (CDMD), and toxic or antiquality constituents. The most widely accepted laboratory procedure in grass breeding programs is near-infrared spectroscopy (NIRS). This procedure is based on the theory that various chemical entities associated with forage quality have distinctive near-infrared absorption properties. It has been successfully used to estimate IVDMD and other measures of forage quality with good precision. Although NIRS instrumentation is relatively expensive, the procedure has several notable advantages. Only a few grams of forage are required per sample, and several samples can be analyzed in a relatively short time. [*See* FORAGES.]

The USDA-ARS at the University of Nebraska has significantly improved the quality of switchgrass (*Panicum virgatum*) through selection for IVDMD. The genetic gains made in IVDMD were accompanied by significant increases in animal performance, based on average daily gain of grazing animals and total beef produced per hectare. This research program led to the release of the improved switchgrass

cultivar, Trailblazer. Similar progress was reported by USDA-ARS scientists at Tifton, Georgia. The bermudagrass (*Cynodon dactylon*) cultivar, Coastcross-1, released from this breeding program, had 12% higher IVDMD and produced 30% more beef per hectare than the cultivar from which it was derived. Breeding to reduce antiquality components in grasses have been successful as well. Reed canarygrass cultivars with lower alkaloid content have been developed, and endophyte-free cultivars of tall fescue have been released. Genetic variability observed in breeding populations indicates that significant genetic progress is imminent in several other forage grasses.

VII. Examples of Genetic Progress

Breeding and selection procedures used to develop improved germplasm and cultivars of perennial grasses for western rangelands can be illustrated through examples of those used in Russian wildrye, interspecific hybrids, and crested wheatgrass.

A. Breeding Russian Wildrye

Russian wildrye is a cross-pollinated perennial bunch grass that is native to the Eurasian interior. It was first introduced into North America in 1927 and has since become a popular grass in the Prairie Provinces of Canada and the Great Plains and Intermountain regions of the United States. It is adapted to loam and clay soils and is moderately tolerant of saline–alkali soils. Although Russian wildrye initiates growth early in the spring, its basal leaves retain their green color and nutritive value later during the summer better than those of most cool season grasses. It is also a valuable forage for grazing animals during the fall and early winter.

The general acceptance of Russian wildrye on semiarid rangelands has been impeded by relatively poor seedling vigor. Stands are difficult to establish, particularly in water-limited environments. The tendency of its seeds to shatter soon after maturity also limits the use of the species. Seed crops are often severely reduced by turbulent weather during seed ripening. Grass tetany is a condition often found in animals grazing forage that has low concentrations of Mg, Ca, and high levels of K. Although this malady has not been a major problem in Russian wildrye, chemical analyses of the forage and results from some grazing trials indicate that it is a potential hazard, particularly early in the grazing season.

Development of cultivars with improved seedling vigor has been a major breeding objective in Russian wildrye. Although various breeding schemes have been used, the most successful breeding approach to date has been to first screen large populations on the basis of seed weight (weight per 100 seeds), coleoptile length, and emergence from deep seedings (5 to 7.5 cm). Final selections are then made on the basis of seedling establishment in the field. Significant genetic variation for these characteristics has been found in Russian wildrye breeding populations, and breeding programs have been effective.

A Russian wildrye breeding program was initiated in 1936 by the USDA-ARS in cooperation with the North Dakota Agricultural Experiment Station. The cultivar Vinall was released from this program in 1960. Vinall, a 5-clone synthetic cultivar, was developed primarily on the basis of improved seed yield and resistance to lodging. Agriculture Canada embarked on a Russian wildrye breeding program in 1947 at Swift Current, Saskatchewan. The first product from this research effort was the cultivar Sawki, a 10-clone synthetic selected largely on the basis of improved seed and forage yield as well as an upright growth habit. The name, Sawki, is a Blackfoot Indian word meaning Great Prairie, which refers to the area of adaptation. The cultivar Mayak was released from the research station in Swift Current in 1971. This cultivar was derived through selection for increased forage and seed yield, and resistance to leaf spot diseases. Final selections were made on the basis of results from open-pollinated (OP) and polycross (PC) progeny tests.

The Russian wildrye cultivar Cabree was released in 1976 by the Agriculture Canada Research Station, Lethbridge, Alberta. This cultivar was derived from first-generation selfed progenies on the basis of seed retention (reduced seed shattering), seedling vigor, forage yield, seed yield, and culm strength. The cultivar Swift was licensed by Agriculture Canada at Swift Current in 1978. Selection criteria used in the development of this cultivar include seedling emergence from deep plantings, disease resistance, seed quality, forage yield, and seed yield.

Bozoisky-select was released in 1984 by the USDA-ARS at Logan, Utah in cooperation with the Utah Agricultural Experiment Station and the USDA-SCS. This cultivar was derived from PI-440627 (Bozoisky), which was originally introduced from the former USSR. Selection criteria included improved vigor, leafiness, seed yield, coleoptile length, and seedling vigor. Bozoisky-select has demonstrated sig-

nificantly improved stand establishment and productivity in evaluation trials on several Intermountain range sites. More recently, the USDA-ARS at Mandan, North Dakota, in cooperation with the North Dakota Agricultural Experiment Station and the USDA-SCS released the cultivar Mankota. Selection criteria used in the development of this 6-clone synthetic included coleoptile length, seedling emergence from deep seedings in the greenhouse, seedling establishment in the field, resistance to leaf-spot diseases, lodging, forage yield, and seed yield. This new cultivar is resistant to drought, persists under heavy grazing, and has demonstrated excellent performance in the northern Great Plains.

Russian wildrye is typically a diploid with $2n=14$ chromosomes. Tetraploid forms ($2n=28$), obtained through treatment with colchicine and nitrous oxide, have larger seeds and have been shown to have better seedling vigor than their diploid counterparts. Tetraploid germplasm has received increased emphasis in Canadian and U.S. breeding programs. The tetraploid cultivar Tetracan was licensed by Agriculture Canada, Swift Current in 1988. This cultivar is reported to have larger spikes and seeds and wider leaves than diploid cultivars. It also has demonstrated improved seedling vigor in evaluation trials conducted in the Northern Great Plains.

The USDA-ARS breeding programs at Mandan, North Dakota and Logan, Utah, also have developed tetraploid breeding populations through chemical treatment and plant introduction.

Genetic variability for Mg, Ca, and K concentration in Russian wildrye has been studied by the USDA-ARS in Utah and Idaho to evaluate the potential of breeding to reduce the risk from grass tetany in Russian wildrye. Based on the magnitude of heritability values and genetic variation found among progeny lines, it was concluded that the grass tetany potential in Russian wildrye could be reduced through breeding. Results from this research did indicate, however, that additional plant exploration would be needed to provide the necessary genetic base.

B. Interspecific Hybridization

Most range-grass cultivars released to date were developed through selection and hybridization within a given species. Interspecific hybridization (hybridization between species) is gaining acceptance as a promising tool for introgressing genes among related species and in some instances creating new species with selected characteristics of the parental species.

Many range grasses, particularly perennial grasses of the Triticeae tribe (wheatgrasses, wildryes, and related species), are well suited to this method of genetic improvement. Intergeneric and interspecific hybridization is a frequent occurrence among these grasses in nature, and many are themselves natural polyploids of hybrid origin. Some wheatgrasses and wildryes have in fact been artificially synthesized from hybrids between their parental species. For example, D. R. Dewey (USDA-ARS, Logan, UT) obtained plants that were morphologically and cytologically identical to Montana wheatgrass (Syn. *Agropyron albicans*) in progenies from crosses between bluebunch wheatgrass (*Pseudoroegneria spicata*) and thickspike wheatgrass (*Elymus lanceolatus*). Western wheatgrass (*Pascopyrum smithii*), a range grass native to North America, is another product of interspecific hybridization. After studying chromosome pairing and morphological variation in various hybrids, Dewey concluded that western wheatgrass emerged from natural hybrids involving thickspike wheatgrass (*Elymus lanceolatus*) and beardless wildrye (*Leymus triticoides*) or their close relatives.

The general acceptance of interspecific hybridization as a breeding tool has been impeded by sterility problems and a preponderance of deleterious characters in hybrid progenies along with additional time and resource requirements. Two basic approaches can be used to incorporate interspecific hybridization in a range-grass breeding program. In some cases, enough fertility exists in the hybrid populations to develop breeding populations without chromosome doubling. Genetic and chromosomal imbalances can generally be overcome in these hybrids by a few generations of selection for improved fertility or seed set. A backcrossing sequence to one of the parental species (recurrent parent) may be used if the objective is to transfer one or two characters from the other species (nonrecurrent parent). This breeding scheme is referred to as genetic introgression or transfer of genetic traits from one species to another.

Other interspecific hybrids are completely sterile. Although complex genetic factors are often involved, sterility problems may be associated with the failure of chromosomes contributed by the parental species to pair during meiosis in the hybrid. Fertility can be at least partially achieved in some of these hybrids through treatment with the drugs colchicine or nitrous oxide to double the chromosome number. But even then, several generations of selection are required to restore the genetic and chromosomal balance necessary for normal meiosis, seed production, and other

metabolic processes in the hybrid plants. When successful, this approach can lead to a "new species" comprising a more or less equal genetic contribution from the two parental species.

Both of these procedures have been used by the USDA-ARS at Utah State University to develop improved range grasses. Examples of these programs are presented below.

C. Genetic Introgression without Chromosome Doubling

The most notable example of this breeding approach is the development of the cultivar NewHy from a hybrid between quackgrass (*Elytrigia repens*) and bluebunch wheatgrass. Quackgrass is a cool-season perennial grass that was probably introduced into North America from Eurasia. It is now one of the most widely distributed temperate grasses in the world. Because of its aggressive rhizomes, it is very difficult to control and often considered to be a troublesome, and even noxious, weed. However, quackgrass does have some good qualities, and it is considered to be a valuable forage in many areas. It produces abundant forage, and it thrives under intense grazing management. Quackgrass also is adapted to adverse soil conditions, and it has excellent resistance to excess soil salinity. From a genetic view point, it is a hexaploid ($2n=42$).

Bluebunch wheatgrass, on the other hand, is a caespitose (bunch type) species, native to the rangelands of western United States and Canada. It has excellent nutritional value, and it is often preferred by grazing animals in mixed stands with other species. The species also is resistant to drought; however, it does not persist well when heavily grazed. Both diploids ($2n=14$) and tetraploids ($2n=28$) occur in nature. The form of bluebunch wheatgrass used in the hybridization program was a tetraploid ($2n=28$).

The first-generation (F_1) hybrid between the quackgrass and bluebunch wheatgrass had an odd chromosome number of $2n=35$ and was meiotically irregular. Although characteristics of both parental species were observed in the hybrid plants, they had relatively poor vegetative vigor, and many were plagued with chlorophyll deficiencies and other morphological irregularities. Although the hybrid population was largely sterile, seed set was adequate to permit selection for improved fertility without modifying the chromosome number. Selection for the next four generations was largely based on seed set, and only plants with characteristics of both parental species were retained each cycle. By the fifth generation, a relatively fertile and meiotically regular population was obtained. This population was then subjected to more intense selection for agronomic performance and adaptation to semiarid range conditions. Progeny testing was initiated and more emphasis was placed on seedling vigor. Seven generations after the initial cross, the hybrid population was fully fertile with $2n=42$ chromosomes.

Five experimental strains of the advanced-generation hybrid were entered in evaluation trials conducted over a 3-year period and several locations. Data were collected on stand establishment, forage yield, forage quality, persistence, and reaction to drought. The superior experimental strain was selected and released as the cultivar, NewHy, more than 20 years after the initial cross was made. This cultivar has only moderate rhizome development and is substantially more drought resistant than the quackgrass parent. The rhizome character proved to be highly heritable and essentially caespitose plants were easily obtained in two or three breeding cycles; however, a limited amount of vegetative spread was found to contribute to the persistency of the hybrid population.

NewHy is adapted to range sites receiving at least 330 mm (13 in.) of annual precipitation or under irrigation. It has demonstrated excellent resistance to soil salinity. In evaluation trials, the salinity tolerance of NewHy approached that of tall wheatgrass (*Thinopyrum ponticum*), which is the most salt-tolerant wheatgrass used in range improvement. NewHy, however, has much better forage quality and leafiness than tall wheatgrass. The new hybrid cultivar also retains its green color later in the growing season than other range grasses. Selection for improved salinity tolerance and other agronomic traits continues in this population and research with more diverse parentage has been initiated. In one particular hybrid, the bluebunch wheatgrass parent has been replaced with its Asian counterpart. This hybrid is more upright and less rhizomatous than NewHy, and it is a potentially valuable grass in pasture mixes with legumes such as alfalfa (*Medicago sativa*).

D. Interspecific Hybridization with Chromosome Doubling

In this breeding approach the first-generation (F_1) hybrid is highly sterile, and it is necessary to double the number of its chromosomes to attain the fertility necessary to initiate a breeding program. This type of hybrid is defined as an induced amphiploid and is exemplified by the hybrid developed by the USDA-ARS, Logan, Utah, between the diploid ($2n=14$) form of bluebunch wheatgrass and tetraploid ($2n=28$) thickspike wheatgrass (*Elymus lanceolatus*). The initial hybrid was triploid ($2n=21$) and almost completely sterile. Hybrid plants were treated with colchicine to double the chromosome number to produce an amphiploid with $2n=42$ chromosomes. The induced amphiploid population was then subjected to three cycles of selection to develop a breeding population with a moderate degree of fertility.

This population was then advanced through six additional cycles. Selection during this phase was based on improved fertility, seed yield, vegetative vigor, reaction to drought stress, leafiness, resistance to insects and disease, seed size, and seedling emergence from deep seedings. A polycross crossing block, consisting of 29 clones selected from the final cycle, then was established to produce seed of an experimental synthetic strain.

The hybrid is morphologically distinct from the parental species; however, characteristics of both are represented in the experimental strain. It has been evaluated in a series of field trials and foundation seed is at present being produced in preparation for its release as a named cultivar. This strain, which is at present designated as the SL-1 hybrid germplasm pool, is essentially a new species since it apparently does not occur in nature.

E. Interspecific Hybridization in Crested Wheatgrass

Crested wheatgrass is a widely adapted cool-season perennial grass that is native to Eurasia. It was introduced into North America in the early 1900s, and has since had a significant impact on the revegetation and stabilization of temperate rangelands of western United States. Although early seedings of crested wheatgrass were made as monocultures, present goals are to include it in an ecosystem with other native and introduced grasses and forbs. As a cool-season grass, it is recommended primarily for grazing during the early spring and summer. It persists under environmental stress, and is relatively easy to establish on harsh range sites; it is best adapted to sites receiving 20 to 36 cm annual precipitation (8 to 14 in.).

Crested wheatgrass is not just one species, but comprises a complex of diploid ($2n=14$), tetraploid ($2n=28$), and hexaploid ($2n=42$) species. The diploid form is known as Fairway and is most prevalent in the Northern Great Plains of the United States and

Canada. Standard and Siberian, the tetraploid forms of crested wheatgrass, are most common in the Intermountain West and Great Plains of the United States. Although hexaploid introductions have shown some promise in breeding nurseries, they have not as yet had an impact on revegetation programs.

Most breeding programs with crested wheatgrass have concentrated on hybridization and selection within species. Difficulty in making interspecific crosses and sterility problems in hybrids when they are obtained have discouraged efforts to transfer genetic traits among species of this complex. Chromosome pairing and fertility in hybrids among the different species indicate, however, that the same basic chromosomes occur at the three ploidy levels. This suggests that with appropriate breeding methods, all species of crested wheatgrass can and should be treated as a single gene pool. All possible crosses now have been made among diploid, tetraploid, and hexaploid forms and breeding schemes have been devised to transfer selected genetic traits across barriers imposed by differences in chromosome numbers.

Genetic transfer from the diploid to the tetraploid level of crested wheatgrass has been a productive breeding approach. Fertile tetraploids have been obtained in the USDA-ARS research program at Logan, Utah, by treating vegetative tillers of diploids with colchicine. These induced tetraploids have been crossed with natural tetraploids (Standard) to produce a relatively fertile hybrid. Selection for improved fertility was effective in the hybrid population, and several hybrid plants were more vigorous than either of the parental species.

This induced tetraploid × natural tetraploid hybrid population was included in a source nursery with several other crested wheatgrass lines obtained through plant exploration and other sources. The source nursery, which consisted of 20,000 spaced plants, was evaluated for vegetative vigor throughout the season, resistance to insects and disease, leafiness, and response to drought. Open-pollinated (OP) seed was harvested from 350 selected clonal lines. These OP progeny lines were evaluated along with their parental lines in replicated trials on representative range sites. In addition, the progeny lines were screened for seedling vigor in the laboratory. Evaluation criteria were in accordance with those described earlier.

Based on the results of these trials, 18 clones were selected and isolated in a polycross crossing block to produce the synthetic-1 generation of an experimental strain. The 18 parental clones were all from the in-duced tetraploid × natural hybrid population. The experimental strain was evaluated along with several other cultivars and strains on several range sites in the Intermountain West, and was subsequently released as the cultivar Hycrest. This new hybrid cultivar has demonstrated superior seedling establishment characteristics on harsh range sites and is widely used in range seeding programs.

VIII. The Future

A. Native vs Introduced Grasses

It has been proposed that only species that are native to North America be used in seeding programs on western rangelands, particularly federally owned lands. The ultimate goal would be to restore these areas to their native or pristine state. Before such a drastic step is taken, some concerns should be addressed. The multiple demands now imposed on western rangelands have significantly altered the environmental forces in the plant community. It is likely that these environmental changes will be reflected in the optimum vegetative climax associated with a particular site. New combinations of grass, forb, and shrub species will be required to meet the needs of user groups for conservation, aesthetics, as well as for habitat and forage for livestock and wildlife.

Introduced grasses have been and will continue to be an essential component of this vegetative complement. It is noteworthy that grasses native to Europe and Asia have evolved under more intense grazing than their native counterparts, and it is not surprising that they are better adapted to more intense management practices on the American continent as well. On the other hand, we should not conclude that native plants do not have a place in the improvement of North American rangelands. Native germplasm has been included in breeding programs and improved cultivars have been developed and released. In some instances, native and introduced grasses are hybridized to combine the most desirable attributes of each in a new species or cultivar.

It is evident that both native and introduced grasses should and will play a major role in the improvement of western rangelands. The future of plant improvement programs will depend on how effectively we can maintain and use the world's genetic resources. We cannot afford to restrict the scope of our gene pool to plants that are native to the North American continent.

Bibliography

Asay, K. H. (1992). Breeding potentials in perennial Triticeae grasses. *Hereditas* **116,** 167–173.

Asay, K. H., Dewey, D. R., Gomm, F. B., Horton, W. H., and Jensen. K. B. (1986). Genetic progress through hybridization of induced and natural tetraploids in crested wheatgrass. *J. Range Manage.* **39,** 261–263.

Asay, K. H., and Johnson, D. A. (1983). Breeding for drought resistance in range grasses. *Iowa State J. Res.* **57,** 441–455.

Berdahl, J. D., and Barker, R. E. (1984). Selection for improved seedling vigor in Russian wild ryegrass. *Can. J. Plant Sci.* **64,** 131–138.

Board on Agriculture National Research Council (1991). "Managing Global Genetic Resources: The U.S. National Plant Germplasm System." National Academy Press, Washington, DC.

Dewey, D. R. (1975). The origin of *Agropyron smithii. Am. J. Bot.* **62,** 524–530.

Hitchcock, A. S. (1951). "Manual of Grasses of the United States," 2nd ed. USDA Misc. Publ. 200.

Lawrence, T. (1963). A comparison of methods of evaluating Russian wild ryegrass for seedling vigor. *Can. J. Plant Sci.* **43,** 307–312.

Sleper, D. A. (1987). Forage grasses. p. 161–208. *In* "Principles of Cultivar Development" (W. R. Fehr, ed.), Vol. 2, pp. 161–208. Macmillan, New York.

Sullivan, C. Y., and Ross, W. M. (1979). Selecting for drought and heat resistance in grain sorghum. *In* "Stress Physiology in Crop Plants" (H. Mussell and R. C. Staples, eds.), pp. 262–281. Wiley, New York.

Turner, N. C. (1979). Drought resistance and adaptation to water deficits in crop plants. *In* "Stress Physiology in Crop Plants" (H. Mussell and R. C. Staples, eds.), pp. 344–372. Wiley, New York.

Vogel, K. P., Haskins, F. A., and Gorz, H. J. (1981). Divergent selection for *in vitro* dry matter digestibility in switchgrass. *Crop Sci.* **21,** 39–41.

Rangeland Grazing

J. L. HOLECHEK, *New Mexico State University*

Glossary

Browsing Consumption of edible leaves and twigs from woody plants (trees and shrubs) by the animal
Deferment Delay of grazing in a pasture until seed maturity of key species
Forage All plant material on a given area potentially edible by livestock and wildlife
Grazing Consumption of standing forage (edible grasses and forbs) by livestock and wildlife
Grazing capacity Maximum animal numbers which can graze each year on a given area of range for a specific number of days without inducing a downward trend on condition
Plant succession Progressive replacement of one plant community by another leading toward a relatively stable plant community (the climax)
Rangeland Uncultivated land that will provide the necessities of life for grazing and browsing animals
Range management Manipulation of rangeland components to obtain the optimum combination of goods and services for society on a sustained basis
Rest Nonuse of a pasture for one year
Retrogression Changes away from the climax or desired plant community that may be caused by overgrazing, drought, fire, flood, etc.

Rangelands typically share the two characteristics of providing forage for large herbivores and being suited for extensive rather than intensive management. Therefore, ecological rather than agronomical principles are the basis for management on most rangelands. The basic tools used in range management involve the control of numbers, kinds, and distribution of grazing animals. The primary objective of range management is to provide the desired combination of products (red meat, fiber, wildlife, water, recreation, amenities, etc.) on a sustained basis.

I. Rangelands Defined

A. Characteristics of Rangelands

Rangelands supply forage for domestic and wild herbivores without annual cultivation and planting (Fig. 1). Much of the world's rangeland is arid or semi-arid (receive under 20 in. average annual precipitation) and has rugged terrain or shallow soils. These limitations make most rangelands unsuited for farming, commercial forestry, or urbanization. [*See* RANGE-LAND.]

The vegetation on rangelands is primarily grasses, forbs, and shrubs. These plants, known as forage, provide food for cattle, sheep, goats, and horses and a variety of wild herbivores such as elk, deer, pronghorn, and bison. Animals like cattle and bison that forage predominately on grasses are called grazers. Those, such as deer and moose, that make considerable use of twigs and leaves from shrubs and trees are referred to as browsers. [*See* FORAGES; RANGELAND GRASS IMPROVEMENT; RANGELAND PLANTS; RANGELAND SHRUBS.]

Historically the most important use of rangelands has been for livestock and wildlife production. Because they require little or no commercial energy input, rangelands are well suited for production of meat, hides, wool, tallow, and milk. In developing countries rangelands are also an important source of fuel-

FIGURE 1 Rangeland in southcentral New Mexico grazed by domestic sheep.

wood, edible plants, and building materials. The manure from range livestock and wildlife is used for both fuel and fertilizer. In affluent countries such as the United States rangelands are becoming increasingly important for recreation, for esthetics, and as a source of ornamental plants. In some parts of the world where human populations are rapidly growing but arid conditions prevail, water is of greater importance than forage as a rangeland product.

B. Types of Rangeland

The basic types of rangelands in the world are grasslands, desert shrublands, and savannah woodlands; forests and tundra have rangeland values but are not classified as rangelands. Variations in climate, soils, topography, and human influences account for differences in the rangeland vegetation types of the world.

Grasslands produce the most forage for wild and domestic herbivores. About 24% of the world's land surface is grassland. They are dominated by plants in the family Gramineae (grasses). Grasses are characterized by having hollow, jointed stems; fine, narrow leaves; and fibrous root systems. Forbs, which are an important component on many grasslands, are distinguished from grasses by having tap roots, broader leaves, and solid, nonjointed stems. Shrubs and trees are a minor component on most grasslands. They differ from grasses and forbs by having woody stems that remain alive during dormant periods. Trees are separated from shrubs by branching from a trunk rather than at ground level. Grasslands are favored by climatic conditions of frequent, light rains during the summer and relatively dry winters. This keeps the soil surface moist during the summer growing period where most grass roots occur.

Desert shrublands account for about 30% of the world's rangeland and they are characterized by the driest climatic conditions. Precipitation is infrequent and varies considerably from year to year. When precipitation does occur it is often from high-intensity storms during a short period. Coarse rooted plants (shrubs) that can collect moisture from a large portion of the soil profile have an advantage under these conditions.

Areas with scattered, low-growing trees with an understory of grasses are referred to as savannah woodlands. They typically occur as a transition zone between forest and grasslands, and are about 8% of the world's rangelands. Without periodic fire many of the wetter grasslands of the world revert to this vegetation type.

Forests are characterized by taller and more closely spaced trees than savannah woodlands. They account for about 30% of the world's rangelands. Generally forests occur under conditions of high rainfall since large amounts of moisture are needed to support the greater biomass of trees compared to grasses and shrubs. Forest soils typically are lower in fertility than grassland soils because of the greater leaching of nutrients that results from wetter conditions. Conversion of tropical forests into grazing land by burning and clearing has been a major environmental concern in recent years. Many ecologists doubt that the soils of the tropical rain forests will sustain production of forage and crop plants.

Cold, treeless areas with low-growing grasses and shrubs in arctic or high-elevation regions are commonly called tundra. This type of rangeland occurs on about 5% of the earth's surface. Trees are restricted on the tundra because the deeper portion of the soil profile remains frozen throughout the year (permafrost).

C. Ownership and Use

About one-third of the rangeland in the United States is publicly owned. The Bureau of Land Management and Forest Service have the responsibility for managing most of the rangeland held by the federal government. Publicly owned rangeland is primarily in the states of Nevada, Wyoming, Utah, New Mexico, Colorado, Idaho, Oregon, Montana, and California.

Public rangelands held by the federal government by law must be managed for multiple use. Thus, these lands are used for wildlife, mineral extraction, recreation, energy extraction, watershed protection, and archeology sites.

Public lands provide forage for about 3% of the nation's beef cattle herd. They are used by about 2% of the country's private ranchers or approximately 7% of the ranchers in the western states.

Ranchers on private rangeland derive most of their income from livestock production. However, income from wildlife (fee hunting) and recreation is becoming increasingly important to private land ranchers. States with large amounts of private rangeland include Texas, New Mexico, Colorado, Montana, Nebraska, South Dakota, North Dakota, Oklahoma, and Kansas.

During each year about 75% of the nation's rangeland is grazed by livestock. About 16% of the feed requirements of the nation's beef cattle herd is provided by rangeland. However, rangelands are critical in U.S. cattle production because most calves are raised on western rangelands or pasturelands. When they near maturity, they are shipped to feedlots where they are fattened on grain prior to slaughter.

II. Rangeland Management Defined

A. Unique Aspects

The distinguishing feature of range management is that it involves the manipulation of the grazing by large herbivores so both plant and animal production will be maintained or increased. Range management as a science and art was born during the early 20th century in the western United States when declining forage production, severe soil erosion, and heavy livestock death losses were recognized as resulting from poorly controlled livestock grazing. Initially range management focused on manipulating the intensity, timing, and frequency of livestock grazing to reduce adverse impacts on soil and vegetation. However, in the last 20 years the scope of range management has broadened to also include manipulation of factors such as fire, wildlife, and human activities. Still the primary aspect of range management is the control of livestock grazing. At present range management is defined as the manipulation of rangeland components to obtain the optimum combination of goods and services for society on a sustained basis. [See RANGELAND MANAGEMENT AND PLANNING.]

B. Basic Concepts

Range management centers around the basic concept that rangeland is a renewable resource that can provide man with food and fiber at very low energy costs compared to those associated with cultivated lands. Ruminant animals are well adapted to use of rangelands because their digestive systems can efficiently break down the fiber which is high in most range plants. [See RANGELAND ECOLOGY.]

Green plants, which are the primary producers on rangelands, use energy from the sun, carbon dioxide from air, and water and nutrients from the soil to produce food for maintenance and growth through the process of photosynthesis. Excessive removal of green plant leaves by grazing destroys the photosynthetic capability of the plant and will ultimately cause its death. However, green plants produce more leaf material than needed for maintenance and growth.

Rangeland productivity is determined by climatic, soil, and topographic characteristics. Generally as precipitation, temperature, and soil depth increase, range

FIGURE 2 A fenceline in southcentral New Mexico that shows the influence of long-term excessive (front) and moderate (back) livestock grazing.

forage productivity increases and more of the forage can be consumed by grazing animals without damage to the range.

III. Grazing Effects on Range Plants

Excessive use of palatable range plants results in their replacement by those that are lower in palatability and often poisonous. Under moderate or light grazing levels the poisonous, unpalatable plants are at a competitive disadvantage because they invest part of their products from photosynthesis in poisonous compounds (alkaloids, oxalates, glycosides, etc.) and appendages (spines, thorns, stickers, etc.) that discourage defoliation rather than contribute to growth. In contrast, the palatable plants use their photosynthetic products mainly for growth in the form of roots, leaves, stems, rhizomes, stolons, seeds, etc. Under excessive defoliation levels the photosynthetic capacity of the palatable plants is reduced to the point that they are unable to produce enough carbon compounds for maintaining root systems, regeneration of leaves, respiration, and reproduction. Over time, they are replaced with unpalatable plants that have had lower rates of defoliation.

As excessive grazing continues the palatable plants tend to be replaced by a succession of plants that increasingly are lower in palatability, lower in pro-

ductivity, and often more poisonous. This process is referred to as retrogression (Fig. 2). When defoliation is reduced to a correct rate, the palatable plants again have the competitive advantage and succession occurs back to the original or climax vegetation.

The driving forces in succession are moisture (rainfall) and temperature. In wet humid range types such as the southern pine forest in the southeastern United States or the tallgrass prairie in the eastern Great Plains, recovery after retrogression is both rapid and predictable. The climax plants will usually again dominate the site within 5 years if severe soil erosion has not occurred. In the drier range types such as the Chihuahuan desert, the palatable plants tend to be less resistant to grazing. Here retrogression can occur within just a few years under excessive grazing but recovery if it occurs is a slow process often requiring 20 or more years. On some sites with serious soil erosion, only minor improvement has been observed after 20 or more years even under complete elimination of livestock grazing. In many areas secondary succession to the original climax does not occur; rather a new and stable climax occupies the site.

IV. Rangeland Condition and Trend

Although range vegetation is seldom stable, it has been well documented that sound grazing management practices will maintain the desirable plants at much higher levels through time than poorly controlled grazing. The group of plants that naturally occupies the site in the absence of severe disturbances such as excessive grazing, cultivation, etc., is referred to as the climax. These plants on grassland ranges are generally the most palatable and productive. In natural forest areas, trees dominate and desirable forage plants are most plentiful when the trees are completely or partially removed by fire or logging.

In the late 1940s E. J. Dyksterhuis developed a system of evaluating the condition of rangelands known as the "quantitative climax approach." This system bases rangeland condition scores on the amount of climax vegetation that remains. Ranges with 76% or more of the climax are considered to be in excellent condition, those with 51 to 75% are in good condition, 26 to 50% are in fair condition, and 0 to 25% are in poor condition. This system has been widely applied by government agencies such as the USDA Soil Conservation Service, and works well on native grasslands. [*See* RANGELAND CONDITION AND TRENDS.]

Under this system plants are categorized into decreasers, increasers, and invaders, depending on their response to grazing (Fig. 3). Highly palatable climax plants that steadily decline under excessive grazing are the decreasers. Increasers are those plants that are part of the climax that initially increase as the decreasers are eliminated. Increaser I plants are distinguished from increaser II plants in that they have a moderate palatability and eventually decline under excessive grazing. Increaser II plants are unpalatable and in many cases poisonous and, therefore, steadily increase under excessive grazing. Invaders are considered to be those that were unimportant or not part of the original climax vegetation. Invader I plants are distinguished from invader II plants by having some

grazing value. Many invader plants are annuals or biannuals while decreasers and increasers are longer lived perennials.

Traditionally range condition has been defined as the state of health of the range. In the past it has commonly been judged on how much climax vegetation remains. In recent years soil stability and how well the existing vegetation is suited for the intended use have also become criteria for assessing range condition. On most ranges limited amounts of increaser and invader plants are advantageous because they provide food and cover for many wildlife species and often grow when decreaser plants are dormant and low in nutritional value. Although if heavily consumed, increaser and invader plants are often toxic, moderate or small amounts in livestock diets can be nutritionally advantageous. Ranges with a mixture of different plant types generally meet the needs of livestock and wildlife better than pure grasslands or shrublands. On most grassland and desert ranges, about 60% remaining climax (good condition) gives an ideal mix of plants for livestock and wildlife and will ensure soil stability. This type of range is sustainable under sound grazing management.

Range trend refers to the rate and direction of change in range condition. Many range managers consider knowledge of trend to be the most important aspect in evaluating the effectiveness of different range management practices. Generally an upward trend is considered to be succession toward the climax or the desired plant community while a downward trend would be retrogression away from the climax or desired plant community.

A problem with evaluating trend is that climatic influences must be separated from those associated with grazing or other management practices. Exclosures (ungrazed areas) are an important tool for making these separations. Generally grazing management is considered to be sound if vegetation improvements on the grazed area equal or exceed those in the exclosure. However, lack of recovery during wet periods and/or more rapid retrogression in drought periods on the grazed area compared to the exclosure would signal improper management.

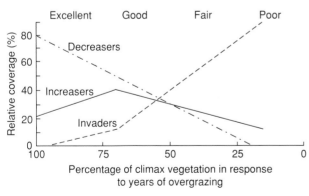

FIGURE 3 Change in proportions of different plant groups with overgrazing.

V. Types of Grazing Animals

A. Comparative Digestive Systems

Large grazing animals (ungulates) can be divided into two groups based on their digestive systems. These systems include those with rumens (cattle, sheep, goats, deer, elk) and those with enlarged cecums (horses, donkeys, elephants, rabbits). Both systems evolved to enable ungulates to digest plant fiber (cell walls) by microbial (bacteria and protozoa) fermentation. Although the digestive processes are similar with both systems, they differ in regard to where fermentation occurs relative to the true stomach. The cecum occurs as an enlarged portion of the large intestine that food enters after passing through the true stomach. The rumen is an enlarged portion of the digestive tract that food must pass through before entering the true stomach. [See ANIMAL NUTRITION, NONRUMINANT; ANIMAL NUTRITION, PRINCIPLES; ANIMAL NUTRITION, RUMINANT.]

In ruminant animals food must be reduced to a fine particle size to pass into the true stomach. This results in more complete digestion of fibrous material which is digested slowly. In cecal digestors fiber can pass easily out of the cecum with no great reduction in particle size. This allows the animal to consume more forage but fiber digestion is less complete than in ruminants.

Although neither type of animal can produce its own protein, microbes in the rumen and cecum have the capability for protein synthesis if a source of nitrogen is available. Microbes are passed from the rumen to the true stomach where they are digested and then absorbed. Little microbial protein is absorbed directly by cecal digestors but these animals will ingest their feces (coprophagy) which enables them to use this source of protein.

B. Forage Selection

Forage selection by range ungulates (large, hooved animals) is governed by four basic factors. These include body size, type of digestive system, size of digestive system relative to body weight, and mouth size and shape. Based on these factors the three basic groups of ungulates are the grazers, the browsers, and the intermediate feeders. They will be discussed.

1. The Grazers

This group of ungulates feeds primarily on grasses and grasslike plants. Cattle, bison, horses, musk oxen, and elk are all grazers. Under certain conditions cattle and elk will consume large amounts of shrubs and forbs. This occurs mostly when green grass is not available. Grasses have lower amounts of protein and higher amounts of fiber than leaves from forbs and shrubs. Because the grazers have a large size and a large digestive system, they require a diet high in

quantity but not high in quality. Therefore, grasses and grasslike plants best meet their needs.

2. The Browsers

This group depends primarily on forbs and leafy material from shrubs and trees as sources of food. Leaves from forbs, shrubs, and trees are high in protein, critical minerals such as phosphorus and calcium, and vitamin A (carotene). These plants have a rapid rate of digestion which compensates for their lower energy concentration compared to grasses. White-tailed deer, pronghorn, mule deer, and moose are examples of browsers. These ungulates may experience digestive problems if forced to consume diets high in mature grass. Selectivity and fine chewing of food are mechanisms used by browsers to avoid toxicity. Fine chewing of food results in the release of some toxic substances as gases minimizing assimilation. More mixing of plant material with saliva occurs with fine-chewing. Proteins in the saliva of some browsers apparently have the capability to detoxify poisonous compounds.

3. The Intermediate Feeders

This group of ungulates by choice tends to select a diet with equal amounts of grasses, forbs, and shrubs. Domestic sheep, goats, and burros are intermediate feeders. These animals will readily use grasses or shrubs depending on what is available. The small, flexible mouth parts, small total size, and large digestive system relative to size give domestic sheep great capability to adjust their diets to available forage resources. However, their large body and short legs make them quite susceptible to predation.

C. Suitability for Different Rangelands

Important considerations in selection of type or types of livestock for a particular range include: water requirements, topography, predators, climate, parasites and disease, poisonous plants, types of forage, and economic costs and returns.

Cattle and horses are considered to be best suited for flat, well-watered areas that are dominated by grasses. European cattle (Bos taurus) such as the Hereford and Angus breeds are well adapted to cooler climates while the African cattle (Bos indicus) such as the Brahman breed do well in hot, humid areas. [See BEEF CATTLE PRODUCTION.]

Sheep and goats are better adapted to hot, humid climates and rugged terrain than cattle or horses. Unlike cattle and horses, they can be watered every other day rather than daily and still perform well. Because of their small size and surefootedness, sheep and goats make better use of steep terrain than large animals. [*See* GOAT PRODUCTION.]

In some areas pests and diseases make it difficult to raise domestic livestock. Large areas in Africa are unsuitable for cattle because of the tsetse fly. Game ranching (raising of native ungulates for meat) shows potential as a sound alternative to domestic livestock based on evaluations by Dr. David Hopcraft in Kenya. [*See* LIVESTOCK PESTS.]

In some parts of the western United States, sheep ranching has been more profitable than cattle ranching. The price of lambs has usually been greater than that for calves, and wool further adds to returns from sheep. However, cattle do have the advantages of reduced susceptibility to predation and they require less labor than sheep. The cowboy persona associated with cattle has given them a special appeal.

Common-use grazing is the intentional use of rangelands by more than one type of animal. This involves some combination of animals such as cattle and sheep in the mountains of Utah or cattle and white-tailed deer in central Texas. Common-use grazing often has a number of advantages such as better distribution of animals, harvesting more of the available plant species, and reduction in risk from diversification. The disadvantages include greater labor requirements, more handling facilities, and conflicts with wildlife when certain combinations of domestic animals are grazed at excessive stocking rates.

VI. Rangeland Management

A. Objectives

Historically the primary goal of range management has been to maximize livestock production without degrading the range. In more recent years wildlife production and water quality and quantity have received greater emphasis than livestock production on some rangelands. Still range management centers around control of livestock. The four basic principles in grazing managements are proper livestock numbers, proper livestock distribution, proper kinds of animals, and proper system of grazing. The other aspect of range management involves manipulation of range vegetation with tools other than large animals. These practices include seeding, fertilization, and control of undesirable plants with fire, herbicides, mechanical means, or insects. This approach is generally much more costly and involves higher risks than vegetation manipulation through control of grazing. However, dramatic increases (2- to 10-fold) in forage

production within short time periods (1 to 5 years) are possible.

B. Determining Grazing Capacity

Grazing capacity is the number of animals which can graze a range each year without inducing a downward trend in condition. Knowledge of the productivity of the vegetation, the forage requirements of the animals, and the degree of use which the vegetation can withstand are required to determine grazing capacity. Adjustments must also be made for distance from water and for topography. Grazing capacity is influenced by range condition, annual climatic conditions, past grazing use, soil characteristics, types of animals, and how long the animals will graze the area. Proper stocking is considered the critical aspect in successful range management (Table I).

TABLE I

Influence of Grazing Intensity on Winter Sheep Production at the Desert Experimental Range in Utah and Cattle Production at the Southern Great Plains Experimental Range in Oklahoma.

	Excessive grazing	Moderate grazing
	Sheep—Desert Experimental Range, Utah[a]	
Utilization of forage (%)	68	35
Ewe weight change (fall to spring) (lb)	+1.1	+9.3
Average fleece weight (lb)	9.68	10.63
Lamb crop (%)	79	88
Death loss (%)	8.1	3.1
Lamb weaned per ewe (lb)	67.0	77.0
Net income (3000 head flock) ($)	5,072	10,390
Net income per ewe ($)	1.69	3.45
	Cow/Calf—Southern Great Plains Experimental Range, Oklahoma[b]	
	Excessive	Moderate
Acres/cow	12	17
Estimated utilization of forage (%)	62	44
Calf crop weaned (%)	81	92
Calf weaning weight/cow (lb)	314	424
Calf weaning weight	388	461
Net returns per cow ($)	9.00	29.44
Net returns per acre ($)	0.70	1.88

[a] *Source.* Hutchings, S. S., and Stewart, G. (1953). Increasing forage yields and sheep production on intermountain ranges *U. S. Dep. Agric. Circ.* **925.**
[b] *Source.* Shoop, M. C., and McIlvain, E. H. (1971). Why some cattlemen overgraze and some don't. *J. Range Management* **24,** 252–257.

Commonly grazing capacity is expressed as the stocking rate which is the number of animal units placed on a given area for a particular time period. An animal unit is a measure of how much forage a mature nonlactating bovine weighing approximately 1000 lbs will remove over a 1-year (365 days) period.[1] Various studies show ruminant animals eat about 2% of their body weight per day on a dry matter basis. Therefore a 1000 lb cow requires approximately 7300 lbs of forage per year, 610 lbs per month, and 20 lbs per day. Horses and donkeys eat about 50% more than ruminants and have dry matter intakes near 3% body weight per day. The following represent some accepted animal unit equivalents:

Animal	Weight (lbs)	Animal unit equivalent
Cow	1000	1.00
Steer	750	0.75
Bull	1200	1.20
Sheep	150	0.15
Goat	100	0.10
Deer	150	0.10
Elk	700	0.70
Bison	1800	1.80
Donkey	700	1.05
Pronghorn	120	0.12
Horse	1200	1.80

Grazing intensities that can be applied to different ranges depend on season, range condition, climate, soil, and topographic features. On flat, wet humid ranges such as the tallgrass prairie where forage production (dry matter basis) exceeds 2000 lbs per acre, 50% of the forage can be removed without damage to the plants. On semi-arid ranges such as the shortgrass prairie, 40 to 45% can be removed. On desert ranges, no more than 30 to 35% removal is recommended. Exceeding these levels not only damages the range under most conditions, but usually reduces livestock productivity and economic returns.

During drought livestock operators need to be prepared to reduce livestock numbers or provide feed supplements to prevent overgrazing. Failure to destock during and following drought periods has been one of the most serious causes of rangeland degradation in the United States and other parts of the world.

C. Improving Distribution

Proper placement and spacing of water have been the primary tool range managers have used to obtain

[1] For a more complete discussion of animal unit equivalents the reader is referred to: Vallentine, J. F. (1990). "Grazing Management." p. 278–284. Academic Press, New York.

uniform grazing over the range. In arid areas water point spacing of about 3 to 4 miles apart has proven most practical when development cost and efficient use of the range are both considered. In the more humid ranges 2- to 3-mile spacings are practical because more grazing capacity is gained relative to watering point cost.

Failure to adjust stocking rates for distance from water has been an important cause of rangeland degradation and poor livestock performance. Various range studies show that with cattle areas over 2 miles from water should be deleted in grazing capacity estimates. A 50% reduction in grazing capacity is required for the zone 1 to 2 miles from water but no adjustment is necessary for the zone within a mile of water.

Salt blocks and supplemental feed in strategic places can also be useful in improving livestock distribution. Another approach is the use of rotation grazing schemes which will be discussed next.

D. Grazing Systems

During the last 20 years specialized grazing systems have been heavily emphasized in range management programs. These systems differ from continuous or season-long grazing in that they involve scheduled moves of livestock from one pasture to another. Deferred-rotation, rest-rotation, and short-duration systems are some of the more popular and widely used specialized grazing strategies. Impressive claims have been made for each system regarding increases in range condition, stocking rate, and livestock production. However, actual research has shown specialized grazing systems gave modest (10 to 30%) to no increase in forage and livestock production compared to continuous or season-long grazing.

The main advantage of rotational grazing schemes is that they provide preferred areas and plants with periodic nonuse that permits recovery from excessive grazing. They have been most useful in rugged terrain, and where climatic conditions favor high forage production and rapid regrowth after grazing (tallgrass prairie, southern pine forest).

1. Continuous Grazing

Under this strategy a given range is grazed year-long or season-long without movement of the livestock. It involves the least cost in terms of fencing and livestock handling. Livestock performance has generally been higher under continuous grazing than under rotational schemes with some exceptions. Both greater handling and the need to adjust to a new pasture increase animal stress under rotational schemes.

One of the big concerns with continuous grazing is that desirable plants will be used excessively. However, actual research from several range types shows that during the critical growing season cattle, sheep, and goats select a variety of plants. Many forbs are highly preferred comprising up to 40% of the overall diet. This reduces grazing pressure on the perennial grasses.

The biggest problem with continuous grazing is that areas near water can receive excessive use. This problem can be reduced by fencing watering points and then regulating access to allow recovery.

Continuous grazing has worked well on flat desert ranges of the southwest, the annual grassland ranges of California, and the shortgrass prairie of the central Great Plains. With judicious application of distribution practices such as fencing, salting, and water development, continuous grazing can work fairly well even on the more rugged range types.

2. Deferred-Rotation Grazing

This system of grazing, developed by Arthur Sampson in the early 1900s, involves dividing the range into two pastures. Every other year grazing is deferred (delay of grazing until set maturity of key forage species) on each pasture. This strategy gives the primary forage plants periodic opportunity to store carbohydrates without defoliation. A number of modifications of this system have been developed involving more than two pastures but they all involve periodic nonuse during active growth.

A problem with deferred-rotation grazing is that stocking pressure is greatly increased in the years when deferment does not occur. The real benefit of deferred rotation grazing is that it gives preferred areas opportunity for recovery. It has given better vegetation response than continuous or season-long grazing in rugged, mountainous areas where meadows, benches, and streamside areas receive heavy use even under light stocking rates.

3. Rest-Rotation Grazing

Rest-rotation grazing involves providing one pasture with a year of nonuse while the other pastures absorb the grazing load. It was developed by Gus Hormay of the U.S. Forest Service in the 1950s and 1960s. Most rest-rotation schemes have involved four pastures with various sorts of rotation schemes applied to the three grazed pastures. There are two main drawbacks to rest-rotation grazing. The first is that

the extra use on the grazed pastures may be more detrimental than the positive effect of periodic rest. The second is that livestock are forced to graze less selectively than under season-long grazing. This can lower the nutritional quality of their diet and, in turn, reduce their weight gains.

On rugged mountainous ranges where livestock distribution problems occur, rest-rotation systems have given better vegetation performance than season-long grazing. When conservative stocking rates were used, cattle weight gains were similar to those under deferred-rotation and season-long systems.

Actually rest-rotation grazing has a number of multiple-use advantages if stocking is kept conservative (around 30% use of key forages). Wildlife are provided with a pasture free from disturbance and defoliation. Many recreationists find it esthetically appealing to see ungrazed ranges. Livestock in the grazed pastures generally more evenly use the grazable area than under season-long grazing.

4. Short-Duration Grazing

This grazing strategy, developed first by A. Voison in France and later refined by Allan Savory in Zimbabwe in the 1960s, involves the greatest division of the range into subunits by fence. When Savory came to the United States in the late 1970s he made numerous refinements in short-duration grazing that have been called the Savory grazing method and most recently, holistic resource management. Other modifications and names for short-duration grazing include rapid-rotation, time-control, cell grazing, and wagon-wheel grazing.

Short-duration grazing has commonly involved subdividing the range into eight or more subunits (paddocks) with water and livestock-handling facilities located in the center (cell). A wagon-wheel design with the paddocks radiating from center is often used although not necessary (Fig. 4). Generally each paddock receives a short grazing period (5 days or less) followed by a much longer period of nonuse (4 weeks or more). During active growth of forage livestock are moved more rapidly than during dormancy. A number of benefits are claimed to result from the high stock density (number of animals per unit area) unique to this system. These include:

1. Improved water infiltration into the soil as a result of hoof action.
2. Increased mineral cycling.
3. Reduced selectivity so that more plants are grazed.

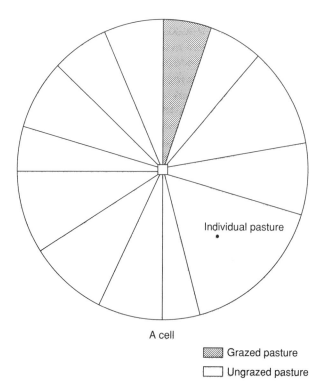

Individual pasture
•

A cell

▦ Grazed pasture
☐ Ungrazed pasture

FIGURE 4 Short-duration grazing system layout on a small cow–calf ranch in southeastern New Mexico. Reprinted with permission from Fowler, J. M., and Gray, J. R. (1986). *New Mexico Agric. Exp. Sta. Bull.* **725.**

4. Improved leaf area index.
5. More even use of the range.
6. Lengthened periods when green forage is available to livestock.
7. Reduced percentage of ungrazed "wolf" plants.

Many ranchers have found short-duration grazing appealing due to claims that stocking rates can be increased substantially (doubled in certain cases), labor costs can be reduced, individual animal performance can be increased, and the range will improve rapidly in condition if degraded. Short-duration grazing has been advocated for range types throughout the world.

One of the biggest benefits from short-duration grazing is better livestock distribution. This appears to account for much of the increase in stocking rates that is possible with this system. When livestock are confined to small areas for short periods of time they are forced to use areas and plants that would be lightly or unused under continuous grazing. A number of studies show the down side of closer confinement and more frequent rotation is reduced animal performance. However, this problem is minimized if livestock can gradually adjust to periodic rotation.

Research casts doubts on claimed benefits to the soil from short-duration grazing. Generally water infiltration into the soil has been reduced and sediment losses increased when short-duration grazing was compared to moderate-continuous grazing. The hoof action resulting from having a large number of animals on a small area seems to have more of a compacting than loosening effect.

In arid areas where forage is sparse, short-duration grazing may be financially impractical due to high fencing costs. The short periods of active forage growth (under 70 days) minimizes the benefits of repeated periods of defoliation and nonuse. Many of the desert grasses recover very slowly from excessive use that can result from failure to move livestock at the right time.

On the other hand, short-duration grazing is theoretically well suited for the flat, more humid range types. Here fencing costs are greatly reduced because much less land is required for a given number of livestock. Partial removal of herbage can potentially increase plant productivity because the high volume of vegetation causes more shading of leaves. Periodic grazing is more likely to prolong the period when forage is nutritious and actively growing and greater opportunity for regrowth exists.

Although much controversy surrounds short-duration grazing, many ranchers have found the holistic research management principles developed and taught by Allan Savory appealing and useful. Rancher ingenuity in applying these principles varies tremendously which explains in part why some claim success and others failure.

E. Stocking Rate versus Specialized Grazing System

Unfortunately managers on both public and private rangelands have held the belief that stocking rate could be largely disregarded if some miracle specialized grazing system was applied. However, actual research has shown overwhelmingly that no specialized grazing system alone will counteract the long-term effects of overstocking. A careful analysis of grazing management research shows stocking rate has had far more impact on range condition, forage production, livestock production, and economic returns than any specialized grazing system.

Across all studies financial returns have averaged about 40% higher under moderate compared to heavy stocking rates. The monetary benefits of moderate grazing have increased as the length of study time has increased. This is because degradation of soil and water resources under excessive grazing occurs gradually.

On the arid western rangelands net profits have consistently been higher under moderate, continuous grazing compared to rotational grazing systems. However, on the more humid ranges of Texas, the Great Plains and the southeastern United States specialized grazing systems have often been advantageous from both vegetation and monetary standpoints.

F. Controlling Vegetation

In many parts of the western United States unpalatable and often toxic woody plants such as sagebrush, mesquite, and juniper have invaded large areas of former grasslands or shrub steppes due to overgrazing and fire suppression. These plants are often well adapted for long-term survival after they have become established. Methods used by range managers to control these plants have included spraying with herbicides, controlled burning, mechanical control, and biological control by releasing animals or insects that will consume the plant.

During the 1950s and 1960s herbicides such as 2,4,5-T were widely used to kill invading woody plants. However, they are now seldom used because of their high cost and possible threats to human health.

Controlled burning has become a more popular method of controlling undesired plants. It is considered to be ecologically natural and involves much lower cost than herbicides. The disadvantages of burning are air pollution, soil erosion, and the possibility of fire getting out of control. Generally burning causes shifts from shrubs and trees to grasses and forbs. When well planned and executed, burns can have minimal negative impacts on the environment, and can be quite beneficial to livestock and wildlife.

Some introduced predatory insects have given impressive biological control of unwanted plants. Goats can be used in suppressing certain shrubs. Biological control is less expensive than other methods, but finding and establishing predatory species on the unwanted plant has been difficult.

Mechanical control of unwanted plants by root plowing, shredding, bulldozing, or chaining has been little used in recent years because of high cost and severe soil disturbance. This approach is most practical in the wetter rangeland types where taller brush species occur and several-fold increases in forage are possible after their removal.

After unwanted plant control, proper management of livestock is essential. Generally 1 to 2 years of nonuse after control is recommended to give desirable plants opportunity to become established.

VII. Rangeland Livestock Production

A. Arid versus Humid Ranges

Grazing on arid and mountainous lands in the West has given lower monetary returns than on the more humid ranges of the Great Plains. This is because more infrastructure such as fences, corrals, roads, watering points, etc., is required to maintain a given number of livestock where forage is sparse and terrain is rugged compared to where the opposite conditions prevail. Most of the arid ranges west of Rocky Mountains produce under 400 lbs of forage per acre per year while those in the Great Plains produce anywhere from 700 to 3000 lbs of forage per acre per year. Flatter terrain, longer periods when the forage is actively growing and nutritious, and the better herd management that is possible when livestock are confined to smaller areas are other economic advantages the Great Plains have over the arid western rangelands. Since the 1980s ranchers in the Great Plains have received a 3 to 5% average annual return on their capital investment compared to a 1 to 2% return for ranchers on arid western lands.

B. Drought

Drought is a serious problem confronting stockmen on arid rangelands. Drought increases in frequency and severity as aridity increases. Coping with drought is the greatest challenge confronting ranchers on western ranges.

Inexperienced ranchers often underestimate the degree to which drought can reduce forage production. Studies from a variety of range types show drought can reduce forage production by more than 50% of the annual average. The best way to deal with drought is to apply conservative stocking rates. Under drought conditions pastures heavily grazed show much greater reductions than those moderately grazed (Table II). Partial confinement and drylot feeding of livestock are often advantageous under drought conditions. This allows ranchers to avoid liquidation of livestock when prices are low followed by repurchase after the drought when prices are high due to restocking by other ranchers. Another problem

TABLE II

Herbage Production (Pounds per Acre) on Heavily and Moderately Grazed Shortgrass Prairie in Colorado during Drought Year Compared to the 5-Year Average

Grazing intensity	Drought year	5-Year average	Drought year as percentage of average
Excessive: 54% use of forage	312	595	52
Moderate: 37% use of forage	577	766	75
Light: 21% use of forage	609	817	75

Source. Klipple, G. E., and Costello, D. F. (1960). Vegetation and cattle responses to different intensities of grazing on shortgrass ranges of the central Great Plains. *U. S. Dept. Agric. Tech. Bull.* **1216.**

with liquidation is that disease-free, high-quality herds are usually difficult to obtain. Animals that have experience on a particular range perform better than those brought in from another area.

C. Poisonous Plants

Annual livestock losses to poisonous plants average between 2 and 5% on western ranges. Heavy livestock losses to poisonous plants are generally linked with excessive grazing that reduces availability on palatable, nonpoisonous species and increases those that are poisonous. Studies from experimental ranges at a variety of locations show annual poisonous plant losses of livestock to be under 2% when stocking rates were moderate.

There are a few poisonous plants such as larkspur that livestock will consume under moderate stocking and when the range condition is good. Although larkspur is quite toxic to cattle, it has little effect on sheep. Delay of grazing until larkspur dries is an effective management practice. Grazing sheep instead of cattle on ranges with larkspur is another approach.

Most poisonous plants are only toxic for a short period of time, usually during active growth. Many poisonous plants initiate growth earlier than nontoxic grasses, and in this period they may be preferred by livestock because they are green and succulent. Delay of livestock grazing until green growth of perennial grasses is readily available is one of the most effective ways to minimize poisonous plant losses on mountain rangelands. Livestock that are born and reared in a particular environment are less prone to ingest poisonous plants than those brought in from another area.

VIII. Future Directions for Rangeland Management

New technologies and range management strategies will be needed to accommodate traditional use by livestock plus the rapidly increasing demand for outdoor recreation, water, wildlife, and open space in the 1990s. In summary range managers will be challenged with producing more goods and services with less land. More productive plants, better herbicides, more efficient machinery, improved grazing methods, and better livestock breeds and management all have potential to meet the above demands.

Computers and biotechnology will undoubtedly play a big role in future range resource management. Plants more resistant to drought, insects, and defoliation will be developed. At the same time livestock breeds will be engineered that are leaner, more efficient in use of forage, and more productive in meat and milk.

Although the future presents formidable challenges it provides great opportunities as well. Historically the problem of increasing demands on limited resources has created incentives for technology that greatly improved the human condition. For this reason I believe the future is bright for rangelands and those who manage them.

Bibliography

Heitschmidt, R. K., and Stuth, J. W. (eds.) (1991). "Grazing Management." Timber Press, Portland, OR.
Holechek. J. L., Pieper, R. D., and Herbel, C. H. (1989). "Range Management Principles and Practices." Prentice-Hall, Englewood Cliffs, NJ.

Savory, A. (1988). "Holistic Resource Management." Island Press, Washington, DC.

Vallentine, J. F. (1989). "Range Development and Improvements." Academic Press, New York.

Vallentine, J. F. (1990). "Grazing Management." Academic Press, New York.

Workman, J. P. (1986). "Range Economics." Macmillan, New York.

Rangeland Management and Planning

REX D. PIEPER, *New Mexico State University*

Glossary

Inventory Measuring or estimating a range attribute at one time to determine range characteristics for management purposes

Monitoring Measuring some attribute (e.g., plant cover, herbage weight, or number of plants) at specified intervals; monitoring is typically used to determine changes in vegetation resulting from climatic patterns such as drought or a management decision such as grazing strategy, stocking rate, or range improvement

Range management Manipulating rangeland systems components such as plants, animals, or soils to meet goals or objectives; sometimes economic restraints are included

Remote Sensing Measuring attributes of an object or phenomenon without physical contact; in natural resources, remote sensing usually refers to measurement information obtained using airplanes or satellites

Sampling Measuring a characteristic of rangeland on a portion of the whole population; a sample is used to characterize a larger portion of the rangeland such as a grazing allotment or pasture

Stocking Rate Number of animals grazing on a rangeland of specific size for a specific time; the stocking rate is the ratio between the number of animals and the size of the area being used, for example, 12 sheep/acre

Rangeland management planning involves setting goals and objectives and formulating a plant to meet these objectives. Because rangelands are diverse and are used for many purposes, the goals and objectives are often multifaceted. Often private owners direct their goals toward economic objectives, while public rangeland administrators may aim their goals toward multiple-use objectives.

I. Introduction

Traditional uses of rangelands included livestock grazing, habitat for animals (both game and non-game), recreational opportunities and game hunting, watersheds, and timber and wood production. As the amount of leisure time, disposable income, and access to rangelands has increased, new demands on rangelands have made range management and planning more complicated. The modern rangeland manager must now accommodate demands on wilderness; increased commercial development for recreation such as ski facilities, seasonal housing, or recreational facilities; trails for hikers, bird watchers, and photographers; power line corridors; energy extraction (coal, geothermal exploration); and a host of other special-interest activities. [See RANGELAND.]

Meeting multiple-use objectives is partly a matter of scale, as it is not possible to manage for all uses on all rangeland areas. Considering multiple use on a "landscape" basis means providing for several uses on large areas of land. For example, large uniform grasslands may yield the most forage for livestock but may be relatively nonproductive for game species that may require areas of woody vegetation interspersed within the grassland. Big game animals require cover, water, and food—all within a particular area. Distributing these essential habitat elements is the key to effective management and must be included in the management plan.

II. Range Management Plan

A sample outline for a range management plan might include:

I. Goals and Objectives
   A. Resources
   B. Economic
II. Inventory
   A. Vegetation
   B. Soils
   C. Topographic features
   D. Physical features
      1. Pastures
      2. Water developments
III. Livestock Management
   A. Number
   B. Type of livestock
   C. Grazing plans
   D. Supplemental feeding
   E. Breeding
   F. Marketing
   G. Management of replacements
IV. Improvements
   A. Brush control
   B. Seeding
   C. Additional water developments and fencing
V. Other Management Concerns
   A. Soil stabilization and watershed objectives
   B. Wildlife habitat
   C. Recreational possibilities
   D. Mineral resources
   E. Timber resources
VI. Monitoring
   A. Objectives
   B. Methods
   C. Timing
   D. Relationship to management

III. Monitoring and Inventorying Rangelands

A survey of current conditions is often the first step in initiating a management plan. Vegetational surveys to determine the current vegetation status are made at this time. A map including broad vegetational types such as grassland, shrubland, savanna (trees with open grassland interspersed throughout), or coniferous forest is developed for the area under consideration. Maps of this type originally were developed from tedious field observation when lines between vegeta-

tion types were drawn on maps as the surveyor stood in the field. [See RANGELAND CONDITION AND TREND.]

Later, the process was greatly simplified when aerial photos became available because the mapper could look at them in the office or lab. These photos have been used as the basis for vegetational mapping since World War I. Nearly all the United States has been covered by aerial photography, and these photos still play an important role in vegetational mapping. Photos have some limitations if detailed mapping is desired as it is difficult to use them to distinguish between different kinds of grasslands and, in some cases, between different kinds of woodlands or forests. These distinctions are even more difficult to make when photos are taken at higher altitudes and photo scales are small. Relatively large-scale color aerial photos, however, have proven useful in making finer distinctions among grassland vegetational types.

With the development of high-speed computers and orbiting satellites, the science of remote sensing advanced greatly. Various systems have been placed on satellites to evaluate reflectance data coming from the earth's surface. Through these systems, spectral characters are changed by the nature of the vegetation intercepting electromagnetic waves from the sun. The initial Landsat spacecraft utilized the Landsat Multiple Scanner (MSS), while later spacecraft utilized the Thematic Mapper (TM), Systeteme Pour l'Ovservation de la Terre (SPOT), and Advanced Very High Resolution Radiometer (AVHRR). These systems have different characteristics, but all have limitations for detailed analysis because they lack high resolution (the ability to render a sharply differentiated image).

Satellite-based mapping systems offer advantages for large-scale surveys and some monitoring work. Software programs are now available for personal computers to use satellite data to produce vegetational maps based on species differences, green herbage weight, or other factors. For example, maps of Africa based on vegetational productivity have been produced using this technology. Obtaining ground-observed data to correlate with satellite data is necessary, in most cases, to obtain maximum value from satellite data.

Physical features such as roads, water developments, pasture fences, livestock handling facilities, or vegetational treatments areas are also included on the base management map. Many land management agencies are now turning to Geographic Information Systems (GIS) to assist in handling all the information needed for managing rangeland resources. GIS involves storing and manipulating spatially oriented information about a specific site in a unified computer system. GIS can be used to advantage for data sets such as vegetational and soils information. This information can be entered separately into the GIS program and then shown as a series of computer-generated overlays. These data can then be analyzed for plant communities occurring on certain soil types. Many other types of information such as pasture fences, past stocking records, physical improvements, or water locations can be included in GIS information.

IV. Estimating Vegetational Characteristics

Most range management plans include a description of the vegetation within the area involved. Vegetation can be characterized by several attributes: weight of individual species, number of individuals of each species, percentage cover of individual species, height of plants, and other properties that are measured less directly. In most range plans, vegetational monitoring and surveys include estimates of vegetation weight or *cover,* sometimes with density (number of plants per unit area) included.

Most techniques developed by ecologists to describe vegetational characteristics of rangelands are applied mainly to small research plots and have limitations on large range areas. Requirements for sampling include low variability among sample units; accurate results, where sample data are close to the range characteristics being estimated; and time efficiency. In either case (research areas or evaluations on large areas), we must resort to sampling where we measure or estimate the characteristic on a fraction of the area and use this sample to infer characteristics about the entire area. On small research plots where time is not a factor, we can use time-consuming techniques if they are both accurate and precise (having low variability among repeated samplings). However, only those techniques considered time efficient for large-scale range surveys and monitoring will be covered in this discussion.

A. Cover

Two methods for large-scale rangeland monitoring or surveying are both time efficient and reliable and can be used to determine either canopy cover or basal cover. *Canopy cover* is defined as the percentage of the soil surface covered by the vertical projection of the crown (aerial portion of plants) on the soil surface. *Basal cover* represents the percentage of the soil surface covered by the base of the plant where it touches the soil surface. The difference between canopy and basal cover for grasses is not as great as that for woody plants, which have large canopies and relatively small stems. Generally, woody species are sampled for canopy cover and herbaceous species are sampled for basal cover. However, when an area contains both woody and herbaceous cover, it is probably most appropriate to estimate canopy cover.

Quadrats are used to estimate cover. A *quadrat* is a small area outlined on the ground using a frame made of light angle iron or metallic rods. In dense vegetation, one end of the frame may be open to place the frame on the soil surface easily. Cover is estimated as the sum of the percentages of species occurring in the quadrat. However, there is no standard for the estimates and there may be large observer differences if observers do not train together to standardize their estimates.

1. Cover Classes
Cover in the quadrat can be estimated by *cover classes.* One system described by Rexford Daubenmire contained six classes:

Class 1: 0–5% cover
Class 2: 5–25% cover
Class 3: 25–50% cover
Class 4: 50–75% cover
Class 5: 75–95% cover
Class 6: 95–100% cover

When using this method, the observer simply records a cover class for each quadrat, as it is easier and more accurate to define cover by one of these cover classes than to estimate absolute cover by species. After cover has been estimated on several quadrats, averages for the entire area are calculated using the midpoints for cover classes in each quadrat. For example, the value used for all quadrats in cover class 2 would be 15%. Many other cover class systems have been proposed by other ecologists, but Daubenmire's system has been used widely in the United States.

A second approach for determining cover, called the *point-step method,* is a modification of the point sampling approach. In point sampling, the sampler imagines covering the entire area to be sampled with a series of dots or points very close together. To determine cover, the sampler counts the dots or points directly over plants and determines cover by dividing the number of dots directly over plants by the total number of dots. Dots not directly over plants would be recorded as bare ground or litter cover. It is obviously not possible to cover an area entirely with points, but using the point-step method, the point of the sampler's toe can be used as the point. The sampler paces across the area and records the plant species directly under the toe whenever the right toe touches the ground. The sampler should look at something on the horizon rather than at the foot to guard against predetermining where the toe will land. Other point sampling techniques involve point frame, line-point, and sighting tube techniques.

One of the problems with point sampling involves point size, as a point that is too large will produce cover estimates that are too high. It is advisable to use a fine wire at the toe of the boot to identify points for cover determinations. This method is fast and many points can be sampled in a day. Sampling is usually conducted by *transects* having 100 points each. Points along the transect are located systematically with equal distances between points, but the transects can be located randomly so that each location in the field has an equal chance of being selected for one transect.

Cover by species and species composition can be calculated from step-point sampling. *Cover* is the number of direct hits for each species divided by the total number of points. *Composition* is calculated by dividing the number of hits for each species by the total number of hits for all species.

B. Herbage Weight (Standing Crop or Biomass)

Herbage weight is an important rangeland characteristic, as it represents the amount of food available for *herbivores* (animals that subsist principally or entirely on plants or plant material, including livestock). Herbage weight is also a good indicator of the ecological importance of the plant species in the community. Many methods have been devised to estimate herbage weight on rangelands, but many are time-consuming and involve destructive sampling. Determining herbage weight is often complicated by several growing season variables.

1. Sampling Time

Herbage weight changes throughout the year. Early in the growing season, above-ground vegetation may consist of dormant plant material with little living material, even for perennial grasses and other herbaceous plants that typically die back to the crown area at the end of the growing season. As the plants begin to grow, new green leaves appear and expand, and then other leaves and stems are produced until herbage weight reaches a peak.

In most range situations, herbage weight reaches a peak once a year, but in some cases there may be more than one growing period. In the Great Basin and Intermountain Region, growth usually occurs in the spring when soil water is available and favorable temperatures stimulate plant growth. Growth is usually completed when soil water is depleted in the summer. Sometimes growth begins again in the fall when temperatures stay high and rainfall replenishes soil water. In the northern Great Plains, cool-season species grow during the early rainfall period while the main warm-season species grow during the late summer when the main rainfall peak occurs. Under all these situations herbage weight declines during the dormant period in the late fall and winter.

In light of these complicating factors, if herbage weight is determined only once during the year, when should it be made? Often herbage weight is determined at the end of the growing season in the fall, or the summer for the Great Basin–Intermountain region, but this practice may underestimate plants growing earlier than in the main growth period. If soil water content is sufficient in the Southwest, many annual forbs grow during the late winter–early spring period. Sampling should be spaced so these species will not be underrepresented through sampling done only at the end of the summer growing season.

2. Sampling Procedures

Herbage weight can be determined directly by clipping plants and weighing them, but such procedures are time-consuming and are not practical for large-scale surveys or monitoring. In these circumstances, estimation or some combination of estimation or clipping and estimation are often used in these circumstances.

The *weight-estimate method* involves estimating the weight of herbage in *quadrats*, generally square, circular, or rectangular areas used for estimating herbage weight. Using squares and circles reduces the necessity to decide which plants are or are not located in the quadrat; using rectangles presents more borderline decisions, but tends to reduce the variation from one quadrat to another. Generally quadrats are small (<1 m²), depending on the vegetation to be sampled.

Calibrating the weight-estimate method requires a training period when the estimator estimates herbage weight in particular quadrats, then clips the plants, weighs the sample, and compares the actual weight to the estimated weight. Once the estimated weight comes close to the clipped weight, the estimator then estimates herbage weight in the relevant quadrats.

Often, the weight-estimate method is supplemented with actual weights from selected quadrats after the weight is estimated. The estimator can then make proportional statistical adjustments in all the estimates. These estimates and clipped weights can be related using correlation and regression analyses and all estimated weights can be adjusted by the regression equation to improve accuracy and precision. Obtaining clipped weights adds to sampling time, but many feel that the gain in accuracy can be worthwhile.

C. Frequency

Some researchers advocate using *frequency* (the absence of presence of a plant species within a sampling unit) as a simpler method of evaluating vegetation. Frequency is sometimes determined on transects or other sampling units, but to be most useful, it should be determined using quadrats. Frequency is calculated by dividing the number of quadrats in which a species occurs by the total number of quadrats sampled, no matter how many individual plants occur in the quadrat and expressing the result as a percentage.

Quadrat size should be determined carefully because frequency is highly sensitive to quadrat size. If the quadrat is too large, all species will have high frequencies; if the quadrat size is too small, all frequencies will be low. Recommendations for using frequency sampling call for adjusting quadrat size so that the frequency of the most abundant species is between 75 and 95%. Some researchers use species : area curves that plot the number of species sampled against quadrat size, for determining optimum quadrat size where the line begins to flatten. In some cases it may be necessary to use more than one quadrat size. For example, in sampling shortgrass vegetation in Colorado, researchers used a small $2 \times 2''$ quadrat to sample blue grama (*Bouteloua gracilis*). The small quadrat was nested in the $14 \times 14''$ quadrat used for all other species.

The main disadvantage in frequency sampling is its lack of sensitivity to distinguish between situations with differing vegetation or treatments; therefore, large sample sizes should always be used with frequency sampling. The main advantages of frequency sampling are speed of use and its usefulness in sampling secondary species that are difficult to detect by other characteristics.

V. Establishing Stocking Rates and Grazing Plans

Stocking rates can be established for both livestock and game animals and should be set to minimize adverse effects on other resources. While livestock operators can control numbers of livestock, game animal populations are more difficult to control because hunting is the main method of regulating game numbers.

In most parts of the western United States, rangelands have been grazed by livestock for decades. Previous stocking rates can be used to determine adjustments based on vegetation changes over time and past herbage and browse utilization. Recent research casts doubt on using utilization to set stocking, but instead *percentage utilization* or *desired residue* can be used as guidelines for adjusting stocking rates and planning moves to new pastures. [See RANGELAND GRAZING.]

Where the stocking history is not known or where a direct estimate is desired, several approaches can be used. Initially, the classic *range reconnaissance method* used plant cover as a basis for setting stocking rates. This method relied on locating comparable areas where stocking was judged satisfactory and using them as comparisons to establish stocking rates on the area in question. This method was appropriate for operators with considerable experience, but it was based on subjective judgment factors. A more direct approach is to determine the herbage available for livestock and game to determine optimal animal consumption, and then estimate the number of animals the area can support. Livestock studies indicate that the daily intake for grazing livestock averages about 2% of body weight and that these ratios can be used to calculate forage demand. Because not all herbage is equally accessible to animals, the operator must make substantial deductions for forage located at considerable distances from water or on step slopes.

A. Animal Numbers

One of the most critical decisions in successful livestock management is the way the operator allows for yearly fluctuations in the amount of herbage produced. Variations in herbage production on arid and semi-arid rangelands can be as great as 100% in successive years, but several approaches can be used to adjust grazing levels for these variations. If an area is stocked at a level appropriate for highest herbage production, grazing will be excessive in all but exceptional years, and both animals and vegetation will suffer. If an area is stocked at a level appropriate for average herbage production, grazing will be excessive during some years, but some feed will not be used in other years. Stocking at a rate appropriate for the lowest herbage production level is a conservative approach that does not maximize livestock production. Varying livestock numbers according to herbage production is ideal, but it is difficult in practice. When forage is low, the operator must sell animals, often when prices are lowest because other operators are reducing their herds. When forage is plentiful, animals are purchased to increase the herd when prices are high.

Some researchers advocate maintaining the herd with one-fourth to one-third as yearlings that can be disposed of more easily than mature animals. If forage conditions are favorable, some young animals can be held over instead of being sold at weaning. If one wishes to keep stable livestock numbers, the prudent approach is to stock at less then capacity and be willing to graze heavily during drought periods. Even with conservative stocking, it might be necessary to reduce numbers during severe drought.

B. Grazing Systems

Another decision concerning livestock management is the type of grazing system to use. The simplest system is continuous grazing where the rangeland is grazed during the entire grazing period. For some ranges, this would mean seasonal grazing, while for others it would mean year-long grazing. Many rangelands cannot be grazed year-long because of climatic limitations (snow, cold weather, poor forage conditions), but in the Southwest, year-long grazing can be practiced. Many kinds of specialized grazing systems have been developed that involve rotating grazing among several pastures or paddocks. *Rotation grazing* involves rotation of the time when each subdivision is again grazed so that each subdivision in the allotment is grazed at a different season until the first subdivision is grazed during the first season. The simplest type of specialized grazing system might be a two-pasture switchback system. In this system the area would be divided into two units. The first unit might be grazed for 3 months, and then livestock moved into the second unit which would be grazed for 6 months after which the first unit would be regrazed for the final 3 months of the year. The sequence would be reversed the second year. More intensive

grazing systems involve more subdivisions and several herds of livestock.

Short-duration or time-controlled grazing systems are highly intensive with many paddocks and only one or a few herds. In these cases all the livestock which might graze the entire area continuously, graze each paddock for a very short time—days or perhaps hours. The paddock is then rested for sufficient time to allow the plants to grow and regain vigor. These specialized grazing systems offer nearly unlimited variety in terms of number of paddocks and herds used.

The idea behind specialized grazing systems is that the deferment or rest from grazing benefits the plants, increases vigor and productivity, and in some cases allows palatable plants to increase in abundance. Thus, specialized grazing systems have been viewed as a means of improving rangeland for grazing without removing cattle or reducing stocking rates. Many studies have been conducted comparing various grazing systems, but the results from these studies are variable. Some have shown an improvement in vegetation (increase in desirable species) and others little or no change.

Stocking rate probably influences vegetational changes more than grazing system. Livestock performance is often reduced under specialized grazing systems probably because livestock are confined to a smaller area and do not have the opportunity to graze selectively for the most nutritious plants or plant parts. However, specialized grazing systems offer the opportunity to improve range conditions, in a modest fashion, while areas are still being grazed. Careful analysis is necessary before specialized grazing systems are implemented because expenses incurred in starting such a program can be considerable. These costs may include fencing, additional water development, and more frequent checking of both animals and range.

C. Wildlife

Judging habitat suitability for wildlife involves evaluating habitat requirements using characteristics of the area under consideration. Habitat diversity is important for most wildlife species, but this diversity is specific for each species. Desirable structural characteristics for pronghorn antelope are different from those of prairie chickens. Wildlife species need palatable plant species to meet their nutritional needs, as well as cover for protection. In addition, often habitat requirements for displaying courtship behaviors (in the case of many upland game birds) or concealment of young are different from those at different times of the year. The manager must know all these requirements to provide suitable habitat for wild animals.

Wildlife scientists Drs. John Kie and Jack W. Thomas outlined a flow diagram for managing vegetation as wildlife habitat. The steps they outlined were as follows:

1. Define management goals.
2. Measure habitat conditions and assess wildlife habitat values.
3. Prescribe and carry out management actions.
4. Monitor results.

If management goals have been achieved, then the cycle is completed. If the goals have not been achieved, then the cycle begins again.

VI. Planning Physical Improvements

Physical improvements on rangeland include water developments, fencing, livestock handling facilities, and trails and driveways. They can often be used to improve livestock handling and to make more uniform use of the range areas. However, these improvements are expensive and should be considered only after detailed economic analysis is made.

A. Fences

Many types of fences can be developed on rangeland depending upon specific needs. Fences are used to control livestock movement and to maintain uniform livestock use for each pasture. In some cases it is important that fences be designed to interfere minimally with wildlife movements. For example, pronghorn antelope are reluctant to jump fences and the bottom wire of the fence must be high enough to allow the pronghorn to crawl under it.

Barbed or smooth wire fences are often used for cattle or larger animals. These can be suspension fences with distances between posts much greater than for traditional fences. With suspension fences, the wire is loosely attached to the posts to allow for movement of the wire. Certain posts act as anchors placed periodically along the fence. These posts are the ones against which the wires are tightened.

For sheep, goats, and small animals, some type of net-wire fence is used most often. If lambs or kids are separated from the mother, they often become orphans. Hence, a tight fence is necessary for these animals. Some type of net-wire fences also has the advantage of limiting access for coyotes or other predators to the sheep or goat flock.

Livestock or game watering facilities are varied on rangelands. These include open water in streams, lakes or ponds, wells, various kinds of dirt tanks, and various catchment systems. Lakes and streams usually provide reliable sources of water, but may be poorly distributed. Often fencing can be adjusted to take advantage of natural water for as many pastures as possible.

Wells can be drilled where desired to provide water near forage supplies provided that adequate groundwater is present within reach of the well. Although wells entail high initial costs and require maintenance, they provide reliable sources of water for livestock and wildlife. Many ranches now have extensive watering system with wells providing water to storage tanks placed on areas of high elevation. Water is then distributed to drinking tubs with plastic pipe by gravity flow. Wells were traditionally equipped with wind mills. After World War II many windmills were replaced with submersible pumps fueled either by gasoline or electricity. However, during the energy crises in the 1970s the use of windmills increased even though many companies had stopped manufacturing the equipment. Solar panels have also been developed to pump water for stock and wildlife.

Dirt tanks are areas in low spots which have been excavated to collect and hold water during rainy seasons for animals. These are often available mainly during the rainy season and may not be reliable sources of water. In some cases these storage catchments may hold water year-long. Dirt tanks require maintenance since they often fill with silt and may require addition of clay to seal the bottom of the tank if they are built on certain soil types. Open reservoirs or tanks also may lose water through evaporation.

Various types of catchments have also been designed to provide water in arid areas. These devices include some collecting area usually lined with material such as butyl rubber sheets, polyethylene or vinyl sheets, concrete, metal roofing, or asphalt material and a storage bag or underground tank. These installations do not provide large amounts of water and may be somewhat undependable during low rainfall periods.

VII. Vegetational Manipulations

In many rangelands in the western United States woody vegetation has increased at the expense of grasslands. Considerable evidence show that pi-ñon–juniper woodlands and mesquite and cresote-bush shrublands have increased their range consid-erably during the last 150 years. Sagebrush grass vegetation has also been modified with increase in density of the shrubs and decrease in abundance of grasses. These changes have generally resulted in de-creased value of these rangelands for livestock grazing and, in some cases, for wildlife as well. During the period from World War II until the 1970s many land management agencies included direct vegetational manipulations as part of range management planning. These manipulations were often aimed at improving livestock grazing but had influences on other resource values as well. [See RANGELAND GRASS IMPROVEMENT; RANGELAND PLANTS; RANGELAND SHRUBS.]

Initially these early vegetational manipulations in-volved brush control and seeding. Brush control was mainly by mechanical and chemical (herbicide) means although fire has been used for a fairly long period. Mechanical brush control usually involved use of large pieces of equipment such as blades and bull dozers to push the trees or shrubs over and to uproot them. Many species such as mesquite, alligator juni-per, and oak resprouted unless their roots were com-pletely exposed. Another common practice was to pull a cable or large anchor chain between two tractors to pull trees and shrubs over and to uproot them. Often it was necessary to pull the cable or chain in two directions to completely uproot the plants. Cabling or chaining often was not effective for small plants and many operations resulted in a greater density of woody plants than previously since the young plants were no longer suppressed by the larger ones. In many cases large-scale projects were planned and conducted without follow-up treatments or evaluation. When fuel prices increased during the 1970s, mechanical treatments were curtailed except for special situations.

With the development of hormone-type herbicides during World War II, the whole field of chemical control of unwanted plants was initiated. Two herbi-cides, 2,4-D and 2,4,5-T proved effective against cer-tain broad-leafed plants and not for grasses. This de-velopment opened the way for selective treatment of unwanted species. The whole field of herbicidal control of expanded greatly during the 1960s until the present. Again herbicides were initially used to control plants to benefit livestock grazing, but have the potential to benefit other uses as well. [See HERBI-CIDES AND HERBICIDE RESISTANCE; PEST MANAGEMENT, CHEMICAL CONTROL; WEED SCIENCE.]

Initially the hormone-type herbicides were applied to the foliage of plants, but later several soil-applied herbicides were developed. Hormone-type herbicides had to be applied during certain periods, usually when plants were growing rapidly and the herbicides could be translocated rapidly. Considerable research was directed at finding the appropriate time and rate of application of various herbicides. Soil-applied herbi-cides had the advantage of being applied at any time and would be dissolved during rainy periods when they were taken up by the roots in the water solution.

During the environmental movement of the 1970s, much concern was expressed about widespread use of pesticides including herbicides. The concern involved impact of herbicides on nontarget species and possible damage by residue if the herbicide was stable or slowly degradable. As a result of the human health concern, the use of herbicides on rangelands has de-creased overall although in some circumstances the selective use of herbicides has increased. Some of the newer herbicides are not persistent in the environ-ment. In addition, a comprehensive study of mesquite control with 2,4,5-T on the Jornada Experimental Range in southern New Mexico showed that there was minimal influence on population of birds, small mammals, insects, and microorganisms.

In most land management plans on public land to-day esthetics and wildlife habitat should be considered before any large vegetational manipulation should be considered. For example, piñon–juniper control is now being considered only on areas where piñon–ju-niper has encroached onto grassland in recent times. Original stands are maintained. The control patterns are designed to blend in with the natural contours of the land instead of being in straight lines. Considering esthetic aspects is difficult because it varies among individuals. Alternate patches of treated areas or strips of treated areas in the background of untreated areas often present better habitat for wildlife than a large treated block adjacent to a nontreated block.

VIII. Economic Analyses

Economic analyses of range management practices are often neglected or conducted in a superficial manner, partly because range economists are scarce and many range conservationists have limited background in economic analysis. In addition, approaches for eco-nomic analyses may differ for public and private rangeland.

Two types of economic analyses are often used for range situations. These are the internal rate of return and the break-even point. The internal rate of return can be calculated several ways: the percentage rate of return from a range improvement having a specified maintenance expense for a given period of time under specified net income, in initial costs, and maintenance cost. Another approach involves the initial investment in a range improvement that can be afforded if the net income, interest rate, and maintenance expense and life of the improvement are known. A third approach is to use amount necessary to cover the initial cost and cost of borrowed funds plus maintenance.

The break-even point is the amount of added income necessary to cover the cost of the improvement or range management practice. Break-even points differ from the internal rate of return in that maintenance costs are not included.

Range management practices to improve livestock production lend themselves to economic analyses more easily than many other types of multiple uses. Esthetic values and even hunting and recreational activities are much more difficult to quantify. Yet these values are becoming much more important as public involvement in land management decisions increases.

Land management decisions often involve trade-offs of one kind or another. For example, studies in eastern New Mexico and Utah have indicated that previous grazing by sheep precludes use of areas by pronghorn antelope. Otherwise these pastures appear to be prime habitat for pronghorn. In other areas there is conflict between elk and cattle use of mountain rangelands. People, wildlife, and livestock often prefer riparian habitats and conflicts arise over use of these sensitive areas. Debates concerning influence of livestock grazing on desert tortoise habitat in the Southwest can become heated. These possible uses of common rangeland need to be considered in plans for the use of rangelands. Research can identify the possible uses for each rangeland, and some techniques are available for optimizing the mix of possible uses, but the ultimate decision concerning which uses to allow on rangeland are made by people. Deciding who makes these decisions is the ultimate question. Now many urban dwellers are involved in these decisions. They greatly out-number the traditional livestock operators who often believe that they have grazing rights on public rangeland. How successful we are in blending the views of the interested and diverse groups of users will ultimately determine the fate of our vast public rangelands.

Bibliography

Harrington, G. N., Wilson, A. D., and Young, M. D. (1984). "Management of Australia's Rangelands." CSIRO Publications, East Melbourne, Australia.

Heitschmidt, R. K., and Stuth, J. W. (eds.) (1991). "Grazing Management: An Ecological Perspective." Timber Press, Portland, OR.

Holechek, J. L., Pieper, R. D., and Herbel, C. H. (1989). "Range Management Principles and Practices." Prentice-Hall, Englewood Cliffs, NJ.

Vallentine, J. F. (1989). "Range Developments and Improvements," 3rd ed. Academic Press, San Diego.

Workman, J. P. (1986). "Range Economics." Macmillan, New York.

Rangeland Plants

JAMES STUBBENDIECK, *University of Nebraska*

Glossary

Browse Parts of woody plants (leaves and twigs) consumed by animals; the act of consuming leaves and twigs

Forage Portions of herbaceous plants consumed by grazing animals

Forbs Herbaceous plants other than grasses and grasslike plants

Grasses Herbaceous plants with hollow (occasionally solid) stems with nodes, and the leaves are two-ranked

Grasslike plants Herbaceous plants (sedges and rushes) that resemble grasses but generally have solid stems without nodes, and the leaves are two- or three-ranked

Palatability Preference for a plant or plant part by animals relative to other plants available for grazing or browsing

Scientific names Binomial name consisting of Latin or Latinized names; the species name formed from a combination of the first part (genus) and second part (specific epithet)

Succulent plants Fleshy plants (such as cacti) that store large amounts of water in the stems (pads) for use during periods of insufficient soil moisture

Woody plants Plants (shrubs and trees) with solid stems containing growth rings and with secondary growth from aerial stems

Rangeland vegetation comprises many individual species of range plants in complex plant communities.

Species respond individually to abiotic and biotic influences making the skill of plant identification a necessity for range managers. Knowledge of individual species and abundance enables managers to determine the condition and trend of the vegetation, select the appropriate type and class of animals, set stocking rates, determine grazing method, and be aware of potential losses to poisonous plants.

I. Units of Vegetation

Range plants are the basic units of rangeland vegetation. A knowledge of the identity, structure, and function of individual plants is necessary for ranchers and other natural resources managers to make sound decisions. A 100-ha unit of rangeland may contain fewer than 100 to as many as 1000 distinct species of plants. It is not practical for most individuals working within the range ecosystem to know the identity of all of these species. Even on diverse rangeland, 10 to 30 species will make up over 90% of the biomass of the area, and knowledge of these species generally will be sufficient. Poisonous plants may be present only in limited numbers and distribution, but knowledge of these species and their management is necessary to prevent livestock losses. [*See* RANGELAND.]

II. Plant Names

A. Scientific Classification and Binomial Names

Plants are described and grouped according to their structure, particularly the flowering or other reproductive parts. The system of classification involves a series of categories arranged to show relationships and similarities between plants. Names are Latin or Latinized. The classification system from most general to most specific for the common range grass little bluestem [*Schizachyrium scoparium* (Michx.) Nash] is as follows:

Kingdom: Plantae
  Division: Embryophyta
    Branch: Angiospermae
      Class: Monocotyledoneae
      Order: Poales
      Family: Poaceae
      Tribe: Andropogoneae
      Genus: *Schizachyrium*
      Species: *Schizachyrium scoparium* (Michx.)
        Nash

Individuals working with or managing range plants are primarily concerned with the final four divisions of the classification system. The family is the basic division of plant orders. It consists of group of closely related genera and is primarily based on flower characteristics. All family names end in -aceae. A tribe is a subdivision of family. Only large, complex families are divided into tribes, and range plant examples include the grass family (Poaceae) and the sunflower family (Asteraceae). Division into tribes is based primarily on flower characteristics. For example, only members of the tribe Andropogoneae have paired spikelets with the lower spikelet being sessile and perfect and a pedicellate upper spikelet that usually does not produce a viable seed. All tribe names end in -eae.

The binomial system of nomenclature is based on the genus and species. Division of plants in a tribe into genera (plural of genus) is based on flower and/or morphological characteristics. An example is *Schizachyrium*. The first letter of the genus is capitalized, and the name is written italics or underlined. The second part of the binomial is the specific epithet, an example is *scoparium*. The specific epithet is written in italics or underlined, and it is never capitalized. The species name comprises the genus and specific epithet, an example is *Schizachyrium scoparium* (Fig. 1). Plants within a species have the ability to interbreed and produce viable offspring. Occasionally, the species is further divided into varieties or subspecies.

The binomial should be followed by an authority or authorities to be complete. The whole name or an abbreviation is used. These individuals described and named the taxon. Sometimes a portion of the authority is parenthetical. Parenthetical authorities recognize work that was later revised by another taxonomist. The authorities for *Schizachyrium scoparium* are (Michx.) Nash. A French botanist, Andrew Michaux (1746–1802), first described and applied the epithet

scoparium to this species. Later, an American agrostrologist, George Nash (1864–1921), transferred the species and its epithet to the genus *Schizacyhrium*.

Only one correct scientific name exists for each species. The International Code of Botanical Nomenclature establishes the rules for naming plants, and names may be corrected or updated by taxonomists. *S. scoparius* (Michx.) Nash was formerly known as *Andropogon scoparius* Michx. It was changed according to the International Code of Botanical Nomenclature due to a consensus of taxonomic opinion favoring Nash over Michaux. *A. scoparius* Michx. is now considered to be a synonym, rather than a correct name.

All scientific names have a meaning. The epithet *scoparium* means broomlike in reference to the shape of the inflorescences. Names may commemorate an individual or may be based on geography. Others may be based on a classical Latin name or describe the growth form, habitat, or a morphological feature.

Scientific names are valuable because only one name is correct, and that name is the same throughout the world, regardless of language. Also, they are organized and evaluated according to a definite system of rules. Unfortunately, scientific names are often long and made up of unusual and unfamiliar syllables. An additional problem is that scientific names are sometimes changed when a species is reclassified or when it is discovered that it was originally improperly named. Learning a new scientific name for an old, familiar species may be difficult.

B. Common Names

Common names, such as little bluestem, are the only names familiar to most people, and they are certainly easier to remember and spell than scientific names. Most common names are simple and are often descriptive of the plant. The common name bluebells instantly brings to mind bell-shaped, blue flowers. Common names have a number of weaknesses. Common names are different in different languages. Use of the Mexican common name, popotillo colorado, for little bluestem in much of the United States would not convey the identity of the plant. A single species may have more than one common name in the same language. Other names used for little bluestem in the United States and Canada include red bunchgrass, prairie beardgrass, prairie bunchgrass, and little bunchgrass. The same common name may be used for more than one species, further confusing the identity of the plant. No standard methods or rules apply

FIGURE 1 Little bluestem [*Schizachyrium scoparium* (Michx.) Nash], a common range grass. [Reprinted with permission from Stubbendieck, J., Hatch, S. L., and Butterfield, C. H. (1992). "North American Range Plants." Copyright 1992 by University of Nebraska Press, Lincoln.]

to formulation of common names. In recent years, common names usually have been reduced to two words. Plants formerly with more words composing the name now have long, complicated names such as serrateleaf eveningprimrose, upright prairiecone-flower, twogrooved posionvetch, and curlleaf moun-tainmahogany. Some common names are ridiculous. For example, peppergrass is not a grass nor is it related to peppers.

Both the scientific and common names should be used wherever possible. This will provide a familiar name to the untrained and a botanically correct name for individuals wanting to obtain additional informa-tion in the literature.

III. Life Span and Season of Growth

Most range plants are classified as perennials or annu-als. Perennials generally live 3 or more years, and some may live for decades or more. Creosotebush [*Larrea tridentata* (DC.) Cov.] (Fig. 2), a common range plant in the Southwest, may live for more than 10,000 years making it the oldest living organism on earth. Perennial herbaceous species die back to near the surface of the soil each year but remain alive under-ground. The aerial portions of woody plants remain alive throughout the year.

Annual plants complete their life span within 1 year, usually within one growing season. Some germinate in the spring, flower, and die before the following winter. Others, winter annual species, germinate in the fall, overwinter in the seedling stage, and complete their life cycle during the following growing season.

Biennial is a third category of life span, and few range plants fit into this category. Biennials germinate during the spring, and remain in a vegetative stage throughout the first growing season. They undergo the process of vernalization during the winter. Vernal-ization stimulates flower initiation, and the plants flower and die during the second growing season. Occasionally, a cold period in the spring following germination will vernalize these plants, and they will flower and die during the first growing season.

Short-term adverse conditions can greatly influ-ence germination and seed production. Therefore, the more reliable and valuable plants in most range-land ecosystems are perennials because their contin-ued productivity does not depend on either of these processes. Conversely, some of the more trouble-some and least desirable plants are perennials because they persist longer and are not susceptible to control

methods designed to interrupt germination or flow-ering.

Range plants, especially grasses, may be separated into two groups based on their growth responses to temperature. Cool-season grasses initiate growth ear-lier in the spring than warm-season grasses. Cool-season plants may become dormant during the warm summer months and regrow in the fall if soil moisture is adequate. Cool-season grasses flower in spring or early summer, while warm-season grasses flower in summer or early fall. During photosynthesis, carbon is initially formed into three-carbon compounds (C_3) in cool-season grasses. Four-carbon compounds (C_4) are initially formed in warm-season grasses.

The balance of warm-season and cool-season grasses in rangeland vegetation at a particular location influences selection of range management practices. For example, rangeland dominated by cool-season species may be best utilized by grazing animals in spring when the plants are more nutritious than in the summer. Range management practices may be designed to alter the balance of cool-season and warm-season plants. Prescribed burning in the spring gener-ally stimulates warm-season plants and reduces cool-season plants, while nitrogen fertilizer applied in the spring will have the opposite effect.

IV. Rangeland Plant Groups

Range plants can be divided into different growth forms. These growth forms, or range plant groups, are grasses, grasslike plants, forbs, shrubs, and succu-lent plants (Fig. 3). Classification into groups is the first step in identifying plants, and members of each group commonly respond in a similar manner to changes in environment, management, and other ex-ternal forces.

A. Grasses

Grasses are usually the most abundant and valuable plants on rangeland. Grasses are herbaceous and have hollow stems with joints (internodes). The leaves attached to two sides of the stem (Fig. 3). Veins in the leaves are parallel. The leaf is divided into the sheath (lower portion) and blade (Fig. 4). Distinguish-ing characteristics between species may be found at the junction of the leaf blade and sheath. Presence or absence of auricles and type and shape of ligules vary between species (Fig. 4). The inflorescence is the

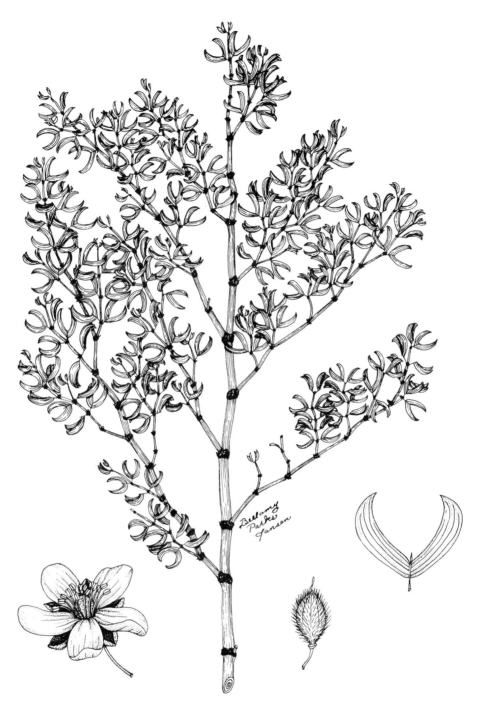

FIGURE 2 Creosotebush [*Larrea tridentata* (DC.) Cov.] a rangeland shrub and the oldest living organism. [Reprinted with permission from Stubbendieck, J., Hatch, S. L., and Butterfield, C. H. (1992). "North American Range Plants." Copyright 1992 by University of Nebraska Press, Lincoln.]

flowering portion of grasses that includes the florets (flowers). The structure of the inflorescence is an identifying characteristic of grasses. The three common types of inflorescences are the panicle, raceme, and spike. The panicle has a main axis and rebranched branches bearing spikelets. Racemes contain spikelets on unbranched pedicels attached to the main axis, and spikes contain spikelets attached directly to the main axis without pedicels. Spikelets consist of two glumes and one or more florets (Fig. 4). Each floret consists

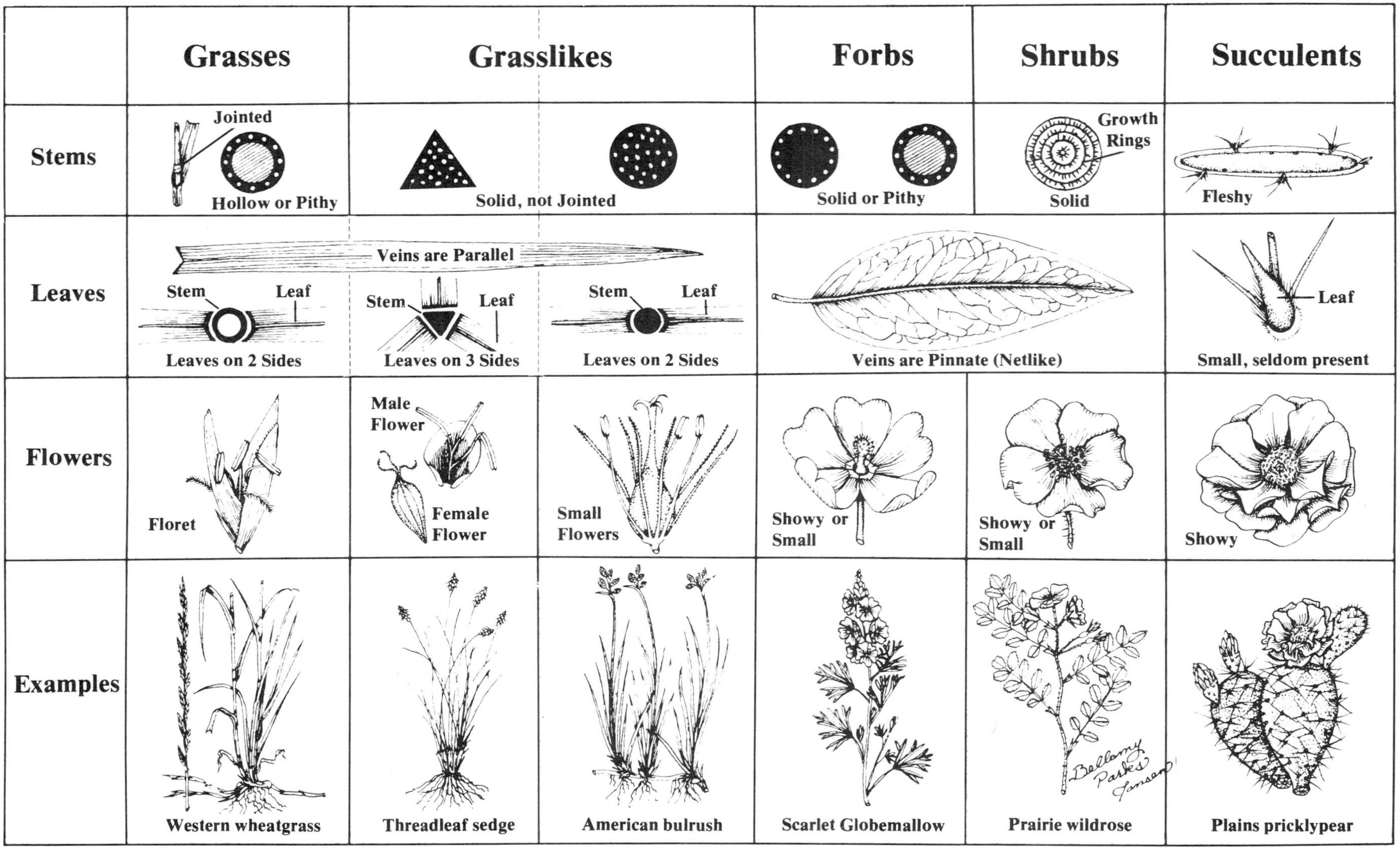

	Grasses	Grasslikes		Forbs	Shrubs	Succulents
Stems	Jointed / Hollow or Pithy / Leaves on 2 Sides	Solid, not Jointed		Solid or Pithy	Growth Rings / Solid	Fleshy
Leaves	Stem / Leaf / Veins are Parallel	Stem / Leaf / Leaves on 3 Sides	Stem / Leaf / Leaves on 2 Sides	Veins are Pinnate (Netlike)		Leaf / Small, seldom present
Flowers	Floret	Male Flower / Female Flower	Small Flowers	Showy or Small	Showy or Small	Showy
Examples	Western wheatgrass	Threadleaf sedge	American bulrush	Scarlet Globemallow	Prairie wildrose	Plains pricklypear

FIGURE 3 Range plant groups. [Reprinted with permission from Stubbendieck, J., Nichols, J. T., and Butterfield. C. H. (1989). "Nebraska Range and Pasture Forbs and Shrubs." Copyright 1989 by University of Nebraska, Lincoln.]

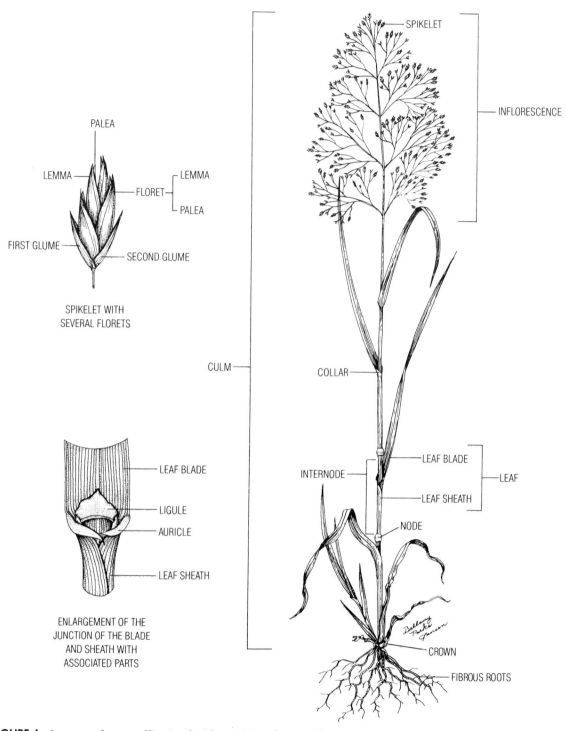

PALEA

LEMMA

FLORET

LEMMA

PALEA

FIRST GLUME

SECOND GLUME

SPIKELET WITH
SEVERAL FLORETS

CULM

LEAF BLADE

LIGULE

AURICLE

LEAF SHEATH

ENLARGEMENT OF THE
JUNCTION OF THE BLADE
AND SHEATH WITH
ASSOCIATED PARTS

SPIKELET

INFLORESCENCE

COLLAR

INTERNODE

LEAF BLADE

LEAF SHEATH

LEAF

NODE

CROWN

FIBROUS ROOTS

FIGURE 4 Structure of grasses. [Reprinted with permission from Stubbendieck, J., Hatch, S. L., and Butterfield, C. H. (1992). "North American Range Plants." Copyright 1992 by University of Nebraska Press, Lincoln.]

of a lemma and palea surrounding a flower or caryopsis (seed). Reproduction by seed is necessary for annual grasses, but it is less important for perennial species. Many perennial grasses reproduce by tillers (shoots arising from adventitious buds at the plant base), rhizomes (horizontal underground stems that emerge to produce a new plant), and/or stolons (horizontal aboveground stems that root at the nodes and produce new plants). Frequently, grasses are grouped by height into the categories of shortgrasses (less than 30 cm tall),

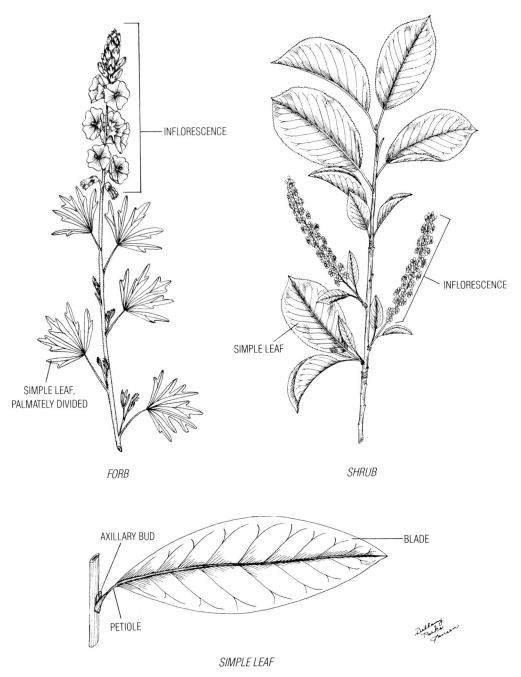

INFLORESCENCE

SIMPLE LEAF,
PALMATELY DIVIDED

FORB

SIMPLE LEAF

INFLORESCENCE

SHRUB

AXILLARY BUD

BLADE

PETIOLE

SIMPLE LEAF

FIGURE 5 Structure of forbs and shrubs and a leaf. [Reprinted with permission from Stubbendieck, J., Hatch, S. L., and Butterfield, C. H. (1992). "North American Range Plants." Copyright 1992 by University of Nebraska Press, Lincoln.]

midgrasses (30 cm to 1 m tall), and tallgrasses (over 1 m tall). [*See* RANGELAND GRASS IMPROVEMENT.]

B. Grasslike Plants

Grasslike plants are generally perennials and resemble grasses. They usually have solid stems without nodes (Fig. 3). The leaves have parallel veins. Leaves arise from the three sides of triangular stems of sedges (Cyperaceae family) and from two sides of round stems of rushes (Juncaceae family). The stems are commonly leafy only at the base. This basal tuft of leaves is sometimes reduced to sheaths. The flowers are in three parts, while grass flowers are in two parts.

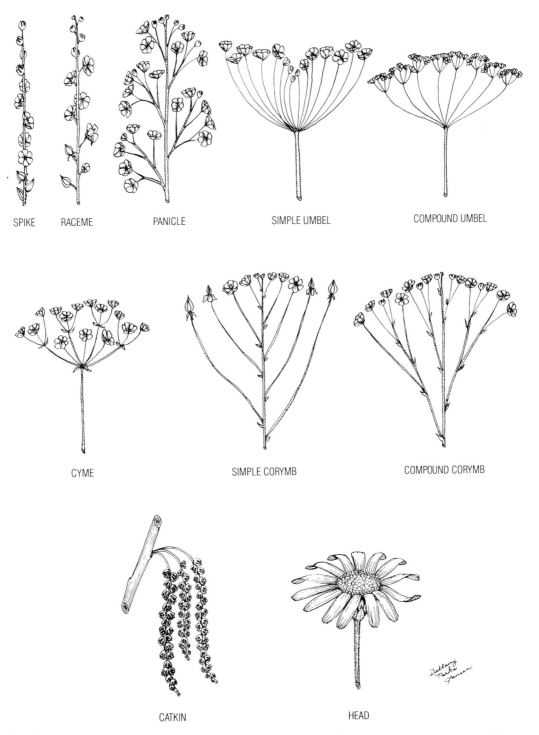

SPIKE   RACEME   PANICLE   SIMPLE UMBEL   COMPOUND UMBEL

CYME   SIMPLE CORYMB   COMPOUND CORYMB

CATKIN   HEAD

FIGURE 6 Inflorescence types. [Reprinted with permission from Stubbendieck, J., Hatch, S. L., and Butterfield, C. H. (1992). "North American Range Plants." Copyright 1992 by University of Nebraska Press, Lincoln.]

Flowers may be solitary or in panicles, corymbs, or heads. The fruit is an achene. Grasslike plants reproduce from achenes, tillers, and/or rhizomes.

Although many grasslike plants grow where moisture is abundant, a few are adapted to semi-arid and arid rangelands. Grasslike plants are most abundant in cool climates. Some are important forage species for both livestock and wildlife. Most rangeland sedges have a high nutrient content. Unfortunately, the task of identification of grasslike plants is formidable. Ex-

FIGURE 7 Honey mesquite (*Prosopis glandulosa* Torr.), a shrub or small tree that has spread rapidly on rangeland in the Southwest and Mexico. [Reprinted with permission from Stubbendieck, J., Hatch, S. L., and Butterfield, C. H. (1992). "North American Range Plants." Copyright 1992 by University of Nebraska Press, Lincoln.]

cept for a few common, distinctive species, most range managers identify grasslike plants only to genus.

C. Forbs

Forbs are herbaceous plants, other than grasses and grasslike plants. They may be annuals, biennials, or perennials. Forbs generally have solid or pithy stems and broad leaves that are pinnately veined (Figs. 3 and 5). The flowers may be small but can be large and colorful. Most prairie wildflowers are forbs. Morphology of forbs is highly variable. Leaves may be simple or compound and attached in several ways. Flower type, color, and size vary. Flower arrange-

FIGURE 8 Plains pricklypear (*Opuntia polycantha* Haw.), a succulent range plant. [Reprinted with permission from Stubbendieck, J., Nichols, J. T., and Butterfield, C. H. (1989). "Nebraska Range and Pasture Forbs and Shrubs." Copyright 1989 by University of Nebraska, Lincoln.]

ments may take many forms (Fig. 6). Forbs reproduce by seed and asexually.

Forbs are second only to grasses in abundance and productivity on rangeland, but more species of forbs than grasses usually are present. Palatability of forbs also is highly variable. Many are valuable sources of nutrients for livestock and wildlife on rangeland and are selectively grazed. Some are consumed only if other plants are not available in adequate quantities. Others will seldom be consumed because they are distasteful or covered with resins or prickles. A few forbs are highly poisonous to animals. Poisonous forbs do not readily fall within families or even within certain genera. For example, the legume family (Fabaceae), and even an individual genus such as *Astragalus* within the legume family, contains highly desirable range species that furnish excellent forage for animals and add atmospheric nitrogen to the soil to benefit all plants while others can cause animal death within a few hours of consumption.

D. Woody Plants

Woody plants have solid stems with growth rings. They may be either shrubs or trees. Secondary growth occurs from aerial stems which live throughout the year, although they may be dormant part of the time. Leaves are often broad and pinnately veined, and flowers may be either showy or small (Fig. 3). Morphology of woody plants is highly variable and plant parts are similar to those of forbs (Fig. 5). Some woody plants furnish browse for range livestock. Generally, shrubs are more valuable to wildlife, especially in the winter months. [*See* RANGELAND SHRUBS.]

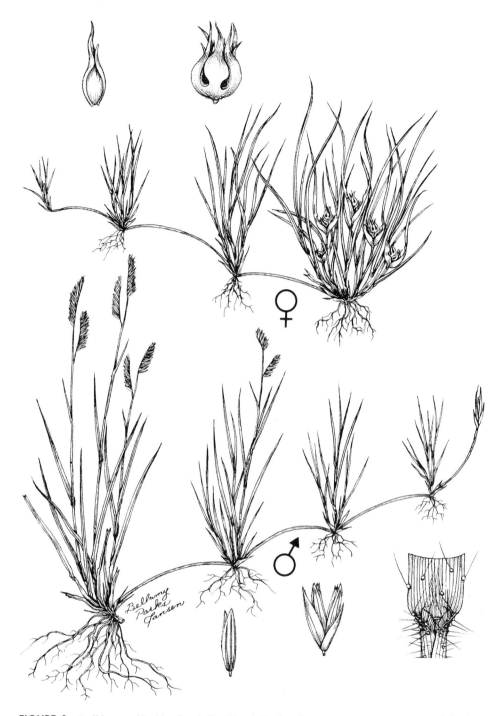

FIGURE 9 Buffalograss [*Buchloe dactyloides* (Nutt.) Englem.], a common range grass used for low-maintence turf. [Reprinted with permission from Stubbendieck, J., Hatch, S. L., and Butterfield, C. H. (1992). "North American Range Plants." Copyright 1992 by University of Nebraska Press, Lincoln.]

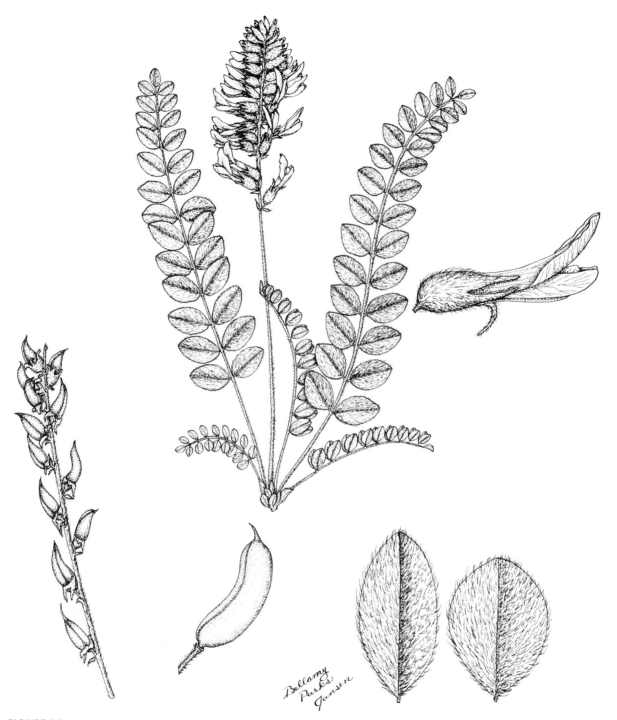

FIGURE 10 Woolly locoweed (*Astragalus mollissimus* Torr.), a common poisonous range forb. [Reprinted with permission from Stubbendieck, J., Hatch, S. L., and Butterfield, C. H. (1992). "North American Range Plants." Copyright 1992 by University of Nebraska Press, Lincoln.]

Saltbushes (*Atriplex* spp.) are examples of woody plants that are important on western rangelands for cattle, sheep, horses, and big game. Saltbush fruits are important food for birds. Many species of sagebrushes (*Artemisia* spp.) are found throughout North American rangeland. Some species are important for livestock and wildlife, but poor management and control of fire have allowed some sagebrushes to spread at the expense of more desirable range grasses and forbs. Few species of woody plants can be placed only in the desirable or only in the undesirable category.

For example, honey mesquite (*Prosopis glandulosa* Torr.) (Fig. 7) infests millions of hectares of rangeland in the southwestern United States and Mexico. These shrubs or small trees are native to that area where they originally were scattered and an integral part of the Savannah vegetation. Honey mesquite is a legume and fixed atmospheric nitrogen. Some of the foliage is utilized as browse, and the legumes (pods) and seeds are an important food for numerous wildlife species and an emergency food for livestock. Honey mesquite rapidly spreads on abused rangeland and where fire was controlled. It replaced more desirable range plants and became the target of a massive brush control effort in the 1950s and 1960s. Despite the efforts, honey mesquite was not controlled, and modern range managers are learning to manage, rather than control, brush.

E. Succulents

Succulent plants are the least abundant range plant group, but many people recognize succulents and associate them with rangeland. Cacti (Cactaceae family) are succulents. The giant saguaro cactus [*Carnegiea gigantea* (Engelm.) Britt. and Rose] is one of the most recognized plants on rangeland although it occurs only in the warm deserts of the southwest.

Large amounts of water may be stored in the fleshy parts of succulents and utilized by the plants during periods when soil moisture is insufficient. Leaves are small, fleshy, and may be present for only a short period each year (Fig. 3). Flattened pads of some species, such as plains pricklypear (*Opuntia polycantha* Haw.) (Fig. 8), are stems and not leaves. Many succulents have showy flowers and are frequently armed with sharp spines. Livestock sometimes utilize succulent plants, and some species of wildlife depend on them. Ranchers in the Southwest may burn spines from pricklypears to allow cattle to utilize the pads as emergency feed during drought.

V. Rangeland Plant Values

A. Forage and Browse

Many range plants provide forage or browse to livestock and wildlife. Many species of native herbivores obtain all of their required nutrients directly from range plants. Most livestock can survive on range plants alone but may be more productive when they receive supplemental feed during periods when range plants are less available or nutritious. [*See* FORAGES; RANGELAND GRAZING.]

Animals prefer some species of range plants. They will seek these preferred plants and consume them first. This preference for a plant species helps to determine its relative palatability. Palatability is relative to the amount available, other species in the plant community, and the kind of animal and its condition. Palatability is related to the flavor, odor, moisture content, texture, and height of the plants. Animals prefer certain plant parts over others and usually prefer the younger, more tender, and more nutritious herbage.

The nutritional value of range plants is measured in relation to the needs of animals. A cow with a young calf requires a higher level of nutrition to maintain her condition and produce milk than does a cow without a calf. The primary nutritional requirements are energy, protein, minerals, and vitamins. Nutrient content of forage and browse varies during the year. Nutrient content is highest during the active periods of plant growth. Nutrient content is lowest during the winter dormant periods when supplemental feed may be required to meet the needs of the animals. [*See* ANIMAL NUTRITION, PRINCIPLES.]

B. Conservation

Most plants protect the soil from erosion losses from wind and/or water. Roots physically bind and hold soil particles while the aerial portions of the plants lift the wind above the soil surface reducing its ability to cause erosion. Foliage intercepts raindrops and breaks their force. Range plants shade the soil, and fallen leaves further protect and cool the soil reducing water evaporation. Range plants add organic matter to the soil improving its fertility. Plants in the legume family (*Fabaceae*) improve soil fertility through a symbiotic relationship with specific bacteria (*Rhizobium* spp.). The rhizobia form nodules on the legume roots. The legumes furnish the necessary food for energy that enables the rhizobia to change nitrogen from the atmosphere to a form that plants can use. This process is called nitrogen fixation. [*See* RANGELAND SOILS.]

C. Wildlife Habitat

Range plant communities furnish both food and cover for wildlife. Some kinds of wildlife, such as predators, do not consume range plants but do consume animals that use range plants for food.

D. Food and Medicine

A largely overlooked value of range plants is a source of food and medicine for humans. One can find numerous references on Native Americans' uses of range plants for these purposes. Common sunflower (*Helianthus annuus* L.) is an example of a range forb that has been selected and bred to produce food and oil for human consumption. Several cultivated fruits and berries can be traced to their rangeland origins. The medicinal properties of range plants have not been thoroughly researched. Numerous range plants are listed in the references *U.S. Pharmacopoeia* and *National Formulary*.

E. Esthetic

Many range plants are attractive and contribute to the esthetics of prairie, desert, and mountain vegetation. Several species of cultivated flowers originated on rangeland. During the past 10–20 years, native plants have been used in landscaping. Buffalograss [*Buchloe dactyloides* (Nutt.) Engelm.] (Fig. 9) and other native grasses are being used increasingly for turf because they require less fertilizer and water than traditional turf grasses.

VI. Poisonous Rangeland Plants

Many range plants have the ability to poison if consumed in large quantities. Poisoning does not automatically mean death. Ingestion of some plants by pregnant female livestock may result in abortions or physical malformation of the offspring. The greatest loss associated with poisonous plants is lowered livestock gain. Some poisonous plants are members of the native plant community. Others have been introduced from other parts of the world. Both types become more abundant with misuse of rangeland increasing the poisoning potential.

Probably the best known poisonous range plant is woolly locoweed (*Astragalus mollissimus* Torr.) (Fig. 10). As is the case with many poisonous plants, woolly locoweed usually is unpalatable to livestock. Generally, it is consumed only when other forage is unavailable. Some animals, especially horses, may become addicted to woolly locoweed and refuse to eat better forage. The toxic substance is locoine, and the effect is accumulative over time. The optic nerve is apparently affected. Animals will leap high over small objects and shy violently from small or imaginary objects. Once animals start to walk, they often continue until walking into an obstruction, hence, the term "loco" which is Spanish for crazy or foolish.

Different parts of poisonous range plants may contain different levels of the poisonous substance. Light grazing of plants with toxic material only in the roots will not adversely affect animals. Likewise, if only the seeds are poisonous, the plants can be grazed after seeds ripen and drop. Some plants retain their poison after they dry in hay, while others are not poisonous upon drying.

Different kinds of animals may have different levels of susceptibility to a particular species of poisonous plant. Generally, sheep are more susceptible to plant poisoning than cattle, but exceptions do occur. For example, larkspurs (*Delphinium* spp.) are highly toxic to cattle, while sheep are less likely to be poisoned.

Wildlife are infrequently poisoned on rangeland. They have evolved with the plants and do not eat them; the poisons are detoxified within their systems, or the poisons pass through their systems without affecting the animals.

It is possible to utilize rangeland containing poisonous plants. Range managers must be able to recognize and identify the poisonous plants. They must know which plant parts are poisonous and at what times of the year. They must know which types of animals are affected. Most importantly, a good cover of suitable forage plants must be maintained so that the animals will not be forced to eat the poisonous plants.

Bibliography

Blackwell, W. H. (1990). "Poisonous and Medicinal Plants." Prentice Hall, Englewood Cliffs, NJ.

Duke, J. A. (1992). "Handbook of Edible Weeds." CRC Press, Boca Raton, FL.

Gilmore, M. R. (1991). "Use of Plants by the Indians of the Missouri River Region." Univ. of Nebraska Press, Lincoln.

James, L. F. (ed.) (1992). "Poisonous Plants, Proceedings of the Third International Symposium, 1989." Iowa State Univ. Press, Ames.

Kindscher, K. (1992). "Medicinal Wild Plants of the Prairie." Univ. of Kansas Press, Lawrence.

Stubbendieck, J., and Conard, E. C. (1989). "Common Legumes of the Great Plains." Univ. of Nebraska Press, Lincoln.

Stubbendieck, J., Hatch, S. L., and Butterfield, C. H. (1992). "North American Range Plants." Univ. of Nebraska Press, Lincoln.

Stubbendieck, J., Nichols, J. T., and Butterfield, C. H. (1989). "Nebraska Range and Pasture Forbs and Shrubs." E. C. 89-118, University of Nebraska, Lincoln.

United States Department of Agriculture (1988). "Range Plant Handbook." Dover, New York.

Rangeland Shrubs

BRUCE L. WELCH, *USDA-Forest Service, Utah*

Glossary

Cultivar Variety of shrub developed by crossing at least two germplasms and selecting offspring which possess the superior characters of the parents

Germplasm Collection of a species or subspecies of shrub from a specific site believed to possess genes different from the same species or subspecies of shrub from other sites

Rangeland Open area with few trees where wild and domestic animals may roam to extract the essence of life needed to complete or partly complete their life cycles. These areas are usually unsuitable for cultivation and receive minimum management

Rangeland shrub improvement Systematic, multidisciplinary approach used for identifying superior germplasms or developing superior cultivars of rangeland shrubs; released, tested germplasms and released cultivars of shrubs are used on rangelands to provide soil and stream stabilization, and to improve and restore animal habitats and esthetics

Released tested germplasm Germplasm found to possess at least one superior characteristic released by an authorized agency for the commercial production and marketing of seeds and plants

Shrub Woody, erect, perennial plant, usually less than 3 to 4 cm tall, with stems branching from or near the soil surface

Rangeland shrubs are a neglected and undervalued dominant plant resource of arid and semi-arid regions of the United States. Their value and use is just now beginning to be understood. Shrubs provide food and habitats for animals, ecosystem enrichment, and low maintenance landscapes. They express much genetic diversity. This diversity provides the opportunity for identifying or developing superior germplasms or cultivars.

I. Nature of Shrubs

A. Distribution of Shrubs in the United States

Shrubs are subordinate or dominant plant-forms in a variety of plant communities. As subordinates, they range as far north as the Arctic tundra where they occur in the zone between trees of the northern coniferous forest and the region of perpetual snow. There they are restricted to protected slopes and valleys. Important shrubs include willow, birch, alder, crowberry, bog rosemary, blueberry, labrador tea, manzanita, cassiope, and rhododendron. Because the Arctic shrub communities border the northern coniferous forest, the two communities share many shrubs. Shrubs of the northern coniferous forest are found mainly in forest openings. Some shade-tolerant shrubs are found as scattered plants in the forest understory. Additional important shrubs found there include hobblebush, cassandra, laurel, chokeberry, hazelnut, spirea, dogwood, and blackberry. Many of these shrubs are also found as subordinate plant-forms in the eastern deciduous forest. Additional important shrubs found there include currant, wahoo, buckthorn, ceanothus, snowberry, baccharis, spicebush, buttonball, rose, sumac, elderberry, leadplant, ash, and laddernut. The range of many of these shrubs extends into the southern coastal plains. This area is not rich in shrubs due to the domination of trees in southern coniferous and hardwood forests. Some shrubs can be found in the central grasslands along streams and slopes that break the generally flat landscape of prairies and plains. Western and southwestern shrubs have extended their range into grasslands where the grass cover has been reduced. It is, however, in the arid and semi-arid regions of the Western United States where shrubs reign supreme.

Because so much of America's grasslands have been used for farming, today, most rangelands are found in the shrublands of the Western United States. These shrublands have been divided into eight provinces based on life-form, geographical features, and ecological relationships. These provinces are: Prairie Brushland, California Chaparral, Intermountain Sagebrush, Mexican Highlands Shrub Steppe, Colorado Plateau, Wyoming Basin, Chihuahuan Desert, and American Desert. While each province has its unique environmental characteristics, all share some characteristics in common. These characteristics are: domination by shrubs, low annual precipitation (12 to 64 cm), low productivity, and, in many areas, poorly developed soils. [See RANGELAND.]

The Prairie Brushland Province covers some 216,500 sq km of west-central and south-central Texas. Major plant communities are: mesquite–buffalo grass with some oak, side-oats grama, and beardgrass; juniper–oak–mesquite with grasses of the previous community present but in lesser amounts; and mesquite–acacia with the least grass, but with prickly pear cactus.

The California Chaparral Province covers some 86,800 sq km of the coastal and southern mountains of California, and the western foothills of Sierra Nevada. Fire was and still is a very important force in the evolution of shrubs there. Most species have developed root crowns that resprout after fire. Fire, while a part of nature in the California Chaparral, is a hazard for homeowners who choose to build in this shrubland province. Important shrubs are: chamise, ceanothus, manzanita, oak, sage, and chinquapin.

The Intermountain Sagebrush Province covers some 526,806 sq km, making it the largest of the United States shrublands. This large province occupies most of southern Idaho and the state of Nevada, the western half of Utah, southeastern and central Oregon, central Washington, and a small part of northern California. This large province is divided into five subprovinces based on vegetation type, precipitation, soil salinity, and geographical features. These subprovinces are: Great Basin Sagebrush–Wheatgrass (rainfall 15 to 40 cm); Great Basin Sagebrush (rainfall 15 to 40 cm); Ponderosa Shrub Forest (rainfall 25 to 45 cm); Lahontan Saltbush–Greasewood (old Lake Lahontan lake bed, Nevada, saline soils, rainfall less than 20 cm); and Bonneville Saltbush–Greasewood (old Lake Bonneville lake bed, Utah, saline soils, rainfall 15 to 30 cm). Important shrubs are: big sagebrush, rubber rabbitbrush, antelope bitterbrush, winterfat, saltbushes, greasewood, and black sagebrush.

The Mexican Highland Shrub Steppe Province covers some 43,325 sq km of southeastern Arizona and southwestern New Mexico. This highly variable province includes four distinct life belts. Important shrubs are: creosote bush, saguaro cactus, mesquite, oak, and juniper.

The Colorado Plateau Province covers some 245,300 sq km. It occupies most of western Colorado, eastern Utah, northern Arizona, northern New Mexico, and many of the desert mountain ranges in the Great Basin. This province includes two subprovinces based on rainfall and vegetation type. The subprovinces are: Juniper–Pinyon Woodland plus Sagebrush–Saltbush Mosaic (rainfall about 50 cm); and Gramma–Galleta Steppe plus Juniper–Pinyon Woodland Mosaic (rainfall about 38 cm, mainly in the summer). Important shrubs are: sagebrushes, saltbushes, mountain mahogany, buckwheat, rabbitbrushes, winterfat, oak, and hollygrape.

The Wyoming Basin Province covers some 109,600 sq km of the north-central and southwestern Wyoming. Precipitation ranges from 12 to 36 cm. The most important shrubs are: sagebrushes, saltbushes, and greasewood.

The Chihuahuan Desert Province covers some 166,800 sq km of south-central New Mexico and southwestern Texas. Precipitation is 12 to 33 cm. Most important shrubs are: creosote bush, flourensia, mesquite, ephedra, acacia, ocotillo, and agave.

The American Desert Province covers some 200,700 sq km. It occupies southern Nevada, southeastern California, southwestern and a small portion of western Arizona, and a very small portion of southwestern Utah. Precipitation is about 5 to 25 cm making this province the driest in the United States. The most important shrubs are: creosote bush, coleogyne, acacia, ambrosia, cercidium, fouquieria, and jatropha.

In summary, shrubs are found from the very moist to the driest ecosystems in the United States. The remainder of this chapter focuses on the drier ecosystems where shrubs dominate.

B. Morphology of Shrubs

Shrubs share two characteristics: woodiness and an extensive and often deep root system. Woodiness saves resources. Shrubs do not have to rebuild conducting and other tissues each year like plants that are not woody. Saved resources are directed toward root growth and maintenance. Also, under stress, shrubs can reproduce with a minimum of growth, whereas plants that are not woody must first build a whole plant. In addition, woody growth is cumulative, allowing shrubs to grow taller and wider so they can intercept more sunlight. Also, many shrubs can use sunlight during winter because they retain chlorophyll in their leaves or stems. For shrubs growing in climates where snow accounts for a large portion of the annual precipitation, woodiness intercepts drifting snow giving shrubs the advantage in water harvesting by as much as 10-fold over plants that are not woody. [*See* RANGELAND PLANTS.]

Shrubs living in dry ecosystems have root to shoot ratios well above 1 and many have ratios as high as 9. The root to shoot ratio increases with dryness so as to maintain plant water potential. It is this commitment to the maintenance and growth of root system that allows shrubs to dominate the drier ecosystems. Not only does the root system give shrubs greater ability to extract water from the soil, it gives them greater ability to extract minerals from deep in the soil. When shrubs lose their leaves, the minerals end up in the top soil where they can be used by shallower-rooted plants. Many might think shrubs have little value. This, however, is not the case.

C. Uses of Shrubs

Shrubs are used for: food for domestic and wild range animals, habitats for wildlife, ecosystem enrichment, and low maintenance landscapes.

1. Food for Domestic and Wild Range Animals

For all animals, survival depends largely on finding the right kind of food and enough of it. Shrubs are food for many animals, sometimes their only food; thus, the nutritional value of shrubs is of major impor-

tance to people who value animals, as well as the animals themselves. [*See* RANGELAND GRAZING.]

The nutrient needs of animals can be divided into four classes: energy, protein, minerals, and vitamins. We measure the nutritive value of shrubs based on their ability to meet these needs.

The nutrient content of shrubs and other range plants, grasses and forbs, is cyclic (Table I). Nutrient content peaks during the spring and gradually decreases, reaching the lowest level in winter. During the spring, shrubs and forbs (small, broad-leaved plants) contain less energy than grasses; shrubs also contain less protein than grasses and forbs. However, shrubs have higher levels of the mineral phosphorus than grasses or forbs. All three have equal amounts of carotene, needed to produce vitamin A. All three forage types can meet the high nutrient needs of females producing milk (lactating) to feed their young. [*See* ANIMAL NUTRITION, PRINCIPLES; FORAGES.]

Nutrient content drops for all three types of forage during summer with grasses and forbs dropping the most. Shrubs still contain less energy, than grasses or forbs. Shrubs and forbs have more protein than grasses. Shrubs have more carotene and phosphorus than grasses or forbs. Both shrubs and forbs still meet the needs of lactating females. Grasses, however, do not meet the needs of lactating females for phosphorus and protein.

Nutrient content continues to decline during the fall. The decline is led by grasses, followed by forbs. Shrubs decline the least. Fall energy content is about equal. Shrubs have more protein, phosphorus, and carotene than grasses and forbs. None of the three can supply enough energy, protein, or phosporus to meet the requirements of lactating females. Shrubs can supply enough carotene. All three can supply enough energy for female range animals with young

TABLE I

Cyclic Nature of the Nutrient Protein in Range Plants

Month/year	Big sagebrush (%)	Antelope bitterbrush (%)	Grass (%)
June 1968	11.8	13.4	13.4
July 1968	12.7	12.8	7.8
September 1968	11.8	9.7	9.6
December 1968	10.5	7.5	2.7
March 1969	14.0	9.9	3.4
May 1969	15.0	11.3	21.3

Source: After Tueller, 1979.
Note: Data are expressed as a percentage of dry matter.

developing (gestating) at this time. Only shrubs and forbs can supply enough phosphorus for gestating females. Only shrubs can supply enough protein and carotene.

Winter is the low point for nutrients. Shrubs can supply enough energy to meet the maintenance needs of range animals (the amount needed to maintain weight and health) and enough protein, carotene, and phosphorus to meet the needs of gestating females. Grasses and forbs can provide enough energy in winters, but otherwise they are unable to meet the maintenance needs of range animals.

Shrubs woodiness and their extensive and deep root system are important to the animals that feed on them. Woodiness allows shrubs to grow tall and helps them to stand up making them more available to feeding animals during periods of snow. The extensive root system allows shrubs to be a more dependable food source during seasonal or extended drought.

2. Habitats for Wildlife

Shrubs are important contributors to wildlife habitats. Nearly half of the 369 mammal species and 58% of the 714 birds native to North America are associated with woody cover of shrubs and trees (less than 4 m tall). One hundred and seventeen species of wildlife native to the United States are known to feed on shrubs during some part of the year. These species include: deer, elk, moose, pronghorn antelope, foxes, coyotes, rabbits, mice, bats, grouse, and songbirds. Shrubs not only provide food or feeding sites for wildlife but also provide areas where they can stay warm in winter, cool in summer, hide from predators, or seek shelter to bear and raise their young. [See WILDLIFE MANAGEMENT.]

3. Ecosystem Enrichment

Shrubs can favorably alter microenvironments by enhancing moisture infiltration, soil moisture-holding capacity, and nutrient concentration, and by moderating wind and temperature. In addition, shrubs can enrich ecosystems by providing sites where grasses and forbs can escape livestock and wildlife grazing, by functioning as biological filters, by preventing soil erosion, and by acting as carbon sinks to help reduce global warming.

Water filters into the soil more quickly under shrubs, allowing more moisture to be stored in the soil there than in adjacent areas where shrubs do not grow. This is partly due to the ability of some shrubs to increase the depth of soil to the caliche layer (hardpan) or to funnel rain to their stem base. But other factors are more important, such as animal holes, decayed root channels, stable aggregates of fine soil particles, and higher soil organic matter. Also, shrubs reduce soil compaction by weakening the impact of raindrops. Shrubs intercept blowing snow, resulting in more snow accumulation and higher soil moisture.

Shrubs accumulate mineral elements or nutrients under their canopies. This is due to the extensive root system which draws nutrients from far away, and to symbiotic fixation of nutrients by shrubs and their associated microorganisms. The moderation of wind and temperature by shrubs helps nutrient accumulation by encouraging animals to use shrubs for nesting, resting, roosting, or feeding. Dead plant material (litter) and wind-blown soil collect at the base of shrubs because they slow the wind. Thus, shrubs create islands of fertility. In areas that are heavily grazed by livestock or wildlife the only seed source for grasses or forbs may be under the protective canopies of shrubs.

Shrubs growing along waterways act as biological filters by reducing the velocity of flowing water. Sediment settles out in the slowed water among the streamside willows, for instance, cleaning the water. In addition, willows shade the water, helping to maintain cool-water fish habitat. Gambel oak clones growing on steep mountain slopes can reduce the velocity of water flowing downhill after high-intensity thunderstorms. Sediment is deposited behind the oak stems. This helps prevent gullies from developing or enlarging and reduces the sediment that washes into waterways. Due to their extensive and deep root system, shrubs hold soil to a greater depth than do grasses or forbs.

Shrubs can reduce the effects of wind erosion by reducing wind velocity. Soil particles accumulate on the leeward side of the shrubs. Since the microclimates at the base of shrubs are more hospitable for other plants, shrubs can become nurse plants.

Rangeland shrubs have great potential for use in landscapes or xeriscapes. Such landscaping relies on plants that require little watering, fertilization, pruning, or pest control. In addition to the diversity of form and function, a major factor in using shrubs is their ability to effectively exploit the limited resources of the environment.

In the Western United States, about 44% of the culinary water is used to irrigate landscapes. Because of the scarcity of water, a number of cities in the Western United States regulate the amount of water that can be used for watering landscapes. Residents may be fined for using too much water. Some cities may not issue building permits until builders have submitted a landscaping plan relying on drought-tolerant plants. Some cities may charge less for building permits or water rates if residents use drought-tolerent plants. Highway departments are using rangeland shrubs for landscaping highway right-of-ways. Benefits are low inputs and diversity of form and function.

Because they are woody, rangeland shrubs tie up carbon for a long time. This helps reduce the build-up of carbon dioxide, a greenhouse gas. To the casual observer, this may seem like "a drop in the bucket," but since shrubs are the dominant plant on billions of hectares worldwide (0.2 billion hectares in Western United States along) they represent a large carbon sink.

II. Rangeland Shrub Improvement

Shrubs contain a rich array of characters not only for shrubs found in different ecosystems, but also within species and subspecies of shrubs in the same ecosystem. This great genetic diversity provides the opportunity for finding (selection) or building (breeding) superior shrubs.

The success of a rangeland shrub improvement program depends on the degree to which superior shrubs meet specific needs. The needs shrubs help to meet are: soil stabilization, animal habitat, and esthetics. [See RANGELAND MANAGEMENT AND PLANNING.]

After a precise description of needs has been formulated, researchers compile a list of shrub species that might meet those needs. For example, one need might be to raise the nutritive content of forage plants on critical mule deer winter ranges. Energy, protein, and phosphorus are commonly the nutrients in short supply. Evergreen shrubs generally supply higher winter levels of these nutrients than deciduous shrubs. Deciduous shrubs generally provide more than dormant grasses and forbs. Therefore, evergreen shrubs would be included on the species list for improving mule deer winter ranges.

Once the species list is complete, the geographical range of each species is defined. The genetic variability of populations of the same species is evaluated. Promising species are selected for intensive evaluation. These evaluations determine the species that will be included in an improvement program.

Selection of germplasm is based on systematic sampling aimed at maximizing the variability in a promising population. Systematic sampling is essential to assure a broad genetic base. Germplasms are established in common gardens from collected seeds, seedlings, or cuttings. The common garden standardizes the environmental effects on character expression. Therefore, any differences in plants grown in common gardens are due to genetic differences among the various germplasms. Germplasms are evaluated based on criteria that help meet the needs established earlier. Some characters require studies in greenhouses or growth chamber rather than common gardens. After the evaluations are complete, promising germplasms are planted at three or four different sites in different environments. Germplasms that show continued superior performance are placed in plant material development programs. A few select germplasms with good market potential become named, tested germplasms released for the commercial production of seeds and plants. [See PLANT GENETIC RESOURCES; PLANT GENETIC RESOURCE CONSERVATION AND UTILIZATION.]

Often, a single germplasm does not express all superior characteristics desired. Such germplasms may be placed in a breeding program aimed at combining superior characteristics. Selected progeny then are released as named cultivars for the commercial production of seeds and plants.

Development of superior germplasms and cultivars of rangeland shrubs is illustrated in the following sections.

III. Released, Tested Germplasms of Sagebrush

There are about 13 species of woody sagebrushes in the Western United States. Only two have been improved: big sagebrush and black sagebrush. Big sagebrush is divided into several subspecies. The subspecies that have been worked on are mountain big sagebrush and Wyoming big sagebrush.

Big sagebrush has been maligned because it is not palatable to cattle. However, a large number of animal species depend on big sagebrush for food, nesting sites, or cover. These include: birds such as sage grouse, sage thrasher, sage sparrow, white crown sparrow, horned lark, Brewer's sparrow, and dark-eyed junco; large mammals such as mule deer, pronghorn antelope, Rocky Mountain elk, and domestic sheep; and small mammals such as pygmy rabbit, cottontail rabbit, black-tailed jack rabbit, and Uinta ground squirrel.

All germplasms of big sagebrushes share at least one common character, superior winter nutrient content (Table II). Big sagebrush ranks number one in digestibility for the winter forages listed in Table II. This

TABLE II
Winter Nutritive Value of Selected Range Plants

Plants	Digestibility (%)	Protein (%)	Phosphorus (%)	Carotene (mg/kg)
Shrubs				
Antelope bitterbrush	23.5	7.6	0.14	—
Big sagebrush	57.8	11.7	0.18	8.0
Birchleaf mahogany	26.5	7.8	0.13	—
Black sagebrush	53.7	9.9	0.18	8.0
Fourwing saltbush	38.3	8.9	—	3.1
Rubber rabbitbrush	44.4	7.8	0.14	—
Winterfat	43.5	10.0	0.11	16.8
Forbs				
Arrowleaf balsamroot	—	3.6	0.06	—
Little sunflower	—	2.8	0.17	—
Grasses				
Bluebunch wheatgrass	45.5	3.2	0.05	0.2
Bluestem wheatgrass	50.2	3.8	0.07	0.2
Bottlebrush squirreltail	42.0	4.3	0.07	1.1
Indian ricegrass	50.5	3.1	0.06	0.4
Needle-and-thread	46.6	3.7	0.07	0.4
Sand dropseed	53.2	4.1	0.07	0.5
Crested wheatgrass	43.7	3.5	0.07	0.2

Note: Data are expressed as a percentage of dry matter, except for carotene, which is expressed as mg/kg of dry matter.

may explain why sage grouse gain weight during the winter. Wintering sage grouse feed almost exclusively on the leaves and shoot tips of big sagebrush. Besides being highly digestible, big sagebrush ranks number one among winter forage in the amount of protein. It is tied with black sagebrush in the amount of winter phosphorus. Although big sagebrush has enough winter carotene to meet the need of animals, winterfat has more.

Some big sagebrush germplasms are preferred by some range animals, are more productive, or have higher nutrient content. These superior germplasms are needed to convert grass plantings back to shrublands or to restore shrublands and wildlife habitats in areas that have suffered overgrazing or other disturbances.

A. Hobble Creek Mountain Big Sagebrush

Hobble Creek mountain big sagebrush is a released, tested germplasm of big sagebrush for improving mule deer and domestic sheep winter ranges, for improving sage grouse habitat, or for restoring damaged habitats. It was released for use on rangelands in the Intermountain Sagebrush, Colorado Plateau, and Wyoming Basin Provinces having 35 to 63 cm of annual precipitation. Because it was to be used as a

forage plant, wildlife and livestock preference, nutrient content (protein, digestibility, and phosphorus), and production were evaluated in common gardens where Hobble Creek was grown with other big sagebrush germplasms.

In common garden studies Hobble Creek was the germplasm most preferred by wintering deer (Table III). Wintering domestic sheep like Hobble Creek, but not as much as some other germplasms. Similar results were obtained for sage grouse. In studies conducted in the states of Colorado, Idaho, Oregon, and Washington, Hobble Creek was preferred by wintering mule deer over the native big sagebrush.

Winter protein content of Hobble Creek was 11% of dry matter (Table IV). Winter digestibility, an indicator of energy, was 52.6% of dry matter. Winter level of phosphorus was 0.21% of dry matter. Its productivity ranked in the top third of the germplasms tested.

Winter forage attributes of Hobble Creek have been compared to antelope bitterbrush, a shrub known to be a desirable winter forage. Wintering mule deer preferred Hobble Creek over antelope bitterbrush. The sagebrush nutrient content was superior to antelope bitterbrush, which had a winter nutrient content of protein 6.0%, digestibility 30.1%, and phosphorus 0.10% (all on a dry matter basis). Hobble Creek had

TABLE III

Preference of Wintering Mule Deer for Germplasms of Big Sagebrush

Germplasm	Growth eaten (%)
Hobble Creek	57.5
Colton	47.0
Petty Bishop's Log	46.0
Indian Peaks	45.6
Sardine Canyon	44.6
Salina Canyon	41.7
Durkee Springs	40.7
Evanston	40.4
Pinto Canyon	40.3
Benmore	39.3
Clear Creek Canyon	38.4
Milford	37.1
Wingate Mesa	36.4
Brush Creek	35.7
Dog Valley	33.8
Kaibab	33.5
Loa	31.6
Dove Creek	30.9
Trough Springs	30.1
Evanston A	28.3

Note: Data are expressed as a percentage of current year's growth eaten.

three times as much vegetative production as antelope bitterbrush.

Hobble Creek big sagebrush can be established on suitable sites by direct seeding or by transplanting bareroot stock or containerized stock. Suitable sites have the following physical characteristics: mean annual precipitation of 35 to 63 cm; deep, well-drained soils with an effective rooting depth of at least 1 m;

TABLE IV

Winter Nutritive Value of 10 Germplasms of Big Sagebrush

Germplasm	Digestibility (%)	Protein (%)
Clear Creek Canyon	64.8	15.3
Dove Creek	64.6	16.0
Loa	57.0	14.5
Indian Peaks	55.8	11.2
Benmore	55.2	10.0
Kaibab	54.9	11.9
Milford	54.6	11.2
'Hobble Creek'	52.6	11.0
Sardine Canyon	48.7	10.5
Trough Springs	44.6	11.0

Note: Data are expressed as a percentage of dry matter.

soil no finer than clay loam (40% clay or less); soil pH between 6.6 and 8.6; and a growing season of 90 days or more.

In 1987, Hobble Creek mountain big sagebrush was formally released as a named tested germplasm for the commercial production and marketing of seeds and plants. It was the first big sagebrush germplasm to be released. The main reason for choosing this germplasm over other big sagebrush germplasms was its high preference ranking.

B. Gordon Creek Wyoming Big Sagebrush

Gordon Creek big sagebrush is a released, tested germplasm of Wyoming big sagebrush. It was released for use on rangelands in the Intermountain Sagebrush, Colorado Plateau, and Wyoming Basin Provinces having 25 to 35 cm of annual precipitation. This precipitation is insufficient to support establishment or growth of Hobble Creek germplasm. Gordon Creek was selected to meet needs similar to those of Hobble Creek. It was to be used for improving mule deer winter ranges, for improving sage grouse habitats, and for restoring damaged habitats. During evaluation, Gordon Creek was compared just to other Wyoming big sagebrush germplasms, whereas Hobble Creek was compared to a broader group of big sagebrush germplasms.

Studies in common gardens showed that wintering mule deer preferred Gordon Creek over the other 12 Wyoming big sagebrush germplasms tested. Sage grouse also ate Gordon Creek. This preference was expected, because at the native site it was browsed evenly by deer. Also, the population was located where the deer could move downslope if they did not like the big sagebrush there. Even during open winters, deer still consumed large quantities of the Gordon Creek Wyoming big sagebrush on its native site. The deer ate this big sagebrush out of choice, not out of necessity.

The winter nutrient content of the Gordon Creek germplasm was: protein 11.9%, digestibility 52.8%, and phosphorus 0.21% (all on a dry matter basis). Gordon Creek was ranked among the most productive Wyoming big sagebrush germplasms tested.

Gordon Creek can be established on suitable sites by direct seeding or by transplanting bareroot stock or containerized stock. Suitable sites have the following physical characteristics: mean annual precipitation of 25 to 36 cm; deep to shallow well-drained soils; clay content up to 55%; soil pH between 6.6 and 8.8; and a growing season of 80 days or more.

In 1992, Gordon Creek Wyoming big sagebrush was formally released as a tested germplasm for the commercial production of seeds and plants. It was the second big sagebrush germplasm to be released. The key characters in choosing this germplasm were wildlife preference and high productivity.

C. Pine Valley Ridge Black Sagebrush

Pine Valley Ridge is a released, tested germplasm of black sagebrush for use as a forage shrub for mule deer, pronghorn antelope, sage grouse, domestic sheep, and cattle, and for use in restoration projects. Black sagebrush is another of the 13 species of woody sagebrushes. It was released for use on rangelands in the Intermountain Sagebrush, Colorado Plateau, and Wyoming Basin Provinces having 15 to 25 cm of annual precipitation. Because this germplasm was to be used as a forage shrub, it underwent evaluations similar to those of the two big sagebrush germplasms. Pine Valley Ridge was compared with other germplasms of black sagebrush.

This black sagebrush was well liked by pronghorn antelope, domestic sheep, and cattle grazing in the Pine Valley area of Utah. Pine Valley Ridge germplasm was preferred by wintering mule deer over all other germplasms tested.

Winter nutrient content of Pine Valley Ridge was: protein 7.3%, digestibility 53.5%, and phosphorus 0.22% (all on a dry matter basis). Its productivity was ranked number two of the germplasms tested.

Pine Valley Ridge black sagebrush can be established on suitable sites by direct seeding or by transplanting bareroot stock or containerized stock. Suitable sites have the following physical characteristics: mean annual precipitation of 15 to 25 cm; deep to shallow soils; clay content up to 18%; soil pH between 7.0 and 8.9; and a growing season of 90 days or more.

In 1994, Pine Valley Ridge was released as a tested germplasm of black sagebrush for commercial production of seeds and plants. The key characters in its selection were its preference for grazing by a number of range animals and its high productivity, and protein and phosphorus content.

D. Changing Attitudes toward Big Sagebrush

Over the past 10 years, some persons have slowly been changing their attitudes toward big sagebrush.

They are beginning to realize that big sagebrush is not a worthless range weed, but a nutritious plant eaten by a number of range animals. In addition, scientific research has shown that the essential oils (monoterpenoids) found in big sagebrush do not adversely affect digestion or the preference of animals that graze it.

In the past, researchers expressed concern that big sagebrush's monoterpenoids might harm microbes living in the digestive tracts of the animals that eat it. Microbes allow range animals to survive on high-fiber forage. Microbes transform energy in the fiber to a form range animals can use. The microbes also manufacture amino acids and vitamins range animals cannot manufacture themselves.

Past concerns about digestibility ignored two points. First wintering wildlife species such as mule deer, pronghorn antelope, pygmy rabbit, and sage grouse eat large amounts of big sagebrush without any apparent harmful effect. Second, monoterpenoids are volatile—they change from a liquid to a gas at relatively low temperatures.

Laboratory tests showed that monoterpenoids were lost after they were placed in digestion flasks. The heat of the water bath drove the monoterpenoids from the flasks. That meant, normal body temperature would be sufficient to volatilize monoterpenoids. Researchers theorized that monoterpenoids could be lost outside the body through chewing, rumination (a special form of digestion), or belching. A study designed to test this theory reported an 80% reduction between the monoterpenoid levels in the ingesta (food that has been chewed and partially digested) of wintering mule deer and the levels expected from the ingested forage. Other studies have shown large reductions in the amount of monoterpenoids in the ingesta of pygmy rabbit, sage groups, greater glider, and brushtail possum. These reductions are large enough to nullify any adverse effects monoterpenoids might have on digestion.

Research has not shown that wildlife avoid eating plants with monoterpenoids. During a study discussed earlier, Hobble Creek mountain big sagebrush, which produces monoterpenoids, was preferred by wintering mule over antelope bitterbrush, which does not produce monoterpenoids.

These results have helped change attitudes, so that some persons now think of big sagebrush as a valuable forage plant.

E. Developing Cultivars of Big Sagebrush

At least two germplasms are crossed when developing cultivars of big sagebrush. Progeny have been produced by crossing Hobble Creek and Dove Creek big sagebrush (Tables III and IV). The progeny have inherited characters from both parents. Through selection, the development of a cultivar sharing the attributes of both parents is feasible. [*See* CULTIVAR DEVELOPMENT.]

IV. Released, Tested Germplasm of Fourwing Saltbush

Rincon, a released tested germplasm of fourwing saltbush, is a facultative evergreen shrub—it is capable of being an evergreen in certain conditions. It is leafier and fuller canopied than other germplasms. The seed source for this germplasm came from the Rincon Blanco area in the Carson National Forest, New Mexico. Its leaves, stems, and utricles (small bladderlike fruits) are browsed in all seasons by livestock and wildlife. Protein varies seasonally from 6.9 to 26.4% of dry matter. In addition to providing forage and cover, Rincon is valuable for rehabilitating overgrazed rangelands, for stabilizing soils, and for restoring damaged habitats. It can be established by direct seeding, by transplanting, and by stem cutting. It is adapted to a wide range of soil textures: sandy areas, gravelly washes, loamy soils, heavy clay soils, and moderately saline soils. It is adapted to sites that vary from 22 to 58 cm annual precipitation. Because it is cold tolerant, Rincon can be grown in the following provinces: Intermountain Sagebrush, Colorado Plateau, and the Wyoming Basin.

V. Released, Tested Germplasm of Antelope Bitterbrush

Lassen is a released, tested germplasm of antelope bitterbrush. It is leafier and larger than other germplasms. The seed source for this germplasm came from stands near Janesville, Lassen County, California. Its leaves and stems provide forage in all seasons for livestock and wildlife. Midwinter protein content was 7.9% and digestibility was 30.6% (both on a dry matter basis). Lassen was highly persistent under moderate to heavy grazing. It can be used for rehabilitating overgrazed rangelands, for stabilizing soils, and for various reclamation projects. It can be established

by direct seeding and by transplanting. It is adapted to deep, coarse, well-drained, neutral to slightly acidic, pumiceous, or granitic soils. It requires sites having 30 to 60 cm of annual precipitation. Lassen can be grown on suitable sites in the Intermountain Sagebrush Province.

VI. Released, Tested Germplasm of Winterfat

Hatch is a released, tested germplasm of winterfat. It is evergreen. Livestock and wildlife prefer it over other winterfat germplasms tested. The seed source for this germplasm came from near the town of Hatch, Utah. Its leaves and stems provide winter forage for livestock and wildlife. Grasshoppers do not feed on this germplasm enough to damage the plants. It can be used for rehabilitating overgrazed rangelands, stabilizing soils, and for various reclamation projects. It can be established by direct seeding and by transplanting. It is adapted to neutral and slightly alkaline sites having 30 to 40 cm of annual precipitation. Soils may vary from sandy to fine-textured. Hatch can be grown in the following provinces: Intermountain Sagebrush, Colorado Plateau, and the Wyoming Basin.

VII. Released, Tested Germplasm of Forage Kochia

Immigrant is a released, tested germplasm of forage kochia. It is a semi-evergreen perennial subshrub or low shrub. This germplasm was introduced from the Tavapol Botanical Gardens, Russia. Its leaves and stems provide forage for livestock and wildlife during all seasons. Immigrant was found to be superior to other germplasms in longevity, forage production, forage quality, palatability, and competitiveness with undesirable annuals such as halogeton or cheatgrass. Protein content varied seasonally from 5.9 to 14.4% of dry matter. Winter digestibility varied from 29.7 to 32.2% of dry matter. Levels of oxalate, a poison, were among the lowest of the germplasm tested. In addition, Immigrant was found to be tolerant of heavy grazing and fire. It can be used for rehabilitating overgrazed rangelands and for various reclamation projects. It can be established by direct seeding and by transplanting. It is adapted to basic, saline soils rang-

TABLE V

Other Releases of Tested Germplasm of Shrubs

Common plant names	Germplasm	Common plant names	Germplasm
Allegheny chinquapin	Golden	Greenleaf manzanita	Altura
Amur honeysuckle	Rem Red	Hardy Privet	Cheyenne
	Cling-Red	Hooker willow	Clatsop
Antelope bitterbrush	Fountain green	Late lilac	ND-83
	Maybell	New Mexico olive	Jemez
Aromatic sumac	Konza	Mongolian cherry	Scarlet
Arroyo willow	Rogue	Mountain mahogany	Montane
Autumn olive	Cardinal	Mountain whitethorn	Maleza
	Ellagood	Pacific willow	Nehalem
Barren ground willow	Oliver	Purpleosier willow	Streamco
Bicolor lespedeza	Natob	Quailbush	Casa
Bladderpod	Dorado	Redosier dogwood	Mason
Bladdersenna	Tierra	Russian almond	ND-283
Bristly Locust	Arnot	Saskatoon serviceberry	9021438
California buckwheat	Duro	Shrub lespedeza	VA-70
Columbia River willow	Multnomah	Silky dogwood	Indigo
Desert willow	Barranco	Silver buffaloberry	Sakakawea
Douglas spiraea	Bashaw	Sitka willow	Plumas
Dwarf willow	Bankers	Skunkbush sumac	Bighorne
Erect willow	Placer	Sulfur-flower buckwheat	Sierra
European cotoneaster	Centennial	Ussurian pear	McDermand
Fourwing saltbush	Marana	Wild plum	Rainbow
Fourwing saltbush	Santa Rita		
	Wytana		

ing from sandy loam to heavy clay on sites receiving from 12 to 68 cm of annual precipitation. Immigrant can be grown in the following provinces: Intermountain Sagebrush, Colorado Plateau, and the Wyoming Basin.

VIII. Other Releases of Tested Germplasms of Shrubs

Agencies that have released other superior tested germplasms of shrubs include: Agriculture Research Service, Soil Conservation Service, Upper Colorado Environmental Plant Center, Utah Division of Wildlife Resources, and various state agricultural experiment stations. Table V lists the germplasms that have been released. For more details the reader is referred to the publication "Conservation Trees and Shrub Cultivars in the United States," Agriculture Handbook Number 692, published by USDA, Soil Conservation Service, Washington DC.

Bibliography

Bailey, R. G. (1980). "Description of the Ecoregions of the United States." United States Department of Agriculture, Forest Service, Miscellaneous Publication Number 1391, Washington, DC.

Carson, J. R., Cunningham, R. A., Everett, H. W., Jacobson, E. T., Lorenz, D. G., and McArthur, E. D. (1991). "Conservation Tree and Shrub Cultivars in the United States." United States Department of Agriculture, Soil Conservation Service, Agriculture Handbook Number 692, Washington, DC.

McArthur, E. D. (1988). New plant development in range management. In "Vegetation Science Application for Rangeland Analysis and Management" (P. T. Tueller, ed.), pp. 81–112. Kluwer, Dordrecht.

McArthur, E. D., Welch, B. L., and Sanderson, S. C. (1988). Natural and artificial hybridization between big sagebrush (Artemisia tridentata) subspecies. J. Heredity 79, 268–276.

McKell, C. M. (1975). Shrubs—A neglected resource of arid lands. Science 187, 803–809.

McKell, C. M. (ed.) (1989). "The Biology and Utilization of Shrubs." Academic Press, San Diego.

McKell, C. M., Blaisdell, J. P., and Goodin, J. R. (eds.) (1972). "Wildland Shrubs—Their Biology and Utilization." United States Department of Agriculture, Forest Service, Intermountain Forest and Range Experiment Station, General Technical Report INT-1, Ogden, UT.

Medin, D. E., and Anderson, A. E. (1979). "Modeling the Dynamics of a Colorado Mule Deer Population." Wildlife Monograph Number 68, The Wildlife Society, Washington, DC.

Tueller, P. T. (1979). "Food Habits and Nutrition of Mule Deer on Nevada Ranges." University of Nevada, Reno.

Welch, B. L., and Pederson, J. C. (1981). *In vitro* digestibility among accessions of big sagebrush by wild mule deer and its relationship to monoterpenoid content. *J. Range Management* **34,** 497–500.

Welch, B. L., Pederson J. C., and Rodriguez, R. L. (1989). Monoterpenoid content of sage grouse ingesta. *J. Chemical Ecology* **15,** 961–969.

Welch, B. L., and Wagstaff, F. J. (1992). "Hobble Creek" big sagebrush vs. antelope bitterbrush as a winter forage. *J. Range Management* **45,** 140–142.

Rangeland Soils

STEPHEN G. LEONARD, *USDI-Bureau of Land Management, Nevada*

GEORGE J. STAIDL, *USDA-Soil Conservation Service, Nebraska*

Glossary

Classification, soil Systematic arrangement of soils into groups or categories on the basis of their characteristics; broad groupings are made on the basis of general characteristics and subdivisions on the basis of more detailed differences in specific properties

Ecological site Kind of land with a specific potential natural community and specific physical site characteristics, differing from other kinds of land in its ability to produce vegetation and to respond to management; it is synonomous with range site when applied to rangelands

Phase, soil Subdivision of a soil series or other unit of classification having characteristics that affect the use and management of soil but do not vary sufficiently to differentiate it as a separate class; a variation in a property or characteristic such as degree of slope, degree of erosion, content of stones, etc.

Physical properties (of soils) Characteristics, processes, or reactions of a soil which are caused by physical forces and which can be expressed in physical terms or equations; examples of physical properties are bulk density, water-holding capacity, hydraulic conductivity, porosity, etc.

Profile, soil Vertical section of the soil through all its horizons and extending into the C horizon

Soil chemistry Division of soil science concerned with the chemical constitution, the chemical properties, and the chemical reactions of soils

Soil survey Systematic examination, description, classification, and mapping of soils in an area; soil surveys are classified according to the kind and intensity of field examination

Soil can be described as a collection of natural bodies occupying parts of the earth's surface that support plants, or even more generally as the superficial unconsolidated and usually weathered part of the earth's surface. Some prefer not to precisely define soil but rather describe individual bodies of the earth's mantle relative to their complex characteristics and properties due to the integrated effects of climate, living organisms, parent material, and relief as they have evolved over time. The soil component of rangeland can, however, be defined specifically in context with the definition of "rangeland." Rangeland soil is the stratum above hard rock or unweathered parent material that supports indigenous vegetation that is predominantly grasses, grasslike plants, forbs, or shrubs and is a component of lands managed as a natural ecosystem. If plants are introduced, they are managed as indigenous species. Soils of rangelands are often characterized by limiting properties or marginal conditions for arable agriculture under natural situations.

I. General Rangeland Soil Characteristics

Approximately 37% of the land area of the United States is arid and semi-arid rangeland situated primarily in the western conterminous states and arctic desert of Alaska. As classified using Soil Taxonomy, Aridisols and Mollisols are the predominant soil orders of native rangelands in the western United States. Aridisols are mostly associated with the desert shrub and grass–shrub ecosystems of the Great Basin and Intermountain ranges while Mollisols are associated with the transition to forests in the mid-elevations of the west, prairie grasslands in the midwest, and transition to forests of the east. Other soil orders are included in less extensive but important rangeland ecosystems such as riparian and wetland areas (the wetter components of almost any soil order) and many flood plains or other recent alluvial landscapes (commonly Entisols in the arid West). [*See* RANGELAND; SOIL GENESIS, MORPHOLOGY, AND CLASSIFICATION.]

Soils of rangelands are extremely varied in their physical and chemical properties. They range from hot to cold climates and can be very shallow to very deep, of varied mineralogy, or nearly any texture or color. However, soils associated with rangelands are most often too dry for nonirrigated agriculture or occasionally too wet without artificial drainage or flood control. Many of these soils are best managed as parts of natural rangeland ecosystems because of physical characteristics such as depth, texture, and stoniness and chemical characteristics such as salinity, sodicity, and availability of soil nutrients that limit other beneficial uses.

Physical characteristics of some soils make them better suited for rangeland even where climatic conditions may be suitable for crops. Shallow soils, depth to bedrock or other restrictive layers can limit moisture storage and availability during the growing season. Rock fragments and stones likewise reduce the volume of soil material available for moisture storage. Native or naturalized range vegetation is most often better adapted to the limited moisture conditions.

Certain textural groups of soils are better suited for management as rangelands because of moisture relationships that would require extensive irrigation practices, erosion hazards, or both. Sandy soils have low water retention. At the other extreme, high moisture tension exhibited by clayey soils can also limit moisture availability to plants. Sandy soils are highly erodible by wind and silty soils are more susceptible to water erosion if vegetation is removed.

Arid and semi-arid rangeland soils are generally slightly acid to alkaline in reaction. Saline soil conditions are a common occurrence resulting from high evapotranspiration rates that exceed yearly precipitation. Accumulation of carbonates, sulfates, and other soil chemicals may also occur. Most soils associated with arid and semi-arid rangelands are characteristically low in organic matter.

II. Availability and Presentation of Rangeland Soil Information in the United States

Soil survey in the United States began with the 1896 U.S. Department of Agriculture appropriations act. The concept of soil survey as we know it today, was initiated by the U.S. Department of Agriculture Soil Conservation Service in the 1930s. Soil surveys have now been completed over a large portion of the United States including private, public, and state lands. [*See* SOIL AND LAND USE SURVEYS.]

All soil surveys are completed within the framework of the National Cooperative Soil Survey (NCSS). Federal leadership is provided by the USDA Soil Conservation Service. Upon completion, each soil survey undergoes a soil correlation and approval process. Soil surveys are then published in a soil survey report for use by public and private entities. In general, a soil survey report contains the soil maps for the area, descriptions of the soils in the area, and all related soil interpretations required to guide the use and management of soils. Where the land use is range and woodland, information about the associated potential native plant community is also included. Pertinent soil and vegetation information contained in recently completed soil survey reports is also being incorporated into the USDA Soil Conservation Service soil survey database. This feature allows electronic access and retrieval of soils data.

A. Soil Survey Publications

Each soil survey that has been correlated and approved is published in a report format. Soils resource data for a specific parcel of land or an entire soil survey area may be obtained at local or state offices of the USDA Soil Conservation Service, USDA Cooperative Extension Service, or any federal agency participating in the National Cooperative Soil Survey. Soil survey reports may also be obtained from Congressional representatives and many libraries. In order to apply soil information to rangeland interpretation and management, we must first know where the data sources exist, in what form, and how to extract the information.

1. Soil Maps

A soil map is a specialized map designed to show the distribution of soils or soil mapping units in relation to other prominent physical and cultural features of the earth's surface. The level of detail shown on the soil maps is determined by principal user objectives identified prior to the start of the soil survey. The objective of soil mapping is not to delineate pure taxonomic classes but rather to separate the landscape into segments that have similar use and management requirements. Boundaries between landscape areas occupied by contrasting soils or groups of soils are delineated in a unique map unit. Each soil will occur on a specific geomorphic position on the landscape. Location and spatial distribution of a soil or groups of soils are linked by a map unit symbol to a map unit description and to ecological site designations (for range and woodlands), tabular data about specific soil properties and interpretations for use and management decisions.

The base maps used in report publication involving rangelands and woodlands are primarily of two kinds: (1) rectified photo base maps (high-altitude photography), and (2) orthophoto base maps (high-altitude photography with the displacement of images removed). Publication is normally on orthophoto quads with black and white images at a 1:24,000 scale.

Map unit symbols are used to identify the map unit assigned to the delineations on the base soil maps. A map unit symbol can be numeric, alphabetic, or a combination of both. The map unit symbol provides the reference to a map unit description and associated information.

2. Map Unit Descriptions

The map unit description is used to characterize the soils, their pattern of occurrence within the map unit, their proportionate extent, and position on the landform. It also provides information about land use, present and potential vegetation, and principal hazards or limitations to be considered in planning. From this information, the relationships and distinctions of one map unit to another can be used to plan rangeland and grazable woodland management alternatives.

Contents of each map unit description are generally systematically arranged. An identifying symbol precedes the map unit name and a general setting is given for the entire map unit. The setting includes the broad landscape characteristics, elevation, and climate. Composition is given for the major components identified in the name of the map unit and for contrasting soil and miscellaneous area inclusions. A specific position within the broad landscape is noted for each major or dominant component. Dominant vegetation at the time of the survey is provided for rangeland and

woodland. A typical profile description for each soil is given as a vertical section of the soil extending from the surface to either a restrictive layer or a depth of 60 in. or more. Additional information about soil properties and qualities may include permeability, available water capacity, hydrologic soil group, erosion hazard, and other data. This is followed by current and readily foreseeable uses. The remainder of the description discusses the soil properties as they affect use and management. Rangeland and grazable woodland management applications and practices are also addressed in this article.

3. Soil Descriptions

Within the soil survey manuscript are representative soil profile descriptions and a range of soil characteristics for each soil taxon identified in the soil survey. The range of characteristics is compiled from many individual field observations and supporting laboratory characterization.

Soil profile descriptions are completed by field soil scientists. The soil is excavated to a limiting layer such as bedrock or to 60 in. or more to expose a vertical section of soil called a profile. A description of the physical and chemical properties of the profile is systematically recorded for each layer. This documentation includes the color, texture, size and shape of soil aggregates, kind and amount of rock fragments, distribution of plant roots, reaction (pH), and other features that enable them to identify and interpret soils. Samples may be taken for further laboratory analysis. The resulting profile description represents a 1 m $\times$ 1 m three-dimensional soil body called a pedon.

Upon completion of the profile description, the soil is classified using Soil Taxonomy and correlated to a soil series. Soil Taxonomy is the hierarchical classification system used in the United States and many other countries. Classification allows each soil to be assigned to a class with precisely defined limits of soil characteristics for interpretation, extrapolation of information, and communication between users.

Soil pedon data used in conjunction with classification and map unit information allow interpretation to proceed from site-specific to landscape map unit delineations for consideration.

4. Tabular Information and Soil Interpretations

Each soil survey report contains soil interpretations based on objectives identified at the start of the survey. Surveys involving rangeland and grazable woodland generally include correlation of soils to ecological sites (or range and grazable woodland sites), soil–vegetation relationships, and other interpretations that support management alternatives. Interpretations are generally located in the back part of the report in a table format by kind of interpretation.

Soil interpretations are developed and field tested through observation of soil properties and soil behavior under different uses and levels of management. Properties and supporting data for each soil are analyzed and rated for various uses such as shallow excavations, pond reservoir areas, excavated ponds aquifer fed, rangeland seeding, and many others. In addition, where the soil is limited for a specific use, the most limiting soil features are identified.

Ratings, limitations, and ecological site designations are given by map unit for each component soil

B. Soil Survey Data Bases

Soil survey information in the form of published reports has been the most common method used to disseminate soils data in the past. An automated soil data system that includes much of the information in a soil survey report is now part of the National Cooperative Soil Survey. The data system is used to store, access, and disseminate soil information. Multiple automated soil applications or modules that stand alone or interact with each other are being developed to provide future soils information.

Four levels of application are available. The broadest is the national level where the data base is designed to include appropriate information for the nation or region. The next includes the most recent data for a state. The soil survey project office and field office levels contain all locally applicable soil information. Technical quality control, management, and maintenance of the soil data system are provided by the USDA Soil Conservation Service.

III. Soil–Vegetation Relationships

Range landscapes are a mosaic of vegetation communities. The patterns may be broad with gradual gradients from one community to the next. Others change abruptly in short distances. Land use such as grazing or disturbances such as fire have a definite effect on these patterns but soil–vegetation relationships are by far the greatest contributor. Soil properties influence the kind, amount, and proportion of plant species capable of growing on a site in pristine conditions.

They also influence community dynamics associated with land use or disturbance. Many relationships are well understood, particularly as they apply to the physiology of individual plants or species. However, the often compensating or cumulative effects of soil properties in the natural setting and their relationship to rangeland vegetation and management still offer great challenges. [See RANGELAND PLANTS; RANGELAND SHRUBS.]

A. Soil–Climate Relationships

Soil moisture and soil temperature are collectively referred to as soil-climate. Soil-climate is related to ambient climate, but is further modified by topographic position, ground water movement and position, overland flow, and vegetation. All soil–plant relationships within rangeland ecosystems are influenced, at least indirectly, by soil-climate. Rooting depth, water potentials, nutrient uptake, and even nutrient distribution are related to the amount of soil moisture available during periods of suitable temperature for root activity.

Figure 1 illustrates the segregation of some major sagebrush (*Artemisia* spp.) habitats in southern Idaho by soil-climate regimes and as modified by other soil properties. Similar segregation of major grasses and pinyon and juniper tree species by soil-climate are also documented. Relationships will vary geographically because of other associated soil properties, climate variables, and ecotypic variation within individual species.

Soil moisture and temperature regimes are defined in soil taxonomy by annual and seasonal distribution of temperature class limits and moisture availability. At the family or higher levels of classification in Soil Taxonomy, combinations of moisture and temperature regimes are incorporated that have proven useful in predicting plant community distribution and, in some cases, individual species distribution on rangelands. Moisture and temperature regimes, however, have specified limits for taxonomic classification that may not always correspond with specific physiologic responses of range plants. Soil Taxonomy recognizes five broad categories for moisture regimes (aquic, udic, ustic, xeric, and aridic), listed from wettest to driest condition. Higher levels of taxa in Soil Taxonomy describe gradations between the categories. Soil temperature classes are: hyperthermic, thermic, mesic, frigid, cryic, and pergelic, listed from hottest to coldest condition. Soil Taxonomy should be consulted for exact definitions.

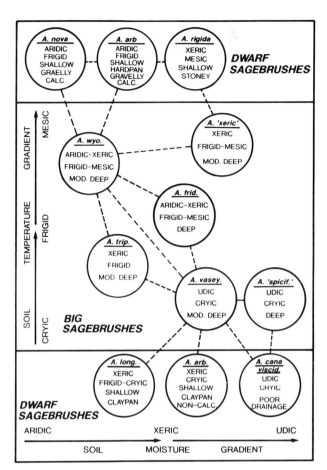

FIGURE 1 Conceptual relationship of sagebrush species and subspecies based on soil moisture, temperature, and other characteristics. [From Hironaka, M., Fosberg, M. A., and Winward, A. H. (1983). Sagebrush-grass habitat types of southern Idaho. Forest, Wildlife and Range Experiment Station. University of Idaho, Moscow.]

An annual pattern of soil water states at incremental depths is often a more sensitive indicator of community differences than broad taxonomic classes. Only a slight lowering of a water table during the growing season of some riparian sites can cause a change in vegetation potential from sedges and rushes to willows. Both situations may still be in the aquic moisture regime, but with dramatic differences in community response.

Dynamics of individual communities is in part due to the annual variance of soil-climate because of weather. Annual fluctuations in soil temperature and moisture favor recruitment of different species between years. Stratification of species response to temperature relative to soil moisture can result in annual differences in species composition based on productivity as well as differences in total productivity. Pro-

ductivity differences as high as 1200% between years have been documented for range ecosystems in the Great Basin.

B. Physical and Morphological Relationships

Soil physical and morphological properties include texture, structure, depth, rock fragment content (stones, cobbles, and gravel), bulk density, consistence, color, and horizonation in relation to these factors. Soil surface features are also included. In Soil Taxonomy, some characteristics become diagnostic at the subgroup or higher level, but more detailed information useful for determining soil–vegetation relationships on rangelands becomes apparent at the family and series levels of classification. Relationships are most often distinct when the landscapes involved contribute to a pattern of soils with distinctly different moisture infiltration and retention characteristics. Some surface soil characteristics such as bulk density, crusting, and microtopography are highly temporal depending on seasonal variation, use, and disturbance. These temporal characteristics can have a tremendous influence on infiltration and localized vegetation response.

Physically different soils (i.e., enough difference to be classified to different soil taxa) are often found to produce extremely similar rangeland vegetation habitats. For instance, similar low sagebrush (*Artemisia arbuscula*) habitats are observed on some soils that are shallow to claypan, some that are shallow to bedrock, and others that are moderately deep or deep but contain over 60% rock fragments. Compensating characteristics within these soils provide similar water supplying capacity during favorable growing conditions for low sagebrush and associated species.

Conversely, soils that are similar in gross morphological characteristics are sometimes found to produce distinctly different rangeland habitats. Although similar enough to be classified in the same series, the soils differ in one or more minor but important characteristics that allow distinctly different vegetation to exist. Recent soil surveys on rangelands "phase" soil series to account for differences that imply a different vegetation potential.

Soil surface characteristics are most susceptible to short-term change from land use or natural disturbances such as fire. Land use leading to compaction of even highly permeable rangeland soils can increase bulk density, reduce infiltration, and increase runoff and erosion. Fire can affect the surface structure as well as the biological and chemical properties of the surface soil. The effects of fire and interactions between soil and vegetation can be extremely variable on rangelands depending on soil type, present vegetation, climate, time and intensity of the fire, etc. Effects can range from temporarily sealing the soil surface or hydrophobic conditions impeding infiltration to improving soil nutrient status.

Both physical and biological surface crusts are common phenomena on rangeland soils. Surface caps associated with durable, cohesive crusts or platy, vesicular crusts limit seedling emergence and establishment. Some plant species (often undesirable ones), however, are better adapted than others for establishment in crusted conditions. Cryptogamic crusts associated with microflora (nonvascular plants such as algae, lichens, and mosses) in and on the soil surface may be either desirable or undesirable.

Some kinds of cryptogamic crusts develop hydrophobic characteristics limiting infiltration; others may have adverse allelopathic relations with other species or otherwise interfere with establishment or growth. In arid and semi-arid rangelands of the Great Basin and Colorado Plateau regions of the western United States, the presence of cryptogamic soil crusts appear to have favorable implications. These crusts, particularly the cyanobacteria (blue-green algae) and lichen components, stabilize the soil, fix nitrogen, and enhance seedling establishment of vascular plants. Population dynamics and effects of microfloral components on soil and vascular vegetation components of rangeland ecosystems are only beginning to be understood. Trying to characterize them as "good" or "bad" as a group is quite possibly like trying to describe all other rangeland communities in a single word.

C. Chemical Relationships

Distribution of chemical characteristics in the soil are influenced by physical characteristics, morphology, parent material chemistry, climate, and vegetation; therefore, interrelationships must be considered or at least concurrently evaluated when considering soil–vegetation relationships. A strong chemical patterning, affecting soil horizonation both vertically and horizontally, often occurs in native rangelands because of these relationships. Vegetation composition, productivity, and distribution are affected by the availability of nutrients, soil water potentials, and sometimes toxic effects of soil chemistry. Conversely, certain vegetation species tend to concentrate soil nutrients, other chemical compounds, and moisture

around themselves. Cryptogamic crusts also influence soil chemistry–vascular vegetation relationships; therefore, evaluation must include the microflora along with the higher plants. [*See* SOIL, CHEMICALS: MOVEMENT AND RETENTION; SOIL CHEMISTRY.]

Increasing levels of salinity and pH often coincide. Soil pH influences nutrient availability (Fig. 2) and the presence of toxic ions. Salinity affects plant growth by decreasing osmotic potentials, producing ion toxicity, and affecting nutrient uptake. Salinity is the total salt content of all salts more soluble than gypsum. As salinity (Table I) and pH levels increase, fewer plants are physiologically adapted to the effects. Different saltbush (*Atriplex* spp.) habitats have been correlated with soil pH, salinity, and sodicity (proportion of exchangeable sodium) in combination with soil properties such as profile depth, surface horizon depth, and particle size. Caution is advised with the extrapolation of localized soil–vegetation relationships involving *Atriplex* species, other members of the Chenopod family, and other genetically active plant families. Hybridization, mutation, and genetic poly-

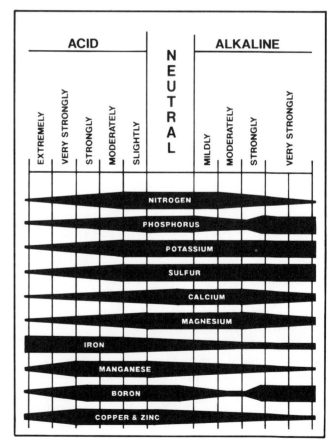

FIGURE 2 Soil reaction and nutrient availability.

TABLE I

Soil Salinity Classes[a]

Electrical conductivity (millimhos/cm)	Salinity	Connotation
0–2	Nonsaline	No effect on plants
2–4	Very slightly saline	Effect on plants is mostly negligible; only most sensitive plants affected adversely
4–8	Sightly saline	Many plants are adversely affected but diversity of adapted species is quite high (e.g., four-wing saltbush, winterfat, galleta)
8–16	Moderately saline	Plant communities are dominated by salt-tolerant species (e.g., greasewood, mat saltbush, western wheatgrass)
>16	Strongly saline	Few plants are adapted (e.g., shadscale, iodine bush, alkali sacaton)

[a] U.S. Soil Conservation Service.

ploidy are common and affect adaptation to soil conditions.

Chemical balance and nutrient ratios influence rangeland vegetation responses. Potassium–magnesium ratios in the soil may in part contribute to differences in shrub- vs grass-dominated sites. Differences in root cation exchange capacity affecting K and Mg absorption suggest a high K–Mg ratio would favor shrubs while low ratios favor grasses. Improvement of nitrogen to sulfur ratios have enhanced nitrogen fixation by subclover (*Trifolium subterranium*) in some California annual ranges and increased productivity. Soils with serpentine parent materials have such an unfavorably high magnesium to calcium ratio that they are characterized by a small number of endemic species evidently able to secure calcium at low degrees of Ca-saturation as well as exhibiting low productivity.

There is considerable microbial activity within rangeland soils in addition to the cryptogamic surface and near-surface components that influence soil–vegetation relationships. Decomposition of organic residue by microbial bacteria affects the availability of organic soil nitrogen and carbon in the soil. Microbial symbiotic nitrogen-fixing relationships are equally as important to sustained rangeland productivity as to arable agriculture. Vesicular–arbuscular mycorrhizal fungi (VAM) have been shown to improve water absorption as well as assist plants in mineral absorption, especially phosphorous, although relationships are variable. VAM levels are affected by both soil disturbance and plant community composition. [*See* SOIL MICROBIOLOGY.]

As range plants are adapted to certain soil environments, plants also create soil environments, both chemically and physically. Creosotebush (*Larrea tridentata*) apparently accumulates nitrogen under its canopy at the expense of interspace areas so that nitrogen may become limiting for grass in the interspace areas. Species such as black greasewood (*Sarcobatus vermiculatus*) and some saltbushes accumulate salts on deciduous leaves, effectively increasing salinity/sodicity under their own canopies. Coppicing around shrub canopies is common in most arid ecosystems. Differential infiltration and evapotranspiration characteristics between the shrub/coppice and interspace areas result in equally different soil chemistry patterns.

D. Soil vs Vegetation Classifications

Even though close relationships exist between soil properties and vegetation communities, soil classification does not necessarily parallel vegetation classification on rangelands or elsewhere. Soil classification must only consider observable or measurable soil properties, while only vegetation characteristics can be considered in vegetation classification. The two may then be compared without the confusion of compensating or cumulative effects on interrelationships. Soils may also vary continuously over the landscape so classification must occasionally set somewhat arbitrary boundaries based on interpretational needs that may or may not include natural vegetation communities. Likewise, vegetation continua are often separated by class limits determined as much by use and management interpretation as by distinct ecological boundaries.

The United States has developed a comprehensive Soil Taxonomy integrating both soil genesis and soil morphology into a hierarchical system of six categorical levels: Order, Suborder, Great Group, Subgroup, Family, and Series. Soil Taxonomy is constantly amended as new information and new needs arise.

Consistency in terminology, classification, and correlation is maintained through the National Cooperative Soil Survey, an organization comprising cooperating agencies and other users of Soil Taxonomy.

No classification or taxonomy of vegetation communities has yet been universally accepted in the United States, although a few major systems have evolved within government agencies and the academic community. Two primary strategies are involved: one based on a taxonomy of present vegetation attributes, the other on a classification of a landscape's ability to produce a potential or climax vegetation community. Both present and potential vegetation communities are related and there is no (scientific) reason that classifications cannot be integrated.

Soil Taxonomic classes are successfully related to ecological site and range site classifications used by many NCSS cooperators when that purpose is specifically identified in the soil survey process. Soil phases are not part of the soil taxonomic hierarchy, but utilitarian distinctions created to accommodate specific interpretational purposes in individual surveys. Soil taxonomic classes are phased in rangeland surveys to account for differences in landscape capability to produce a characteristic potential or climax plant community. Other present plant communities are then linked by virtue of this distinct soil–site correlation for interpretation.

IV. Application of Soils Information to Rangeland Planning and Management

When soil–site correlation is accomplished as described, soil survey map units and associated information provide a reliable basis for extrapolating rangeland vegetation ecology and management interpretations as well as soil interpretations. Range management is dependent on the inherent natural productivity of the land because land treatments such as leveling, irrigation, or even fertilization are usually not economically feasible. Mechanical or chemical treatments, seeding, and management facilities such as fences or water developments are often feasible only because of an expected long-term benefit. Soil information and soil–vegetation relationships are essential knowledge for planning and implementing successful range management actions with some predictability. Soil information is also useful for

monitoring and evaluating the subsequent results of those actions. [See RANGELAND MANAGEMENT AND PLANNING.]

A. Present and Desired Conditions

There are many reasons that the present conditions of rangelands may not be what individuals or society in general want now or in the future. Settlement and past use have sometimes resulted in the loss of productivity or undesirable plant community changes. Our values and needs also change over time. Catastrophic events such as fire, flood, prolonged drought, and others can often result in undesirable conditions even though they are quite natural phenomena. As assessment of goods, values, and uses expected from the land, however, should precede any determination of the capability of the soil to produce vegetation. [See RANGELAND CONDITION AND TREND.]

Once society's needs are identified, resource inventories provide information about the capability of the land to provide those values. Properly designed rangeland soil surveys not only provide an inventory of soil properties, but also provide a map base on which vegetation resource inventories and uses can be associated for interpretation. Assessments of condition for some values such as watershed quality or sustainable production are necessarily a function of both soil and vegetation attributes (as well as other factors) while others appear to be primarily a function of vegetation. Regardless of a present vegetation community's value for livestock forage, wildlife habitat characteristics, or other individual uses, the question remains whether there is a better alternative for use and management of the area.

The answer is at least partially dependent on the soil capability to produce different plant communities. Correlated soil surveys and vegetation inventories provide an observed range of properties associated within ecosystems perceived as possessing similar ecological potential. Desirable present vegetation characteristics and interrelated soil characteristics at one location can then be used to formulate desired characteristics in other areas using a known comparison.

Present and desired conditions for soils are often described using interrelationships between calculated soil factors such as erodibility (K factor), soil loss tolerance (T factor), and wind erodibility group (WEG) with vegetation cover and degree of surface disturbance. Models such as the Revised Universal Soil Loss Equation (RUSLE), Water Erosion Prediction Project (WEPP) model, and Wind Erosion Prediction (WEP) model are often helpful tools in establishing relationships between present and desired conditions. Hydrologic relationships between water table and annual pattern of soil water states must also be considered relative to frequency, extent and duration of flooding, channel characteristics, and vegetation in riparian–wetland areas.

B. Management Actions

Range management actions to achieve or maintain desired conditions nearly always include grazing prescriptions. Spatial distribution of soils and associated ecological sites presented on soil maps help determine the suitability of different practices within a management unit. Susceptibility of soils to compaction or erosion may limit seasonal use or stocking densities. Access to certain areas can be limited by seasonal saturation or flooding. Choice of grazing prescription may also be influenced by the physical and economic capability to implement vegetation treatments and support facilities. [*See* RANGELAND GRAZING.]

Mechanical and chemical treatments or fire control of unwanted or competitive plant species is often necessary to achieve desired plant community characteristics. Reseeding of desired species may also be required in conjunction with these practices.

Mechanical control of undesirable species consists of practices such as bulldozing, various kinds of plowing, chaining and cabling, and various cutting, mowing, beating, or shredding treatments. The methods used depend upon soils as well as the kind of vegetation to be controlled. Equipment limitations associated with soil topography, depth, amount and size of rock, and compaction must be considered in choosing the appropriate method. Erodibility must also be considered relative to soil disturbance and exposure.

Depending on application method, herbicidal plant control can reduce many equipment limitations and soil disturbance and exposure of bare soil is reduced or eliminated. Soil properties do affect application rates and effectiveness of many herbicides, however. Excessive losses of soil-applied herbicides can occur in some areas due to leaching or absorption on soil colloids. At the other extreme, low infiltration rates can inhibit penetration of herbicides into the soil where rainfall is marginal. Soil moisture and soil temperature affect translocation and effectiveness of foliar herbicides through effects on rate of plant growth. Rapid growth associated with favorable soil texture,

structure, and fertility can also increase herbicide effectiveness.

Soil moisture status and related fuel moisture are at times factors in predicting prescribed fire success. However, postfire effects on the soil should be a consideration. Fire can affect the soil through direct action of heat, removal of vegetation and litter cover, and changes in soil chemistry including redistribution of nutrients. Except where accelerated soil erosion is initiated as a result of reduced cover, the effects are apparent but short lived (a few months to a year or so). Soil water is usually decreased following fire. Decreased soil water content may persist for a while because of increased solar radiation and evapotranspiration. Infiltration may or may not be affected depending on surface soil characteristics and topography. Increases in surface pH, total phosphorous and nitrogen, exchangeable bases, and total soluble salts, and decreases in carbon to nitrogen ratio are often apparent following fire. Effects on plant growth are usually negligible 2 to 3 years after a fire.

Rangeland seeding suitability guides are available from agriculture and land management agencies or university extension offices in most if not all states with appreciable amounts of native range. Table II provides an example of a soil properties and ratings guide for rangeland seeding. Criteria may change slightly by state or region and additional criteria such as slope, erodibility, and others may be incorporated.

Standard soil interpretations in the Soil Conservation Service's State Soil Survey Area Database and published soil surveys provide considerable information regarding soil suitability and limitations associated with range management support facilities. Many interpretations are directly applicable such as for local roads, pond reservoir areas, and certain equipment. Others, such as for cattleguard installation and other kinds of facilities, can be readily interpolated from building site interpretations or determined directly from estimated soil properties.

C. Monitoring

Rangeland monitoring is performed to evaluate effects of management and determine adjustments that may be needed to achieve objectives. Monitoring objectives must relate to overall land use objectives, but are usually more site or activity specific. Some monitoring objectives pertain directly to effects on soils or soil attibutes such as erosion, compaction, or physical disturbance. More often rangeland monitoring focuses on vegetation. Soil information is an im-

TABLE II

Example Soil Suitability Rating Guide for Rangeland Seeding

Soil property	Good	Fair	Poor
Moisture regime (soil taxonomy)	Aquic, xeric, ustic, and xeric and ustic bordering on aridic or torric	Aridic and torric bordering on aquic, xeric, or ustic	Aridic and torric
Effective moisture[a]	>10 in.	7–10 in.	<7 in.
Available water capacity	Surface 10 in. >1.25 in. Profile >4 in.	Surface 10 in. 0.75–1.25 in. Profile 2.5–4 in.	Surface 10 in. <0.75 in. Profile <2 in.
Texture, surface 7 in.	LVFS, COSL, SL, FSL, VFSL, L SIL, SCL, and CL SICL with <35% C.	VFS, LFS, SC, SIC, C, and CL and SICL with >35% C.	LS, LCOS, FS, COS
Rock fragments, surface 7 in.	GR < 35%; CB < 15%; ST < 3%. Total rock fragments <35%	GR < 35%; CB 15–35%; ST 3–15%. Total rock fragments <35%	GR > 35%; CB > 35%; ST > 15%. Total rock fragments >35%
Depth, abrupt A–B text. boundary[b]	>10 in.	>10 in.	<10 in.
Depth, bedrock, or hardpan	>20 in.	10–20 in.	<10 in.
Elec. conductivity saturation-ext. −25°C.	<2 mmhos/cm in upper 20 in.	2–4 mmhos/cm, upper 10 in., and 4–8 mmhos/cm 10–20 in.	>4 mmhos/cm, upper 10 in. and/or >8 mmhos/cm 10–20 in.
Sodium adsorption-ratio	<8, upper 20 in.	8–13, upper 10 in. and <20 10–20 in.	>13, upper 10 in. and/or >20 10–20 in.

[a] Moisture from precipitation, run-on, and groundwater budgeted to actual evapotranspiration.
[b] Rate Vertisols and Vertic subgroups as poor.

portant consideration in locating, evaluating, and extrapolating vegetation monitoring activities and results.

Short-term monitoring includes utilization of forage plants relative to numbers, distribution, and period of use by grazing animals, livestock, and/or wildlife. Because plant growth is variable between years, evaluation of utilization data may include normalizing productivity over the period of observation. Models such as the Ekalaka Rangeland Hydrology and Yield model (ERHYM) and Simulation of Production and Utilization (SPUR) model assist in the evaluation process and require soil information for calculations. Utilization mapping may also be included. Soil survey maps and associated map unit descriptions and soil information such as slope breaks, surface cobbles, and other factors affecting animal distribution can be helpful to mappers in locating utilization class breaks.

Trend studies are longer-term monitoring to detect plant community changes over a period of years. Reinventory of vegetation resources provides the most accurate assessment of trend but is seldom feasible because of cost. In lieu of re-inventory, key areas are selected that represent the management unit or specific values within the unit. Tabulations of soils and associated ecological sites provide an indication of importance, either by their extent or relative value (i.e., productivity). Soil survey information also provides a basis for extrapolating information from a key area. Vegetation response can be expected to be similar throughout the distribution of a particular soil under similar use and management (barring other changes such as weather variation, etc.).

Bibliography

Branson, F. A., Gifford, G. F., Renard, K. G., and Hadley, R. F. (1981). "Rangeland Hydrology." Society for Range Management, Kendall/Hunt, Dubuque.

Jenny, H. (1983). The soil resource, origin and behavior. *Ecol. Studies* **37.**

Soil Survey Staff (1992). "Keys to Soil Taxonomy." SMSS Technical Monograph No. 19. USDA, Soil Conservation Service. Pocahontas Press, Blacksburg.

Tueller, P. T. (ed.) (1988). Vegetation science applications for rangeland analysis and management. *In* "Handbook of Vegetation Science," Vol. 14. Kluwer, Boston.

United States Department of Agriculture (1983). "National Soils Handbook." Soil Conservation Service, Washington DC.

United States Department of Agriculture (in press). "Soil Survey Manual." Soil Conservation Service, Washington DC.

Vallentine, J. F. (1980). "Range Development and Improvements." Brigham Young Univ. Press, Provo.

Rangeland Watershed Management

LINDA H. HARDESTY, *Washington State University*

HUGH BARRETT, *Bureau of Land Management*

Glossary

Ground water Water that exists below the level of the water table

Hydrologic cycle Process in which water circulates through the environment from the atmosphere to the earth's surface by precipitation, through the soil and biota, and back to the atmosphere by evaporation, transpiration, and sublimation

Infiltration Movement of water into the soil profile

Litter Recently dead plant material on the soil surface

Percolation Downward movement of water within the soil profile

Riparian area Ecotone between aquatic and terrestrial systems characterized by plants rooted in or below the water table

Runoff Volume of water discharged from a watershed

Soil moisture Water that exists in the soil above the level of the water table

Watershed (syn. catchment, hydrologic basin, drainage basin) Area of land, all of which drains to a single point

Water table Zone beneath the surface of most watersheds below which all soil pore spaces are completely filled with water

A watershed is an area of land that contributes to the flow of water at a given point. The size and boundaries of a watershed are established by the specific point used to define the watershed. If the point selected is at the mouth of the Columbia River, the watershed includes an area of nearly 260,000 square miles in Oregon, Washington, Idaho, Montana, Nevada, and British Columbia. Within the Columbia River watershed are many tributary streams, each supported by its own, smaller watershed. All lands exist within a watershed. Rangeland watersheds are generally characterized by arid and semi-arid climates. In many cases rangelands make up only a portion of a watershed that may also include forest, agricultural, and urban land.

I. Natural Watershed Functions

Rangelands occupy 40% of the earth's surface, giving them ecological and economic significance on a global scale. Rangeland watersheds contribute in many ways to ecosystem health and function and to the economic and social well-being of those communities within their embrace and downstream. Fully functioning watersheds can provide sustained supplies of water for plant growth, fish and wildlife habitat, domestic and municipal use, industrial and agricultural use, hydroelectric power generation, recreation, and esthetics. [*See* RANGELAND.]

The basic functions of any watershed relate to its role in the hydrologic cycle: the capture, storage, and release of water that falls in the watershed (Fig. 1). These functions, operating in concert, govern the kinds and amounts of plants that grow in the watershed, the diversity and health of wildlife populations, the quality of water and duration of streamflow the watershed produces, and the uses to which the lands are put. This article focuses on these functions and the factors which affect them, as these are the principles that underlie effective rangeland watershed management.

A. Capture Water

The initial step in the terrestrial phase of the water cycle is the capture of water. Moisture in the form

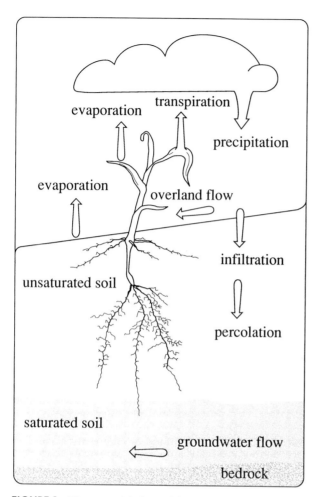

FIGURE 1 The terrestrial phase of the hydrologic cycle.

of rain, snow, hail, sleet, or fog can follow several possible pathways. When intercepted by vegetation, rocks, plant litter, or other materials, water may return to the atmosphere by evaporation and sublimation, or it may flow downward to the soil surface.

Moisture that reaches the soil surface is available for infiltration through the soil surface and into the soil profile. The rate of infiltration is influenced by a number of factors. Coarse textured, sandy surfaces can have very rapid infiltration rates, while finer materials like silt and clay have slower infiltration rates. Fine textured soils low in organic matter may lose their capacity to take in water after surface disturbance. The result, "puddling," occurs when aggregations of soil particles are broken down and redistributed as random plates that resist the passage of moisture. Plants and plant litter that absorb the shock of raindrop impact protect the soil surface from the dislocation of soil particles and maintain its ability to absorb water. Water runs rapidly off steep slopes,

limiting the time available for infiltration. Broken topography, microrelief, and litter detain overland flow, increasing opportunities for infiltration. [*See* SOIL DRAINAGE; SOIL–WATER RELATIONSHIPS.]

As infiltration commences, moisture begins to percolate through the soil profile. If the supply of moisture is adequate and its downward movement is not impeded, all the pores in the upper soil profile will become saturated. There will be continued downward movement as gravity drains the larger pore spaces. Eventually only the smaller micropores will contain water—a condition referred to as field capacity.

Often the movement of water in the soil is restricted by changes in soil texture or density at horizon boundaries. Dense clay underlying a loamy upper horizon will drastically slow downward water movement. Discontinuities in the profile such as compaction layers, mineral pans, or ice lenses have the same effect. These restrictions can result in percolation rates slower than infiltration rates, resulting in a modified terminal infiltration rate.

When the rate of precipitation or snowmelt exceeds the terminal infiltration rate, overland flow results. If overland flow is not detained, it can dislodge and transport soil particles and organic material on the soil surface. If this eroded material is transported into open water, it is termed sediment.

B. Store Water

Soil texture and depth delimit the volume of water that can be stored in the soil profile. Other factors affecting soil water storage are the kinds and amounts of plants that occupy a specific site—their rates of water uptake and transpiration, plants and litter that limit evaporation, and soil surface texture.

Not all of the moisture held within the root zone of plants is available for plant use. Available moisture refers to soil moisture that can be used by plants. Some water is held so tightly by soil particles that it is unavailable to plants.

Moisture in excess of field capacity percolates deeply into the soil profile and may eventually reach the water table to be stored as ground water. In some instances downward movement is stopped or slowed by less permeable layers, producing a perched water table, or inducing horizontal flow giving rise to seeps or springs. Deep percolation into subterranean reservoirs or aquifers can result in storage of large volumes of water for long periods of time. [*See* GROUND WATER.]

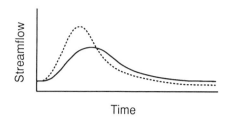

FIGURE 2 A comparison of hypothetical stream flow regimes from healthy and degraded watersheds, demonstrating the healthy watershed's more moderate flow. Solid line is streamflow curve of a healthy watershed; dashed line is streamflow curve of a deteriorated watershed. (Adapted, with permission, from Oregon Watershed Enhancement Coalition. "Watersheds, Their Importance and Functions." Department of Rangeland Resources, Oregon State University, Corvallis, OR.)

C. Release Water

Runoff is the water released from a watershed. Runoff may occur as overland flow, streamflow or underground water flow. The timing and duration of runoff and the quality of the water leaving the watershed depend on the capture and storage of precipitation. Ideally, infiltration rates are adequate to meet precipitation inputs and soil moisture supports plant cover sufficient to protect the soil surface, promote infiltration, and detain occasional overland flows. Riparian vegetation will flourish, protecting streambanks from erosion and, in all likelihood, both vertical and horizontal stream channel stability can be maintained. Corresponding stream hydrographs commonly portray long-term flows of moderate volume with few, if any, "spikes" of discharge (Fig. 2).

In contrast to this ideal situation are those cases where infiltration rates are inadequate—because of past erosion, lack of plant cover and organic matter, recent severe fire, soil restrictions, or any combination of these and other factors—to capture all precipitation on the site. Moisture lost at the soil surface is not available to recharge the soil. Overland flow occurs frequently and destructively. The high-flow/no-flow regime in the stream channel commonly results in a high degree of channel instability, stream entrenchment, and bank erosion.

II. Rangeland Watershed Characteristics

Although all watersheds share the functions of water capture, storage, and release, each is unique; its performance, products and management opportunities proscribed by interacting physical and biological characteristics.

A. Watershed Components

Watersheds generally consist of uplands, riparian zones, and the aquatic zone. Uplands form the largest spatial component of the watershed. Here is where most of the water received during precipitation events is captured and stored, or lost through interception, evaporation, overland flow, or transpiration. Many range watersheds are sufficiently arid that surface water does not occur, or occurs only intermittently.

When surface water does occur, the riparian zone is the transition between the terrestrial and aquatic ecosystems. Here, plants respond to increased subsurface moisture by producing distinctive vegetation often recognized from a distance as a green ribbon through the landscape. Floodplains are components of riparian zones associated with moderate and low gradient streams. These low-lying areas often have deep, productive alluvial soils supporting lush, diverse vegetation ranging from meadows of forbs, grasses, and grasslike plants to gallery forests (Fig. 3).

By virtue of their position in the landscape and the kinds of plants they support, riparian areas contribute in many ways to the function of the watershed and the quality of water issued by the watershed. Properly functioning riparian areas dissipate floodwater energy as high flows spread across accessible floodplains, reducing the concentration of energy in the stream channel. When floodplains are inundated during peak flows, their soils store water for later, gradual release. In an intact riparian area underlain by 20 feet of alluvium, an area one-quarter mile wide and a mile long is capable of storing up to 1500 acre feet of water. Riparian vegetation slows down and filters turbulent waters, removing much of the suspended sediment. Sediment deposition builds streambanks and narrows stream channels with the attendant benefits of increased water storage capacity and improved instream habitat. Riparian vegetation also shades the stream, cooling water during hot weather, and moderating extreme cold, reducing the build-up of destructive ice formations. Riparian vegetation is a foundation for both aquatic and terrestrial food chains.

While riparian zones constitute only 1–2% of the land in most rangeland watersheds, their vegetative structure and spatial distribution support diverse plant and animal life within the riparian area and in adjacent uplands. Uplands and associated riparian zones are functionally linked to the degree that they should be

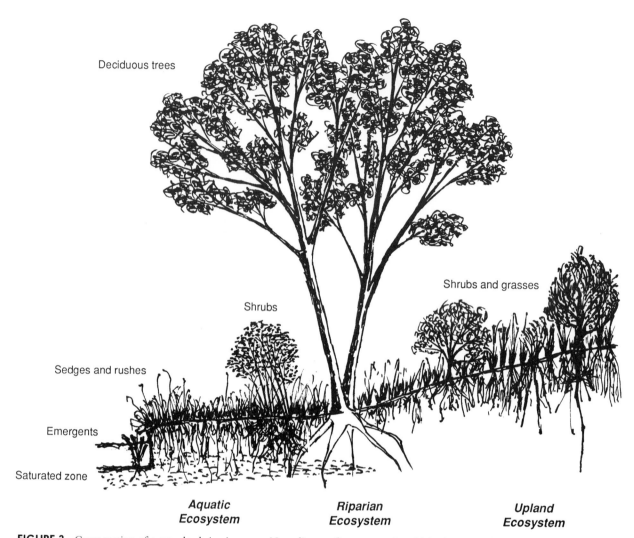

Deciduous trees

Shrubs and grasses

Shrubs

Sedges and rushes

Emergents

Saturated zone

Aquatic Ecosystem

Riparian Ecosystem

Upland Ecosystem

FIGURE 3 Cross-section of a rangeland riparian zone. Note distance from stream in which plant roots have access to saturated soil.

managed as a single system. The quality of management of the entire watershed and the integrity of its function are often reflected in the condition of the riparian zone. This system is influenced by physical and biological factors including climate, geology, topography, soils, vegetation, fauna, management history, and time.

B. Climate

Climate describes the long-term weather patterns that prevail in a specific region during which considerable year to year variation occurs. By dictating the availability, timing, and effectiveness of water in an ecosystem, climate defines the potential productivity of a watershed. Many rangelands exist in extreme climatic zones: alpine and arctic tundras, deserts, and semiarid shrublands.

Precipitation on rangelands is usually limited; hence, precipitation amounts and distribution over time are key determinants of watershed characteristics and function. Precipitation influences the rate of weathering of rock and of soil development. A watershed in a hot, arid region receiving 200 mm of rainfall annually is harsh and barren compared to a more temperate region receiving 500 mm. The source of storms determines the intensity of precipitation events. Intense tropical rains and convection storms produce large volumes of precipitation in short periods of time, often resulting in overland flow. Gentler storms allow more time for water to infiltrate the soil. Snow can be stored on site, gradually infiltrating as the soil thaws and snow melts. Rapid snowmelt or warm rain over a snow mantle, however, can lead to overland flow. In arid regions, much precipitation is lost as evaporation before it ever infiltrates the soil.

Effectiveness of precipitation refers to how much of the precipitation received is useful to plants.

Temperature influences watershed function by controlling the dominant form of precipitation received in the watershed (rain or snow), establishing runoff periods and the timing of moisture entry into the soil. In cold climates ice in the upper soil profile can restrict percolation such that snowmelt exceeds terminal infiltration rates. Ideally, plant litter helps insulate the soil, reducing the frequency, duration, and severity of winter freezing and the roots of vigorous plants help maintain soil permeability by breaking down restrictions in the soil profile.

The combined effects of temperature and precipitation lay the foundation for ecosystem characteristics. The cold wet winters and warm dry summers typical of much of western North America support cool season vegetation that requires stored winter moisture for its growth in spring and early summer. In regions where rainfall occurs during the warmer months, humidity determines how effective that moisture will be in supporting plants and higher life forms dependent upon them. More water evaporates and is transpired when humidity is low; hence, less water enters or remains in the soil to support plant growth and stream flow. Precipitation is often more effective in humid regions where evaporation and transpiration losses are reduced. Temperature affects humidity because as air cools, water condenses out of it.

Wind delivers storm systems, distributes snow upon the landscape, and can cause significant soil erosion. Wind is also an agent of desiccation on rangelands, drawing moisture from plants and soils into the atmosphere.

C. Geology/Topography

The nature of the geologic material and geologic processes in which it develops determine many of the features of a watershed. Dense crystalline material like granite weathers to coarse, infertile sands that exhibit rapid infiltration rates and low water-holding capacity. Rising granitic mountain ranges are characterized by steep droughty uplands and high gradient streams. The alluvial fans and valley bottoms derived from these materials also have rapid infiltration rates and low water-holding capacity. Igneous materials like basalt tend to weather to fine textured, nutrient-rich soils with moderate to slow infiltration rates and high water-holding capacity. Alluvial soils derived from igneous parent material exhibit these properties as well and are potentially very productive. Many rangeland soils developed from sediments of mixed

origin deposited in shallow inland seas and lakes. These fine silts and clays have slow infiltration rates and high water-holding capacity. Where these conditions occur in basins that lack external drainage, salts and alkali often accumulate in the soils.

The size, shape, steepness, and aspect (compass orientation) of the watershed determine how it intercepts storms. Watersheds on the lee side of mountain ranges are drier than their counterparts to windward which are able to milk storms of their moisture due to condensation as moist air rises and cools over the mountain crest. This orographic effect accounts for the aridity of the Great Basin of North America relative to conditions on the windward side of the Sierra Nevada Range. Both steepness and aspect affect the watershed's angle to the sun, influencing soil surface temperatures, evaporation rates, vegetation, and many other features. In the northern hemisphere, south slopes capture the greatest amounts of solar radiation and are generally warmer, drier, and more sparsely vegetated than cooler, moister, more shaded north slopes.

Natural, or geologic erosion by wind and water is part of the continuing process of landscape development. Most landscapes we see today are still undergoing geologic change, although the slow rate may lead to false assumptions of stability. Accelerated erosion, the result of human activity in the watershed, is an amplification of the natural process that, while sometimes difficult to distinguish from natural erosion, indicates the need to adjust management expectations to those the watershed can support over time.

Channel cutting results when large amounts of water flow at high speed into the channel, perhaps as a result of a 100-year precipitation event, vegetation removal upstream, or the compounding effects of both. Conversely, aggradation tells of diminishing energy and the reduced capacity of flowing water to carry suspended matter.

Water acts upon the land to sculpt it, but water's action is also influenced by existing topographic features. Water flows more rapidly on long, steep, uninterrupted slopes, building greater erosive power. Even small-scale topographic features such as a hoofprint or protruding pebble may divert overland flow into a growing channel or gully, or slow the water, allowing it drop suspended sediments. Interpretation of topographic features allows us to predict the behavior of water on a site.

D. Soils

An understanding of soil characteristics and their influence on watershed function is essential. Soil texture, the size of the mineral particles in the soil, is one determinant of how fast water can infiltrate and move through the soil profile and how much pore space and surface area exist to hold that water within the soil. Finer textured soils can hold relatively more water than coarser soils, but less of it may be available to plants (Fig. 4). The arrangement of these particles, their structure, determines how easily particles can be broken away and transported as sediment. Aggregates of small particles may act as larger particles, resulting in more stable soil. Decaying organic matter and soil-dwelling organisms produce compounds that bind soil particles, helping develop and maintain soil structure. The mineral content of the soil and the availability of those minerals affect the degree to which soil moisture is able to support plant growth. [*See* RANGELAND SOILS.]

E. Vegetation

Vegetation has both a physical and biological role in the watershed. A tree branch, whether it is dead or alive, may break the fall of a raindrop, reducing its erosive force, but only a live tree yields water from deep within the soil profile back into the atmosphere through transpiration. The influence of vegetation on a watershed is determined not only by the species present and their growth form and function, but their requirements for water, water use efficiency, nutrient needs, productivity, and potential human uses (Fig. 5). The vegetation currently present is of immediate importance, but so is knowledge of what has been there and what will occur in the future under different management scenarios. A perennial grass dominated range site that is being invaded by long-lived shrubs is one in which water relations, soil stability, and water yield may be expected to change. Knowledge of successional patterns and the influence of management upon them allows range scientists to determine the current and future health of an entire ecosystem and if the trend in its condition is towards or away from desired states. [*See* RANGELAND ECOLOGY; RANGELAND PLANTS; RANGELAND SHRUBS.]

Plant communities are rarely stable as they respond to variable environments and human activities. Most species and communities have a range of tolerance for environmental fluctuations, but beyond this threshold, dramatic changes may occur rapidly. Stream channels and riparian zones that take the brunt of the energy generated by intense storms and runoff events, and steep slopes subject to high gravitational forces are particularly unstable even under undisturbed conditions.

Plants shelter the soil surface from the force of falling rain, insulate it from solar radiation and desiccating wind, and reduce potential evaporation. Plant litter on the soil surface functions similarly and contributes humus that binds soil aggregates and maintains soil fertility. The roots of living and dead plants provide physical strength, holding the soil together. As roots decompose, they provide channels through the soil that enhance aeration and infiltration. Deep rooted plants take up nutrients in the soil profile and, through their incorporation into plant tissue that eventually is consumed or dies, make those nutrients available closer to the soil surface.

F. Fauna

All forms of life that exist in the watershed depend on it for sustenance and shelter, and in turn, contribute to the watershed's function and development. Microorganisms, insects, and burrowing rodents work within the soil converting organic matter into plant nutrients and organic compounds that aid in water storage, soil aggregation, and nutrient retention. Faunal movement in the soil creates pore spaces and cavities that enhance aeration and drainage. Rodent excavations break down restrictive layers in the soil and bring subsoil nutrients to the upper soil profile where they are available to plants.

When feeding, large herbivores break down plant material into nutrients available for future plant growth and distribute these nutrients across the landscape. Appropriate levels of herbivory stimulate some

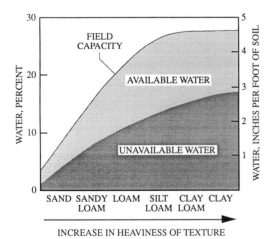

FIGURE 4 Curves representing the relationship between soil texture and soil moisture. [Reprinted, with permission, from Buckman, H. O., and Brady, N. C. (1960). "The Nature and Properties of Soils," 6th ed. Macmillan, New York.]

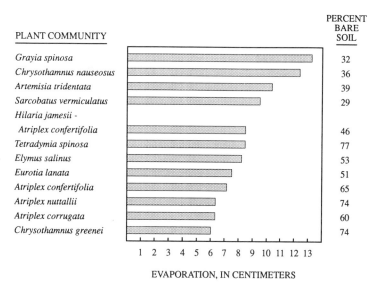

PLANT COMMUNITY — PERCENT BARE SOIL

Plant Community	Percent Bare Soil
Grayia spinosa	32
Chrysothamnus nauseosus	36
Artemisia tridentata	39
Sarcobatus vermiculatus	29
Hilaria jamesii -	
Atriplex confertifolia	46
Tetradymia spinosa	77
Elymus salinus	53
Eurotia lanata	51
Atriplex confertifolia	65
Atriplex nuttallii	74
Atriplex corrugata	60
Chrysothamnus greenei	74

EVAPORATION, IN CENTIMETERS

FIGURE 5 Relative amount of water evapotranspired and percent bare soil for 12 different plant communities. Reprinted, with permission, from F. A. Branson, R. F. Miller, and I. S. McQueen (1970). "Plant Communities and Associated Soil and Water Factors on Shale Derived Soils in Northeastern Montana," *Ecology*, **51**, 391–407. [Copyright © 1970 by Ecological Society of America. Reprinted by permission.]

plant species to sprout vigorously, enhancing their contribution to the mechanical stability of the watershed. Beaver (*Castor canadensis*) dams can reduce stream gradients, slowing water movement and encouraging sediment deposition and development of riparian zones. Riparian zones often support the majority of the wildlife species found within a watershed and constitute critical habitat for many of them. [*See* RANGELAND GRAZING.]

Increasingly, sensitive wildlife and insect species are being used to diagnose water quality and watershed conditions. Examples range from using changing aquatic invertebrate populations to detect water pollution, to viewing anadromous fish populations as indicators of conditions over huge areas such as the Columbia–Snake–Salmon river system in the northwestern United States.

G. Current and Historical Management

The effect of human activities in watersheds ranges from minor, reversible impacts to irretrievable devastation. Improper management of uplands or riparian zones can cause undesirable changes in the vegetation and bank structure of a stream as well as increase sedimentation and water temperature. Proper management can eventually return conditions to a desirable state if degradation is not too advanced. When substantial degradation has occurred, a return to pre-

disturbance conditions may not be feasible, although the watershed may be stabilized. This is the case in areas that have experienced desertification, a process of degradation by which the site becomes effectively much drier and less productive than would be expected given its climate and original soil and vegetation. [*See* RANGELAND MANAGEMENT AND PLANNING; WETLANDS AND RIPARIAN AREAS: ECONOMICS AND POLICY.]

Past management often constrains future options to the point where the potential of the watershed to perform each of its functions must be reassessed. In general, vegetation recovers more rapidly than soils from disturbance or damage. Activities with significant long-term impacts include surface and subsurface mining, heavy vehicle traffic, improper grazing, mismanaged fire, intensive recreational use, agricultural activity, timber harvest, brush control, and any other activities that significantly alter the soil or vegetation. [*See* RANGELAND CONDITION AND TREND.]

H. Time Scales

Activities within watersheds occur on multiple time scales. Geologic time measures the rise of mountains, the wearing away of uplands, the carving of great canyons, and the filling of valleys. Natural erosion of relatively level sites occurs at rates of about 50 mm in 1000 years, a rate not easily conceived of by humans. Yet a single event within these centuries, a 100-year's storm, may gully uplands, scour out stream channels, undercut banks, topple trees, and deposit thick layers of sediment in flood plains, all of which will be evident for centuries into the future. Occurrences on these time scales are beyond human control, but must be anticipated in order that extreme natural events do not become natural disasters.

Annual cycles are more recognizable. In mountainous regions, snow will accumulate over the winter, and the water will be released rapidly in annual high water events as the weather warms. Such events are sufficiently reliable to allow aging of lake sediments by counting their annual layers. Many watershed management activities aim at ameliorating the flood potential of peak flows and trapping water in the soil profile, or in natural or engineered bodies, for use during drier months.

There are other cycles. Prolonged droughts are particularly prevalent in the world's drier regions. When annual rainfall of less than 75% of the average occurs in a year, or a sequence of years, we must revise expectations formulated during years of average precipitation. Abandoned farms and villages testify to the human consequences of past drought cycles, particularly when combined with destructive management practices.

III. Managing Rangeland Watersheds

A. The Concept of Proper Management

Management is a human activity, motivated by a desire for resource characteristics that favor intended uses or products. Water, wood, and forage are some products of rangeland watersheds, as are wildlife habitat, recreation, solitude, and esthetics. Different actions are needed to realize each objective, or as is more often the case, to jointly produce multiple benefits and values. Insuring the long-term quality, productivity, and economic value of rangeland watersheds requires that watershed function be maintained, or where impaired, restored. Proper management is based upon sufficient knowledge of the various factors operating within the system to insure a reasonable chance of achieving management objectives within the limits of that system. The basic principle for managing all rangeland watersheds is encouraging the watershed's ability to capture, store, and release water.

B. Anticipate Drought

Rangelands are some of the world's most arid regions, and as a general rule, the less average annual precipitation a region receives, the less reliable that rainfall is

from year to year or within a season each year. Thus, arid ecosystems are subject to wide fluctuations in annual water flow and plant growth that must be considered in formulating management objectives and strategies. It may take decades for range vegetation to recover from misuse during periods of drought stress. Consequent soil loss may cause an irreversible decline in site productivity (Fig. 6). A wise course is to formulate management plans in anticipation of drought and to temper expectations to the lower ranges of production cycles. [See DESERTIFICATION OF DRYLANDS.]

C. Soil is the Bottom Line

Although soil is technically a renewable resource, the hundreds of years needed to replace lost soil make it nonrenewable within the human frame of reference. It is imperative to limit accelerated erosion to the greatest degree possible in order to maintain the watershed's ability to regulate its own functions.

Guidelines for protecting soil integrity include minimizing vehicle and animal traffic on wet, fragile, or steep soils, avoiding creating channels on slopes, and working within the limits imposed by soil texture, structure, and topographic characteristics. Maintaining adequate vegetative and litter cover to protect the soil surface from the forces of wind and water and enhance infiltration is critical (Fig. 7). The amount of

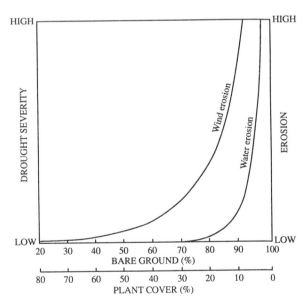

FIGURE 6 Relationships between drought, plant cover, and erosion by wind and water. [Reprinted, with permission, from Marshall, J. K. (1973). *In* "The Environmental, Economic, and Social Significance of Drought" (J. V. Lovett, ed.). Angus and Robertson.]

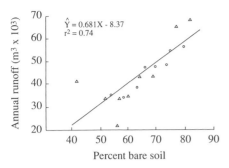

FIGURE 7 Regression between percent bare soil and average annual runoff from 17 watersheds in western Colorado. [Adapted, with permission, from Branson, F. A., and Owen, J. R. (1970). Plant cover, runoff, and sediment yield relationships on rangelands of the western United States. *Water Resources Res.*, **6,** 783–790. © American Geophysical Union.]

vegetation needed to insure these functions depends upon numerous site characteristics. Soil surveys prepared by the USDA Soil Conservation Service can be extremely useful in interpreting soil characteristics and formulating appropriate management strategies.

D. Vegetation—The Most Directly Managed Resource

The vegetation of a watershed is one of the attributes most directly subject to management. As a living resource, it offers more opportunities for management than climate, topography, and soils, that must generally be understood but can rarely be directly manipulated.

Volumes have been written on proper management of grazing including selection of the correct livestock species, number of animals, and season of use. This approach has limited application to management of wild herbivores whose impact may equal or exceed that of livestock. Riparian zones offering water, shade, and succulent forage are especially attractive to herbivores and often need particularly attentive management. Range developments such as watering facilities, fencing, and vegetation manipulation can be used to influence the distribution of both livestock and wildlife.

Vegetation management objectives should include maintaining sufficient amounts of vegetative cover to protect the soil surface and maximize infiltration, protecting the vigor and reproductive capacity of desirable plant species, and preventing encroachment of undesirable species that are often deleterious to watershed function.

E. Maintain Water Quantity and Quality

Water may be the highest value product of many range watersheds and will continue to increase in value to a growing human population. Management may be implemented to maximize streamflow in some cases and in others to reduce peak flows and diminish flood hazards. Vegetation cycles substantial water back into the atmosphere by evaporation from its surface and transpiration. When maximum water yield is desired, vegetation may be manipulated to favor water-efficient species and reduce vegetative cover to the minimum needed to optimize infiltration and provide soil stability.

Geochemical factors determine natural water quality in a watershed. Movement of water through vegetation and soil helps to remove some contaminants and improve water quality. As natural variability in water quality is better understood, changes in water quality can be used to evaluate management. When water is a primary product for domestic, agricultural, or industrial uses, quality standards may be well defined. Water quality standards for wildlife and, particularly, fish habitat are currently being identified. Standards can be established for water chemistry, temperature, sediment load, and microbiology, as well as for acceptable variation in these parameters. To meet these standards, sediment, pathogens, excess organic matter, and other pollutants should be prevented from entering open water or ground water. Modifications of riparian vegetation or channels that would cause increased temperature and decreased dissolved oxygen are also undesirable.

F. Restoring Watershed Functions

Proper management is not limited to maintaining watersheds, but often consists of attempts to stabilize or restore degraded watersheds. Over time, concepts of watershed management change; hence, what was "state of the science" at some time in the past may today be a site in need of restoration. An extreme example is current consideration of returning some reservoirs to free flowing streams in order to protect or restore indigenous fish populations.

Restoration measures such as placing structures in streams to stabilize banks or alter channels are often suggested, but experience has shown that better results are often achieved at lower cost by addressing improper land use in the uplands. Restoring natural watershed functions, allowing vegetation to regenerate, and allowing banks to stabilize as a result is a slower, but more reliable process in many cases.

The current and future value of rangeland watersheds are dependent upon the basic functions that integrate the many interacting factors that make each watershed unique in both its ecological characteristics and potential value to society.

Bibliography

Bedell, T. E. (ed.) (1991). "Watershed Management Guide for the Pacific Northwest," EM 8436. Oregon State Univ. Extension Service, Corvallis, OR.

Branson, F. A., Gifford, G. F., Renard, K. G., and Hadley, R. F. (1981). "Rangeland Hydrology," 2nd ed. Kendall/Hunt, Dubuque, IA.

Gresswell, R. E., Barton, B. A., and Kershner, J. L. (eds.) (1989). "Practical Approaches to Riparian Resource Management." U.S. Government Printing Office, Washington, DC.

Johnson, R. R., Ziebell, C. D., Patton, D. R., Ffolliot, P. F., and Hamre, R. H. (eds.) (1985). "Riparian Ecosystems and Their Management: Reconciling Conflicting Uses." USDA Forest Service, Fort Collins, CO.

Satterlund, D. R., and Adams, P. W. (1992). "Wildland Watershed Management," 2nd ed. Wiley, New York, NY.

Rice Genetics and Breeding

GURDEV S. KHUSH, *International Rice Research Institute, Manila, Philippines*

Glossary

Abiotic stress Yield reduction caused by abiotic factors such as drought or salinity
Alien addition lines Organism with one extra chromosome of a wild species
Embryo rescue Culturing of a hyrid embryo on artificial medium to save it from death due to lack of food
Genome Set of basic number of chromosomes
Harvest index Proportion of grain to total biomass
Hybrid rice Rice grown from first generation hybrid
RFLP markers DNA fragments separated with restriction enzymes
Trisomic Organism with one normal extra chromosome
Yield potential Yielding ability of a variety under optimum management

Rice is the world's single most important food crop and a primary food source for more than a third of the world's population. More than 90% of the world's rice is grown and consumed in Asia where about 60% of the earth's people live. Rice accounts for 35 to 60% of the calories consumed by 2.8 billion Asians. Rice is planted on about 146 million hectares annually, or 11% of the world's cultivated land. Wheat covers a slightly larger land area, but much of the wheat crop is fed to animals. Rice is the only major cereal crop that is consumed almost exclusively by humans. World's rice production was 523 million tons in 1991.

China, the largest producer, produced 188 million tons followed by India (110 million tons), Indonesia (47 million tons), Bangladesh (29 million tons), Thailand (20 million tons), and Vietnam (19 million tons). Only about 12 million tons or 3.5% of the world's rice production is traded internationally.

I. Introduction

Thailand is the world's leading rice exporter, selling about 4–6 million tons annually. The United States is the second-largest exporter, even though it ranks only 11th in production. It produces about 6 million tons of rice annually and exports about 40% of it. Pakistan and Vietnam each export about 1 million tons annually. Both Italy and Myanmar export about 0.5 million tons every year.

Iran, Iraq, and Saudi Arabia are the major importers, taking in about 0.9, 0.7, and 0.5 million tons per year, respectively. African countries, where demand for rice is increasing at the rate of about 2% annually buy around 3 million tons or about 25% of the total world imports each year.

The importance of rice in the diet varies among countries. It accounts for over 70% of the daily calorie intake in countries such as Bangladesh, Cambodia, Laos, and Myanmar but drops to 35% in countries such as China and India whose northern areas consume more wheat. Rice is also an important staple in Latin America, Africa, and most of the middle East. Health food advocates in Western countries pay a premium price for brown or unpolished rice, but rice is polished wherever it is a staple food. Why? Because its bran layers contain oils (free fatty acids) that turn rancid if the surface is scarred. Thus, brown rice cannot be stored for more than a few weeks. In fact, brown rice may be less nutritious than white rice, because the body's digestion and absorption of it is

lower. However, brown rice does have more B vitamins, and 1% more protein. The difference in fiber
and minerals is insignificant. Rice is 7–8% protein
(wheat is 11–12%). [*See* RICE PROCESSING AND
PRODUCTS.]

Rice is probably the world's most diverse crop. It
is grown as far north as Manchuria in China (50°N)
and as far South as Uruguay and New South Wales,
Australia (around 35°S). It grows at elevations of more
than 3000 m in Nepal and Bhutan and 3 m below sea
level in Kerala, India. Rice is cultivated under five
major ecosystems namely irrigated, rainfed lowland,
upland, deepwater, and tidal wetlands. About 80 million hectares or 55% of the world's rice land is irrigated and has adequate water supply throughout the
growing season. In much of this area, rainfall supplements irrigation water. Perhaps 75% of world rice
production comes from irrigated areas.

About one-fourth of the world's rice area is rainfed
lowland. Rice is grown in puddled and bunded fields
which are like irrigated areas, but the crop is dependent upon rainfall. Thus, there may be periods of
drought or excessive water during the growing season. Rice grown in rainfed, naturally well-drained
unbunded fields without surface water accumulation
is called upland rice and is planted to about 20 million
hectares annually. Largest areas of upland rice are in
Brazil, India, and Indonesia. Deepwater or floating
rices grow in 1–4 m of water in the deltas of mighty
rivers of Asia. About 7 million hectares of deepwater
rice is located in Vietnam, Cambodia, Thailand,
Burma, Bangladesh, and India. Tidal wetland rices
are planted to about 5 million hectares along sea coasts
and inland estuaries (Fig. 1).

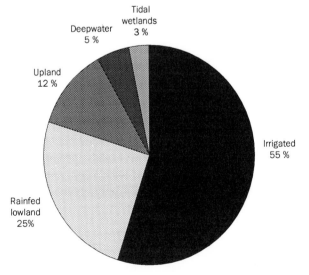

FIGURE 1   Distribution of world rice area in different ecologies.

II. The Origin and Diversity of Rice

Like wheat, corn, rye, oats, and barley, rice belongs
to Gramineae or the grass family. The genus *Oryza*
probably originated at least 130 million years ago and
spread as a wild grass in Gondwanaland, the super
continent that eventually broke and drifted apart to
become Asia, Africa, the Americas, Australia, and
Antarctica. Today's species of genus *Oryza* are distributed in all of these continents except Antarctica.
The cultivated rice *Oryza sativa* was perhaps domesticated about 10,000 years ago. The domestication may
have occurred independently at several locations in a
broad belt extending from the foothills of the Himala-
yas to Vietnam and Southern China. This dispersal,
along with farmer selection over several millennia,
has led to about 120,000 different varieties of *O. sativa*,
the Asian cultivated rice. In Western Africa, another
cultivated species of rice, *O. glaberrima*, is grown on
a small scale which was domesticated much later than
O. sativa. Besides these two cultivated species, about
20 wild species of *Oryza* are recognized (Table I).

Three morphologically distinct forms of Asian rice,
namely indica, japonica, and javanica, were thought
to exist. However, on the basis of recent investigations on their genetic affinity using isozymes, rice
varieties are classified into six groups. Group I varieties are the typical indicas with worldwide distribution. Group II comprises short duration, drought-
tolerant upland rices called Aus and distributed in
Indian subcontinent. Groups III and IV consist of
deepwater rices of India and Bangladesh. Aromatic
rices from Indian subcontinent such as Basmati 370
belong to Group V. Group VI comprises typical japonicas and javanicas. The former are now referred
to as temperate japonicas and the latter as tropical
japonicas. Crosses between varieties belonging to the
same group are fertile whereas crosses between varieties belonging to different groups show varying levels
of sterility.

III. Rice Genetics

Cultivated rice is a diploid with 24 or 12 pairs of
chromosomes. These chromosomes differ in size and
arm ratio. They have been numbered according to
the decreasing order of length at the pachytene stage
of sexual cell division. Thus, the longest chromosome
is number 1, second longest number 2, and the shortest number 12. Chromosomes of both the cultivated

TABLE I

Chromosome Numbers and Genomic Constitutions of Cultivated and Wild Species of *Oryza*

Species	2*n*	Genome	Distribution
O. sativa L.	24	AA	Worldwide
O. nivara Sharma et Shastry	24	AA	India
O. perennis Moench			
Asiatic (ssp. *rufipogon*)	24	AA	Asia
American (ssp. *cubensis* Ekman)	24	$A^{cu}A^{cu}$	West Indies
African (ssp. *barthii* A. Chev.)	24	A^bA^b	Africa
O. glaberrima Steud.	24	A^gA^g	Tropical West Africa
O. breviligulata A. Chev. et Roehr.	24	A^gA^g	Tropical West Africa to Sudan
O. longistaminata A. Chev. et Roehr.	24	A^gA^g	Tropical West Africa to Madagascar
O. punctata Kotschy ex Steud.	24, 48	BB, BBCC	Tropical northeast Africa, Madagascar
O. minuta J.S. Presi, ex C.B. Presi	48	BBCC	Malaysia, Philippines, Indonesia
O. Officinalis Wall. ex Watt	24	CC	India, Bangladesh, Burma, Thailand
O. eichengeri A. Peter	24, 48	CC, BBCC	Tanzania, Uganda
O. latifolia Desv.	48	CCDD	Central and South America, West Indies
O. alta Swallen	48	CCDD	Honduras, Brazil, Paraguay, South America
O. grandiglumis (Doell) Prod.	48	CCDD	Brazil, South America
O. australiensis Domin.	24	EE	Western and Northern Australia
O. brachyantha A. Chev. et Roehr.	24	FF	West tropical and Central Africa
O. schlechteri Pilger	—	Unknown	New Guinea
O. meyeriana (Zoll. et Mor. ex Steud.) Baili.	24	Unknown	Indonesia, Philippines, Thailand
O. granulata Nees et Arn. ex Watt	24	Unknown	India, Bangladesh, Sri Lanka, Burma, Thailand, Indonesia
O. ridleyi Hook. f.	48	Unknown	Malaysia, Thailand, New Guinea
O. longiglumis Jansen	48	Unknown	New Guinea

species and closely related wild species are similar and their genomes (a basic complement of 12 chromosomes) are designated as AA. The chromosomes of other wild species, however, differ from those of cultivated rice and they belong to genomes designated as BB, CC, DD, EE, and FF. A few of the tetraploid species have BBCC or CCDD genomes. Genomes of a few of the wild species have not as yet been designated (Table I).

A. Marker Genes

Nearly 400 mutant genes affecting morphological, physiological, biochemical, disease and insect reactions, abiotic stresses, and coloration of plant parts have been described. Many of these were identified as naturally occurring variations of spontaneous origin in varietal populations. Others were induced through mutagenic treatments. As a result of painstaking efforts of numerous rice geneticists over the last 75 years, inheritance of most of the marker genes has been studied. [See PLANT GENE MAPPING; PLANT GENETIC ENHANCEMENT.]

A vast majority of the mutants affect morphological traits. Among them 11 govern anthocyanin coloration, 8 are inhibitors of anthocyanin coloration, 12 affect hull and pericarp color, 28 for awn and spikelet traits, 10 for grain size and shape, 20 for endosperm traits, 22 for panicle characters, 9 for culm characters, 27 for leaf characters, 11 for duration to heading, 30 for reproductive traits, and as many as 52 have been reported for short stature. About 50 mutants for chlorophyll deficiency have been identified. These include albino, chlorina, fine striped, virescent, and zebras. Genes for physiological traits include those with brittle culms and spotted leaves.

Genes for disease and insect resistance are undoubtedly of greatest value for rice improvement. About 40 genes for resistance to rice diseases are known of which 18 are for resistance to bacterial blight, 15 for resistance to blast, 2 each for resistance to cercospora and stripe virus, and one each for resistance to grassy stunt, hoja blanca, and yellow dwarf virus. Many of these genes have been utilized for developing resistant varieties. Thirty genes for resistance to insects such as brown planthopper, green leafhopper, whitebacked planthopper, zigzag leafhopper, and gall midge are known and have been incorporated into improved varieties.

Isozymes are important biochemical markers and are useful in genetic and breeding research. Fifty isozyme loci have been identified through starch gel electrophoresis and 33 of them have been assigned to 9 of the 12 rice chromosomes. Most of the isozyme loci have multiple alleles.

Mutant genes have been assigned to different linkage groups through genetic analysis and 12 linkage groups corresponding to the 12 chromosomes have been established. These linkage groups were associated with respective chromosomes through the genetic analysis using primary trisomics.

Primary trisomics have been established in several varieties of rice and are being used to determine the chromosome locations of the newly discovered genes. Primary trisomics in the genetic background of IR36 developed at International Rice Research Institute (IRRI) have been most extensively used. Secondary trisomics of rice have also been produced and are being utilized for determining the arm location of the genes. A few monosomics of rice have been reported. However, the monosomic condition is not inherited to the next generation. Because rice is a diploid plant, the chromosomal deficiencies are tolerated at the diploid level only and the gametes with chromosomal deficiencies are inviable.

Alien addition lines having full chromosome complement of rice and an alien chromosome of a wild species have been produced. The alien chromosomes have been introduced from *O. officinalis*, *O. australiensis*, *O. latifolia*, and *O. brachyantha*. The alien addition lines closely resemble the primary trisomics, suggesting similar gene contents of the homoeologous chromosomes.

B. Molecular Markers and Gene Tagging

The morphological markers, although useful in genetic studies, are of limited value in rice improvement. The second category of markers are isozymes, which are more useful than the morphological markers. They are codominant and thus all genotypes can be distinguished in segregating populations. They have no deleterious effect on plant phenotype and large number of samples can be scored in the laboratory at very early stages of growth. However, only limited number of isozymes are known and the number is not enough to saturate the genetic maps. The third category of markers is the restriction fragment length polymorphism or the RFLP markers. These molecular markers have all the advantages of isozyme mark-

ers but are numerous in number and saturated maps can be prepared. The first RFLP map of rice consisting of 250 markers was published in 1988. Since then additional markers have been added by various workers and the RFLP map now consists of more than 1000 markers. These markers are now being used to tag genes of economic importance such as those for resistance to diseases and insects. Some of the genes for resistance to blast, bacterial blight, hoja blanca, and whitebacked planthopper have already been tagged.

IV. Rice Breeding

In the late 1950s and early 1960s, populations in Asian countries were increasing faster than food production. Because of these trends, several authorities such as Paddock and Paddock brothers and Borgstrom predicted large-scale famines in the late 1970s. However, recent advances in rice breeding have produced varieties with a yield potential two to three times higher than that of traditional varieties. These improved varieties respond better to modern agronomic practices. Wide-scale adoption of these varieties and proper agronomic practices have led to major increases in rice production and launched the green revolution in rice farming. [*See* CULTIVAR DEVELOPMENT.]

A. Role of IRRI

The International Rice Research Institute (IRRI) was established in 1960 by the Rockefeller and Ford Foundation in cooperation with the Government of the Philippines to develop technologies for increasing rice production. A group of relatively young and well-trained scientists was assembled at the new center, with well-equipped laboratories and a well laid out experimental farm. The staff, charged with the mission of increasing the productivity of rice farms, soon realized that the tropical rice plant had to be improved to make it more responsive to inputs. The plant breeding efforts at IRRI and elsewhere have resulted in a series of high yielding Green Revolution rice varieties which are now planted to 70% of the world's riceland and have led to major increases in rice production. The following improvements were made in the Green Revolution varieties.

1. Yield Potential
Pre-Green Revolution varieties were tall and leafy with weak stems and had a harvest index (ratio of dry grain weight to total dry matter) of 0.3. When

FIGURE 2 Different plant types of rice. Left; tall conventional plant type. Center; improved high-yielding, high-tillering plant type. Right; proposed low-tillering ideotype with higher yield potential.

nitrogenous fertilizers were applied at rates exceeding 30–40 kg/hectare, these varieties tillered profusely, grew excessively, lodged early, and yielded less than they would with lower fertilizer inputs. To increase the yield potential of tropical rice, it was necessary to improve the harvest index and increase nitrogen responsiveness by increasing lodging resistance. This was accomplished by reducing plant height through the incorporation of a recessive gene for short stature from a Chinese variety, Dee-geo-woo-gen. The first short statured variety, IR8, developed at IRRI also had a combination of other desirable features such as profuse tillering, dark green and erect leaves, and sturdy stems (Fig. 2). It responded to fertilizer inputs much better than traditional varieties such as Peta. It had a harvest index of 0.5 and double the yield potential of traditional rice. Being photoperiod insensitive it could be planted at any time of the year in the tropics.

The value of the IR8 plant type in raising the yield potential was so convincing that rice breeders the world over initiated hybridization programs to develop short statured varieties. Since then more than 800 short statured varieties have been developed by IRRI and national rice improvement programs and are now planted to most of the irrigated and favorable rainfed lowlands.

2. Shorter Growth Duration

Most of the traditional rice varieties in tropical and subtropical countries matured in 160–170 days and many were photoperiod sensitive. These were suitable for producing only one crop a year during the rainy season and were not suitable for multiple cropping systems. IR8 and subsequent varieties mature in about 130 days. However, if the farmers plant these varieties, it is not possible to grow a second crop of rice or another crop after rice in one rainy season. Therefore, varieties with even shorter growth duration were developed. For example, IR28, IR30, and IR36 mature in 110 days and IR50 and IR58 are even earlier. These earlier maturing varieties have the same yield potential as medium duration varieties. They were selected for faster growth rates at earlier stages and are able to produce approximately the same total biomass in 85–90 days as medium duration varieties do in 110–115 days. Because the short duration varieties produce the same amount of grain in fewer days than medium duration varieties, their per day productivity is much higher.

Short duration varieties are excellent for input economy. Because they grow rapidly during the vegetative period and are thus more competitive with weeds, weed control costs are reduced. They utilize less irrigation water. Short duration varieties developed at

IRRI and by National Rice Improvement programs are now widely grown in Asia. IR64 and IR72 are the most popular. Former is now planted to about 8 million hectares of rice land and is probably the most widely grown variety of rice in the world today. The availability of short duration varieties has resulted in major changes in cropping patterns in Asia and area under double cropping has rapidly increased. These changing cropping patterns have resulted in increased food supplies, greater food security, more opportunities for on-farm employment, and higher income for Asian farmers.

3. Multiple Disease and Insect Resistance

Rice is a host of numerous diseases and insects. Five diseases—blast, bacterial blight, sheath blight, tungro, and grassy stunt—and four insects—brown planthopper, green leafhopper, stem borers, and gall midge—are of common occurrence in most countries of tropical and subtropical Asia. The improved varieties are grown with better management and higher inputs such as fertilizers. These practices also favor the build up of diseases and insects. Therefore, incidence of diseases and insects increased after the introduction of improved varieties. Chemical control of diseases and insects for prolonged periods in a tropical climate is very expensive and impractical. The use of host resistance for disease and insect control is the logical approach to overcome these production constraints. Therefore, IRRI breeding program has placed major emphasis on developing germplasm with multiple resistance to major diseases and insects. Many national programs have similarly given priority to developing varieties with multiple resistance to diseases and insects.

Fortunately, large germplasm collections are maintained by IRRI and national rice improvement programs. These germplasm collections have been screened for disease and insect resistance and donors for resistance have been identified. These donors have been utilized in the hybridization programs and varieties with multiple resistance to as many as four diseases and four insects have been developed (Table II). These varieties can be grown without any insecticide and fungicide use and thus affect enormous savings to rice growers. Moreover, they have greater yield stability. As Fig. 3 shows, the yield of susceptible IR8 fluctuates from year to year. If there is a disease and/or an insect attack, the yield is drastically reduced. However, if the disease or insect incidence is low, the yield is high. On the other hand, varieties with multiple resistance such as IR36 and IR42 show only minor fluctuations in yield from year to year and thus have greater yield stability. [See PLANT GENETIC RESOURCES; PLANT GENETIC RESOURCE CONSERVATION AND UTILIZATION.]

4. Improved Grain Quality

Grain quality preferences vary from region to region. Most consumers in tropics and subtropics prefer long or medium long and slender translucent grains. However, in temperate areas preference is for short, bold, and roundish grains. Cooking quality is largely determined by amylose content (a component of starch) and gelatinization temperature of rice starch. In tropics and subtropics, varieties with intermediate amylose content (20–23%) and intermediate gelatinization temperature are preferred. In temperate areas of China, Korea, and Japan, however, consumers prefer rices with low amylose content and gelatinization temperature.

Earlier improved varieties such as IR5 and IR8 had poor grain quality. They cook dry because of high amylose content and have bold chalky grains which frequently break during milling. However, later varieties were selected for slender and translucent grains and high milling recovery. Improvements in cooking quality were only slowly achieved because all the donors of disease and insect resistance used in the hybridization programs had high amylose content and low gelatinization temperature. IR64 was the first improved variety with a desirable combination of long slender and translucent grains with intermediate amylose content and intermediate gelatinization temperature. It has been widely adopted in Asia because of its superior grain quality and high yield potential. Aroma is another component of superior grain quality but very few high-yielding varieties with aroma have been developed.

5. Tolerance to Abiotic Stresses

In many rice lands, crops suffer from nutritional deficiencies and toxicities. Large tracts of land suitable for growing rice remain unplanted because soils have high salinity or acidity. Efforts have been made to incorporate moderate to high level of tolerance for several nutritional deficiencies and toxicities. IR36 for example has tolerance to salinity, alkalinity, peatiness and iron and boron toxicities. It is also tolerant of zinc deficiency. Similarly, IR42 and more recent varieties such as IR64 and IR72 have a broad spectrum of tolerance for many problem soils. Varieties tolerant of these deficiencies and toxicities have a more stable

TABLE II

Disease and Insect Reactions to IR Varieties of Rice

Variety	Disease or insect reactions[a]							
	Blast	Bacterial blight	Tungro	Grassy stunt	Green leaf hopper	Brown plant hopper	Stem borer	Gall midge
IR5	MR	S	S	S	R	S	MR	S
IR8	S	S	S	S	R	S	S	S
IR20	MR	R	S	S	R	S	MR	S
IR22	S	R	S	S	S	S	S	S
IR24	S	S	S	S	R	S	S	S
IR26	MR	R	MR	S	R	R	MR	S
IR28	R	R	R	R	R	R	MR	S
IR32	MR	R	R	R	R	R	MR	R
IR36	MR	R	R	R	R	R	MR	R
IR38	MR	R	R	R	R	R	MR	R
IR42	MR.	R	R	R	R	R	MR	R
IR46	MR	R	R	R	R	R	MR	R
IR50	S	R	R	R	R	R	S	—
IR54	MR	R	R	R	R	R	MR	—
IR58	MR	R	R	R	R	R	S	—
IR60	MR	R	R	R	R	R	MR	—
IR62	MR	R	R	R	R	R	MR	—
IR64	MR	R	R	R	R	R	MR	—
IR66	MR	R	R	R	R	R	MR	—
IR68	MR	R	R	R	R	R	MR	—
IR72	MR	R	R	R	R	R	MR	—
IR74	MR	MS	R	R	R	R	MR	—

[a] R, resistant; MR, moderately resistant; MS, moderately susceptible; S, susceptible; —, not known.

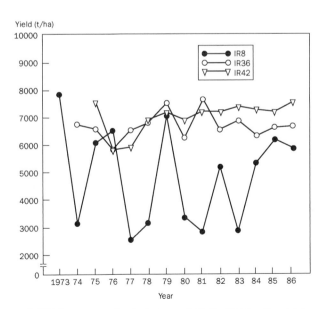

FIGURE 3 Yields of IR8, IR36, and IR42. Yields of multiple resistant IR36 and IR42 show little year-to-year variation; yield of susceptible IR8 fluctuates widely. Dry season replicated yield trials at IRRI.

yield performance and do well across several locations.

6. Combining Desirable Traits

A popular variety must have a favorable combination of adaptive traits such as high yield potential, appropriate growth duration, multiple resistance to diseases and insects, and superior grain quality. Many such varieties have been developed at IRRI and by national rice improvement programs. IR36 released in 1976 was the first such variety. It has a harvest index of 0.55 and excellent yield potential. It matures in 110 days and has excellent long slender grains, multiple resistance to major diseases and insects and tolerance to problem soils. Because of a combination of desirable attributes, it was accepted widely and became the most widely planted variety of rice or any other crop the world has ever known. In the early 1980s it was planted on 11 million hectares worldwide. Later releases such as IR64 and IR72 have all the desirable attributes of IR36 but outyield it by 20%. In addition, IR64 released in 1985, has superior grain quality and has replaced IR36 in most areas.

B. Impact of Modern Rice Breeding

About 800 improved rice varieties developed at IRRI and by the national rice improvement programs are now planted to about 65% of the world's riceland. In China, Korea, Philippines, and Sri Lanka, more than 90% of the rice area is now planted to improved varieties (Table III). In India, Indonesia, Pakistan, Burma, Malaysia and Vietnam more than 60% of the area is planted to such varieties. Because of large-scale adoption of improved varieties and associated management practices, rice production has dramatically increased in most of the major rice growing countries. The Asian rice belt, which used to be a rice deficit area, now has more than 8 million tons of exportable surplus. Indonesia which used to import up to 2 million tons of rice, became self-sufficient in 1984 and had an exportable surplus in 1985 and 1986. Indonesian rice production increased from 12.9 million tons in 1965 to 47 million tons in 1991. Average rice yield in Indonesia increased from 1.76 tons in 1965 to 4.2 tons in 1990. In India, rice production increased from 45.9 million tons in 1965 to 110 million tons in 1991.

Worldwide rice production doubled in a 25-year period from 257 million tons in 1965 to 523 million tons in 1990. Due to similar increases in wheat production, there are almost 400 million tons of food grain reserves worldwide. Instead of the food scarcity and famines forecast by Paddock brothers, there are problems associated with overproduction such as storage of surplus grain in several countries such as Indonesia and Vietnam. In addition to food security,

increased rice production resulted in a substantial decline in rice prices, particularly from the late 1970 to the present. Lower real prices clearly benefit the poor segments of society who spend 60–70% of their incomes on food.

V. Future Challenges in Rice Breeding

The population of rice eaters continues to increase at a frightening rate. Each year there are 50 million more rice eaters to feed. Although the rate of annual population increase worldwide has fallen from 2.0 to 1.7%, the population in Asia, where 90% of the rice is grown and consumed, continues to increase at 2.0% annually. It is estimated that 370 million tons of additional rice will be needed to feed the rice consumers in the year 2020. During the 1960s and 1970s, increases in rice production were due to increase in area planted to rice and increased productivity. There has been no increase in the area planted to rice since 1980. Further increases in area planted to rice are unlikely as there is very little land left unplanted which is suitable for growing rice. Thus, the increases in production of this staple will have to come from the increased productivity of the existing land.

As mentioned earlier most of the favorable rice lands (irrigated and favorable rainfed lowlands) are now planted to improved varieties. However, improved varieties for the unfavorable environments (rainfed lowlands, uplands, deepwater, and tidal wetlands) are still not available and there has been very

TABLE III

Total Area Planted to Rice and to Improved Varieties and Rice Production Increases in Selected Asian Countries, 1990

Country	Total area (million ha)	Area planted to improved varieties (%)	Rice production (million tons)		% increase in 1990 over 1965
			1965	1990	
Bangladesh	10.6	35	15.7	29.4	187.2
Burma	4.6	53	8.0	14.5	181.2
China	32.9	98	92.0	188.3	204.6
India	41.8	60	45.9	109.5	238.5
Indonesia	10.0	85	12.9	46.8	339.5
Korea	1.2	95	4.8	8.1	168.7
Malaysia	0.6	70	1.2	1.8	150.0
Nepal	1.4	35	2.2	3.5	159.0
Pakistan	2.1	51	1.9	5.2	273.6
Philippines	3.4	90	4.0	9.6	240.0
Thailand	10.2	20	11.1	20.4	183.0
Vietnam	5.9	60	9.8	19.1	194.0
Sri Lanka	0.7	91	0.8	2.3	287.5

little improvement in rice yields under unfavorable environments. The challenges of rice improvement therefore are twofold.

1. Develop varieties for the favorable environments with higher yield, greater yield stability, and superior grain quality.
2. Develop improved varieties with higher yield potential for the unfavorable environments.

A. Increasing the Yield Potential of Irrigated Rice

Since the initial breakthrough in yield potential of irrigated rice when IR8 was developed, there has been only a marginal increase in the yield potential of rice, although per day productivity of the short duration varieties is much higher. Therefore, the major objective of rice improvement now is to develop rice varieties with a quantum jump in yield over the existing varieties. Yield is a function of total dry matter and harvest index. Therefore, yield can be increased by increasing either the total dry matter or harvest index or both. Efforts are now under way to improve the harvest index from 0.5 to 0.6. For this purpose, architecture of the rice plant has been redesigned (Fig. 2).

Existing high-yielding varieties have 20–25 tillers of which only 14–15 produce small panicles (with 100–120 grains) and rest of the tillers remain unproductive. A plant with reduced tiller number but with all productive tillers and with large panicles (with 200–250 grains) should have higher harvest index. Other desired features of such a plant are: short stature (80–90 cm), sturdy stems, dark green erect and thick leaves, and vigorous root system. Such plants have already been developed and are being evaluated for harvest index and yield potential. The target is to raise the yield potential of rice from 10 tons per hectare to 12.0 tons in dry season and from 6 tons to 7 tons per hectare in wet season.

The second approach for increasing the yield potential of rice is through hybrid breeding. The first generation offspring of a cross between genetically different parents of a species are called hybrids. Chinese rice scientists were the first to demonstrate the successful development and use of hybrid rices in the mid-1970s after 12 years of intensive research. These hybrids give about 15% more yield than the best improved rice varieties. To produce commercial hybrid rice crop, farmers need to purchase fresh seeds every season. Specific genetic tools such as cytoplasmic male sterility are required to produce hybrid seeds. Such

tools have been identified and hybrid rice seed production practices have been developed. Currently about 50% of the ricelands in China are planted to hybrid rice. However, these hybrids are not adapted to tropical environments. Hybrid rices developed at IRRI are suitable for tropics and are being evaluated in several countries. Several national rice improvement programs are also developing hybrid rices. Hybrid rices when widely adopted should help increase the yield potential by about 15%.

B. Increasing the Yield Stability

Resistant varieties when widely grown for several years become susceptible due to the development of new races or biotypes. Therefore, new genes for resistance must be identified and incorporated into improved varieties. At IRRI, there is a strong emphasis to genetically analyze donors for major diseases and insects to identify new genes for resistance and as mentioned in an earlier section, several genes have been identified for resistance to blast, bacterial blight, brown planthopper, and green leafhopper.

Genes for disease and insect resistance are also being transferred from wild species to cultivated rice. Wild species are highly resistant to diseases and insects. However, they are difficult to cross with cultivated rice as the hybrid embryos abort a couple of weeks after fertilization. If the embryos are rescued and grown on culture medium, F_1 hybrids can be obtained. Using this embryo rescue technique, hybrids between 10 wild species and cultivated rice have been obtained and useful genes for disease and insect resistance have been transferred to cultivated germplasm. Three varieties with genes for resistance to brown planthopper from wild species, *O. officinalis,* have been released in Vietnam. Alien genes for resistance are also being transferred to cultivated rice through genetic engineering. For example, Japanese scientists have transferred *Bt* gene from the bacterium, *Bacillus thuringiensis* into rice and the transgenic plants are resistant to stem borer. Other alien genes for resistance such as protease inhibitors are being transferred to rice germplasm at IRRI. Such genetic diversity when incorporated into improved varieties should lead to more durable resistance.

C. Developing Germplasm with Superior Grain Quality

Aromatic rices command higher prices in the domestic and international markets. However, few improved rices with aroma have been developed. Basmati rices from India and Pakistan with strong aroma have been utilized in the hybridization programs but have proved to be poor combiners. However, through backcrossing of derived lines, improved germplasm with aroma has been developed. Several other aromatic donors such as Khao Dawk Mali 105 from Thailand are better combiners and have been extensively used in the hybridization programs. Improved germplasm with aroma is now being evaluated in several countries.

D. Developing Improved Germplasm for Unfavorable Environments

Unfavorable environments are characterized by variable water regimes, occurrence of floods, drought, water logging, and soil toxicities and deficiencies. Progress in developing improved varieties for these environments has been extremely slow and farmers in these areas continue to grow traditional varieties which have low yields but adaptability traits which impart yield stability at low yield levels. Breeding strategies aim at retaining these adaptability traits but improving the plant architecture to make them more productive. Such plant types have been defined for each of the unfavorable ecologies which are likely to have higher yield potential. Efforts to develop germplasm with these traits are under way.

One of the reasons for slow progress in developing improved germplasm for the unfavorable environments is the lack of appropriate sites for evaluating the segregating breeding materials. Research centers and experiment stations are generally located in favorable environments with good water control, and breeding materials for unfavorable environments cannot be exposed to appropriate selection pressures if grown on experiment stations. In recent years appropriate sites in target environments have been identified

and breeding materials when grown at these locations are exposed to appropriate stresses. This approach should help breed materials with improved and stable yields for unfavorable environments.

Bibliography

David, C. C. (1991). The world rice economy: Challenges ahead. *In* "Rice Biotechnology." (G. S. Khush and G. H. Toenniessen eds.). pp. 1–18. C. A. B. International Wallingford, England, and International Rice Research Institute, P.O. Box 933, Manila, Philippines.

Glaszman, J. C. (1987). Isozymes and classification of Asian rice varieties. *Theor. Appl. Genet.* **74,** 21–30.

Jena, K. K., and Khush, G. S. (1990). Introgression of genes from *Orzya officinalis* Well ex Watt to cultivated rice, *O. sativa* L. *Theor. Appl. Genet.* **80,** 737–745.

Khush, G. S. (1989). Multiple disease and insect resistance for increased yield stability of rice. *In* "Progress in Irrigated Rice Research." pp. 79–92. International Rice Research Institute, P.O. Box 933, Manila, Philippines.

Khush, G. S. (1990). Rice breeding—accomplishments and challenges. *Plant Breeding Abstracts* **60,** 461–567.

Khush, G. S. and Kinoshita, T. (1991). Rice karyotype, marker genes and linkage groups. *In* "Rice Biotechnology." (G. S. Khush and G. H. Toenniessen, eds.). pp. 83–108. C. A. B. International Wallingford, England and International Rice Research Institute, P.O. Box 933, Manila, Philippines.

McCouch, S. R., Kochert, G., Yu, Z. H., Wang, Z. Y., Khush, G. S., Coffman, W. R., and Tanksley, S. D. (1988). Molecular mapping of rice chromosomes. *Theor. Appl. Genet.* **76,** 815–829.

Morishima, H. (1984). Wild plants and domestication. *In* "Biology of Rice." (S. Tsunoda and N. Takahashi eds.). pp. 3–30. Japan Scientific Societies Press Tokyo, Elsevier, Amsterdam.

Oka, H. I. (1988). Origin of cultivated rice. Japan Scientific Societies Press Tokyo and Elsevier, Amsterdam.

Rice Processing and Utilization

BOR S. LUH, *University of California, Davis*

Glossary

Amylograph Instrument used by the baking industry and research laboratories to determine the starch characteristics of cereals important to baking applications; amylograph pasting curves of 10% flour slurries (50 g rice flour and 450 ml water) give useful information on rice starch characteristics important in baking applications; tests on 20% flour slurries (100 g rice flour and 400 ml water) can give estimates of initial gelatinization temperature of rice starch within ±0.5°C when gelatinization temperature is taken as the point of initial increase in viscosity; these values correlate significantly ($r = 0.94$) with BEPT values determined microscopically

BEPT Birefringence end-point temperature (BEPT) of rice flour can be determined on a 1% slurry using a polarizing microscope equipped with a Kofler hot stage and set at a heating rate of 5°C/min; BEPT is usually recorded at 95–98% loss of birefringence

Freeze-drying Freeze-drying process consists of removing moisture from foods at the frozen state by sublimation under high vacuum; the low temperature used in the process inhibits undesirable chemical and biochemical reactions and minimizes the loss of volatile aromatic compounds; the dried product can be stored in air-tight containers for long periods without refrigeration; for institutional feeding and for markets where low-temperature facilities are absent, freeze-dried foods are even more desirable than frozen and dehydrated foods; because of the high vacuum and low temperature requirements, freeze-drying is an expensive method of dehydration; but, for high-valued items, freeze-drying is a desirable process, yielding high-quality products

Parboiled rice Made by a hydrothermal process in which the crystalline form of starch present in the paddy rice (the rice grain from the field) is changed into an amorphous one as a result of the irreversible swelling and fusion of starch; this is accomplished by soaking, steaming, drying, and milling the rice; the parboiling process produces physical, chemical, and organoleptic modifications in the rice, with economic and nutritional advantages

Retort pouch Flexible, heat sealable, flat container capable of withstanding the high temperature (121°C) required for pressure processing rice and other low-acid foods; the most commonly used retort pouch is a three-ply laminate of polyester, aluminum foil, and polypropylene

Starch Starch is a homopolymer of α-D-glucopyranoside of two distinct types. The linear polysaccharide, *amylose,* has a degree of polymerization on the order of several hundred glucose residues connected by α-D-(1,4)glucoside linkages. *Amylopectin,* a branched polymer has a degree of polymerization on the order of several hundred thousand glucose residues; the segments between the branched points average about 25 glucose residues linked by α-D-(1,4)glucoside bonds, while the branched points are linked by α-D-(1,6) bonds

Tempering After drying, the parboiled paddy rice must be allowed to rest for a time before milling; this time interval is called the tempering period; tempering period of about 48 hr is needed for the product to dissipate the heat it received during drying. Also, the moisture content inside each grain must become uniform throughout

Viscosity Property of fluids which appears as a dissipative resistance to flow; fluid subject to external forces can be in static equilibrium only if the forces are derivable from a potential function and the usual result is steady flow resisted by viscous stresses set up by distortion of fluid elements; the mechanisms and nature of the viscous effect may be very different in gases and in liquids

Waxy-rice flour Waxy-rice flour, or sweet-rice flour, contains less than 2% amylose in the starch and an appreciable amount of the amylase enzyme; this flour has a lower peak viscosity than some of the short-grain rices, probably because of its amylolytic activity, and it has practically no setback viscosity. Waxy-rice flour is more resistant to liquid seperation (syneresis) during freezing and thawing treatments

Rice is considered a semiaquatic annual plant, although it could survice as a perennial in the tropics or subtropics. Cultivars of the two cultivated species, *Oryza sativa L.* and *O. glaberrima* Steud., can grow in a wide range of water–soil regimes, from a prolonged period of flooding in deep water to dry land on hilly slopes. Today rice is grown in more than 100 countries, extending from 53°N latitude to 40°S and from sea level to an altitude over 3000 m. *Oryza glaberrima* is grown only in Africa on a limited scale. The production practices for rice in various countries vary from extremely primitive to highly mechanized operations. Harvesting, drying, milling, and packaging are important procedures for successful handling of rice grains. Due to limitation of space we will not discuss them here. The readers may refer to Luh, B. S. (1991), "Rice Production," Vol I, 2nd ed. Van Nostrand Reinhold, New York, NY. for detailed information.

I. Introduction

A. Importance of Rice as Food

Rice is one of the world's most important cereals for human consumption. In the densely populated countries of Asia, especially Bangladesh, China, India, Indonesia, Iran, Japan, Korea, Pakistan, and Sri Lanka, rice is the most important staple food. As much as 80% of the daily caloric intake of people in these Asiatic countries is derived from rice. [*See* RICE GENETICS AND BREEDING.]

Rice is also consumed in the form of noodles, puffed rice, fermented sweet rice, and snack foods made by extrusion cooking. It is used in making beer, rice wine, and vinegar. Some oriental desserts require the use of glutinous or sweet rice, which consists entirely

of amylopectin in the starch, in contrast to the nonglutinous rice that contains both amylopectin and amylose (10–30%).

Rice oil extracted from the bran is rich in vitamin E and has received considerable attention by researchers as a source of oil for the developing countries. Up to this time, the use of rice oil has lagged behind its potential value because of activation of lipase and lipoxygenase enzymes during milling, which caused rancidity and development of off-flavor. Extrusion heating of rice bran immediately after milling to inactivate the enzymes has improved the stability of rice oil. As the new technology in oil extraction and refining becomes available to the developing countries, the consumption of rice oil should gradually increase.

When the actual extraction rates of the cereals (the fraction of each grain utilized as food) are considered, rice is calculated to produce more food energy per hectare than the other cereals. When the superior quality of rice protein is considered, the yield of utilizable protein is actually higher for rice than for wheat.

B. Rice in World Trade

Over the period studied, rice traded on the world markets ranged between 11.8 and 12.7 million tons per annum. The volume was 4.5% of the total production. Asia remained the largest rice-importing region, followed by Africa (over 3 million tons in 1985–1986), Europe (2 million tons in 1980–1984), and South America (1.6 million tons in 1986).

Thailand remained the top rice-exporting country (2.8–4.8 million tons/annum), followed by the United States (2.3–3.1 million tons); China (0.7–1.4 million tons); Pakistan (0.7–1.3 million tons); Burma and Italy (0.5–0.8 million tons); and Australia, Uruguay, and Japan (0.1–0.7 million tons).

Among the rice-importing countries, Indonesia ranked first, followed by Saudi Arabia, Nigeria, the USSR, Senegal, and the Ivory Coast. Brazil showed highly variable imports during the period.

C. Utilization of Rice and Its By-products

Milled rice and parboiled rice are consumed mainly as boiled rice. Different rice varieties with specific amylose–amylopectin ratios are used in specific rice products. Waxy (glutinous) rice is the staple food in China, Laos, and Thailand and is usually prepared by steaming milled rice previously soaked in water. Waxy rice is used also in sweets and desserts. In the United States, medium-grain low-amylose rice

(12–20% amylose) is used in making baby foods, beer, and breakfast cereals. Rice in temperate countries (Japan, Korea, and China) are low-amylose varieties. Intermediate-amylose (20–25%) rice is used mainly for fermented rice cakes and in making canned soups. Intermediate-amylose rice is preferred over high-amylose rice in China, Japan, Philippines, Indonesia, Thailand, Malaysia, and Vietnam. In the United States, the short- and medium-grain varieties have low amylose content, and the long-grain varieties have intermediate amylose content. High-amylose (>25%) rices are used for making extruded rice noodles. In some countries, high-amylose rice is a staple food.

II. Rice Flours

There are two types of commercial rice flour available in the United States. The first is produced from waxy or glutinous rice, used as a thickening agent for white sauces, gravies, and puddings and in oriental snack foods. It can prevent liquid separation (syneresis) when these products are frozen, stored, and subsequently thawed. The other type of rice flour is prepared from broken grains of ordinary raw or parboiled rice. The flour prepared from parboiled rice is essentially a precooked flour. It differs from wheat flour in baking properties because it does not contain gluten, and its doughs do not readily retain gases generated during baking. There is, however, a steady basic demand for rice flours for use in baby foods, breakfast cereals, and snack foods; for separating powders for refrigerated, preformed, unbaked biscuits, dusting powders, and breading mixes and for formulations for pancakes and waffles. These uses are sufficient to sustain a market for rice-flour production.

Rice is consumed largely as a whole grain. Some breakage, however, is unavoidable. Some of the rice kernels are cracked while still in the husk, and some breakage occurs during harvesting, handling, drying, and milling.

The larger pieces of broken rice sell for a little more than half the price of whole rice, whereas the smaller fragments sell for less than half of comparable whole grains. Therefore, these smaller pieces are used for grinding into rice flour or for brewing.

About a third of the broken rice produced in the United States is used by brewers and a small portion by cereal manufacturers and baby-food formulators.

A. Properties of Rice Flours

Rice flours made from long-, medium-, and short-grain and waxy rice are available commercially. There are varietal differences in protein, lipid, and starch content and the amylose and amylopectin ratios in the starch. The proximate analyses of some rices are presented in Table I. Composition differences contribute to the diversity of chemical and physical properties of various rice flours, such as viscometric properties, starch gelatinization temperatures, birefringence end-point temperatures, water absorption, and other characteristics. Differential scanning calorimetry is routinely used now to measure gelatinization temperature of starch in rice flour.

Rice flours from each variety, with the exception of the waxy type, have charactristic viscosity patterns during the heating and cooling cycles of their pastes. The changes in viscosity depend largely on the composition of the starch and, to a lesser extent, on the protein and oil components. Usually, long-grain rice containing starch with an amylose content of over 22% has a relatively low peak viscosity and forms a rigid gel on cooling (high setback viscosity). Those with starches low in amylose have high peak and low setback viscosities.

B. Determination of Rice-Flour Properties

Birefringence end-point temperature (BEPT) of rice flour can be determined on a 1% slurry using a polarizing microscope equipped with a Kofler hot stage and set at a heating rate of 5°C/min. BEPT is usually recorded at 95–98% loss of birefringence.

Amylograph pasting curves of 10% flour slurries (50 g of rice flour to 450 ml of water) give useful information on rice starch characteristics important in baking applications. A test on 20% flour slurries (100 g of rice flour to 400 ml of water) can give estimates of initial gelatinization temperature of rice starch within ±0.5°C, when gelatinization temperature is taken as the point of initial increase in viscosity. These values correlate significantly ($\gamma = 0.94$), with BEPT determined microscopically.

These techniques are used to characterize rice varieties that give contrasting texture results in yeast-leavened rice breads. Figure 1 shows representative results with four flours. The short curves (20% slurries) indicate distinctly different temperatures for initial viscosity increase (i.e., estimate of gelatinization temperature) among the rices. The full curves (10% slurries) show initial viscosity increases occurring

TABLE I

Proximate Analyses of Four Rice Varieties Grown in California

Variety	Grain type	Amylose, %	Amylopectin, %	Protein, %	Fat, %	Ash, %
S–6	Short	19.0	56.5	6.50	2.05	1.22
M–5	Medium	22.5	54.7	6.60	1.50	1.07
74–y–52	Medium	24.0	56.2	6.75	1.82	1.55
72 3764	Long	27.8	54.2	7.07	1.88	1.65

much later in the heating cycle; this is because the amylograph lacks the sensitivity to detect small increments in viscosity in less concentrated slurries. However, the 10% slurries do provide a complete pasting history that can be correlated with the textural properties of the baked products. For example, sample B (IR8 variety) had a low initial gelatinization temperature, which is favorable for baking, but a high final viscosity and high positive setback (viscosity at 50°C minus peak viscosity), which adversely affect bread texture. Sample B had a high amylose content, which results in increased retrogradation of starch. Bread made from this variety had a harsh, dry, crumbly texture within 24 hr.

C. Waxy-Rice Flour

Waxy-rice flour, or sweet-rice flour, contains less than 2% amylose in the starch and an appreciable amount of amylase. This flour has a lower peak viscosity than some of the short-grain rices, probably because of its amylolytic activity, and it has practically no setback viscosity.

Waxy-rice flour is different from other rice flours in its resistance to liquid separation (syneresis) during freezing and thawing. The remarkable stability of cooked waxy-rice flour pastes after repeated freeze–thaw cycles was observed during a study of factors determining the stability of white sauces and gravies commonly used with frozen, precooked meat and vegetables.

Waxy-rice flours are superior to other grain flours because they are more stable under freeze–thaw treatment than any other flours or starches. This behavior may be attributable to the virtual lack of amylose starch in waxy-rice. This flour, mixed with wheat flour even at 40–60% levels, stabilizes sauces and gravies held at 0°C for 5–6 months. When waxy-rice flours are used as a sole source of starch, stability can be maintained for a year or longer. The unusual stability of waxy-rice flour over other flours may be due to its special chemical structure or to the small size of the starch granules.

D. Composite Flour for Baking

Composite flours for baking are blends of nonwheat flours with or without wheat. They are blended after milling or during dough preparation at a bakery. Such mixtures are made as adjustments to crop shortages or surpluses.

Experimental reports on the use of rice flour in wheat-based baked products are worldwide. In Italy, additions of 15% rice flour in bread and 60% in pastries are known. Brown-rice flour was preferred for the better nutritive value needed by yeasts during fermentation. Increases in yeast level and fermentation time, along with 3% shortening in the formula, alleviated most of the problems due to substitution of rice.

Composite flours and baked products were extensively developed to increase the nutritional and caloric intake of people in developing countries. Small amounts of rice flour from indica or japonica types can be added to wheat flour if dough improvers are included to compensate for the dilution of the wheat gluten protein. These include ascorbic acid, potassium bromate, lipid-based surfactants, solid shortenings, and vegetable oils. These additives strengthen and enhance the carrying capacity of the wheat flour for the nongluten flour diluent.

Application of 10, 15, and 20% substitutions of rice flours in a wheat-flour pan-bread formula containing 4% shortening can be made. Untreated rice flour from brown or white milled rice showed a typical dilution effect at 20% substitution. Loaf volumes were lower, and farinograph and extensigraph curves showed slightly less strength. Amylograph viscosities were higher presumably because of the lower amylase activity from 80% wheat than from 100% wheat. Emulsified lipids rather than α-amylase suppress amylograph viscosity of waxy rice flour as compared to

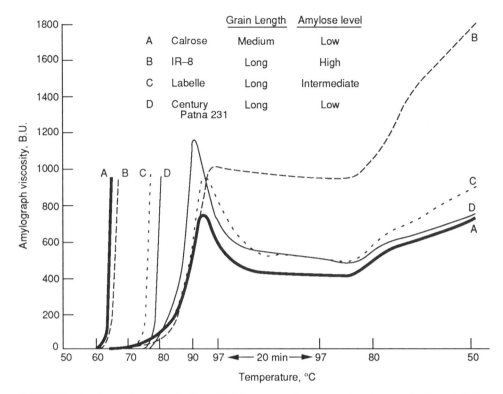

FIGURE 1 Amylograph curves of roller-milled flours from four contrasting rice varieties. Sample B (IR-8) is a Phillippine variety, the others are U.S. varieties. Short curves, 20% slurries; full curves, 10% slurries. [*Source:* from Nishita and Bean (1979) *Cereal Chem.* **56**, 185–189.]

waxy rice starch, because the effect is not reduced by boiling of the aqueous extract (Merca and Juliano, 1981).

E. Rice Starch

Rice starch is a special product used as a major component of face powder. The fine particle size of rice starch makes it especially suitable for cosmetic use. Chemical treatment is necessary to disperse the rice proteins. For example, broken rice may be steeped at ambient temperature for 24 hr in five times its weight of a 0.3% caustic soda solution as the first step in rice-starch production. The soaking solution may be heated to 48.8°C to speed up the extraction process.

The caustic-treated granules are washed and then dried before being ground into flour. The flour is then mixed with 10 times its weight of 0.3% caustic soda solution and stirred for 24 hr. The starch is allowed to settle, and the supernatant solution, which contains most of the rice protein, is removed. Washing with water, settling, and decanting are employed to remove most of the soluble materials from the starch

granules. The washed starch is dewatered by filtering or centrifuging; complete removal of residual alkali is a very important step, which must be carefully controlled. The washed starch is dried in ovens or rotary-drum driers; the rice-starch cake is ground to the desired particle size and sieved.

The rice protein in the combined effluent is precipitated by the addition of hydrochloric acid. The supernatant fluid is discarded, and the precipitated material is partially dewatered in a filter press and finally dried in a rotary drier. The product can be used as a protein supplement for cattle feed.

F. Rice Bread

The problem associated with rice-bread formulation is the absence of gluten in rice flour. The manufacture of rice bread without gluten presents considerable technological difficulties because gluten is the important structure-forming protein. Research has been done using gum or suitable surfactants such as glyceryl monostearate (GMS) as the binding agent. A yeast-leavened rice-bread formula consists of modifying a typical wheat-bread formula in which wheat flour is completely replaced by rice flour. The effects of hydroxypropylmethylcellulose, locust bean and guar, sodium carboxymethylcellulose, carrageenan, and xanthan gum on the loaf volume of rice bread were reported by Delgado 1977; Nishita and Bean 1979; and Nishita et al. 1976.

The bread formula consists of 100 parts rice flour, 75 parts water, 7.5 parts sugar, 6 parts oil, 3 parts fresh compressed yeast, 3 parts hydroxypropylmethylcellulose, and 2 parts salt. These ingredients are mixed thoroughly, panned or shaped as rolls, fermented to the desired volume, and baked.

Several gums were tested in the formula presented above as gluten substitutes, including xanthan gum, which was successfully used in wheat-starch breads. Hydroxymethylcellulose provided the proper dough viscosity and film-forming characteristics so that the rice-flour dough would retain fermentation gases during proofing and expand during baking to produce a crumb grain similar to that of typical white pan bread. Surfactants that normally improve the texture of wheat breads had negative effects on rice doughs and bread because they interfere with the gum–water–rice–flour complex such that no fermentation gases were retained and thus no leavening occurred.

Only the short- and medium-grain rice flour had the necessary physicochemical properties to give a soft-textured bread crumb. The long-grain type yielded sandy, dry crumb characteristics.

The short- and medium-grain rice kernels have stickly properties when cooked in the traditional manner as raw milled rice. Their starches have low gelatinization temperatures (below 70°C), with amylose contents at 20% or lower. In contrast, the long-grain types produce fluffy cooked rice, with starches that gelatinize above 70°C, and the amylose content is above 23%. The cooking characteristics of the long-grain types appear to be directly related to the unacceptable crumb grain properties of baked products made from the flour.

G. Rice Cakes

Rice cakes have a wide range of processing and product characteristics. In the United States, "cake" generally refers to angel, sponge, or layer types in a variety of sizes and shapes. Angel and sponge cakes depend on egg whites or whole eggs for the major structural foundation, with flour supporting and strengthening the foam like structure, and are leavened primarily by steam and air. Layer cakes can be made without eggs; they depend on flour as the main structural component and are leavened primarily with carbon dioxide provided by baking powder. Layer cakes characteristically contain high levels of sugar, shortening, butter, or oil.

Layer cakes containing 100% rice flour were developed, using 100 parts rice flour, 80 parts sugar, 80 parts water, 15 parts oil, and 5–7 parts double-acting baking powder. Rice flours from short- and medium-grain rices having low amylose contents and low gelatinization temperatures were preferred over those from long-grain rices.

Layer cake formulas typically contain a high level of sugar, which markedly increases the gelatinization temperature of the starch. In a batter, the ratio of sugar to water should be such that starch gelatinization and granule swelling can occur and set the structure during baking when the leavening action has expanded the batter to its maximum volume.

III. Parboiled Rice

Parboiling is a hydrothermal process in which the crystalline form of starch present in the paddy rice is changed into an amorphous one. This is accomplished by soaking, steaming, drying, and milling the rice. The parboiling process produces physical, chemical, and organoleptic modifications in the rice, with economic and nutritional advantages. The major objectives of parboiling are to: (1) increase the total and head yield of the paddy, (2) prevent the loss of nutrients during milling, (3) salvage wet or damaged paddy, and (4) prepare the rice according to the requirements of consumers. The changes occurring in the parboiling process are as follows:

1. The water-soluble vitamins and mineral salts are spread throughout the grain. The riboflavin and thiamin contents are four times higher in parboiled rice than in milled rice. The thiamin is more evenly distributed in the parboiled rice, and the niacin level in this rice is eight times greater.

2. The moisture content is reduced to 10–11% for better storage.

3. The starch grains embedded in a proteinaceous matrix are gelatinized and expanded until they fill up the surrounding air spaces.

4. The protein substances are separated and sink into the compact mass of gelatinized starch, becoming less liable to extraction.

After rice is parboiled, the milling yield is higher because there are fewer broken grains. The milled parboiled rice keeps longer and better than in the raw state since germination is no longer possible. The nutritional value of parboiled rice is greater because of the higher content of vitamins and mineral salts that have spread into the endosperm. Parboiled and prepared rices are not always accepted by raw-rice consumers, because they are prone to oxidative rancidty. Good packaging material will help to minimize or retard oxidation. In some parts of India, Brazil, and other countries, parboiled rice is consumed in significant amounts.

A. Cleaning

The impurities present in paddy rice are varied; weeds, animals used for threshing, and natural drying all account for the extraneous materials found in the paddy. Impurities and seeds other than rice are usually removed during milling. Some are removed before shelling, and others after polishing along with the broken and damaged grains.

B. Grading

Sorting on the basis of kernel thickness is essential for good-quality parboiled rice. This is done by means of grading reels fitted with a steel sheet with rectangular slots or with wire netting. Further grading may complete the selection according to the length and bulk weight of the grain to obtain a final product of improved and uniform quality. Sorting by bulk weight, if necessary, is done by specific-gravity separators.

C. Steeping

Different varieties of paddy rice have their own steeping characteristics. An efficient steeping process in medium-temperature water (65°C) should be used for the production of parboiled rice. The treatment must be done quickly to avoid fermentation.

The methods used to achieve steeping include: (1) the use of high- and medium-temperature water; and (2) the addition of wetting agents to the steeping water. These systems have been used either alone or in conjunction with one another to increase water penetration and to reduce steeping time. Steeping is needed to provide the starch with a sufficient amount of water for gelatinization. A moisture content of not less than 30% is required to fully gelatinize the starch in the caryopsis.

In the United States, parboiling of rice has been fully mechanized. The facilities include six to eight steeping vessels with built-in steam coils, hot-water tank, and boiler; rotary-drum driers with steam-heat exchangers or husk-fired furnaces for drying paddy; mechanical handling equipment; rotary-hot-air driers; bin driers; milling equipment; and packaging machinery. The steeping vessels are fed with hot water at 80–85°C, and the raw paddy is transferred into them from an overhead surge bin. The water is circulated for 15 min and then maintained at 65°C for 4–5 hr, after which it is drained off. Steam is let into the built-in steam coils, and the paddy is steamed for 10–20 min and then moved to the driers through belt conveyors.

The discoloration of the parboiled milled rice increases with the duration of steeping and the temperature of the water, subsequent steaming doing the same in both cases. The color becomes much deeper once the limit of 70°C is exceeded.

The color of the parboiled rice varies with the pH of the steeping water. If the pH is close to 5, coloring is minimal. The color deepens as the pH rises.

Occasionally, both vacuum and hydrostatic pressure methods are used to reduce steeping time, keeping the temperature of the water within limits that do not adversely affect the quality of the final product. By removing interstitial air and by applying hydrostatic pressure to the steeping water, the steeping time can be reduced.

D. Steaming

The purpose of steaming is to increase the milling yield and to improve storage characteristics and eating quality. Steaming improves the firmness after cooking and achieves better vitamin retention in the milled rice.

If the starch in the endosperm is not fully gelatinized, there will be white cores present in the parboiled product. The time exposed to steam must therefore be long enough to gelatinize the whole kernel completely.

The quantity of water to be absorbed, the time of exposure to steam, and the temperature or pressure of the steam itself provide the parameters that will decide the quality of the parboiled rice.

Through a variation of these factors, parboiled rice possessing particular characteristics and degrees of gelatinization can be obtained.

Within defined limits of temperatures and pressure of the steam used, the milled parboiled rice shows differences in: (1) color, (2) volume after exposure to air heated to 121^0C, and (3) soluble-starch content.

E. Drying

The objective of drying parboiled rice is to reduce the moisture content to an optimum level for milling and subsequent storage and to obtain the maximum milling yield.

The drying process is stopped for a while when the moisture content reaches 16% and then drying is resumed using the appropriate temperature and drying time. This interval is called "conditioning." The optimum temperature and time needed for final drying are related to the temperature of the paddy after conditioning. Generally, slow and prolonged drying is essential in the final stage to ensure a maximum yield of whole grains.

Rice kernel breakage is related to the moisture content and the condition period between the first and second drying stages. After steeping and steaming, a sample was dried to the critical moisture content of 16%. A part of it was then given a further drying while the rest was put aside and the second stage begun after periods varying from 2 to 48 hr. The percentage of broken grains after milling decreased with the increase of tempering time and with a reduction in moisture content.

F. Driers

Various types of vertical-column driers and horizontal, continuous-flow, rotating-cylindrical hot-air driers have been developed in modern plants. The vertical-cylindrical driers are preferred where low-temperature drying air is used and the rice is exposed to the drying air for a long time. Horizontal-rotating, continuous-flow, cylindrical driers are normally used when rapid drying at high temperatures is required.

In rotary-cylindrical driers, hot air is used, and heat is also applied to the cereal by fitting an external steam jacket to the drier and a tube nest inside it. High-temperature drying (80–100°C) with a horizontal-cylindrical drier is used for reducing the moisture content to 16–18%. It is followed by further drying at lower temperatures in a conventional column drier.

G. Tempering

After drying, the parboiled paddy must be allowed to rest for a time before milling. This time interval is called the tempering period. A tempering period of about 48 hr is needed for the product to dissipate the heat it received during drying. Also, the moisture content inside each grain must become uniform throughout.

Tempering must be done to ensure dissipation of heat without speeding up the cooling by artificial means. Milling is done only when the rice has become stabilized at an ambient level and the grains have hardened and become glassy in texture.

The moisture content of parboiled milled rice may be brought up to 12–14%, even if that of the raw paddy used for the process is below these percentages.

H. Milling

Parboiled rice, when properly prepared and milled, gives the maximum yield of edible rice with a minimum amount of broken grains. Parboiling gives hardness and seals any cracks in the caryopsis. Any breakages are caused only by mechanical action of the milling machines. Good results from the treatment depend to a great extent on the drying process.

It is necessary to pass the product through a cone-type whitening machine abrasive or a horizontal cylinder covered with abrasive material in order to remove the pericarp, the perisperm, and the layer of aleuronic cells. Polishing is done in a fraction machine.

In many cases, parboiled rice is undermilled and still carries most of the aleuronic cells and traces of the perisperm, as well as the germ at one end.

When the paddy is put directly into the huller without prior shelling, the hull, which came off the caryopsis during the first stage, acts as an abrasive and, at the same time, absorbs some of the fatty substances, thus facilitating polishing.

The bran from parboiled rice has prolonged resistance to the formation of free fatty acids. This makes it better and easier to use for the extraction of edible oil. The bran obtained from raw paddy has a fat content of 12–14%, and the bran from parboiled rice may contain 16–22%. In modern milling plants, the bran from milled rice is immediately passed through an extruder under 2.8–7.0 kg/cm^2 pressure at 138°C for 5–15 sec to inactivate the lipase activity, thus preventing the formation of free fatty acid due to enzymic hydrolysis of the rice oil present in the rice bran. The bran so treated is more stable during storage.

I. Color Sorting

The parboiled rice must be sorted to remove discolored grains. A flat conveyor belt about 0.9 m wide is used. The speed of the belt is adjustable as desired by the operator. The rice is spread on the belt in a thin layer and inspected as it moves along by sorters who pick out the discolored grains by suction, using a plastic or rubber tube connected to a centrifugal air pump. The grains thus sucked up are deposited inside a cyclone separator through which the flow of air passes before reaching the pump.

Automatic machines based on photoelectric devices have been used to sort the rice by color. The existence of such machines enables rapid sorting of parboiled rice.

The automatic sorting machines have the following advantages over the manual sorting belts: (1) the speed is faster and the rice passes through the machine at the same speed, irrespective of its content of discolored grains, and (2) sorting is more efficient as the grains are checked from all angles.

In modern sorting machines, the rice presented for scanning is made to slide in line down a straight slope from which it reaches the scanning area at a previously calculated speed and curve. The scanning unit, the photodetectors, and the impulse amplifier are the essential components of the machine. Transistors have completely replaced electron tubes, and plug-in circuit boards are now commonly used.

Newer machines use a stream of pressurized air that knocks the grain aside from its path. Improvements made on these machines have brought about a continuous rise in sorting speeds and output. Running costs of both power consumption and maintenance are low.

J. The Parboiling Processes

Steam is required to gelatinize the starch in the rice grain during parboiling.

About 600 kg of water is needed to steep 1000 kg of rice paddy. During steeping, proper amount of water is absorbed by the paddy.

The quantity of steam needed to parboil paddy rice is a mere fraction of that required for the whole process because steam is also needed for heating the steeping water and the air where the paddy is artificially dried. The steam required to produce parboiled rice in modern plants is supplied by high-pressure boilers sent to the various points at which steam is needed at a low pressure. In some plants, steam is produced at high pressure for power in turbines or engines before it enters the heating system of the parboiling plant at a low pressure.

The parboiling process includes cleaning and grading the paddy, parboiling, steam production, drying, milling, color sorting, and packaging. Between stages, bins are used for storing the products or by-products so that the various operations are kept flexible.

The plants may operate under continuous or batch processes. Some use a long steeping and steaming cycle with low temperature, and others use short cycles with high temperatures and pressures.

When a parboiling plant is built, the continuous production system is preferable if the paddy consists of only a few varieties grown on a large scale. A batch production system is more suitable where the paddy is of many different varieties and characteristics. The steeped rice should be moved from the soaking tanks to the steaming autoclave and from there to the driers by gravity.

A highly mechanized plant with automated processes may be suitable in a country where labor costs are high.

IV. Quick-Cooking Rice

Ordinary milled rice requires 20–30 min to cook to a satisfactory culinary acceptability. In some instances, the rice is soaked, washed, and steamed, requiring a total attention time of about 1 hr. Thus, effort has been directed toward development of quick-cooking rice to reduce the cooking time.

Various rice varieties yield cooked rice of different textural characteristics. Variations in recipes also have a significant effect on the texture, flavor, and acceptability of the cooked rice.

Quick-cooking rice should be cooked within 5 min, and the cooking method should be simple. After cooking, the product should match the characteristic flavor, taste, and texture of conventionally cooked rice. The rice must be easily processed in mass quantities and must possess good keeping quality.

The Nissin Food Company in Osaka, Japan, has developed an instant "Cup Rice," which can meet most of the conditions mentioned above. The rice is precooked under high pressure and temperature and then dehydrated. The product can be reconstituted with boiling water within 5 min in a polystyrene cup.

Quick-cooking rice is precooked and gelatinized to some extent in water, steam, or both. The cooked or partially cooked rice is usually dried in such a manner as to retain the rice grains in a porous and open-structured condition. The finished product should consist of dry, individual kernels, free of lumps or aggregates, and have approximately 1.5–3.0 times the bulk volume of the raw rice.

Many quick-cooking rice products, although varying in texture, bulk volume, appearance, taste, and performance qualities, are designed specifically for certain consumer uses. Some quick-cooking rice for special applications, such as in dry soup mixes, casseroles, or other dry food mixtures that have certain rehydration time requirements, were designed to be compatible with the other ingredients in the mix.

A. Types of Quick-Cooking Rice

Difference in moisture levels, precooking times and temperatures, drying conditions, and other processing variables can produce various types of quick-cooking rice. These range from relatively undercooked rice, requiring 10–15 min of "cooking" or a good-quality "table" rice requiring a 5-min preparation time, such as Minute Rice, to a variety that can be hydrated in several seconds to a minute or two. The last type yields a fairly mushy product when boiled. Some of these are marketed as ready-to-eat breakfast cereals.

Some consumers prefer long-grain, light, fluffy, or slightly dry individual kernels of rice, with typical cooked-rice flavor, having essentially no gritty or hard, uncooked centers. This has been the target for most quick-cooking rice development. A notable exception to this is that, in Japan and China, people prefer short-grain rice, which is somewhat pasty and sticky when cooked. In fact, as rice gradually becomes "drier" (when cooked) with time in storage after harvesting and milling, the short-grain type may become too dry and nonpasty for Japanese textural preference, to the extent that "waxy" or "sweet" rice may be added in small amounts to increase the pastiness.

Among the types of products, completely precooked rice should be used in cup and standing form because no further cooking of these products is necessary during preparation, whereas the other types could use either completely or partially precooked rice products. Except for the standing-type commercial product, all others usually contain seasoning mix. As for quality consistency, cup and standing types give consistent quality in the finished dish because no real cooking is involved, whereas the others may vary, especially simmering and sauté/simmering types. The sauté/simmering type gives a Chinese fried-rice texture and appearance, whereas boil-in-bag products eliminate the problem of boiled rice sticking to the pot during regular preparation. Microwave type products are most convenient because of their short preparation time (microwave heating time is about one-fourth of regular stove-top heating), but this type of product also has its shortcomings, among them:

1. Because of variations in power, type, and structure of microwave ovens used (for example, some have a rotaing feature), the cooking results vary, and adjustment of cooking conditions is needed.

2. Limited amounts of rice (and water) should be cooked because of the size limit of cooking dishes used in microwave ovens. Larger quantities of rice can be cooked by standing or simmering types without significant increase in cooking time. Thus, microwave type products are usually limited to single-serving portions, with cooking time less than 7 min.

3. A cooking dish with optimum size, depth, and shape is needed to accommodate the amount of rice and water used; this ensures surface coverage of the rice by a minimum amount of water, as well as the shortest heating time. An oval or round dish offers more uniform cooking than a square one and a shallow, large surface area is better than a dish with a deep, small area. The dish material is usually polyester-coated paperboard, which can be used in both microwave and regular electric ovens (so-called dual-ovenable).

4. Foaming and spillage of products sometimes occur during microwave heating; addition of butter, margarine, oil, seasoning mix, and an optimum amount of water and rice can reduce spillage.

Practically all quick-cooking rice processes described in patents emphasize the treatment of the rice. Efforts have been made to improve milling characteristics and yields, to remove surface fats, to improve storage stability, and to enhance flavor by parching the grain. Some of these processes improve nutritional quality by infusing the surface vitamins from the bran and aleurone layers into the endosperm. This latter treatment has been developed to form products now commonly known as parboiled rice.

B. Quick-Cooking Processes

Many quick-cooking rice products and processes have been developed and patented during the past years. Among the processes and products developed in the past, the following are the commercially useful quick-cooking processes.

1. Soak–boil–steam–dry methods: Raw milled white rice is soaked in water to 30% moisture and cooked in hot water to 50–60% moisture, with or without steaming. The product is further boiled or steamed to

increase the moisture content to 60–70% and then dried carefully to 8–14% moisture to maintain a porous structure. A significant modification of the procedure is a dry-heat pretreatment to fissure the grains prior to cooking and drying.

2. Expanded and pregelatinized rice: Rice is soaked, boiled, steamed, or pressure-cooked to gelatinize the grain thoroughly, dried at a low temperature to yield fairly dense glassy grains, and then expanded or puffed at a high temperature to produce the desired porous structure.

3. The rolling or "bumping" treatment: Rice is pregelatinized, rolled, or "bumped" to flatten the grains and dried to a relatively hard, glassy product.

4. Dry heat treatment: Rice is exposed to a blast of hot air at 65–82°C for 10–30 min, or at 272°C for 18 sec, to dextrinize, fissure, or expand the grains. No boiling or steaming is applied. The product cooks in less time than untreated grains.

5. The freeze–thaw process: Rice is precooked, then frozen, thawed, and dried. This procedure combines the hydration and gelatinization steps 1–3, plus the critical steps of freezing and thawing before drying.

6. Gun puffing: Gun puffing is a combination of preconditioning the rice to 20–22% moisture followed by steam cooking in a retort at $3.5–5.5 \ kg/cm^2$ for 5–10 min. Then the product is puffed to atmospheric pressure or into a vacuum. The optimum terminal condition is at 165°C at 20–25% moisture levels.

7. Freeze-drying: To freeze-dry cooked rice, the cooked rice is cooled and then frozen in a blast freezer. The water is removed by sublimation.

8. Chemical treatments: Chemical treatments with sodium chloride, disodium phosphate, or food-grade surfactant reduce clumping.

V. Canning, Freezing, and Freeze-Drying

A. Introduction

The types of canned rice products on the market include soups with rice, meat and rice dinners, casseroles, Spanish rice, unflavored cooked rice, fried rice, and rice pudding.

Rice with a high protein content takes a long time to cook because of the physical barrier to water absorption created by the protein matrix around starch granules. Low-protein rice is more tender, more cohesive, and sweeter than high-protein rice. Parboiled rice is often used in preparing canned-rice products because of the stability of the kernel and the retention of its shape without disintegration under rigid retorting and heating conditions.

Other factors that may affect the quality of canned rice include pH, fat content, salt concentration, and blanching time. For example, alkaline solutions cause rice to develop a yellow color.

B. Canning

Various methods have been studied for making canned rice more acceptable. These fall into two categories: wet pack and dry pack. A product in which there is an excess of liquid, such as in soup media, is termed wet pack. Proper density is the prime objective with these types of products. The rice is precooked or blanched sufficiently to promote buoyancy in the product and prevent settling and matting but not to the point that kernel texture is degraded. The parboiled rice is cooked slowly in an excess of water, followed by draining and washing in cold water. This washing process removes excess surface starch. The rice is put into cans, together with the sauce. The cans are sealed, and the product is retorted to sterilize the product. [*See* THERMAL PROCESSING: CANNING, AND PASTEURIZATION.]

A canned product, such as Chinese-style fried rice, in which the grains are separate and devoid of free or excess moisture, is called the dry pack. The prime objective is to provide enough moisture for gelatinization of the starch during retorting without causing pastiness or cohesiveness in the kernels. Cooking oil as an ingredient helps to minimize grain cohesion. The usual procedure involves slowly precooking parboiled rice in an excess of water. The rice is subsequently washed in cold water and then mixed with the other ingredients. After filling and sealing, the canned product is slowly heated and then retorted.

A process for canning white rice is as follows. The rice is washed and soaked in cold water for 30–45 min and boiled for 2–4 min or until the moisture content is approximately 55%. Limiting the moisture to this level minimizes kernel disintegration. The partially cooked rice is put into cans, sealed under 71.1 cm of vacuum, and then retorted. The canned product is prepared for serving after being heated in boiling water. The grains remain white and well separated.

Oil emulsions applied in the rinse step following soaking and cooking of the rice have caused a significant reduction in the kernel cohesion. Only a few of the surfactants tested have given a substantial reduction in clumping; sorbitan mono-oleate has been found to be effective.

The process of post-can-freezing may be used for both regular and parboiled rice. However, the canned parboiled rice flows with greater ease. In addition, it has a better appearance, stands up better in soups and casseroles, and has better keeping quality in the unused portion after being taken out of the can.

There are relatively few commercially canned rice products on the market, primarily because of the lack of stability of the rice grain. Most canned products that could use rice in the formulation require processing for approximately 60 min at 115.6°C or a shorter process at 121.1°C in conventional retorting equipment.

1. Rice in Retort Pouches

The retort pouch is a flexible, heat-sealable, flat container capable of withstanding the high temperature (121°C) required for pressure-processing rice and other low-acid foods. This container represents an alternative packaging system to conventional metal cans and glass jars. Over 750 million pouches of foods have been consumed annually in Japan alone.

The most commonly used retort pouch is a three-ply laminate of polyester, aluminum foil, and polypropylene. The polyester, usually 0.0005 in. (0.0127 mm) thick, serves as the outer protective layer because of its strength and resistance to scuffing and flexing. A middle layer of aluminum foil is used to increase product shelf life by acting as a barrier to passage of light, water, and oxygen. The foil layer may vary in thickness (0.00035–0.0007 in., or 0.0089–0.0179 mm), according to its intended application. The inner layer of polypropylene provides an excellent food contact surface because of its inert properties. This material was approved by the FDA in 1977 for use only at retort temperatures of 121°C or less. Pouch film is supplied to the processor either as roll stock or as preformed pouches. If roll stock is used, the pouch-forming operation is accomplished in the food plant by continuous form-fill-seal equipment. Present commercial pouches are sized to hold 5–12 oz of product, an amount generally sufficient for a single serving. Also available are larger institutional-size pouches, which hold a quantity comparable to that of a No.10 (603 × 700) can.

Rice in retort pouches was made by hermetically sealing cooked rice in laminated plastic and aluminum-laminated plastic pouches and heat processing at 120°C. Its shelf life is 6 months at room temperature. Sekihan (steamed waxy rice with red beans) accounts for 80% of all retort rice in Japan. Retort rice made from nonwaxy rice includes chicken and rice (Japanese style). The consumer soaks the intact pouch in hot water for 10–15 min (or transfers the contents from the aluminum pouch to a plate and heats it in a microwave oven for 1–2 min). Boiling for 10 min increases the degree of gelatinization of retort rice from 55% before boiling to 90%. Freshly cooked japonica rice has 92% gelatinization.

2. Rice in Canned Soups

Canned condensed soup is one of the products in which rice stability is most important to overall product quality. The process includes, first of all, a precooking or blanching treatment of the rice in boiling water for 15–18 min. The rice, almost completely cooked after the blanching treatment, is blended with the other ingredients, and the soup is canned, sealed and heat-processed. During heat processing, the cans may be heated to 121.1°C and held for 30 min or longer. It is important to remember that there is an excess of water present. The water uptake and expansion of the rice kernels is not restricted by the lack of water. Rice showing kernel instability is not suitable for use. Upon heating, the ends of the kernels are split, the surface is ragged, and sloughing of fragments increases the turbidity of the surrounding liquid.

3. Rice Puddings

There is growing trend toward using the aseptic canning process for rice puddings. The pudding is sterilized and cooled separately from the container, thus avoiding the slow heat-penetration problems inherent in the in-container canning process. The sterilized and cooled product is filled into presterilized containers and sealed in a sterile atmosphere with a sterile cover.

The two components of the rice pudding, the rice kernels in a small amount of liquid and the sauce, are sterilized individually and then combined in the can. This step is necessary because of the different sterilization treatments required by the two components. The sauce can be quickly sterilized in swept-wall heat exchangers. Sterilization time of the rice is much longer because of the greater time interval required for the heat to penetrate the kernels completely. If the sauce and grains are heated together until the rice is sterilized, there is a tendency for the sauce to become overheated, with excess browning and off-flavors as the result. The pudding may be sterilized at 137.8°C for 30–60 sec. At this temperature, a F_0 of 20–30 is reached.

C. The Freezing of Rice

Frozen cooked rice, like canned rice, is convenient to use since it requires less time to prepare than raw rice.

The rice may be frozen plain or in combination with other foods. Rice is an integral part of Chinese frozen dinners. Recently, microwave heating of precooked frozen rice in plastic containers has been a new quick method for serving rice.

1. Freezing Technology

There are a number of excellent frozen rice products on the market today. Some of these are combination dishes that can be reheated by the boil-in-bag method or in a microwave oven.

1. Place rice in an excess of water at 54.4–60°C, which contains enough citric acid to reach a pH of 4.0–5.5. Enough water should be used to cover the rice after it has soaked for 2 hr.

2. After 2 hr, drain off the soak water, and rinse with more of the same pH-adjusted water to remove fines.

3. Drain thoroughly, tapping the screen to shake loose the adhering water or blowing the rice layer with air.

4. Place a small volume of water in the bottom of a pressure cooker, and heat to boiling with the cover on to heat up the apparatus. The soaked drained rice is placed in layers 5 cm deep or less over a screen supported above the water in the vessel. Close the vessel, and heat with the vent open until steam is emitted; then close off the vent, raise the pressure to 2.09 kg/cm^2, and hold for 12–15 min. Then, blow off steam gradually enough to prevent violent boiling and flashing of the hot water.

5. Place the hot steamed rice in an excess of hot water at 93.3–98.9°C without stirring. The rice will imbibe water until the grains are large, tender, and quite free. Stirring will cause the rice to become sticky. The rice should be held in a perforated vessel so that water may circulate freely through it.

6. Cooking should require only 10–15 min following the method described in step 4. Drain off the hot water, and rinse twice with cold water that has the pH adjustment described in step 1.

7. Tap and shake to remove free water, or suck off the free water over a vacuum filter.

8. Convey the cooked rice on a stainless steel mesh belt through an air-blast cooler to reduce it to room temperature, and then package in cartons or plastic pouches. Freeze the rice in air-blast freezers. The rice may be frozen as individually quick-frozen (IQF) products prior to packaging in a fluidized bed freezer.

Both boiled and steamed white rice that have been frozen and reheated are virtually indistinguishable from their unfrozen counterparts. Frozen storage at -18.8°C up to 1 year appears to have no deleterious effects on quality.

Prechilling the rice results in removing most of the surface moisture that may be present after cooking and, at the same time, permits quicker freezing. Be-fore freezing takes place, the individual grains are separated and maintained out of contact with one another during the freezing part of the process. The product is then frozen solid. Any appropriate freezing temperature may be used, but good results have been obtained by subjecting the rice to a moving air blast at -34.4°C. After the separate grains are solidly frozen, they can be brought together in a mass and packed in any desirable manner.

D. Freeze-Drying

The freeze-drying process consists of removing moisture from foods at the frozen state by sublimation under high vacuum. The low temperature used in the process inhibits undesirable chemical and biochemical reactions and minimizes the loss of volatile aromatic compounds. The dried product can be stored in airtight containers for long periods without refrigeration. For institutional feeding and for markets where low-temperature facilities are absent, freeze-dried foods are even more desirable than frozen or dehydrated foods.

Because of the high-vacuum and low-temperature requirements, freeze-drying is an expensive method of dehydration. But, for highly valued items, freeze-drying is a desirable process, yielding high-quality products. The freeze-dried rice should be packaged in proper containers for protection against moisture, light, and oxygen attack.

VI. Rice Breakfast Cereals and Baby Foods

A. Rice Breakfast Cereals

Rice breakfast cereals may be divided into two major categories: those that require cooking or adding boiling water before serving (hot cereals) and those that are fully cooked and ready to eat directly from the package.

1. Hot Cereals

There are two types of hot rice cereals: (1) precooked or instantized products, such as quick-cooking rices and instant rice gruels; and (2) granular products produced from granulated white milled rice.

2. Ready-to-Eat Breakfast Cereals

The use of rice in prepared cereals, both alone and in combination with products of other cereal grains, has assumed considerable importance in recent years.

The cooking time, steam pressure, temperature of the basic meterials, and toasting procedure greatly influence the quality of the final product. Flavoring materials, vitamin and mineral premixes, and protein-containing ingredients are added to improve the nutritive value.

3. Puffed Rice

The puffing processes may be divided into two types: (1) atmospheric pressure procedures, which rely on the sudden application of heat to obtain the necessary rapid vaporization of water, and (2) pressure-drop processes, which involve sudden transferring of superheated moist particles into a space at lower pressure. In the latter case, the pressure drop may be achieved by releasing the seal on a vessel containing a product that has been equilibrated with high-temperature steam, or it may be secured by transferring the hot material from the atmosphere into an evacuated chamber. The former process is much more widely used.

The puffing phenomenon results from the sudden expansion of water vapor (steam) in the interstices of the granule. Gun puffing may result in an increase of apparent volume (bulk density decrease) of six- to eightfold. Oven puffing causes a lesser increase, about three- to fourfold.

Puffed products must be maintained at about 3% moisture in order to achieve the desired crispness.

a. Oven-Puffed Rice Oven-puffed rice is prepared from whole kernels of California Pearl (short-grain) rice. Frequently, the rice is parboiled. Each batch consists of 635 kg of rice and 202 liters of sugar syrup. Some salt may be added. The mixture is cooked in a retort for 5 hr under 100–150 kPa steam pressure.

The cooked rice is broken up and dried to approximately 25–30% moisture content in a rotating drier. The partially dried product is stored in stainless steel bins for about 15 hr to equilibrate the moisture. This reduces the stickiness and toughens the kernel so that it is in perfect condition for bumping. Lumps may form during the tempering process. They must be broken up before being sent to the flaking rolls.

The individual kernels are dried so that a moisture content of 18–20% is reached. The kernels are passed under a radiant heater, which brings the external layers of the rice to 82.2°C. The outside layers of the kernel are plasticized by the heat so they do not split when the grain is run through the flaking rolls. The rolls used in preparation of oven-puffed rice are set relatively far apart so that they contact only the central part of the rice kernel. The bumped grains are again tempered, this time for about 24 hr.

To secure the puffed effect, the cooled and tempered rice is passed through rotating toasting ovens, which are usually gas-fired. The moist flake is tumbled through the perforated drums and passed within a few inches of the gas flames. Treatments are at 232.2–301.7°C for 30–45 sec. The oven-puffed rice emerges from the oven with less than 3% moisture. It is then carried by belts to expansion bins, where it is permitted to cool to room temperature before packaging.

The processing of a cereal called Special K, manufactured by Kellogg Corp., is similar to that of oven-puffed rice. The rice kernels are cooked and then coated, while in a moistened condition, with wheat gluten, wheat-germ meal, dried skim milk, debittered brewer's yeast, and other nutritional adjuncts. Following partial drying and tempering, the grain is run through steel flaking cylinders revolving at a speed of about 180–200 rpm. Hydraulic controls maintain a pressure of over 40 tons at the point of contact of the rolls. The rolls are cooled by internal circulation of water.

The flaking process presses the rice kernels into thin flakes. The product is still rather flexible at this time. The flakes are toasted in the same manner as oven-puffed rice. In addition to being thoroughly dehydrated by the process, the flakes are toasted and blistered.

b. Gun-Puffed Rice Rice puffing is a relatively simple process. It consists of essentially three steps: (1) heating the cleaned rice, (2) cooking with steam at high pressure in a sealed chamber or gun, and (3) suddenly releasing the pressure.

Short-grain rice is preferred for gun-puffing. California Pearl rice is generally used. The clean milled rice is introduced into the gun manually by a swing spout, and the gun is then closed. With the gun rotating, the heating phase of the process is started. After a period of preheating, superheated steam at $15.1 \, \text{kg/cm}^2$ is introduced into the gun. It is important that the steam be dry. Free water will cause clumping, pitting, and lower uneven expansion. Sufficient time is allowed for the superheated steam to cook the rice to a semiplastic state. Finally, the pressure in the gun is suddenly released by manually triggering the end gate, and the puffed rice is caught in a cage and then dried to 3% moisture before packaging.

Satisfactory puffing depends on attaining grain temperatures at which starch exhibits plastic flow characteristics under pressure. The time and temperature at which the rice is preheated are critically important. In general, the required temperature should be reached as quickly as possible without scorching the grain. The rate of steam flow to the puffing gun is very important and must be controlled precisely.

c. Puffing by Extrusion Puffed ready-to-eat breakfast cereals are being made by extruding superheated and pressurized doughs through an orifice into the atmosphere. Either single-screw or twin-screw extruders can be used. The sudden expansion of water vapor as the excess pressure is released increases the volume several times. Apparent specific volumes can reach or exceed those attained by gun puffing, and the process seems to have several advantages over gun puffing, such as higher production rates, greater versatility in product shape, and easier control of product density.

The rice premix containing 60–75% expandable starch base is moisturized with water or steam. The resultant mash is compacted by a screw revolving inside a barrel, which may be heated by steam. The thread of the screw has a progressively closer pitch as it approaches discharge. In some extruder designs, the rice premix is fed directly to the extruder. The water and/or steam are injected into the barrel and mixed with the premix. The pressurizing, shearing, and steam heating bring the dough to a temperature of around 150–175°C and a pressure of 5–10 MPa at the die head. Under these conditions, the dough is quite flexible and easily adapts to complex orifice configurations.

The dough pieces expand very rapidly as they leave the die orifice, but the expansion may continue for a few seconds since the dough is hot and still flexible and water continues to boil off. The moisture content of pieces is on the order of 10–15% and is too high for satisfactory crispness and stability. Thus, the pieces are flash-dried or dried on vibrating screens in hot-air ovens to a final moisture content of 3–5%. The products may be coated with sugar syrup and flavoring if desired, dried again, cooled, and packaged.

3. Shredded Rice Cereal

Shredded rice cereal is a very popular ready-to-eat breakfast cereal. Whole-kernel or broken rice can be used as the starting material. The rice is washed and cooked in a rotary cooker with sugar, salt, and malt

syrup under 100–150 kPa steam pressure for a period of 1–2 hr or until the rice is uniformly cooked throughout, with no free white centers, and the kernels are soft and pliable but still individual particles. The cooked particles are then discharged at a moisture content of about 40% and are partially dried to a moisture content of about 25–30%. The dried kernels are tempered to ensure a uniform moisture distribution and form a hard, glazed surface. This allows the rice kernels to flow freely through the process and reduces hang-ups and choking problems.

The shredding rolls are from 15.2–20.3 cm in diameter and as wide as 60 cm or more. They are much smaller than flaking rolls. On one of the pair of rolls is a series of about 20 shallow corrugations running around the periphery. In cross section, these corrugations may be square, rectangular, or a combination of these shapes. The other roll of the pair is smooth. Soft and cooked rice is drawn between these rolls as they rotate and issues as continuous strands of dough.

Rice Chex and Crispix, manufactured by Ralston Purina, use two pairs of shredding rolls. Rice Chex uses rice as the sole cereal ingredient, while Crispix uses both rice and corn as the cereal components. The dough sheet formed from the first pair of rolls is placed on a moving belt. The dough sheet from the second pair of rolls is then laid on top of the first sheet on the same moving belt. The layered sheets can be cut by one or two pairs of cutting rolls, which fuse a thin line of the dough sheets into a solid mass at regular intervals to form a continuous matrix of biscuits. The wet biscuits are transferred to a metal belt moving through a gas-fired oven. The biscuits are toasted, cooled, and broken apart from each other through a vibratory conveyor before packaging.

4. Fortification with Vitamins and Minerals

Fortification of ready-to-eat rice breakfast cereals with vitamins and minerals is now a very common practice. The usual approach is to add the minerals and more stable vitamins such as niacin, riboflavin, and vitamin B-6 to the basic formula mix and then spray the more labile vitamins such as vitamin A and thiamin on the product after processing. The nutrients to be added must be stable during processing and storage, and a sufficient amount is added to compensate for losses in processing and storage.

B. Rice in Baby Foods

Rice in the form of rice flour or as granulated rice is used in the formulation of many strained baby foods, particularly in meat and vegetable combinations. Rice flour, waxy-rice flour, parboiled rice, rice polishings, and rice oil are used in baby foods. The largest use of rice in the baby-food industry is in the manufacture of precooked infant rice cereal.

1. Precooked Rice Cereal

Precooked rice cereal is frequently prescribed as the infant's first solid food. The cereal must be easily reconstituted with milk or formula with a minimum of lumps.

The process for making precooked baby cereals consists of preparing and cooking a cereal slurry. The slurry is dried on a double-drum atmospheric drier, and the dried cereal is flaked and packaged. Each baby-food manufacturer has his own formulation and process for the manufacture of precooked rice cereal. The ingredients used in the formulation of precooked rice cereals are rice flour, rice polishings, sugar, dibasic calcium phosphate, iodized salt, sodium iron pyrophosphate, glycerol monostearate (emulsifier), rice oil, thiamin hydrochloride, riboflavin, and niacin or niacinamide.

The precooked rice slurry is dried on an atmospheric drum drier. The thickness of the film on the drier surface, the spacing between the drums, the temperature of the drum surface, the drum speed, and the flowing properties of the slurry are controlled. The bulk density of the cereal is related to the thickness of the sheet coming off the drier and the size distribution of the flakes. Dry rice cereal does not flake well and, if the cereal process is not controlled, an excess weight of cereal must be packaged in order to meet the package headspace requirement. Because of the high starch content in rice, the apparent viscosity of the cooked cereal slurry is markedly affected by slight variations in the solids content. The solids, drum speed, drum temperature, etc., are adjusted to obtain a finished product of good quality.

A precooked rice cereal product with strawberries appears to be gaining favorable consumer acceptance. These products are prepared in a manner similar to that employed for regular precooked infant cereals. Cereal ingredients, fruit, sugar, oil, vitamins, and minerals are cooked, dried on an atmospheric drum drier, flaked, and packaged. Because of the hygroscopicity related to the fruit and sugar, fruit cereals require moisture-proof packages. An antioxidant is incorporated in the packaging material to prolong shelf life.

Sometimes, the enzyme diastase can lower the liquid requirement for reconstitution by hydrolyzing part of the starch. This is a very sophisticated process. The temperature, time of digestion, enzyme activity, and solids concentrations must be closely controlled to obtain a satisfactory product.

Rice cereal will become rancid if packaged in a hermetically sealed container. The package material most suitable for rice cereal is one that allows transmission of both moisture vapor and gas. Most precooked infant cereals are packaged in paperboard cartons. A bleached manila liner on the interior of the carton is occasionally used. The carton is overwrapped with a glue-mounted, preprinted paper label. The tight wrap offers sifting and insect protection to the package.

Dry cereal products other than precooked cereals are used in infant feeding. These are manufactured by dry-blending cereal grains, including rice, sugar, and milk powder. These blends require cooking and are fed as a gruel.

2. Extrusion-Cooked Baby Foods

In addition to drum drying, extrusion cooking is another method used in the preparation of precooked rice-based baby foods. The type of extruders used, the particle size of the rice flour, and the extrusion conditions are some of the major factors influencing the properties of extruded baby foods. Examples of extrusion-cooked rice-based foods are Kaset infant food, with 71.7% rice flour and 12.5% full-fat soy flour and Babymate, 75% milled rice and 25% dehulled mung bean [Juliano, B. O., and Sakunari, J. (1985). Miscellaneous rice products. In "Rice Chemistry and Technology" (B. O. Juliano, ed.), Chap. 16. American Association of Cereal Chemists, Inc., St. Paul, MN].

3. Formulated Baby Foods

Rice cereal products are customarily used in the preparation of soups and casserole dishes. Similar products designed for infant feeding also contain these ingredients. Not only is rice a food ingredient, but its use in baby foods has a significant role in the consistency of the product. The variety of rice used in these products is very important. Long-grain rice, because of its amylose content, causes the product to thicken during storage (retrograde) and eventually to produce a very rigid gel and water separation. The presence of free liquid in a product packaged in a glass container is a serious defect. The strong gel associated with water separation is also undesirable. Waxy-rice flour is a good stabilizer for canned and frozen food products. The stability is achieved as a result of a reduction in the amylose/amylopectin ratio.

"Junior" baby foods have a coarser texture. They help the baby acquire the "mouth feel." To produce the "junior particle," granulated rice is incorporated into the formulation of many junior vegetable and meat items. Care must be taken in the formulation of junior baby foods because the consistency can thin upon cooking and the particle may settle out and form a mat in the bottom of the jar. To ensure uniform distribution of the junior particle, modified waxy-maize starch is frequently incorporated into the product.

C. Packaging

Packaging of rice cereals is very important in rice processing. Many packaged ready-to-eat cereals are stored for 6 months or longer before they are actually consumed. Thus, it is necessary to use special packaging materials to protect the crispness and storage stability of rice cereals. [*See* FOOD PACKAGING.]

The cereals must be acceptable in flavor, aroma, and texture at the time of consumption. For this reason, many cereal manufacturers adjust their production schedules to meet the demand of the market so that the products will not stay longer than 6 months on the shelves of supermarkets and stores.

Crispness of breakfast cereals depends on the moisture content of the products. Puffed rice, for example, is more hygroscopic than other rice cereals. It will rapidly lose its crispness unless a good moisture barrier is provided in the package. Optimum crispness can be kept if the moisture content of the puffed rice is 3–4%.

VII. Rice Snack Foods

Recent studies have shown rice bran to be a highly effective means of reducing serum cholesterol levels in hamsters and also in human beings. As a result, food manufactures are using rice and oat brans as ingredients in such snack foods as crackers, cookies, breads, side dishes, hot and cold breakfast cereals, and pancake mixes. Damardjati and Luh (1989) in "Proceedings of the 7th World Congress on Food Science and Technology," Vol. VII, p. 12, reported on physicochemical properties of extrusion-cooked rice cereals from medium-grain rice flour enriched with rice bran.

A. Rice Snacks in America

1. Rice Cakes

Rice cake is a relatively new snack food. It is a disk-shaped puffed product, low in calories (35–40 kcal per cake). The main ingredient is long- or medium-grain brown rice. Other minor ingredients such as sesame seed, millet, and salt may be added. Brown rice is milled from paddy rice by removing the hulls and retaining the bran and polish layers (mainly aleurone), which have higher levels of nutrients and dietary fiber than conventional white rice. Because of consumer interest in low-calorie and dietary fiber-containing foods, rice cake is rapidly gaining widespread consumer acceptance.

Although rice cakes are a puffed product, they are unique in that no added binder is used to hold the individually puffed rice kernels together.

The procedures for making rice cake are as follows. Water is added to the long- or medium-grain raw brown rice to adjust its moisture content to 14–18%. The added water and brown rice are mixed and tempered in a liquid–solids blender and tumbled for a selected time (1–3 hr) at room temperature. The moistened rice is then introduced to a rice cake machine that has been preheated to 200°C or higher. An example of the rice cake machine is the Lite Energy Rice Cake Machine (Real Foods Pty, Ltd., St. Peters, Australia) shown in Fig. 2.

The mold in this type of rice cake machine consists of three parts, a ring-shaped side piece and upper and lower platens, which can be moved up or down to adjust the gap between them. The rice is then pressed

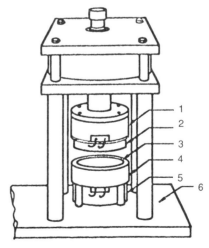

FIGURE 2 Rice-puffing parts of the rice-cake machine: (1) insulation block, (2) upper platen, (3) lower platen, (4) ring, (5) insulation block, (6) base plate.

between the movable upper and lower platens. At the end of a prescribed heating time, the upper platen is lifted and stopped at the upper edge of the ring. The heat-softened rice kernels are puffed because of the sudden release of water vapor as a result of moisture flash vaporization and are fused together to form the rice cake. Each cake is 10 cm in diameter and 1.7 cm high and weighs approximately 10 g. The cakes are then discharged and cooled in air before packaging.

F. Hsieh *et al.* [1989, *J. Food Sci.* **54**(5), 1310–1312] investigated the effects on rice cake volume of raw rice tempering conditions (time and moisture level) and heating conditions (temperature and time) immediately before puffing. In general, a lower moisture level (14% vs 16–20%) in raw rice and a longer tempering time (5 hr vs 1–3 hr) resulted in higher specific volumes in the rice cakes. Higher heating temperature (230°C vs 200–220°C) and an 8-sec heating time produced rice cakes of higher specific volumes. Darker cakes were obtained from combinations of high temperature and long tempering time.

Riceland Foods, Inc. (Stuttgart, Arkansas), utilizes an extrusion process to produce an American-style product similar to rice cake. Extruded pellets about the size of puffed breakfast rice are produced. The pellets are pressure-formed into approximately 10-cm-diam × 1.3-cm-thick disks or cakes.

The advantages of the extrusion process are flexibility in selecting ingredients to produce the pellets and easy control of the parameters affecting product acceptability compared to traditional methods.

Many types of cereal grains and combinations have been used in preparing extruded cakes from rice, wheat, corn, rye, buckwheat, oats, and barley. Either whole-grain or selected grain components are used, including high-fiber bran portions. A limitation of the extrusion process is the presence of oil-rich cereal components. Oil content must be less than 3% in order to obtain satisfactory expansion of the pellets and to maintain a desired texture in the cake.

2. Extruded Rice Cakes

Another category of rice cakes is represented by the square Crispy Cakes, which are similar in shape to the flat bread or crispbread type of product so popular in Europe. Rice flour is the major ingredient. Extruded rice cakes are usually produced in self-wiping twin-screw extruders. In addition to rice flour, rice bran, malt, or other minor ingredients such as sugar or salt may be added. A narrow slit die or dies are used. Hot and expanded extrudate ribbon about 7.5 cm wide is gently pulled away by a pair of rollers, which control the final thickness (0.7–0.8 cm) of the rice cake. Because of evaporative cooling, the temperature of the extrudate drops quickly after exiting from the die and, hence, the ribbon does not stick to the rollers. The traveling ribbon from the rollers is continuously cut into 7.5-cm-square cakes, which are then conveyed to a drying oven. After drying to 2–4% moisture, the cakes are cooled, stacked, and packaged. In many cases, apple, strawberry, and cinnamon extracts in vegetable oil are sprayed onto the cakes to improve the aroma of the product.

3. Granola Snacks

The most recent form pertinent to the snack food industry is granola bars, which compete for consumers of many snack and confectionery products.

Compared with other snack foods, granola bars have added appeal among consumers as "natural," and wholesome.

The important ingredient in granola bars is crisp rice or Rice Krispies. It contributes to the desirable eating characteristics, such as increased crispness and reduced roughness, and is a bulking agent. The slightly toasted note and relatively bland flavor of crisp rice also blends well with other ingredients in the granola snack bars.

Crisp rice can be manufactured by either oven puffing or a high-pressure extrusion puffing process. Kellogg's Rice Krispies are manufactured by the oven-puffing process. More often than not, crisp rice used in granola bars is manufactured by the high-pressure extrusion puffing process.

4. Rice Fries

Rice Fries are a snack food using rice as a basic ingredient. It first appeared in retail supermarkets in 1975, produced by American Frozen Foods. First, the rice is fully cooked in a broth containing butter, salt, and selected seasonings. After cooking, the rice is compressed and pumped or extruded through a 1.3-cm-square die to form ribbons. The ribbons are passed through a cutter, where they are cut into 7.6-cm-long units. The units are then fried in a hot (177–204°C) vegetable oil for about 1 min or until proper crust is formed on their surfaces. Finally, the products are cooled and quick-frozen, packaged, and ready for shipment.

For food service institutions, the rice fries can be served immediately after frying. The product has a crisp exterior crust while retaining the fluffy, light interior.

5. Rice Pudding

In Europe and in the United States, rice is frequently used in making pudding.

There are a variety of rice types, and these require consideration for cooking. Pastry chefs cook the rice in boiling water and then strain the product. It is then mixed with milk before cooking is completed. Rice must be handled carefully during cooking to prevent lumps from forming and rice kernels from breaking. Egg yolks, sugar, vanilla, and light cream are other ingredients. Rice pudding, with a variety of fruit combinations, serves as a popular dessert. It is preserved in enameled 5-oz aluminum cans by the high-temperature short-time aseptic canning process.

B. Flavor of Snack Foods

The cereal-processing industry is placing more emphasis on the variety and flavor quality of snack foods. Also, as in most of the food industries, health and fitness are important considerations.

Extruded foods are typically flavored after extrusion. This is done by dusting or spraying a flavoring mix on the product. If there is inadequate surface oil to cause good adhesion, the product must be sprayed with edible oil before the flavor is dusted on. If the product has been fried in oil, it is unnecessary to add additional oil to facilitate the adherence of the flavoring mix.

Some problems associated with flavoring an extruded product by surface coating are:

1. High dosage levels of flavor are required.
2. Flavoring may be unevenly distributed on the product.
3. Flavor may be lost from the product during packaging and shipping.
4. Once the surface coating is gone, the product has little flavor.
5. The potential for microbial contamination is increased.

The method does, however, permit the manufacturer to make more than one flavor of product from the same base material, and products can be flavored as a batch just prior to shipping, resulting in fresher-flavored products.

The quality of snack foods would be enhanced if the product base itself could be either totally or partially flavored.

For internal flavoring to be cost effective, it would be desirable to obtain better retention of flavors during the extrusion process. Encapsulation can help protect the flavor during extrusion.

Bibliography

Damardjati, D. S., and Luh, B. S. (1989). Physicochemical properties of extrusion-cooked rice breakfast cereals. *In* "Proceedings of the 7th World Congress on Food Science and Technology," Vol. VII. p.12. The Singapore Institute of Food Science and Technology, Singapore. Held in October, 1987.

Delgado, C. (1977). Improvement of rice bread. M. S. Thesis. University of California, Davis.

Gerdes, D. L., and Burns, E. E. (1982). Techniques for canning instant parboiled rice. *J. Food Sci.* **47**, 1734–1735.

Hsieh, F., Huff, H. E., Peng, I. C., and Marek, S. W. (1989). Puffing of rice cakes as influenced by tempering and heating conditions. *J. Food Sci.* **54**(5), 1310–1312.

Juliano, B. O. (1985)." Rice: Chemistry and Technology." The American Association of Cereal Chemists, Inc., St. Paul, MN.

Juliano, B. O. (1990). Rice grain quality: Problems and challenges. *Cereal Foods World* **35**(2), 245–253.

Juliano, B. O., and Sakurai, J. (1985). Miscellaneous rice products. *In* Rice Chemistry and Technology" (B. O. Juliano, ed.), Chap. 16. American Association of Cereal Chemists, Inc., St. Paul, MN.

Luh, B. S. (1991). "Rice I. Production." Van Nostrand Reinhold, New York.

Luh, B. S. (1991). Canning, freezing and freeze-drying. *In* "Rice" II. "Utilization" (B. S. Luh, ed.), Chap. 7. Van Nostrand Reinhold, New York.

Luh, B. S. (1991). Quick cooking rice. *In* "Rice." II. "Utilization" (B. S. Luh, ed.), 2nd Ed., Chap. 6, pp. 121–140. Van Nostrand Reinhold, New York.

Luh, B. S. (1991). "Rice." II. "Utilization" (B. S. Luh, ed.), 2nd Ed. Van Nostrand Reinhold, New York.

Merca, F. E., and Juliano, B. O. (1981). Physicochemical properties of starch of intermediate-amylose and waxy rices differing in grain quality. *Staerke* **33**, 253–260.

Nishita, K. D., and Bean, M. M. (1979). Physicochemical properties of rice in relation to rice bread. *Cereal Chem.* **56**, 185–189.

Nishita, K. D., Roberts, R. L., Bean, M. M., and Kennedy, B. M. (1976). Development of a yeast-leavened rice bread formula. *Cereal Chem.* **59**, 626–635.

Root and Tuber Crops

STEPHEN K. O'HAIR, *University of Florida*

Glossary

Annual Completing life cycle during one growing season

Apical Tip of the shoot (stem)

Biennial Requiring two growing seasons to complete life cycle; after a required cool period, flowering normally occurs during the second season

Bulb Compressed modification of the stem surrounded by fleshy, leaflike structures (scales)

Bulbil Bulblike structures formed on aerial stems

Corm Very compressed stem with nodes, internodes, and a large terminal bud

Cormel Fleshy, underground cormlike shoot arising from the lower portions of the corm; cormels are generally club-shaped, with a large terminal bud

Curing Subjecting harvested storage organs to high humidity and temperature for a few days after harvest to help extend shelf life; the result is healing of wounds and thickening of skin

Node Enlarged portion of the stem giving rise to leaves

Internode Stem section between nodes

Perennial Growing year after year, not necessarily dying after flowering

Propagule Vegetative structure that propagates a plant; sometimes misnamed "seed" as in potato "seed"

Rhizome Underground fleshy stem with several normally compressed internodes, growing parallel to the soil surface; during active growth, rhizomes continue to grow in length by producing new nodes and internodes and radial growth is slower than longitudinal

Root Portion of a plant that is normally underground; over one-half of a plant's biomass is root tissue, the major function of which is absorption of moisture and nutrients; however, some plants produce modifications which become fleshy edible storage organs for starches and sugars

Stolon Modified horizontal stem which grows near the ground

Tuber Fleshy modification of a stolon or stem; during growth phases, they enlarge radially and longitudinally; tubers have nodes, internodes, and a terminal bud

Tuberous root Modified storage root, resembling a tuber

The term "root crops" refers to a diverse grouping of food crops, known for their ability to store starches and sugars in specialized structures at or below the soil surface. These structures arise from either stem or root tissue and are divided into three main groups: (1) roots, (2) tubers, or (3) rhizomes and corms. Although many are biennial or perennial in growth habit, most are cultivated as annuals. Root crops specialize in concentrating either starches or sugars, but seldom both, in specialized parenchymatous cells. The deposition of concentrated carbohydrates normally begins as stem and leaf biomass approach their optimum and continues during the life of the plant. A review of plants producing enlarged storage organs at or below the soil surface would include a large number of plant species. However, only those few which are cultivated as food crops will be discussed.

TABLE I

Primary Information on the Major Root and Tuber Crops

Name		Family	Region of origin	Major producers	Edible portion	Propagules	Days to harvest	Shelf life under ambient conditions (days)	Methods of extending shelf life	Extended shelf life under controlled conditions (months)
Common	Scientific									
Cassava	*Manihot esculenta* Crantz	Euphorbiaceae	Brazil	Brazil, Thailand, Zaire	Root	Stem cuttings	240–730	2–3	Waxing roots	1–2
Sweet potato	*Ipomoea batatas* L.	Convolvulaceae	Peruvian Andes	China, Japan	Tuberous root	Stem cuttings and sprouts from roots	90–240	5–10	Curing roots	1–4
Tannia	*Xanthosoma* spp (L.) Schott.	Araceae	Venezuela	Caribbean Islands, W. Africa, Florida	Lateral corms	Corm pieces	240–420	56	Refrigeration	4–5
Taro	*Colocasia esculenta* (L.) Schott	Araceae	Bay of Bengal	W. Africa, S. Pacific	Corm	Corm pieces	180–540	14–30	Refrigeration	6
Yam	*Dioscorea* sp.	Dioscoreaceae	W. Africa, S. E. Asia, Tropical America	Nigeria	Tuber	Tuber pieces	240–330	180	Curing	2–3
Potato	*Solanum tuberosum* L.	Solanaceae	Peruvian Andes	Europe, China, N. America	Tuber	Tuber pieces and botanical seed	90–210	30	Refrigeration	3–8
Radish	*Raphanus sativus* L. fam.	Cruciferae	Europe, Asia	N. America, Europe, Asia	Root	Botanical seed	22–60	2–4	Refrigeration and plastic closure	1–2
Carrot	*Daucus carota* L.	Umbelliferae	Asia, Near East	Europe, Asia	Root	Botanical seed	60–85	2–3	Refrigeration and plastic closure	2–3
Beet	*Beta vulgaris* L.	Basellaceae	Mediterranean, Near East	Europe, Asia	Root	Botanical seed	50–80	2–3	Refrigeration	2–3

These can be divided into major (Table I) and less well-known or minor (Table II) crops. A few of the minor root crops are grown primarily as condiments. Members of the Allium family produce bulbs and will not be discussed.

I. Description of Root Crops

A. Crops Producing Storage Roots

For plants in this large group, the storage organ is formed as an enlargement of specialized roots. Unfortunately, there is no unique name for this morphologically distinct structure, which is often wrongly described as a "tuber." A more appropriate term is tuberous root. Cassava, sweet potato, beet, carrot, turnip, and radish are the best known members of this group.

B. Crops Producing Tubers

This group has the fewest members with the potato being the most notable, followed by yams (*Dioscorea* sp.). Tubers typically have a dormant period that coincides with adverse growing conditions associated with cold or dry seasons. [*See* POTATO.]

C. Crops Producing Rhizomes or Corms

There are only a few crops in this group that are grown extensively. Edible aroids (edible plants in the family Araccac) are the most common. Of these, the most popular are taro, which was the staple food in Polynesia, and tannia, also known as malanga, yautía, and cocoyam. Other edible crops in this family include giant taro, elephant yam, and swamp taro. Edible canna and arrowroot also produce edible rhizomes. Dormancy occurs in elephant yam, and to some extent in tannia, cormels. [*See* DORMANCY.]

II. Origins and Domestication

Most of the root crops originated and were domesticated in tropical areas in the Americas, Europe, or Asia (Table I and II). They were selected and domesticated as food crops that required little input and could be left unharvested for extended periods of time. A major factor for selection of the tropical root crops included vegetative propagation, which is generally less complicated than planting from seed. During this same period, it is suspected that the ability to flower and set seed were diminished in these crops. Since flowering and seed set were not necessary, the saved energy that would be otherwise allocated to physical support structures could be channeled to storage organs. In contrast the crops domesticated in temperate regions were selected for their ability to produce seed to facilitate overwintering.

III. Importance

Most root crops, with the exception of a few, are staple foods, being a source of low-cost carbohydrates. Many of the minor crops are limited to specific market or use niches or are valued as traditional foods and thus show little promise for expanded production. A few that are considered underexploited may have potential for expanded production. The remainder are utilized as condiments or for their specialized starches or sugars. In this case the market limits production. St. Vincent arrowroot is known for its fine starch and Jerusalem artichoke for its inulin, and indigestible sugar.

A. Major Producers

Since most tropical root crops require a long growing season, their production is limited to the tropics and subtropics (Tables I and II). Potato and sweet potato are the exceptions. They have been selected for short season production during temperate summers. In this case, special structures must be constructed to protect the planting material during periods when freezing temperatures are likely to occur. The temperate root crops are limited in production to regions where night temperatures are cool enough or night length is short enough to keep respiration at a value that is less than photosynthate accumulation.

B. Major Consumers and Trade

The main tropical root crops are grown as subsistence crops, being consumed by the growers and the local community. Because of their high water content, they are more costly to transport on a unit biomass basis than grains and legumes. Consequently, they are less popular as items for trade. The potato, which is marketed throughout the world, is a major exception. Most can be processed to extract starch and dried for milling into flours. The market for these products is rather limited, and is far surpassed by starches and

TABLE II

Primary Information on Minor Root and Tuber Crops

Name					
Scientific	Common	Family	Common region	Adaptation[a]	Main use	Other uses
Alocasia macrorrhiza (L.) Schott.	Giant taro	Araceae	S. Pacific, Asia	Lowland	Vegetable	
Alpinia conchigera Griff.		Zingiberaceae	Malaysia, China	Lowland	Condiment	
Alpinia galanga (L.) Wild.	Langwas, greater galangal	Zingiberaceae	Tropical Asia	Lowland	Condiment	Essential oil
Alpinia officinarum Hance.	Lesser galangal	Zingiberaceae	S. E. Asia	Lowland	Condiment	Stimulant, carminative
Alpinia pyramidata Blume.		Zingiberaceae	Tropical Asia	Lowland	Condiment	Intoxicating beverage
Amorphophallus campanulatus (Roxb.) Blume ex Dene	Elephant yam, whitespot giant arum, oroy	Araceae	Tropical Asia	Lowland	Vegetable	
Amorphophallus harmandii Engl. & Gehr.		Araceae	Tonkin	Lowland	Vegetable	
Amorphophallus prainii Hook. f.	Lekir, begung	Araceae	S. W. Malaya	Lowland	Vegetable	Source of an arrowposion
Amorphophallus rivieri Dur. (*A. konjac* Koch.)	Elephant yam, konjak	Araceae	Tropical Asia	Lowland	Vegetable	Source of konjaku flour
Arracacia xanthorrhiza Bancroft.	Arracacha, apio	Umbelliferae	Andes	Highland	Vegetable	
Brassica rapa L.	Turnip	Cruciferae	Europe, United States	Temperate	Vegetable	Oil seed
Brassica napobrassica (L.) Mill.	Rutabaga, swede	Cruciferae	Europe, United States	Temperate	Vegetable	
Calathea allouia (Aubl.) Lindl.	Topee tambo, sweet corn root	Marantaceae	Caribbean, America	Tropical Lowland	Vegetable	
Canna edulis Ker-Gawl.	Queensland arrowroot	Cannaceae	S. Pacific, Africa, Tropical America	Lowland	Starch	Leaves used as feed
Cochlearia armoracia L.	Horseradish	Cruciferae	N. America, Europe, N. Asia	Temperate	Condiment	
Curcuma aeruginosa Roxb.	Amada, mango ginger	Zingiberaceae	Burma	Lowland	Starch	
Curcuma amada Roxb.		Zingiberaceae	India	Lowland	Condiment	Carminative, stimulant
Curcuma angustifolia Roxb.	East Indian arrowroot	Zingiberaceae	Himalaya	Highland	Starch	
Curcuma heyneana Valeton		Zingiberaceae	Java	Lowland	Starch	
Curcuma longa L.	Turmeric	Zingiberaceae	Asia	Lowland	Condiment	Carminative

640

Scientific name	Common name	Family	Region	Climate	Use	Notes
Curcuma xanthorrhiza Roxb.		Zingiberaceae	Indonesia	Lowland	Starch	Stimulant, carminative, cosmetic
Curcuma zedoaria (Berg.) Roscoe	Shoti, zedoary	Zingiberaceae	S. E. Asia, India	Lowland	Starch, condiment	Beverage, oil
Cyperus aristatus Rottb.		Cyperaceae	N. America	Temperate	Vegetable	
Cyperus esculentus L.	Chufa, yellow nut grass	Cyperaceae	Pan tropic	Lowland	Vegetable	
Cyrtosperma chamissonis (Schott.) Merr. (C. edule Schott, C. merkussi (Hasskarl) Schott)	Swamp taro	Araceae	S. Pacific	Lowland bogs	Vegetable	
Eleocharis dulcis Trin. (E. tuberosa Schultes)	Water chestnut	Cyperaceae	Asia, S. Pacific	Lowland	Vegetable	
Eutrema wasabi (Sieb.) Maxim.	Wasabi, Japanese horseradish	Cruciferae	Japan	Temperate	Condiment	
Helianthus maximiliani Schrad		Compositae	N. America	Temperate	Vegetable	
Helianthus tuberosus L.	Jerusalem artichoke	Compositae	N. America	Temperate	Vegetable	Inulin
Hitchenia caulina (Grah.) Baker	Chavar	Zingiberaceae	India	Lowland	Starch	
Icacina senegalensis A. Juss.	False yam	Icacinaceae	Tropical Africa	Lowland	Famine food	Edible seeds
Lepidium meyenni Walpers	Maca	Cruciferae	Andes	Highland	Vegetable	
Maranta arundinacea L.	St. Vincent arrowroot	Marantaceae	Caribbean	Lowland	Fine starch	
Nelumbium luteum (Willd.) Pers.	Water chinquapin, water nut	Nymphaeaceae	S. United States	Temperate	Vegetable	Entire plant edible
Nelumbium speciosum Willd. (Nelumbo nucifera Gaertn.)	Lotus	Nymphaeaceae	S. E. Asia	Lowland	Vegetable	Edible seeds
Oxalis cernua Thumb.		Oxalidaceae	S. Africa	Temperate	Vegetable	
Oxalis tuberosa Molina (O. crenata Jacq.)	Oca	Oxalidaceae	Andes	Highland	Vegetable	
Oxalis deppei Lodd.		Oxalidaceae	Mexico	Temperate	Vegetable	
Pachyrrhizus ahipa (Wedd.) Parodi	Ahipa	Leguminosae	S. America	Highland	Vegetable	
Pachyrrhizus angulatus Rich.	Wayaka yam bean	Leguminosae	Tropical Asia	Lowland	Vegetable	
Pachyrrhizus bulbosus Spreng.	Yam bean	Leguminosae	Pan tropic	Lowland	Vegetable	
Pachyrrhizus erosus (L.) Urban	Yam bean	Leguminosae	Pan tropic	Lowland	Vegetable	

TABLE II (*continued*)

Primary Information on Minor Root and Tuber Crops

| Name | | Family | Common region | Adaptation[a] | Main use | Other uses |
Scientific	Common					
Pachyrrhizus palmatilobus Benth. & Hook	Jicama	Leguminosae	Mexico	Lowland	Vegetable	
Pachyrrhizus tuberosus (Lam.) Spreng. (possibly same as *P. palmatilobus*)	Yam bean	Leguminosae	Caribbean, America	Tropical Lowland	Vegetable	Young pods edible
Pastinaca sativa L.	Parsnip	Umbelliferae	Worldwide	Temperate	Vegetable	
Polymnia sonchifolia Poepp & Endl. (*P. edulis* Wedd.)	Yacon	Compositae	Andes	Lowland	Vegetable	Inulin, fodder
Pueraria lobata (Wild.) Ohwi (*P. thunbergiana* Benth.)	Kudzu	Leguminosae	Asia	Subtropics	Starch	Fodder
Sagittaria latifolia Willd.	Arrowleaf, duck potato	Alismaceae	N. and Centeral America	Lowland	Vegetable	
Sagittaria sagittifolia L.	Arrowhead	Alismaceae	S. E. Asia, Pacific	Lowland	Vegetable	
Solenostemon rotundifolius (Poir.) J. K. Morton	Hausa potato	Labiaceae	S. E. Asia, Africa	Lowland	Vegetable	
Sphenostylis stenocarpa (Hochst. ex Rich.)	African yam bean	Leguminosae	Tropical Africa	Lowland	Vegetable	Edible seeds
Tacca hawaiiensis Limpr. f.	Hawaiian arrowroot	Taccaceae	Hawaii	Lowland	Vegetable	
Tacca leontopetaloides (L.) Kuntze (*T. pinnatifida* Forst.)	East Indian arrowroot	Taccaceae	Africa, Asia, S. Pacific	Lowland	Starch, vegetable	
Tacca umbrarum Jum. & Perr.	Kobitsondolo	Taccaceae	W. Madagascar	Lowland	Vegetable	
Tragopogon porrifolius L.	Salsify, oyster plant	Compositae	Mediterranean, N. America and Europe	Temperate	Vegetable	
Tropaeolum tuberosum Ruiz & Pav.	Añu, ysaño	Tropaeolaceae	Andes	Highland	Vegetable	
Ullucus tuberosus Caldas.	Ullucu	Basellaceae	Andes	Highland	Vegetable	

[a] Tropical lowland or tropical highland or temperate regions.

flours from grains. The minor crops which are valued as condiments or for their special starches are marketed worldwide.

IV. Ranges of Production

The ecological requirements of the root crops are extremely diverse, ranging from tropical swamps to alpine meadows. Although most root crops are limited to specific ecosystem, others like potato and sweet potato are widely adapted. The latter two were carried throughout the world by early explorers, which placed them under diverse regimes of natural selection. Potato, sweet potato, and cassava have been the subject of extensive breeding to increase their ranges of adaptation. Of these, potato breeding has been in progress for the longest period and production ranges reflect these inputs. Most of the production of crops originating in temperate regions has remained in that region, possibly because of the high input requirements. The remaining tropical crops have been highly neglected and occupy relatively narrow ranges of production. Swamp taro, found in South Pacific bogs, and maca from the high Andes are examples of extremes in this group. Lack of genetic variability and appropriate selection by growers are reasons for their limited range.

A. Humid Tropics

Crops in this group are typified by a long growing season requirement of at least 8 months of moist, warm, frost-free weather. Taro and tannia are adapted to a tropical rain forest climate, while cassava, yam, and sweet potato are more common in regions which experience a monsoon climate.

B. High Altitude Tropics

All crops discussed in this niche are found in parts of the Andes (Tables I and II). Potato and its related tuber bearing species are the best known. The others are rarely grown outside this ecological zone since they require cool nights and a relatively long growing season. Additionally, the temperate region short night lengths may encourage production of foliage and retard or inhibit storage organ formation. In the absence of selection and breeding, these qualities limit expansion of production.

C. Temperate

Crops adapted to temperate regons are typified being able to produce a crop in a short period of time

(1–4 months) and having tolerance to temperatures at or slightly below freezing (Tables I and II). Parsnips are the only root crop that can survive a hard freeze. Carrots, beets, rutabaga and celery root can be overwintered in the ground in mild climates. Mulching can be used to protect some root crops in the soil or they can be harvested and stored in a cool environment. Most are biennials, producing an abundance of seed which is used for crop propagation. They are most productive in environments where day temperatures are warm and night temperatures are cool and are usually less productive in environments with high night temperatures. In these latter environments respiration rates negate previous photosynthate accumulation.

V. Productivity and Maturity

The productivity or yield of root crops can be presented in several ways. This is due in part to the storage organs' ability to continue growth as long as the plant is actively accumulating photosynthate. Thus, crop maturity is a vague term for these crops. It relates more to the time when the appropriate percentage of the storage organs has reached an acceptable marketable size than to a specific physiological stage of growth. In many cases there is no distinct event or characteristic that indicates "maturity." Usually, production is timed, with harvesting scheduled to start a specified number of days after planting. Delays in harvesting can result in oversizing of the storage organs to the point that they are no longer marketable. In many instances the qualities of the storage organs change over time, with a decrease in palatability after a certain period of growth.

A. Yield Per Crop

Although yield is usually reported as a weight of the storage organs harvested after a specified growth period, it is not necessarily the best or fairest. As mentioned above, the size and weight of the individual storage organs increase in a curvilinear manner over time. With crops like cassava, the crop can be left unharvested for more than one season. During subsequent seasons, the roots continue to enlarge during periods favorable for growth. Thus, the yields could be twice as large by the end of the second growing season and cannot be fairly compared with yields of a crop harvested at any other time. Yields should be compared using a unit of production per unit of time such as days. Additionally, yields are often reported on a fresh weight basis. This too can be deceiving, since the dry matter content is the main quality of concern. Although the percentage of dry weight does not change dramatically over time, cultivars can vary significantly. When the yield of carbohydrate or low cost calories is the concern, dry weight is a more meaningful measure of yield.

VI. Propagation

Most tropical root crops are propagated vegetatively, while the temperate crops are propagated from botanical seed. Vegetative propagation results in rapid crop establishment, while botanical seed can survive long-term storage. [See PLANT PROPAGATION.]

A. Methods

Propagation of crops normally grown from botanical seed is similar to that of other vegetables. For those propagated vegetatively, propagules comprise portions of healthy, mature stems (cuttings) or tubers (seed pieces) of an adequate size to produce nourishment for rapid early growth. Vegetative propagules have a shorter shelf life and a narrower range of environmental conditions than most botanical seeds.

B. Problems

Propagation problems when botanical seed is utilized are similar to those for other crops grown from seed. On the other hand vegetative propagation is a problem for crops in temperate regions, since protection during periods when temperatures are close to or below freezing is an added expense. Potato and yam propagation present problems similar to those of grain crops, in that the plant part consumed as food is also used for propagation purposes (Table I). This presents a greater problem in regions where food supplies are limited. On the other hand, cassava is grown from stem cuttings, sweet potatoes can be grown from vine cuttings, and cocoyam is propagated from the unpalatable main corm. In all instances, cut surfaces offer sites for pathogen attack. Other problems for vegetative propagules relate to transmission of pests and systemic diseases from one generation to the next. Some pests and diseases carried on or in propagules are minor, while others can result in complete crop loss. The latter can demand expensive control programs as is used for potatoes via official seed crop certification. In addition, vegetative propagules are fresh plant material. Consequently, they are (1) bulky,

requiring large storage areas and difficult to transport and (2) extremely perishable, requiring special storage environments to maintain quality.

C. Remedies for Vegetative Propagation Problems

Tissue culture and development of techniques utilizing botanical seed in propagation of crops that are normally propagated vegetatively are options which show promise as remedies for problems related to: (1) pest and disease transmission, (2) reduction of food supplies from the use of the same plant part for food and propagation, (3) bulkiness of cuttings, and (4) perishability. Both will require considerable scientific input before they can be used in commercial settings. Potato will likely be the first crop to benefit from this advanced technology since it is receiving the majority of the attention. [See PLANT TISSUE CULTURE.]

VII. Storage

When tropical root crops are grown in their regions of origin or similar ecological niches, harvest schedules can be very flexible, allowing for convenient in field storage. Temperate root crops, on the other hand have various tolerances for temperatures at or near freezing. Usually mulching or special protection is required for in field storage. Otherwise the crops must be harvested and stored in a cool location. Like other harvested fresh vegetables, root crops are perishable commodities. Shelf life varies among crops and species, ranging from a few days to several months. Some, like potato and yam, have a natural dormancy period of several weeks to months, which controls shelf life. Careful handling at all stages of harvest and marketing, curing, refrigeration, and minimizing moisture loss are methods of extending shelf life of these crops. [See GRAIN, FEED, AND CROP STORAGE; POSTHARVEST PHYSIOLOGY.]

VIII. Utilization

A. Special Nutritional Qualities

Several of the root crops are known to contain various quantities of potentially toxic compounds. For this reason, traditional preparation methods may involve extensive processing. Undomesticated genotypes and species may have quantities which could be harmful when unprocessed storage organs are consumed. The quantities that must be consumed before adverse reactions occur can very widely, depending on plant genotype and environmental conditions during production and after harvest. The extent of harm is also determined by body weight and general health of the individual.

Potatoes are a classic example, containing varying amounts of solanine, which is found primarily near the peel and in the sprouting "eyes." Most modern clones have been selected to contain minimal amounts of solanine and are safe to consume.

Since cassava is widely grown, the presence of cyanogenic glucosides (linamarin and lotaustralin) and their effects on human metabolism have been well documented. These glucosides change to hydrocyanic acid (HCN) when they come in contact with the naturally occurring enzyme linamarase. This enzyme helps to liberate the cyanide (CN^-), once tissues are disturbed. In large quantities, unprocessed cassava could be lethal when consumed at a rate of 0.5–3.5 mg HCN/kg body weight. Fresh unprocessed root tissue of high HCN clones can contain as much as 250 mg HCN/kg, while most clones have less than 50 mg HCN/kg.

Sweet potato, taro, swamp taro, and giant taro contain various amounts of proteinase inhibitors, which are destroyed in the cooking process. All of the edible aroids contain an acrid factor that causes itching and swelling of the skin. Although the acridity is mostly destroyed in cooking, extensive washing and soaking are required to remove it from the more acrid clones. Alkaloids, tannins, and saponins are found in yams. The main edible species have safe, low levels of these antinutritional factors. Some undomesticated species are highly poisonous, requiring extensive soaking and washing to make them edible.

B. Uses

There are numerous uses for the root crops which revolve around the starches found in the storage organs. The starch qualities vary considerably according to the size and uniformity of the grains and the ratio of amylose to amylopectin. These qualities make the crops ideal for specialized uses. Cassava starch has about 20% amylose and can be used as a substitute for waxy-corn starch. The starches of all crops can be converted to dextrose and be used as a sweetening agent. They can also be used in fermentation and other industrial processes. All starchy roots and tubers can be dried, milled, and used as partial or total substitutes for wheat flour.

C. Processing

With the exception potato, processing technologies and the number of industries utilizing root crops have been minimal. Nevertheless, many of the same products made from potato could be made from other starchy root crops. The lack of adequate financial input, absence of infrastructure to organize large-scale processing, and undocumented demand for processed products are reasons for the low incidence of commercial processing.

IX. Prominent Centers for Research

The volume of root crop research has been small in comparison to that of other crops valued for trade. Until recently, many root crops were considered to be neglected and underexploited. Participation of the more affluent countries in funding research to reduce world hunger has led to the organization of specialized, international research centers under the auspices of the Consultative Group for International Agricultural Research (CGIAR). Their mandates are to address improvement of specific underexploited crops. Centers headquartered in Nigeria, Colombia, and Peru address root crop problems. In addition, organizations such as the United Nations, national governments, and governmental consortiums fund projects dealing with specific root crop problems for specific regions or nations. There are several North American and European programs addressing the needs of potato. In contrast, the number and size of programs covering the other crops are dwarfed. [*See* CONSULTATIVE GROUP FOR INTERNATIONAL AGRICULTURAL RESEARCH.]

Bibliography

Coursey, D. G. (1967). "Yams." Longmans, London.
Kay, D. E. (1973). "TPI Crop and Product Digest, 2. Root Crops." Trop. Prod. Inst., London.
National Research Council (1989). "Lost Crops of the Incas." National Academy Press, Washington, DC.
Onwueme, I. C. (1978). "The Tropical Tuber Crops: Yams, Cassava, Sweet Potato, and Cocoyams." Wiley, New York.
O'Hair, S. K., and Asokan, M. P. (1986). Edible aroids: botany and horticulture. *Hort. Rev.* **8,** 43–99.
Purseglove, J. W. (1974). "Tropical crops: Monocotyledons." Longmans, London.
Purseglove, J. W. (1984). "Tropical crops: Docotyledons." Longmans, London.

Rural Development, International

MIGUEL A. ALTIERI, *University of California, Berkeley*

Glossary

Agroecology Ecological approach to the study and management of agricultural systems

Agroecosystem Collection of organisms that interact with the physical environment and other management factors to produce food and fiber

Biodiversity Set of biological species that play key ecological roles in agroecosystems

Ethnoecology Study of the ecology of a particular ethnic grouping

Ethnoscience Ecological knowledge systems of an ethnic group

Genetic conservation Preservation of native plant and animal gene pools in agroecosystems

Grassroots From the rural people

Indigenous knowledge Knowledge that has been passed down from generation to generation and relates to all aspects of survival

Input Economic term for materials brought into a system, in this case, fertilizers, seeds, etc.

Rural development Increase in the economic, social, and political status of rural populations

Socioeconomic Involving both social and economic factors

Sustainability Ability to maintain productive agricultural lands over time

Rural development is an evolving process in the betterment of the quality of life of rural populations involving changes in the social organization, technological innovation and adoption, conservation of natural resources, promotion of income generating activities, and enhancement of food security and provision for health, education, and general infrastructural support.

I. Introduction to Rural Development

Food scarcity, malnutrition, and absolute poverty are widespread problems in developing countries. These problems are generally perceived as the result of high population growth and low agricultural yields. Agricultural and economic development aimed at the rural population has been proposed as the answer. International- and state-sponsored agricultural research and development agencies have offered strategies of rural development for expanded food production and improved productivity to generate economic surplus for the poor. After 3 decades of technological innovations in agriculture, rural poverty and low land and labor productivity still persist in the developing world. Clearly then, topdown development strategies have been fundamentally limited in their ability to promote equitable and sustainable development. Distribution of benefits has been extremely uneven, favoring larger, better-off farmers who control optimum lands. In certain areas this has resulted in increased concentration of land ownership, peasant differentiation, and loss of land. [*See* WORLD HUNGER AND FOOD SECURITY.]

New technologies are distributed unevenly and biased toward modern, high-input farming. They are also channeled through institutions whose policies perpetuate conditions of land tenure, credit, and technical assistance which favor large-scale farmers. These technologies are not made available to small farmers on favorable terms and often are not suited to the agroecological conditions of small-scale farmers. Furthermore, agricultural modernization has proceeded in the absence of effective land distribution and has emphasized high-input technologies, all factors contributing to environmental problems in the developing world. As the interrelated issues of environmental quality, food security, and rural poverty are further analyzed and understood in the Third World, the sustainability of agriculture has emerged as a central development theme. The goal is to maintain agricultural productivity with minimal environmental impacts, with adequate economic returns, while providing for the social needs of the total population. There are no recipes on how to achieve sustainability, however. In fact, policy makers and development practitioners alike seem caught up in a series of rhetorical discussions:

- should agricultural production for export or for domestic consumption be emphasized?
- should efforts be devoted to raising agricultural productivity in optimal or marginal areas?
- should high-input or low-input technologies be used?
- should high-yielding varieties or traditional ones be emphasized?
- should local farmers be included in research and development efforts?
  [See FARMING SYSTEMS; SUSTAINABLE AGRICULTURE.]

Dilemmas abound. Meanwhile, top-down development sponsored by the state and/or international organizations has yet to address the pressing economic, environmental, and production needs of the developing countries and to provide viable options to the large masses of resource-poor peasants. [See INTERNATIONAL AGRICULTURAL RESEARCH.]

Agricultural production challenges can no longer be considered separately from social and environmental concerns, and therefore approaches to agricultural development must provide for the needs of present and future generations without compromising the natural resource base. There is a growing need to build new research and extension capabilities which effectively promote rural development. During the past 10–15 years in many developing countries, hundreds of grassroots and nongovernment organizations (NGOs) have become the new actors in rural development, focusing their attention on neglected crops, lands, and people. Their approach has been to question "conventional wisdom" in technological design through the search for new kinds of agricultural development and resource management that, based on local participation, skills, and resources, enhance produc-

tivity while conserving the local resources. Local farmers' knowledge about plants, soils, and ecological processes regain unprecedented significance within this new "agroecological" paradigm.

By focusing on the roots of poverty and low land productivity, NGOs, together with the peasants, are also attempting to change the socioeconomic and political environment where these systems operate. This work is providing the basis for a new approach to rural development and is helping to legitimize the role of agroecology in the development of low-input and regenerative technological innovations that serve the needs of the rural poor.

NGO efforts are not free of obstacles and limitations. Most NGOs involved in the implementation of agroecological proposals are aware of their knowledge and technical gaps and therefore are interested in strengthening the technical capabilities of their personnel. NGOs are also faced with the need to promote productive alternatives that are not only ecologically sound, but also economically profitable. It is important to realize that profitability at the household level depends not only on what the peasants can do, but mainly on the macro conditions under which the peasant production operates. It will be crucial, therefore, to create the right socioeconomic conditions needed for massive replicability of agroecological strategies.

II. Agroecology and Rural Development

In developing countries, purely technological approaches to agricultural development have ignored the enormous variations in ecology, population pressures, economic relations, and social organizations prevailing in the region; therefore, agricultural development has not been matched with the needs and potentials of local peasants. This mismatch has been characterized by three aspects:

a. Technological change has been concentrated mainly in temperate and subtropical areas where physical and socioeconomic conditions are similar to those in the industrial countries and/or experiment stations.

b. Technological change mostly benefits the production of export and/or commercial crops produced primarily for the large farmer, marginally affecting the production of food crops, which are largely grown by the peasant sector.

c. Many countries have become net importers of chemical inputs and agricultural machinery, increasing government expenditures and exacerbating technological dependence.

Agroecology has emerged as a new agricultural development approach more sensitive to the complexities of local agricultures, by broadening its performance criteria to include properties of sustainability, food security, biological stability, resource conservation, and equity along with the goal of increased production.

Due to its novel approach to peasant agricultural development, agroecology has heavily influenced the agricultural research and extension work of many institutions and farmers' organizations. Several characteristics of the agroecological approach to technology development and diffusion make it especially compatible with new forms of rural development:

a. Agroecology, with its emphasis on reproduction of the household and regeneration of the agricultural resource base, provides an agile framework for analyzing and understanding the diverse factors affecting small farms. It also provides methodologies that allow the development of technologies closely tailored to the needs and circumstances of specific peasant communities.

b. Low-input and regenerative agricultural techniques and designs proposed by agroecology are socially activating since they require a high level of popular participation.

c. Agroecological techniques are culturally compatible since they do not question peasants' rationale, but actually build upon traditional farming knowledge, combining it with the elements of modern agricultural science.

d. Techniques are ecologically sound since they do not attempt to radically modify or transform the peasant ecosystem, but rather identify management elements that, once incorporated, lead to optimization of the production unit.

e. Agroecological approaches are economically viable since they minimize costs of production by enhancing the use efficiency of locally available resources.

In practical terms the application of agroecological principles in rural development programs has translated into a variety of research and demonstration programs on alternative systems of production aimed at:

a. Improving the production of basic foods at the farm level to enhance the family nutritional intake, including the valuation of traditional food crops (*Amaranthus,* quinoa, lupine, etc.) and conservation of native crop germplasm.

b. Rescuing and reevaluation of peasants' knowledge and technologies.

c. Promoting an efficient use of local resources (i.e., land, labor, agricultural subproducts, etc.).

d. Increasing crop and animal diversity to minimize risks.

e. Improving the natural resource base through water and soil conservation and regeneration, emphasizing erosion control, water harvesting, reforestation, etc.

f. Reducing the use of external inputs to reduce dependency, but sustaining yields with appropriate technologies, through testing and implementation of organic farming and other low-input techniques.

g. Ensuring that alternative systems have an empowering effect not only on individual families but on the total community. To achieve this the technological process is complemented by programs of popular education that tend to preserve and strengthen "peasant's productive rationale" while supporting peasants in the process of technological adaptation, linkage to markets, and social organization.

The various examples of grassroots rural development programs currently functioning in developing countries suggest that the process of agricultural betterment must: (a) utilize and promote autochthonous knowledge and resource-efficient technologies; (b) emphasize the use of local agricultural diversity, including indigenous crop germplasm as well as essentials like firewood resources and medicinal plants; and (c) be a self-contained, village-based effort with the active participation of peasants. The evaluation of projects in Latin America suggests that promoted methods represent important alternatives for better water-use efficiency, environmentally sound pest control, effective soil conservation, and fertility management that subsistence farmers can afford.

III. Rescuing and Applying Traditional Farmers' Technical Knowledge

Traditional agricultural systems are the product of centuries of accumulated experience. By mimicking natural ecological processes, farmers have evolved complex "agroecosystems," the sustainability of which has stood the test of time. Moreover, unlike modern monocultures, traditional agroecosystems reflect the priorities of peasant farmers; they produce a varied diet, achieve a diversity of sources of income, use locally available resources, minimize the risk to farmers from crop losses, protect against the incidence of pests and disease, and make efficient use of available labor. Such multiple cropping methods are estimated to provide as much as 15–20% of the world's food supply. Throughout Latin America, farmers grow from 70–90% of their beans with maize, potatoes, and other crops. Maize is intercropped on 60% of the region's maize-growing area.

The development of agroecosystems has not been a random process: on the contrary, intercropping (the growing of two or more crops on the same land at the same time), agroforestry (intercropping systems which include trees), shifting cultivation, and other traditional farming methods are all based on a thorough understanding of the elements and the interactions between vegetation and soils, animals, and climate. Indeed, the ethnobotanical knowledge of many traditional farmers is prodigious: the Tzeltals Mayans of Mexico, for example, can recognize more than 1200 species of plants, while the P'urepechas recognize more than 900 species and the Yucatan's Mayans some 500. Such knowledge enables peasants to assign specific crops to the areas where they will grow best.

Although traditional agroecosystems vary as a result of different historical and geographical circumstances, they share the following structural and functional features:

- they contain high numbers of species;
- they exploit the full range of microenvironments differing in characteristics such as soil, water, temperature, altitude, slope, or fertility, whether within a single field or a region;
- they rely on local resources plus human and animal energy which utilize low levels of input technology;
- they rely on local varieties of crops and the use of wild plants and animals. Production is usually for local consumption.

Clearly then, traditional farming systems exhibit two salient features: a high degree of biodiversity and a complex system of indigenous technical knowledge or ethnoscience. Both elements are obviously interrelated as the maintenance of biodiversity is dependent upon local farmers' knowledge about their environments. Despite acknowledgement of the importance of both elements, agriculturalists have yet to take full advantage of the benefits of biodiversity and ethnoscience in the implementation of rural development projects.

One of the features of the new kinds of agricultural development and resource management strategies is that local farmers' knowledge about the environment, plants, soils, and ecological processes is regarded as significant. Understanding the ecological and cultural features of traditional agriculture, such as the ability to bear risks, knowledge of biological folk taxonom-

ies, the production efficiencies of symbiotic crop mixtures, and the use of local plants for pest control, is crucial for obtaining useful and relevant information to guide the development of appropriate agricultural strategies. These strategies are more sensitive to the complexities of peasant agriculture and are also tailored to the needs of specific peasant groups and regional agroecosystems.

Agricultural research and development should operate on the basis of a "bottom up" approach, starting with what is there already: local people, their needs and aspirations, their farming knowledge, and their autochthonous natural resources. In practice the approach consists of preserving and strengthening peasants' productive rationale through programs of popular education and training, using demonstration farms that incorporate both traditional peasant techniques as well as new viable alternatives. In this way farmers' knowledge and environmental perceptions are integrated into schemes of agricultural innovation that attempt to link resource conservation and rural development.

An example of the application of this approach is the work of several groups in Puno, Peru, which are rescuing an ingenious system of raised fields that evolved on the high plains of the Peruvian Andes about 3000 years ago. These "waru-warus," which consisted of platforms of soil surrounded by ditches filled with water, were able to produce bumper crops in the face of floods, droughts, and the killing frosts common at altitudes of almost 4000 meters. Around Lake Titicaca, remnants of over 80,000 hectares of them can still be found.

Technicians have assisted local farmers in reconstructing some 10 hectares of the ancient farms, with encouraging results. They have found, for instance, that yields of potatoes from waru-warus can outstrip those from chemically fertilized fields. Recent measurements indicate yields from waru-warus of 10 tons per hectare compared with an average in the Puno region of 1–4 tons per hectare. The combination of raised beds and canals has proved to have remarkably sophisticated environmental effects. During droughts, moisture from the canals slowly ascends to the roots by capillary action, and during floods, the furrows drain away excess runoff. Waru-warus also reduce the impact of extremes of temperature. Water in the canals absorbs the sun's heat by day and radiates it back by night, thereby helping protect crops against frost. On the raised beds, nighttime temperatures can be several degrees higher than in the surrounding region. The system also maintains its own soil fertility. In the canals, silt, sediment, algae, and plant and animal remains decay into a nutrient-rich muck which can be dug out seasonally and added to the raised beds.

This ancient technology is proving so productive and inexpensive that it is actively being promoted throughout the Altiplano, in preference to modern agriculture. It requires no modern tools or fertilizers; the main expense is for labor to dig canals and build up the platforms.

Nowhere is the potential for biodiversity and ethnoscience more applicable than in the realm of pest management and soil conservation. The regulating effects of vegetational diversity and the importance of vegetational cover on soil conservation are well known. Certain ethnic groups of farmers have a thorough knowledge of the biology of a variety of insect pests and weeds and also possess complex classification systems of soils and plants which guide their management practices. The challenge is to find creative ways to use biodiversity and ethnoscience to design production methods tailored to the socioeconomic circumstances and cultural needs of small farmers throughout the developing world.

In Honduras, NGOs are combining local farmers' knowledge of soil conservation with those of external agents to develop site-specific and socioeconomically adapted farming techniques for hillsides. The program has revived and/or introduced soil conservation practices such as drainage and contour ditches, grass barriers and rock walls, and organic fertilization methods such as using chicken manure and intercropping leguminous plants. In the first year, yields tripled or quadrupled from 400 kg per hectare to 1200–1,600 kg. In the next 5 years, 40 other villages requested training in the soil conservation practices.

It is important to understand that for a resource conservation strategy to be compatible with a production strategy and to succeed among small farmers, the process must be linked to rural development efforts that give equal importance to local resource conservation and food self-sufficiency and/or participation in local markets. Any attempt at soil, forest, or crop genetic conservation must struggle to preserve the agroecosystems in which these resources occur. It is clear that preservation of traditional agroecosystems cannot be achieved isolated from maintenance of the ethnoscience and sociocultural organization of the local people. It is for this reason that many researchers and development workers emphasize an agroecological–ethnoecological approach as an effective mechanism to link farmers' knowledge with Western scien-

tific approaches in the design of agricultural development that matches local needs, constraints, and resource-bases (Fig. 1).

IV. Socioeconomic Aspects of Rural Development

Despite the many advances, bottom-up grassroots development efforts in poverty alleviation have met with mixed success. A key reason is that they are attempting to counteract an environment in which their constituents have little access to political and economic resources and in which institutional biases against peasant production prevail. Grassroots development is difficult to implement where landownership is very skewed or where institutional arrangements (i.e., credit, technical assistance, etc.) and markets favor the large farm sector.

All the groups involved in the implementation of agroecological proposals are faced with the need to promote productive alternatives that are not only ecologically sound but also economically profitable. In other words, the diffusion of agroecological agriculture will be only possible if it is a good business for small producers and takes into account peasants' criteria for taking risks.

There are many policy obstacles that prevent peasants from a fair competition in the market, thus limiting the chances for any agroecological strategy to be assumed at the household level. Removal of policy constraints must occur in at least three areas:

a. Elimination of anti-peasant institutional biases in access to credit, research, and technical advice.
b. Elimination of the perennial social underinvestment in peasant communities in education, health, and infrastructure.
c. Elimination of subsidies to capital intensive and agrochemical agriculture.

In addition, it is important to create the right policy climate to improve the terms of trade for peasant production by providing competition to local monopolistic intermediaries and allow peasants to capture the externalities that a peasant sustainable agriculture might produce. This will require a definition of adequate tax policies to charge "free riders" who take advantage of peasants' efforts. This kind of economic policy could help create subsidies to encourage peasants to assume sustainable agricultural practices.

Up to now, the macro perspectives for a more sustainable agriculture in developing countries have

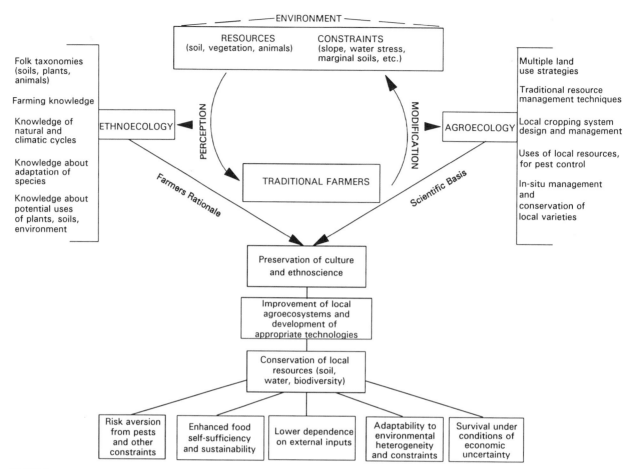

FIGURE 1 An agroecological and ethnological paradigm to rescue and validate peasants' farming knowledge for its application in sustainable agriculture development.

been uncertain. On the one hand it is possible to observe that high real exchange rates push for a more local resource oriented agriculture, since labor costs have gone down and imports have become more expensive. On the other, the strength of export-oriented economic approaches dominated by large multinational companies are preventing the emergence of a technological option based on regional resources.

Defining a sustainable rural development agenda will ultimately depend on the analysis of the economic, social, political, and institutional conditions determining poverty and environmental degradation in the developing world, and on the types of structural changes that will be promoted to affect such conditions. Several tasks lie ahead: (a) determining ways in which agricultural and environmental national and international policies can be changed to promote rural development in the region; (b) fostering research on specific policy change that removes biases against rural development and assessing the impacts of such

changes on socioeconomic and environmental parameters within a natural resource accounting, as well as a social welfare, framework; (c) defining the elements of an appropriate agricultural strategy that enhances local development and the viability of nonconventional land-use strategies; (d) reorienting and redesigning university curricula to more effectively address the problems of agricultural development, environmental degradation, and rural poverty; their connections; and viable solutions.

V. Promoting Agroecological Strategies

Today there is great concern about the process of systematic impoverishment that peasant agriculture is undergoing, with populations rising, landholdings becoming smaller, environments degrading and per capita food production remaining static or declining. In view of this deepening crisis, a major goal of rural development programs should be to prevent the collapse of peasant agriculture by transforming it into a more sustainable and productive one. Such transformation can only occur if projects are capable of realizing the potential contributions of agroecology and incorporating them into rural development strategies that:

a. Improve the quality of life of those peasants farming small landholdings and/or marginal lands through the development of ecologically based subsistence strategies.

b. Raise land productivity of those peasants competing in the market through the design and promotion of low-input technologies that reduce production costs.

c. Promote income and job generation through the design of appropriate technologies targeted at food-processing activities that enhance the aggregate value of what is produced within the peasant units.

It is now increasingly recognized that agroecological techniques can produce high yields of varied crops, while maintaining soil fertility and reducing farmers' reliance on expensive chemical inputs and unstable markets. Cropping systems and techniques specially tailored to the needs of specific agroecosystems result in a more fine-grained agriculture, based on a mosaic of appropriate traditional and improved genetic varieties, local inputs, and techniques with each combination fitting a particular ecological, social, and economic niche.

It is clear that improving the access of peasants to land, water, other natural resources, as well as to equitable credit, markets, appropriate technologies, etc. is crucial to ensure sustainable development. Securing control and access to resources can only be assured by political reforms or well-organized, community-based actions. Given these structural limitations, agroecology can only hope to provide the ecological basis to manage the resources once they become available to the rural poor. In other words, as an agricultural development approach, agroecology cannot confront the structural and economic factors that underlie rural poverty. This will require a much broader development approach that strongly emphasizes the social organization of the peasantry. In this regard, technological issues must assume their corresponding role within a development strategy that incorporates social and economic dimensions.

Bibliography

Altieri, M. A. (1987). "Agroecology: The Scientific Basis of Alternative Agriculture." Westview Press, Boulder, CO.

Altieri, M. A., and Hecht, S. B. (1990). "Agroecology and Small Farm Development." CRC Press, Boca Raton, FL.

Brokensha, D., Warren, D. M., and Werner, D. (1980). "Indigenous Knowledge Systems and Development." Univ. Press of America, Lanham.

Chambers, R. (1983). "Rural Development: Putting the Last First." Longman, London.

Harwood, R. R. (1979). Small Farm Development: Understanding and Improving Farming Systems in the Humid Tropics." Westview Press, Boulder, CO.

Richards, P. (1985). "Indigenous Agricultural Revolution." Westview Press, Boulder, CO.

Rural Sociology

FREDERICK H. BUTTEL, *University of Wisconsin*

Glossary

Adoption/diffusion of technology Processes by which individual farmers (or other prospective users of technology) make decisions about the use or nonuse of a new technology

Rural–urban continuum Notion that the nature of social structure, social relationships, values, and so on vary systematically across rural and urban communities

Sociology of (agricultural) science Study of the processes through which scientists and scientific institutions discover new knowledge

Technology assessment Study of the social and environmental implications of emerging technologies

Treadmill of technology Process of adoption of new technology in which the market effects of the technology cause the use of the technology eventually to become obligatory for all producers

Social stratification Distribution of social resources such as wealth, income, status, power, and prestige among a social grouping such as a community, formal organization, or society and the processes that generate relative equality or inequality of that distribution

Rural sociology has traditionally been defined as the study of social organization and social processes that are characteristic of geographical zones where population sizes and densities are relatively low. Thus, rural sociology can be defined as the sociological study of rural societies. In practice, however, modern rural sociology has had to become considerably more comprehensive than the study of the social structures and processes of rural societies. Rural societies do not exist in isolation or a social vacuum. The "rural" is, in part, a reflection of the larger processes of the regional differentiation and allocation of populations, economic activities, and other human activities within a society as a whole (or, increasingly, within the global economy and society). Science and technology are particularly good examples of social forces that have metropolitan origins (e.g., in the laboratories of land-grant universities, ARS stations, and agribusiness firms) and that in turn have major impacts on rural people, communities, and regions.

I. The Development of American Rural Sociology

As much as contemporary rural sociologists can justifiably say that the field has made major strides in understanding rural social processes over the past four or five decades, rural sociology, like the larger discipline, still has very strong roots in nineteenth-century social thought. As G. F. Summers has noted, many of the major issues of concern to the giants of nineteenth-century social theory are still important a century later. Among the most important of the preoccupations of nineteenth-century social theory was whether village and farm life was morally or socially superior to metropolitan life, and whether it would be resilient in the face of urban-industrialism. The classical sociologists Emile Durkheim (1858–1917) and Karl Marx (1818–1883), though they agreed on little else, both argued that urban-industrial capitalism and modern technologies and organizational practices were inexorable—and, on balance, progressive—social forces that would eventually supersede or supplant rural social forms. Ferdinand Toennies (1855–1936), by contrast, saw that urbanization and industrial capitalism were leading to the decline of intimate communal bonds that are characteristically rural and intrinsic to healthy social life. Max Weber (1864–1920) had a more variegated view of rural societies and social structures. Weber recognized the dynamism of the various forces (bureaucratization, industrialization, urbanization) that were marginalizing traditional village and farming structures. But at the same time Weber felt that bureaucratization and the Industrial Revolution (what he referred to collectively as an "iron cage") could be expected to lead to social movements, religious ethics, political ideologies, and other forms of resistance and challenge to the forces of rationalization and uniformity.

For most of its history, rural sociology has reflected these nineteenth-century debates over the desirability and resilience of traditional rural social organization. Two general orientations have alternated in prominence over time: (1) the view that rural society, on account of its more intimate social bonds, lower incidence of social pathologies, stronger religious institutions, and so on, is socially or morally superior to urban society and therefore deserves to be preserved, and (2) the view that traditional rural beliefs, social structures, technologies, practices, and institutions can and must be superseded for the quality of rural life to be enhanced.

A notable historical benchmark in the debate between these two views of rural society was the publication of P. A. Sorokin and C. C. Zimmerman's *Principles of Rural–Urban Sociology* in 1929, which synthesized the dominant thrust of rural sociological thought of the preceding 15 years and which for another generation was an influential treatise in the field. The Sorokin and Zimmerman book drew primarily on Ferdinand Toennies' (1887) analysis of how urbanization and industrial capitalism had led to the undermining of the primary social bonds of community. Accordingly, rural sociology prior to World War II was largely devoted to rural community studies and often had overtones of seeing rural life as being socially richer or morally superior to metropolitan life.

The Toennies–Sorokin–Zimmerman perspective, usually referred to as the theory of rural–urban continuum, was ultimately superseded, however, for several reasons. Shortly after *"Principles"* was published, rural sociology was confronted with the Great De-

pression and rural squalor, and rapidly embraced New Deal reformism. During the New Deal period, rural sociology increasingly reflected the view that rural society stood in need of the types of legal, social, and structural reforms that were then under way in nonrural arenas such as labor relations, state regulation of economy and society, old age assistance, and so on. In other words, there was an emerging perspective, that the traditional structures and practices of rural society were not necessarily socially desirable. The rural–urban continuum perspective was also slowly but surely undermined empirically (e.g., by studies showing that communitylike primary social bonds persist even within large metropolitan places and that social beliefs of rural and urban people exhibit few differences).

Rural sociology was also shaped, albeit less directly, by America's agrarian politics. Though the expression "rural sociology" was not widely employed until the second decade of this century, the first American rural sociological studies were initiated in the 1890s in the midst of a decade of Populism and other forms of agrarian unrest. It is now generally agreed that the pioneering rural sociological studies in the United States were those initiated in the late 1890s by W. E. B. DuBois, a black sociologist who was then on the staff of the U.S. Department of Labor (and whose later writings are now considered to be a precursor of African-American separatist thought). DuBois' studies emphasized how the postbellum crop-lien system in Southern plantation agriculture had had the effect of tying black farmers to the plantation system and subordinating them to the power of the planter class, thus reinforcing black poverty. Numerous studies of agricultural communities in the Northeast were subsequently undertaken by F. H. Giddings and associates at Columbia University. In these early days, land-grant universities had little or no presence in the area of scholarship we now call rural sociology.

Although the radical Populist critique of industrial capitalism was the most dramatic phenomenon in rural America during the earliest years of rural sociology, Populism did not directly influence early rural sociology. During the first full decade of American sociology there was widespread suspicion among university administrators and their patrons that sociology was suspect and should be scrutinized to ensure that it would concern itself with empirical research rather than radical politics. This political environment led most sociologists of the era to distance themselves from radical movements and social theories; thus, for example, there is no indication that any American "rural sociologists" in these early years were active supporters of Populism.

The indirect influences of Populism on rural sociology, however, were important. The Populists' occasional (and mostly unsuccessful) attempts to bring black farmers into their movement contributed to the questioning of the sharecropping system and Southern rural social structure by DuBois and others. Even more important was the fact that Populist radicalism was a source of concern to urban industrialists who benefited from existing arrangements that provided a stable supply of cheap food for their workers. Although William Jennings Bryant's defeat in the Presidential election of 1898 signaled a decisive defeat of Populism in electoral politics, fear of a resurgence of Populist radicalism remained for more than a decade.

The aim of providing a moderate alternative to Populism was integral to the establishment of the Country Life Movement at the turn of the century. Founded and funded by industrial and other elites, the Country Life Movement promoted the view that the problems of rural society were not due to the shortcomings of industrial capitalism as the Populists had held. These problems were rather caused by a lack of organization, poor infrastructure, and technological backwardness in rural areas, which in turn led rural society to lag behind urban America. In 1908 President Theodore Roosevelt recognized this reform movement by appointing a Presidential Commission on Country Life, chaired by Liberty Hyde Bailey, dean of agriculture at Cornell University. The next six decades of American rural sociology were largely prefigured by the Country Life Commission Report, which was published in 1909. The Report did acknowledge many of the social problems of rural America (e.g., the inequities of the crop-lien system, widespread tenancy, land speculation), but the dominant thrust was the case it made that an expanded effort to modernize rural America technologically, while simultaneously restructuring rural society to facilitate this process of technological upgrading, was integral to improving the condition of rural America. The Country Life Commission recommended the establishment of what is now called the Extension Service to speed the technological modernization of rural America (which was accomplished in 1914 by the passage of the Smith-Lever Act). [See COOPERATIVE EXTENSION SYSTEM.]

The Country Life Commission also recommended the harnessing of the social sciences, particularly agricultural economics and rural sociology, to support the technological modernizationist vision of rural America. Rural sociological studies in land-grant colleges of agriculture were seen to be important in helping to remove social barriers to technological modernization and to stabilize rural communities. The establishment of rural sociology, however, was slow and uneven, particularly compared to the field of agricultural economics. Only a few land-grant universities—generally the larger ones with the greatest scientific capacity in the Northeast and Midwest—established major rural sociology programs. And it was not until the Purnell Act was signed into law in 1925 that federal funds were available to support rural sociological research. For all practical purposes, the land-grant colleges at which rural sociology was present in the 1930s are the same 25 or so land-grant colleges where rural sociology exists in the early 1990s. [See EDUCATION: UNDERGRADUATE AND GRADUATE UNIVERSITY.]

Three other historical factors were crucial in shaping and institutionalizing rural sociology over time. First, as noted earlier, the Great Depression and the fashioning of the Depression-era New Deal opened up vast opportunities for rural sociological scholarship aimed at rural reform and relief. By the mid-1930s, sociologists in the USDA Bureau of Agricultural Economics (BAE) and elsewhere in the USDA and other federal agencies had initiated an impressive program of empirical research on rural communities, rural population, and the structure of agriculture and had linked this research with an agenda for far-reaching rural reforms (e.g., reduction of the power of Southern landlords, land tenure reform, encouragement of cooperative forms of production).

Second, the course of the development of rural sociology was decisively shaped when, in 1936–1937, rural sociologists started their own journal (*Rural Sociology*) and broke from the American Sociological Society (later the American Sociological Association) to establish the Rural Sociological Society. The organizational break with the larger discipline of sociology led rural sociology to become even more completely identified with and institutionalized within the land-grant and USDA complex.

Third, from the late 1930s through the mid-1940s, the active reformism of the New Deal came under assault by conservative members of Congress. The crackdown on New Deal reformist rural sociology (and agricultural economics) was completed when the USDA's BAE was disbanded in 1944. For two decades after the dismantlement of the BAE, the newer generations of rural sociologists were dissuaded from embracing the scholarly and political reform orientations of the New Deal rural sociologists.

Though rural sociology is now undertaken in dozens of nations and there are rural sociology professional associations in many countries and on all continents, rural sociology is essentially an American phenomenon. The very category rural sociology was created in association with the American land-grant system. Approximately half of the active professional rural sociologists in the United States are university faculty appointed in land-grant colleges of agriculture. Though in many other countries there had been some tradition of scholarship similar to rural sociology well before their rural sociological associations were established, the global spread of rural sociology was due mainly to the missionary zeal of American rural sociologists from the 1940s through the 1960s. Funded mainly through the rapid expansion of American foreign aid in the post-World War II period, many of the luminaries of American rural sociology traveled extensively and sowed the seeds of American-style rural sociology across an impressive range of countries. As late as the 1960s, rural sociology in other (industrial and Third World) countries was very similar to the American variety. Over the past two decades, however, rural sociology has become increasingly diverse. U.S. rural sociology is not nearly so homogeneous or globally influential as it once was and it increasingly borrows from the ideas of European and Third World scholars. Even so, the historical importance of American rural sociology can be gauged by the fact that in some overseas rural sociological circles the content of American rural sociology is stereotyped (as being similar to what is was in the 1960s) and subjected to critique, and the stereotypical elements (e.g., excessive reliance on quantitative methods, a lack of a critical view of technology and technology-development institutions) are studiously avoided.

II. The Major Foci of Rural Sociology

Modern rural sociology has six major branches: rural population, rural community, rural social stratification, natural resources and environment, agriculture, and science and technology. Each of these areas has active theoretical and empirical research wings as well as applied research and practice wings (e.g., community development, technology assessment, social impact assessment, rural poverty alleviation). This review will give principal stress to science and technology and secondarily to agriculture. Before doing so, however, a brief description of each of the major specialties in American rural sociology will be provided.

From the very inception of rural sociology, sociological analysis of rural population and rural community dynamics through census and social survey data has been central to the field. DuBois' early work on black farming, for example, was largely based on population census data. The work of the Division of Farm Population and Rural Life of the BAE was pivotal in providing basic descriptive data about the rural population and in establishing rural population studies as one of the pillars of the field. To this day there is not a major rural sociology program that lacks a specialist in rural population and in rural community studies. In the early days of rural sociology, these two areas of study were very closely articulated, with the same people often doing work in both. In large part this was because census of population data for counties and rural places were the most accessible data sources for characterizing rural communities and farm structures. Community studies using survey methods often began with developing a censuslike morphology of the rural community. Even today there remains a significant relationship between rural population and community studies.

Although there has been a continuing basis for articulation between rural population and community studies (and, to a lesser extent, the sociology of agriculture), these areas have become increasingly differentiated so that they are now seen as quite distinct specialties. The use of census data to document rural–urban differences, à la Sorokin and Zimmerman, fell into disrepute. Rural population research has followed the larger subdiscipline of demography in becoming more quantitative, whereas the rural sociological analysis of communities, though it has a strong quantitative wing, has a significant cadre of researchers who use qualitative or field methodologies.

Since the time of ancient societies and empires, there has been a tendency for poverty, disadvantage, and political subordination to be exhibited disproportionately among rural people. Preindustrial rural social structures (e.g., feudalism, the antebellum and postbellum Southern plantation systems in the United Sates, landlord-dominated rural economies in much of contemporary South and Southeast Asia, and the latifundia–minifundia complex of Latin America) tended to promote extraordinary social inequality. In modern societies, rural people in aggregate have tended to be poorer than urban–metropolitan people.

For example, in the United States, the average annual incomes of nonmetropolitan households are only slightly more than 70% of those in metropolitan counties. The equation of rural with poverty and inequality has thus made analyses of rural social stratification and inequality an important dimension of rural sociology.

The analysis of rural social stratification, however, has long been sensitive and controversial. Scrutiny of sociology by the administrators of turn-of-the-century universities, for example, served to steer sociological attention away from class and other inequalities and toward perspectives, such as population analysis and the rural–urban continuum, that deemphasized social class and inequality. The New Deal milieu of state-led reformism created more space for sociological and rural sociological attention to social class and poverty, though this space would ultimately shrink as the New Deal was rolled back, the BAE was dismantled, and East–West rivalry intensified. Rural social stratification and related analyses enjoyed a considerable renaissance in tandem with the War on Poverty and Great Society era of the 1960s. The study of rural social stratification continues to be a lively, and increasingly more diversified, arena of inquiry up to the present time. Studies of regional and labor market inequalities, rural gender inequality, and rural racial inequality have enriched the analysis of rural social inequality. These analyses have been extended to the global level and to metropolitan–corporate forces, such as international competition among private multinational agricultural input firms.

Though there has been a long and significant tradition of rural sociological scholarship on the relations between people, communities, and natural resources, the sociology of natural resources and environment in the contemporary sense (of analyzing "environmental problems") did not emerge until the early 1970s. However, it is an enduring feather in the cap of rural sociology that its analyses of environmental problems were at the leading edge of the larger subdiscipline of environmental sociology. Rural sociology has also continued to provide strong leadership in applied areas of environmental sociology such as social impact assessment.

As noted earlier, many of the most prominent American rural sociologists of the post-World War II period devoted major segments of their careers to encouraging the diffusion of rural sociology across the globe, particularly in the developing world. The impulse for elaborating international rural sociology (or the sociology of rural development) was partly the

intellectual one of wanting to promote a comparative approach to understanding rural social organization, but even more important was the fact that the post-War period was an emerging era of "developmentalism" (i.e., an era of faith in the efficacy of planned social change and development in the decolonizing world). Rural sociologists of this era were increasingly confident that they had a great deal to offer in promoting socioeconomic development in the preindustrial world. Rural sociologists trained in the adoption and diffusion of agricultural technology (see Section III) were particularly well represented among the first cohort of international rural sociologists. This modernizationist/developmentalist tradition of rural sociology would hold sway for nearly three decades, but ultimately it would be displaced intellectually by critiques of its lack of efficacy and of having ignored how development is constrained and blocked by the global and national dynamics of capitalism and the international political economy. The previous tradition of international rural sociology would be undermined, in particular, through critical assessments of the Green Revolution experience and the role of rural sociology in the Green Revolution.

The sociology of international rural development has contributed enormously to the development of rural sociology as a whole. International rural sociological studies have arguably helped to temper the tendency for this land-grant-bound discipline to become parochial. International rural sociology scholarship has also contributed to important theoretical trends in the field. For example, the period from the middle to the late 1970s is generally regarded as a period of ascendance of the "new rural sociology," whose principal theoretical underpinnings were largely worked out initially, or most thoroughly, through the critique of international development programs and the role of developmentalist/modernizationist rural sociology in contributing to these programs.

In the early 1990s, however, three major social forces are combining to threaten the future of international rural sociology. First, land-grant universities are faced with growing fiscal pressures, and more often than not their international development programs suffer disproportionately as budgets are cut. Second, the past 15 or so years have witnessed a profound crisis of whether "development interventions" (e.g., World Bank loans, U.S. Agency for International Development programs) can make a positive difference in the developing world now that global recession, international mobility of finance and industry, and imposition of structural adjustment programs threaten to foreclose the possibility of vibrant, balanced Third World growth. The current era is often characterized as being a "postdevelopmentalist" one in which there has been erosion of faith that meaningful "planned development" can occur in the late twentieth-century political–economic environment of globalization and fiscal austerity. Third, the past half dozen years have witnessed a significant decline in U.S. development assistance programs, which had traditionally been the major source of funding for international development sociology.

Of all the major rural sociological specializations, the sociology of agricultural/rural science and technology is, in one sense, one of its oldest and, in another sense, rural sociology's newest major speciality. As noted in Section III, there were significant studies of technological change and labor displacement in Southern plantation agriculture from the 1930s through the 1950s. The sociological study of the diffusion of agricultural innovations began during World War II. Shortly thereafter and until the late 1960s, the adoption–diffusion perspective was the single most important and dynamic area of rural sociological research. But as some critics of the diffusion of innovation have held, the diffusion tradition tended to ignore the social origins of science and technology and the social consequences of technology. Serious study of the processes by which agricultural science is practiced and new agricultural technologies are developed did not emerge until Lawrence Busch and William Lacy began to draw from the sociology of science to develop the field now generally referred to as the sociology of agricultural science. Sociological study of the social consequences of agricultural technology did not become a serious and enduring area of scholarship until Busch and Lacy, K. F. Goss, W. H. Friedland and colleagues, and G. M. Berardi, and C. C. Geisler, among others, propelled it to prominence.

III. Rural Sociology and Technological Change: The Diffusion Era

The rural sociological embracement of the adoption and diffusion of technology perspective was one of the most pivotal turns in the development of rural sociology. Prior to the rise of the diffusion perspective, the rural population, community studies, and rural–urban continuum traditions had been domi-

nant, and attention to technological change had been mainly confined to a few studies, such as that by B. O. Williams on the mechanization of cotton production in the plantation South.[1]

The adoption and diffusion of innovations perspective is essentially a social-psychological approach to explaining how and why farmers adopt new technological innovations. Farmers were conceptualized as *social actors* with varying *cognitive states* (due to variation in education and other factors) and various *value-orientations* and predispositions (e.g., different levels of modernity, innovativeness, risk-taking, and so on) who respond to *stimuli* such as the charastics of new agricultural technologies, mass media messages, and influences from "reference groups." Unlike the earlier traditions in rural sociology in which community and place were stressed, the diffusion perspective almost totally deemphasized the spatial aspect of rural society.

The rise of the adoption-diffusion perspective, which came to dominate rural sociology for about two decades, was a natural outcome of many trends of the time. The rural–urban continuum perspective was essentially being exhausted and its role in data reconnaissance to support New Deal-type reform interventions had withered. The early 1940s were also a time when there was much excitement about and fascination with hybrid corn, and it is no accident that the pioneering adoption studies focused on this technology. The social climate of the time, in other words, was conducive to shifting rural sociology to a new position of more strongly embracing the technologically driven modernization of rural society as the best means of improving the rural quality of life. Adoption research was also a practical way to make rural sociology appear useful to experiment station directors in the new post-BAE climate. And in sociology at large, the 1940s was a period of increased emphasis on social psychology, social statistics, and survey research, each of which was integral to executing adoption research.

The adoption tradition has had an impressive history, an admirable overview of which has re-

[1] B. O. Williams documented how the mechanization of cotton had led to the mass dislocation of sharecroppers (particularly black sharecroppers) and how the benefits of mechanization accrued largely to plantation owners. It should also be noted that even during the height of diffusion research in rural sociology, the dislocations caused by mechanization of Southern agriculture led many rural sociologists in the region to be more questioning of the implications of new technology than was typical elsewhere. [Williams, B. O. (1939). The impact of mechanization of agriculture on the farm population of the South. *Rural Sociol.* **4,** 300–311.]

cently been provided by F. C. Fliegel. It has produced a number of generalizations about the process of agricultural technology diffusion that remain relevant to this day. For example, it was repeatedly found that plots of cumulative adoption over time tend to take the form of a logistical growth (or "S") curve. It was also found that earlier adopters of new technology tend to have higher levels of education, to be younger, and to have more "modern" and "cosmopolitan" (or less "traditional") value-orientations than do later adopters or nonadopters. In addition, it was also found that correlations between farm size and farm household wealth, on one hand, and adoption of new commerical innovations, on the other, had tended to increase over time (notable summaries of this literature were written by F. H. Buttel and colleagues, F. C. Fliegel, and E. M. Rogers).

Some researchers have stressed, however, that one of the shortcomings of the adoption-diffusion perspective was that it tended to be promotional toward new technology and tended to see new technologies as being unambiguous improvements. Farmers who did not adopt new technology were referred to with the pejorative term "laggards." Adoption researchers seldom considered how technologies might affect the structure of agriculture or the environment. Despite the fruitfulness of the adoption-diffusion perspective, its influence began to decline in the late 1960s and early 1970s in tandem with a growing skepticism toward modern agricultural technologies among sociologists and society at large.

Although adoption-diffusion research no longer has the prominent position it enjoyed in the 1950s and 1960s, in updated forms it remains viable and insightful. Modified diffusion theories and methodologies lend themselves to understanding the processes of technological change in two important respects. First, by rigorously incorporating the biophysical setting and geographical space into diffusion models, there have been significant advances in understanding the interplay between social factors and the environment in technological decision making. Second, the past half dozen years or so have witnessed growing public interest in *ex ante* agricultural technology assessment, and diffusion knowledge and research methods are often integral in generating primary data (survey data on intentions to adopt a new technology) that can assist in *ex ante* assessment research.

IV. The "New Rural Sociology" and the Social Impacts of Agricultural Technology

Though there are a good many differences in the theories and methods used by various groups of sociologists who study science and technology, they agree that the social context of science, broadly construed, affects the practice of, and the knowledge that is generated from, science in both obvious and not-so-obvious ways. This observation is actually quite apt in the case of the science of rural sociology during the 1970s and 1980s. Rural sociology was decisively shaped by several phenomena in the larger society. Chief among the influences on rural sociology were five separate but interrelated social movements: the 1970s critique of the Green Revolution; the 1970s critique of the land-grant system and of modern mechanization and chemical technologies (especially J. Hightower's *"Hard Tomatoes, Hard Times,"* published in 1973); public interest group resistance to the rise of biotechnology in the 1980s; the sustainable agriculture movement; and the rise of global environmentalism in the middle to late 1980s. By no means did all or even most sociologists embrace these movements and their views fully. But many sociologists placed considerable emphasis on explaining the origins of these movements and how they have shaped research policies and the content of new technologies. Most importantly, however, disagreements over new technologies in society at large have prompted sociologists to ponder whether there are intellectually rigorous and socially constructive ways to take a neutral, detached, or agnostic position toward the practice of science and the development of new technologies.

It was noted at the outset that rural sociology's history has been one of alternating between opposing views about the distinctiveness and resilience of traditional rural social structures. Rural sociology came into its own during the post-World War II era. Post-War rural sociology tended to reject the rural romanticism of 1920s rural sociology and embraced a variety of technological-modernizationist assumptions: that traditional rural social structures were destined to disappear, that there was nothing intrinsic to rural America that was suffficiently socially or morally superior so that it ought to be preserved, and that technological modernization was the most desirable future for rural America. In the 1970s and early 1980s, however, the balance between these two views would be equalized to a degree. Emerging controversies over agricultural technologies and their social and environmental impacts removed some of the luster from the technological-modernizationist view. The 1970s were a period of "rural renaissance," as the population of nonmetropolitan counties grew faster than that of metropolitan counties as a result of net metro-to-nonmetro migration. And the 1970s were also an era in which the rapid declines in farm numbers that characterized virtually all of the post-War period began to level off.

These new realities of rural America during the 1970s, along with the growing popularity within the sociology of "critical" theories, led to what I have elsewhere referred to as the "new rural sociology." These new theories were actually quite diverse, ranging from various forms of neo-Marxism and neo-Weberianism to "critical theories" drawn from hermeneutics and related traditions. To the degree there was a coherence among these theories, it was due to their common critique of "modernizationist" rural sociology. Though the "new rural sociology" was most prominent in the sociological literature on agriculture, it had a major influence on the rural sociological study of agricultural science and technology as well.

In the early "post-Hightower" period, rural sociologists were unable to develop a distinctly sociological theory of technological change, but they did creatively borrow from and build on the "treadmill of technology" notion of the noted agricultural economist W. W. Cochrane. Cochrane's theory was based partly on an observation from diffusion research—that better-educated, less risk-averse farmers with larger, better-capitalized farm operations tend to be the earlier adopters of new technology. The early adopters, who receive "innovator's rents" (because of lower per unit production costs) for the period during which relatively few farmers use the technology, tend to receive the lion's share of the benefits of new technology that accrue to farmers.[2] Because most agricultural commodities have low price and income elasticities of demand and because new technology tends to lead to increases in output, adoption of new technology can be expected to contribute to overproduction and declining farm product prices. At the point at which there is sufficient additional production to depress product prices, innovator's rents disappear, and economic pressure ("the treadmill") is then placed on

[2] Cochrane and most economists who study technological change agree, however, that over the long term the bulk of the benefits of technological change accrue to consumers rather than farmers.

those who have not yet adopted. The choice that the nonadopting farmer faces would typically be either to adopt in order to stay in business, or not to adopt and eventually be forced out of business. Cochrane argued that there was a growing tendency for the lands of those farmers forced out of business by technological change to be consolidated into larger units (what he referred to as "cannibalism"). He also noted that when farm commodity programs place a floor under prices, the treadmill of technology receives state-subsidized acceleration and causes the benefits of technology to become capitalized in farm asset values (with these benefits accruing to large landowners in the form of capital gains).

Sociologists appropriated this Cochrane account of technological change and its relationship to farm structural change and gave it sociological embellishment. This genre of scholarship would become particularly important during the late 1970s as there emerged strong Congressional and USDA interest in the social consequences of new agricultural technologies. But as useful as the Cochrane treadmill of technology perspective was in synthesizing a range of scholarship on the social impacts of new technology, a number of scholars have identified significant limitations. As E. P. LeVeen has argued persuasively, the treadmill theory is more valid with respect to family farming systems than to large-scale corporate-industrial agriculture. The Cochrane perspective may also exaggerate the obligatory nature of technological change in family-farm agricultures. Cochrane's analysis assumes that most farmers are engaged full time in agriculture, yet about two-thirds of contemporary farmers have significant off-farm work, see their main occupation as being a nonfarm one, and have only modest dependence on the farm enterprise for the household's income. Accordingly, a part-time farming household will have more autonomy concerning adoption decisions than is depicted in the Cochrane account. Cochrane's theory of technological change relies mainly on reasoning about mechanization technology, which in the late twentieth century is far less important in farm technological change than it was at midcentury. Finally, and perhaps most importantly, treadmill of technology reasoning does not address the origins of new technologies, a limitation that would shortly be redressed through the development of a rural sociology of agricultural science.

V. The Sociology of Agricultural Science

One of the most important breakthroughs in the history of rural sociological thought occurred during the late 1970s when Lawrence Busch and William Lacy began to grapple with how rural sociology could meaningfully engage public debates over agricultural research and technology. They recognized that critiques of technology can be sterile unless one can demonstrate that there are meaningful technological alternatives (or, in other words, that a change in the structure of research institutions or research policy can affect the content and impacts of new technology). And to understand whether there are alternatives to existing or emerging technologies, one needs to study science and research itself rather than just the social impacts of technology. Busch and Lacy would thereby pioneer the field of the sociology of agricultural science. Their first major works were a 1978 article by Busch that was a theoretical critique of the implicit sociology of science and knowledge within the adoption-diffusion tradition, and a coauthored paper applying recent theory in the sociology of science to how one can conceptualize the sources of influence on the research of agricultural scientists.

The most well-known work of the Busch–Lacy group was their *"Science, Agriculture, and the Politics of Research,"* a monograph that reported results from a national survey of agricultural scientists. The book demonstrated the important impacts that scientists' social backgrounds (e.g., farm vs. nonfarm) and institutional nexus (SAES vs. ARS) have on the research goals of scientists. The book was therefore powerful testimony to the fact that science is a social product, and that social factors help to determine which of several alternative priorities and approaches will be stressed in scientific institutions.

The sociology of agricultural science remains a minority activity in rural sociology. Even so, the Busch and Lacy volume has had a broad impact in rural sociology and elsewhere in land-grant circles. Their work has provided a way for rural sociology to find some middle ground between ritualistic Hightower-style criticism of new technology on one hand, and unqualified defense of land-grant technology on the other. The impact of *"Science, Agriculture, and the Politics of Research"* has been far broader than one might expect from a few dozen rural sociologists interested in agricultural science and technology. The book has contributed to a better understanding of how land-grant and other public agricultural research institutions function and has also helped to galvanize rural sociological interest in contributing to public research policy (e.g., through Experiment Station Committee on Policy and National Association of State Universities and Land-Grant Colleges committees).

It was significant that the Busch–Lacy volume appeared just as interest was growing in the conflict over agricultural biotechnology. Their volume provided a general theoretical and empirical template for the numerous studies that would be done over the next decade to explain the course of the development of agricultural biotechnology. Although some of these studies of biotechnology took exception with one or another aspect of Busch and Lacy's work, it was recognized as well that *"Science, Agriculture, and the Politics of Research"* was the point of departure for what has become a sizable literature. Their own book on biotechnology has been a significant contribution to the study of plant biotechnology, and their innovative theoretical position, which combines the insights of actor–network theory from the sociology of science and induced innovation theory from economics, will likely become influential as well.

Rural sociological studies of biotechnology were wide-ranging. Some of these were aimed at placing modern biotechnology in a historical context of the changing division of labor between the state and private capital in developing and promoting new technology. Other scholars have stressed the new political-economic environment of biotechnology (e.g., new intellectual property restrictions such as patents), the importance of international competition in biotechnology, and how modern biotechnology could be expected to alter the dynamics of agrarian change in developed and developing countries. One of the major focal points of social science research on new technology has been recombinant bovine growth hormone (rBGH). rBGH has been the object of numerous studies of its political-economic and ideological context and of projections of farmer and consumer adoption patterns and their social consequences.

Rural sociologists have also pioneered in studies of several areas of agricultural science other than biotechnology, such as developing methodologies for "commodity systems" or "commodity subsector" analysis. These studies have demonstrated that the forces that shape public and private agricultural research are, more often than not, relatively specific to the commodity sector involved. For example, whether the benefits from a new technology such as the FLAVRSAVR tomato or high-lauric-acid canola varieties will accrue mainly to input manufacturers, farmers, processors, or consumers is shaped by the rules, structures, and institutions of a particular commodity sector.

Another important area of research has concerned sustainable agriculture, which has included research aimed at facilitating agricultural sustainability as well

as research on the origins and implications of the sustainable agriculture movement. Several rural sociologists have also made significant advances in our understanding of the relations between indigenous farmer knowledge and crop plant genetic resources, for example, how plant genetic resources struggles have been structured and why a North–South struggle emerged over these resources in the 1980s. J. Kloppenburg has demonstrated that indigenous technical knowledge exists and remains useful, even in the advanced countries, and he uses the example of intensive rotational grazing to suggest that farmer-developed knowledge has its own methods of discovery and may be as efficacious as technologies from knowledge developed within traditional experimental science. [See SUSTAINABLE AGRICULTURE.]

VI. Nature, Technology, and Agriculture

For most of the history of rural sociology, farm numbers have been in rapid decline and the scale of commerical farms has progressively increased. This trend was interrupted during the 1970s, which became the first post-World War II decade in which there was not a significant decline in farm numbers. The stabilization of farm numbers during the 1970s suggested that increased scale and industrialization of agriculture are not inexorable tendencies, and stimulated interest in the overall issue of what is the structural basis of the persistence of family farms. [See U.S. FARMS: CHANGING SIZE AND STRUCTURE.]

Agriculture is one of the few sectors of the advanced economies in which there are millions of firms, and in which the centralization of productive capital has yet to lead to the obliteration of small firms. Considerable research turned to determining whether there are intrinsic attributes of agriculture that lead to the persistence of family farming. A variety of explanatory schemes were proposed. It was argued, for example, that agriculture is distinctive because of the fact that land is its essential production input. Land is fixed in quantity and cannot be manufactured, which slows the pace of concentration of assets. The tie of agriculture to land also leads the farm labor process to be dispersed across large expanses and to be incompatible with supervision of large labor forces. It has also been argued that the tie of agriculture to the seasons makes it difficult to recruit a hired labor force, prevents agriculture from being fully industrialized, and, given the

idleness of labor in slack seasons, causes agriculture to be unprofitable. Finally, others have suggested that the family farm persists because of its "flexibility." The family farm enterprise need not earn the average rate of profit to remain in business, and the household is able to adapt to economic downturns by squeezing consumption, whereas prolonged downturns will ordinarily cause large corporate businesses to liquidate their assets. Finally, one influential account has claimed that agriculture is a "natural production process" whose very biology causes agriculture to resist full industrialization.[3] Thus, while agriculture has been penetrated by industrially produced inputs and agricultural products are marketed through large-scale corporate channels, agriculture by and large remains relegated to household producers because the natural production process inhibits the scale economies and production structures integral to large-scale enterprise.

Although there has been general agreement that one or another of the processes just noted has contributed to the persistence of family farming, several analysts have stressed the possibility that technological change could counter or override these barriers to "full industrialization" of production agriculture. There are historical precedents in which new technologies have made possible agricultural industrialization, most notable being the cases of U.S. poultry production and cattle feedlots, which are made possible through a combination of mechanization, confinement, and veterinary technologies. Industrial forms of production have also begun to take hold in pork and dairy production, two sectors that were once thought to be the exclusive preserve of diversified family farming. But these precedents basically involve the separation of livestock from crop production, which is of limited significance because it has been recognized that agriculture's distinctiveness lies particulary in crop production. Also, industrialization of animal agriculture has appeared to be due more to policy and institutional factors than to technological change per se.

There is now emerging a rich literature on the social and technical forces that are restructuring American and world agricultures. A variety of analysts have made the case that the post-World War II development of Western agricultures was a particular one, in

[3] The criteria for "full industrialization" would include most or all work being done by hired, nonfamily labor and continuous year-round production that is largely insulated from natural vagaries. [D. Goodman, B. Sorj, and J. Wilkinson. (1987). "From Farming to Biotechnology. Basil Blackwell, Oxford.]

that diverse farming structures became homogenized through the subordination of farming to the logics of mass production and technological standardization, though there is variation in the degree to which theory and evidence suggest that science and technology have played an autonomous role in this process. However, this perspective has been challenged in recent years, mainly on the grounds that it has exaggerated the homogeneity and sociotechnical coherence of post-War agriculture. There is also vigorous debate as to whether the technological and broader social changes that are emerging in the late twentieth century will override the natural production process of agriculture, and if so what the consequences of this will be. Some research on the implications of modern biotechnology has suggested that it will increasingly lead to a shift of food and fiber production from farm to factory and relegate farmers to being producers of low-value-added, undifferentiated commodities. Other research suggests that "identity-preserved" biotechnologies (e.g., high-lauric-acid canola varieties) will predominate in biotechnology R&D for the foreseeable future and that these technologies will reinforce the heterogeneity of agricultural systems.

VII. Summary

In this review the contributions of rual sociology to understanding the role and significance of agricultural science and technology have been stressed. Rural sociologists of science and technology are forging increasingly stronger linkages with colleges of agriculture, the land-grant system, and the agricultural sciences. But as noted earlier, modern rural sociology is extremely diverse and has several other areas of theoretical, empirical, and applied work, and most have larger cadres of scholars than does the sociology of agricultural science and technology. This diverse field is a vibrant one. Its relevance continues to increase in tandem with the growing importance of rural issues such as agricultural research, farm commodity programs, rural development, and related policy areas.

Bibliography

Busch, L., and Lacy, W. B. (1983). "Science, Agriculture, and the Politics of Research." Westview Press, Boulder, CO.

Busch, L., Lacy, W. B., Burkhardt, J., and Lacy, L. R. (1991). "Plants, Power, and Profit." Basil Blackwell, Oxford.

Buttel, F. H., Larson, O. F., and Gillespie, G. W., Jr. (1990). " The Sociology of Agriculture." Greenwood Press, Westport, CT.

Cochrane, W. W. (1979). "The Development of American Agriculture." University of Minnesota Press, Minneapolis.

Fliegel, F. C. (1993). "Diffusion Research in Rural Sociology." Greenwood Press, Westport, CT.

Garkovich, L. (1989). "Population and Community in Rural America." Greenwood Press, Westport, CT.

Goodman, D., and Redclift, M. (1991). "Refashioning Nature." Routledge, New York.

Havens, A. E. (1972). Methodological issues in the study of development. *Sociologia Ruralis* **12,** 252–272.

Hightower, J. (1973). "Hard Tomatoes, Hard Times." Schenkman, Cambridge, MA.

Kloppenburg, J., Jr. (1988). "First the Seed." Cambridge University Press, New York.

Sorokin, P. A., and Zimmerman, C. C. (1929). "Principles of Rural–Urban Sociology." Henry Holt, New York.

Summers, G. F. (1992). Rural sociology. *In* "Encyclopedia of Sociology." (Borgatta, E. F., and Borgatta, M. L. eds.), Macmillan, New York.

Wilkinson, K. (1991). "The Community in Rural Society." Greenwood Press, Westport, CT.

ISBN 0-12-226673-0

90065